Comprehensive Analytical Chemistry
Volume 68

Advanced Mass Spectrometry for Food Safety and Quality

Advisory Board

Comprehensive Analytical Chemistry

Volume 68

Advanced Mass Spectrometry for Food Safety and Quality

Edited by

Yolanda Picó

Food and Environmental Safety Research Group
Faculty of Pharmacy
University of Valencia
Valencia, Spain

ELSEVIER

AMSTERDAM • BOSTON • HEIDELBERG • LONDON
NEW YORK • OXFORD • PARIS • SAN DIEGO
SAN FRANCISCO • SINGAPORE • SYDNEY • TOKYO

Elsevier
Radarweg 29, PO Box 211, 1000 AE Amsterdam, Netherlands
The Boulevard, Langford Lane, Kidlington, Oxford OX5 1GB, UK
225 Wyman Street, Waltham, MA 02451, USA

Notices
Knowledge and best practice in this field are constantly changing. As new research and experience broaden our understanding, changes in research methods, professional practices, or medical treatment may become necessary.

Practitioners and researchers must always rely on their own experience and knowledge in evaluating and using any information, methods, compounds, or experiments described herein. In using such information or methods they should be mindful of their own safety and the safety of others, including parties for whom they have a professional responsibility.

To the fullest extent of the law, neither the Publisher nor the authors, contributors, or editors, assume any liability for any injury and/or damage to persons or property as a matter of products liability, negligence or otherwise, or from any use or operation of any methods, products, instructions, or ideas contained in the material herein.

ISBN: 978-0-444-63340-8
ISSN: 0166-526X

Contents

Contributors to Volume 68 xiii
Series Editor's Preface xv
Preface xvii

Part I
Advanced Mass Spectrometry Approaches and Platforms

1. Mass Spectrometry in Food Quality and Safety: An Overview of the Current Status

Yolanda Picó

1. **Introduction** 3
2. **Current Instrumentation and Operation in Mass Spectrometry** 5
3. **Matrix-Assisted Laser Desorption Ionization Coupled to Time-Of-Flight Mass Spectrometry (MALDI-TOF)** 13
4. **Inductively Coupled Plasma Mass Spectrometry (ICP-MS)** 21
5. **Mass Spectrometry (MS) Combined with Chromatography or Related Techniques** 26
   - 5.1 Liquid Chromatography–Mass Spectrometry (LC-MS) 26
   - 5.2 Gas Chromatography–Mass Spectrometry (GC-MS) 48
   - 5.3 Capillary Electrophoresis–Mass Spectrometry (CE-MS) 61
6. **Conclusions and Future Trends** 65

References 66

2. Advanced Mass Spectrometry

Yolanda Picó

1. **Introduction** 77
2. **Mass Spectrometric Resolution and Accuracy** 78
   - 2.1 Magnetic Sector 82
   - 2.2 Time-of-Flight (TOF) 84
   - 2.3 Fourier Transform Ion-Cyclotron Resonance (FT-ICR) 95
   - 2.4 Orbitrap 97
3. **Obtaining and Processing Information** 100
   - 3.1 Data-Dependent Acquisition (DDA) or Information-Dependent Acquisition (IDA) 105
   - 3.2 Data-Independent Acquisition (DIA) 108

3.3 Preprocessing and Postprocessing Steps 111
4. **Imaging Mass Spectrometry (IMS)** 112
5. **Conclusions** 115
References 116

3. Elemental and Isotopic Mass Spectrometry

Constantinos A. Georgiou and Georgios P. Danezis

1. **Introduction** 132
1.1 Mass Spectrometry 132
1.2 Toxic Elements 133
1.3 Nutritive Elements 135
2. **Theoretical Aspects** 139
A Elemental Mass Spectrometry 139
2.1 ICP-MS 139
B Isotopic Mass Spectrometry 174
2.2 IRMS 174
2.3 TIMS 184
2.4 MC-ICP-MS 187
3. **Applications** 193
3.1 ICP-MS 194
3.2 IRMS 213
3.3 TIMS 220
4. **Conclusion and Future Outlook** 222
References 224

4. Ambient Ionization Techniques

Marinella Farré and Damià Barceló

1. **Introduction** 245
2. **Principles and Classes of Ambient Ionization MS Techniques** 248
2.1 Momentum Desorption: Solid–Liquid Extraction Ambient Ionization Techniques 249
2.2 Techniques Based on the Thermal Desorption and Chemical Ionization 254
2.3 Techniques Based on Laser Ablation and Ionization 257
3. **Application of Ambient Ionization MS Techniques in Food Analysis** 259
4. **Conclusions** 267
Acknowledgments 268
References 268

5. High-Performance Ion Mobility Spectrometry

Wenjie Liu and Herbert H. Hill Jr

1. **Introduction** 276
2. **Ion Mobility Spectrometer** 276

2.1 Classification of Ion Mobility Spectrometers 276
2.2 Sample Introduction and Ionization Method of Ion Mobility Spectrometer 277
2.3 The Performance of IMS: Resolving Power and Peak Capacity 279
2.4 Hyphenation of Ion Mobility Spectrometry and Mass Spectrometry 280
3. **Analysis of Food Flavors and Composition with Ion Mobility Spectrometry** 282
3.1 Food Flavor Compounds 282
3.2 Foodborne Microbial Detection with Ion Mobility Spectrometry 285
3.3 Biogenic Amines 286
3.4 Other Flavor-Related Compounds 288
4. **Determination of Food Compositions and Food Additives with IMS** 290
4.1 Amino Acids 290
4.2 Flavonoids 290
4.3 Caffeine and Theophylline 290
4.4 Food Additives and Nutritional Supplements 292
4.5 Detection of Food Adulterants from Dietary Supplements and Other Food with IMS 294
5. **Analysis of Food Contaminants and Residue with Ion Mobility Spectrometry** 297
5.1 The Detection of Pesticides 297
5.2 Analysis of Veterinary Drugs 301
5.3 Analysis of Nitrite and Nitrate in Food 302
5.4 Analysis of Microbial Toxins in Food 302
6. **Conclusions** 303
References 303

Part II
Mass Spectrometry Applications within Food Safety and Quality

6. Food Proteins and Peptides

Roberto Samperi, Anna Laura Capriotti, Chiara Cavaliere, Valentina Colapicchioni, Riccardo Zenezini Chiozzi and Aldo Laganà

1. **Introduction** 309
2. **Proteomic Techniques** 313
2.1 Top-Down Proteomics 315
2.2 Bottom-Up Proteomics 316
3. **Food Security** 326
3.1 Plant Proteomics 326
3.2 Stress Effects on Gene Expression 327
3.3 What about Livestock? 330

4. **Food Safety, Quality, and Authentication** 332
4.1 Genetically Modified Organisms 332
4.2 Food Processing and Quality Control 332
5. **Food Peptidomics and Digestomics** 341
5.1 Bioactive Peptides 341
6. **Conclusions and Outlook** 346
References 348

7. Mass Spectrometry in Food Allergen Research

Linda Monaci, Rosa Pilolli, Elisabetta De Angelis and Gianfranco Mamone

1. **Food Allergy** 359
1.1 Overview 359
1.2 IgE versus Non-IgE-Mediated Food Allergy 360
1.3 Food Proteins Triggering Allergic Reactions 362
1.4 A Non-IgE-Mediated Food Allergy: The Case of Celiac Disease 363
1.5 Food Allergy Prevalence and Legislative Frame Worldwide 365
2. **The Role of Mass Spectrometry Applied to Food Allergen Research** 367
2.1 Mass Spectrometry in Allergen Characterization 367
2.2 Mass Spectrometry in Celiac Disease 370
2.3 Features of Different Mass Spectrometers Applied to Food Allergen Detection 372
3. **Advances in MS Methods for Multiple Detection of Food Allergens** 375
3.1 Detection of Intact Food Allergens 375
3.2 Detection of Peptides Tracing for Allergenic Proteins 376
4. **Concluding Remarks** 386
References 387

8. Lipidomics

Paola Donato, Francesco Cacciola, Marco Beccaria, Paola Dugo and Luigi Mondello

1. **Introduction and Overview** 396
2. **Lipids: Definitions and Classification** 397
3. **Structures, Functions, and Occurrence of Lipids** 398
3.1 Fatty Acids 399
3.2 Glycerolipids 401
3.3 Glycerophospholipids 402
3.4 Sphingolipids 404
3.5 Sterol Lipids 406
3.6 Prenol Lipids 406
3.7 Saccharolipids 407
3.8 Polyketides 407

4. **Extraction Methods of Lipids** 408
4.1 Sampling and Sample Preparation 408
4.2 Soxhlet Extraction 410
4.3 Method of Folch, Lees, and Stanley 410
4.4 Method of Bligh and Dyer 410
4.5 Supercritical Fluid Extraction (SFE) 411
4.6 Accelerated Solvent Extraction (ASE) 412
4.7 Microwave-Assisted Extraction (MAE) 412
4.8 Solid Phase Extraction (SPE) 412
5. **Mass Spectrometry in Lipidomics Research** 413
5.1 Ionization Modes 413
5.2 Advances in Mass Spectrometry for Lipidomics 419
6. **Chromatographic Separations in Lipidomics Research** 420
6.1 Thin-Layer Chromatography (TLC) 421
6.2 High-Performance and Two-Dimensional TLC 422
6.3 High-Performance Liquid Chromatography 422
6.4 Recent Advances in LC/MS Techniques for Lipidomics 425
6.5 Multidimensional Liquid Chromatography (MDLC, LC × LC) 428
6.6 Gas Chromatography 429
7. **Conclusions and Future Perspectives** 433
References 434

9. **Food Forensics**

Maurizio Aceto

1. **Introduction** 442
1.1 Definitions 445
1.2 The Role of Chemometrics in Food Forensics 448
1.3 The Principal Foods Subjected to Fraud 449
2. **Mass Spectrometric Analysis for Food Forensics** 460
2.1 Analytical Methods in Food Forensics 460
2.2 Isotope Ratios as Chemical Markers 461
2.3 Trace Elements as Chemical Markers 476
2.4 Molecular Ions as Chemical Markers 486
3. **Conclusions and Future Trends** 498
Acknowledgments 499
References 499

10. **Emerging Contaminants**

Julián Campo and Yolanda Picó

1. **Introduction** 515
2. **Emerging Contaminants** 518
3. **Extraction Techniques** 521
4. **Mass Spectrometry Analyzers** 524

5. **Gas Chromatography-Mass Spectrometry (GC-MS)** 528
   5.1 GC-MS/MS ($GC-MS^2$) 539
   5.2 Comprehensive GC (GC × GC) 543
6. **Liquid Chromatography-Mass Spectrometry (LC-MS)** 545
   6.1 LC-MS/MS ($LC-MS^2$) 555
   6.2 LC-High Resolution Mass Spectrometry (LC-HRMS) 558
   6.3 Comprehensive LC (LC × LC) 561
7. **Other Methods** 563
8. **Conclusions** 568
   Acknowledgments 569
   References 569

11. Engineered Nanomaterials in the Food Sector

Ralf Greiner, Volker Gräf, Anna Burcza, Birgit Hetzer, Johanna Milsmann and Elke Walz

1. **Introduction** 580
2. **Food Nanotechnology** 581
3. **Naturally Occurring Nanomaterials and Those Derived from Conventional Food Processing** 582
4. **Fabrication of Nanomaterials** 583
5. **Food Contact Materials** 584
   5.1 Food Packaging Systems 584
   5.2 Coatings for Kitchen Utensils and Processing Equipment 588
   5.3 Nanofiltration Membranes 589
6. **Nanosized Food Ingredients and Nanosized Delivery Systems** 589
   6.1 Nanosized Food Ingredients 589
   6.2 Nanosized Delivery Systems 590
7. **Structuring of Foods** 592
8. **Nanosensors** 593
9. **Safety Aspects of Nanomaterials in Food** 593
10. **Legal Obligation** 595
11. **Analytical Approaches to Characterize Nanoparticles in Food** 595
   11.1. FFF-ICP-MS 600
   11.2. Single-Particle ICP-MS 602
12. **Conclusions** 606
   References 608

12. Food Pathogens

Isin Akyar

1. **Introduction** 618
2. **Common Foodborne Pathogens (Spoilage and Pathogenic Bacteria)** 618
   2.1 Spoilage Bacteria 618
   2.2 Pathogenic Bacteria 619

3. **Diagnosis of Microorganisms** 620
4. **Mass Spectrometry in Microbial Analysis** 620
4.1 Ionization Source 622
4.2 Mass Analyzer 623
4.3 Ion Detector 624
5. **Identification of Food Pathogens by Mass Spectrometry** 624
5.1 Matrix-Assisted Laser Desorption Ionization Time-of-Flight-Mass Spectrometry (MALDI-TOF-MS) 624
5.2 Electrospray Ionization Mass Spectrometry (ESI-MS) 636
5.3 PCR Coupled to Mass Spectrometry Using Electrospray Ionization (PCR/ESI-MS) 638
5.4 Bioaerosol Mass Spectrometry 640
5.5 Fractionation Methods Coupled with Mass Spectrometry: GC-MS and Pyrolysis–GC-MS 641
5.6 Capillary Electrophoresis-MS 642
5.7 Liquid Chromatography MS 642
5.8 Surface Enhanced Laser Desorption Ionization-Mass Spectrometry (SELDI-MS) 644
5.9 MS Data Analysis Methods 645
6. **Conclusions and Future Aspects** 646
References 648

13. Foodomics

Koichi Inoue and Toshimasa Toyo'oka

1. **Introduction** 654
2. **Definition of Foodomics** 654
3. **General Flowchart on Foodomics of MS** 655
4. **Food Matrices** 656
4.1 Food Metabolites Related to Genes 657
4.2 Food Metabolites Related to the Impact of Stress 658
4.3 Food Metabolites Related to Nutritional Status 658
4.4 Food Metabolites Related to Quality Grade 659
4.5 Food Metabolites Related to Environmental Conditions 659
5. **Sample Preparations** 660
5.1 Liquid–Liquid Extraction, Solid-Phase Extraction, and Dispersive Solid-Phase Extraction 660
5.2 SPME and Similar Methods 661
5.3 Centrifugal Ultrafiltration 662
6. **Separation** 663
6.1 GC Techniques 664
6.2 LC Techniques 664
6.3 CE Techniques 666
6.4 Others Separating Techniques 667

7. MS Detection 669
7.1 MALDI-MS Techniques 669
7.2 Direct MS Techniques 670
7.3 MS Combined with Separation Techniques 670
8. Data Analysis 671
9. Data Assessment 674
10. Advanced Future Foodomics Approach with MS 676
10.1 Advanced Foodomics Approach for Human Response by Foods 676
10.2 Advanced Foodomics Approach for Food Microbiology 677
10.3 Advanced Foodomics Approach for Food Safety and Quality 677
List of Abbreviations 679
References 680

Index 685

Contributors to Volume 68

Maurizio Aceto, Dipartimento di Scienze e Innovazione Tecnologica, Università degli Studi del Piemonte Orientale "Amedeo Avogadro", Alessandria, Italy

Isin Akyar, Department of Medical Microbiology & Acibadem Labmed Clinical Laboratories, Acibadem University School of Medicine, Istanbul, Turkey

Elisabetta De Angelis, Institute of Sciences of Food Production (ISPA-CNR), Bari, Italy

Damià Barceló, Institute of Environmental Assessment and Water Research (IDAEA-CSIC), Barcelona; and Catalan Institute of Water Research (ICRA), Girona, Spain

Marco Beccaria, "Scienze del Farmaco e Prodotti per la Salute" Department, University of Messina, Messina, Italy

Anna Burcza, Department of Food Technology and Bioprocess Engineering, Federal Research Institute of Nutrition and Food, Max Rubner-Institut, Karlsruhe, Germany

Francesco Cacciola, "Scienze dell'Ambiente, della Sicurezza, del Territorio, degli Alimenti e della Salute" Department, University of Messina, Messina, Italy

Julián Campo, Environmental Forensic and Landscape Chemistry Research Group. Desertification Research Centre - CIDE (Spanish Council for Scientific Research, University of Valencia, Generalitat Valenciana) Valencia, Spain

Anna Laura Capriotti, Dipartimento di Chimica, Università di Roma "La Sapienza", Rome, Italy

Chiara Cavaliere, Dipartimento di Chimica, Università di Roma "La Sapienza", Rome, Italy

Riccardo Zenezini Chiozzi, Dipartimento di Chimica, Università di Roma "La Sapienza", Rome, Italy

Valentina Colapicchioni, Dipartimento di Chimica, Università di Roma "La Sapienza", Rome, Italy

Georgios P. Danezis, Food Science & Human Nutrition, Agricultural University of Athens, Athens, Greece

Paola Donato, "Scienze dell'Ambiente, della Sicurezza, del Territorio, degli Alimenti e della Salute" Department, University of Messina, Messina, Italy

Paola Dugo, "Scienze del Farmaco e Prodotti per la Salute" Department, University of Messina, Messina, and Integrated Research Center, University Campus Bio-Medico of Rome, Roma, Italy

Marinella Farré, Institute of Environmental Assessment and Water Research (IDAEA-CSIC), Barcelona, Spain

Constantinos A. Georgiou, Food Science & Human Nutrition, Agricultural University of Athens, Athens, Greece

Volker Gräf, Department of Food Technology and Bioprocess Engineering, Federal Research Institute of Nutrition and Food, Max Rubner-Institut, Karlsruhe, Germany

Ralf Greiner, Department of Food Technology and Bioprocess Engineering, Federal Research Institute of Nutrition and Food, Max Rubner-Institut, Karlsruhe, Germany

Birgit Hetzer, Department of Food Technology and Bioprocess Engineering, Federal Research Institute of Nutrition and Food, Max Rubner-Institut, Karlsruhe, Germany

Herbert H. Hill Jr., College of Life Science, Tarim University, Alar, Xinjiang, China and Department of Chemistry, Washington State University, Pullman, WA, USA

Koichi Inoue, Laboratory of Analytical and Bio-Analytical Chemistry, School of Pharmaceutical Sciences, University of Shizuoka, Suruga-ku, Shizuoka, Japan

Aldo Laganà, Dipartimento di Chimica, Università di Roma "La Sapienza", Rome, Italy

Wenjie Liu, College of Life Science, Tarim University, Alar, Xinjiang, China and Department of Chemistry, Washington State University, Pullman, WA, USA

Gianfranco Mamone, Institute of Food Sciences (ISA-CNR), Avellino, Italy

Johanna Milsmann, Department of Food Technology and Bioprocess Engineering, Federal Research Institute of Nutrition and Food, Max Rubner-Institut, Karlsruhe, Germany

Linda Monaci, Institute of Sciences of Food Production (ISPA-CNR), Bari, Italy

Luigi Mondello, "Scienze del Farmaco e Prodotti per la Salute" Department, University of Messina, Messina; Integrated Research Center, University Campus Bio-Medico of Rome, Roma, and Chromaleont s.r.l., c/o "Scienze del Farmaco e Prodotti per la Salute" Department, University of Messina, Messina, Italy

Yolanda Picó, Food and Environmental Safety Research Group, Faculty of Pharmacy, University of Valencia, Valencia, Spain

Rosa Pilolli, Institute of Sciences of Food Production (ISPA-CNR), Bari, Italy

Roberto Samperi, Dipartimento di Chimica, Università di Roma "La Sapienza", Rome, Italy

Toshimasa Toyo'oka, Laboratory of Analytical and Bio-Analytical Chemistry, School of Pharmaceutical Sciences, University of Shizuoka, Suruga-ku, Shizuoka, Japan

Elke Walz, Department of Food Technology and Bioprocess Engineering, Federal Research Institute of Nutrition and Food, Max Rubner-Institut, Karlsruhe, Germany

Series Editor's Preface

I am delighted to introduce this new volume on Food Safety and Quality in the *Comprehensive Analytical Chemistry* Series edited by my friend and colleague, Professor Yolanda Picó. Professor Picó previously edited Volume 58 on a similar topic, Food Contaminants and Residue Analysis, published in 2008. Then, the first lines of my Preface already indicated that food safety has become a key issue in our society, due to globalization where food is traveling around the world. Under these circumstances quality issues are always on the agenda and their importance has steadily increased since then.

In the series we have devoted much attention to Food Quality. We published a volume on pesticide residues in food in Volume 43, edited by my old friend and colleague Professor Amadeo R Fernández-Alba, already 10 years ago, back in 2004. Amadeo was also the editor of Volume 58, in 2012, on Pesticides and other Food Contaminants. Volumes 63 and 64 on the Fundamentals and Applications of Advanced Omic Technologies, edited by Carolina Simó, Alejandro Cifuentes, and Virginia García Cañas, reported on the applications of these technologies to food science and food safety among other related applications to cancer and clinical research.

The volume that you have now in your hands is an excellent addition to the series, reinforcing our interest in keeping you updated on this essential research topic. The book includes chapters on the most advanced mass spectrometric techniques, such as different types of ionization forms and analyzers, like isotopic, ambient ionization, and ion mobility as well as novel applications in the broad field of food safety. There is a comprehensive, state-of-the-art list of emerging contaminants and materials in the application chapters that include food allergen, lipidomics, food forensics, engineered nanomaterials, food pathogens, and foodomics. In total the book comprises 13 chapters with a good balance between the description of the advanced mass spectrometric techniques and new applications.

Finally I would like to thank Yolanda for the amount of work, time, and expertise she has devoted as editor of and contributor to this book. I would like to acknowledge as well the other well-known authors for their contributions in compiling such a world-class and timely book that will be of help to newcomers, PhD students, and senior researchers. The book should serve as a tool to improve food quality in our globalized market of goods exchange.

D. Barceló
IDAEA-CSIC, Barcelona and ICRA, Girona, February 5, 2015
Editor in Chief of the *Comprehensive Analytical Chemistry* Series, Elsevier

Preface

The scope, relevance, and degree of food safety and quality have never been higher than in today's global world. Confidence in the safety and integrity of the food supply is an important requirement for consumers. Recent advances in mass spectrometry technologies provide tremendous opportunities for a range of food-related applications. However, the distinctive characteristics of food, such as the wide range of different components and their extreme complexity, give rise to enormous difficulties. In this book, recent advances in mass spectrometry-based techniques and their applications in food safety and quality, as well as the major challenges associated with implementing these technologies for more effective identification of unknown compounds, food profiling or candidate biomarker discovery are summarized and discussed. Recent developments in sample preparation, matrix treatment, and traditional biochemical techniques, in combination with substantial improvements in MS platforms, have enabled the study of new areas within food quality and safety, such as food pathogens, biopolymers (lipids or proteins), components produced by genetically modified organisms, naturally occurring antibiotics or allergens, food authenticity and adulteration, emerging contaminants and nanomaterials. Despite these significant advances and efforts, major challenges associated with the dynamic range of measurements and extent of food coverage, confidence of compound identifications, quantitation accuracy, analysis throughput, and the robustness of present instrumentation must be addressed before mass spectrometry platforms that are suitable for efficient food quality and safety applications can be routinely implemented.

The book is structured in two parts; the first describes the basic and essential topic—the current status of mass spectrometry within food safety and quality—and will identify those issues that are promising but not fully resolved yet. The chapters of this part introduce a number of hot topics, including high-resolution mass spectrometry, imaging mass spectrometry, isotope ratio-mass spectrometry, ambient ionization techniques, and high-performance ion-mobility spectrometry. The second part describes the main and most innovative applications in food safety and quality that are current in mass spectrometry, and the difficulties in porting them to this technique. These applications include proteomics, lipidomics, and emerging contaminants with special detail to nanomaterials, food forensics and foodomics. The upcoming developments of mass spectrometry in food quality and safety are presented and the future of this line of work is discussed.

The proposed book has some idiosyncratic features that make it more interesting to read, such as new topics in food quality and safety and descriptions of the latest advances in mass spectrometry. This book aims to serve as a general reference for postgraduate students, who are not expert in many of the emerging mass spectrometry platforms and their applications, as well as a practical reference for a wide range of specialists: biologists, biochemists, microbiologists, food chemists, toxicologists, chemists, agronomists, food hygienists, and everybody who need to use mass spectrometry for evaluating food quality and safety. The different advances and applications described here are not only emerging now but will also be critical in the future for assuring an affordable, safe, and available food supply.

I thank all the authors who have contributed to the successful completion of this book for their labor and dedication in producing high quality work. Time is a precious resource and all of us are beyond our capacity, and writing a chapter is not usually high on the priority list. Therefore, the remarkable effort of the contributors, all of them specialists in their fields, is greatly appreciated. I also acknowledge the Elsevier staff for their helpful assistance through the development of this project, especially to Derek Coleman for his incredible patience when I missed nearly every deadline—it has been a pleasure to work with all of you. Prof. Dr Damià Barceló of CSIC-Spain and editor of this Elsevier series deserves special mention for the opportunity to publish this work and his encouragement. Many, many thanks!

Last but not least, I hope that you find reading and studying the chapters as exciting as I found the writing and editing process. I trust that you will find this book an invaluable source of information.

Yolanda Picó
University of Valencia, Spain
February 2015

Part I

Advanced Mass Spectrometry Approaches and Platforms

Chapter 1

Mass Spectrometry in Food Quality and Safety: An Overview of the Current Status

Yolanda Picó
Food and Environmental Safety Research Group, Faculty of Pharmacy, University of Valencia, Valencia, Spain
E-mail: Yolanda.Pico@uv.es

Chapter Outline

1. Introduction 3
2. Current Instrumentation and Operation in Mass Spectrometry 5
3. Matrix-Assisted Laser Desorption Ionization Coupled to Time-Of-Flight Mass Spectrometry (MALDI-TOF) 13
4. Inductively Coupled Plasma Mass Spectrometry (ICP-MS) 21
5. Mass Spectrometry (MS) Combined with Chromatography or Related Techniques 26
5.1 Liquid Chromatography–Mass Spectrometry (LC-MS) 26
5.1.1 Target Analysis 40
5.1.2 Nontarget Analysis 44
5.2 Gas Chromatography–Mass Spectrometry (GC-MS) 48
5.2.1 Target Analysis 48
5.2.2 Target and Nontarget Metabolomics 57
5.2.3 Profiling of Volatile Compounds 58
5.3 Capillary Electrophoresis–Mass Spectrometry (CE-MS) 61
6. Conclusions and Future Trends 65
References 66

1. INTRODUCTION

According to the World Health Organization (WHO) [1], the terms food safety and food quality can sometimes be confusing. Food safety refers to all those hazards, whether chronic or acute, that may make food injurious to the health of the consumer. It is not negotiable. Quality includes all other attributes that influence a product's value to the consumer. This comprises negative attributes such as spoilage, contamination with filth, discoloration, off-odors and positive attributes such as the origin, color, flavor, texture, processing method of the food.

Comprehensive Analytical Chemistry, Vol. 68. http://dx.doi.org/10.1016/B978-0-444-63340-8.00001-7

The importance of food safety and quality today is clear considering the food and drink industry is a fundamental pillar of the European Union (EU) economy with 500 million of consumers. The sector is among the top three manufacturing industries in terms of turnover (1016 € millions in 2011–2012) and employment (4 million people in approximately 280,000 companies in 2011–2012) in several Member States. It ranks first in France, Spain, the UK, Denmark and Belgium [2]. Transforming today's challenges and demands within the sector (e.g., higher quality and safety standards at affordable prices to the consumer, sustainable crop management, environmental compatibility, etc.) into opportunities is required to increase the competitiveness of the EU food and drink industry.

Quality and safety testing of food is an area of growing importance, where accurate analytical results are critical, be that for exporters, importers, or government bodies [3]. As a result of the rising globalization levels, food suppliers, producers, manufacturers, and regulatory agencies are all facing greater pressure than ever before to test more food products for more contaminants and attributes and quantify their presence at lower levels with greater accuracy—and in less time [4–6]. The utilization of mass spectrometry has grown rapidly within the sector because it is now recognized as an extremely specific and exceptionally sensitive technique for testing food products with superior accuracy and higher throughput [7,8]. The ability to simultaneously monitor a range of contaminants and/or natural components greatly improves throughput, and new levels of sensitivity allow more simplified sample preparation protocols [9,10].

Several reviews give an idea of the complexity of some aspects of the applications of mass spectrometry (MS) in food. Some of them presented the most recent applications of MS-based metabolomic approaches for food quality, safety, and traceability assays [11,12]. Ibañez et al. [13] reported an overview of the current developments and applications of capillary electrophoresis–mass spectrometry (CE-MS) as an analytical platform for Foodomics. The highest number of reviews covers the field of liquid chromatography–mass spectrometry (LC-MS) in food analysis [14] and food safety [3]. Mohamed et al. [15] evaluated the contribution and the potentialities of MS-based techniques to ensure the absence of chemical contaminants in food-based products. Recently, the state of the art of MS within food quality and safety was already reviewed, including matrix-assisted laser desorption ionization coupled to time-of-flight mass spectrometry (MALDI-TOF-MS) and ambient ionization MS for direct food analysis [16]. The revolution of "omic" sciences introduced within MS integrated high-throughput approaches to address the understanding of the biochemical systems and their dynamic evolution as well [17]. Applications of modern MS-based techniques in proteomics, all ergonomics, glycomics, metabolomics, lipidomics, food safety and traceability were also surveyed [18]. These advances in MS provide the capacity to screen for more analytes at lower levels, with greater accuracy, and in less time.

This chapter presents an overview of the current status of MS techniques applied in food safety and quality, as well as their future prospects that will be developed in other chapters of the book. The general characteristic steps of MS-based techniques and the main chemical analytical procedures applied in the field are introduced. Furthermore, applications of the different techniques and ionization sources are discussed.

2. CURRENT INSTRUMENTATION AND OPERATION IN MASS SPECTROMETRY

MS appeared at the beginning of the twentieth century but current advances in this technique are the result to face the inherent challenge in it—interfacing atmospheric pressure ionization sources (760 torr) to analyzers maintained at 10^{-6} to 10^{-11} torr, a remarkable pressure differential of more than nine orders of magnitude [19,20]. Nowadays most common methods of sample introduction are direct insertion with a probe or plate commonly used with MALDI, direct infusion or injection into the ionization source such as electrospray ionization (ESI) or electron impact (EI) ionization [21]. These ionization sources are schematized in Figure 1.

Prior to the 1980s, EI was the primary ionization source for MS analysis. EI energetic electrons interact with gas phase atoms or molecules to produce ions. The ionization process often follows predictable cleavage reactions that give rise to fragment ions which, following detection and signal processing, convey structural information about the analyte [22,23]. However, EI limited analyst to small molecules well below the mass range of common food components, such as proteins, lipids and polysaccharides. This limitation motivated scientists during 1980s such as John B. Fenn, Koichi Tanaka, Franz Hillenkamp, Michael Karas, Graham Cooks, and Michael Barber to develop ionization techniques, including fast atom/ion bombardment (FAB), particle beam (PB), MALDI, and ESI [23–25]. Although FAB and PB have already fallen into disuse, these techniques have revolutionized food analyses, especially for large molecules such as protein. Among them, ESI and MALDI have clearly evolved to be the methods of choice for food analysis and they are still evolving to the ambient ionization and imaging techniques [25]. In ESI, high voltage is applied to a liquid supplied through an emitter (usually a glass or metallic capillary). The charged droplets, generated at the exit of the electrospray tip, pass down a pressure gradient and a potential gradient toward the analyzer region of the mass spectrometer. MALDI is a soft ionization technique in which a short laser pulse, instead of continuous laser, of nitrogen gas usually around 237 nm is used to ionize molecules. ESI is also a soft ionization technique that is typically used to determine the molecular weights of either small molecules or biological macromolecules [25,26].

During the 1990s, research moved from the ionization source to the mass analyzer. Quadrupoles (single and triple), three-dimensional ion traps (^{3}DIT), linear ion traps (LTQ), time of flight (TOF), quadrupole time of flight (QqTOF), and different types of orbitrap mass analyzers have been developed and/or undergone numerous modifications/improvements [4,26].

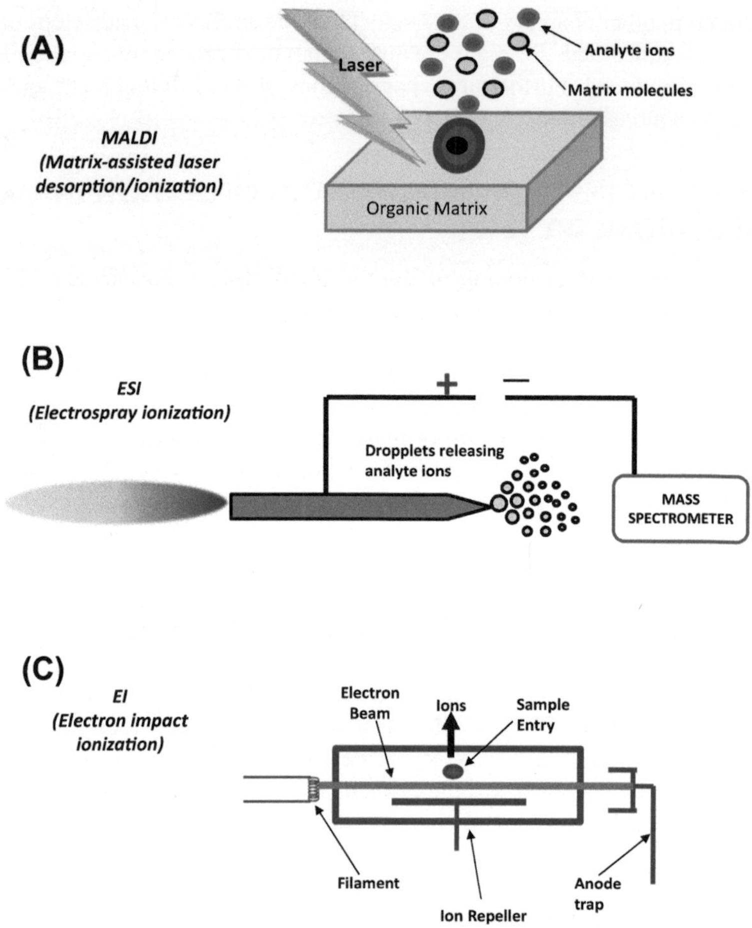

FIGURE 1 Scheme of the most common ionization sources (A) MALDI; (B) ESI; (C) EI ionization.

These mass analyzers can be distinguished into (1) quadrupole-base, (2) ion traps, and (3) TOF. The fundamentals of the different mass spectrometers is beyond the scope of this chapter, and the reader is referred for a good description of the theory behind them [8,11,12,14,16,27]. However, schemes of the most used ones are outlined in Figure 2 Quadrupoles consists of four cylindrical rods, set parallel to each other. In a quadrupole mass spectrometer the quadrupole is the component of the instrument responsible for filtering sample ions, based on their mass-to-charge ratio (*m*/*z*). Ions are separated in a quadrupole based on the stability of their trajectories in the oscillating electric fields that are applied to the rods. Only ions of a certain *m*/*z* will reach the detector for a given ratio of voltages: other ions have unstable trajectories and will collide with the rods [14]. These mass analyzers are little sensitive in scan mode (MS is obtained for a full range of *m*/*z*) and

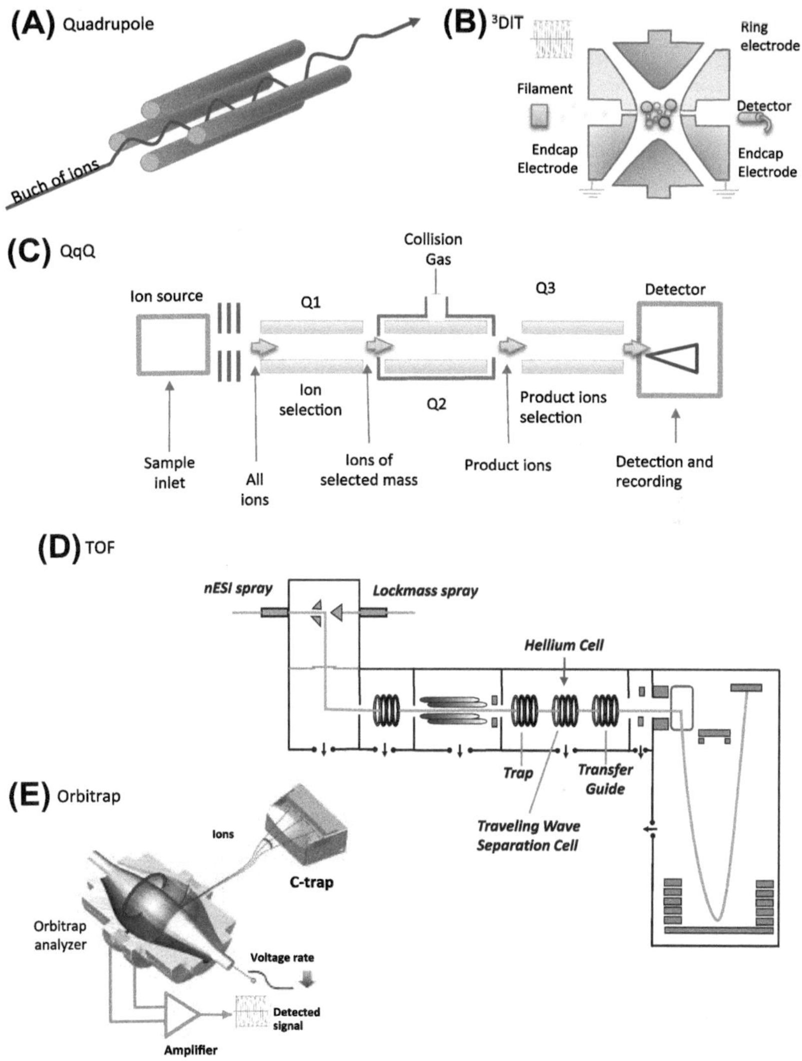

FIGURE 2 Scheme of several mass analyzers (A) Quadrupole, (B) ^{3}D ion trap, (C) triple quadrupole, (D) quadrupole time-of-flight and (E) Orbitrap.

then, commonly they work in selected ion mode (SIM). A linear series of three quadrupoles is known as a triple quadrupole (QqQ) mass spectrometer. The first (Q1) and third (Q3) quadrupoles act as mass filters, and the middle (Q2) quadrupole is employed as a collision cell. This mass analyzer is nowadays widely used to carry out a number of determinations within food safety and quality [15,28,29].

There are several trapped-ion mass analyzers: ^{3}DIT ("dynamic" traps), LTQ, and Fourier transform ion cyclotron resonance (FT-ICR) mass spectrometers ("static" traps) [18]. All these instruments operate by storing ions in the trap and manipulating the ions by using direct current (DC) and radio frequency (RF) electric fields in a series of carefully timed events. In a ^{3}DIT, ions are dynamically stored in a three-dimensional quadrupole ion storage device. The RF and DC potentials can be scanned to eject successive *m/z* from the trap into the detector (mass-selective ejection). In a LTQ, ions are confined radially by a two-dimensional RF field, and axially by stopping potentials applied to end electrodes. These instruments can work as an either QqQ or ion trap [30]. Working as QqQs are today's most sensitive and selective method. Furthermore, in ion-trap operation, different scan modes can be used to get complementary, qualitative information about the sample [21,22]. The LTQ scan modes used third quadrupole as an ion trap; ions are stored in the trap being scanning out, giving increased sensitivity to full scan experiments [e.g., enhanced MS, enhanced multicharge, enhanced product ion (EPI), time delayed fragmentation, etc.]. MS can be performed in the ion trap, allowing to get more information on the sample. Other interesting feature is the ability of the instrument to work in information-dependent acquisition (IDA) that automatically runs experiments based on results obtained from previous experiments. The Fourier transform MS or more precisely FT-ICR MS is based on ions that move in a circular path in a magnetic field. The cyclotron frequency of the ion's circular motion is mass-dependent. By measuring the cyclotron frequency, the ion's mass can be determined. Orbitrap instruments are similar to FT-ICR mass spectrometers. Ions are electrostatically trapped in an orbit around a central, spindle shaped electrode. The electrode confines the ions so that they both orbit around the central electrode and oscillate back and forth along the central electrode's long axis. This oscillation generates an image current in the detector plates which is recorded by the instrument [31–34]. The stand-alone orbitrap provides high mass resolution (>15,000 FMWH) and high mass accuracy (<2 ppm) without mass selection [35]. The LTQ–Orbitrap instruments have significant features, such as a dual linear ion trap, and a more efficient higher energy collisional dissociation (HCD) cell interfaced directly to the C-trap. These instruments offer a wide range of fragmentation modes that can help to solve different analytical problems [36,37]. More recently, the combination of a quadrupole with an orbitrap (Q-Orbitrap) also showed the possibility to perform fast HCD fragmentation [37].

A TOF-MS measures the mass-dependent time needed for the ions of different masses to move from the ion source to the detector [22]. This is the simplest mass spectrometer since it consists of a tube where the lighter ions travel faster than the heavier. The ions are accelerated with a fixed amount of kinetic energy in order to travel down the flight tube. The ions are separated according to their *m/z* ratio [33]. This requires a well-defined starting time (time at which the ions leave the ion source). Recent improvements in TOF technology, including orthogonal acceleration, ion mirrors (reflection), and high speed electronics, have significantly improved TOF resolution. Today, QqTOF is the most popular hybrid instrument. The configuration can be regarded either as the addition of a mass-resolving quadrupole and collision cell to an ESI-TOF, or as the replacement of the Q3 in a QqQ by a TOF mass spectrometer. From any viewpoint, the benefits are high sensitivity, mass resolution, and mass accuracy of the resulting tandem mass spectrometer in both precursor (MS) and product ion (MS/MS) modes [23,24,33].

In that context, MS-based techniques coupled or not to chromatography offer a vast panel of features such as sensitivity, selectivity, quantification at trace levels, and/or structural elucidation. There are a large number of chromatographic techniques, interfaces, or ionization sources and mass analyzer that can be combined among them helping to configure food quality and safety analysis. Table 1 summarizes these possible combination as well as their main applications within the field. The versatility of mass spectrometry has created many different ways to work.

A key distinction in food safety and quality analysis platforms is between nontargeted identification and targeted quantitative methods [38–41]. Targeted analysis is based on determining a list of known analytes, often using reference standards. Typical methods are validated prior to analysis of real samples [42–44]. Target analysis is excellent for providing sensitive and selective results for known compounds [45]. Although essentially any MS instruments that use chromatographic separation coupled to MS, such as gas chromatography–quadrupole MS (GC–qMS) or high-resolution MS (GC–HRMS), liquid chromatography–QqQ MS (LC–MS/MS), are popular choices for target analysis; particularly, QqQ and LTQ MS instruments are most common choices [46–48]. Both use the Q1 to select targeted precursor ions for further fragmentation in a collision cell. Specified fragment ions then are detected either by the Q3 or by the LTQ. Although both instruments can execute full scan detection of all product ions, the most sensitive and repeatable measurements are achieved when the Q3 targets selected product ions in multiple selected reaction monitoring (SRM) [39,49,50]. However, by design, target analysis filters out all ions not corresponding to the target analyte(s), making it inappropriate for new compound identification. The advantages of QqQ and LTQ mass spectrometers provide measurements of a wide dynamic range ($>10^5$), high sensitivity, and low variation for the SRM transitions. A limitation of these instruments is the relatively low resolution of precursor m/z measurements, which may allow interference from nominally isobaric background compounds in food. Some QqQ mass spectrometers as well as the LTQ have hyperbolic quadrupoles able to work in enhanced mass resolution mode (mass resolution at $0.1 m/z$ FWHM, full with half maximum). However, this mode can only be used when a short number of compounds are simultaneously analyzed [51,52]. QqQ instruments are also limited by their duty cycle, the rate at which transitions can be sampled with an acceptable signal-to-noise ratio [43,53].

On the contrary, nontarget analysis offers the possibility of detecting both unexpected and unknowns. Commonly, nontarget analysis is based on the use

TABLE 1 Different mass Spectrometric Approaches and Applications within Food Quality & Safety

Previous Chromatographic Techniques	Source of Ionization	MS	Applications
LC	EI	Q	• Target analytes (natural components and toxics) • Detection of food adulterations • Characterization of food allergens • Structural modifications of food constituents • Identification and characterization of microorganisms • Determination of mineral and trace elements • Determination of heavy metals • Determination of geographical origin by multi-element fingerprinting • Isotope analysis • Profiling determination of volatile compounds • Nontarget metabolomics • Widely target metabolomics • Food fingerprinting • proteomics and peptidomics • Lipidomics • Foodomics • Identification and characterization of nontarget compounds
	CI	QqQ	
	MALDI	LTQ	
GC	ESI	IT	
	APCI	TOF	
	APPI	QqTOF	
CE	ICP	Orbitrap	
	DART	Q-Orbitap	
	DESI	LTQ-Orbitrap	

APCI, atmospheric pressure chemical ionization; APPI, atmospheric pressure photo-ionization; CE, capillary electrophoresis; CI, chemical ionization; DART, direct analysis in real time; DESI, desorption electrospray; EI, electron impact ionization; ESI, electrospray; GC, gas chromatography; LC, liquid chromatography; LTQ, linear ion trap; MALDI, matrix assisted laser desorption/ionization; Q, single quadrupole; QqTOF, quadrupole time-of-flight; QqQ, triple quadrupole; TOF, time-of-flight.

of HRMS, such as TOF or Orbitrap, for food analysis. These analyses are largely driven by the advantages of using the full-scan acquisition mode with high sensitivity, combined with high resolving power (>30,000 FWHM) and accurate mass measurement (1–5 ppm). This powerful analytical tool allows the development of analytical strategies that combine (1) target analysis (determination of specific priority analytes with available standards and for which accurate mass, retention time window, isotopic pattern, and fragments are reliable identification tools); (2) nontarget screening analysis based on an accurate-mass customized database of known precursor molecules, some diagnostic fragment or product ions and isotopic pattern; and (3) nontarget with no selection of analytes to be searched. Here, the ion chromatograms of the individual posttarget analytes are extracted from the raw data of the entire full-scan spectra and then verified. The hyphenation of these instruments to achieve MS/MS is of help in this difficult task. The QqTOF, Q-Orbitrap, and LTQ-Orbitrap are able to obtain product ion mass spectrum. Furthermore, the possibility of work using data- or information-dependent acquisition (DDA or IDA)—different names for the same feature—improve the possibilities of these equipments, because the instrument uses the information in a dependent experiment to make decisions about the next step of the experiment automatically, without the input of the analyst.

To identify unexpected or unknown compounds, full-scan MS spectra must be examined by molecular feature algorithm for masses whose abundances build chromatographic and mass spectrometric peak profiles. Given the exact mass, along with the isotope pattern, it is then possible to use a formula generator to compute a molecular formula of the feature. The allocation of the molecular structure of the compound (feature) can be determined with the help of databases. Database hits are then verified by means of reference substances and MS/MS experiments. For compounds, not being identified in this way, the structure elucidation is commonly done by sophisticated MS scan techniques (e.g., MS^n experiments) supplemented by derivatization reactions and hydrogen–deuterium exchange combined with information from other analysis methods. Due to the hard work required for identification, the selection of those compounds recognized by nontarget screening to be submitted for further identification is usually based on signal intensity. The signal intensity represents the product of the concentration and the ionization efficiency of the compound, which is perhaps why relevant compounds with high concentration but small ionization efficiency are often not considered. Due to the number of compounds that are usually detected in most food samples, however, it is necessary to focus on the identification of only relevant compounds [40].

The nontarget screening and nontarget identification are also linked to the “omic” concept and to different ways in order to analyze or characterize the sample, such as metabolomics, profiling, or fingerprinting. By definition, metabolomics describes the scientific study of small molecules, the metabolites, of a biological system based on comprehensive chemical analysis (omics technologies) with the aim to detect and identify as many substances as possible

[40]. Food profiling focuses on the analysis of a group of metabolic products in combination with a certain metabolic pathway or a class of compounds (multicomponent analysis). This strategy is based on the prior knowledge of the analyst concerning the biological system (commonly obtained by nontarget analysis) and, hence, is rather performed by targeted analysis and optional subsequent multivariate data analysis. In contrast, food fingerprinting techniques do not deal with the identification of all metabolites, but on the recognition of patterns, the so-called fingerprints of the matrix.

MS has capacity to provide valuable information in diverse aspects related to food safety and quality, such as metabolites identification, contaminants determination, proteomics, lipidomics, glycomics, and foodomics. The generation of large volumes of data possesses the challenge of processing the information into a meaningful and practical way. To process data efficiently, new software packages and algorithms are continuously being developed to improve protein identification and characterization in terms of high-throughput and statistical accuracy. Different bioinformatics tools useful for interpretation of the mass spectra information are being introduced, especially in the analysis of peptide mixtures for protein identification by MALDI-MS and ESI-MS/MS. A list of these tools, as well as their references, is provided in a recent review of MS/MS analysis and validation strategies [54]. Bioinformatics toolbox provides a set of functions for mass spectrometry data analysis. These functions enable preprocessing, classification, and marker identification from MALDI, LC-MS, GC-MS, and CE-MS data and include baseline correction, smoothing, calibration, and resampling. Raw spectral data can be aligned using the *m/z* axis and perform retention-time alignment on LC-MS and GC-MS data (Figure 3). Computational proteomics for protein quantitation is one of the most exciting areas for bioinformatics researchers and it is witnessing rapid developments in computational frameworks, data standardization, and software development [55] that is also being exported to other areas of food safety and quality.

Once a data matrix has been produced from raw data, subsequent steps usually involve different forms of statistical analysis and data mining to allow the identification of samples or variables (metabolites) that capture the bulk of variation between data sets and that may represent candidates for biologically meaningful variables. Typical analyses of metabolomic data consist of two phases: initially an overview of the given data sets is generated using multivariate analysis, and then individual peaks are subsequently graded by univariable analysis. Several univariable and multivariate analyses and classification and assessment methods are widely used in analyzing MS-based metabolomics data sets.

Future trends in the area of MS that are already envisaged now are: (1) the shift from low-resolution to (ultra)high-resolution tandem mass analyzers providing high-mass accuracy, (2) the addition of a second dimension in mass spectrometry as, in the case of ion mobility (IM) MS, (3) the ambient ionization techniques, and (4) the imaging mass spectrometry (IMS). The enhancement of HRMS/MS is clear considering the explosion of "omic" techniques previously presented and the role play by the nontarget analysis. The IM separation

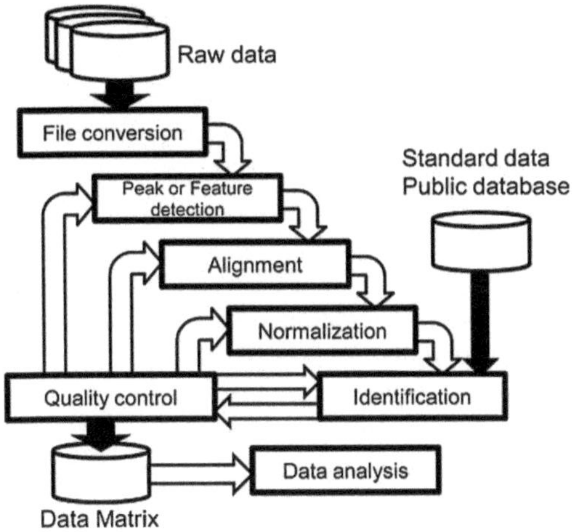

FIGURE 3 Typical processing flow of MS data in the field of metabolomics. Raw data are sequentially processed in multiple phases, including file conversion, feature detection, alignment, and normalization. Standard data and public databases that include metabolite information, such as mass spectrometric data, are used for subsequent feature identification. These processes are then assessed using quality control criteria and the previous phase is repeated if necessary. Once calibrated, the data matrix (aligned detected features across multiple data sets) can be transferred for subsequent data analysis phases. *Reproduced from Ref. [55] with permission of Bentham Science Publishers.*

is different compared to chromatographic and mass spectrometric separations, and the combination of these separation modes provides a better resolution for complex samples. Three-dimensional data set consists of retention times, drift times, and *m/z* values. Direct mass spectrometric analysis at ambient conditions without the chromatographic separation has recently been introduced. The main advantage of such approaches is the fast analysis without any (or minimum) sample preparation, which significantly increases the sample throughput. The main group is a family of ambient desorption ionization techniques, such as desorption ESI (DESI), desorption atmospheric pressure chemical ionization (APCI), desorption atmospheric pressure photo-ionization, and the direct analysis in real time (DART). IMS is designed for the determination of spatial distribution of analyte molecules on the surface, which is typically used for the spatial imaging of biomolecules in biological tissues. All these techniques will be outlined in detail in the following chapters.

3. MATRIX-ASSISTED LASER DESORPTION IONIZATION COUPLED TO TIME-OF-FLIGHT MASS SPECTROMETRY (MALDI-TOF)

MALDI—MS has a considerable impact on food safety and quality because it covers from proteomics to lipidomics [56], including polymer analysis [57] and, more recently, even low molecular weight analytes due to the introduction of matrix-less ionization techniques [e.g., desorption ionization on porous silicon (DIOS)] or new matrices such as ionic liquids, proton sponges, and metal nanoparticles. However, protein identification by peptide mass fingerprint still remains the main routine application [8]. This identification has played a relevant role in food chemistry especially in detection of food adulterations [30], characterization of safety, quality and food allergens, and investigation of protein structural modifications induced by various industrial processes that could be detrimental for food quality and safety [58–60]. Sample handling and pretreatment is a crucial step in these cases that can be very different depending on the physical state of the sample. Two major factors restricting the wider application of this technique have been throughput and cost per data point. Classic approaches are based on long, tedious, and expensive procedures (e.g., SDS gel electrophoresis separation followed by in-gel digestion). However, in recent years, novel molecular tools, the re-invention/re-application of mature technologies, and miniaturization of technology have both increased throughput and significantly reduced these costs (e.g., faster in-solution digestion of whole samples) [61–64]. Table 2 outlines a list of relevant applications recently published.

The capacity of MALDI-TOF-MS of identify proteins also offers a safe, cost-effective, and adaptable system for rapid identification of bacteria and fungi. Species differentiation of food pathogenic and spoilage bacteria is important to ensure food quality and safety. The growth of microbial contaminants in industrially produced beverages can cause turbidity, haze, and off-flavors resulting in quality loss often rendering the product undrinkable. MALDI-TOF-MS could deliver discriminative peptide mass fingerprints within minutes and could thus be a rapid and reliable tool for identification and differentiation [65]. To this end, different sample preparation methods ranging from plain cell smears to more elaborate extraction procedures including mechanical and enzymatic disruption of cells have been investigated [66,67]. Methods can be based on the analysis of whole bacterial cells that were suspended in an organic solvent or applied directly to the sample target. In a different sample preparation technique, cell extracts were obtained from intact bacterial cells by a dissolution/centrifugation step. The protocol applied for the study of cell extracts was shown to be very fast and simple, allowing the standardization of sample preparation. Furthermore, the analysis of cell extracts had several advantages with respect to the analysis of suspensions of whole bacterial cells. Thus, spectral profiles obtained from cell extracts showed less noise and more reproducible peaks as compared to spectra obtained from intact cells [68].

MALDI-TOF can also be used for DNA determination. Recently, MALDI-TOF-MS has been applied to single nucleotide polymorphism (SNP) detection by analysis of the mass of single base extension oligonucleotides specific to the SNP of interest. Sequenom® MassARRAY® MALDI-TOF-MS were tested as a medium-throughput platform for cereal varietal identification and found that MALDI-TOF-MS could produce SNP profiles specific to barley varieties [63]. A representative mass spectrum barley variety Baudin (Multiplex C) is shown in Figure 4.

TABLE 2 Selected Applications of MALDI for Food Safety and Quality

Analytes	Matrix	MS Technique	Additional Techniques/Main Results	References
Small Molecules				
PFOS and PFOA	Drinking water	DIOS-TOF-MS	Detect PFOS but not PFOA. In DIOS, only analytes are deposited on DIOS chips without using organic matrix for ionization. It was demonstrated that DIOS can be are suitable for the analysis of PFOS with the high sensitivity (1 ppt) and the quantitative analysis from 2.5 to 10 ppb, compared to the traditional MS methods such as MALDI-MS and ESI-MS.	[58]
Polymer Analysis				
Thermomechanical degradation of PET	Food contained plastic bottles	MALDI-MS	MALDI-MS data coupled to PCA analysis monitor the wt% of PETpc-btg present in PET-btg blends and its overall quality as a function of intrinsic viscosity due to thermomechanical degradation. Moreover, subtle changes in oligomeric composition were also detected.	[57]
Lipids Analysis				
Phospholipids	Olive oil fruits	MALDI-TOF/MS	TiO_2 NP was utilized as MSPD sorbent and chloroform and methanol (1/2, v/v) as eluent. The efficiency was compared to that of the classic Bligh & Dyer method. Fast lipidomic fingerprinting of olive fruits and oils for quality control.	[56]

Continued

TABLE 2 Selected Applications of MALDI for Food Safety and Quality—cont'd

Analytes	Matrix	MS Technique	Additional Techniques/Main Results	References
Protein Analysis				
Simple protocols for protein extraction	Liquid (milk) and solid (hazelnuts)	MALDI-MS	Sample handling and pretreatment can be very different depending on the physical state. Comparison of traditional SDS gel electrophoresis and in-gel digestion and fast in-solution digestion.	[8]
Water soluble proteins	Pork meat (longissimus thoracis)	SELDI-TOF-MS	Identification of candidate protein markers for meat quality. Associations between protein peaks obtained with SELDI-TOF-MS and meat quality traits, mainly water holding capacity, texture, and skatole were observed.	[59]
Protein profiling	Liver	DIOS-TOF-MS NI-TOF-MS	The DIOS™ substrate includes a thick nano-sized porous layer, a high surface concentration of fluorocarbon and silicon oxides, and superhydrophobicity. In contrast, the QuickMass™ substrate consisted of a nonporous germanium thin-film. Higher ionization efficiency is obtained by DIOS™ substrate. However, the use of germanium (QuickMass™) suppresses better background interference of mass spectra.	[60]
Drugs, lipids, proteins, and small peptides	Skin	MALDI-IMS	Curcumin can be used as a novel MALDI matrix. Versatility of curcumin is also demonstrated by the possibility to image a.	[61]
Spatial distribution of GABA	Eggplant	MALDI-IMS	The distribution of GABA, one of the functional food factors, some other amino acids, and carbohydrates in eggplant sections.	[62]

DNA Analysis				
Variety identification SNP-based markers	Cereals	MALDI-TOF-MS	MALDI-TOF applied to SNP detection by analysis of the mass of single base extension oligonucleotides specific to the SNP of interest. Accurate DNA sequence data is an important prerequisite for MALDI-TOF SNP assay design and genotyping. Sequenom® MassARRAY® MALDI-TOF-MS as a medium-throughput platform for cereal varietal identification.	[63]
Bacterial Identification				
Characterization of moderately halophilic lactic acid bacteria	Tetragenococcus halophilus and Tetragenococcus muriaticu	MALDI-TOF-MS	Cells washed with ethanol and treated with formic acid, then ionized using x-cyano-4hydroxycinnamic acid as matrix.	[64]
Spoiling yeasts	Beverages	MALDI-TOF-MS	Discriminative peptide mass fingerprints within minutes. Optimize sample preparation and measurement parameterization using three spoilage yeasts (*Saccharomyces cerevisiae* var. *diastaticus*, *Wickerhamomyces anomalus*, and *Debaryomyces hansenii*).	[65]
Identification of spoilage bacteria	Beverages	MALDI-TOF-MS	Generation of peptide mass fingerprints, which form a distinctive protein peak pattern. The influence of environmental or physiological parameters including oxygen availability, different nutrients, cell density, and growth phase were analyzed and revealed small differences in mass fingerprints.	[66]

Continued

TABLE 2 Selected Applications of MALDI for Food Safety and Quality—cont'd

Analytes	Matrix	MS Technique	Additional Techniques/Main Results	References
Spore analysis	22 Fusarium species	MALDI-IC/IS-TOF-MS	A Critical point for analysis is a proper sample preparation of spores, which increases the quality of mass spectra with respect to signal intensity and *m/z* value variations. Results show a potential to build a database on Fusarium species for accurate species identification, for fast response in the case of infections in the cornfield.	[67]
Seafood-borne pathogenic and spoilage bacteria	Seafood	MALDI-TOF-MS MALDI-IC-TOF-MS	Different sample preparation protocols were applied and compared to each other. The analysis of cell extracts by MALDI-TOF-MS was also applied to create a mass spectral library of the main pathogenic and spoilage bacteria potentially present in seafood.	[68]

DIOS, desorption/ionization on porous silicon; GABA, γ-aminobutiric acid; IC/IS, intact cell/intact spore; IMS, imaging mass spectrometry; MALDI, matrix-assisted laser desorption ionization; MSPD, matrix solid-phase dispersion; NI, nanoinitiator; PCA, principal component analysis; PFOA, perfluorooctanoic acid; PFOS, perfluoroctane sulfonate; PTE, poly(ethylene terephthalate); PETpc-btg, post-consumption bottle-grade PET; PET-btg, bottle-grade PET; SNP, single nucleotide polymorfins; SELDI, surface-enhanced laser desorption; TOF-MS, time-of-flight mass spectrometry.

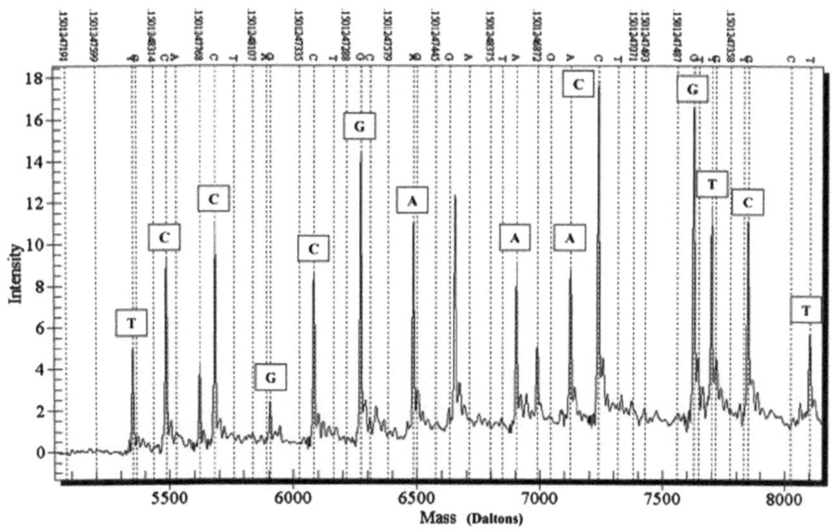

FIGURE 4 MALDI-TOF SNP mass spectrum for barley variety Baudin, Multiplex C. *Reproduced from Ref. [63] with permission of Elsevier.*

The latest developments in MALDI-MS are focused on two aspects: (1) the elimination of the matrix of ionization that often interferes with the analyte and prevents the detection of small molecules, and (2) the IMS for in situ determination as well as the direct spatial mapping of peptides in cells and tissues. Surface-enhanced laser desorption/ionization time-of-flight mass spectrometry (SELDI-TOF-MS) substitutes the chemical matrix of MALDI for an active surface, which means that matrix interference can be eliminated. SELDI combined with Protein Chip was used to obtain the soluble protein profiles of fresh pork longissimus thoracis for meat quality. The developed models explained a limited proportion of the variability; however, they point out interesting relationships between protein expression and meat quality [59]. Recent advances in matrix-free laser desorption techniques, mostly by the employment of nanochips, emerged as nano–initiator mass spectrometry (NIMS) and include commercial platforms such as DIOS and nano-assisted laser desorption ionization (NALDI™) that have already found application within food safety and quality [58,60]. In DIOS-MS, analytes are deposited onto the porous silicon surface of a chip and desorbed/ionized by the irradiation of a pulsed UV laser (337 nm), which succeeded at almost eliminating the background ion interference and offered a new method for high-speed analysis of low mass compounds [60]. Several other matrix-free laser desorption/ionization substrates—as NALDI™—have also been developed and/or commercialized.

The imaging capability of MALDI-MS represents another significant advance in the field of mass spectrometry. MALDI-IMS generates molecular profiles of analytes (commonly proteins) directly from thin food sections. Therefore, with IMS it is possible to determine the distribution of hundreds of unknown compounds with a single measurement while maintaining the cellular and molecular integrity within the tissue [61]. IMS might be a powerful tool for exploring functional food factors, investigating the specific distribution of nutrients in unused natural resources, and evaluating the quality of functional foods [62]. As an example, Figure 5(A) shows optical images of eggplant, vertically sliced eggplant, and eggplant in round

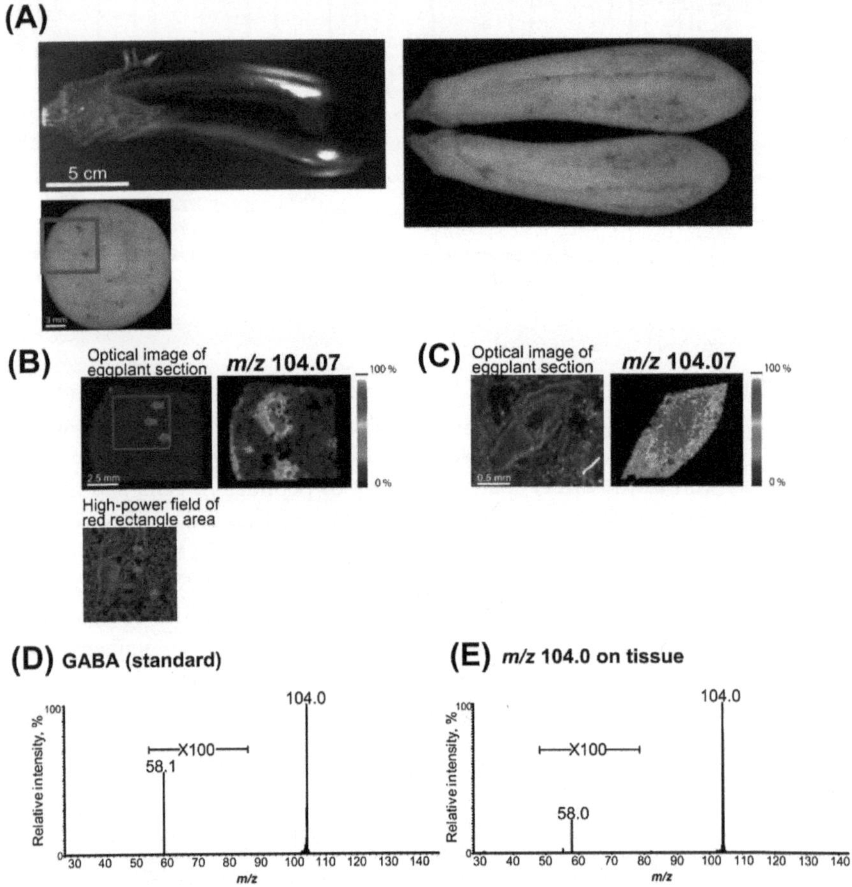

FIGURE 5 Optical images of eggplant, the results of IMS and tandem mass analyses. (A) Optical images of eggplant, vertically cut eggplant, and round-cut eggplant. A gray rectangle in a round-cut image shows the region of analyses by IMS. (B) Optical image of eggplant section and ion image of the *m/z* values at 104.07 (red (dark gray in print versions) arrows in the optical image show seed locations). Scale bar: 2.5 mm. Reproducibility was confirmed (n = 3). (C) Optical image of eggplant section and ion image of the *m/z* values at 104.07 with higher spatial resolution at 25 μm on a seed. Scale bar: 0.5 mm. (D) The tandem mass spectrum of standard GABA, and (E) *m/z* 104.0 on eggplant tissue. *Reproduced from [62] with permission of The Japan Society for Analytical Chemistry.*

slices. Figure 5(B) shows the ion image of *m/z* 104.07 [γ-aminobutiric acid (GABA)] that was detected intensely in seeds. To validate the accurate distribution of GABA, next higher-resolution imaging of a seed on the serial section was tried by setting the spatial resolution at 25 μm in positive ion mode (Figure 5(C)) [62].

4. INDUCTIVELY COUPLED PLASMA MASS SPECTROMETRY (ICP-MS)

ICP-MS is used for elemental determinations. The technique was commercially introduced in 1983 and has gained general acceptance in food quality and safety laboratories. These instruments hyphenate a high-temperature ICP source with a mass spectrometer. The multi-elemental and multi-isotopic nature of the method offers the potential to analyze a whole suite of elements in a single run, saving considerable time and money and allowing faster and more cost-effective decision-making. The ICP source converts the atoms of the elements in the sample to ions. These ions are then separated and detected by the mass spectrometer. Three main types of mass spectrometers are used in commercial ICP-MS systems: quadrupole, TOF, and magnetic sector. For overall performance and economic value, most laboratories choose an ICP-MS with a quadrupole mass spectrometer.

Nowadays, the use of ICP-MS is becoming common in food analysis laboratories. Compared to graphite furnace atomic absorption (GF-AAS) or inductively coupled plasma optical emission spectrometry (ICP-OES), this technique has some distinct advantages, including simultaneous multi-element measurement capability coupled with very low detection limits. Moreover, it offers a wider linear dynamic range which allows the determination of major and trace elements at same sample injection. Additionally, compared to ICP-OES, ICP-MS provides simpler spectral interpretation and isotopic information [69]. Other important benefits of ICP-MS include increased sensitivity, a high signal-to-noise ratio, and the flexibility to analyze almost any element in the periodic table. The method offers much lower detection limits and less interference than GF-AAS and ICP-OES.

However, ICP-MS has some limitations for food sample analysis, such as the high concentration of organic matrix that often results in matrix interferences and/or spectral interferences from polyatomic ions. The organic matter of the food samples is commonly oxidized with appropriate methods to reduce the amount of carbon. Sample digestion is still a critical step for routine determination of chemical elements in food samples that incorporates some form of "ashing." Classical dry ashing procedures require the sample to spend long times in a crucible furnace at elevated temperatures, in most cases following evaporation of the water in the sample and/or carbonizing on a hot plate. Wet digestion procedures in open vessels require the use of concentrated acids and careful monitoring of digestion for varying periods. Both methods are time-consuming and losses of analytes by volatilization are common. Alternatively, closed-vessel acid decomposition in microwave oven systems may provide faster and more efficient sample digestion. Moreover, the risk of sample contamination and losses of analytes by volatilization are practically eliminated. The determination of some elements by ICP-MS suffers from polyatomic isobaric interference. For example, the scandium signal at *m/z* 45 is affected by the possible presence of $^{13}C^{16}O^{2+}$ and $^{29}Si^{16}O^{+}$ or the chromium signal at *m/z* 53 by $^{40}Ar^{13}C^{+}$ and $^{37}Cl^{16}O^{+}$. These effects may be eliminated or minimized by the use of alternative isotopes and/or interference correction equations [70]. However, more effective systems as the dynamic reaction cell have been introduced in order to successfully eliminate those interferences. This cell is placed before the quadrupole also has a quadrupole that is filled up with reaction (or collision) gases (ammonia, methane, oxygen or hydrogen), which reacts with the introduced sample, eliminating some of the interference [70].

Table 3 briefly summarizes most remarkable types of applications of ICP-MS within food safety and quality. Therefore, it is important that all food types are monitored for contaminants and toxic elements. It is also equally important to monitor levels of nutritionally significant elements, which are present at elevated levels. Much work has been done on the analysis of inorganic elements in food with ICP-MS [69,71–74]. Discriminate the geographical origin of food based on multi-element fingerprinting (e.g., pork provenance and origin of different fruit and vegetables) can be attained by the use of ICP-MS multi-elemental analysis and the discriminatory power of inter-element association [70,75,76]. As a multi-isotopic technique, ICP-MS can also provide accurate and precise isotope ration information, important for authenticity studies or for pinpointing contamination, by verifying the origin of the foodstuff [75,76].

ICP-MS can easily be coupled with separation techniques like LC and GC, resulting in a literally matrix-independent method capable of performing dependable speciation analyses for toxicological or bioavailability studies [77,78]. Because of these benefits, ICP-MS is mandated in standard operating procedures for analyzing heavy metals in foodstuffs.

Single particle (SP)-ICP-MS is an emergent ICP-MS method for detecting, characterizing, and quantifying nanoparticles, which is one of the recent challenges between food safety and quality [79]. Asymmetrical flow field-flow fractionation (AsFFFF or ^{4}F) separated particles according to their hydrodynamic diameters. ^{4}F has already been combined with SP-ICP-MS to successfully characterize and quantified TiO_2NPs using a method based on focused sonication for preparing NPs dispersion. Figure 6 shows the particle size distribution pattern of moisturizing cream. Nowadays, SP-ICP-MS is the way to expand out from research laboratory.

TABLE 3 Selected Applications of ICP-MS for Food Safety and Quality

Research Aim	Matrix	MS Technique	Additional Techniques/Main Results	References
Evaluation of the Mineral (Nutrients and Contaminants) Content in the Diet				
As, Ba, Cd, Co, Cr, Cu, Fe, Hg, Mn, Mo, Ni, Pb, Sb, Se, Sn, V and Zn	Gluten-free food	ICP-MS	High sensitivity that improved the limits of quantification levels for some elements that are present at low quantities in some samples. Determination of different element content.	[71]
Ca, Cu, Fe, K, Mg, Mn, Mo, P, and Zn	Peanuts	ICP-MS	Comparison of ICP-MS and ICP-OES	[69]
Cd, Pb, As	Infant formula	ICP-MS	LOQs (μg/kg): As 6.2, Cd 1.2, and Pb 4.5. Recovery rates: As 105%, Cd 98%, and Pb 108%. Expanded uncertainties were Cd 13% and Pb 19%. The LOQ and the uncertainty for Pb meet the requirements of Commission Regulation (EC) No. 333/2007.	[74]
Total I	Infant formula and adult nutritional products	ICP-MS	To prevent loss of iodine, ammonium hydroxide solution was added to the samples immediately after digestion.	[72]

Continued

TABLE 3 Selected Applications of ICP-MS for Food Safety and Quality—cont'd

Research Aim	Matrix	MS Technique	Additional Techniques/Main Results	References
Metal Speciation in Food				
As (III), As (V), MA, DMA, AsB, AsC, TMAO and TMAs.	Seafood (mussels, oysters, shrimps, and different types of fish).	IC–ICP-MS	Development of an MAE procedure. Using only H_2O as extractant and a nitric acid gradient as an eluent is most compatible with the long-term stability of both IEC separation and ICP-MS detection. Determination of bioaccessibility of As species.	[77]
iAs, MA, and DMA	Infant food-based cereals	IC-ICP-MS	Arsenic speciation	[78]
Discrimination of the Geographical Origin				
^{7}Li, ^{11}B, ^{49}Ti, ^{51}V, ^{53}Cr, ^{59}Co, ^{60}Ni, ^{71}Ga, ^{74}Ge, ^{75}As, ^{82}Se, ^{85}Rb, ^{88}Sr, ^{98}Y, ^{90}Zr, ^{93}Nb, ^{98}Mo, ^{101}Ru, ^{105}Pd, ^{109}Ag, ^{111}Cd, ^{115}In, ^{120}Sn, ^{123}Sb, ^{126}Te, ^{133}Cs, ^{138}Ba, ^{139}La, ^{140}Ce, ^{141}Pr, ^{146}Nb, ^{152}Sm, ^{153}Eu, ^{158}Gd, ^{159}Tb, ^{162}Dy, ^{165}Ho, ^{166}Er, ^{169}Tm, ^{172}Yb, ^{175}Lu, ^{178}Hf, ^{181}Ta, ^{182}W, ^{202}Hg, ^{205}Tl, ^{208}Pb, ^{209}Bi, ^{232}Th, ^{238}U	Pig	ICP-MS	Comparison to ICP-AES LDA was used to assess the variation in the multi-element profile of pigs from different geographic regions.	[76]

^{10}B, ^{11}B, ^{55}Mn, ^{59}Co, ^{63}Cu, ^{85}Rb, ^{86}Sr, ^{95}Mo, ^{111}Cd, ^{133}Cs, ^{137}Ba, ^{206}Pb, ^{207}Pb, ^{208}Pb (*low resolution*) ^{26}Mg, ^{27}Al, ^{47}Ti (*medium resolution*) ^{51}V, ^{52}Cr, ^{54}Fe, ^{56}Fe, ^{57}Fe, ^{60}Ni, ^{66}Zn (medium–high resolution) ^{75}As, ^{82}Se (high resolution)	Rice	HR-ICP-MS (magnetic sector)	Discrimination of geographical origin of rice based on multi-element fingerprinting. The classification of rice samples was carried out based on elemental composition by a radar plot and multivariate data analysis, PCA and DA.	[75]
Al, As, Ba, Be, Ca, Cd, Ce, Cu, Dy, Fe, K, La, Lu, Mg, Mn, Na, Nd, Pb, Rb, Sm, Sr, Th, U, V, Zn	Tomatoes and triple concentration tomato paste	DRC-ICP-MS	The origin of tomato fruits and the areas of production as "Italy" and "non-Italy" of the triple concentrated pastes were evaluated by three supervised pattern recognition procedures, LDA, SIMCA and KNN	[70]
Characterization of Nanoparticles				
TiO2NPs	Sugar glass and coffee cream	AsFlFFF–ICP-MS	TEM for visualization of the particles For accurate characterization of TiO2NPs, fractions containing NPs were collected from the AsFlFFF system, concentrated, and Analyzed by TEM TiO2 was determined by ICP-MS after MAE digestion using On-line off-line and all the possible combinations	[79]

AES, atomic emission spectroscopy; AsB, arsenobetaine; AsC, arsenocholine; AsFlFFF, asymmetrical flow field-flow fractionation; DA, discriminat analysis; DRC, dynamic reaction cell; DMA, dimethylarsinic acid; HR, High resolution; iAs, inorganic arsenic; ICP, inductively coupled plasma; IC, ion chromatography; IEC, ion exchange chromatography; KNN, K-nearest neighbors; LDA, linear discriminant analysis; LOQs, limits of quantification; MA, methylarsonic acid; MAE, microwave-assisted extraction; MS, mass spectrometry; NPs, nanoparticles; OES, optical emission spectroscopy; PCA, principal component analysis; SIMCA, soft independent modeling of class analogy; TEM, transmission electron microscopy; TMAO, trimethylarsine oxide; TMAs, tetramethylarsonium ion.

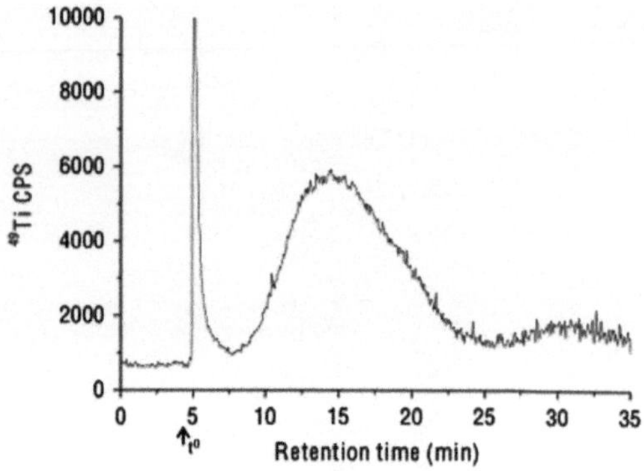

FIGURE 6 AsFlFFF-ICP-MS fractograms showing the detection of TiO_2NPs in moisturizing cream. The black arrow (t^0) indicates the void peak. *Reproduced from Ref. [79] with permission of Elsevier.*

5. MASS SPECTROMETRY (MS) COMBINED WITH CHROMATOGRAPHY OR RELATED TECHNIQUES

5.1 Liquid Chromatography–Mass Spectrometry (LC-MS)

LC-MS is one of the most used combinations of chromatography and mass spectrometry. There are already several reviews that highlight the role LC-MS is playing within the field of food safety and quality [3,14]. LC-MS is one of the fastest developing fields in science and industry. Table 4 shows some of the most important application within food safety and quality. Analytical results are grouped together based on the type of chemicals analyzed (lipids, carbohydrates, glycoproteins, vitamins, flavonoids, mycotoxins, pesticides, allergens, and food additives) [4–6,9,30,39,42,44,46,47,50–53,80–123]. Results are also shown for various types of food (ham, cheese, milk, cereals, olive oil, and wines). Although it is not an exhaustive list, it illustrates the main current directions of LC-MS applications within target and non-target analysis:

- Target analysis
  - Determination of the content of some compounds
  - Multitarget screening
  - Target metabolomic profile
- Nontarget analysis
  - Nontarget metabolomics
  - Nontarget fingerprinting.

The LC separation is important to ensure the highest quality data. Reversed-phase LC (RP-LC) using or not ion pairs in the mobile phase covers more than 95% of applications. Recently, hydrophilic interaction chromatography (or hydrophilic interaction liquid chromatography, HILIC)—a variant of normal phase liquid chromatography is emerging as a promising alternative for some very polar compounds. HILIC uses hydrophilic stationary phases with reversed-phase type eluents [81,88,114,115]. LC column miniaturization—reducing the

TABLE 4 Application of LC-MS for Food Safety and Quality

Research Aim	Matrix	MS Technique	Additional Techniques/Main Results	References
Target Analysis				
Food Components				
Macronutrients Amino acids	Traditional food *Potentilla anserina* L. root	UHPLC-ESI-QqTOF-MS	The resultant data were statistically processed by partial least squares-discriminant analysis (PLS-DA). Various metabolites, including amino acids, small peptides, nucleosides, urea cycle intermediates, and organic acids, which are responsible for the unique taste and nutritional and functional quality of fermented soy foods, were clearly altered by increasing the fermentation period.	[80]
Amino acids	*Ginkgo biloba*	HILIC-UHPLC-QqQ-MS/MS PI-SRM	The profiles of the amino acids in *G. biloba* leaves would be extraordinarily helpful for improving their potential values as dietary supplements and also may be used as markers for their quality control.	[81]
Monoacylglicerols (MAGs) and free fatty acids (FAs)	Fats and oils	LC-ESI-QqQ-MS/MS PI-SRM Two transitions	Various animal fat and vegetable oil samples were characterized for their MAG/FA profile indicating the usefulness of the method to address quality and authenticity of fats and oils.	[82]
Phospholipids	Soybean	UPLC-ESI-LTQ-MS/MS PI-SRM	Variation in the composition and content of PLs in soybeans following genetic modification, due to their beneficial effects on human health.	[83]

Continued

TABLE 4 Application of LC-MS for Food Safety and Quality—cont'd

Research Aim	Matrix	MS Technique	Additional Techniques/Main Results	References
Detection of traces of wheat gluten	Food and consumer products	HPLC-ESI-LTQ-MS/MS PI-SRM	This approach used LC-ESI-Ion Trap-MS^n to identify and characterize potential immunogenic peptides. About 25 potential wheat gluten peptide sequences were thus identified that could potentially serve as markers for gluten content in food. Samples were treated with proteases using an in vitro procedure that utilized conditions and enzymes that model gastric and duodenal protein digestion in humans.	[84]
Micronutrients Folates	Food-based standard reference materials	LC-ESI-QqQ-MS/MS PI-SRM Two transitions	Methods that include trienzyme digestion and isotope dilution.	[85]
Antioxidants Flavonoids	Vegetable (*Lophatherum gracile*)	UHPLC-ESI-QqQ-MS/MS NI-SRM	The method was validated using calibration curves, limits of detection and quantification, precision, repeatability, stability, and accuracy.	[5]
Isoflavones	Wild and cultivated soybeans, bean products	HPLC-ESI-^{3}DIT PI and NI Full scans 50–1000 *m/z*	Malonylglucosides and glucosides were the chemical markers in the two soybean species, while aglycones and glucosides were increased in products. Samples grouping based on PCA and cluster analysis coincided nicely with their species and geographical origin.	[86]
Identification of flavonol and triterpene glycosides	Luo-Han-Guo extract	UHPLC-ESI-QqTOF-MS PI Full scan *m/z* 100–1600 Target MS/MS	Two flavonol glycosides and 20 triterpene glycosides were identified, and in addition the profiles of Luo-Han-Guo samples collected at different growing periods (15-, 40-, and 80-days) were compared using PCA.	[87]

Phenolic compounds	Vegetables	LC × LC-ESI-QqTOF-MS Mass range *m/z* 200–2000 MS^E (two full scan functions at low and high collision energy)	Off-line HILIC × RP-LC revealed information which could not be obtained by one-dimensional LC methods, while the structured elution order for the anthocyanins simplifies compound identification and facilitates the comparison of anthocyanin content of natural products by means of contour plots.	[88]
Polyphenols	Chokeberry (*Aronia melanocarpa*)	LC-QqQ-MS/MS NI-SRM Quantified by LC-UV	Lyophilized leaves (0.5 g) were finely ground and submerged in 70% aqueous methanol (30 mL). The antioxidant activity of the three different mature leaf extracts was determined.	[6]
Other healthy substances Ginsenosides	American ginseng functional foods and ginseng raw plant materials	LC-ESI-QqQ-MS/MS PI-SRM	Ten selected analytes comprised five PPD group and four PPT group ginsenosides and one pseudoginsenoside, some of which were characteristic compounds in specified type of ginsengs. Combining with principal component analysis results, ginseng source could be roughly decided by plotting PCs score.	[89]
Active components and antidiabetic drugs	Propolis	LC-ESI-QqQ-MS/MS NI-SRM	The samples were extracted by ultrasound extraction with methanol. The insoluble residue of extract was removed by freeze-centrifuging.	[90]
Organic acids	Fruit juices	LC-ESI-QqQ-MS/MS NI two SRM	A Method based on stable isotope dilution liquid chromatography-tandem mass spectrometry is described for citric, malic, quinic, and tartaric acids in fruit juices.	[91]
β-sitosterol	*Saraca asoca* and its preparations	LC-ESI-QqTOF-MS/(MS)	Different concentrations of beta-sitosterol and crude extracts were estimated by LC and targeted mass spectrometry.	[38]

Continued

TABLE 4 Application of LC-MS for Food Safety and Quality—cont'd

Research Aim	Matrix	MS Technique	Additional Techniques/Main Results	References
Quantitative and transformation product analysis of major Active physalins	*Physalis alkekengi* var. *franchetii* (Chinese Lantern)	UHPLC-ESI-QqTOF-MS PI Full scan *m/z* 100 to 1000 UHPLC-ESI-QqQ-MS/MS PI-SRM	Extraction with 60% of methanol.	[92]
Cholina	Infant formulas	UHPLC-ESI-QqQ-MS/MS PI two SRM	Sample preparation was adopted from AOAC Official Method(SM) 999.14.	[93]
Isoflavones	Radix Puerariae	LC-DAD-^{3}DIT-TOF-MS^n With post-column derivatization PI-ESI Full scan *m/z* 100 to 1000	LC fingerprints and the structural elucidation of isoflavonoids in *Radix puerariae*.	[94]
Nucleotides, nucleosides, and nucleobases	Ziziphus plants	HILIC-UHPLC-ESI-QqQ-MS/MS PI-SRM	The analysis results showed that the fruits and leaves of Ziziphus plants are rich in nucleosides and nucleobases, and could be selected as the healthy foods.	[44]
Nucleotides	*Mactra veneriformis*	LC-DAD-ESI-qMS PI-SIM	The validated method was successfully applied to identifying 10 nucleosides and nucleobases in 48 *M. veneriformis* samples. PCA was used to classify the 48 samples based on the contents of the nucleosides and nucleobases.	[95]

Catechins	Tea	UHPLC–ESI-QqQ-MS/MS PI-SRM NI-SRM	This method guarantee broad spectrum of research applications in the analysis of catechins in dietary supplements, tea extracts, tea infusions, and also in very complex matrices such as biological fluids and tissues studies involving the administration of tea flavonoids.	[96]
Food Contaminants and Residues				
Additives Additives and additive degradation products	Solutions in contact with the plasticizers	HPLC-ESI-QqQ-MS/MS HPLC-ESI-QqQ-MS/MS	A Comparison of ESI and APCI interfaces and the optimization of the extraction procedure were discussed. The extraction procedure was checked for two types of sorbents: a polymeric sorbent (Strata-X) and asilica sorbent (Strata-C18). The impact of the cartridge nature on the analytical method was shown by comparing two types of cartridges: Polypropylene and Teflon®.	[97]
Preservative and antimicrobial compounds	Fish	UHPLC-ESI-QqQ-MS/MS NI two SRM	The estimated dietary exposure values of the four parabens in the Philippines through fish is four orders of magnitude lower than the ADI of 10 mg/kg/day, but the values of antimicrobials are just half of the ADI of triclosan.	[98]
Preservatives	Cheeses	LC-ESI-^{3}DIT-MS/MS PI-SRM	The method consists of a simple extraction procedure of the preservatives from the cheese, an RP-LC separation of the preservatives, an ESI-MS/MS revelation.	[99]
Cyanidin-3-glucoside (red kernel color, RKC)	Beverages	UHPLC-ESI-TOF-MS/MS	Study of their degradation and a new component was identified when RKC degraded under various pH conditions and a degradation pathway is proposed. The results will assist in understanding the degradation mechanism of the colorant Cyd-3-Glu.	[100]

Continued

TABLE 4 Application of LC-MS for Food Safety and Quality—cont'd

Research Aim	Matrix	MS Technique	Additional Techniques/Main Results	References
Synthetic sweeteners	Wine	UHPLC-ESI-QqQ-MS/MS PI-SRM	This method is rapid, accurate, highly sensitive, and suitable for the quality control of low concentration of the synthetic sweeteners, which are illegally added to wines and other foods with complex matrices.	[101]
Natural Toxins Mycotoxins	Herbal medicine	UHPLC-QqTOF-MS Mass range *m/z* 50–1000 MS^E (two full scan functions at low and high collision energy)	Each of the accurate masses of the markers obtained was searched in a mycotoxins/fungal metabolites database. The molecular formulas with relative mass error were then applied to mass fragment analysis for further confirmation of their structures. With the use of this approach, five mycotoxins that have never been reported were identified in *Imperatae Rhizoma*.	[102]
Mycotoxins	Cereals	UPLC-ESI-QqQ-MS/MS PI-SRM	The extraction procedures considered were a QuEChERS-like method and one using PLE.	[103]
Aflatoxins	Nuts and nut products	LC-ESI-QqQ-MS/MS PI two SRM	All four aflatoxins of interest (B_1, B_2, G_1, and G_2) were quantified using aflatoxin M_1 as the internal standard.	[104]
Marine lipophilic toxin	Mussels, oysters, and cockle	LC-ESI-QqQ-MS/MS PI-SRM NI-SRM	Thirteen laboratories participated in an interlaboratory study to evaluate the method performance characteristics of an LC-MS/MS for marine lipophilic shellfish toxins.	[105]
Plant and fungal metabolites	Wheat, maize, and animal feed	LC-Orbitrap-MS (Q Exactive) Multiscreening against a database	On-line cleanup using turbo-flow chromatography. This screening method is using a database of over 600 metabolites to establish contamination profiles in food and feed.	[42]

Residues Pesticides	Food (apple, lemon, lettuce, wheat grain)	LC-ESI-QqQ-MS/MS PI-SRM	Three new different versions of the QuEChERS method using more volatile salts (ammonium chloride and ammonium formate and acetate buffers).	[106]
Pesticides	Food	LC-ESI-QqQ-MS/MS PI-SRM	QuEChERS was used as an extraction method. Due to the large number of %R data, the complexity of the matrix effect, and the uncertainty associated with the separation parameters, PCA was used to explore the property of these data.	[107]
Pesticides	Fruits and vegetables	LC-ESI-LTQ-MS/MS PI-SRM-triggered EPI (qualitative) PI-SRM (quantitative/ confirmatory)	QuEChERS method with acetate buffering (AOAC Official Method 2007.01) was used for sample preparation, which has been previously shown to yield high-quality results for hundreds of pesticide residues in foods.	[108]
Veterinary drugs (Glucocorticoid residues)	Edible tissues of swine, cattle, sheep, and chicken	LC-ESI-QqQ-MS/MS NI-SRM	After deconjugation in alkali media, samples were extracted with ethyl acetate for glucocorticoids followed by SPE cleanup.	[109]
Veterinary drugs (carbadox and metabolites of carbadox and olaquindox)	Meat muscle	LC-APCI-QqQ-MS/MS PI-SRM	Carbadox, its major metabolite (QCA), and the major metabolite of Olaquindox (MQCA) in animal muscle. Application of the method to test samples showed no false negative or false positive results even after the analysis of a significant number of samples from different matrices.	[110]
Forming during processing Acrylamine	Potato chips	LC-ESI-QqQ-MS/MS PI-SRM	Extract produced from matrices was directly analyzed without derivatization, the use of isotopically labeled acrylamide or the use of clean up cartridges.	[111]

Continued

TABLE 4 Application of LC-MS for Food Safety and Quality—cont'd

Research Aim	Matrix	MS Technique	Additional Techniques/Main Results	References
Biogenic amines	Fish	UHPLC-ESI-LTQ-MS/MS PI-SRM	The biogenic amines were dansylated and separated on a C8 column under LC gradient of 7 min duration.	[112]
Four biogenic and three volatile amines	Anchovy	LC-ESI-QqQ-MS/MS PI-SRM	Other amines, such as spermidine and spermine, which are used to determine the chemical quality index, could be included.	[47]
Formetanate hydrochloride	Food	LC-ESI-q-MS PI-SIM UHPLC-ESI-QqQ-MS/MS PI-SRM	QuEChERS version AOAC method 2007.01 was used as extraction method. The method was used to provide residue data for dietary exposure determinations of formetanate hydrochloride.	[113]
Migrating Bisfenols (A, F, E, B, S)	Soft drink	LC-ESI-QqQ-MS/MS NI-SRM	An automated online solid-phase extraction coupled to fast liquid chromatography–tandem mass spectrometry (online SPE fast LC-MS/MS).	[52]
UV ink photoinitiators	Packaged food	LC-ESI-QqQ-MS/MS PI-SRM PI H-SRM	In H-SRM mode, a mass resolution of 0.1 *m/z* FWHM on Q1 and a scan width of 0.01 *m/z* were employed, while the other quadrupole operated at low resolution (0.7 *m/z* FWHM).	[51]
Bisphenol A and alkylphenols	Cereals	LC-ESI-QqQ-MS/MS NI-SRM	Automated method based on online SPE-LC-MS/MS for the simultaneous determination of bisphenol A, nonylphenol and octylphenol in cereals.	[46]

Multitargeted				
Pesticides Multiresidue	Fruits	LC × LC-ESI-QqQ-MS/ MS PI-SRM	The combination of HILIC and RP chromatography provides a high orthogonality and, hence, a good matrix separation.	[114]
Pesticides Multiresidues	Fruits and vegetables	LC × LC-ESI-QqQ-MS/ MS PI-SRM	LC × LC include HILIC and RP modes. Comparison of the three of the most. Widely used methods (1) DFG S19 (extraction with acetone), (2) ChemElut method (extraction with methanol), and (3) QuEChERS. Permanent post-column infusion to evaluate matrix effects.	[115]
Pesticides, veterinary drugs, and mycotoxin	Bakery ingredients and food	UHPLC-ESI- Orbitrap-MS PI *m/z* 150 to 1000 100.000 FWHM	Targeted screening The natural coexistence of many compounds of interest requires the development of multiresidue analytical strategies for fast, reliable information at low concentration levels.	[116]
Target Metabolomics				
Biomarkers of *Vitex agnus-castus* fruits	Food supplements	LC-ESI-LTQ-MS PI-SRM Full scan *m/z* 100 to 1000 (nontarget) LC-ESI-QqQ-MS/MS PI-SRM (target)	PCA approach was applied to the whole spectroscopic data generated by LC-ESI-LTQ-MS analysis (positive ion mode) of *V. agnus-castus* crude extract and mother tinctures. PCA was also performed in the target screening by applying the peak heights of the total peaks present in the LC-MS data set (excluding the noisy), thus a matrix was obtained by using these heights (variables), and the columns of the matrix were different analyzed samples.	[50]

Continued

TABLE 4 Application of LC-MS for Food Safety and Quality—cont'd

Research Aim	Matrix	MS Technique	Additional Techniques/Main Results	References
Detection of adulteration with other milk species	Buffalo mozzarella	LC-ESI-QqTOF-MS PI Full scan *m/z* 100 to 1000 ddMS/MS LC-ESI-QqQ-MS/MS PI-SRM	A Novel specific and proteotypic peptide marker derived from beta-casein, highly diagnostic for bovine and buffalo species, has been identified by accurate LC-ESI-QqTOF-MS/MS. Then, a sensitive and selective methodology has been developed for the rapid detection and quantification of the fraudulent addition of bovine milk to buffalo milk for the 'Mozzarella di Bufala Campana' production. The targeted quantitative analysis was performed by monitoring specific transitions of the phosphorylated beta-casein f33-48 peptide, identified as a novel species-specific proteotypic marker.	[53]
Botanicals	Food supplements	LC-ESI-LTQ-MS/MS PI-SRM and IDA-EPI	For SRM transition >1500 cps, three EPIs were generated at three different collision energies, namely low (20 eV), medium (35 eV) and high energy (50 eV).	[39]
Nontarget and Unknown Identification				
Flavonoid oxidation	Rutin	LC-Orbitrap-MS (LTQ Orbitrap) Unknown identification	High resolution LC–MS analysis of the mixture of oxidized products of rutin revealed the presence of rutin dimer.	[117]
Fermentation	Production of meju-related foods	UHPLC-ESI-QqTOF-MS PI Full scans 50–1000 *m/z* MS/MS spectra	Metabolomic study of water-soluble extracts of meju during various times in the fermentation process. Resultant data were statistically processed by PLS-DA. Various metabolites contribute to both the nutritional and sensory quality of fermented soy products were remarkably altered during the fermentation period.	[118]

Quantitative determination of AMX qualitative analysis of metabolomic profiles	Chicken tissues	LC-ESI-LTQ-Orbitrap MS	Medicated and non-treated animals were clearly clustered in distinct groups. The multivariate analysis revealed some relevant mass features contributing to this separation. Biomarkers were not identified.	[4]
Detection of horse and pork in Halal beef	Meat	nLC-ESI-LTQ Orbitrap or nLC-ESI-QqTOF (identification) UHPLC-ESI-LTQ-MS/MS PI-SRM^3	Biomarker peptides were identified by a shotgun proteomic approach using tryptic digests of protein extracts and were verified by the analysis of 21 different meat samples from the 5 species. Potential biomarker peptides were searched against the UniProt knowledge database.	[9]
Characterisation of the phytochemical compounds	*Phaseolus vulgaris* L.	LC-ESI-TOF-MS PI and NI Full scan *m/z* 50–1100	Hydromethanol extracts from green beans were analyzed. The compounds were characterised based on interpreting their mass spectrum provided by the TOF–MS as well as by comparison with information from the literature (some compounds have been described previously in Fabaceae).	[119]
Assessment of toxins	Amanita spp.	UHPLC-UV-TOF-MS PI Full scan *m/z* 100–1600	Nontarget analysis Estraction with acidified methanol followed by SPE.	[120]
Muscle proteome	Atlantic cod (*Gadus morhua*)	UHPLC-ESI-LTQ-Orbitrap-MS PI Full scans 200–2000 *m/z* ddMS/MS	A large-scale proteomics approach has been used. One-dimensional polyacrylamide gel electrophoresis, nanoflow liquid chromatography peptide separation, and LTQ were used to identify 4804 peptides, which retrieved 9113 cod expressed sequence tags, which in turn were mapped to 446 unique proteins. Each raw file was analyzed using the Proteome Discoverer 1.0.	[121]

Continued

TABLE 4 Application of LC-MS for Food Safety and Quality—cont'd

Research Aim	Matrix	MS Technique	Additional Techniques/Main Results	References
Specific volatile and aminothiols	Wine	UHPLC-ESI-TOF-MS	A Novel foodomics assay of thiol-containing compounds, such as free aminothiols and related conjugate. FL-specific derivatization was applied along with multivariate statistical analysis.	[122]
Plant metabolites (Glycosylated phenolic acids)	Broccoli	LC-ESI-QqTOF-MS	With LC-MS and NMR, profiling metabolites in complex extracts is feasible at high throughput. However, the identification of key metabolites remains a limitation given the analytical effort necessary for traditional structural elucidation strategies. The hyphenation of LC-SPE-NMR is a powerful analytical platform for isolating and concentrating metabolites for unequivocal identification by NMR measurements.	[123]
Fingerprinting	*G. biloba* leaves	LC-ESI-QqTOF-MS	By both peak analysis and similarity analysis of the fingerprint chromatograms, variation of constituents was easily observed in the leaves from different sources. By comparison with batches of authentic leave, the authenticity, and quality consistency of related health foods in different matrixes were effectively estimated.	[30]

[3]DIT, ion trap; ADI, admissible daily intake; AOAC, association of official analytical chemists; APCI, atmospheric pressure chemical ionization; DAD, diode array detector; EPI, enhanced product ion; FAs, free fatty acids; ESI, electrospray ionization; FL, fluorescent; FWHM, full width at half maximum; HILIC, hydrophilic interaction liquid chromatography; HPLC, high performance liquid chromatography; IDA, information-dependent acquisition; LC, liquid chromatography; LTQ, linear trap quadrupole; MAG, monacylglicerols; MQCA, methyl-3-quinoxaline-2-carboxylic acid; MS, mass spectrometry; MS/MS, tandem mass spectrometry; NI, negative ion; NMR, nuclear magnetic resonance; PCA, principal component analysis; PCs, principal components; PI, positive ion; PLE, pressurized liquid extraction;
PLs, phospholipids;PLS-DA, partial least squares discrimatory analysis; PPD, protopanaxadiol; PPT, protopanaxatriol; QCA, Quinoxaline-2-carboxylic acid; q-MS, quadrupole mass spectrometry; QqQ, triple quadrupole mass spectrometer; QqTOF, quadrupole time of flight; QuEChERS, quick, easy, cheap, effective, rugged, and safe; RKC, red kernel color; RP, reversed phase; SIM, selective ion monitoring; SPE, solid-phase extraction; SRM, selected reaction monitoring; TOF, time of flight; UHPLC, ultra-high performance liquid chromatography.

inner diameter or the particle size—increased sensitivity, peak resolution, and efficiency, together with shorter analysis times than those achieved with conventional LC. The literature shows a clear shift toward smaller diameters in the use of normal-bore columns—from 4.6 to 2.1 mm that offer a better ESI-MS compatibility and lower mobile phase consumption. On the particle size, the introduction of ultra-high performance liquid chromatography (UHPLC) columns, with smaller particle size (1.7 and 1.8 μm), has allowed improved peak resolution and, therefore, increased sensitivity in chromatographic separations [5,96,100]. Although UHPLC technology has been launched recently and its separation capabilities when combined with MS are not fully exploited, it is used in more than 50% of the applications reported in Table 4. However, there are still some key challenges of LC that are difficult to overcome even by UHPLC, such as peak coelution and poorly retained peaks eluting in the solvent front. Multidimensional chromatography is getting interest to solve the separation challenges that cannot be achieved by one-dimension technique. Different coupling techniques with various interface solutions have been developed taking into account that one of the principal requirements of an efficient LC × LC system is that the separation modes employed be orthogonal. This implies that the separation mechanisms to be used in each dimension be carefully selected in order to minimize the retention correlation between dimensions. The combination of HILIC and RP-LC provides the appropriate degree of orthogonality and the desire compatibility with LC-MS [88,114,115] because these LC approaches does not need to use high salt content.

The most common ionization source used to couple LC and MS is ESI. However in some cases, APCI is also used. If possible it is always interesting to check ionization sources other than ESI that could result in higher efficiency, as for example for carbadox and the acidic metabolites quinoxaline-2-carboxylic acid and methyl-3-quinoxaline-2-carboxylic acid, APCI resulted in higher efficiency [110]. LC is available combined almost with any type of MS being the most common QqQ, QTRAP, TOF, QqTOF, and the different orbitraps. Other mass analyzers can also be combined but they have falling into disuse.

The major drawback in quantitative analysis with LC-MS is the matrix effect—unexpected suppression or enhancement of the analyte response due to coeluting matrix constituents that influence the ionization process. The matrix effect is related to the ionization source of the mass spectrometer, and then all mass analyzers are susceptible to suffer matrix effects. Matrix effect can seriously degrade the accuracy of LC-MS analysis results. Method performance parameters such as limit of detection, limit of quantification, linearity, accuracy and precision, are affected by the loss of sensitivity and selectivity. Taking matrix effect into account (without actually reducing its influence) and thus improving the accuracy of the results can be done via standard addition [5], echo-peak technique [89], isotope dilution [42], postcolumn standard infusion [100], internal standard usage [52], and matrix matched calibration [106,107,108].

5.1.1 Target Analysis

The target analysis is done with analytical standards that are needed to quantify. Before entering the description of the different applications that exist of these methods, it is important to note that for the majority of food quality and safety control determinations, target analysis has been traditionally the most basic system. Normally the target-quantitative methods were optimized and validated for some few compounds. However, these analyses were expensive and long and thus, little viable, especially for the control laboratories that had to process a high number of samples, of which only a small percentage present any compound of interest. For this reason, target analysis has evolved toward the increasing application of more general extraction techniques, which simultaneously extracted a wide range of compounds (e.g., different contaminants, several vitamins or other nutritional compounds), and the techniques of multiscreening and multisearching, which allow to identify a large number of possible analytes in the sample. The ultimate advance in these analyses is to incorporate unknown or unexpected compounds identification (meaning nontarget analysis).

Target analysis for food analysis has been applied to determine the content of polyphenols [6], additives (and their degradation products) authorized in plastic products such as pharmaceutical packaging (e.g., antioxidants, release agents, and light absorbers) [51,97], food additives [99,101,124], free amino acids [80,81], mycotoxins [45,102,104,125], nucleobases, nucleosides and nucleotides [44,95], monoacylglycerols and free fatty acids [82], β-sitosterol [38], pesticides [103,115,126–128], residues of veterinary medicines [110], choline [93], folic acid and folates [85], ginsenosides [89], physalis [92], bisphenol A and alkylphenols [46,52], biogenic and volatile amines [47,112], isoflavones, flavonols, and triterpene glycosides [86,87,100], organic acids [91], marine lipophilic toxins [105] and tea catechins [96]. The state-of-the-art LC-MS/MS instrumentation has superb sensitivity and allowed the development of mega LC-MS/MS methods that achieved the simultaneous analysis of at least 100 different substances with superior data quality and efficiency. This, concomitant with the implementation of artificial intelligence-based data acquisition software, such as, scheduled SRM, dynamic SRM, and timed SRM, allowed the development of LC-MS/MS methods for the effective determination of many pesticides.

Multitargeted screening including multiclass and multiresidue methods of analysis tends to provide greater overall laboratory efficiency than the use of multiple methods. This approach can be conceived as the analysis of a sample for the presence or absence of wide range known compounds, and it is often assumed as semi-quantitative, using standard solutions of commonly encountered compounds, allowing the analyst to identify signals that can be associated with values close to, or exceed relevant legislative levels. Targeted analysis with LC-MS/MS operated in SRM mode enables specific and sensitive detection of targeted molecules. A survey study of fruits and vegetables and their products for detection of 300 pesticides has been published recently. Initially, the measurement was carried out as an aliquot of a raw acetonitrile extract. The subsequent cleanup was carried out fully automated by a multidimensional LC. Matrix compounds and analytes are separated in the first dimension on a hydrophilic interaction LC column [114]. In a second step, this new approach was compared with the classical methods. Sample extracts, obtained by three different methods, were injected directly into the LC × LC-MS/MS system. In total 44 targeted compounds could be identified and confirmed using this methodology [115].

With thousands of organic compounds present in complex food matrices, the development of new analytical solutions leaned toward simplified extraction/clean-up procedures and chromatography coupled with MS. Efforts must also be made regarding the instrumental phase to overcome sensitivity/selectivity limits and interferences. For this purpose, high-resolution full-scan analysis in MS is an interesting alternative to the traditional tandem approach. Modern HRMS instruments can be operated at resolutions up to R = 240,000 using Orbitrap technology with mass accuracy <5 ppm in full-scan mode. Another advantage of HRMS is that the acquired data can even be evaluated retrospectively for additional compounds or reprocessing of raw data by another compound class database. Parallel to full-scan detection, fragmentation experiments using HCD enables higher specificity by detection of fragment ions for compound confirmation where analytical standards are available. An example of using HRMS as a screening tool for the analysis of plant and fungal toxins and metabolites has been described by using a database of over 600 metabolites to establish contamination profiles in food and feed. Extracts were injected directly into an automated turbulent flow sample clean-up system, coupled to an LC-Q-Orbitrap-MS. The wide applicability of the method was demonstrated by the validation of 15 fungal and plant metabolites in maize, wheat, and animal feed samples [42]. A fast method for extracting and purifying bakery matrices was also developed and combined with the exploitation of UHPLC coupled to an Orbitrap Exactive(TM) HRMS [116] to determine several contaminants (e.g., pesticides, mycotoxins, veterinary drugs). The study demonstrates the advantages of analyzing a large variety of food contaminants.

However, within the "omic" sciences the workflow has been inversed. The blooming of the "omics" techniques, mostly metabolomics and proteomics, was based on obtaining the spectra of whole range of masses and the subsequent identification and characterization of molecules or proteins responsible for a certain characteristics. Nontarget methods were the basis of the primitive discoveries within "omics." However, in a second stage, the nontarget methods, once identified the compounds responsible for a particular feature, are going to become target methodologies, given the need to quantify the compounds mostly because the difference may be related to concentration. Nowadays, mass spectrometry-targeted approach is considered one of the most powerful way to obtain metabolite profiling of biological samples, thus it represents an ambitious tool in metabolomic quantitative analysis field.

The LC-MS in SRM is particularly suitable for the detection of food adulteration and the verification of authenticity to assure food safety and quality, both recognized as top priorities by the EU Commission. Using this system, the accidental or fraudulent blending of meat from different species can be easily detected [9], demonstrating the specificity of previously identified biomarker peptides against chicken, lamb, and beef. Biomarker peptides were previously identified by a shotgun proteomic approach using tryptic digests of protein extracts and were verified by the analysis of 21 different meat samples from five species. For the most sensitive peptides, SRM method was developed that allows for the detection of 0.55% horse or pork in a beef matrix. To enhance sensitivity, SRM experiments were applied and were able to detect down to 0.13% pork contamination in beef. Sensitive and selective detection of buffalo mozzarella adulteration can be carried out by monitoring specific transitions of the phosphorylated beta-casein f33-48 peptide [53]. This peptide is a novel specific and proteotypic peptide marker derived from beta-casein, highly diagnostic for bovine and buffalo species, that has been previously identified by accurate LC-ESI-QqTOF-MS/MS. The high sensitivity of SRM-based mass spectrometry and the wide dynamic range of QqQ spectrometers provide a valuable tool for the analysis of complex matrices such as dairy products.

Efficient multitargeted methods have been developed to screen selected botanicals of interest in herbal food using both QqQ and QTRAP. A validated analytical method for the quantitative determination of seven chemical markers occurring in a hydroalcoholic extract of *Vitex agnus-castus* fruits by LC-ESI-QqQ-MS/MS is reported. To carry out a comparative study, five commercial food supplements corresponding to hydroalcoholic extracts of *V. agnus-castus* fruits were analyzed under the same chromatographic conditions of the crude extract. Principal component analysis (PCA) based only on the variation of the amount of the seven chemical markers was applied in order to find similarities between the hydroalcoholic extract and the food supplements [50]. A second PCA analysis was carried out considering the whole spectroscopic data deriving from LC-ESI-LTQ-MS analysis. High similarity between the two PCA was observed, showing the possibility to select one of these two approaches for future applications in the field of comparative analysis of food supplements and quality control procedures.

A generic method was developed for multitargeted screening of biomarkers with the aim of characterizing plant species in herbal food supplements [39]. Botanicals were characterized by means of appropriate biomarkers, which were unambiguously identified by mass spectrometry using an IDA experiment which combined an SRM survey with dependent EPI scans. During this procedure, product ion scans of targeted analytes were generated at three collision energies and compared with an in-house library of MS/MS spectra acquired from reference standards of all biomarkers. This generic method enables detection, identification, and quantification of 98 biomarkers intended to characterize 79 selected plants. Figure 7 compares EPI spectra from an extract of *V. agnus-castus* L. with those from the library based upon pure standards. Box (A) shows

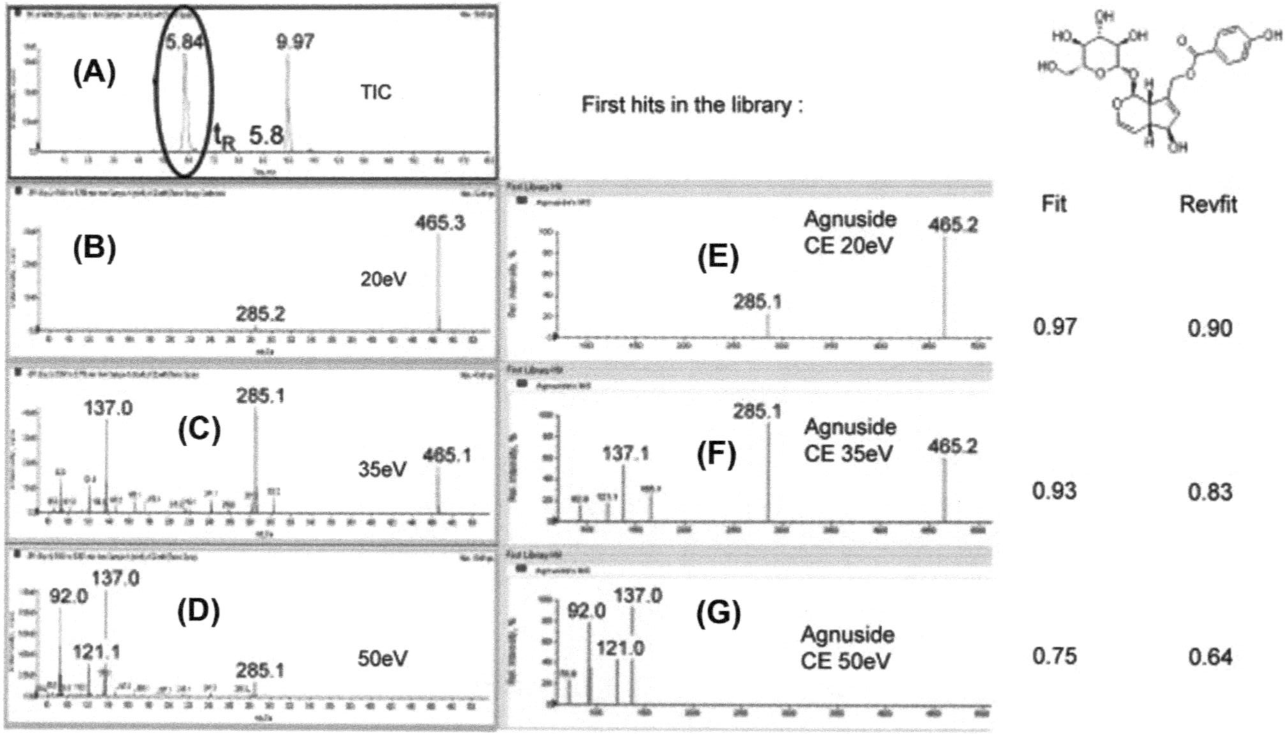

FIGURE 7 Identification of agnuside in an extract of *Vitex agnus-castus* L. based on MS/MS spectra comparisons. (A) total ion chromatogram; (B–D) MS/MS spectra at 20, 35, and 50 eV, respectively; (E–G) first hits of library search. *Reproduced from Ref. [39] with permission of Elsevier.*

the total ion chromatogram (TIC) in the negative ionization mode, where two peaks can be observed. The first one at 5.8 min corresponds to the biomarker-agnuside and the second one at 9.9 min to the internal standard (fipronil). For proper identification, the relative retention time (biomarker/internal standard ratio) must be within a time window of ±0.05 min compared to that of the standard solution. Boxes (B), (C), and (D) depict the three EPI spectra generated at 20, 35, and 50 eV, respectively. Boxes (E), (F), and (G) show the first library hit with the best match at each CE. In all cases, the first hit was agnuside with Fits higher than 0.75 and Revfits above 0.64. These results enabled the unambiguous identification of agnuside and thus in turn identified the plant as *V. agnus-castus* L. Likewise, other 97 biomarkers could be identified in the same way by their specific EPI spectra, which were eventually used to characterize 79 botanicals. In terms of identification capabilities, EPI spectra are structurally much more informative than a combination of conventional SRMs. Mass transitions and retention times are indeed excellent identification indicators to characterize a compound, but in the case of coeluting isobaric compounds, the classical SRM detection mode may not guarantee an unambiguous identification. Thus, in cases of isomers, which are often encountered in botanicals, it is necessary to obtain additional information by generating EPIs. Besides, if EPIs acquired at different collision energies show complementary fragmentation patterns, the identification performance is significantly improved due to the additional characterization criteria (Figure 7, boxes (B–G)).

5.1.2 Nontarget Analysis

The nontarget analysis can be divided in nontarget metabolomics and nontarget fingerprinting.

Metabolomics is defined as technology designed to give us the broadest, least biased insight into the richly diverse population of small molecules present in living things. Changes in the water-soluble metabolites of meju during fermentation were analyzed by UPLC-QqTOF-MS, and the resultant data were statistically processed by partial least squares-discriminant analysis (PLS-DA). Various metabolites, including amino acids, small peptides, nucleosides, urea cycle intermediates, and organic acids, which are responsible for the unique taste and nutritional and functional quality of fermented soy foods, were clearly altered by increasing the fermentation period. Changes in these metabolites allowed discrimination among meju samples with different fermentation periods (0, 10, 20, 40, and 60 days) on a PLS-DA score plot, and the fermentation was mainly processed between 10 and 40 days of fermentation. Twenty-two metabolites (phenylalanine, glutamic acid, leucine, adenine, citrulline, arginine, glutamine, gamma-aminobutyric acid, proline, acetylornithine, valine, pipecolic acid, methionine, citric acid, xanthine, tyrosine, isoleucine, Glu–Tyr, Ser–Pro, tryptophan, Glu–Phe, and Leu-Val-Pro-Pro) with high PLS-DA values of over 1.00 were determined as the major compounds contributing to the discrimination of meju samples. These metabolites, which were positively related to the sensory quality of meju, can be used as fermentation biomarkers for the production of meju and to construct the meju fermentation metabolic pathway. Therefore, our results indicate that monitoring the changes in metabolites during meju fermentation might be important for producing meju-related foods with good nutritional and sensory quality [118].

The identification of metabolites in biochemical studies is a major bottleneck in the proliferating field of metabolomics. In particular in plant metabolomics, given the diversity and abundance of endogenous secondary metabolites in plants, the identification of these is not only challenging but also essential to understand their biological role in the plant and their value to quality and nutritional attributes as food crops. With the new generation of analytical technologies, in which LC-MS and nuclear magnetic resonance play a pioneering role, profiling metabolites in complex extracts is feasible at high throughput. However, the identification of key metabolites remains a limitation given the analytical effort necessary for traditional structural elucidation strategies [123].

An interesting example was the comprehensive characterization of the phytochemical compounds found in three varieties of green bean [119]. For this, hydromethanol extracts from green beans were analyzed by LC-ESI-TOF-MS in order to characterize phytochemical compounds from different varieties. The compounds were characterized based on interpreting their mass spectrum provided by the TOF-MS as well as by comparison with information from the literature (some compounds have been described previously in *Fabaceae*). In this work, 72 phytochemical compounds were tentatively characterized and classified as, 10 phenolic acids, 59 flavonoids, two lignans and an iridoid. Notably, out of the 72 compounds, 54 are new and their isomers have been characterized for the first time in green beans. These results highlight the influence that variety can exert on the quality of phytochemicals. Given that new phytochemical compounds have been characterized, this study offers a useful approach for improving and updating the food-composition tables.

Detection of unexpected contaminants in complex matrices such as food is fraught with challenges. Scientists undertaking targeted analysis will normally have sufficient time, well characterized methods at their disposal, and adequate supplies of calibration standards and quality control materials. However, when faced with an unusual request, for example, to rapidly detect compounds for which calibration standards are not readily available, a more generic approach may have to be used. Modern analytical approaches such as LC-TOF have potential to undertake nontargeted analyses and a remarkable example is the semiquantitative measurement of selected mushroom toxins. Using authenticated mushrooms, a chromatographic profile for death cap mushrooms *Amanita phalloides* has been established and a consistent profile noted across five UK sampling sites. Representative specimens of *Amanita muscaria*, *Amanita rubescens*, and *Amanita citrina* were found to be free of these specific 22 toxins (<c. 1 mg/kg). About 80% of the mushroom toxicity is associated with the

amatoxins. The extraction and detection method can be used to determine the accidental or malicious adulteration of food with mushroom tissues containing, amatoxins and phallotoxins, but the approach has wider applications for the detection of unexpected chemicals in complex matrices and in assessing their impact on food safety [120].

The presence of specific volatile and aminothiols in wine is associated with quality, worth, price, and taste. Their identification in various wines has been reported using a novel foodomics assay of thiol-containing compounds, such as free aminothiols and related conjugates (Figure 8). Fluorescent specific derivatization was applied along with multivariate statistical analysis. First, the optimal experimental conditions were studied using representative thiols, such as L-cysteine, N-acetyl-L-cysteine, cysteamine, and L-glutathione, and then the UHPLC-fluorescent derivatization and separation steps were fixed for the subsequent screening of unknown thiol-containing compounds. The screening assay consisted of monitoring the UHPLC-TOF-MS peaks of unknown thiols, which decreased due to the derivatization as compared to the nonderivatized thiols. The PCA of the UPLC-TOF-MS data could be well-differentiated and categorized into two groups. The orthogonal signal correction PLS-DA the so-called S-plot, showed that the quality differentiation is directly related to the decrease of native thiols and increase of derivatized thiols. With this strategy, the mass difference from the derivatization reagent (+*m/z* 198) could be utilized for the identification of these thiols using the fluorescent derivatized peaks retention time and metabolomic databases. The presence of L-glutathione in rice

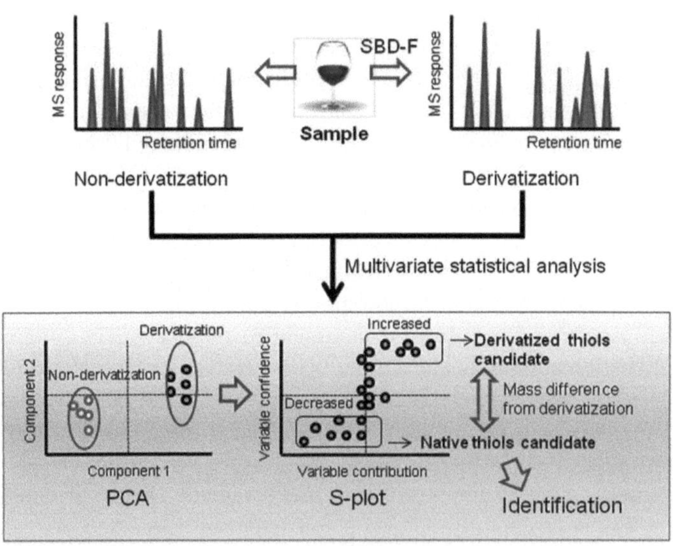

FIGURE 8 Proposed foodomics platform for the novel screening assay of thiol-containing compounds in wine by UPLC-FL/TOF/MS with derivatization and multivariate statistical analysis. *Reproduced from Ref. [122] with permission of the American Chemical Society.*

wine was for the first time reported on the basis of the available metabolomic databases and standard matching. This novel concept based on foodomics could be applied in food analysis for the ready screening of specific functional compounds by exploiting the various derivatization modes available [122].

Traditional shotgun proteomics has been widely used to detect thousands of proteins in biological samples using data-dependent LC-MS/MS methods. A typical proteomics study involves polyacrylamide-based fractionation of proteins extracted from biological samples followed by MS analyses of the peptides to acquire protein structural information for use in protein identification and characterization. The sensitivity of MS analyses can be greatly increased when a second round of fractionation is applied to the mixture of peptides using nano-flow LC before introduction to the MS. De novo sequencing of peptides using MS/MS provides an invaluable tool for identifying proteins on the basis of amino acid sequence homology searches from available databases. Currently, only five ray-finned fish species have complete (or nearly complete) genome sequences deposited in databases, including zebrafish, fugu, medaka, tetradon, and three-spined stickleback, which can be used for identification of proteins from proteomic approaches. Because Atlantic cod (*Gadus morhua*) has high economic value and its protein-rich muscle tissue is a food source, an increased understanding of the effects and consequences of environmental, nutritional, biological, and industrial factors on meat quality is necessary. To gain insight into cod muscle tissue's protein composition, a large-scale proteomics approach has been used. One-dimensional polyacrylamide gel electrophoresis, nanoflow LC peptide separation, and LTQ-MS were used to identify 4804 peptides, which retrieved 9113 cod expressed, sequence tags, which in turn were mapped to 446 unique proteins. The same data set identified 3924 proteins from the zebrafish protein database, which highlights the complementary value of the two approaches. The generated data sets will act as a foundation for studies related to physiological status assessment of cod under different environmental conditions, screening for diseases, and biomarker identification for assessment of fish quality during industrial processing and preservation [121].

Fingerprint refers to the characteristic spectrum of a food (recognition pattern), which can be related to its properties and thus to its authenticity in the same way as a human fingerprint is specific of a certain person and unequivocally identifies him/her. The term thus recalls a comprehensive description of a test material that is carried out in a nonselective (or untargeted) way. A fingerprint analysis method was developed for *Ginkgo biloba* leaves and was successfully used for quality evaluation of related health foods by HPLC-ESI-MS. Fifteen samples of *G. biloba* leaves, which were collected from 15 different locations in China, were analyzed and identified by both peak analysis and similarity analysis of the fingerprint chromatograms, variation of constituents was easily observed in the leaves from different sources. By comparison with batches of authentic leaves, the authenticity and quality consistency of related health foods in different matrixes were effectively estimated. It is important to mention that studying a wide range of authentic leaves from various habitats made the quality evaluation of commercial products more convincing and reasonable. The fingerprint-based strategy of the developed method should provide improved quality control of *G. biloba* leaves and products [30].

5.2 Gas Chromatography–Mass Spectrometry (GC-MS)

In comparison to LC, GC is faster, provides higher separation efficiency and has better properties for combination with mass spectrometric identification. Virtually the only limitation is that this technique is only applicable for volatile or semivolatile and thermally stable analytes. If not, possible solution is the use of chemical derivatization where the polar groups of the target molecules are converted into less polar moieties, which favorably affects the vapor pressure and the adsorption characteristics. The drawback of derivatization reactions is that they are time-consuming, and intensive and tedious labor. Table 5 summarizes the most important applications of this technique within food safety and quality [7, 41, 106, 129–157].

The great analytical strength of GC lies in its high resolution. This means that the average capillary column (30 m long) has approximately 100,000 theoretical plates. GC-MS has been widely used within food quality and safety for:

- analysis of target compounds,
- target and nontarget metabolomics, and
- importantly, profiling of volatile compounds.

5.2.1 Target Analysis

This technique is still one of the answers to screen/identify and quantify many groups of moderately polar and nonpolar food components, contaminants, and residues [129]. The most traditional way to perform GC-MS is using capillary columns—available in different dimensions with various stationary phases—and single quadrupole mass spectrometer EI ionization. Columns with different selectivity are available including the possibility of enantiomer separations. The successful development of a method for the chiral speciation of simeconazole based on the combination of such modified QuEChERS–enantioselective GC procedure with ^{3}DIT-MS/MS detection has been worked out [144]. In addition to EI ionization, positive and negative chemical ionization (PCI and NCI) can be used particularly for those molecules that provide low-mass fragment ions with similar *m/z* values, such as pyrethroid insecticides [146].

Conventional applications in food quality and safety include single compounds that need to be separated from matrix compounds as well as complex component mixtures that require the separation power of GC. Successful example of high sensitive and selective application within the first group is the determination of furan in food by solid-phase microextraction (SPME) combined with GC-MS [129]. Another example is to quantify the formation of ethyl carbamate

TABLE 5 Application of GC–MS for Food Safety and Quality

Research Aim	Matrix	MS Technique	Additional Techniques/Main Results	References
Target Analysis				
Forming during processing Furan	Chinese foods	GC-qMS	*Target analyte* SPME online desorbed in the GC *Estimation of the daily intake*	[129]
Ethyl carbamate, chloropropanols, and acrylamide	Fermented products, flavoring, and related foods	GC-QqQ-MS/MS SRM transitions (target)	The sample was extracted and cleaned in one single step by a combined clean-up column packed with Extrelut-NT and PSA.	[130]
Ethyl carbamate, chloropropanols, and acrylamide	Fermented products, flavoring, and related foods	GC-QqQ-MS/MS SRM transitions (target)	The sample was extracted and cleaned in one single step by a combined clean-up column packed with Extrelut-NT and PSA.	[130]
Ethyl carbamate	Brazilian spirit	GC-qMS-SIM *m/z* 62, 74 and 89	Study formation of ethyl carbamate during fermentation.	[131]
Acrylamide	Sudanese foods	GC-qMS SIM *m/z* 44, 55, 71	Extraction by QuEChERS followed by dSPE methyltrimethoxysilane-tetraethoxysilane.	[132]
Acrylamide	Cereal-based foods and potato chips	GC-IT-MS Full scan *m/z* 50–400	Before analysis acrylamide was efficiently derivatized with trifluoroacetic anhydride	[133]
PAHs	Edible oil	RTL-GC-qMS-SIM	Extraction was performed with acetonitrile by ultrasonication, and subsequent cleanup was applied using narrow GPC column	[134]
Industrial PBDEs and PCBs	Fishery and aquaculture products	LVI-GC–QqQ-MS/MS SRM transitions	Samples were extracted by an accelerated solvent extraction system and cleaned up by sequential SPE including dSPE and tandem SPE.	[135]

Continued

TABLE 5 Application of GC–MS for Food Safety and Quality—cont'd

Research Aim	Matrix	MS Technique	Additional Techniques/Main Results	References
Polybrominated biphenyls	Fish	GC-^{3}DIT-MS-MS Full product ion scan	Extraction by PLE. The optimization of the MS/MS parameters allowed maximum sensitivity and selectivity in the determination of PBBs, providing low limits of detection and good precision.	[136]
Migrating Bisphenol A	Large set of food items	GC-QqQ-MS/MS SRM 357.2 > 191.2, 372.2 > 357.2, 357.2 > 207.2, 357.2 > 341.1	A standard operating procedure, based on the combination of two successive SPE, was developed for various liquid and solid foodstuffs. The derivatization step consisted in a silylation reaction of bisphenol A using MSTFA.	[137]
Phthalate esters	Hydroalcoholic food beverages	GC-^{3}DIT-MS Full scan *m/z* 50–400	Extraction based on SPE with Amberlite XAD-2 adsorbent used as stationary phase.	[138]
Residues Pesticide	Fruits and vegetables	LP-GC-TOF-MS (low resolution)	GC-MS exhibited larger matrix effects, but as in LC-MS/MS, the differences were reasonably consistent among the 20 samples tested.	[139]
Pesticide	Seaweed	GC-qMS-SIM	MSPD was developed to extract 17 pesticides from seaweed samples using Florisil and GBC as clean-up adsorbents. The extraction has been optimized by a Box–Behnken design.	[140]
Pesticide	Oil seed	GC × GC-TOF-MS	An oil-absorbing MSPD. The quality of present screening method was evaluated by the Document No. SANCO/10684/2009.	[141]

Pesticides	Fruits and vegetables	LPGC-QqQ-MS/MS (target) SRM transitions	Evaluate three new different versions of the QuEChERS method using more volatile salts (ammonium chloride and ammonium formate and acetate buffers)	[106]
Pesticide	Fresh produce	GC-qMS/SIM or GC-QqQ-MS/MS SRM	• Several QuEChERS platforms were tested • Although a QqQ mass spectrometer is generally more expensive than a single quadrupole, its specificity makes it more reliable for pesticide identification.	[142]
Pesticide	Fruits and vegetable	GC-qMS Scan mode *m/z* 40 to 550 Identification in AMDIS database	• Independent evaluation of a commercial deconvolution reporting software • The detection of pesticides is based on fixed retention times using RTL and full scan mass spectral comparison with a partly customer built AMDIS database.	[143]
Triazole fungicide simeconazole	Vegetables, fruits and cereals	GC-^{3}DIT-MS/MS Two SRM transitions (*m/z* 121–101 and 195–153)	The enantiomers were resolved by capillary GC using a commercial chiral column (BGB-172)	[144]
Pyrethroids	Apple juice	GC-qMS (target) SIM	Extraction with Cl_2CH_2 and SPE Quantitation was based on the measurement of concentration ratios of the natural and isotope analogues in the sample and calibration blends.	[145]
Pyrethroid insecticides	Food	GC-QqQ-MS/MS NCI SRM transitions	First, a comparison of two different ionization modes, EI and NCI was carried out using MS and MS/MS. NCI-MS/MS provided the best results in terms of selectivity and sensitivity	[146]

Continued

TABLE 5 Application of GC–MS for Food Safety and Quality—cont'd

Research Aim	Matrix	MS Technique	Additional Techniques/Main Results	References
Solvents BTEXs	Olives and olive oil	GC-qMS-SIM	Sample pretreatment or clean-up stages were not necessary because samples are almost put directly into an HS vial, automatically processed by HS and then injected in the GC-MS for chromatographic analysis.	[147]
Toluene and other residual solvents	Food Packaging materials	GC-qMS Full scan 20–200 *m/z*	Automated HS sampling Levels of toluene and other residual solvents were evaluated in food packaging materials as a preventive action against hazard occurrence in foods via migration and to confirm food safety	[148]
Target and Nontarget Metabolomics				
Metabolic profiling	*Saccharomyces cerevisiae*	GC-qMS (nontarget) GC-QqQ-MS/MS (target)	Target metabolomics Trimethylsilyl derivatives of 110 metabolites	[7]
Metabolomic profiling	Millet (*Panicum miliaceum* L.)	GC-TOF-MS	A total of 48 metabolites were identified from millet, including 43 primary metabolites and five phenolic acids. The metabolite profiles were subjected to PCA and PLS-DA to evaluate the differences among varieties. PCA and PLS-DA fully distinguished the three varieties tested.	[149]
Pseudo-targeted profiling	Typical tobacco leaf extract	RTL-GC-MS-SIM	The established GC-MS-SIM method was compared with GC-MS-full scan (the total ion current and extracted ion current, TIC and EIC) methods	[41]

Metabolic profiling	Soy sauce	GC-TOF-MS Identification using SIMCA, PCA and PLS	Derivatization with MSTFA This new approach which combines metabolite profiling with QDA is applicable to analysis of sensory attributes of food as a result of the complex interaction between its components.	[150]
Profiling of Volatile Compounds (VOCs)				
VOCs	New Zealand Greenshell mussels	GC-qMS Full scan 40–500 amu Comparison with NIST library	Sample preparation by HS-SPME Of the 34 VOCs identified 20 were reliably identified throughout the storage treatment and 9 were found to change in relative concentration in homogenized mussel meat.	[151]
Fingerprinting and profiling	Food	GC-qMS Full scan 50–500 *m/z*	For fingerprinting studies all observed mass features, here called mass spectral tags, are quantified in a nontargeted and comprehensive approach.	[152]
VOCs	Chinese soy sauce aroma type liquor	GC-qMS Full scan 35–350 *m/z*	Extraction based on SBSE A Total of 76 volatile compounds were identified from 14 baijiu samples. PCA was employed to group of these baijiu liquors and to detect the components that contributed most to differentiating the liquor samples.	[153]
Profiling food volátiles Fingerprinting	Roasted hazelnuts from different origins	GC × GC-qMS	Analytes were extracted by HS-SPME The resulting patterns were processed by: (1) "Chromatographic fingerprinting," and (2) "comprehensive template matching." Fingerprint analysis showed to be effective for sample comparison and classification of roasted hazelnuts.	[154]

Continued

TABLE 5 Application of GC–MS for Food Safety and Quality—cont'd

Research Aim	Matrix	MS Technique	Additional Techniques/Main Results	References
Assessment of moisture damage	Cocoa beans	GC × GC-TOF-MS Full scan 40–250 *m/z* NIST library search	• Moisture damage to cocoa beans alters the volatile chemical signature of the beans in a way that can be tracked quantitatively over time. • 29 analytes that change in concentration levels via the time-dependent moisture damage process were measured using chemometric software.	[155]
Evaluation of the pervaporation process	Kiwifruits juices	GC-^{3}DIT-MS (target)	Volatile profile established by HS-SPME The most representative volatile compounds of the kiwifruit volatile fractionwere chosen for evaluating the pervaporation process.	[156]
Sugar compounds	Cereal and pseudocereal flour	GC-qMS Full scan *m/z* 50–400 WILEY 275 library for MS analysis	Compare sugar components of tested samples of flour of cereals bread wheat and spelt and pseudocereals (amaranth and buckwheat). Results were analyzed using descriptive statistics (dendrograms and PCA).	[157]

AMDIS, automated mass spectral deconfolution and identification system; ^{3}DIT, ion trap; dSPE, dispersive solid-phase extraction; EI, electron impact; EIC, extracted ion chromatogram; GBC, graphitized black carbon; GC, gas chromatography; GC × GC, comprehensive 2D gas chromatography time-of-flight mass spectrometry; GPC, gel permeation chromatography; HS, head space; LC, liquid chromatography; LP, low pressured; LVI, large-volume injection; MS, mass spectrometry; MS/MS, tandem mass spectrometry; MSPD, matrix solid-phase dispersion; MSTFA, N-methyl-N(trimethylsilyl)-trifluoroacetamide; NCI, negative chemical ionization; NIST, national institute of standards and technology of USA; PBBs, polybrominated biphenyls; PCA, principal componenet analysis; PLE, pressurized liquid extraction; PLS-DA partial least square discriminate analysis; PSA, primary secondary amine; QDA, Quantitative descriptive analysis; q-MS, quadrupole mass spectrometry; QqQ, triple quadrupole; QuEChERS, quick, easy, cheap, effective, rugged, and safe; RTL, retention time lock; SBSE, stir-bar sorptive extraction; SIM, selected ion monitoring; SIMCA, soft independent modelling of class analogy; SPE, solid-phase extraction; SPME, solid-phase micro extraction; SRM, selected reaction monitoring; TIC, total ion chromatogram; TOF, time of flight; VOCs, volatile organic compounds.

in the fermentation processes or in different distilled fractions [131]. Acrylamine, bisphenol A, and their corresponding diglycidyl ethers are also widely determined individually or covering a few metabolites [130,132,133,137,158,159]. The second group includes several applications to determine a broad range of pollutants in complex matrices. The detection is commonly operated in selective ion-monitoring mode (SIM) because of its increased sensitivity. Although GC-MS/SIM provides qualitative and quantitative MS information on the presences of contaminants in foods, there are difficulties with this approach. The detection can also be based on fixed retention times using retention time locking (RTL) and full-scan mass spectral comparison with a partly customer built automated mass spectral deconvolution and identification system (AMDIS) database. The deconvolution software is very important in those cases [143]. The determination of multiresidue pesticides in foods is still widely performed by conventional GC-MS for qualitative and quantitative purposes [140–143,145]. Polycyclic aromatic hydrocarbons (PAHs) [134,160], polychlorinated biphenyls (PCBs) [135], polybrominated compounds [135,136], phtalates [138,159], chloropropanols [130], and mono aromatic hydrocarbons such as benzene, toluene, ethylbenzene, and xylene [160] are the most common food safety applications of GC-MS. Currently, the top GC applications within food quality involve lipids [161–163] and carbohydrates [157], even though most of the applications are within the metabolomic field.

Innovation strategies that configure the current approaches are mostly in two fields (1) mass analyzers and (2) separation strategies. Within the former, quadrupole mass spectrometer is still the general work horse but GC has coupled depending on the specific analytical aims to other mass analyzers including those that provide nominal mass such as ^{3}DIT or QqQ as well as those that provide high resolution and mass accuracy (e.g., TOF). To improve the selectivity of GC-MS detection, GC equipped with QqQ or MS/MS (GC-MS/MS) can operate in SRM. Several studies already developed methods to brominated flame retardants and PCBs [135], bisphenol A [137], ethyl carbamate, chloropropanols and acrylamide [130], pesticide [106,142,146], using GC-MS/MS for a variety of food products. As an example, GC-MS/MS chromatograms of detected pesticides in fruits are shown in Figure 9. Organochlorine, organophosphorus, and pyrethroid pesticides were found in these samples with concentrations ranging from 5 (1 μg/kg p,p′-DDT in spinach) to 180 (5 μg/kg chlorothalonil in tomato). GC-TOF-MS has also several advantages, including high peak capacity, excellent retention time repeatability, and readily available compound libraries, so that compounds in a sample can be usually identified without using standard compounds [149,150].

GC often suffers from long analysis times. Speed of analysis is important to many of today's GC analysts as they look for ways to improve sample throughput. Fast GC used smaller diameter columns to help to reduce run times. The reason that this works is due to the increased efficiency of smaller ID columns which allow the use of shorter columns [106,139].

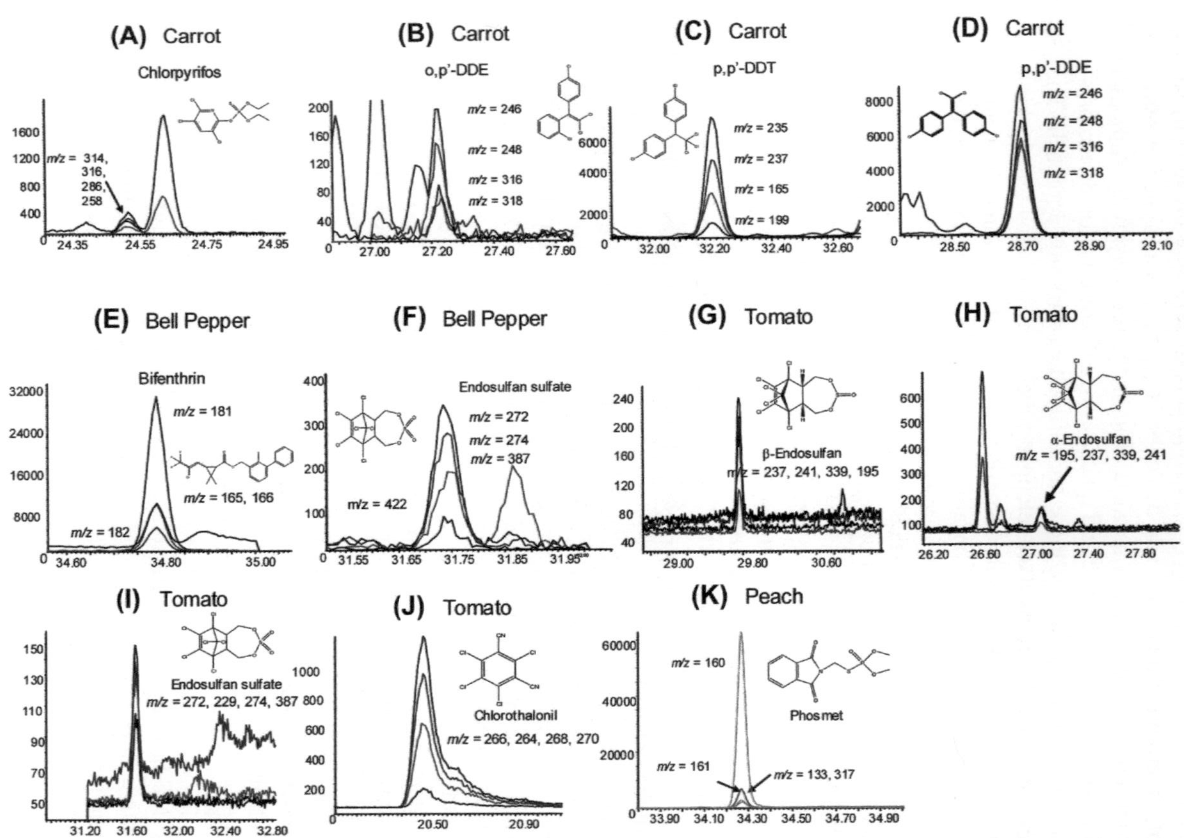

FIGURE 9 Reconstructed GC-MS/MS chromatograms of several commodities containing various pesticides including chlorpyrifos (A), o,p0-DDE(B), p,p0-DDT (C),and p,p0-DDE (D) present in carrot; bifenthrin (E) and endosulfan sulfate (F) present in bell pepper; β- (G) and R- (H) endosulfan, endosulfan sulfate (I) and chlorothalonil (J) present in tomato; and phosmet (K) in peach. Included are the transitions from precursor to product ions and the relative ion ratios between the two transitions, primary (top) and secondary (bottom), which are used for pesticide identification. *Reproduced from Ref. [142] with permission of the American Chemical Society.*

Comprehensive two-dimensional gas chromatography (GC × GC) consists of two columns connected serially, such that all sample portions emerging from the first column enter the second and are analyzed sequentially. Wang et al. [141] described the development of an oil-absorbing matrix solid-phase dispersion extraction with comprehensive GC × GC-TOF-MS suitable for screening of 68 pesticide residues in peanut, soybean, rape seed, sesame, and sunflower seed. A 35-min orthogonal separation was performed using a nonpolar–polar columns set. Identification of pesticide residues in the extract was finished by using the TOF-MS in the assistance of an automated peak-find and spectral deconvolution software. Cordero et al. [154] carried out a comparative analysis of the volatile fraction of roasted hazelnuts (*Corylus avellana* L.) from different origins by GC × GC-qMS.

The GC can also be combined to other less conventional detectors, such as ICP-MS. GC-ICP-MS provides a powerful laboratory technique for analyses, with the capability to separate and quantitate ultra-trace levels of metals and organometallic compounds with the possibility of speciation. Bruno et al. [164] already described the development and validation of an analytical method for the determination of monomethylmercury and monoethylmercury at parts-per-billion level simultaneously in various types of food items.

Compound-specific isotope analysis by GC combustion isotope ratio mass spectrometry is a powerful technique for the sourcing of food, such as determination of the geographic origin of food and food adulteration [165].

5.2.2 Target and Nontarget Metabolomics

In metabolomics studies, GC-MS has frequently been used to analyze low-molecular-weight metabolites because of the high equipment stability and availability of user-friendly tools for data analysis. Nontargeted metabolic profiling is the most widely used method for metabolomics. In metabolomic studies, GC-TOF or q-MS has frequently been used for the nontargeted analysis of hydrophilic metabolites. An interesting example to illustrate GC-TOF-MS application within the field was the establishment of the diversity among primary metabolites and phenolic acids in three varieties of millet (*Panicum miliaceum* L.). A total of 48 metabolites were identified from millet, including 43 primary metabolites and five phenolic acids. The PCA and PLS-DA of the metabolic profile fully distinguished the three varieties tested showing that this approach is a viable alternative method for evaluating food quality [149]. Other approach combined GC-TOF-MS metabolite profiles—primarily concerning low-molecular-weight hydrophilic components—with quantitative descriptive analysis (QDA), obtained from well-trained sensory panelists, to comprehend the relationship between components and the sensory attributes of soy sauces produced from different ingredients and brewing processes [150]. QDA data were accurately predicted by projection to latent structure, with metabolite profiles serving as explanatory variables and QDA data set serving as a response variable. It was indicated that sugars are important components of the tastes of soy sauces demonstrating that it is effective to search important compounds that contribute to the attributes. Commonly, the analytical platform employs the deconvolution method to extract single-metabolite information from coeluted peaks and background noise.

Therefore, there is a need of highly sensitive and selective methods capable of pure peak extraction without the need of any complicated mathematical techniques. For this purpose, a novel analytical method using GC-QqQ-MS in SRM was developed to analyze the trimethylsilyl derivatives of 110 metabolites, using EI. This methodology enables to utilize two complementary techniques nontargeted and widely targeted metabolomics in the same sample preparation protocol, which would facilitate the formulation or verification of novel hypotheses in biological sciences. The GC-QqQ-MS/MS analysis can accurately identify a metabolite using multichannel SRM transitions and intensity ratios in the analysis of food. In addition, the methodology offers a wide dynamic range, high sensitivity, and highly reproducible metabolite profiles, which will contribute to the biomarker discoveries and quality evaluations in biology, medicine, and food sciences [7].

A novel approach was also established to transform a nontargeted metabolic profiling method to a pseudo-targeted method using RTL-GC-MS/SIM. To achieve this transformation, an algorithm based on the AMDIS, GC-MS raw data, and a bi-Gaussian chromatographic peak model was developed. The established GC-MS/SIM method was compared with GC-MS-full-scan [TIC and extracted ion current (EIC)] methods, it was found that for a typical tobacco leaf extract, 93% components had their relative standard deviations of relative peak areas less than 20% by the SIM method, while 88% by the EIC method and 81% by the TIC method. 47.3% components had their linear correlation coefficient higher than 0.99, compared with 5.0% by the EIC and 6.2% by TIC methods. Multivariate analysis showed the pooled quality control samples clustered more tightly using the developed method than using GC-MS-full-scan methods, indicating a better data quality. With the analysis of the variance of the tobacco samples from three different planting regions, 167 differential components ($p < 0.05$) were screened out using the RTL-GC-MS/SIM method, but 151 and 131 by the EIC and TIC methods, respectively. The results show that the developed method not only has a higher sensitivity, better linearity and data quality, but also does not need complicated peak alignment among different samples. It is especially suitable for the screening of differential components in the metabolic profiling investigation [41].

5.2.3 Profiling of Volatile Compounds

Most of the methods applied in the analysis of volatile compounds in food rely on GC-MS procedures, which provide enormous separation power, adequate resolution, and high sensitivity. These compounds have an important contribution to the aroma and flavor of food. Aroma compounds can be found in food, wine, spices, perfumes, fragrance oils, and essential oils. For example, many form biochemically during ripening of fruits and other crops. The identification of the complicated aroma compounds' pattern is useful to establish food quality and safety. Compounds present in trace-level concentrations may be responsible for the flavor of some foods. Therefore the isolation and concentration of the analytes is a crucial step. The large diversity of components belonging to several chemical classes and their differences in volatility may further complicate the sample preparation process. The instability of certain components is another important aspect as well. Diverse concentrations and isolation methods can be combined to obtain a profile which is representative of the sample. The most popular one is the combination of head space isolation with SPME (HS-SPME) and/or stir-bar sorptive extraction (HS-SBSE) concentrations.

Quality control of cocoa beans is a significant issue in the chocolate industry. Moisture damage to cocoa beans alters the volatile chemical signature of the beans in a way that can be tracked quantitatively over time before the visible molds were presented. To monitor this, the chemical signature of the beans was established via HS-SPME and GC × GC-TOF-MS. Twenty-nine analytes that were independent of the origin and change in concentration levels via the time-dependent moisture damage process were measured using chemometric software. Furthermore, prediction algorithms were used to demonstrate that moisture damage could be verified before there were visible signs of mold by analyzing subsets of the 29 analytes [155].

The processing of kiwifruit for obtaining products with a higher sensory quality, to be used in the food industry, is associated with the use of techniques which can limit the physical and chemical losses of aroma compounds. Pervaporation represents an alternative to the techniques based on distillation/evaporation or partial condensation to concentrate the aroma compounds preserving the molecule integrity (mild operational conditions used), having a high selectivity toward the organic volatile compounds, and respecting also the environment. The most representative volatile compound of the kiwifruit aroma was chosen for evaluating the pervaporation process. SPME-GC-^{3}DIT-MS method was exploited to determine the amounts of these compounds. The approach was based on chemical ionization acquisition with isobutane as reagent gas and 1-heptanol and (Z)-3-hexen-yl acetate as internal standards. In these conditions, the calibration curves were satisfactory as demonstrated by the R^2 values of the straight lines (0.9937–0.9999). A kiwifruit fresh juice was processed by pervaporation through the composite commercial membrane (GFT1070) and the one self-prepared, made of styrene butadiene co-styrene, at three different feed temperatures. Marked effects for both membranes were observed in the total and partial fluxes of aroma compounds as the temperature was increased. The studied aroma compounds were differently affected by the temperature changes during pervaporation process, with the result of a change in concentration in the permeate of the aroma compounds recovered [156].

The volatile fractions of roasted hazelnuts (*Corylus avellana* L.) from nine different geographical origins, comparably roasted for desirable flavor and texture, were sampled by HS-SPME and then analyzed by GC × GC-qMS. The resulting patterns were processed by: (1) "chromatographic fingerprinting," i.e., a pattern recognition procedure based on retention time criteria, where peaks correspondences were established through a comprehensive peak pattern covering the chromatographic plane; and (2) "comprehensive template matching" with reliable peak matching, where peak correspondences were constrained by retention time and MS fragmentation pattern similarity criteria. Fingerprinting results showed how the discrimination potential of GC × GC can be increased by including all the detected components in sample comparisons and correlations and, in addition, providing reliable results in a comparative analysis by locating compounds with a significant role. Results were completed by a chemical speciation of volatiles, and sample profiling was extended to known markers whose distribution can be correlated to sensory properties, geographical origin, or the effect of thermal treatment on different classes of compounds. Figure 10 illustrates a GC × GC plot of Italian hazelnuts from Piedmont (i.e., Piemonte), with the locations of all 422 peaks in the consensus template. The subset of 196 template peaks with matches in all nine chromatograms are shown with white filled circles. The comprehensive approach for data interpretation proposed here may be useful to assess product specificity and quality, through measurable parameters strictly and consistently correlated to sensory properties and origin [154].

SBSE is a fast, simple, effective, and reliable method for extracting volatile compounds from food. Considering the importance of aroma compounds on the quality of baijiu (a Chinese liquor aromatized with soy sauce), SBSE

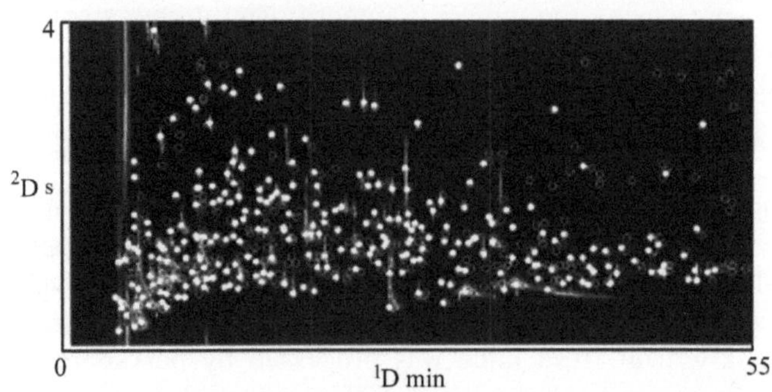

FIGURE 10 GC × GC-qMS plot of Italian hazelnuts from Piedmont (i.e., Piemonte). Circles indicate the 422 peaks in the consensus template. The subset of 196 template peaks with matches in all nine chromatograms is shown with white filled circles. *Reproduced from Ref. [154] with permission of Elsevier.*

coupled with GC-MS was applied. Based on the results of optimization, a total of 76 volatile compounds were identified from 14 Chinese liquors, including 25 esters, 10 alcohols, 9 aldehydes and ketones, 8 aromatic compounds, 5 furans, 3 nitrogen-containing compounds, 6 acids, 4 phenols, 3 terpenes, 1 sulfide-containing compound, 1 lactone, and 1 acetal. Quantification results indicated that the volatile profiles of different aroma types of Chinese liquor and the concentrations of some aroma compounds were quite distinct. PCA was employed to find the natural grouping of these baijiu liquors and to detect the components that contributed most to differentiating the liquor samples. According to the results from PCA, it was discovered that liquors with soy sauce aroma exhibited more prominent sauce flavor through extending the time for storage [153].

5.3 Capillary Electrophoresis–Mass Spectrometry (CE-MS)

Coupling CE-based separation techniques to MS creates a powerful platform for analysis of a wide range of biomolecules from complex samples because it combines the high separation efficiency of CE and the sensitivity and selectivity of MS detection. ESI and MALDI, as the most common soft ionization techniques employed for CE and MS coupling, offer distinct advantages for biomolecular characterization [25]. CE-MS has been applied in food proteomics, peptidomics, metabolomics, and foodomics [13]. CE-MS has also proved to be a powerful alternative to chromatographic techniques in the analysis of drugs and metabolites in foods, such as betains, isoflavones, antihelmintics, etc. Particularly, direct analysis of amino acids and their analogs is routinely performed using strongly acidic buffer conditions under positive-ion mode ESI-MS [166]. Figure 11 shows the extracted ion electropherogram (EIE) for a representative sample with a low content in D-Carnitine (0.16 l g/mL). Despite the low concentration of D-Carnitine present in the sample, its unequivocal identification was possible through its MS/MS spectrum. These results corroborated the high potential of the developed method in the control of the quality and safety of foods containing Carnitine in their composition.

Table 6 summarizes some of the relevant recent applications of this technique within food safety and quality [167–174]. On their instrumental setup, CE has been mainly coupled to single quadrupole, ^{3}DIT, or TOF mass analyzers and the most common way to interface both systems is the sheath interface. In the context of quantitative analysis, a significant increase of the signal-to-noise ratio can be obtained using MS/MS experiments.

New advances in online sample preconcentration, chemical derivatization, and/or ESI interface designs can further improve assay performance allowing for resolution with low nanomolar detection limits.

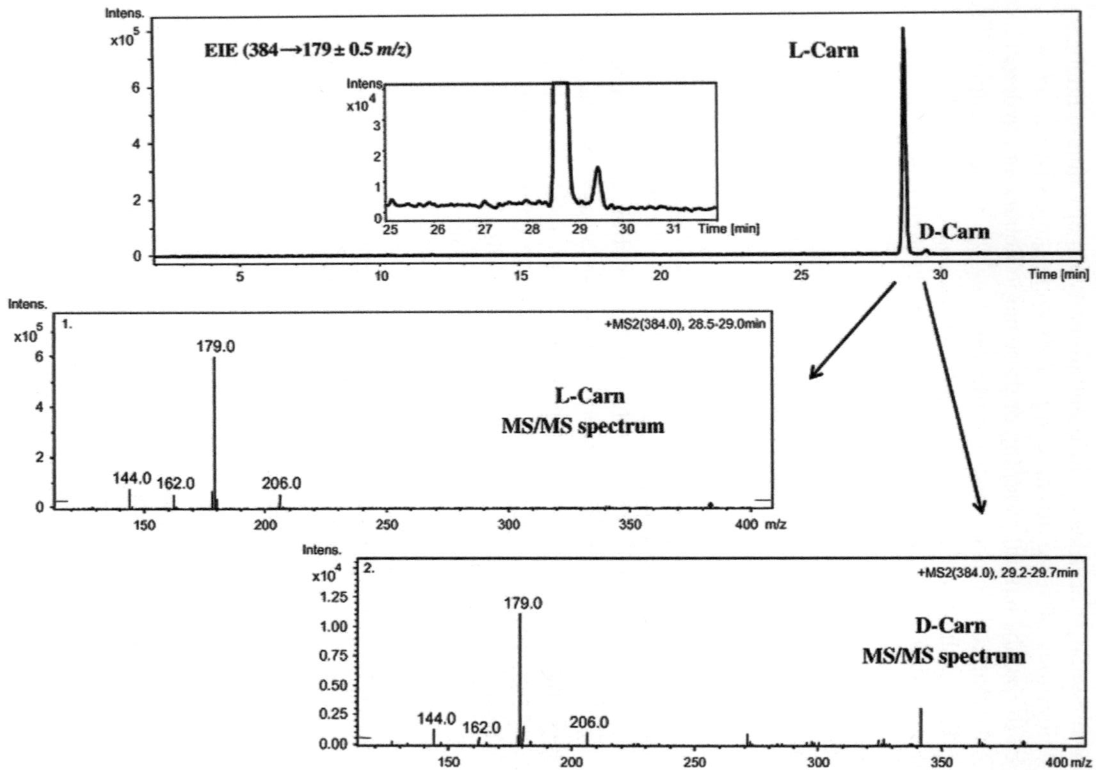

FIGURE 11 CE-MS/MS EIE for a biscuit (Biscuit#2) and the corresponding MS/MS spectra for the peaks of L- and D-Carnitine. CE conditions: BGE, 0.5M ammonium formate buffer (pH 2.5) with 0.2% (m/v) succ-c-CD (4 succinyl groups/CD ring); uncoated fused-silica capillary, 50 lmID × 100 cm; injection by pressure at 50 mbar × 12 s; applied voltage, 25 kV; temperature, 25 °C. ESI conditions: positive ion mode; spray voltage, 4.5 kV; sheath liquid, isopropanol/water (50/50 v/v) with 0.1% formic acid at 3.3 lL/min; drying gas flow, 5 L/min; drying temperature, 300 °C; nebulizer pressure, 2 psi; compound stability, 50%. MS/MS conditions: MS/MS transitions, 384→179*m/z*; width, 4*m/z*; fragmentation amplitude, 1.20 V. *Reproduced from Ref. [168] with permission of Elsevier.*

TABLE 6 Selected Application of CE-MS for Food Safety and Quality

Resarch Aim	Matrix	MS Technique	Additional Techniques/Main Results	References
Betaines	Vegetable oil	CE-ESI-^{3}DIT-MS/MS Via sheath liquid sprayer	Butyl ester derivatization of betaines was carried out. MS^2 experiments provided the fingerprint fragmentation for the betaines identified in vegetable oils. In extra virgin olive oils, carnitines were not detected, making it possible to propose them as a feasible novel marker for the detection of adulterations of olive oils.	[167]
L and D-carnitine	Food supplement	CE-ESI-^{3}DIT-MS/MS Via sheath liquid sprayer	Derivatization with FMOC. The composition of the sheath liquid (isopropanol:water (50:50, v:v) containing 0.1% of formic acid).	[168]
Vitamins B	Dietary supplements	CE-ESI-QqQ-MS/MS Via sheath liquid sprayer	Background electrolyte: 50 mmol/L formic acid with pH 2.05, ESI ionization (sheath liquid: 0.1% v/v formic acid in 50% v/v aq methanol) and QqQ detection (100–160 V fragmentor voltage, 15–25 eV collision energy, depending on the type of analyte).	[169]
Free amino acids	Dietary supplements containing royal jelly	CE-ESI-^{3}DIT-MS/MS Via sheath liquid sprayer	The CE separation was achieved in an uncoated fused-silica capillary using a 1 M formic acid solution (pH 1.8) as the electrolyte, followed by MS/MS detection after mixing with a sheath liquid comprising 50% (v/v) methanol.	[170]
Isoflavones	Soy-based foods	CE-ESI-qMS Via sheath liquid sprayer PI mode	The composition and the flow rate of the sheath liquid have strong effects on the ionization efficiency of analytes in MS and on spray stability. Different isopropanol:water mixtures were tested and it was found that increasing the proportion of isopropanol to 90% provided the highest responses in MS.	[171]

Continued

TABLE 6 Selected Application of CE-MS for Food Safety and Quality—cont'd

Resarch Aim	Matrix	MS Technique	Additional Techniques/Main Results	References
Anthelmintic benzimidazoles	Eggs	CE-ESI-qMS Via sheath liquid sprayer PI mode	Four different solvents and mixtures were assayed as sheath liquids: pure isopropyl alcohol and methanol, and mixtures of isopropyl alcohol/water 50% (vol.) and methanol/water 50% (vol.).	[172]
Endocrine disruptor	Honey	CE-PNP-ESI-qMS NI mode	For sample pretreatment, different LLE procedures were assayed and compared to the QuEChERS (c) methodology prior to electrophoretic analysis.	[173]
Aminoacids	Food safety	CE-ESI-MS Different approaches	Direct analysis of amino acids and their analogs is routinely performed using strongly acidic buffer conditions under positive-ion mode ESI-MS with a coaxial sheath liquid interface. New advances in online sample preconcentration, chemical derivatization, and/or ESI interface designs can further improve assay performance allowing for resolution of amino acid stereoisomers and labile aminothiols with low nanomolar detection limits.	[166]
Pesticide residues	Fruits	CE-ESI-qMS Via sheath liquid sprayer PI mode	The best results were obtained using 35 mM ammonium formiate (pH 9.7) as separation buffer, 20° C as capillary temperature and 23 kV as applied voltage in an uncoated fused-silica capillary with 50 cm total length, 25 cm thermostated length, 25 cm at room temperature length, connected to an MS detector by an ESI sprayer kit.	[174]

CE, capillary electrophoresis; ^{3}DIT, three-dimensional ion trap; ESI, electrospray ionization; FMOC, 9-Fluorenylmethoxycarbonyl; LLE, liquid/liquid extraction; MS, mass spectrometry; MS/MS, tandem mass spectrometry; PI, positive ionization; PNP, Programed nebulizing-gas pressure mode; QuEChERS, quick, easy, cheap, effective, rudge and safe; qMS, quadrupole mass spectrometry; QqQ, triple quadrupole.

6. CONCLUSIONS AND FUTURE TRENDS

Mass spectrometry has ideal features to be used in most areas of food safety and quality including food component analysis, food processing, food quality/authenticity assessment, and food additives, and contaminant analysis. The actual and potential contributions of the mass spectrometry to these areas are unlimited. However, this impact is still restrained by the instrumental cost as well as the technology and software/database availability. In this sense, recent developments in mass spectrometry have provided new options to ensure food quality and safety based mainly on profiling/fingerprinting and metabolomic approaches.

In the light of current knowledge, there are two ways to work with different combinations of mass spectrometry—target and nontarget analysis. The former has the advantage of allowing the quantification with accuracy and precision of the analytes and the inconvenience of any related molecule not considered a priori remains invisible for the analyst. The latter has the benefit of being able to detect any substance present in the sample and that can be detected by the selected technique but the drawback that identification is as much semiquantitative.

The determination of individual compounds began working with specific and quantitative methods and now as it is moving to multiscreening and wide-coverage methods at the same time as in the "omics," the process is backward, i.e., is moving from techniques that are capable of detecting a wide range of compounds to those that detect a few but quantitatively. The above fact is a good indicator of the interest of both types of systems. Each one has advantages and limitations that make them complementary.

Mass spectrometry can be combined with different ionization sources and/or chromatographic techniques which may be complementary or alternative depending on the situation. Each one is certainly more appropriate to cover a specific aspect. MALDI ionization offers unique versatility to identify compounds of high molecular weight such as proteins and lipids. ICP is the interface of choice for the determination of the different chemical elements. ESI works for any organic molecule easily ionizable while EI is the technique of choice for volatile nonpolar molecules. This set of techniques allows to tackle almost any problem within the food quality and food safety.

The full possibilities of MS within this field are far away to be reached. Future trends in MS are the wider implementation of (ultra)high-resolution tandem mass analyzers providing high-mass accuracy, the test IMS (considering it as a second MS dimension) as well as the establishment of the ambient ionization techniques and the IMS. As recent progress demonstrates, chemometrics can be used in parallel with data collected from food samples to test food associations, to explore biological mechanisms, and to determine origin and authenticity. This will be also an area of future research within MS.

REFERENCES

[1] WHO, Assuring Food Safety and Quality: Guidelines for Strengthening National Food Control Systems, vol. 79, Food and Agriculture Organization of the United Nations, World Health Organization, Rome, 2003.

[2] FoodDrink Europe, Priorities for the development of an EU industrial policy for food, Competitiveness Report (2012) pp.1–44 (available at http://www.fooddrinkeurope.eu).

[3] A.K. Malik, C. Blasco, Y. Pico, Liquid chromatography-mass spectrometry in food safety, J. Chromatogr. A 1217 (25) (2010) 4018–4040.

[4] M.P. Hermo, et al., High-resolution mass spectrometry applied to the study of metabolome modifications in various chicken tissues after amoxicillin administration, Food Chem. 153 (2014) 405–413.

[5] Y. Shao, et al., Determination of flavones in *Lophatherum gracile* by liquid chromatography tandem mass spectrometry, Instrum. Sci. Technol. 42 (2) (2014) 173–183.

[6] J.E. Lee, et al., Determination of chokeberry (*Aronia melanocarpa*) polyphenol components using liquid chromatography-tandem mass spectrometry: overall contribution to antioxidant activity, Food Chem. 146 (2014) 1–5.

[7] H. Tsugawa, et al., Highly sensitive and selective analysis of widely targeted metabolomics using gas chromatography/triple-quadrupole mass spectrometry, J. Biosci. Bioeng. 117 (1) (2014) 122–128.

[8] C.D. Calvano, C. De Ceglie, C.G. Zambonin, Proteomic analysis of complex protein samples by maldi-tof mass spectrometry, Methods Mol. Biol. 1129 (2014) 365–380 Clifton, N.J.

[9] C. von Bargen, et al., New sensitive high-performance liquid chromatography tandem mass spectrometry method for the detection of horse and pork in Halal beef, J. Agr. Food Chem. 61 (49) (2013) 11986–11994.

[10] P. Cheng, W. Fan, Y. Xu, Quality grade discrimination of Chinese strong aroma type liquors using mass spectrometry and multivariate analysis, Food Res. Int. 54 (2) (2013) 1753–1760.

[11] C. Hu, G. Xu, Mass-spectrometry-based metabolomics analysis for foodomics, Trac-Trends Anal. Chem. 52 (2013) 36–46.

[12] M. Castro-Puyana, M. Herrero, Metabolomics approaches based on mass spectrometry for food safety, quality and traceability, Trac-Trends Anal. Chem. 52 (2013) 74–87.

[13] C. Ibanez, et al., Metabolomics, peptidomics and proteomics applications of capillary electrophoresis-mass spectrometry in Foodomics: a review, Anal. Chim. Acta 802 (2013) 1–13.

[14] V. Di Stefano, et al., Applications of liquid chromatography-mass spectrometry for food analysis, J. Chromatogr. A 1259 (2012) 74–85.

[15] R. Mohamed, P.A. Guy, The pivotal role of mass spectrometry in determining the presence of chemical contaminants in food raw materials, Mass Spectrom. Rev. 30 (6) (2011) 1073–1095.

[16] X. Wang, S. Wang, Z. Cai, The latest developments and applications of mass spectrometry in food-safety and quality analysis, Trac-Trends Anal. Chem. 52 (2013) 170–185.

[17] G. Picariello, et al., Novel mass spectrometry-based applications of the 'Omic' sciences in food technology and biotechnology, Food Technol. Biotechnol. 50 (3) (2012) 286–305.

[18] D. Aiello, et al., Multistage mass spectrometry in quality, safety and origin of foods, Eur. J. Mass Spectrom. 17 (1) (2011) 1–31.

[19] M. Thaysen-Andersen, N.H. Packer, Advances in LC-MS/MS-based glycoproteomics: getting closer to system-wide site-specific mapping of the N- and O-glycoproteome, Biochim. Biophy. Acta-Proteins and Proteomics 1844 (9) (2014) 1437–1452.

[20] R.N. Rao, Emerging liquid chromatography-mass spectrometry technologies improving dried blood spot analysis, Expert Rev. Proteomics 11 (4) (2014) 425–430.

[21] M. Pernemalm, J. Lehtio, Mass spectrometry-based plasma proteomics: state of the art and future outlook, Expert Rev. Proteomics 11 (4) (2014) 431–448.

[22] H. Awad, M.M. Khamis, A. El-Aneed, Mass spectrometry, review of the Basics: ionization, Appl. Spectrosc. Rev. 50 (2) (2015) 158–175.

[23] O. Gonzalez, et al., Bioanalytical chromatographic method validation according to current regulations, with a special focus on the non-well defined parameters limit of quantification, robustness and matrix effect, J. Chromatogr. A 1353 (2014) 10–27.

[24] H. Faber, M. Vogel, U. Karst, Electrochemistry/mass spectrometry as a tool in metabolism studies-A review, Anal. Chim. Acta 834 (2014) 9–21.

[25] X.F. Zhong, et al., Recent advances in coupling capillary electrophoresis- based separation techniques to ESI and MALDI-MS, Electrophoresis 35 (9) (2014) 1214–1225.

[26] R. Ramautar, G.J. de Jong, Recent developments in liquid-phase separation techniques for metabolomics, Bioanalysis 6 (7) (2014) 1011–1026.

[27] S. Taira, et al., Mass spectrometry imaging: applications to food science, Anal. Sci. 30 (2) (2014) 197–203.

[28] L. Gao, et al., Precision and accuracy in the quantitative analysis of biological samples by accelerator mass spectrometry: application in microdose absolute bioavailability studies, Analytical Chem. 83 (14) (2011) 5607–5616.

[29] I. Ortea, B. Canas, J.M. Gallardo, Selected tandem mass spectrometry ion monitoring for the fast identification of seafood species, J. Chromatogr. A 1218 (28) (2011) 4445–4451.

[30] J. Song, et al., Fingerprint analysis of *Ginkgo biloba* leaves and related health foods by high-performance liquid chromatography/electrospray ionization-mass spectrometry, J. AOAC Int. 93 (6) (2010) 1798–1805.

[31] C. Junot, et al., High resolution mass spectrometry based techniques at the crossroads of metabolic pathways, Mass Spectrom. Rev. 33 (6) (2014) 471–500.

[32] E.B. Erba, Investigating macromolecular complexes using top-down mass spectrometry, Proteomics 14 (10) (2014) 1259–1270.

[33] K.P. Law, Y.P. Lim, Recent advances in mass spectrometry: data independent analysis and hyper reaction monitoring, Expert Rev. Proteomics 10 (6) (2013) 551–566.

[34] S.G. Ma, S.K. Chowdhury, Data acquisition and data mining techniques for metabolite identification using LC coupled to high-resolution MS, Bioanalysis 5 (10) (2013) 1285–1297.

[35] A. Makarov, M. Scigelova, Coupling liquid chromatography to Orbitrap mass spectrometry, J. Chromatogr. A 1217 (25) (2010) 3938–3945.

[36] R.A. Zubarev, A. Makarov, Orbitrap mass spectrometry, Anal. Chem. 85 (11) (2013) 5288–5296.

[37] A.A. Makarov, Orbitrap mass spectrometry in proteomics: past, present and future, FEBS J. 280 (2013) 632.

[38] A. Gahlaut, et al., β-sitosterol in different parts of *Saraca asoca* and herbal drug ashokarista: quali-quantitative analysis by liquid chromatography-mass spectrometry, J. Adv. Pharm. Technol. Res. 4 (3) (2013) 146–150.

[39] C. Mathon, et al., Multi-targeted screening of botanicals in food supplements by liquid chromatography with tandem mass spectrometry, Food Chem. 138 (1) (2013) 709–717.

[40] K. Fraser, et al., Analysis of low molecular weight metabolites in tea using mass spectrometry-based analytical methods, Crit. Rev. Food Sci. Nutr. 54 (7) (2014) 924–937.

[41] Y. Li, et al., A novel approach to transforming a non-targeted metabolic profiling method to a pseudo-targeted method using the retention time locking gas chromatography/mass spectrometry-selected ions monitoring, J. Chromatogr. A 1255 (2012) 228–236.

[42] E. Ates, et al., Screening of plant and fungal metabolites in wheat, maize and animal feed using automated on-line clean-up coupled to high resolution mass spectrometry, Food Chem. 142 (2014) 276–284.
[43] K. Kalachova, et al., Gas chromatography-triple quadrupole tandem mass spectrometry: a powerful tool for the (ultra)trace analysis of multiclass environmental contaminants in fish and fish feed, Anal. Bioanal. Chem. 405 (24) (2013) 7803–7815.
[44] S. Guo, et al., Hydrophilic interaction ultra-high performance liquid chromatography coupled with triple quadrupole mass spectrometry for determination of nucleotides, nucleosides and nucleobases in Ziziphus plants, J. Chromatogr. A 1301 (2013) 147–155.
[45] L. Liu, et al., Determination of aflatoxins in medicinal herbs by high-performance liquid chromatography-tandem mass spectrometry, Phytochem. Anal. 23 (5) (2012) 469–476.
[46] Y. Niu, et al., Analysis of bisphenol A and alkylphenols in cereals by automated on-line solid-phase extraction and liquid chromatography tandem mass spectrometry, J. Agr. Food Chem. 60 (24) (2012) 6116–6122.
[47] R. Romero-Gonzalez, et al., Simultaneous determination of four biogenic and three volatile amines in anchovy by ultra-high-performance liquid chromatography coupled to tandem mass spectrometry, J. Agr. Food Chem. 60 (21) (2012) 5324–5329.
[48] F. Wieland, et al., Online monitoring of coffee roasting by proton transfer reaction time-of-flight mass spectrometry (PTR-ToF-MS): towards a real-time process control for a consistent roast profile, Anal. Bioanal. Chem. 402 (8) (2012) 2531–2543.
[49] C. Zhen, et al., Differentiation of Chinese liquors by using ambient glow discharge ionization mass spectrometry, Analyst 138 (13) (2013) 3830–3835.
[50] A. Mari, et al., Liquid chromatography tandem mass spectrometry determination of chemical markers and principal component analysis of *Vitex agnus-castus* L. fruits (Verbenaceae) and derived food supplements, J. Pharm. Biomed. Anal. 70 (2012) 224–230.
[51] H. Gallart-Ayala, et al., Analysis of UV ink photoinitiators in packaged food by fast liquid chromatography at sub-ambient temperature coupled to tandem mass spectrometry, J. Chromatogr. A 1218 (3) (2011) 459–466.
[52] H. Gallart-Ayala, E. Moyano, M.T. Galceran, Analysis of bisphenols in soft drinks by on-line solid phase extraction fast liquid chromatography tandem mass spectrometry, Anal. Chim. Acta 683 (2) (2011) 227–233.
[53] R. Russo, et al., Detection of buffalo mozzarella adulteration by an ultra-high performance liquid chromatography tandem mass spectrometry methodology, J. Mass Spectrom. 47 (11) (2012) 1407–1414.
[54] A.I. Nesvizhskii1, O. Vitek, R. Aebersold, Analysis and validation of proteomic data generated by tandem mass spectrometry, Nat. Methods 3 (2007) 10.
[55] M. Sugimoto, et al., Bioinformatics tools for mass spectroscopy-based metabolomic data processing and analysis, Curr. Bioinform. 7 (1) (2012) 96–108.
[56] Q. Shen, et al., Lipidomic study of olive fruit and oil using TiO_2 nanoparticle based matrix solid-phase dispersion and MALDI-TOF/MS, Food Res. Int. 54 (2) (2013) 2054–2061.
[57] W. Romao, et al., Fingerprinting of bottle-grade poly(ethylene terephthalate) via matrix-assisted laser desorption/ionization mass spectrometry, Polym. Degrad. Stabil. 95 (4) (2010) 666–671.
[58] K.P. Law, Laser desorption/ionization mass spectrometry on nanostructured semiconductor substrates: DIOS™ and QuickMass™, Int. J. Mass Spectrom. 290 (2–3) (2010) 72–84.
[59] B. Marcos, et al., Surface-enhanced laser desorption/ionisation time-of-flight mass spectrometry: a tool to predict pork quality, Meat Sci. 95 (3) (2013) 688–693.

[60] H. Kawasaki, et al., Desorption/ionization on porous silicon mass spectrometry (DIOS-MS) of perfluorooctane sulfonate (PFOS), Colloids Surf. A Physicochem. Eng. Asp. 347 (1–3) (2009) 220–224.

[61] S. Francese, et al., Curcumin: a multipurpose matrix for MALDI mass spectrometry imaging applications, Anal. Chem. 85 (10) (2013) 5240–5248.

[62] N. Goto-Inoue, M. Setou, N. Zaima, Visualization of spatial distribution of gamma-aminobutyric acid in eggplant (*Solanum melongena*) by matrix-assisted laser desorption/ionization imaging mass spectrometry, Anal. Sci. 26 (7) (2010) 821–825.

[63] J.A. Pattemore, et al., Cereal variety identification using MALDI-TOF mass spectrometry SNP genotyping, J. Cereal Sci. 52 (3) (2010) 356–361.

[64] T. Kuda, et al., Rapid identification of *Tetragenococcus halophilus* and *Tetragenococcus muriaticus*, important species in the production of salted and fermented foods, by matrix-assisted laser desorption ionization-time of flight mass spectrometry (MALDI-TOF MS), Food Control 35 (1) (2014) 419–425.

[65] J.C. Usbeck, et al., Optimization of experimental and modelling parameters for the differentiation of beverage spoiling yeasts by Matrix-Assisted-Laser-Desorption/Ionization Time-of-Flight Mass Spectrometry (MALDI-TOF MS) in response to varying growth conditions, Food Microbiol. 36 (2) (2013) 379–387.

[66] C.C. Kern, et al., Optimization of matrix-assisted-laser-desorption-ionization-time-of-flight mass spectrometry for the identification of bacterial contaminants in beverages, J. Microbiol. Methods 93 (3) (2013) 185–191.

[67] M. Marchetti-Deschmann, et al., Using spores for *Fusarium* spp. Classification by MALDI-based intact Cell/Spore mass spectrometry, Food Technol. Biotechnol. 50 (3) (2012) 334–342.

[68] K. Boehme, et al., Comparative analysis of protein extraction methods for the identification of seafood-borne pathogenic and spoilage bacteria by MALDI-TOF mass spectrometry, Anal. Methods 2 (12) (2010) 1941–1947.

[69] P.-T. Kim-Yen, G.C. Wright, N.A. Lee, Inductively coupled plasma-mass spectrometry (ICP-MS) and -optical emission spectroscopy (ICP-OES) for determination of essential minerals in closed acid digestates of peanuts (*Arachis hypogaea* L.), Food Chem. 134 (1) (2012) 453–460.

[70] G. Lo Feudo, et al., Investigating the origin of tomatoes and triple concentrated tomato pastes through multielement determination by inductively coupled plasma mass spectrometry and statistical analysis, J. Agr. Food Chem. 58 (6) (2010) 3801–3807.

[71] S. Orecchio, et al., Determination of trace elements in gluten-free food for celiac people by ICP-MS, Microchem. J. 116 (0) (2014) 163–172.

[72] L.H. Pacquette, A.M. Levenson, J.J. Thompson, Determination of total iodine in infant formula and nutritional products by inductively coupled plasma/mass spectrometry: single-laboratory validation, J. AOAC Int. 95 (1) (2012) 169–176.

[73] S. Millour, et al., Internal quality controls applied in inductively coupled plasma mass spectrometry multi-elemental analysis in the second French Total Diet Study, Accredit. Qual. Assur. 15 (9) (2010) 503–513.

[74] A. Sorbo, et al., Development and validation of an analytical method for the determination of arsenic, cadmium and lead content in powdered infant formula by means of quadrupole Inductively Coupled Plasma Mass Spectrometry, Food Control 44 (0) (2014) 159–165.

[75] P. Cheajesadagul, et al., Discrimination of geographical origin of rice based on multi-element fingerprinting by high resolution inductively coupled plasma mass spectrometry, Food Chem. 141 (4) (2013) 3504–3509.

[76] N.M. Kreitals, R.J. Watling, Multi-element analysis using inductively-coupled plasma mass spectrometry and inductively-coupled plasma emission spectroscopy for provenancing of animals at the Continental scale, Forensic Sci. Int. 244 (0) (2014) 116–121.

[77] A. Leufroy, et al., Determination of seven arsenic species in seafood by ion exchange chromatography coupled to inductively coupled plasma-mass spectrometry following microwave assisted extraction: method validation and occurrence data, Talanta 83 (3) (2011) 770–779.

[78] T. Llorente-Mirandes, et al., A need for determination of arsenic species at low levels in cereal-based food and infant cereals. Validation of a method by IC–ICPMS, Food Chem. 147 (0) (2014) 377–385.

[79] I. López-Heras, Y. Madrid, C. Cámara, Prospects and difficulties in TiO_2 nanoparticles analysis in cosmetic and food products using asymmetrical flow field-flow fractionation hyphenated to inductively coupled plasma mass spectrometry, Talanta 124 (0) (2014) 71–78.

[80] L. Xia, J. You, The determination of amino acids composition of the traditional food *Potentilla anserina* L. root by high-performance liquid chromatography via fluorescent detection and mass spectrometry, Int. J. Food Sci. Technol. 46 (6) (2011) 1164–1170.

[81] X. Yao, et al., Direct determination of underivatized amino acids from *Ginkgo biloba* leaves by using hydrophilic interaction ultra high performance liquid chromatography coupled with triple quadrupole mass spectrometry, J. Sep. Sci. 36 (17) (2013) 2878–2887.

[82] B.-S. Chu, K. Nagy, Enrichment and quantification of monoacylglycerols and free fatty acids by solid phase extraction and liquid chromatography-mass spectrometry, J. Chromatogr. B Analyt. Technol. Biomed. Life Sci. 932 (2013) 50–58.

[83] S.J. Lee, et al., Determination of phospholipids in soybean (Glycine max (L.) Merr) cultivars by liquid chromatography-tandem mass spectrometry, J. Food Composit. Anal. 23 (4) (2010) 314–318.

[84] J.A. Sealey-Voyksner, et al., Novel aspects of quantitation of immunogenic wheat gluten peptides by liquid chromatography-mass spectrometry/mass spectrometry, J. Chromatogr. A 1217 (25) (2010) 4167–4183.

[85] J.E. Camara, M.S. Lowenthal, K.W. Phinney, Determination of fortified and endogenous folates in food-based Standard Reference Materials by liquid chromatography-tandem mass spectrometry, Anal. Bioanal. Chem. 405 (13) (2013) 4561–4568.

[86] J.-L. Hong, et al., Comparative study of isoflavones in wild and cultivated soybeans as well as bean products by high-performance liquid chromatography coupled with mass spectrometry and chemometric techniques, Eur. Food Res. Technol. 233 (5) (2011) 869–880.

[87] H. Zhang, et al., Identification of flavonol and triterpene glycosides in Luo-Han-Guo extract using ultra-high performance liquid chromatography/quadrupole time-of-flight mass spectrometry, J. Food Compos. Anal. 25 (2) (2012) 142–148.

[88] C.M. Willemse, et al., Comprehensive two-dimensional liquid chromatographic analysis of anthocyanins, J. Chromatogr. A 1359 (0) (2014) 189–201.

[89] Y. Shi, et al., Simultaneous determination of ten ginsenosides in american ginseng functional foods and ginseng raw plant materials by liquid chromatography tandem mass spectrometry, Food Anal. Methods 6 (1) (2013) 112–122.

[90] M.-D. Zhu, et al., Simultaneous determination of active components and anti-diabetic drugs in propolis health foods by ultra performance liquid chromatography tandem mass spectrometry, Chinese J. Anal. Chem. 38 (2) (2010) 169–174.

[91] S. Ehling, S. Cole, Analysis of organic acids in fruit juices by liquid chromatography-mass spectrometry: an enhanced tool for authenticity testing, J. Agr. Food Chem. 59 (6) (2011) 2229–2234.

[92] Y. Zheng, et al., Quantitative and transformation product analysis of major active physalins from *Physalis alkekengi* var. *franchetii* (Chinese Lantern) using ultraperformance liquid chromatography with electrospray ionisation tandem mass spectrometry and time-of-flight mass spectrometry, Phytochem. Anal. 23 (4) (2012) 337–344.

[93] S. Fu, et al., Determination of total choline by liquid chromatography-electrospray ionization-tandem mass spectrometry in infant formulas, J. AOAC Int. 95 (1) (2012) 157–162.

[94] Y. Niu, et al., Identification of isoflavonoids in Radix Puerariae for quality control using on-line high performance liquid chromatography-diode array detector-electrospray ionization-mass spectrometry coupled with post-column derivatization, Food Res. Int. 48 (2) (2012) 528–537.

[95] R. Liu, et al., Characterisation of nucleosides and nucleobases in *Mactra veneriformis* by high performance liquid chromatography coupled with diode array detector-mass spectrometry (HPLC-DAD-MS), Food Chem. 135 (2) (2012) 548–554.

[96] Z. Spacil, L. Novakova, P. Solich, Comparison of positive and negative ion detection of tea catechins using tandem mass spectrometry and ultra high performance liquid chromatography, Food Chem. 123 (2) (2010) 535–541.

[97] C. Pouech, et al., Monitoring the extraction of additives and additive degradation products from polymer packaging into solutions by multi-residue method including solid phase extraction and ultra-high performance liquid chromatography-tandem mass spectrometry analysis, Anal. Bioanal. Chem. 406 (5) (2014) 1493–1507.

[98] B.R. Ramaswamy, et al., Determination of preservative and antimicrobial compounds in fish from Manila Bay, Philippines using ultra high performance liquid chromatography tandem mass spectrometry, and assessment of human dietary exposure, J. Hazard. Mater. 192 (3) (2011) 1739–1745.

[99] F. Fuselli, et al., Multi-detection of preservatives in cheeses by liquid chromatography-tandem mass spectrometry, J. Chromatogr. B Anal. Technol. Biomed. Life Sci. 906 (2012) 9–18.

[100] J. Song, et al., Determination of cyanidin-3-glucoside (red kernel food colour) in beverages by high performance liquid chromatography and a study of its degradation by quadruple time-of-flight mass spectrometry, Food Addit. Contam. Part A Chem. Anal. Control Expo. Risk Assess. 28 (12) (2011) 1645–1656.

[101] C. Ji, et al., Determination of five synthetic sweeteners in wines using high performance liquid chromatography-tandem mass spectrometry, Se pu Chin. J. Chromatogr./Zhongguo hua xue hui 28 (8) (2010) 749–753.

[102] L-x. Fang, et al., A strategy for screening and identifying mycotoxins in herbal medicine using ultra-performance liquid chromatography with tandem quadrupole time-of-flight mass spectrometry, J. Sep. Sci. 36 (18) (2013) 3115–3122.

[103] A. Desmarchelier, et al., Development and comparison of two multiresidue methods for the analysis of 17 mycotoxins in cereals by liquid chromatography electrospray ionization tandem mass spectrometry, J. Agr. Food Chem. 58 (13) (2010) 7510–7519.

[104] Y.H. Leong, et al., Determination of aflatoxins in commercial nuts and nut products using liquid chromatography tandem mass spectrometry, World Mycotoxin J. 4 (2) (2011) 119–127.

[105] H.J. van den Top, et al., Quantitative determination of marine lipophilic toxins in mussels, oysters and cockles using liquid chromatography-mass spectrometry: inter-laboratory validation study, Food Addit. Contam. Part A Chem. Anal. Control Expo. Risk Assess. 28 (12) (2011) 1745–1757.

[106] M.Á. González-Curbelo, et al., Use of ammonium formate in QuEChERS for high-throughput analysis of pesticides in food by fast, low-pressure gas chromatography and liquid chromatography tandem mass spectrometry, J. Chromatogr. A 1358 (0) (2014) 75–84.

[107] J. Wong, et al., Development and interlaboratory validation of a QuEChERS-based liquid chromatography-tandem mass spectrometry method for multiresidue pesticide analysis, J. Agr. Food Chem. 58 (10) (2010) 5897–5903.

[108] B. Kmellar, et al., Routine approach to qualitatively screening 300 pesticides and quantification of those frequently detected in fruit and vegetables using liquid chromatography tandem mass spectrometry (LC-MS/MS), Food Addit. Contam. Part A Chem. Anal. Control Expo. Risk Assess. 27 (10) (2010) 1415–1430.

[109] D. Chen, et al., Development of a liquid chromatography-tandem mass spectrometry (LC-MS/MS) method for the quantification of glucocorticoid residues in edible tissues of swine, cattle, sheep, and chicken, Food Addit. Contam. Part A Chem. Anal. Control Expo. Risk Assess. 27 (10) (2010) 1363–1371.

[110] A. Merou, G. Kaklamanos, G. Theodoridis, Determination of Carbadox and metabolites of Carbadox and Olaquindox in muscle tissue using high performance liquid chromatography-tandem mass spectrometry, J. Chromatogr. B Anal. Technol. Biomed. Life Sci. 881 (82) (2012) 90–95.

[111] F. Tateo, M. Bononi, F. Gallone, Acrylamide content in potato chips on the Italian market determined by liquid chromatography tandem mass spectrometry, Int. J. Food Sci. Technol. 45 (3) (2010) 629–634.

[112] S. Lee, et al., Determination of biogenic amines in Cheonggukjang using ultra high pressure liquid chromatography coupled with mass spectrometry, Food Sci. Biotechnol. 20 (1) (2011) 123–129.

[113] L.V. Podhorniak, A. Kamel, D.M. Rains, Determination of formetanate hydrochloride in fruit samples using liquid chromatography-mass selective detection or-tandem mass spectrometry, J. Agr. Food Chem. 58 (10) (2010) 5862–5867.

[114] S. Kittlaus, et al., Development and validation of an efficient automated method for the analysis of 300 pesticides in foods using two-dimensional liquid chromatography–tandem mass spectrometry, J. Chromatogr. A 1283 (0) (2013) 98–109.

[115] S. Kittlaus, G. Kempe, K. Speer, Evaluation of matrix effects in different multipesticide residue analysis methods using liquid chromatography-tandem mass spectrometry, including an automated two-dimensional cleanup approach, J. Sep. Sci. 36 (13) (2013) 2185–2195.

[116] E. De Dominicis, I. Commissati, M. Suman, Targeted screening of pesticides, veterinary drugs and mycotoxins in bakery ingredients and food commodities by liquid chromatography-high-resolution single-stage Orbitrap mass spectrometry, J. Mass Spectrom. 47 (9) (2012) 1232–1241.

[117] S. Savic, et al., Enzymatic oxidation of rutin by horseradish peroxidase: kinetic mechanism and identification of a dimeric product by LC-Orbitrap mass spectrometry, Food Chem. 141 (4) (2013) 4194–4199.

[118] H.J. Kang, et al., Metabolomic analysis of meju during fermentation by ultra performance liquid chromatography-quadrupole-time of flight mass spectrometry (UPLC-Q-TOF MS), Food Chem. 127 (3) (2011) 1056–1064.

[119] I.M. Abu-Reidah, et al., Phytochemical characterisation of Green beans (*Phaseolus vulgaris* L.) by using high-performance liquid chromatography coupled with time-of-flight mass spectrometry, Phytochem. Anal. 24 (2) (2013) 105–116.

[120] D.B. Clarke, A.S. Lloyd, P. Robb, Application of liquid chromatography coupled to time-of-flight mass spectrometry separation for rapid assessment of toxins in Amanita mushrooms, Anal. Methods 4 (5) (2012) 1298–1309.

[121] M. Gebriel, et al., Cod (Gadus morhua) muscle proteome cataloging using 1D-PAGE protein separation, nano-liquid chromatography peptide fractionation, and linear trap quadrupole (LTQ) mass spectrometry, J. Agr. Food Chem. 58 (23) (2010) 12307–12312.

[122] K. Inoue, et al., Foodomics platform for the assay of thiols in wines with fluorescence derivatization and ultra performance liquid chromatography mass spectrometry using multivariate statistical analysis, J. Agr. Food Chem. 61 (6) (2013) 1228–1234.

[123] S. Moco, J. Vervoort, Chemical identification strategies using liquid chromatography-photodiode array-solid-phase extraction-nuclear magnetic resonance/mass spectrometry, Methods Mol. Biol. 860 (2012) 287–316 Clifton, N.J.

[124] F. Feng, et al., Simultaneous screening of 14 illegal food additives in wines using ultra performance liquid chromatography tandem mass spectrometry, Chinese J. Anal. Chem. 39 (11) (2011) 1732–1737.

[125] L.C. Jackson, M.B. Kudupoje, A. Yiannikouris, Simultaneous multiple mycotoxin quantification in feed samples using three isotopically labeled internal standards applied for isotopic dilution and data normalization through ultra-performance liquid chromatography/electrospray ionization tandem mass spectrometry, Rapid Commun. Mass Spectrom. 26 (23) (2012) 2697–2713.

[126] E. Fornal, A. Stachniuk, Application of a truly one-point calibration for pesticide residue control by liquid chromatography-mass spectrometry, J. Chromatogr. B Anal. Technol. Biomed. Life Sci. 901 (2012) 107–114.

[127] G. Chen, P. Cao, R. Liu, A multi-residue method for fast determination of pesticides in tea by ultra performance liquid chromatography-electrospray tandem mass spectrometry combined with modified QuEChERS sample preparation procedure, Food Chem. 125 (4) (2011) 1406–1411.

[128] M. Garcia-Chao, et al., Validation of an off line solid phase extraction liquid chromatography-tandem mass spectrometry method for the determination of systemic insecticide residues in honey and pollen samples collected in apiaries from NW Spain, Anal. Chim. Acta 672 (1–2) (2010) 107–113.

[129] S. Wu, E. Wang, Y. Yuan, Detection of furan levels in select Chinese foods by solid phase microextraction-gas chromatography/mass spectrometry method and dietary exposure estimation of furan in the Chinese population, Food Chem. Toxicol. 64 (2014) 34–40.

[130] W-m. Mo, et al., Simultaneous determination of ethyl carbamate, chloropropanols and acrylamide in fermented products, flavoring and related foods by gas chromatography–triple quadrupole mass spectrometry, Food Control 43 (0) (2014) 251–257.

[131] J.C. Baffa Junior, et al., Ethyl-carbamate determination by gas chromatography-mass spectrometry at different stages of production of a traditional Brazilian spirit, Food Chem. 129 (4) (2011) 1383–1387.

[132] M.M.A. Omar, W.A. Wan Ibrahim, A.A. Elbashir, Sol–gel hybrid methyltrimethoxysilane–tetraethoxysilane as a new dispersive solid-phase extraction material for acrylamide determination in food with direct gas chromatography–mass spectrometry analysis, Food Chem. 158 (0) (2014) 302–309.

[133] M.V. Russo, et al., Rapid and simple determination of acrylamide in conventional cereal-based foods and potato chips through conversion to 3-[bis(trifluoroethanoyl)amino]-3-oxopropyl trifluoroacetate by gas chromatography coupled with electron capture and ion trap mass spectrometry detectors, Food Chem. 146 (0) (2014) 204–211.

[134] J.-H. Wang, C. Guo, Ultrasonication extraction and gel permeation chromatography clean-up for the determination of polycyclic aromatic hydrocarbons in edible oil by an isotope dilution gas chromatography-mass spectrometry, J. Chromatogr. A 1217 (28) (2010) 4732–4737.

[135] D. Lu, et al., Determination of polybrominated diphenyl ethers and polychlorinated biphenyls in fishery and aquaculture products using sequential solid phase extraction and large volume injection gas chromatography/tandem mass spectrometry, J. Chromatogr. B 945–946 (0) (2014) 75–83.

[136] J. Malavia, F.J. Santos, M.T. Galceran, Simultaneous pressurized liquid extraction and clean-up for the analysis of polybrominated biphenyls by gas chromatography-tandem mass spectrometry, Talanta 84 (4) (2011) 1155–1162.

[137] Y. Deceuninck, et al., Development and validation of a specific and sensitive gas chromatography tandem mass spectrometry method for the determination of bisphenol A residues in a large set of food items, J. Chromatogr. A 1362 (1) (2014) 241–249.

[138] G. Cinelli, et al., Study of XAD-2 adsorbent for the enrichment of trace levels of phthalate esters in hydroalcoholic food beverages and analysis by gas chromatography coupled with flame ionization and ion-trap mass spectrometry detectors, Food Chem. 146 (0) (2014) 181–187.

[139] H. Kwon, S.J. Lehotay, L. Geis-Asteggiante, Variability of matrix effects in liquid and gas chromatography-mass spectrometry analysis of pesticide residues after QuEChERS sample preparation of different food crops, J. Chromatogr. A 1270 (2012) 235–245.

[140] D. Garcia-Rodriguez, et al., Analysis of pesticide residues in seaweeds using matrix solid-phase dispersion and gas chromatography-mass spectrometry detection, Food Chem. 135 (1) (2012) 259–267.

[141] X. Wang, et al., Screening for pesticide residues in oil seeds using solid-phase dispersion extraction and comprehensive two-dimensional gas chromatography time-of-flight mass spectrometry, J. Sep. Sci. 35 (13) (2012) 1634–1643.

[142] J.W. Wong, et al., Multiresidue pesticide analysis in fresh produce by capillary Gas chromatography-mass Spectrometry/Selective ion monitoring (GC-MS/SIM) and-tandem mass spectrometry (GC-MS/MS), J. Agr. Food Chem. 58 (10) (2010) 5868–5883.

[143] H.R. Norli, A. Christiansen, B. Holen, Independent evaluation of a commercial deconvolution reporting software for gas chromatography mass spectrometry analysis of pesticide residues in fruits and vegetables, J. Chromatogr. A 1217 (1) (2010) 2056–2064.

[144] J. Li, et al., Enantioselective determination of triazole fungicide simeconazole in vegetables, fruits, and cereals using modified QuEChERS (quick, easy, cheap, effective, rugged and safe) coupled to gas chromatography/tandem mass spectrometry, Anal. Chim. Acta 702 (1) (2011) 127–135.

[145] S-k. Wong, K-c. Yu, C-h. Lam, Isotope dilution gas chromatography/mass spectrometry method for determination of pyrethroids in apple juice, Anal. Bioanal. Chem. 396 (5) (2010) 1877–1884.

[146] M.L. Feo, E. Eljarrat, D. Barcelo, Performance of gas chromatography/tandem mass spectrometry in the analysis of pyrethroid insecticides in environmental and food samples, Rapid Commun. Mass Spectrom. 25 (7) (2011) 869–876.

[147] B. Gilbert-Lopez, et al., Rapid determination of BTEXS in olives and olive oil by headspace-gas chromatography/mass spectrometry (HS-GC-MS), Talanta 83 (2) (2010) 391–399.

[148] I. Seo, H.-S. Shin, Determination of toluene and other residual solvents in various food packaging materials by Gas Chromatography/Mass spectrometry (GC/MS), Food Sci. Biotechnol. 19 (6) (2010) 1429–1434.

[149] J.K. Kim, et al., Metabolic profiling of millet (*Panicum miliaceum*) using gas chromatography-time-of-flight mass spectrometry (GC-TOFMS) for quality assessment, Plant Omics 6 (1) (2013) 73–78.

[150] S. Yamamoto, et al., Metabolite profiling of soy sauce using gas chromatography with time-of-flight mass spectrometry and analysis of correlation with quantitative descriptive analysis, J. Biosci. Bioeng. 114 (2) (2012) 170–175.

[151] N.P.L. Tuckey, J.R. Day, M.R. Miller, Determination of volatile compounds in New Zealand Greenshell (TM) mussels (*Perna canaliculus*) during chilled storage using solid phase micro-extraction gas chromatography-mass spectrometry, Food Chem. 136 (1) (2013) 218–223.

[152] A. Luedemann, et al., TagFinder: preprocessing software for the fingerprinting and the profiling of gas chromatography-mass spectrometry based metabolome analyses, Methods Mol. Biol. 860 (2012) 255–286 Clifton, N.J.

[153] W. Fan, H. Shen, Y. Xu, Quantification of volatile compounds in Chinese soy sauce aroma type liquor by stir bar sorptive extraction and gas chromatography-mass spectrometry, J. Sci. Food Agr. 91 (7) (2011) 1187–1198.

[154] C. Cordero, et al., Profiling food volatiles by comprehensive two-dimensional gas chromatography coupled with mass spectrometry: advanced fingerprinting approaches for comparative analysis of the volatile fraction of roasted hazelnuts (*Corylus avellana* L.) from different origins, J. Chromatogr. A 1217 (37) (2010) 5848–5858.

[155] E.M. Humston, et al., Quantitative assessment of moisture damage for cacao bean quality using two-dimensional gas chromatography combined with time-of-flight mass spectrometry and chemometrics, J. Chromatogr. A 1217 (12) (2010) 1963–1970.

[156] A. Figoli, et al., Evaluation of pervaporation process of kiwifruit juice by SPME-GC/ion trap mass spectrometry, Desalination 250 (3) (2010) 1113–1117.

[157] M.M. Ačanski, D.N. Vujić, Comparing sugar components of cereal and pseudocereal flour by GC–MS analysis, Food Chem. 145 (0) (2014) 743–748.

[158] M.L. Oca, et al., Optimum pH for the determination of bisphenols and their corresponding diglycidyl ethers by gas chromatography–mass spectrometry. Migration kinetics of bisphenol A from polycarbonate glasses, J. Chromatogr. A 1360 (0) (2014) 23–38.

[159] A.K. Sakhi, et al., Concentrations of phthalates and bisphenol A in Norwegian foods and beverages and estimated dietary exposure in adults, Environ. Int. 73 (0) (2014) 259–269.

[160] P. Poinot, et al., Study of current analytical strategies for the dual investigation of polycyclic aromatic hydrocarbons and benzene, toluene, ethylbenzene and xylene in apples, J. Food Compos. Anal. 35 (2) (2014) 83–93.

[161] D.R. Briskey, et al., Optimized method for quantification of total F2-isoprostanes using gas chromatography–tandem mass spectrometry, J. Pharm. Biomed. Anal. 90 (0) (2014) 161–166.

[162] G.A. Bataglion, et al., Comprehensive characterization of lipids from Amazonian vegetable oils by mass spectrometry techniques, Food Res. Int. 64 (0) (2014) 472–481.

[163] J. Ecker, G. Liebisch, Application of stable isotopes to investigate the metabolism of fatty acids, glycerophospholipid and sphingolipid species, Prog. Lipid Res. 54 (0) (2014) 14–31.

[164] B.L. Batista, et al., Mercury speciation in seafood samples by LC–ICP-MS with a rapid ultrasound-assisted extraction procedure: application to the determination of mercury in Brazilian seafood samples, Food Chem. 126 (4) (2011) 2000–2004.

[165] Y. Zhang, et al., Calibration and data processing in gas chromatography combustion isotope ratio mass spectrometry, Drug Test. Anal. 4 (12) (2012) 912–922.

[166] P. Britz-McKibbin, New advances in amino acid profiling by capillary electrophoresis-electrospray ionization-mass spectrometry, Methods Mol. Biol. 828 (2012) 83–99 Clifton, N.J.

[167] L. Sanchez-Hernandez, et al., Determination of betaines in vegetable oils by capillary electrophoresis tandem mass spectrometry - application to the detection of olive oil adulteration with seed oils, Electrophoresis 32 (11) (2011) 1394–1401.

[168] L. Sanchez-Hernandez, et al., Determination of L- and D-carnitine in dietary food supplements using capillary electrophoresis-tandem mass spectrometry, Food Chem. 120 (3) (2010) 921–928.

[169] K. Marakova, et al., Simultaneous analysis of vitamins B in pharmaceuticals and dietary supplements by capillary electrophoresis hyphenated with triple quadrupole mass spectrometry, Pharmazie 69 (9) (2014) 663–668.

[170] S. Akamatsu, T. Mitsuhashi, Development of a simple analytical method using capillary electrophoresis-tandem mass spectrometry for product identification and simultaneous determination of free amino acids in dietary supplements containing royal jelly, J. Food Compos. Anal. 30 (1) (2013) 47–51.
[171] M. Bustamante-Rangel, et al., QuEChERS method for the extraction of isoflavones from soy-based foods before determination by capillary electrophoresis-electrospray ionization-mass spectrometry, Microchem. J. 108 (2013) 203–209.
[172] J. Dominguez-Alvarez, et al., Capillary electrophoresis coupled to mass spectrometry for the determination of anthelmintic benzimidazoles in eggs using a QuEChERS with preconcentration as sample treatment, J. Chromatogr. A 1278 (2013) 166–174.
[173] J. Dominguez-Alvarez, et al., Programed nebulizing-gas pressure mode for quantitative capillary electrophoresis-mass spectrometry analysis of endocrine disruptors in honey, Electrophoresis 33 (15) (2012) 2374–2381.
[174] A. Juan-Garcia, et al., Pressurised liquid extraction and capillary electrophoresis-mass spectrometry for the analysis of pesticide residues in fruits from Valencian markets, Spain. Food Chem. 120 (4) (2010) 1242–1249.

Chapter 2

Advanced Mass Spectrometry

Yolanda Picó
Food and Environmental Safety Research Group, Faculty of Pharmacy, University of Valencia, Valencia, Spain
E-mail: Yolanda.Pico@uv.es

Chapter Outline

1. Introduction 77
2. Mass Spectrometric Resolution and Accuracy 78
2.1 Magnetic Sector 82
2.2 Time-of-Flight (TOF) 84
2.3 Fourier Transform Ion-Cyclotron Resonance (FT-ICR) 95
2.4 Orbitrap 97
3. Obtaining and Processing Information 100
3.1 Data-Dependent Acquisition (DDA) or Information-Dependent Acquisition (IDA) 105
3.2 Data-Independent Acquisition (DIA) 108
3.2.1 MS^E 109
3.2.2 All-Ion Fragmentation (AIF) 109
3.3 Preprocessing and Postprocessing Steps 111
4. Imaging Mass Spectrometry (IMS) 112
5. Conclusions 115
References 116

1. INTRODUCTION

Since the 1980s decade, mass spectrometry (MS) has been increasingly used in the food sciences to investigate the presence of target organic pollutants, micro and macronutrients, and other food components [1–4]. MS has been widely coupled with chromatographic techniques including gas chromatography (GC), liquid chromatography (LC), and capillary electrophoresis (CE) because of their complementary nature when facing a broad range of compounds of different polarity and volatility. Structural determination and identification are strengths of modern MS [1,3,5].

Rapid advances have been made toward the design of analytical instruments with improved sensitivity, delivering greater resolution to control food safety and quality [2,6]. The strategic separation techniques—ultrahigh

Comprehensive Analytical Chemistry, Vol. 68. http://dx.doi.org/10.1016/B978-0-444-63340-8.00002-9

performance liquid chromatography (UHPLC), GC, and the two-dimensional approach of both techniques (GC×GC or LC×LC)—may be used alone or, more commonly, in combination with MS and significantly increase the food control capability through a more integrated and comprehensive analytical approach (e.g., chemical fingerprinting) [7–9]. MS is acknowledged to be the third most powerful spectral technique in the structural elucidation of unknown organic compounds, trumped only by X-ray diffraction and NMR spectroscopy [10,11].

All these advances have been possible because MS instruments are becoming increasingly powerful. The high mass resolution has significantly enhanced the selectivity of the ions. The sensitivity of the systems has also increased considerably. Another noticeable direction of development includes the coupling or incorporation of ion mobility (IM) spectroscopy to the mass spectrometers that allow separation of ions in the gas-phase orthogonal to other separation techniques [12,13]. The combination of these advances has already permitted researchers to identify and/or quantify small and large molecules in complex biological systems (including foods) several more times than what it used to be [14,15]. The emerging methods of data acquisition and associated data minding or interrogation approaches are only possible with the improved performance of mass spectrometers [16]. MS in combination with separation methods and bioinformatic tools is the key analytical technique on which the emerging "omic" technologies such as proteomics, glycomics, lipidomics, transcriptomics, and metabolomics are based [6,17–19]. In particular, in the field of food sciences, MS-based techniques have rapidly evolved for determining quality, authenticity, functionality, and safety issues. The increased availability of high resolution mass spectrometry (HRMS) in chemical analysis has dramatically improved the detection and identification of compounds in food. The ultimate advance of HRMS has been the imaging mass spectrometry (IMS), which encompasses a wide variety of HRMS techniques that enable the chemical analysis of the analytes directly from food matrices [17].

This has opened up new research opportunities in food sciences, demonstrated in Figure 1, by over 100 research papers per year in the last 2 years and the strong increase during the last decade [20].

In this chapter, the principles and techniques of the highest resolution analytical mass spectrometers [time-of-flight (TOF), Fourier transform ion cyclotron resonance (FT-ICR), and Orbitrap mass analyzers] are reviewed and some representative HRMS applications in food safety and quality are described, mainly in terms of their applicability in routine analysis. This chapter will introduce also IMS and explains how it presents a paradigm shift in mass spectrometric detection, and finally presents a brief selection of applications.

2. MASS SPECTROMETRIC RESOLUTION AND ACCURACY

Since many years ago, HRMS has been used in food analysis, but until recently it was associated with elevated instrumental intricacy (magnetic

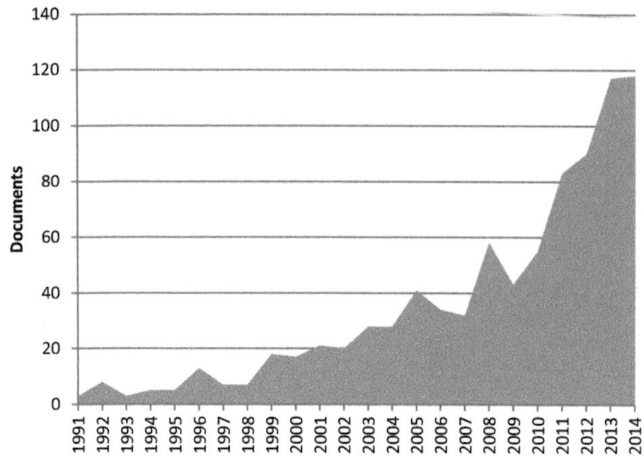

FIGURE 1 Number of articles published per year since 1990. *Scopus, keywords "high resolution mass spectrometry" and "food" using as search field abstract, title and keywords.*

sectors and double-focus magnetic sectors) [15,21]. Therefore, it was limited to the most demanding applications, such as the analysis of natural organic matter or dioxin-related compounds. The appearance of modern HRMS instruments such as TOF and Orbitrap has changed the situation. The use of HRMS in many fields of food safety and quality, including metabolite identification and structure elucidation, food contaminants, or food control, is notably increasing [22–25].

Apart from the fundamentals behind HRMS, detailed description of general MS topics including ionization processes and mass analyzers are found elsewhere [26–29]. Table 1 summarizes a comparison of the characteristics of commonly used mass analyzers. As could be observed, it is important to keep in mind two important basic concepts that are key points in HRMS. The performance of mass analyzers is typically quantitated in terms of resolution and mass accuracy. According to the nomenclature rules of the International Union of Pure and Applied Chemistry (IUPAC) [30] and the American Society for Mass Spectrometry (ASMS) [31], resolution or mass resolving power measures the ability to distinguish two peaks of slightly different mass-to-charge ratios and can be calculated using the formula:

$$mass\ resoving\ power = \frac{M}{\Delta M}$$

The resolving power can be calculated either from the peak width of the peak in the mass spectrum or from the valley between two continuous peaks of equal height (Figure 2).

The mass accuracy is the difference between the mass measured by the mass spectrometer and theoretical value. The mass accuracy can be expressed as an absolute value, for example in millidalton (mDa),

TABLE 1 A General Comparison of High Resolution Mass Analyzers

	Accuracy (ppm)	Resolving Power (FWHM)	*m/z* Range	Scan Speed	Linear Dynamic Range	Sensitivity	Tandem MS	Tandem MS Comments	Other Comments
Magnetic sector	<1	100,000	10,000	~1 s	1×10^9	10^6–10^9	MS^2	Limited resolution, high-energy collisions	High cost and large space requirements
TOF	3–5	30,000	>300,000	Milliseconds	10^6	10^6	MS		Moderate cost and space requirements
QqTOF	3–5	30,000	10,000	~Milliseconds	10^3–10^4	10^6	MS^2	Excellent accuracy, good resolution, low-energy collisions, high sensitivity	High cost
IT-TOF	3–5	100,000		~0.1 s	10^3–10^4	10^5	MS^n	Excellent accuracy, good resolution, low-energy collisions, high sensitivity	High cost

TOF–TOF	3–5	100,000	>500,000	~0.1 s	10^3–10^4	10^3–10^4	MS^2	Excellent accuracy, good resolution, low-energy collisions, high sensitivity	Very high cost and space requirements
FT-ICR	<1	1,000,000	10,000	~1 s	10^3–10^4	10^3–10^4	MS^n	Excellent accuracy and resolution of product ions	High cost and large space requirements
Orbitrap	<3	100,000	4000	0.1 s	>5000	10^6	MS		Moderate cost and space requirements
LTQ-Orbitrap	<3	100,000	4000	0.1 s	>5000	10^6	MS^n	$n=1$ to 10 Multiple dissociation techniques: CID, ETD, and HCD	Parallel MS and MS^n analysis
Q-Orbitrap	<3	140,000	6000	0.1 s	>5000	10^6	MS^2	Excellent accuracy, good resolution, high sensitivity	Spectral multiplexing for enhanced duty cycle

CID, collision induced dissociation; ETD, electron transfer dissociation; FT-ICR, fourier transform ion cyclotron resonance; HCD, higher energy collisional dissociation; IT-TOF, ion-trap time-of-flight; LTQ-Orbitrap, linear ion trap Orbitrap; Q-Orbitrap, single quadrupole Orbitrap; QqTOF, quadrupole time-of-flight; TOF, time-of-flight.

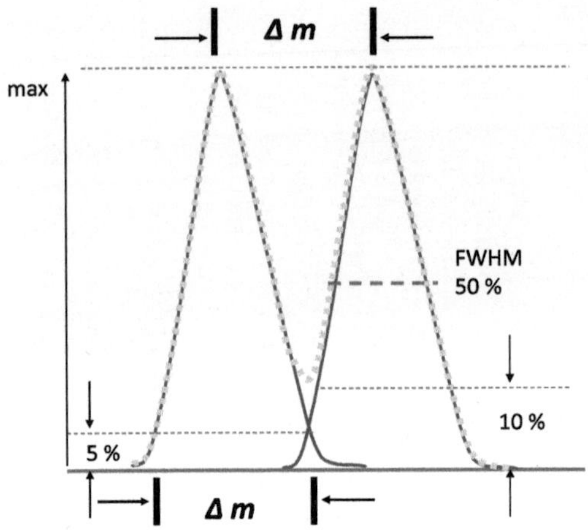

FIGURE 2 Methods of calculating mass resolving power.

$$mass\ accuracy\ (mDa) = \left(m/z_{(exp)} - m/z_{(theor)}\right) \cdot 1000$$

Or as a relative value, for example, in parts per million (ppm),

$$mass\ accuracy\ (ppm) = \frac{\frac{m}{z}_{(exp)} - \frac{m}{z}_{(theor)}}{\frac{m}{z}_{(theor)}} \cdot 10^6$$

2.1 Magnetic Sector

Until recently, magnetic sector and double-focusing instruments were the only ones that provided exact mass measurements. HRMS using double-focusing mass analyzers was established in chemistry laboratories in the 1960s. Coupled to a direct insertion probe or to GC, it revolutionized the elemental analysis of organic compounds [32,33].

The fundamentals of magnetic sectors are based on the curved path taken by ions moving in a magnetic field, and the relation of that path to *m/z*. The relation between *m/z* and the curvature of the ion's path can be derived from basic physics. A magnetic analyzer alone does not provide impressive resolution because of the distribution of initial kinetic energies of ions. Kinetic energies can be unified by the use of the second sector analyzer with electrostatic focusing. This is called double-sector or two-sector instruments. The electric sector serves as a kinetic energy focusing element allowing only ions of a particular kinetic energy to go through its field irrespective of their *m/z*. Thus, the addition of an electric sector allows only ions of uniform kinetic energy to reach the detector, thereby decreasing the kinetic energy spread, which in turn increases resolution. It should be noted that the corresponding increase in resolution is parallel to a decrease in sensitivity. These double-focusing (Figure 3) mass analyzers are mostly used with electrospray (ESI) and electron impact ionization. The

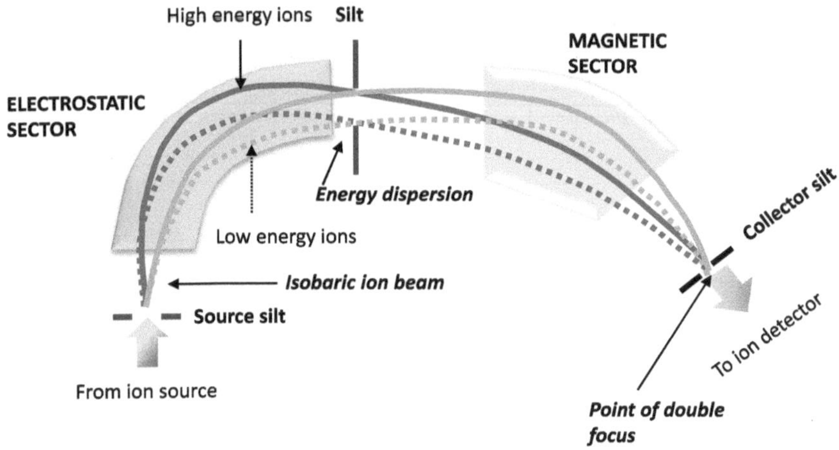

FIGURE 3 Scheme of a double focusing sector mass spectrometer.

coupling of magnetic analyzer with LC or GC is possible, but not so common because this analyzer has higher vacuum requirements and a slower scanning speed.

Sectors deliver the ultimate in high resolution GC-MS for ultra-trace quantification of environmental contaminants like dioxins and polychlorinated biphenyls (PCBs) in food, but it is almost the only application for which it is still used. Magnetic sector mass spectrometers are significantly more sensitive than the quadrupole. For example, magnetic sector spectrometers detect dioxins at parts per trillion (ppt) levels in soil and parts per quadrillion (ppq) in water. However, a search on Scopus as that previously described in the introduction, for the last 5 years and introducing "magnetic sector" and "food" in the fields "article title" and "article title, abstract, and keyword," respectively as keywords, only gives a study [32]. Background levels of polychlorinated dibenzo-*p*-dioxins, polychlorinated dibenzofurans, PCBs, and related persistent organic pollutants are determined. This study takes as a premise that GC × GC can enhance the detectability limit up to 2–20 times better than single dimensional GC due to the signal-to-noise ratio enhancement related to thermal modulation. Therefore, modulators can also have some potential applications as signal enhancers, as well as peak capacity enhancers. A liquid nitrogen jet-cooled thermal modulator has been mounted in a GC oven coupled to a high resolution magnetic sector MS instrument. The data acquisition parameters of the slow double-focusing magnetic sector MS instrument needed optimization to accommodate it to the narrow modulated GC peaks. Acquisition rates were increased to 20 Hz, while maintaining high mass resolution. Selected ion monitoring was limited to one or two to ensure enough MS cycle time. For maximization of the sensitivity enhancement due to cryogenic zone compression, the entire GC peak of interest was trapped and remobilized in one event. Optimization of the method resulted in the ability to detect low attogram (ag) amounts, a level that had not yet been attained using classical GC-HRMS. This study has been detailed because it gives a clear picture of the advantages and disadvantages of the double focusing magnetic sector. It makes clear that this instrument is in use and that it is still possible to improve their detection limits and its scope.

2.2 Time-of-Flight (TOF)

TOF mass analyzers utilize an electric field to accelerate a pulse of ions toward the detector. The *m/z* is determined by the time required by the ion to reach the detector. At higher masses, resolution is difficult because flight time is longer. Also at high masses, not all of the ions of the same *m/z* values reach their ideal TOF velocities. To solve this, a reflectron—series of ring electrodes of very high voltage placed at the end of the flight tube—is added to the analyzer. The high voltages reflect the ions in the opposite direction increasing resolution by narrowing the broadband range of flight times for a single *m/z* value. Faster ions travel further and slower ones travel less into the reflector, making that both of them reach the detector at the same time rather. The basic principles of operation are shown in Figure 4.

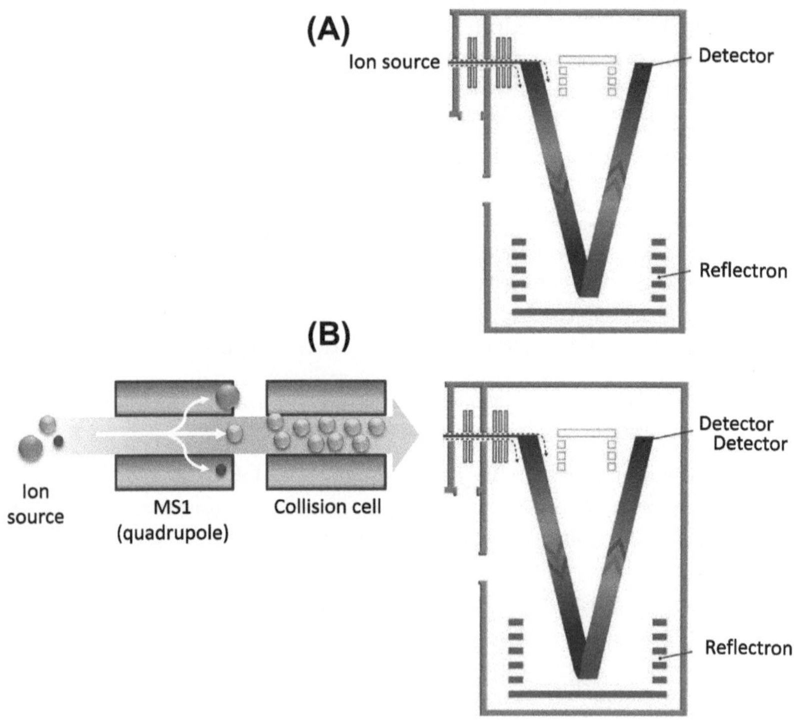

FIGURE 4 Mass spectrometers incorporating a TOF analyzers. (A) The stand-alone TOF instrument; and (B) hybrid QqTOF.

Today's TOF mass spectrometers offer excellent sensitivity and high mass range as well as high-speed analysis [34–41]. This has made TOF-MS an essential instrument for food analysis applications—typically with matrix-assisted laser desorption/ionization (MALDI) and ESI ionization—particularly with the development of high-resolution and hybrid instruments [quadrupole time-of-flight (QqTOF), ion-trap time-of-flight (IT-TOF), and TOF-TOF configurations]. These hybrid instruments benefit from the versatile ability of quadrupoles to carry out various tasks and from the high performance of TOF spectrometers in both MS and MS/MS modes [42–47]. The QqTOF is a hybrid quadrupole TOF mass spectrometer with MS/MS capability. The most common configuration is also illustrated in Figure 4. The quadrupole is operated as an ion guide in MS mode and as mass selection device in MS/MS mode. A reflectron TOF analyzer is placed orthogonally to the quadrupole and serves as a mass resolving device for both MS and MS/MS modes [48–55]. A collision cell is located between the quadrupole and the TOF analyzer to induce fragmentation in MS/MS experiments [56–61]. The QqTOF is almost as used as TOF mass analyzers (Table 2).

TABLE 2 Applications of TOF Analyzers in Food Safety and Quality

Type	Application	References
Target Analysis		
ICP-TOF	Multielement analysis of food and beverages	[57]
QqTOF	Ni speciation in tea infusions	[62]
ICP-TOF	Arsenic species in methanolic extracts of cod liver	[63]
ICP-MS	Analysis of heavy metal content in *Indocalamus tessellatus* samples	[56]
LC-TOF	Simultaneous determination of B complex vitamins in nutritional yeast	[64]
MALDI-TOF	Monitoring of barley starch amylolysis	[65]
LC-TOF	Determination of free and bound phenolic compounds in buckwheat spaghetti. Effect of thermal processing	[66]
LC-TOF	Analysis of pesticides in fruits and vegetables	[67]
GC-TOF	Quantitative analysis of acetaldehyde in foods consumed	[68]
LC-TOF	Screening natural antioxidants in peanut shell	[69]
UHPLC-TOF	Hexabromocyclododecane diastereomers	[70]

Continued

TABLE 2 Applications of TOF Analyzers in Food Safety and Quality—cont'd

Type	Application	References
UHPLC-TOF	19 antibiotics in dairy products	[71]
MALDI-TOF	Microscale strategy to detect spermine and spermidine in food	[72]
LC-TOF	Determination of phthalates in edible salts from food markets in Korea	[73]
LC-QqTOF	Simultaneous determination five major antioxidant compounds	[74]
MALDI-TOF	Monitoring of she-donkey's milk contamination or adulteration	[75]
MALDI-TOF	Quantitating curcumin in food condiments and dietary supplements and screening of acrylamide-induced ROS protein indicators	[76]
LC-QqTOF	Quantitating curcumin in food condiments and dietary supplements and screening of acrylamide-induced ROS protein indicators	[45]
UHPLC-TOF	Analysis of illicit dietary supplements sold in the Italian market: Identification of a sildenafil thioderivative as adulterant	[46]
LC-TOF	Simultaneous investigation of sesquiterpenes, pyrrolizidine alkaloids, and N-oxides in butterbur (*Petasites hybridus*)	[77]
UHPLC-QqTOF	Determination of 141 pesticides in tea	[78]
UHPLC-QqTOF	Determination of tadalafil and N-desmethylsibutramine in health and dietary supplements	[79]
LC-QqTOF	Seven fluorescent whitening agents in food contact paper package materials	[80]
Lipidomics		
GC × GC-TOF	Unsaponifiable fraction of lipids belonging to various types of milk	[22]
UHPLC-QqTOF	Rapid and highly accurate detection of steryl glycoside	[81]
MALDI-TOF	Rapid lipid profiling of bacteria	[82]
MALDI-TOF	Detection of glycated phospholipids in heat-treated food samples	[83]

TABLE 2 Applications of TOF Analyzers in Food Safety and Quality—cont'd

Type	Application	References
MALDI-TOF	Triacylglycerol profile in cocoa liquors	[84]
MALDI-TOF	Influence of *Pseudomonas fluorescens* on phospholipids in salmon fillet	[85]
MALDI-TOF	Extra-virgin olive oil adulteration with hazelnut oil by analysis of phospholipids using an ionic liquid as matrix and extraction solvent	[86]
MALDI-TOF	Determination of intact phospholipids as markers of illegal bovine milk adulteration of high-quality milk	[87]
Metabolomics		
LC-QqTOF	Nontarget analysis for varietal separation of onions	[88]
LC × LC-IT-TOF	Comprehensive identification of pigments in food	[35]
LC-QqTOF	The comparative research on constituents of *Radix aconite*	[89]
UHPLC-QqTOF	Nonvolatile migrants from new active packaging materials	[59]
MALDI-TOF	Caffeine-precipitated SII thearubigins from 15 commercial black teas	[90]
UHPLC-QqTOF	Comprehensive characterization of an *Eryngium bourgatii* extract and their antioxidant and anti-inflammatory activities	[91]
MALDI-TOF	NMR, HPLC-ESI-MS, and MALDI-TOF-MS analysis of condensed tannins from *Delonix regia* (Bojer ex Hook.) Raf. and their bioactivities	[92]
LC-QqTOF	Comprehensive characterization of bioactive phenolic compounds in cucumber whole fruit extract	[93]
LC-QqTOF	Migration of nonintentionally added substances from adhesives and the role of EVOH to avoid migration in multilayer packaging materials	[94]
SELDI-TOF	Analysis of raw hams to predict the final quality of dry-cured hams	[43]

Continued

TABLE 2 Applications of TOF Analyzers in Food Safety and Quality—cont'd

Type	Application	References
GC-TOF CE-TOF	Metabolic profiling of cheonggukjang (fast-fermented bean paste) during fermentation and its correlation with metabolic pathways	[95]
GC-APCI-TOF	Metabolomic analysis of avocado fruits. Effects of ripening degrees and fruit varieties	[96]
GC-APCI-QqTOF	Validation of a qualitative screening method for pesticides in fruits and vegetables	[48]
UHPLC-QqTOF	Metabolomic analysis of meju during fermentation	[97]
LC-TOF	Determination of phenolic compounds of "Sikitita" olive leaves	[98]
MALDI-TOF	Cocoa content influences chocolate *molecular profile*	[99]
PTR-TOF	Characterization of food and tasters: the case study of coffee	[100]
PTR-TOF	Isobaric compounds in food-flavor applications	[101]
LC-IT-TOF	Profiling and identification of the metabolites of (+)-catechin	[34]
GC-APCI-TOF	Quantitative characterization of important metabolites of avocado fruit	[102]
TOF-SIMS	Characterization of antimicrobial textiles for the food processing industry	[38]
UHPLC-QqTOF	Analyses of products of aflatoxin B1 after ozone treatment	[103]
GC-TOF	Metabolomics for comparison of caffeinated and decaffeinated coffee	[104]
UHPLC-TOF	Metabolite profiling of Brazilian Lippia species	[105]
Profiling		
MALDI-TOF	Discrimination of freshwater fish species	[106]
MALDI-TOF	Fingerprinting as a new tool for quality control and health risk assessment of virgin olive oil	[107]
LC-TOF-MS	Activity guided isolation of isoflavones and metabolome profiling of *Azorella madreporica* Clos. from northern Chile	[108]

TABLE 2 Applications of TOF Analyzers in Food Safety and Quality—cont'd

Type	Application	References
LC-QqTOF	Profiling the ginsenosides of three ginseng product	[109]
UHPLC-TOF	Evaluation of the effect of extraction methodology on the yield of 6-shogaol, a biomarker of *Zingiber officinale*	[110]
LC-IT-TOF	Profiling of polyphenolic compounds extracted from green lettuce (*Lactuca sativa* L.; var. Maravilla de Verano)	[49]
GC × GC-TOF	Characterization and authentication of four important edible oils using free phytosterol profiles	[111]
PTR-TOF	Volatile compound changes during shelf life of dried *Boletus edulis*	[112]
LC-TOF	Evaluation of the composition of vine shoots and oak chips for enological purposes	[113]
Proteomics and Peptidomics		
MALDI-TOF	Proteomic approach to detect powdered milk in fresh cow's milk	[114]
MALDI-TOF-TOF	Proteins as targets of ferrous-catalyzed oxidation: identification of protein carbonyls	[37]
LC-TOF	Evaluation of the immunostimulating activity of soybeans via the determination of the functional peptide soymetide	[115]
SELDI-TOF	Effect of NaCl and sucrose tastants on protein composition of oral fluid	[44]
MALDI-TOF	Hydrolysis of β-lactoglobulin by trypsin under acidic pH	[116]
UHPLC-TOF	Quantification of Se-methylselenocysteine and its γ-glutamyl derivative from naturally Se-enriched green bean (*Phaseolus vulgaris vulgaris*)	[50]
MALDI-TOF	Detection of sheep and goat milk adulterations by direct MALDI-TOF-MS analysis of milk tryptic digests	[117]
CE-TOF	Identification of bioactive peptides in hypoallergenic infant milk formulas	[58]
UHPLC-QqTOF	Differentiation bovine and porcine gelatins in food products	[118]

Continued

TABLE 2 Applications of TOF Analyzers in Food Safety and Quality—cont'd

Type	Application	References
LC-TOF	Identification of modified lysozyme peptides upon photooxidation	[119]
MALDI-TOF	Identification and characterization of a new IgE-binding protein in mackerel (*Scomber japonicus*)	[120]
MALDI-TOF	Quality control of high protein content sport supplements	[60]
MALDI-TOF	Proteomic analysis of complex protein sample	[121]
MALDI-TOF	Identifying glycation inhibitors	[122]
Tanscriptomics		
MALDI-TOF	Gene analysis using mass spectrometric cleaved amplified polymorphic sequence (MS-CAPS)	[123]
Microorganisms Identification		
MALDI-TOF-TOF	Rapid identification of food-borne pathogens by top-down proteomics	[36]
MALDI-TOF	Rapid species identification of seafood spoilage and pathogenic Gram-positive bacteria	[124]
MALDI-TOF	Characterization of *Staphylococcus aureus* strains isolated from Italian dairy products by mass fingerprinting	[125]
MALDI-TOF	Detection and quantification of bacterial spoilage in milk and pork meat	[126]
MALDI-TOF	Classification and identification of lactic acid bacteria, with a focus on isolates from traditional fermented foods in Northern Vietnam	[127]
MALDI-TOF	Characterization of the cultivable microbial community in a spinach-processing plant	[128]
MALDI-TOF	Identification of newly described *Aeromonas* spp.	[129]
MALDI-TOF	Detection of *Listeria monocytogenes* from selective enrichment broth	[130]
MALDI-TOF	Identification and differentiation of food-related bacteria	[131]
MALDI-TOF	Identification of lactobacilli isolated	[132]

TABLE 2 Applications of TOF Analyzers in Food Safety and Quality—cont'd

Type	Application	References
GC × GC-TOF	Multivariate analysis for the characterization of food-borne pathogen bacteria profiles	[133]
MALDI-TOF	Detection of enterotoxins produced by pathogenic strains of the *Bacillus cereus* group	[134]
UHPLC-QqTOF	Analysis on crude and wine-processed *Siegesbeckia pubescens*	[135]
MALDI-TOF	Optimization of experimental and modeling parameters for the differentiation of beverage spoiling yeasts	[136]
MALDI-TOF	Identification and differentiation of brewery isolates of *Pectinatus* sp.	[137]
MALDI-TOF	Characterization of different food-isolated *Enterococcus* strains by fingerprinting	[138]
MALDI-TOF	Rapid identification of microorganisms isolated from pickled vegetables	[139]
MALDI-TOF	Characterization and profiling of *Bacillus subtilis*, *Bacillus cereus*, and *Bacillus licheniformis* by mass fingerprinting	[140]
MALDI-TOF	Rapid and accurate bacterial identification in probiotics and yogurts	[141]
MALDI-TOF	Species identification of *Aspergillus* section *Flavi* isolates from Portuguese almonds using phenotypic	[142]
MALDI-TOF	Identification and classification of seafood-borne pathogenic	[143]
MALDI-TOF	Assessment of seafood safety and quality by rapid identification of pathogenic and spoilage bacterial species by fingerprinting	[144]
MALDI-TOF	Rapid identification of *Tetragenococcus halophilus* and *Tetragenococcus muriaticus*, important species in the production of salted and fermented food	[145]
Volatile Compounds		
PTR-TOF	Rapid characterization of dry cured ham produced following different PDOs	[146]
PTR-TOF	Linking GC-MS and PTR-TOF-MS fingerprints of food samples	[147]

Continued

TABLE 2 Applications of TOF Analyzers in Food Safety and Quality—cont'd

Type	Application	References
PTR-TOF	In vitro and in vivo flavor release in cereal bars with varying sugar composition	[148]
PTR-TOF	Online monitoring of coffee roasting	[149]
PTR-TOF	Monitoring the effect of high pressure and transglutaminase treatment of milk on the evolution of flavor compounds during lactic acid fermentation	[54]
PTR-TOF-MS	Effect of the pig rearing system on the final volatile profile of Iberian dry-cured ham	[55]
GC × GC-TOF	Sensomics analysis of key hazelnut odorants (*Corylus avellana* L. "Tonda Gentile")	[150]
GC × GC-TOF	Capture and identification of the volatile components in crude and processed herbal medicines	[151]
GC × GC-TOF	Analysis of roasted and unroasted *Pistacia terebinthus* volatiles	[152]

Much less used, the IT-TOF combines ion trap and TOF technologies. When combining IT and TOF, the greatest technological challenges were the efficient introduction of ions into the IT and the simultaneous ejection of trapped ions to the TOF. The insertion of a quadrupole before the ion trap can introduce ions from the source to the ion trap by compressed ion injection method efficiently. The IT gives an efficient MS^n capability. In addition, this system provides high resolution/accuracy for either MS or MS^n modes. There are applications to profiling and identification of catechin metabolites [34] and to the comprehensive identification of carotenoids in fruits [35].

TOF-TOF-MS is targeted for proteomics, microbiology, and tissue imaging. This system offers flexibility for both profiling and identification applications in a single instruments that has good sensitivity, mass accuracy, and resolution. TOF-TOF instrument can operate in both MS and MS/MS. In the past few years, MALDI-TOF/TOF MS/MS has been shown to provide extensive structural information. MALDI high-energy collision induced dissociation (CID) spectra of both positive and negative molecular ions provide sequence, branching, and linkage information [36,37].

Time-of-flight secondary ion mass spectrometry (TOF-SIMS) is a surface analytical technique that focuses a pulsed beam of primary ions onto a sample surface, producing secondary ions in a sputtering process. Analyzing these secondary ions provides information about the molecular and elemental species present on the surface. For example, if the TOF-SIMS analysis were of organic contaminants, such as oils adsorbed on the surface, TOF-SIMS would reveal this information, whereas other techniques may not. Since TOF-SIMS is a survey technique, all the elements in the periodic table, including H, are detected [38].

The TOF-MS suitability for food analysis has been widely demonstrated. An analysis of the published references during the last 5 years (source: Scopus, keywords "time-of-flight" and "food" using as search field article title and abstract, title, and keywords, respectively) [20] showed 143 studies published, 42 using QqTOF, three using IT-TOF and three using TOF-TOF. It should be taken into account that this search is not extensive because there are many studies that use other words in the title, such as HRMS and apply any type of TOF that are not reflected in this search.

MALDI-TOF is by far the principal application of TOF-MS. It is widely used in the characterization of peptides, proteins, nucleotides, and microorganisms. MALDI-TOF MS is a very sensitive technique that permits a fast sampling of chemical and biological compounds at abundances of sub-femtomole primarily with single charge. Unlike LC methods, MALDI-TOF-MS does not require a stationary phase for sample separation or a mobile phase for sample elution. Hence, because it can be performed relatively quickly and easily, MALDI-TOF-MS has greater potential use as a high-throughput analytical platform compared to LC-based methods. Since the advent of the use of MALDI-TOF-MS as a tool for microbial characterization, efforts to increase the taxonomic resolution of the approach have been made. These efforts are one of the most widely covered fields in several review articles in which recent finding and new approaches are discussed [39–42]. An alternative developed two decades ago is surface enhanced laser desorption/ionization time-of-flight mass spectroscopy (SELDI-TOF-MS). This proteomic technology combines MALDI-TOF-MS with surface chromatography utilizing sample chips, which display various chemically enriched and active surfaces to bind proteins. The technique enables rapid and high-throughput detection of proteins and peptides directly from crude mixtures. For this reason, SELDI-TOF-MS has been used in a number of studies related to screening identification, biomarkers discovery in food proteomics [43,44].

The use of LC-TOF-HRMS methods can also be found in a variety of analytical applications [45,46]. Their major disadvantage in comparison to MALDI is the time required for the chromatographic separation. LC has been indistinctly used as either classical LC or recently developed UHPLC. The major advantages of UHPLC (1.7 μm of particle size in the stationary phase) over classical LC (between 3 and 5 μm) include improved resolution within a shorter retention time and higher analytical sensitivity [47]. LC-TOF is not only constrained to peptide and protein applications but has also a wide field of application to a wide range of molecules. The TOF and QqTOF hybrid instruments used with LC are able to identify and elucidate the structure of small molecule (mostly bioactive compounds and contaminants) [48]. The use of high resolution and mass accuracy

LC-MS systems as provided by TOF-MS is a very valuable solution to cover these demands in a cost-effective way. In the past, the application of TOF instruments to the field of food analysis was not been frequent as consequence of some limitations in performance, but nowadays, its use is extending increasingly due to the improvements achieved in the new equipment. Their characteristic features, such as high resolving power, accurate mass measurement, and high full-scan sensitivity make LC-(Qq)TOF-MS an attractive tool for identifying and quantifying both target and nontarget compounds in vegetable matrices. LC-(Qq)TOF-MS has the ability to record an unlimited number of compounds operating in full-scan mode, which makes this technique very convenient for the development of screening strategies based on the use of accurate mass databases with automated library-based detection. Another application of LC-TOF-MS and LC-QqTOF-MS/MS for pesticide residues analysis in fruits and vegetables is the identification of pesticide transformation products and metabolites, making use of the great advantage of having the possibility to identify and confirm with one single system [49,50].

Ions generated after ionization (ESI, MALDI, etc.) may be further separated by exploiting their different behaviors across a tube filled with an inert gas. IM refers to the velocity of an ion across a drift tube filled with a neutral gas, often nitrogen, under the influence of a weak electric field. In a drift tube, smaller molecules experience lesser collisions with the gas and thus travel faster than larger molecules [51]. This added orthogonal separation technique can therefore make structural elucidation of molecules more efficient.

GC-TOF-MS can screen hundreds of compounds at high sensitivity within one run. It is a powerful analytical technique with excellent capabilities due to its high sensitivity in full-spectrum-acquisition mode together with its resolving power and accurate-mass measurements. These features make this technique very attractive in qualitative analysis, especially for wide-scope screening of a large number of organic contaminants and residues at trace levels as well as for lipidomics. The recently developed TOF that can be combined with GC×GC preserving the HRMS characteristics open new prospects within this field of application. Hernandez et al. [52] discussed in detail the characteristics and the potential of GC-HR-TOF-MS describing different analytical strategies from wide-scope target screening to investigate unknowns in food safety. Recent instrumental developments (e.g., high-speed analog-to-digital converter or soft ionization sources) and advances in software for processing the huge amount of data available open up new prospects to investigate the presence of organic compounds in different fields. One of the possible solutions to advance in the possibilities of GC-full scan HRMS is a new atmospheric pressure chemical ionization (APCI) source that allows coupling GC to analyzers designed for LC. APCI is a very promising alternative source to couple GC with TOF and QTOF providing highly sensitive and selective detection. Its use has been recently reported in GC-MS based methods for metabolomic screening of avocado fruits [53] and wide coverage search of pesticides [48].

Recently the coupling of proton transfer reaction ionization with a TOF mass analyzer (PTR-TOF-MS) has been proposed to realize a volatile organic compound (VOC) detector that overcomes the limitations in terms of time and mass resolution of the previous instrument based on a quadrupole mass analyzers (PTR-Quadrupole-MS). This opens new horizons for research and allows for new applications in fields where the rapid and sensitive monitoring and quantification of VOCs are crucial in food sciences [54]. Very briefly, PTR-MS is based on the reaction of protonated water with VOC molecules with a proton affinity higher than that of water and the successive detection by a mass spectrometer [55].

The introduction of ICP-TOF mass spectrometers to the analytical practice has brought many practical advantages, e.g., enhanced sample throughput and elemental coverage, enhanced resolution, unlimited use of internal standards without performance compromises, improved isotope ratio precision capability owing to the high correlation of the noise sources for all isotopes, which occur with simultaneous sampling and measurement of transient signals [56]. It seems that just these advantages open new opportunities for ICP-MS analysis also in the field of food analysis because of the growing trend toward increasing consumer awareness [57].

CE has also been applied, but to a lower extent. CE-MS may be regarded as an excellent complement to LC-MS, because the charge-to-radius ratio mechanism governing electrophoretic migration can be considered orthogonal to the partitioning mechanism of reverse phase LC retention, which is mainly based on the hydrophobic character of the molecules. CE-TOF-MS has already attained the tentative identification of a large set of low molecular mass bioactive peptides [58].

The range of TOF applications within food analysis is impressive, TOF has been used for the analysis of traces of organic compounds in food safety [52] including pesticides, compounds coming from food contact materials [59], heavy metals [56], etc. There are also a number of applications to lipidomics [22], glycomics, metabolomics, and proteomics [60]. Table 2 summarizes the most important applications reported in the last 5 years. There are a number of reviews that highlight the latest technical advances within the instrumentation [34–36, 61]. Special attention is paid on the continuous internal calibration of the mass axis, the introduction of the reference mass ions, and the adjustment of the ionization source for robust operation [61].

2.3 Fourier Transform Ion-Cyclotron Resonance (FT-ICR)

Mass analysis by FT-ICR was first described in 1974, and the methods has grown to became the state of the art in terms of mass resolving power and mass accuracy capabilities. FT-ICR-MS is a type of mass analyzer (or mass spectrometer) for determining *m/z* of ions based on their cyclotron frequency in a fixed magnetic field. The ions are trapped in a Penning trap (a magnetic field with electric trapping plates) where they are excited (at their resonant cyclotron frequencies) to a larger cyclotron radius by an oscillating electric field orthogonal to the magnetic field. After the excitation field is removed, the ions are rotating at their cyclotron frequency in phase (as a "packet" of ions). These ions induce a charge (detected as an image current) on a pair of electrodes as the packets of ions pass close to them. The resulting signal is called a free induction decay, transient, or interferogram that consists of a superposition of sine waves. The useful signal is extracted from this data by performing a Fourier transform to get the mass spectrum. The basics of the technique are illustrated in Figure 5.

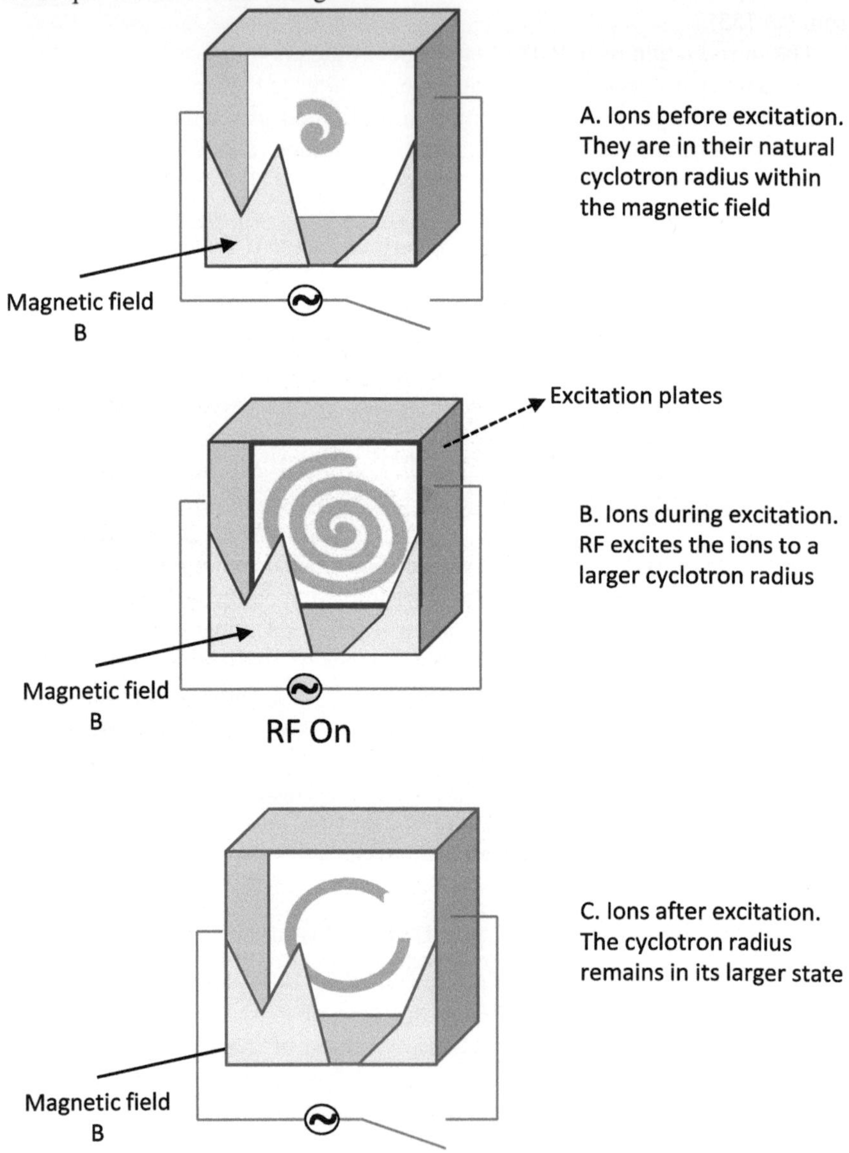

FIGURE 5 The operating principles of FT-ICR mass analysis.

FT-ICR-MS is able to achieve higher levels of resolution than other forms of mass spectrometer, in part, because a superconducting magnet is much more stable than radio frequency (RF) voltage. Currently, commercially available instruments readily provided resolving powers higher than 100,000 (FWHM). However, the same search carried out for other mass analyzers clearly shows that its use within food safety is not common (only three studies appear). These few applications pointed out that FT-ICR MS allows for the identification of complex organic mixtures without the need for any prior extraction or separation steps. FT-ICR MS is applied in all the "omics" sciences (i.e., metabolomics, proteomics, and lipidomics), thus enabling molecular-level analyses of complex mixtures. FT-ICR-MS allied to an ESI source enables the characterization of in situ chemical compounds produced during the natural aging of coconut water [153], to analyze samples of raw fermented cocoa beans [154] and of aged macadamia nut oils [155]. Whereas ESI-FT-ICR-MS offers unprecedented resolving power and mass accuracy, the technique suffers from the disadvantage that higher molecular mass ions escape detection because they appear as double- or multiple-charged ions in the mass range of monocharged ions.

2.4 Orbitrap

The Orbitrap mass analyzer was described for the first time in the year 2000 by Makarov [156]. These instruments, like the FT-ICR and the quadrupole ion traps, function as trapping devices. However, Orbitraps do not require magnetic or RF fields to carry out their trapping functions. Ions are introduced into the Orbitrap in small packets. A strong electrical field inside the analyzer initiates axial oscillation. Ions oscillate harmonically with a period proportional to (*m/z*) 1/2 and produce an image current on split outer electrodes. Subsequently, the signal obtained is converted from time-domain into a mass spectrum by Fourier transformation. The fundamentals associated with Orbitrap mass analyzers are illustrated in Figure 6.

The Orbitrap was introduced onto the market in 2005 [157]. Now spectrometers equipped with Orbitrap analyzers are available in diverse configurations—as a single-stage mass spectrometer and as a hybrid instrument [with a quadrupole mass filter or a linear ion trap (LTQ)]. The stand-alone instrument, which provides high mass resolution (>15,000 FMWH) and high mass accuracy (<2 ppm) but without mass selection, is already successfully used in routine for determining pesticide residues [158–162]. Single-stage Orbitrap-MS was demonstrated to be a very useful tool with a high potential for efficient multiresidue analysis in the field of food safety. This alternative to the traditional tandem MS approach can be considered also flexible in postacquisition data processing and potentially allows retrospective data mining. Identification and characterization of chloramphenicol is illustrated in Figure 7.

In proteomics and related fields, this combination of a low resolution LTQ with the high resolution Orbitrap analyzer—termed "LTQ Orbitrap"—has now become widespread. The LTQ Orbitrap instruments represent a multistage trap

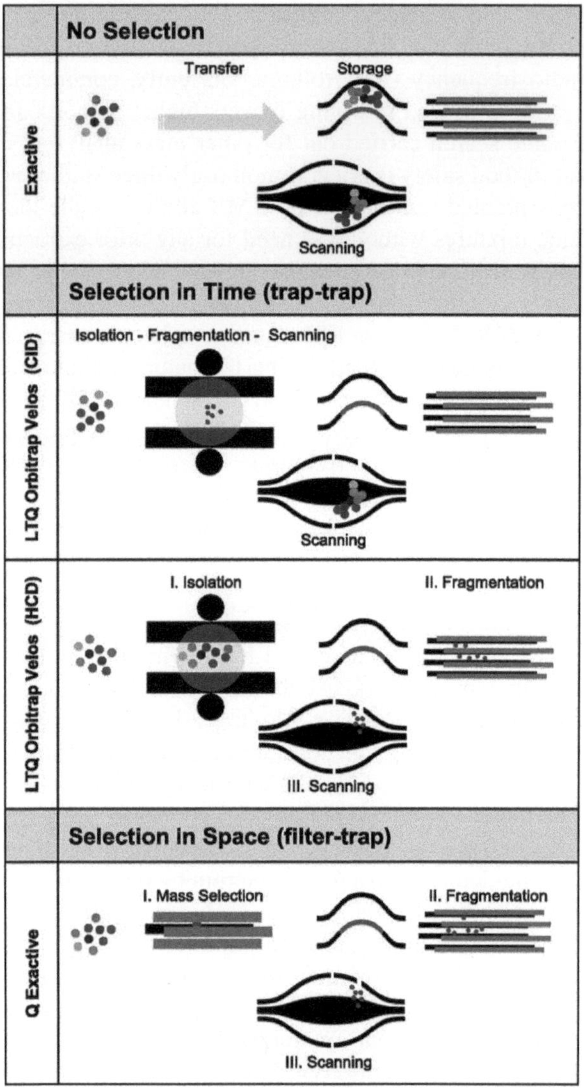

FIGURE 6 Mass spectrometers incorporating an Orbitrap analyzer. The Exactive is a stand-alone instrument without mass selection. The total ion population is collected in the C-trap and injected into the Orbitrap analyzer. In the LTQ Orbitrap Velos combination, ions can be selected "in time" by mass selective scans in the linear ion trap. In CID mode, the LTQ and Orbitrap operate as separate mass spectrometers. In HCD mode its function is to isolate a particular precursor, which is then fragmented in the HCD cell. In contrast, in the Q Exactive mass selection is "in space" as ions of only a specified *m/z* range have stable trajectories and are transferred to the storage or fragmentation devices before Orbitrap analysis. *Reproduced from Ref. [232].*

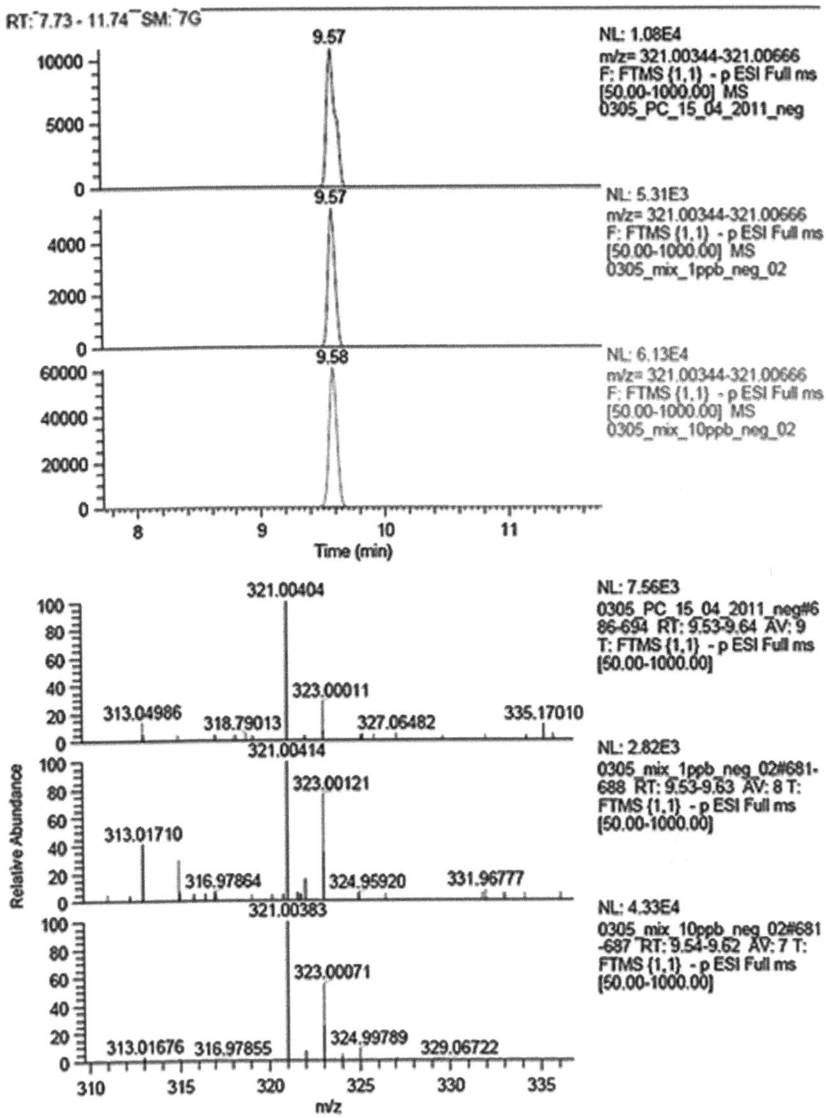

FIGURE 7 Minicake black sample containing chloroamphenicol at 5 mg/kg (upper chromatogram) juxtaposed with a minicake matrix spiked with chloramphenicol at 1 mg/kg (middle chromatogram) and 10 mg/kg (lower chromatogram). Analytical data collected using the present developed full scan mode Exactive (Orbitrap) approach: it can be also observed the presence of both [M(^{35}Cl)-H]- and [M(^{37}Cl)-H]-isotopes either in the black sample or in the two minicake matrix-matched recovery tests. *Reproduced from Ref. [178] with permission of John Wiley & Sons, Ltd.*

combination. In MS mode, the LTQ performs the function of collecting the ion population, passing them on to an intermediate C-trap for injection and analysis in the Orbitrap analyzer at high resolution. In MS/MS mode, the LTQ only retains a chosen mass window, which is activated by a supplemental RF field leading to fragmentation of the trapped precursor ions, and records the signal of a mass-dependent scan at low resolution. Note that the high resolution MS scan can be performed at the same time as the low resolution MS/MS scans in the LTQ. The current generation of LTQ-Orbitrap instruments has significant features of an S-lens with up to 10-fold improved ion transmission for the atmosphere, a dual LTQ, and a more efficient higher energy collisional dissociation (HCD) cell interfaced directly to the C-trap. The HCD method was introduced as a fragmentation process that overcomes two of the shortfalls when performing CID in the LTQ ion trap: (1) low mass resolution and accuracy and (2) losses due to "low mass cutoff." HCD fragmentation is similar to the fragmentation in triple quadrupole or QqTOF instruments and its products are analyzed with high mass accuracy in the Orbitrap. Moreover, other fragmentation methods, such as electron transfer dissociation (ETD) have recently been incorporated to the LTQ-Orbitrap [163,164]. In ETD, an anion donates an electron to a cationic peptide, which then causes the peptide backbone to fragment [165]. ETD is of significant usefulness in proteomics because of its ability to preserve most post-translational modifications of general interest. Thus, the LTQ-Orbitrap instruments offer versatile fragmentation modes depending on the analytical problem.

Recently, the combination of a quadrupole with an Orbitrap (Q-Orbitrap), which also showed high ion currents because of the S-lens, and fast high energy collision-induced dissociation fragmentation, was also successfully applied to determine pesticide residues in fruits and vegetables. Although sensitivity of the Q-Orbitrap is not limiting, it only can perform HCD fragmentation, making its fragmentation less versatile than in the LTQ-Orbitrap. The Q-Orbitrap is the last addition to the family of hybrid instrument. However, it is increasingly applied, particularly, in the field of profiling and fingerprinting metabolites [18,70,161,166,167].

The Orbitrap is the second most used high resolution mass analyzers. An analysis of the published references during the last 5 years (source: Scopus, keywords "orbitrap" and "food" using as search field article title and abstract, title and keywords, respectively) [20] showed 54 studies published. Unlike TOF that has been combined with almost all types of separation techniques, Orbitrap is only commercially combined with LC or with ambient ionization MS. They are now being connected to GC [168,169]. The applications reported within the field of food safety and quality are summarized in Table 3.

3. OBTAINING AND PROCESSING INFORMATION

The latest generations of high resolution instruments (TOF-MS or Orbitrap TM-MS) with improved resolution and stability of accurate mass measurements showed great quantitative potential, and various investigations have assessed their

TABLE 3 Applications of Orbitrap Analyzers in Food Safety and Quality

Type	Application	References
Metabolomics		
UHPLC-LTQ-Orbitrap	Characterization of polyphenols in beer	[170]
UHPLC-Q-Orbitrap	Analysis of pesticide and veterinary drug residues in baby food	[161]
UHPLC-Q-Orbitrap	Analysis of hexabromocyclododecane diastereomers	[70]
UHPLC-Q-Orbitrap	Analysis of phthalates in milk and milk products	[18]
UHPLC-Q-Orbitrap	Simultaneous identification, confirmation, and quantitation of illegally adulterated antidiabetics in herbal medicines and dietary supplements	[167]
UHPLC-Orbitrap	Targeted metabolite profile of food bioactive compounds	[171]
UHPLC-Orbitrap	Identification and quantification of the main isoflavones and other phytochemicals in soy-based nutraceutical products	[172]
UHPLC-Q-Orbitrap	Rapid screening, confirmation, and quantification of illegally adulterated phosphodiesterase-5 inhibitors in herbal medicines and dietary supplements	[173]
LC-LTQ-Orbitrap	High-throughput screening and confirmation of 22 banned veterinary drugs in feedstuffs	[174]
UHPLC-Orbitrap	Multiresidue method for the detection of veterinary drugs in distillers grains	[13]
UHPLC-Orbitrap	Long-term stability of natural and synthetic glucocorticoids in livestock urine	[175]
LC-Orbitrap	Detection of type A trichothecene di-glucosides produced in corn	[176]
UHPLC-LTQ-Orbitrap	Determination of 166 pesticides in fruits and vegetables	[164]
UHPLC-Orbitrap	Determination of polar organophosphonate pesticides	[177]
UHPLC-Q-Orbitrap	Targeted screening of pesticides, veterinary drugs, and mycotoxins in bakery ingredients and food commodities	[178]

Continued

TABLE 3 Applications of Orbitrap Analyzers in Food Safety and Quality—cont'd

Type	Application	References
UHPLC-Orbitrap	Comprehensive qualitative and quantitative determination of pesticides and veterinary drugs in honey	[162]
UHPLC-Orbitrap (HCD cell)	Qualitative aspects and validation of a screening method for pesticides in vegetables and fruits	[160]
UHPLC-Orbitrap (HCD cell)	Determination of deoxynivalenol, T-2, and HT-2 toxins in a bread model food	[179]
UHPLC-LTQ-Orbitrap	Determination of triazine herbicides in rice	[163]
LC-Orbitrap	Screening of plant toxins in food, feed, and botanicals	[180]
LC-Orbitrap	Quantification of Se-methylselenocysteine and its γ-glutamyl derivative from naturally Se-enriched green bean (*Phaseolus vulgaris vulgaris*)	[50]
UHPLC-LTQ-Orbitrap	Metabolic profiling of *Vitex agnus-castus* leaves, fruits, and sprouts: analysis	[181]
LC-Q-Orbitrap	Pesticide multiresidue screening	[159]
LTP-Orbitrap	Rapid and quantitative analysis of pesticides in fruits	[182]
UHPLC-Q-Orbitrap	High-throughput screening of pesticide and veterinary drug residues in baby food	[183]
UHPLC-Q-Orbitrap	Multimycotoxin analysis in dairy products	[184]
UHPLC-Q-Orbitrap	Analysis of additives in dairy products	[185]
UHPLC-LTQ-Orbitrap	Qualitative and quantitative assessment of pesticide residues	[186]
UHPLC-Q-Orbitrap	Targeted and untargeted data-dependent experiments for characterization of polycarbonate food-contact plastics	[187]
MALDI-Orbitrap	Hop bitter acids in beer	[188]

TABLE 3 Applications of Orbitrap Analyzers in Food Safety and Quality—cont'd

Type	Application	References
UHPLC-LTQ-Orbitrap	Determination of acrylamide in Sudanese food	[47]
LC-LTQ-Orbitrap	Identification of the environmental neurotoxins annonaceous acetogenins in an *Annona cherimolia* Mill	[189]
UHPLC-Orbitrap	Validation of a confirmatory method for lipophilic marine toxins in shellfish	[190]
LC-Q-Orbitrap	Determination of dyes in wines	[191]
LC-LTQ-Orbitrap	Enzymatic oxidation of rutin by horseradish peroxidase: kinetic mechanism and identification of a dimeric product	[192]
LC-Q-Orbitrap	Exploration of olive oil main metabolites	[193]
LC-Orbitrap	Detection of masked mycotoxins derived from type A trichothecenes in corn	[194]
LC-Orbitrap (HCD cell)	Analysis of antimicrobial agents in pig feed	[53]
LC-Q-Orbitrap	Application to determination of synthetic hormones in animal urine	[166]
LC-Orbitrap	Confirmation of anabolic steroids in meat	[195]
DART-LTQ-Orbitrap	Direct peel monitoring of xenobiotics in fruit	[196]
LC-Orbitrap	Rapid and semiautomated method for the analysis of veterinary drug residues in honey	[33]
LC-LTQ-Orbitrap	Determination of sedatives and adrenergic blockers in blood meal	[197]
LC-Orbitrap	Determination of eight biogenic amines in selected seafood	[198]
UHPLC-Orbitrap	Comprehensive qualitative and quantitative determination of pesticides and veterinary drugs in honey	[162]
LC-LTQ-Orbitrap	Rapid identification of antifungal compounds produced by lactic acid bacteria	[165]
DART-Orbitrap	Evaluating a direct swabbing method for screening pesticides on fruit and vegetable surfaces	[199]

Continued

TABLE 3 Applications of Orbitrap Analyzers in Food Safety and Quality—cont'd

Type	Application	References
DART-Orbitrap	Rapid qualitative analysis of phthalates added to food and nutraceutical products	[200]
LC-Orbitrap	Pesticide residues in fruits and vegetables	[201]
LC-Orbitrap	Quantification of selected mycotoxins and the simultaneous screening of fungal metabolites in food	[202]
LC-Orbitrap	Detection of a new *Fusarium*-masked mycotoxin in wheat grain	[203]
LC-Orbitrap	Quantification of residual milk allergens in caseinate-fined white wines	[204]
LC-Orbitrap	Rapid screening for pesticides using automated online sample preparation	[158]
UHPLC-Orbitrap (HCD cell)	Simultaneous quantification of eight biogenic amine compounds in tuna by matrix solid-phase dispersion	[205]
LC-Orbitrap	Screening of lipophilic marine toxins in shellfish and algae	[206]
UHPLC-LTQ-Orbitrap	In-depth analysis of the chicken egg white proteome	[207]
Proteomics and Peptidomics		
LC-LTQ-Orbitrap	Comparison of five methods for direct extraction of surface proteins from *Listeria monocytogenes* for proteomic analysis	[208]
UHPLC-LTQ-Orbitrap	Multiallergen screening in wine	[209]

effectiveness in quantitative and qualitative analyses of food relevance. This can be achieved by performing full MS or full MS/MS acquisition analysis in a selected mass range and by applying postacquisition mass filtering on the total ion current on the basis of the accurate masses expected for the ions of interest [209,210].

The hybrid instruments can exploit different features depending on the specific configuration of the MS system, taking advantage of the information provided by the tandem configurations, which can be operated under different schemes giving valuable contribution to the structural elucidation studies, untargeted screening, or multitargeted experiments. Hence, the selection of markers (peptides, proteins, or small molecules) can be either dependent or independent of the data (or information). In the data-dependent acquisition (DDA) mode, a precursor ion is selected and fragmented for sequencing, which is inherently biased due to the existence of many compounds in the sample. On the other hand, in the data-independent acquisition (DIA) mode, rather than selecting a precursor, a non-overlapping *m/z* window is selected and all the precursor ions within the window are cofragmented yielding more fragment ions per samples, which increases the analyte identification score [118]. A summary of the different approaches supported by the various instrument manufacturers and their advantages, limitations, and potential impact to proteomics are given in Table 4.

3.1 Data-Dependent Acquisition (DDA) or Information-Dependent Acquisition (IDA)

The most common way to work with hybrid mass spectrometers has been DDA or IDA. Both refer to the same way to work, but the nomenclature depends on instrument. Commonly, DDA is used for Orbitrap and IDA for QqTOF. In this mode, the HRMS/MS first performs a survey scan, from which certain ions chosen "a priori" by the analysts or "nonchosen" but with the intensity above a predefined threshold value, are selected, isolated, fragmented, and sequenced by product ion scanning in the instruments.

A full MS/ddaMS^2 experiment with "a priori" selection of the precursors consists of a full scan event followed by MS/MS scan events of top N precursors in the inclusion list. The use of an inclusion list with the monoisotopic mass and specific collision energy of the precursors provided sensitive products as precursors are selected in the quadrupole. Full scan MS/MS of some peptides are illustrated in Figure 8.

However, the most common way to perform a DDA experiment is when the instruments use the information obtained without the input of the analyst to make decisions about the next step of the experiment. In this nontarget MS/ddaMS^2 (TopN), which is used for confirmatory purpose, the mass analyzer performs data-dependent scans. The simplest experiment is a full scan followed by a data-dependent MS^2. The software commonly selects the most intense ion from the scan as the candidate for the data-dependent MS^2. Other additional selection criteria, such as dynamic exclusion, background subtraction, charge state selection, etc., are also used to prevent redundant acquisition of the most abundant ions, or to avoid acquiring product ion spectra of the interferences. Using this strategy, co-eluting matrix compounds from the matrix or noisy peaks can be easily excluded, facilitating the identification and quantification of known or novel analytes in a single-run analysis. Figure 9 shows the identification of the nontarget pesticide cymoxanil and two metabolites of Alternaria fungus mycotoxins. The compounds were identified by comparison with literature data, from retention times, MS and MS^2 analysis. The identity was confirmed based on the mass accuracy of the molecular ion (≤5 ppm), the appropriate RDB, the isotopic profile confirmed by simulation, and the coincidence of the characteristic MS^2.

TABLE 4 A Summary of Selected Emerging Data Analysis/HSM Approaches and Their Respective Pros and Cons

Method	Instrument Type	Brief Description	MS^1 Isolation Window	Duty Cycle, *m/z* Range	Advantages	Disadvantages
DDA						
Previous selection of the precursor ion	QqTOF LTQ-Orbitrap	The *m/z* of the precursor ions is selected "a priori" by the analyst.	1	Millisecond	Sensitive product ion spectrum as the precursors are selected in the quadrupoles.	Only MS/MS spectrum of the target precursor is obtained.
Precursor ions not selected a priori	QqTOF LTQ-Orbitrap	The precursor is selected by the introduction of parameters based on intensity, dynamic exclusion, background substraction, etc.	1	0.1–0.01 s	Need no prior knowledge of the sample components.	Large duty cycles. Intensity of the precursor ions is one of the limiting factors.
DIA						
MS^E or AIF	QqTOF Q-Orbitrap LTQ-Orbitrap	QqTOF: The collision cell switches rapidly and continuously between low and high collision energy. Orbitrap: The system performs alternating MS scans of the precursor ions and AIF scans in which all the precursors are transferred to the HCD cell and fragmented. Complex sample requires efficient separation using UPLC or further with IM separation.	300–2000 Th (filter low mass ions)	1 s (0.5 s for intact/precursor ions scan and 0.5 s for fragment ions scan), *m/z* 300–1990	Requires a minimum method development. Need no prior knowledge of the sample components. Unbiased data acquisition, regardless of the ion intensity. Closes to 100% duty cycle and potentially allow a higher number of compounds to be identified.	The method is not a replacement of DDA. The assignation of fragment ions to precursors is not always correct. Identity of the compounds may require further validation with directed DDA.

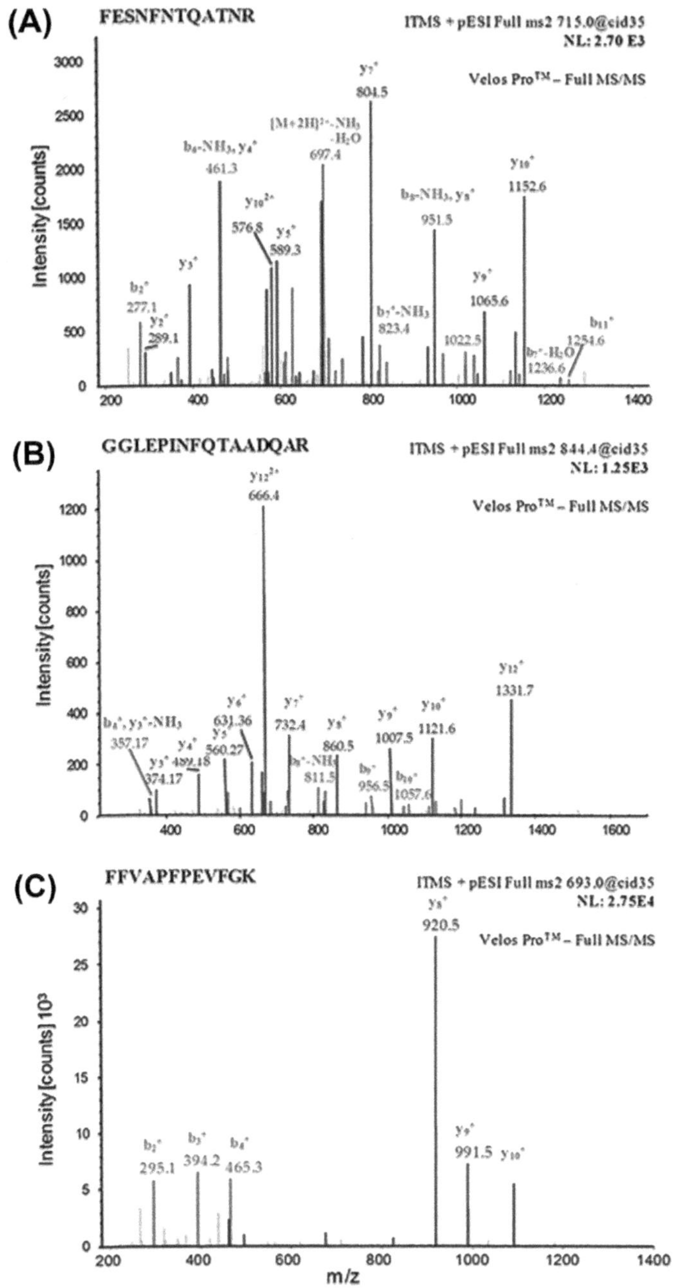

FIGURE 8 Overlay of typical full MS/MS spectrum matches acquired with Velos ProTM in DDA acquisition mode, referred to (A) marker 714.8 (LYS, FESNFNTQATNR), (B) 844.4 (OVA, GGLEPINFQTAADQAR), and (C) 692.9 (αS1-CN, FFVAPFPEVFGK), respectively. Transition assignments were achieved via software. *Reproduced from Ref. [209] with permission of John Wiley & Sons, Ltd.*

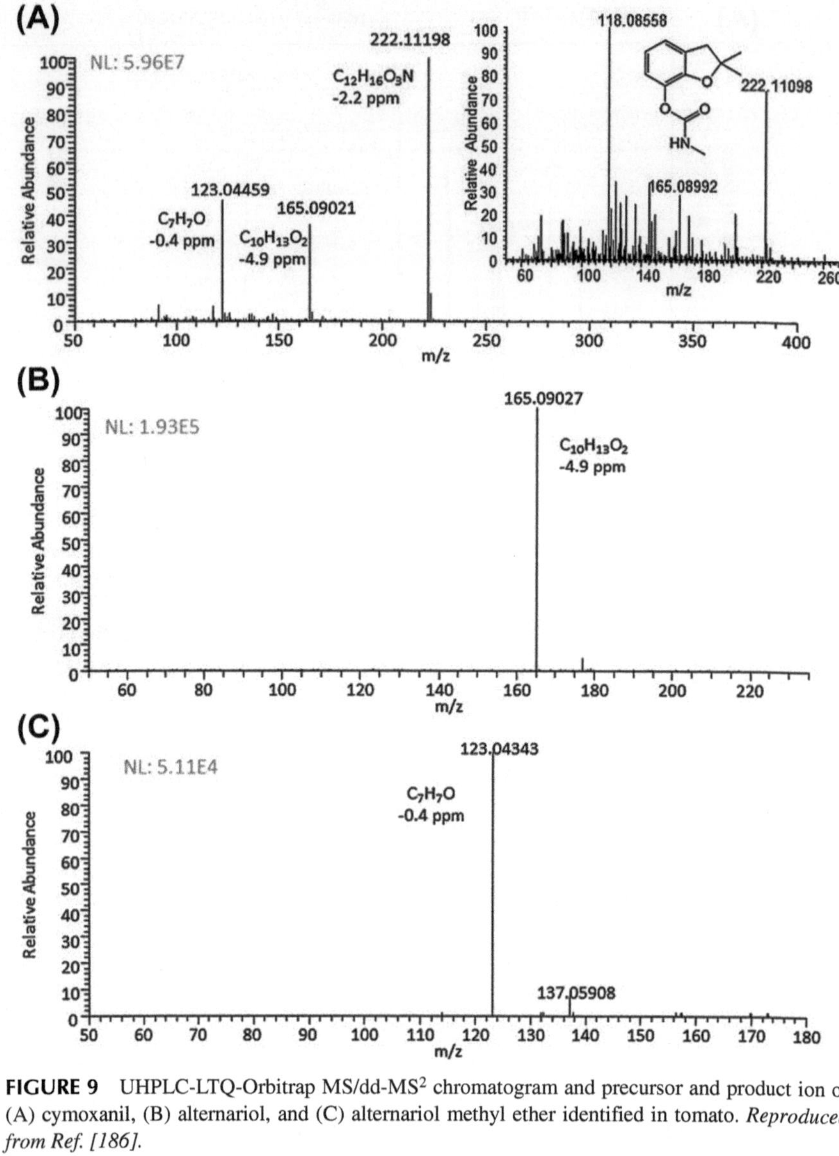

FIGURE 9 UHPLC-LTQ-Orbitrap MS/dd-MS^2 chromatogram and precursor and product ion of (A) cymoxanil, (B) alternariol, and (C) alternariol methyl ether identified in tomato. *Reproduced from Ref. [186].*

The elucidation of the possible structure was carried out using chemical databases as ChemSpider, and a search in the scientific literature related to the compound.

3.2 Data-Independent Acquisition (DIA)

DIA experiments are conducted on a hybrid system with full scan data being recorded in a TOF, Orbitrap, or other high resolution mass analyzers. A major difference in these methods lies on the width (or the absent) of the precursor windows at the first stage of ion isolation. This in turn determines the range of ions being transmitted, fragmented, and sequenced. A consequence of having a wide precursor isolation window in DIA approaches is that it produces a very complex data structure and requires coherent and intricate data processing.

3.2.1 MS^E

MS^E is the most common DIA approach. It acquires MS^1 and MS^2 mass spectra in an unbiased and parallel manner. As such, it increases both the number of compounds detected and the reproducibility. During data acquisition, the energy of the gas-filled traveling-wave collision cell is dynamically switched between a low-energy and an elevated-energy status. This produces alternating composite mass spectra of all intact molecular ions, followed by hypothetic fragmented mass spectra of all precursors. Ion detection and charge/isotopic deconvolution are performed in data preprocessing. The "reconstructed" DDA-like data is then evaluated by database searching algorithm prior to the compound identification. MS^E method is based on parallel CID of molecules and provides masses of the precursor and fragment ions in parallel, which is advantageous for complex mixture analysis [118].

MS^E has been a popular choice among authors and has found applications both in proteomics and metabolomics. The method is compatible with multivariate data analysis, such as principal component analysis. A number of label-free quantitative studies have successfully applied MS^E, including identification of differentially expressed proteins in rat frontal cortex following subchronic treatment with haloperidol or olanzapine, profiling the serum [54] and postmortem brain tissues of schizophrenia patients [53], and analyzing the human pituitary proteome [55]. MS^E has also been employed in the studies of the proteins from plants [56,57] and microorganisms [40,48,58,59]. Figure 10 shows the MS^E analysis of flavonoids. Parent ions were detected when using a low voltage of 6 V. Fragmentation was observed at higher voltages, while at 45 V, all relevant product ions were observed. Fragmentation of flavonoids after high energy CID in MS^E mode were found to yield similar diagnostic ions as reported, which indicates that the method is indeed able to structurally characterize flavonoid glycosides.

3.2.2 All-Ion Fragmentation (AIF)

AIF is the system for the Orbitrap and was primarily developed for applications to small molecules. AIF, despite being named differently, is a similar acquisition mode to MS^E in which all precursor ions are fragmented without a preselection by the quadrupole. Fragmentation, however, is obtained with the HCD cell, located at the far side of the C-trap. During filling of the HCD collision cells, the energy can be set to step between values at specified percent values around

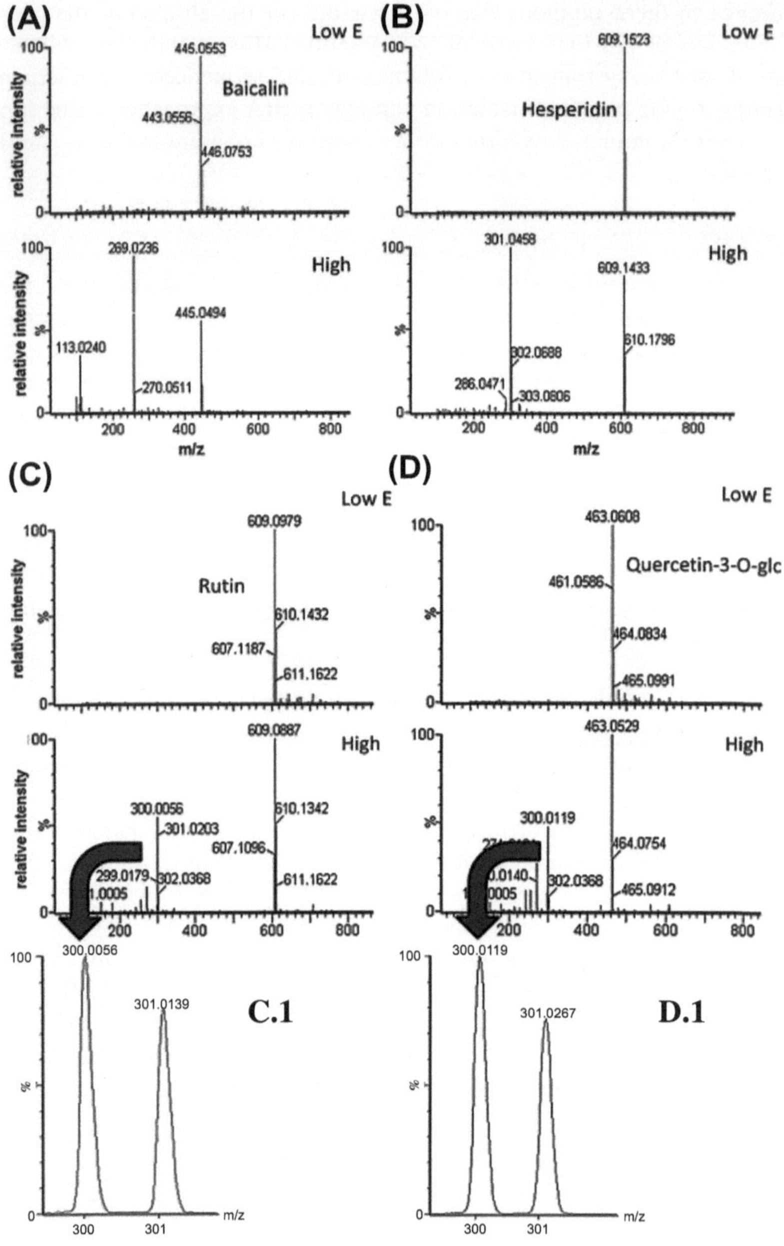

FIGURE 10 Fragmentation of flavonoid mono-O-glycosides in MSE (low energy 6V; high energy 45 V in 2-kV negative ionization mode). (A) Baicalein-7-O-glucuronide; (B) Hesperitin-7-O-rutinoside; (C) Rutin (quercetin-3-O-rhamnosylglucoside); (D) Quercetin-3-O-glucoside; (C.1, D.1) relative abundance of [Y0–H]•– (*m/z* = 300) to [Y0]– (*m/z* = 301). *Reproduced from Ref. [51] with permission of Elsevier.*

the chosen middle energy regardless of the ion's characteristics. This is similar to the energy ramping normally employed in MS^E.

3.3 Preprocessing and Postprocessing Steps

The mathematical processing of MS signals can be roughly divided into preprocessing and postprocessing. First, in the "preprocessing" step, the data are recovered from the original signal as accurately as possible, the "true" signal reflecting the *m/z* distribution of the ions originating from the sample. The preprocessing step includes registration, denoising, baseline correction, peak selection, and across samples peak alignment. In this step, these operations are performed independently of any biological information one seeks to extract from the data. The second type of processing attempts to represent the data in a form that facilitates the extraction of biological information. This step involves operations, such as dimension reduction, feature selection, clustering, and pattern recognition for classification.

The full methodology, summarized in Figure 11, can be applied directly to standard mass spectra generated from commercial instruments, i.e., from spectra to data visualization/exploration and application of up-to-date data mining methods. Beside improvements brought in the fields of screening and multiplexed quantification of analytes in food, HRMS methods have also been

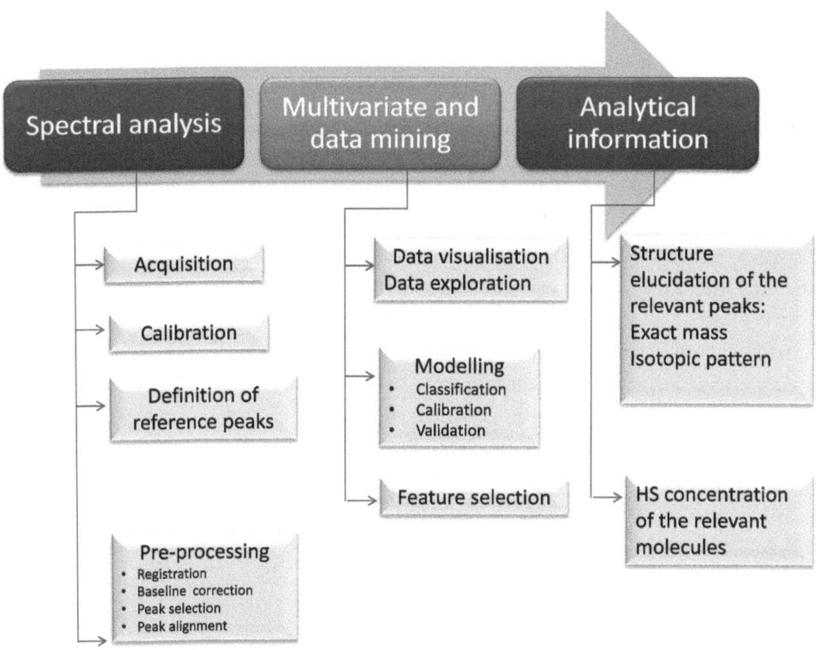

FIGURE 11 Schematics of the data analysis methodology.

coupled to untargeted data processing approaches or to effect directed analysis studies to highlight untargeted beneficial or hazardous chemicals in a variety of matrices. Furthermore, HRMS is nowadays the most used tool for "omics," which deals with the large-scale detection and quantification of any type of compounds in food.

4. IMAGING MASS SPECTROMETRY (IMS)

IMS encompasses a variety of techniques that enable the chemical imaging of analytes, which range in size from atoms and small molecules to intact proteins directly in the matrix [211]. It offers a unique combination of chemical and spatial information. IMS as an imaging technique presents as main advantage its ability to simultaneously detect and identify multiple known and unknown compounds without labeling the analytes because the mass of the analytes is a universal parameter [212]. The isotopic labeling of specific compounds can be beneficial but is not necessary [211]. A major drawback of the system is that it cannot be used easily in vivo and consumes the sample during the analysis [213].

In IMS, the analyte is transferred to the gas phase by (1) laser irradiation or high-energy ion beams and (2) ESI. There are several modes of imaging acquisition. In the most common, a laser or ion microprobe (micron or submicron dimensions) sequentially analyzes an array of isolate points as the microprobe beam is scanned across the sample surface in a raster pattern. Reconstruction of mass imaging is given after two-dimensional MS measurements on a sample section. Individual MS are acquired for each point and stored digitally. The software (specially designed for these applications) allows the selection of an analyte signal from the array of mass spectra and plots the intensity of the signal for each individual point in a 2D array. The signal intensity is plot in a color scale, creating a map of the analyte distribution [213,214]. The acquisition of chemical images by MS involves multiple, often elaborate steps beginning with sample preparation and ending with the generation of the chemical imaging. They are schematized in Figure 12.

Alternatively, the image can also be captured in the ion microscope mode by detecting desorbed (sputtered) ions. These ions are detected by special ion optics that keeps their relative position in the plane of the sample surface. The chemical imaging is generated by detecting these ions in a spatial resolving manner. This system is more similar to the optical imaging [215].

IMS is performed in the same mass spectrometers used for standard MS analysis. However, the most commonly applied are TOF instruments.

Three ionization methods have been used in IMS such as (1) laser desorption using both MALDI and laser desorption/ionization (LDI), (2) ion beam-induced desorption SIMS, and (3) ESI [216,217]. Although, the two first are already well-established techniques, the third one has test as a promising

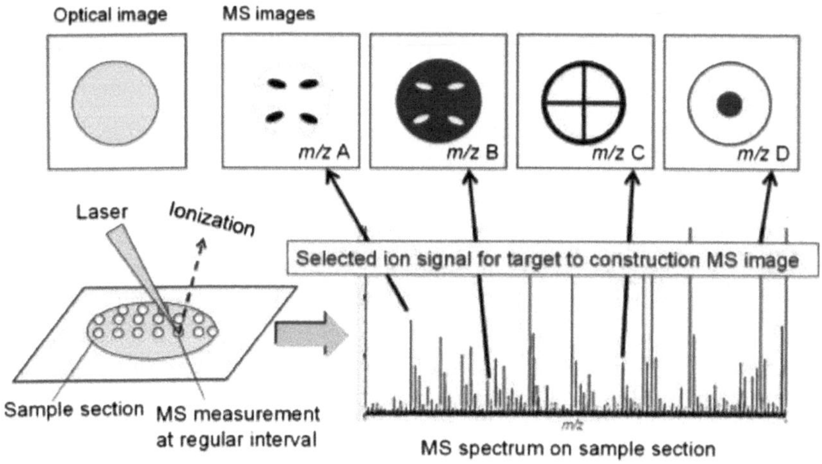

FIGURE 12 Schematic illustration of IMS. *Reproduced from Ref. [222] with permission of the Japan Society for Analytical Chemistry.*

approach for the future, MS for food imaging has increased significantly over the past decade, aided by the emergence of various ambient ionization techniques that allow direct and rapid analysis in ambient environment. The current status of food imaging with IMS has already been reviewed [218]. The described approaches include MALDI, but emphasize desorption atmospheric pressure photoionization, electrospray-assisted laser desorption/ionization, probe electrospray ionization, surface desorption atmospheric pressure chemical ionization, and laser ablation flowing atmospheric pressure afterglow.

IMS was initially developed for diagnostic applications in the medical and biomedical field. However, nowadays, it is starting to play an important role in food quality and safety [219,220]. To the moment, there are already some applications of IMS to locate biomolecules of interest and contaminants mostly in plant tissues. These applications are reported in Table 5. One of the most interesting example was presented by de Oliveira et al. [221], who carried out direct and simultaneous screening of aflatoxins and a stilbenoid (resveratrol) in peanuts (*Arachis hypogaea* L.) skin and kernel using silica plate imprinting as a sorptive tape-like extraction method followed by LDI-IMS. A silica gel (60 Å) plate is used as a molecular trapping surface for the samples. The greatest advantage associated with these methods is that they do not require chromatographic separation or many steps of sample preparation. This is also the first work that assesses both health hazardous and beneficial compounds to humans in a single sample, at the same time. Figure 13 shows the distribution of several mycotoxins in peanut kernel.

As can be observed, IMS is gaining attention as an important technique in its own right, with unique capabilities and requirements.

TABLE 5 Applications of IMS in Food Safety and Quality

Type	Application	References
MALDI-TOF-IMS	Analysis and localization of bioactive nutrients (ginsenosides and capsaicin)	[222]
MALDI-TOF-IMS	Visualization of anthocyanin species in rabbiteye blueberry *Vaccinium ashei*	[223]
MALDI-TOF-IMS	Different localization patterns of anthocyanin species in the pericarp of black rice	[224]
TLC-Blot-MALDI-TOF-IMS	Selective analysis of lipids	[225]
MALDI-TOF-IMS	*Oryza sativa* (rice)	[226]
MALDI-TOF-IMS	Visualization of spatial distribution of γ-aminobutyric acid in eggplant (*Solanum melongena*)	[227]
MALDI-TOF-IMS	Unambiguous characterization and tissue localization of Pru P 3 peach allergen	[228]
LA-ICP-TOF-IMS	Imaging of nutrient elements in the leaves of *Elsholtzia splendens*	[229]
LAESI-TOF-IMS	Macroscopic and microscopic imaging of pesticides, mycotoxins, and plant metabolites on rose leaves, orange and lemon fruits, ergot bodies, cherry tomatoes, and maize kernels	[230]
SPI-LDI-IMS	Simultaneous in situ assessment of aflatoxins and stilbenes	[221]
LDI-SIMS-IMS	Comparison of sample pretreatments for the imaging of *Miscanthus × giganteus*	[220]
LDI-SIMS-IMS	Localization of flavonoids in seed sections of peas (*Pisum sativum*) and *Arabidopsis thaliana*	[231]

LA-ICP-MS, laser ablation inductively coupled plasma mass spectrometry; LAESI-TOF-MS, laser-ablation electrospray ionization time of flight mass spectrometry imaging; LDI-SIMS-IMS, laser desorption/ionization secondary ion mass spectrometry imaging; MALDI-TOF-IMS, matrix-assisted laser desorption/ionization time of flight imaging mass spectrometry; SPI-LDI-QqTOF-IMS, silica plate imprinting laser desorption/ionization imaging mass spectrometry; TLC-Blot-MALDI-TOF-IMS, thin-layer chromatography blotting matrix-assisted laser desorption/ ionization time of flight imaging mass spectrometry.

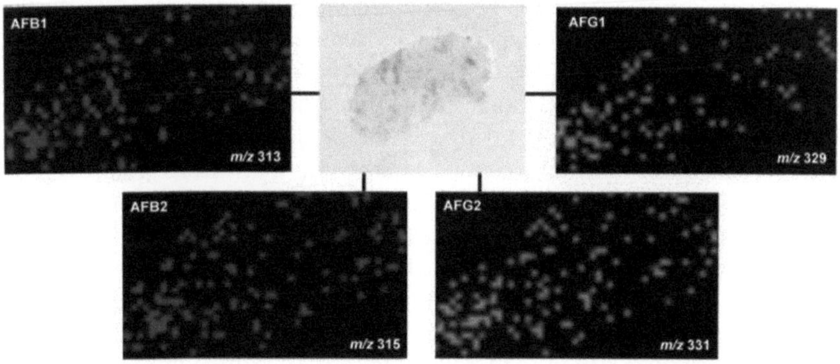

FIGURE 13 Example of SPILDI-MS images of the peanut skin: aflatoxins B1, B2, G1, and G2 are noted in their characteristic spatial distributions in positive ion mode. *Reproduced from Ref. [221].*

5. CONCLUSIONS

Control of food safety and quality is been transformed by mass spectrometers of increasingly high mass-resolving power. TOF and Orbitrap mass spectrometers including their multiple hybrid instruments are the key tools that have expanded the use of HRMS to transform it in a routine methodology within food analysis. For small molecules up to ~400 Da (e.g., contaminants, metabolites, and various natural organic mixtures), it is possible to determine elemental compositions of thousands of chemical components or contaminants simultaneously from accurate mass measurements (the same can be done up to 1000 Da if additional information is included). At higher mass, it becomes possible to identify proteins (including posttranslational modifications) from proteolytic peptides as well as lipids, glycoconjugates, and other food components. At even higher mass, it is possible to characterize posttranslational modifications of intact proteins and to map the binding surfaces of large biomolecule complexes. Fingerprint analysis is an important tool to extend the informative potential of HRMS. The fingerprint-assisted investigation of the distributions of known and unknown markers of a food matrix can be very useful for the definition of the so-called product signature in terms of sensory properties, botanical/geographical origin, and/or to study the modifications induced by thermal treatments on primary and secondary metabolites.

The high amount of information provided by HRMS is not only dependent on the resolving power of the mass spectrometer but also dealt with the high amount of data obtained. In practice, a food experiment can generate an MS data set that contains tens of thousands of spectra, and each spectrum contains tens of thousands sampling pairwise data points (*m/z* location and the corresponding intensities). Hence, the necessity to develop new procedures to extract manageable data sets arises, which can be employed in preliminary data visualization and analysis or as inputs for data mining procedures.

Few people predicted the explosion of proteomics research a decade ago that paved the way for MS usage today, including the wide application of HRMS. It is just as difficult to forecast the major use of MS instrumentation 10 years from now. However, it is probable that further improvements in separation devices, ionization techniques, and mass analyzers will foster advancements in the quantitative aspects of MS that will result in better performance specifications and improved precision. However, nowadays is already patented that IMS is a rapidly developing field of analytical science that has already demonstrated promising results in industry and biomedicine and very few insides within food quality and safety. Although limitations still exist on sample preparation, MS tissue imaging and profiling has the potential to become an important tool in food composition mapping within the next decade.

REFERENCES

[1] K. Klepárník, Recent advances in combination of capillary electrophoresis with mass spectrometry: methodology and theory, Electrophoresis 36 (2015) 159–178.

[2] M. Castro-Puyana, M. Herrero, Metabolomics approaches based on mass spectrometry for food safety, quality and traceability, TrAC-Trends Anal. Chem. 52 (2013) 74–87.

[3] X. Wang, S. Wang, Z. Cai, The latest developments and applications of mass spectrometry in food-safety and quality analysis, TrAC-Trends Anal. Chem. 52 (2013) 170–185.

[4] C. Hu, G. Xu, Mass-spectrometry-based metabolomics analysis for foodomics, TrAC-Trends Anal. Chem. 52 (2013) 36–46.

[5] A. Masiá, C. Blasco, Y. Picó, Last trends in pesticide residue determination by liquid chromatography-mass spectrometry, Trends Environ. Anal. Chem. 2 (2014) 11–24.

[6] S.J. Hird, et al., Liquid chromatography-mass spectrometry for the determination of chemical contaminants in food, TrAC-Trends Anal. Chem. 59 (2014) 59–72.

[7] N. Arroyo-Manzanares, et al., Holistic approach based on high resolution and multiple stage mass spectrometry to investigate ergot alkaloids in cereals, Talanta 118 (2014) 359–367.

[8] A.D. Troise, A. Fiore, V. Fogliano, Quantitation of acrylamide in foods by high-resolution mass spectrometry, J. Agric. Food Chem. 62 (1) (2014) 74–79.

[9] B.D. Eitzer, W. Hammack, M. Filigenzi, Interlaboratory comparison of a general method to screen foods for pesticides using QuEChERs extraction with high performance liquid chromatography and high resolution mass spectrometry, J. Agric. Food Chem. 62 (1) (2014) 80–87.

[10] S.V. Malysheva, et al., Identification of novel metabolites from Aspergillus flavus by high resolution and multiple stage mass spectrometry, Food Addit. Contam. Part A Chem. Anal. Control Expo. Risk Assess. 31 (1) (2014) 111–120.

[11] J. Xu, et al., Chemical composition, sources, and processes of urban aerosols during summertime in northwest China: insights from high-resolution aerosol mass spectrometry, Atmos. Chem. Phys. 14 (23) (2014) 12593–12611.

[12] E. Ates, et al., Screening of plant and fungal metabolites in wheat, maize and animal feed using automated on-line clean-up coupled to high resolution mass spectrometry, Food Chem. 142 (2014) 276–284.

[13] G. Kaklamanos, U. Vincent, C. von Holst, Multi-residue method for the detection of veterinary drugs in distillers grains by liquid chromatography-orbitrap high resolution mass spectrometry, J. Chromatogr. A 1322 (2013) 38–48.

[14] P. Cheajesadagul, et al., Discrimination of geographical origin of rice based on multi-element fingerprinting by high resolution inductively coupled plasma mass spectrometry, Food Chem. 141 (4) (2013) 3504–3509.

[15] T. Rezanka, et al., Identification of plasmalogen cardiolipins from pectinatus by liquid chromatography-high resolution electrospray ionization tandem mass spectrometry, Lipids 48 (12) (2013) 1237–1251.

[16] M. Mattarozzi, et al., Liquid chromatography-full scan-high resolution mass spectrometry-based method towards the comprehensive analysis of migration of primary aromatic amines from food packaging, J. Chromatogr. A 1320 (2013) 96–102.

[17] H. Gallart-Ayala, et al., Potential of mass spectrometry metabolomics for chemical food safety, Bioanalysis 7 (1) (2015) 133–146.

[18] W. Jia, et al., Analysis of phthalates in milk and milk products by liquid chromatography coupled to quadrupole orbitrap high-resolution mass spectrometry, J. Chromatogr. A 1362 (2014) 110–118.

[19] V. Cunsolo, et al., Mass spectrometry in food proteomics: a tutorial, J. Mass Spectrom. 49 (9) (2014) 768–784.

[20] SCOPUS, 2015. Available from: www.scopus.com.

[21] A. Rizzuti, et al., A combined approach for characterisation of fresh and brined vine leaves by X-ray powder diffraction, NMR spectroscopy and direct infusion high resolution mass spectrometry, Food Chem. 141 (3) (2013) 1908–1915.

[22] P.Q. Tranchida, et al., Analysis of the unsaponifiable fraction of lipids belonging to various milk-types by using comprehensive two-dimensional gas chromatography with dual mass spectrometry/flame ionization detection and with the support of high resolution time-of-flight mass spectrometry for structural elucidation, J. Chromatogr. A 1313 (2013) 194–201.

[23] G. Pierri, et al., Analysis of bovine milk caseins on organic monolithic columns: an integrated capillary liquid chromatography-high resolution mass spectrometry approach for the study of time-dependent casein degradation, J. Chromatogr. A 1313 (2013) 259–269.

[24] P. Faber, et al., Anthropogenic sources of aerosol particles in a football stadium: real-time characterization of emissions from cigarette smoking, cooking, hand flares, and color smoke bombs by high-resolution aerosol mass spectrometry, Atmos. Environ. 77 (2013) 1043–1051.

[25] L. Monaci, et al., Multi-allergen quantification of fining-related egg and milk proteins in white wines by high-resolution mass spectrometry, Rapid Commun. Mass Spectrom. 27 (17) (2013) 2009–2018.

[26] P. Donato, et al., Mass spectrometry detection in comprehensive liquid chromatography: basic concepts, instrumental aspects, applications and trends, Mass Spectrom. Rev. 31 (5) (2012) 523–559.

[27] F.G. Strathmann, A.N. Hoofnagle, Current and future applications of mass spectrometry to the clinical laboratory, Am. J. Clin. Pathol. 136 (4) (2011) 609–616.

[28] J. Koch, D. Günther, Review of the state-of-the-art of laser ablation inductively coupled plasma mass spectrometry, Appl. Spectrosc. 65 (5) (2011) 155A–162A.

[29] H. Awad, M.M. Khamis, A. El-Aneed, Mass spectrometry, review of the basics: ionization, Appl. Spectrosc. Rev. 50 (2) (2015) 158–175.

[30] IUPAC, Compendium of Chemical Terminology, second ed., Blackwell Scientific Publications, Oxford, 1997.
[31] P. Price, Standard definitions of terms relating to mass spectrometry: a report from the committee on measurements and standards of the American society for mass spectrometry, J. Am. Soc. Mass Spectrom. 2 (4) (1991) 336–348.
[32] D.G. Patterson Jr., et al., Cryogenic zone compression for the measurement of dioxins in human serum by isotope dilution at the attogram level using modulated gas chromatography coupled to high resolution magnetic sector mass spectrometry, J. Chromatogr. A 1218 (21) (2011) 3274–3281.
[33] M.M. Aguilera-Luiz, et al., Rapid and semiautomated method for the analysis of veterinary drug residues in honey based on turbulent-flow liquid chromatography coupled to ultrahigh-performance liquid chromatography-orbitrap mass spectrometry (TFC-UHPLC-Orbitrap-MS), J. Agric. Food Chem. 61 (4) (2013) 829–839.
[34] J. Liang, et al., The profiling and identification of the metabolites of (+)-catechin and study on their distribution in rats by HPLC-DAD-ESI-IT-TOF-MS^n technique, Biomed. Chromatogr. 28 (3) (2014) 401–411.
[35] P. Donato, et al., The beauty and the health: comprehensive LC×LC/PDA/IT-TOF of pigments in food, Food Sci. Technol. 25 (2) (2011) 46–47.
[36] C.K. Fagerquist, Rapid identification of food-borne pathogens by top-down proteomics using MALDI-TOF/TOF mass spectrometry, in: ACS Symposium Series, 2011, pp. 99–120.
[37] M. Pazos, et al., Fish proteins as targets of ferrous-catalyzed oxidation: identification of protein carbonyls by fluorescent labeling on two-dimensional gels and MALDI-TOF/TOF mass spectrometry, J. Agric. Food Chem. 59 (14) (2011) 7962–7977.
[38] C. Brunon, et al., ToF-SIMS and XPS characterization of antimicrobial textiles for the food processing industry, Surf. Interface Anal. 43 (1–2) (2011) 604–608.
[39] T.R. Sandrin, J.E. Goldstein, S. Schumaker, MALDI TOF MS profiling of bacteria at the strain level: a review, Mass Spectrom. Rev. 32 (3) (2013) 188–217.
[40] C.D. Doern, Charting uncharted territory: a review of the verification and implementation process for matrix-assisted laser desorption ionization-time of flight mass spectrometry (MALDI-TOF MS) for organism identification, Clin. Microbiol. Newslett. 35 (9) (2013) 69–78.
[41] M. Pavlovic, et al., Application of MALDI-TOF MS for the identification of food borne bacteria, Open Microbiol. J. 7 (1) (2013) 135–141.
[42] M. Welker, MALDI-TOF MS for identification of microorganisms: a new era in clinical microbiological research and diagnosis, in: The Role of New Technologies in Medical Microbiological Research and Diagnosis, 2012, pp. 56–71.
[43] B. Marcos, et al., Analysis of raw hams using SELDI-TOF-MS to predict the final quality of dry-cured hams, Meat Sci. 93 (2) (2013) 233–239.
[44] E. Silletti, J.H.F. Bult, M. Stieger, Effect of NaCl and sucrose tastants on protein composition of oral fluid analysed by SELDI-TOF-MS, Arch. Oral Biol. 57 (9) (2012) 1200–1210.
[45] C.W. Lim, et al., Quantitative assessment of moniliformin in cereals via alternative precipitation pathways, aided by LC-LIT-MS and LC-Q-TOF-MS, Food Chem. 174 (2015) 372–379.
[46] F. Damiano, et al., Analysis of illicit dietary supplements sold in the Italian market: identification of a sildenafil thioderivative as adulterant using UPLC-TOF/MS and GC/MS, Sci. Justice 54 (3) (2014) 228–237.
[47] M.M.A. Omar, A.A. Elbashir, O.J. Schmitz, Determination of acrylamide in Sudanese food by high performance liquid chromatography coupled with LTQ Orbitrap mass spectrometry, Food Chem. 176 (2015) 342–349.

[48] T. Portolés, et al., Validation of a qualitative screening method for pesticides in fruits and vegetables by gas chromatography quadrupole-time of flight mass spectrometry with atmospheric pressure chemical ionization, Anal. Chim. Acta 838 (2014) 76–85.

[49] G. Pepe, et al., Evaluation of anti-inflammatory activity and fast UHPLC-DAD-IT-TOF profiling of polyphenolic compounds extracted from green lettuce (*Lactuca sativa* L.; Var. Maravilla de Verano), Food Chem. 167 (2015) 153–161.

[50] S. Shao, et al., Quantification of Se-methylselenocysteine and its γ-glutamyl derivative from naturally Se-enriched Green bean (*Phaseolus vulgaris vulgaris*) after HPLC-ESI-TOF-MS and orbitrap MS^n-based identification, Food Anal. Methods 7 (5) (2014) 1147–1157.

[51] G.B. Gonzales, et al., Ultra(high)-pressure liquid chromatography–electrospray ionization-time-of-flight-ion mobility-high definition mass spectrometry for the rapid identification and structural characterization of flavonoid glycosides from cauliflower waste, J. Chromatogr. A 1323 (0) (2014) 39–48.

[52] F. Hernandez, et al., Gas chromatography coupled to high-resolution time-of-flight mass spectrometry to analyze trace-level organic compounds in the environment, food safety and toxicology, TrAC-Trends Anal. Chem. 30 (2) (2011) 388–400.

[53] K. George, U. Vincent, C. von Holst, Analysis of antimicrobial agents in pig feed by liquid chromatography coupled to orbitrap mass spectrometry, J. Chromatogr. A 1293 (2013) 60–74.

[54] M. Tsevdou, et al., Monitoring the effect of high pressure and transglutaminase treatment of milk on the evolution of flavour compounds during lactic acid fermentation using PTR-ToF-MS, Food Chem. 138 (4) (2013) 2159–2167.

[55] J. Sánchez del Pulgar, et al., Effect of the pig rearing system on the final volatile profile of Iberian dry-cured ham as detected by PTR-ToF-MS, Meat Sci. 93 (3) (2013) 420–428.

[56] Y. Qin, et al., Inductively coupled plasma orthogonal acceleration time-of-flight mass spectrometry (ICP-oa-TOF-MS) analysis of heavy metal content in *Indocalamus tessellatus* samples, Food Chem. 141 (3) (2013) 2154–2157.

[57] L. Husáková, et al., Analytical capabilities of inductively coupled plasma orthogonal acceleration time-of-flight mass spectrometry (ICP-oa-TOF-MS) for multi-element analysis of food and beverages, Food Chem. 129 (3) (2011) 1287–1296.

[58] S. Català-Clariana, et al., Identification of bioactive peptides in hypoallergenic infant milk formulas by CE-TOF-MS assisted by semiempirical model of electromigration behavior, Electrophoresis 34 (13) (2013) 1886–1894.

[59] M. Aznar, et al., UPLC-Q-TOF-MS analysis of non-volatile migrants from new active packaging materials, Anal. Bioanal. Chem. 404 (6–7) (2012) 1945–1957.

[60] C. De Ceglie, C.D. Calvano, C.G. Zambonin, MALDI-TOF MS for quality control of high protein content sport supplements, Food Chem. 176 (2015) 396–402.

[61] C. Ferrer, et al., Application of HPLC-TOF-MS and HPLC-QTOF-MS/MS for pesticide residues analysis in fruit and vegetable matrices, in: Comprehensive Analytical Chemistry, 2012, pp. 1–60.

[62] J. Ščančar, et al., Ni speciation in tea infusions by monolithic chromatography—ICP-MS and Q-TOF-MS, Anal. Bioanal. Chem. 405 (6) (2013) 2041–2051.

[63] U. Arroyo-Abad, et al., Determination and identification of hydrophilic and hydrophobic arsenic species in methanol extract of fresh cod liver by RP-HPLC with simultaneous ICP-MS and ESI-Q-TOF-MS detection, Food Chem. 141 (3) (2013) 3093–3102.

[64] K. Hälvin, T. Paalme, I. Nisamedtinov, Comparison of different extraction methods for simultaneous determination of B complex vitamins in nutritional yeast using LC/MS-TOF and stable isotope dilution assay, Anal. Bioanal. Chem. 405 (4) (2013) 1213–1222.

[65] K. Mazanec, F. Dycka, J. Bobalova, Monitoring of barley starch amylolysis by gravitational field flow fractionation and MALDI-TOF MS, J. Sci. Food Agric. 91 (15) (2011) 2756–2761.

[66] V. Verardo, et al., Determination of free and bound phenolic compounds in buckwheat spaghetti by RP-HPLC-ESI-TOF-MS: effect of thermal processing from farm to fork, J. Agric. Food Chem. 59 (14) (2011) 7700–7707.

[67] I. Ferrer, E.M. Thurman, J. Zweigenbaum, LC/TOF-MS analysis of pesticides in fruits and vegetables: the emerging role of accurate mass in the unambiguous identification of pesticides in food, in: Methods in Molecular Biology, 2011, pp. 193–218.

[68] K. Kaseleht, T. Paalme, I. Nisamedtinov, Quantitative analysis of acetaldehyde in foods consumed by children using SPME/GC-MS(Tof), on-fiber derivatization and deuterated acetaldehyde as an internal standard, Agron. Res. 9 (Sppl. iss. 2) (2011) 395–401.

[69] J. Qiu, et al., Screening natural antioxidants in peanut shell using DPPH-HPLC-DAD-TOF/MS methods, Food Chem. 135 (4) (2012) 2366–2371.

[70] D. Zacs, et al., Ultra high performance liquid chromatography-time-of-flight high resolution mass spectrometry in the analysis of hexabromocyclododecane diastereomers: method development and comparative evaluation versus ultra high performance liquid chromatography coupled to orbitrap high resolution mass spectrometry and triple quadrupole tandem mass spectrometry, J. Chromatogr. A 1366 (2014) 73–83.

[71] J. Zhang, et al., Simultaneous analysis of 19 antibiotics in dairy products using ultra-performance liquid chromatography coupled with high resolution time-of-flight mass spectrometry, Se Pu 30 (10) (2012) 1031–1036.

[72] H.H. Su, et al., Micro-scale strategy to detect spermine and spermidine by MALDI-TOF MS in foods and identification of apoptosis-related proteins by nano-flow UPLC-MS/MS after treatment with spermine and spermidine, J. Chromatogr. B: Anal. Technol. Biomed. Life Sci. 978–979 (2015) 131–137.

[73] W. Dirwono, et al., LC-TOF/MS determination of phthalates in edible salts from food markets in Korea, Food Addit. Contam. Part B Surveill. 6 (3) (2013) 203–208.

[74] J. Wang, et al., Antioxidant activity of leaf extracts from different *Hibiscus sabdariffa* accessions and simultaneous determination five major antioxidant compounds by LC-Q-TOF-MS, Molecules 19 (12) (2014) 12226–12238.

[75] V. Cunsolo, et al., MALDI-TOF mass spectrometry for the monitoring of she-donkey's milk contamination or adulteration, J. Mass Spectrom. 48 (2) (2013) 148–153.

[76] Y.S. Huang, T.J. Hsieh, C.Y. Lu, Simple analytical strategy for MALDI-TOF-MS and nanoUPLC-MS/MS: quantitating curcumin in food condiments and dietary supplements and screening of acrylamide-induced ROS protein indicators reduced by curcumin, Food Chem. 174 (2015) 571–576.

[77] A.A. Aydin, T. Letzel, Simultaneous investigation of sesquiterpenes, pyrrolizidine alkaloids and N-oxides in butterbur (*Petasites hybridus*) with an offline 2D-combination of HPLC-UV and LC-MMI-ToF-MS, J. Pharmaceut. Biomed. Anal. 85 (2013) 74–82.

[78] J. Wang, W. Chow, D. Leung, Applications of LC/ESI-MS/MS and UHPLC/Qq-TOF-MS for the determination of 141 pesticides in tea, J. AOAC Int. 94 (6) (2011) 1685–1714.

[79] S.H. Roh, et al., Determination of tadalafil and N-desmethylsibutramine in health and dietary supplements using ultra-performance liquid chromatography (UPLC) coupled with quadrupole-time-of-flight mass spectrometry (Q-TOF MS), Food Addit. Contam. Part A Chem. Anal. Control Expo. Risk Assess. 28 (11) (2011) 1475–1482.

[80] Z.-F. Du, et al., Simultaneous determination of seven fluorescent whitening agents in food contact paper package materials by high performance liquid chromatography tandem triple TOF-high resolution mass spectrometry, Chin. J. Anal. Chem. 42 (5) (2014) 636–642.

[81] S.R. Oppliger, L.H. Münger, L. Nyström, Rapid and highly accurate detection of steryl glycosides by ultraperformance liquid chromatography-quadrupole time-of-flight mass spectrometry (UPLC-Q-TOF-MS), J. Agric. Food Chem. 62 (39) (2014) 9410–9419.

[82] X. Shu, et al., Rapid lipid profiling of bacteria by online MALDI-TOF mass spectrometry, Int. J. Mass Spectrom. 321–322 (2012) 71–76.

[83] C.D. Calvano, C. De Ceglie, C.G. Zambonin, Development of a direct in-matrix extraction (DIME) protocol for MALDI-TOF-MS detection of glycated phospholipids in heat-treated food samples, J. Mass Spectrom. 49 (9) (2014) 831–839.

[84] L. Bono, et al., Triacylglycerol profile in cocoa liquors using MALDI-TOF and LC-ESI tandem mass spectrometry, J. Mass Spectrom. 49 (9) (2014) 894–899.

[85] Q. Shen, Q. Yang, H.Y. Cheung, Hydrophilic interaction chromatography based solid-phase extraction and MALDI TOF mass spectrometry for revealing the influence of Pseudomonas fluorescens on phospholipids in salmon fillet, Anal. Bioanal. Chem. 407 (5) (2014) 1475–1484.

[86] C.D. Calvano, et al., MALDI-TOF mass spectrometry detection of extra-virgin olive oil adulteration with hazelnut oil by analysis of phospholipids using an ionic liquid as matrix and extraction solvent, Food Chem. 134 (2) (2012) 1192–1198.

[87] C.D. Calvano, et al., MALDI-TOF mass spectrometric determination of intact phospholipids as markers of illegal bovine milk adulteration of high-quality milk, Anal. Bioanal. Chem. 405 (5) (2013) 1641–1649.

[88] J. Lee, J. Zweigenbaum, A.E. Mitchell, Nontargeted unknown LC(ESI-)-Q/TOF MS approaches for food verification, in: ACS Symposium Series, 2013, pp. 17–29.

[89] J. Wu, et al., The comparative research on constituents of Radix Aconiti and its processing by HPLC quadrupole TOF-MS, Biomed. Chromatogr. 26 (11) (2012) 1301–1307.

[90] J.W. Drynan, et al., MALDI-TOF mass spectrometry: avoidance of artifacts and analysis of caffeine-precipitated SII thearubigins from 15 commercial black teas, J. Agric. Food Chem. 60 (18) (2012) 4514–4525.

[91] M.D.L.L. Cádiz-Gurrea, et al., Comprehensive characterization by UHPLC-ESI-Q-TOF-MS from an *Eryngium bourgatii* extract and their antioxidant and anti-inflammatory activities, Food Res. Int. 50 (1) (2013) 197–204.

[92] W.M. Chai, et al., NMR, HPLC-ESI-MS, and MALDI-TOF MS analysis of condensed tannins from *Delonix regia* (Bojer ex Hook.) Raf. and their bioactivities, J. Agric. Food Chem. 60 (19) (2012) 5013–5022.

[93] I.M. Abu-Reidah, et al., HPLC-ESI-Q-TOF-MS for a comprehensive characterization of bioactive phenolic compounds in cucumber whole fruit extract, Food Res. Int. 46 (1) (2012) 108–117.

[94] F. Isella, et al., Migration of non intentionally added substances from adhesives by UPLC-Q-TOF/MS and the role of EVOH to avoid migration in multilayer packaging materials, J. Mass Spectrom. 48 (4) (2013) 430–437.

[95] J. Kim, et al., GC-TOF-MS- and CE-TOF-MS-based metabolic profiling of cheonggukjang (fast-fermented bean paste) during fermentation and its correlation with metabolic pathways, J. Agric. Food Chem. 60 (38) (2012) 9746–9753.

[96] E. Hurtado-Fernández, et al., Metabolomic analysis of avocado fruits by GC-APCI-TOF MS: effects of ripening degrees and fruit varieties, Anal. Bioanal. Chem. 407 (2) (2014) 547–555.

[97] H.J. Kang, et al., Metabolomic analysis of meju during fermentation by ultra performance liquid chromatography-quadrupole-time of flight mass spectrometry (UPLC-Q-TOF MS), Food Chem. 127 (3) (2011) 1056–1064.

[98] N. Talhaoui, et al., Determination of phenolic compounds of "Sikitita" olive leaves by HPLC-DAD-TOF-MS. Comparison with its parents "Arbequina" and "Picual" olive leaves, LWT-Food Sci. Technol. 58 (1) (2014) 28–34.

[99] C.C. Bonatto, L.P. Silva, Cocoa content influences chocolate molecular profile investigated by MALDI-TOF mass spectrometry, J. Sci. Food Agric. (2015). http://dx.doi.org/0.1002/jsfa.6740.
[100] A. Romano, et al., Nosespace analysis by PTR-ToF-MS for the characterization of food and tasters: the case study of coffee, Int. J. Mass Spectrom. 365–366 (2014) 20–27.
[101] E. Zardin, et al., Performance assessment of proton-transfer-reaction time-of-flight mass spectrometry (PTR-TOF-MS) for analysis of isobaric compounds in food-flavour applications, LWT-Food Sci. Technol. 56 (1) (2014) 153–160.
[102] E. Hurtado-Fernández, et al., Quantitative characterization of important metabolites of avocado fruit by gas chromatography coupled to different detectors (APCI-TOF MS and FID), Food Res. Int. 62 (2014) 801–811.
[103] X. Luo, et al., Analyses by UPLC Q-TOF MS of products of aflatoxin B1 after ozone treatment, Food Addit. Contam. Part A Chem. Anal. Control Expo. Risk Assess. 31 (1) (2014) 105–110.
[104] K.L. Chang, P.C. Ho, Gas chromatography time-of-flight mass spectrometry (GC-TOF-MS)-based metabolomics for comparison of caffeinated and decaffeinated coffee and its implications for Alzheimer's disease, PLoS One 9 (8) (2014).
[105] C.S. Funari, et al., High resolution ultra high pressure liquid chromatography-time-of-flight mass spectrometry dereplication strategy for the metabolite profiling of Brazilian Lippia species, J. Chromatogr. A 1259 (2012) 167–178.
[106] P. Volta, et al., Discrimination of freshwater fish species by matrix-assisted laser Desorption/Ionization-time of flight mass spectrometry (MALDI-TOF MS): a pilot study, J. Limnol. 71 (1) (2012).
[107] C.D. Calvano, et al., Matrix-assisted laser desorption/ionization time-of-flight (MALDI-TOF) mass spectrometry fingerprinting as a new tool for quality control and health risk assessment of virgin olive oil, in: Olive Consumption and Health, 2012, pp. 131–154.
[108] J. Bórquez, E.J. Kennelly, M.J. Simirgiotis, Activity guided isolation of isoflavones and hyphenated HPLC-PDA-ESI-ToF-MS metabolome profiling of *Azorella madreporica* Clos. from northern Chile, Food Res. Int. 52 (1) (2013) 288–297.
[109] C. Chu, et al., Profiling the ginsenosides of three ginseng products by Lc-Q-Tof/Ms, J. Food Sci. 78 (5) (2013) C653–C659.
[110] U.V. Mallavadhani, R. Panigrahi, UPLC-Q-TOF-MS based studies to evaluate the effect of extraction methodology on the yield of 6-shogaol, a biomarker of *Z. officinale*, Ind. Crops Prod. 50 (2013) 821–827.
[111] B. Xu, et al., Characterization and authentication of four important edible oils using free phytosterol profiles established by GC-GC-TOF/MS, Anal. Methods 6 (17) (2014) 6860–6870.
[112] E. Aprea, et al., Volatile compound changes during shelf life of dried *Boletus edulis*: comparison between SPME-GC-MS and PTR-ToF-MS analysis, J. Mass Spectrom. 50 (1) (2014) 56–64.
[113] M.P. Delgado de la Torre, F. Priego-Capote, M.D. Luque de Castro, Evaluation of the composition of vine shoots and oak chips for oenological purposes by superheated liquid extraction and high-resolution liquid chromatography-time-of-flight/mass spectrometry analysis, J. Agric. Food Chem. 60 (13) (2012) 3409–3417.
[114] C.D. Calvano, et al., Proteomic approach based on MALDI-TOF MS to detect powdered milk in fresh cow's milk, J. Agric. Food Chem. 61 (8) (2013) 1609–1617.
[115] M. Guijarro-Díez, et al., LC-ESI-TOF MS method for the evaluation of the immunostimulating activity of soybeans via the determination of the functional peptide soymetide, J. Agric. Food Chem. 61 (15) (2013) 3611–3618.

[116] S.C. Cheison, et al., Hydrolysis of β-lactoglobulin by trypsin under acidic pH and analysis of the hydrolysates with MALDI-TOF-MS/MS, Food Chem. 125 (4) (2011) 1241–1248.

[117] C.D. Calvano, et al., Detection of sheep and goat milk adulterations by direct MALDI-TOF MS analysis of milk tryptic digests, J. Mass Spectrom. 47 (9) (2012) 1141–1149.

[118] M.T. Yilmaz, et al., A novel method to differentiate bovine and porcine gelatins in food products: NanoUPLC-ESI-Q-TOF-MSE based data independent acquisition technique to detect marker peptides in gelatin, Food Chem. 141 (3) (2013) 2450–2458.

[119] B. Kerkaert, et al., Identification of modified lysozyme peptides upon photo-oxidation by LC-TOF-MS, J. Agric. Food Chem. 61 (51) (2013) 12727–12736.

[120] B. Wang, et al., Identification and characterization of a new IgE-binding protein in mackerel (*Scomber japonicus*) by MALDI-TOF-MS, J. Ocean Univ. China 10 (1) (2011) 93–98.

[121] C.D. Calvano, C. De Ceglie, C.G. Zambonin, Proteomic analysis of complex protein samples by MALDI-TOF mass spectrometry, Meth. Mol. Biol. (Clifton, N.J.) 1129 (2014) 365–380.

[122] Q. Zhang, et al., A high throughput screening assay for identifying glycation inhibitors on MALDI-TOF target, Food Chem. 170 (2015) 160–168.

[123] H. Kajiwara, Gene analysis using mass spectrometric cleaved amplified polymorphic sequence (MS-CAPS) with matrix-assisted laser desorption ionization time-of-flight mass spectrometry (MALDI-TOF), in: Methods in Molecular Biology, 2015, pp. 205–214.

[124] K. Böhme, et al., Rapid species identification of seafood spoilage and pathogenic gram-positive bacteria by MALDI-TOF mass fingerprinting, Electrophoresis 32 (21) (2011) 2951–2965.

[125] K. Böhme, et al., Characterization of *Staphylococcus aureus* strains isolated from Italian dairy products by MALDI-TOF mass fingerprinting, Electrophoresis 33 (15) (2012) 2355–2364.

[126] N. Nicolaou, Y. Xu, R. Goodacre, Detection and quantification of bacterial spoilage in milk and pork meat using MALDI-TOF-MS and multivariate analysis, Anal. Chem. 84 (14) (2012) 5951–5958.

[127] N.T.L. Doan, et al., Validation of MALDI-TOF MS for rapid classification and identification of lactic acid bacteria, with a focus on isolates from traditional fermented foods in Northern Vietnam, Lett. Appl. Microbiol. 55 (4) (2012) 265–273.

[128] L. Hausdorf, et al., Characterization of the cultivable microbial community in a spinach-processing plant using MALDI-TOF MS, Food Microbiol. 34 (2) (2013) 406–441.

[129] A. Vávrová, et al., Evaluation of the MALDI-TOF MS profiling for identification of newly described *Aeromonas* spp., Folia Microbiol. (2015). http://dx.doi.org/10.1007/s12223-014-0369-4.

[130] S. Jadhav, et al., Detection of Listeria monocytogenes from selective enrichment broth using MALDI-TOF mass spectrometry, J. Proteomics 97 (2014) 100–106.

[131] M. Wenning, et al., Identification and differentiation of food-related bacteria: a comparison of FTIR spectroscopy and MALDI-TOF mass spectrometry, J. Microbiol. Meth. 103 (2014) 44–52.

[132] M. Dušková, et al., Identification of lactobacilli isolated from food by genotypic methods and MALDI-TOF MS, Int. J. Food Microbiol. 159 (2) (2012) 107–114.

[133] J.Y. Gardner, et al., The use of GC×GC/TOF MS with multivariate analysis for the characterization of foodborne pathogen bacteria profiles, J. Sep. Sci. 34 (2) (2011) 176–185.

[134] V. Tsilia, et al., Application of MALDI-TOF mass spectrometry for the detection of enterotoxins produced by pathogenic strains of the *Bacillus cereus* group, Anal. Bioanal. Chem. 404 (6–7) (2012) 1691–1702.

[135] W.G. Ren, et al., Analysis on crude and wine-processed *Siegesbeckia pubescens* by UPLC-Q-TOF/MS, Chin. Tradit. Herb. Drugs 45 (2) (2014) 181–187.

[136] J.C. Usbeck, et al., Optimization of experimental and modelling parameters for the differentiation of beverage spoiling yeasts by matrix-assisted-laser-desorption/ionization-time-of-flight mass spectrometry (MALDI-TOF MS) in response to varying growth conditions, Food Microbiol. 36 (2) (2013) 379–387.

[137] C.C. Kern, R.F. Vogel, J. Behr, Identification and differentiation of brewery isolates of *Pectinatus* sp. by matrix-assisted-laser desorption-ionization time-of-flight mass spectrometry (MALDI-TOF MS), Eur. Food Res. Technol. 238 (5) (2014) 875–880.

[138] M. Quintela-Baluja, et al., Characterization of different food-isolated *Enterococcus* strains by MALDI-TOF mass fingerprinting, Electrophoresis 34 (15) (2013) 2240–2250.

[139] Z.L. Tan, et al., Rapid identification of microorganisms isolated from pickled vegetables by MALDI-TOF mass spectrometry, in: Advanced Materials Research, 2013, pp. 494–497.

[140] I.C. Fernández-No, et al., Characterisation and profiling of *Bacillus subtilis*, *Bacillus cereus* and *Bacillus licheniformis* by MALDI-TOF mass fingerprinting, Food Microbiol. 33 (2) (2013) 235–242.

[141] E. Angelakis, et al., Rapid and accurate bacterial identification in probiotics and yoghurts by MALDI-TOF mass spectrometry, J. Food Sci. 76 (8) (2011) M568–M572.

[142] P. Rodrigues, et al., Species identification of *Aspergillus* section *Flavi* isolates from Portuguese almonds using phenotypic, including MALDI-TOF ICMS, and molecular approaches, J. Appl. Microbiol. 111 (4) (2011) 877–892.

[143] K. Böhme, et al., Identification and classification of seafood-borne pathogenic and spoilage bacteria: 16S rRNA sequencing versus MALDI-TOF MS fingerprinting, Electrophoresis 34 (6) (2013) 877–887.

[144] K. Böhme, et al., Assessment of seafood safety and quality by rapid identification of pathogenic and spoilage bacterial species by MALDI-TOF mass fingerprinting, in: 6th International CIGR Technical Symposium—towards a Sustainable Food Chain: Food Process, Bioprocessing and Food Quality Management, 2011.

[145] T. Kuda, et al., Rapid identification of *Tetragenococcus halophilus* and *Tetragenococcus muriaticus*, important species in the production of salted and fermented foods, by matrix-assisted laser desorption ionization-time of flight mass spectrometry (MALDI-TOF MS), Food Control 35 (1) (2014) 419–425.

[146] J.S. Del Pulgar, et al., Rapid characterization of dry cured ham produced following different PDOs by proton transfer reaction time of flight mass spectrometry (PTR-ToF-MS), Talanta 85 (1) (2011) 386–393.

[147] L. Cappellin, et al., Linking GC-MS and PTR-TOF-MS fingerprints of food samples, Chemometr. Intell. Lab. Syst. 118 (2012) 301–307.

[148] S. Heenan, et al., PTR-TOF-MS monitoring of in vitro and in vivo flavour release in cereal bars with varying sugar composition, Food Chem. 131 (2) (2012) 477–484.

[149] F. Wieland, et al., Online monitoring of coffee roasting by proton transfer reaction time-of-flight mass spectrometry (PTR-ToF-MS): towards a real-time process control for a consistent roast profile, Anal. Bioanal. Chem. 402 (8) (2012) 2531–2543.

[150] J. Kiefl, G. Pollner, P. Schieberle, Sensomics analysis of key hazelnut odorants (*Corylus avellana* L. "Tonda Gentile") using comprehensive two-dimensional gas chromatography in combination with time-of-flight mass spectrometry (GC×GC-TOF-MS), J. Agric. Food Chem. 61 (22) (2013) 5226–5235.

[151] G. Cao, et al., Capture and identification of the volatile components in crude and processed herbal medicines through on-line purge and trap technique coupled with GC×GC-TOF MS, Nat. Prod. Res. 28 (19) (2014) 1607–1612.

[152] F. Gogus, et al., Analysis of roasted and unroasted *Pistacia terebinthus* volatiles using direct thermal desorption-GC×GC-TOF/MS, Food Chem. 129 (3) (2011) 1258–1264.

[153] H.B. Costa, et al., Monitoring the physicochemical degradation of coconut water using ESI-FT-ICR MS, Food Chem. 174 (2015) 139–146.

[154] B.P. Milev, et al., Fourier transform ion cyclotron resonance mass spectrometrical analysis of raw fermented cocoa beans of Cameroon and Ivory Coast origin, Food Res. Int. 64 (2014) 958–961.

[155] N.W. Proschogo, et al., Aging effects on macadamia nut oil studied by electrospray ionization fourier transform ion cyclotron resonance mass spectrometry, J. Agric. Food Chem. 60 (8) (2012) 1973–1980.

[156] A. Makarov, et al., Injection of Sample Ions into Electrostatic Trap by Supplying Trapping and Release Voltages to Storage Device with Sample Ions, and Trapping the Received Sample Ions in Electrostatic Trap by Applying Potential to Electrodes. Thermo Finnigan Llc; Thermo Masslab Ltd; (A. Makarov, M.E. Hardman, J.C. Schwartz, M.W. Senko).

[157] M. Thevis, et al., Mass spectrometry of stanozolol and its analogues using electrospray ionization and collision-induced dissociation with quadrupole-linear ion trap and linear ion trap-orbitrap hybrid mass analyzers, Rapid Commun. Mass Spectrom. 19 (22) (2005) 3369–3378.

[158] Y. Shi, et al., Rapid screening for pesticides using automated online sample preparation with a high-resolution benchtop orbitrap mass spectrometer, Food Addit. Contam. Part A Chem. Anal. Control Expo. Risk Assess. 28 (10) (2011) 1383–1392.

[159] T. Rajski, M.D.M. Gómez-Ramos, A.R. Fernández-Alba, Large pesticide multiresidue screening method by liquid chromatography-orbitrap mass spectrometry in full scan mode applied to fruit and vegetables, J. Chromatogr. A 1360 (2014) 119–127.

[160] H.G.J. Mol, P. Zomer, M. de Koning, Qualitative aspects and validation of a screening method for pesticides in vegetables and fruits based on liquid chromatography coupled to full scan high resolution (Orbitrap) mass spectrometry, Anal. Bioanal. Chem. 403 (10) (2012) 2891–2908.

[161] M.L. Gómez-Pérez, et al., Analysis of pesticide and veterinary drug residues in baby food by liquid chromatography coupled to orbitrap high resolution mass spectrometry, Talanta 131 (2015) 1–7.

[162] M.L. Gómez-Pérez, et al., Comprehensive qualitative and quantitative determination of pesticides and veterinary drugs in honey using liquid chromatography-orbitrap high resolution mass spectrometry, J. Chromatogr. A 1248 (2012) 130–138.

[163] R.-X. Mou, et al., Simultaneous determination of triazine herbicides in rice by high-performance liquid chromatography coupled with high resolution and high mass accuracy hybrid linear ion trap-orbitrap mass spectrometry, Anal. Chim. Acta 706 (1) (2011) 149–156.

[164] J. Wang, et al., Application of ultrahigh-performance liquid chromatography and electrospray ionization quadrupole orbitrap high-resolution mass spectrometry for determination of 166 pesticides in fruits and vegetables, J. Agric. Food Chem. 60 (49) (2012) 12088–12104.

[165] B. Brosnan, et al., Rapid identification, by use of the LTQ Orbitrap hybrid FT mass spectrometer, of antifungal compounds produced by lactic acid bacteria, Anal. Bioanal. Chem. 403 (10) (2012) 2983–2995.

[166] P. Kumar, et al., Targeted analysis with benchtop quadrupole-orbitrap hybrid mass spectrometer: application to determination of synthetic hormones in animal urine, Anal. Chim. Acta 780 (2013) 65–73.

[167] C. Guo, et al., Simultaneous identification, confirmation and quantitation of illegal adulterated antidiabetics in herbal medicines and dietary supplements using high-resolution benchtop quadrupole-orbitrap mass spectrometry, J. Chromatogr. B-Anal. Technol. Biomed. Life Sci. 967 (2014) 174–182.

[168] A.C. Peterson, et al., Development of a GC/quadrupole-orbitrap mass spectrometer, part I: design and characterization, Anal. Chem. 86 (20) (2014) 10036–10043.

[169] A.C. Peterson, et al., Development of a GC/quadrupole-orbitrap mass spectrometer, part II: new approaches for discovery metabolomics, Anal. Chem. 86 (20) (2014) 10044–10051.

[170] P. Quifer-Rada, et al., A comprehensive characterisation of beer polyphenols by high resolution mass spectrometry (LC-ESI-LTQ-Orbitrap-MS), Food Chem. 169 (2015) 336–343.

[171] A.D. Troise, et al., Targeted metabolite profile of food bioactive compounds by orbitrap high resolution mass spectrometry: the "FancyTiles" approach, Food Res. Int. 63 (2014) 139–146.

[172] N. Lopez-Gutierrez, et al., Identification and quantification of the main isoflavones and other phytochemicals in soy based nutraceutical products by liquid chromatography-orbitrap high resolution mass spectrometry, J. Chromatogr. A 1348 (2014) 125–136.

[173] F. Shi, et al., Application of a high resolution benchtop quadrupole-orbitrap mass spectrometry for the rapid screening, confirmation and quantification of illegal adulterated phosphodiesterase-5 inhibitors in herbal medicines and dietary supplements, J. Chromatogr. A 1344 (2014) 91–98.

[174] X. Wang, et al., High-throughput screening and confirmation of 22 banned veterinary drugs in feedstuffs using LC-MS/MS and high-resolution orbitrap mass spectrometry, J. Agric. Food Chem. 62 (2) (2014) 516–527.

[175] N. De Clercq, et al., A validated analytical method to study the long-term stability of natural and synthetic glucocorticoids in livestock urine using ultra-high performance liquid chromatography coupled to orbitrap-high resolution mass spectrometry, J. Chromatogr. A 1301 (2013) 111–121.

[176] H. Nakagawa, et al., Detection of type a trichothecene di-glucosides produced in corn by high-resolution liquid chromatography-orbitrap mass spectrometry, Toxins 5 (3) (2013) 590–604.

[177] J. Antonio Padilla-Sanchez, et al., Innovative determination of polar organophosphonate pesticides based on high-resolution orbitrap mass spectrometry, J. Mass Spectrom. 47 (11) (2012) 1458–1465.

[178] E. De Dominicis, I. Commissati, M. Suman, Targeted screening of pesticides, veterinary drugs and mycotoxins in bakery ingredients and food commodities by liquid chromatography-high-resolution single-stage orbitrap mass spectrometry, J. Mass Spectrom. 47 (9) (2012) 1232–1241.

[179] L. Monaci, E. De Angelis, A. Visconti, Determination of deoxynivalenol, T-2 and HT-2 toxins in a bread model food by liquid chromatography-high resolution-orbitrap-mass spectrometry equipped with a high-energy collision dissociation cell, J. Chromatogr. A 1218 (48) (2011) 8646–8654.

[180] H.G.J. Mol, et al., Screening of plant toxins in food, feed and botanicals using full-scan high-resolution (Orbitrap) mass spectrometry, Food Addit. Contam. Part A Chem. Anal. Control Expo. Risk Assess. 28 (10) (2011) 1405–1423.

[181] A. Mari, et al., Metabolic profiling of *Vitex agnus castus* leaves, fruits and sprouts: analysis by LC/ESI/(QqQ)MS and (HR) LC/ESI/(Orbitrap)/MS^n, J. Pharmaceut. Biomed. Anal. 102 (2015) 215–221.

[182] A. Albert, et al., Rapid and quantitative analysis of pesticides in fruits by QuEChERS pretreatment and low-temperature plasma desorption/ionization orbitrap mass spectrometry, Anal. Methods 6 (15) (2014) 5463–5471.

[183] W. Jia, et al., High-throughput screening of pesticide and veterinary drug residues in baby food by liquid chromatography coupled to quadrupole orbitrap mass spectrometry, J. Chromatogr. A 1347 (2014) 122–128.

[184] W. Jia, et al., Multi-mycotoxin analysis in dairy products by liquid chromatography coupled to quadrupole orbitrap mass spectrometry, J. Chromatogr. A 1345 (2014) 107–114.

[185] W. Jia, et al., Analysis of additives in dairy products by liquid chromatography coupled to quadrupole-orbitrap mass spectrometry, J. Chromatogr. A 1336 (2014) 67–75.

[186] M. Farré, Y. Picó, D. Barceló, Application of ultra-high pressure liquid chromatography linear ion-trap orbitrap to qualitative and quantitative assessment of pesticide residues, J. Chromatogr. A 1328 (2014) 66–79.

[187] C. Bignardi, et al., Targeted and untargeted data-dependent experiments for characterization of polycarbonate food-contact plastics by ultra high performance chromatography coupled to quadrupole orbitrap tandem mass spectrometry, J. Chromatogr. A 1372 (2014) 133–144.

[188] B. Ivanova, M. Spiteller, Solid-state determination of hop bitter acids in beer by UV–MALDI–Orbitrap mass spectrometry, J. Food Meas. Charact. 8 (4) (2014) 343–355.

[189] J. Le Ven, et al., Identification of the environmental neurotoxins annonaceous acetogenins in an *Annona cherimolia* Mill. alcoholic beverage using HPLC-ESI-LTQ-Orbitrap, J. Agric. Food Chem. 62 (34) (2014) 8696–8704.

[190] G. Orellana, et al., Validation of a confirmatory method for lipophilic marine toxins in shellfish using UHPLC-HR-Orbitrap MS, Anal. Bioanal. Chem. 406 (22) (2014) 5303–5312.

[191] W. Jia, et al., Simultaneous determination of dyes in wines by HPLC coupled to quadrupole orbitrap mass spectrometry, J. Sep. Sci. 37 (7) (2014) 782–791.

[192] S. Savic, et al., Enzymatic oxidation of rutin by horseradish peroxidase: kinetic mechanism and identification of a dimeric product by LC-Orbitrap mass spectrometry, Food Chem. 141 (4) (2013) 4194–4199.

[193] P. Kanakis, et al., From olive drupes to olive oil. An HPLC-orbitrap-based qualitative and quantitative exploration of olive key metabolites, Planta Med. 79 (16) (2013) 1576–1587.

[194] H. Nakagawa, et al., Detection of masked mycotoxins derived from type A trichothecenes in corn by high-resolution LC-Orbitrap mass spectrometer, Food Addit. Contam. Part A Chem. Anal. Control Expo. Risk Assess. 30 (8) (2013) 1407–1414.

[195] L. Vanhaecke, et al., High resolution orbitrap mass spectrometry in comparison with tandem mass spectrometry for confirmation of anabolic steroids in meat, Anal. Chim. Acta 767 (1) (2013) 118–127.

[196] M. Farré, Y. Picó, D. Barceló, Direct peel monitoring of xenobiotics in fruit by direct analysis in real time coupled to a linear quadrupole ion trap-orbitrap mass spectrometer, Anal. Chem. 85 (5) (2013) 2638–2644.

[197] J.H. Choi, et al., Determination of sedatives and adrenergic blockers in blood meal using accelerated solvent extraction and orbitrap mass spectrometry, J. Chromatogr. A 1260 (2012) 111–119.

[198] R.L. Self, W.H. Wu, Determination of eight biogenic amines in selected seafood products by MSPD extraction followed by UHPLC-Orbitrap MS, J. Food Compos. Anal. 27 (2) (2012) 169–173.

[199] E. Crawford, B. Musselman, Evaluating a direct swabbing method for screening pesticides on fruit and vegetable surfaces using direct analysis in real time (DART) coupled to an exactive benchtop orbitrap mass spectrometer, Anal. Bioanal. Chem. 403 (10) (2012) 2807–2812.

[200] R.L. Self, W.H. Wu, Rapid qualitative analysis of phthalates added to food and nutraceutical products by direct analysis in real time/orbitrap mass spectrometry, Food Control 25 (1) (2012) 13–16.

[201] L. Alder, A. Steinborn, S. Bergelt, Suitability of an orbitrap mass spectrometer for the screening of pesticide residues in extracts of fruits and vegetables, J. AOAC Int. 94 (6) (2011) 1661–1673.

[202] S.M. Lehner, et al., Evaluation of LC-high-resolution FT-Orbitrap MS for the quantification of selected mycotoxins and the simultaneous screening of fungal metabolites in food, Food Addit. Contam. Part A Chem. Anal. Control Expo. Risk Assess. 28 (10) (2011) 1457–1468.

[203] H. Nakagawa, et al., Detection of a new Fusarium masked mycotoxin in wheat grain by high-resolution LC-Orbitrap™ MS, Food Addit. Contam. Part A Chem. Anal. Control Expo. Risk Assess. 28 (10) (2011) 1447–1456.

[204] L. Monaci, et al., Towards the quantification of residual milk allergens in caseinate-fined white wines using HPLC coupled with single-stage orbitrap mass spectrometry, Food Addit. Contam. Part A Chem. Anal. Control Expo. Risk Assess. 28 (10) (2011) 1304–1314.

[205] R.L. Self, W.H. Wu, H.S. Marks, Simultaneous quantification of eight biogenic amine compounds in tuna by matrix solid-phase dispersion followed by HPLC-orbitrap mass spectrometry, J. Agric. Food Chem. 59 (11) (2011) 5906–5913.

[206] A. Gerssen, P.P.J. Mulder, J. de Boer, Screening of lipophilic marine toxins in shellfish and algae: development of a library using liquid chromatography coupled to orbitrap mass spectrometry, Anal. Chim. Acta 685 (2) (2011) 176–185.

[207] K. Mann, M. Mann, In-depth analysis of the chicken egg white proteome using an LTQ Orbitrap Velos, Proteome Sci. 9 (2011).

[208] H.K. Tiong, S. Hartson, P.M. Muriana, Comparison of five methods for direct extraction of surface proteins from Listeria monocytogenes for proteomic analysis by orbitrap mass spectrometry, J. Microbiol. Meth. 110 (2015) 54–60.

[209] R. Pilolli, et al., Orbitrap™ monostage MS versus hybrid linear ion trap MS: application to multi-allergen screening in wine, J. Mass Spectrom. 49 (12) (2014) 1254–1263.

[210] L. Cappellin, et al., On data analysis in PTR-TOF-MS: from raw spectra to data mining, Sens. Actuator B Chem. 155 (1) (2011) 183–190.

[211] B. Spengler, Mass spectrometry imaging of biomolecular information, Anal. Chem. 87 (1) (2015) 64–82.

[212] N. Winograd, Imaging mass spectrometry on the nanoscale with cluster ion beams, Anal. Chem. 87 (1) (2015) 328–333.

[213] M.M. Gessel, J.L. Norris, R.M. Caprioli, MALDI imaging mass spectrometry: spatial molecular analysis to enable a new age of discovery, J. Proteomics 107 (2014) 71–82.

[214] B. Cillero-Pastor, R.M.A. Heeren, Matrix-assisted laser desorption ionization mass spectrometry imaging for peptide and protein analyses: a critical review of on-tissue digestion, J. Proteome Res. 13 (2) (2014) 325–335.

[215] M.L. Kraft, H.A. Klitzing, Imaging lipids with secondary ion mass spectrometry, Biochim. Biophys. Acta-Mol. Cell Biol. Lipids 1841 (8) (2014) 1108–1119.

[216] N. Bjarnholt, et al., Mass spectrometry imaging of plant metabolites-principles and possibilities, Nat. Prod. Rep. 31 (6) (2014) 818–837.

[217] D.S. Urgast, J.H. Beattie, J. Feldmann, Imaging of trace elements in tissues: with a focus on laser ablation inductively coupled plasma mass spectrometry, Curr. Opin. Clin. Nutr. Metab. Care 17 (5) (2014) 431–439.

[218] E. Handberg, et al., Mass spectrometry imaging for visualizing organic analytes in food, Mass Spectrom. Rev. (2014).

[219] K.H. Ko, et al., Application of matrix-assisted laser desorption/ionization time-of-flight imaging mass spectrometry (MALDI-TOF IMS) for premalignant gastrointestinal lesions, Clin. Endosc. 46 (6) (2013) 611–619.

[220] Z. Li, P.W. Bohn, J.V. Sweedler, Comparison of sample pre-treatments for laser desorption ionization and secondary ion mass spectrometry imaging of *Miscanthus × giganteus*, Bioresour. Technol. 101 (14) (2010) 5578–5585.

[221] D.N. de Oliveira, M.S. Ferreira, R.R. Catharino, Rapid and simultaneous in situ assessment of aflatoxins and stilbenes using silica plate imprinting mass spectrometry imaging, PLoS One 9 (3) (2014).

[222] S. Taira, et al., Mass spectrometry imaging: applications to food science, Anal. Sci. 30 (2) (2014) 197–203.

[223] Y. Yoshimura, et al., Visualization of anthocyanin species in rabbiteye blueberry *vaccinium ashei* by matrix-assisted laser desorption/ionization imaging mass spectrometry, Anal. Bioanal. Chem. 403 (7) (2012) 1885–1895.

[224] Y. Yoshimura, et al., Different localization patterns of anthocyanin species in the pericarp of black rice revealed by imaging mass spectrometry, PLoS One 7 (2) (2012).

[225] N. Zaima, et al., Selective analysis of lipids by thin-layer chromatography blot matrix-assisted laser desorption/ionization imaging mass spectrometry, J. Oleo Sci. 60 (2) (2011) 93–98.

[226] N. Zaima, et al., Application of imaging mass spectrometry for the analysis of *Oryza sativa* rice, Rapid Commun. Mass Spectrom. 24 (18) (2010) 2723–2729.

[227] N. Goto-Inoue, M. Setou, N. Zaima, Visualization of spatial distribution of γ-aminobutyric acid in eggplant (*Solanum melongena*) by matrix-assisted laser desorption/ionization imaging mass spectrometry, Anal. Sci. 26 (7) (2010) 821–825.

[228] V. Cavatorta, et al., Unambiguous characterization and tissue localization of Pru P 3 peach allergen by electrospray mass spectrometry and MALDI imaging, J. Mass Spectrom. 44 (6) (2009) 891–897.

[229] B. Wu, et al., Imaging of nutrient elements in the leaves of *Elsholtzia splendens* by laser ablation inductively coupled plasma mass spectrometry (LA-ICP-MS), Talanta 78 (1) (2009) 132–137.

[230] M.W.F. Nielen, T.A. van Beek, Macroscopic and microscopic spatially-resolved analysis of food contaminants and constituents using laser-ablation electrospray ionization mass spectrometry imaging, Anal. Bioanal. Chem. 406 (27) (2014) 6805–6815.

[231] A. Sever, et al., Localization of flavonoids in seeds by cluster time-of-flight secondary ion mass spectrometry imaging, Anal. Chem. 82 (6) (2010) 2326–2333.

[232] A. Michalski, et al., Mass spectrometry-based proteomics using Q exactive, a high-performance benchtop quadrupole orbitrap mass spectrometer, Mol. Cell. Proteomics 10 (9) (2011).

Chapter 3

Elemental and Isotopic Mass Spectrometry

Constantinos A. Georgiou* **and Georgios P. Danezis**
Food Science & Human Nutrition, Agricultural University of Athens, Athens, Greece
**Corresponding author: E-mail: cag@aua.gr*

Chapter Outline

1. Introduction **132**
1.1 Mass Spectrometry 132
1.2 Toxic Elements 133
1.2.1 Contamination 133
1.2.2 Toxicity between Chemical Forms 134
1.2.3 Health Problems by Toxic Elements 134
1.2.4 Legislation on Toxic Elements 135
1.3 Nutritive Elements 135
1.3.1 Biological Function, Deficiency Problems and Bioavailability 135
1.3.2 Worldwide Situation, Guidelines—Reference Intakes 138
2. Theoretical Aspects **139**
A Elemental Mass Spectrometry 139
2.1 ICP-MS 139
2.1.1 Throwback—Historical Overview 139
2.1.2 Main Features and Operation Principles 140
2.1.3 Sample Introduction 140
2.1.4 Ion Source 141
2.1.5 Ion Formation 141
2.1.6 Ion Optics 144
2.1.7 Mass Analyzer 145
2.1.8 Mass Resolution and Abundance Sensitivity 146
2.1.9 Ion Detection Systems 154
2.1.10 Interferences 154
2.1.11 Sample Preparation—Pretreatment 159
2.1.12 Microwave-Assisted Digestion 161
2.1.13 Direct Solid Sampling Analysis 161
2.1.14 Hyphenated Techniques 164
2.1.15 Conclusion, Strengths and Limitations 171
2.1.16 ICP-MS versus Other Techniques 172
B Isotopic Mass Spectrometry 174
2.2 IRMS 174
2.2.1 Isotopes 174

Comprehensive Analytical Chemistry, Vol. 68. http://dx.doi.org/10.1016/B978-0-444-63340-8.00003-0

2.2.2 Expression of Stable Isotope Abundances and δ-Notation 174
2.2.3 Causes of Stable Isotopes Variations 176
2.2.4 Isotopic Variations in Nature 176
2.2.5 Isotopic Variations in Plants and Animals 177
2.2.6 Instrumentation 178
2.2.7 Inlet System 178
2.2.8 The Ion Source: Electron Impact Ion Production 182
2.2.9 Separation and Detection of Ions in the Mass Spectrometer 182
2.2.10 Isobaric Interferences 184
2.2.11 Conclusions 184
2.3 TIMS 184
2.3.1 Instrumentation 184
2.3.2 Strengths 186
2.3.3 Limitations 186
2.3.4 Comparison with MC-ICP-MS 186
2.4 MC-ICP-MS 187
2.4.1 Instrumentation 187
2.4.2 Fundamentals of MC-ICP-MS 190
2.4.3 Strengths 190
2.4.4 Limitations 191
2.4.5 Isotope Dilution 191
3. Applications 193
3.1 ICP-MS 194
3.1.1 Trends 194
3.1.2 Toxic and Nutritive Elements 195
3.1.3 Authentication 205
3.1.4 Migration Studies—Food Nanomaterials—Others 210
3.2 IRMS 213
3.2.1 Trends 213
3.2.2 Authentication 214
3.2.3 Nutrient Intake and Bioavailability 218
3.2.4 Other 220
3.3 TIMS 220
4. Conclusion and Future Outlook 222
References 224

1. INTRODUCTION

Food safety and quality are important and have attracted more public concern in recent years. Consumers' confidence in food safety and quality is an undisputed priority worldwide. The presence of undesired chemicals, basically residues and contaminants, in food as well as the lack of essential chemical substances at the required concentrations can pave the way to very serious consequences for human health. Therefore, consumers increasingly demand reassurance regarding the content and origin of their foods, fearing of safety and adulteration issues. Many of them are also interested in the relationship between diet and health, so they utilize nutrient content and health claim information from food labels to make the best choices for purchasing. Consequently, manufacturers need to be able to confirm the authenticity of components of their products and comply with legislation.

1.1 Mass Spectrometry

An important task of modern analytical chemistry is to provide accurate, reliable methodologies for the assurance of food safety and quality, particularly including identifying, monitoring, and quantifying newly recognized food contaminants and components. Mass spectrometry (MS) has played and will play a crucial role in this task. Modern mass spectrometry provides unique, reliable, and affordable methodologies in order to approach almost any problem, which may be posed in the field of food science. Mass spectrometry has been considered one of the most suitable techniques that is widely used in food safety and quality analysis due to its advantages of high sensitivity, selectivity, and throughput. Recent developments and improvements in mass spectrometry provide more accurate, precise, and faster analysis of harmful or dangerous compounds in food at very low concentrations, the acceptable residue levels are easily achieved with MS instruments. Moreover, applications in this area are rapidly growing with a current focus on environment-friendly, cost-effective, faster, and fit for purpose methods with a trend toward multicompound and multielement analysis. For these reasons, mass spectrometry is used as a frontline technology as it replaces other analytical technologies. In recent years, expansion of mass spectrometry across medium and small labs is becoming a reality, while MS instruments occupy the main stream of new instrument purchases, regarding food analysis laboratories. Besides, nowadays, mass spectrometry is considered as something obvious and essential for advanced research in food safety and quality. Consequently, it can be easily deduced that mass spectrometry plays a significant and established role in all aspects of food safety and quality [1,2].

Determination of elemental and isotope content are two very important types/categories of analyses, which are performed on foodstuffs, in terms of food safety and quality. Especially elemental and isotope mass spectrometry techniques show great convenience and suitability for food analysis due to their sensitivity, efficiency, high-throughput capability, accuracy, precision, robustness, and ease of automation. Inductively Coupled Plasma-Mass Spectroscopy (ICP-MS) is used for multielement measurements. Respectively, Isotope Ratio Mass Spectrometry (IRMS), Thermal Ionization Mass Spectrometry (TIMS), and Multi Collector-Inductively Coupled Plasma-Mass Spectroscopy (MC-ICP-MS) are used for isotope ratio measurements.

In this chapter, the contribution of isotopic and elemental mass spectrometry techniques in food quality and safety is presented. It aims at providing clues on the fundamental role of the basic principles, characteristics, perspectives, and applications of these techniques in food field. Applications to various aspects such as toxic and nutritive elements determination, food authentication, food nanomaterials, metallomics, speciation, and migration have also been surveyed.

1.2 Toxic Elements

1.2.1 Contamination

Concerning food safety, there is a plethora of contaminants which are arising during food processing or they derive from the environment or from the packaging materials. Toxic elements and especially heavy metals constitute a significant class of these contaminants. Contamination of food products by heavy metals is becoming a frequent problem these days. Heavy metals are found everywhere in nature, and cannot be degraded or destroyed. Air, soil, and water pollution are contributing to the presence of harmful elements in foodstuff. The occurrence of heavy metals enriched by ecosystem components, firstly arise from rapid industrial growth, advances in agricultural chemicalization, and the urban activities of humans. Also, possible contact between food and coat metal surface of packing containers or the processing equipment is a significant source of toxic metals' contamination in food. Consequently, sources of food contamination include environmental and industrial pollution, agricultural practices, food processing, and migration from packaging [3–5].

1.2.2 Toxicity between Chemical Forms

As previously told metals occur in many foods, either naturally or as a result of pollution and/or processing. While trace amounts of some metals (e.g., Fe, Mg, Mn, Mo, Zn, Co, Ni, Cr) are important nutrients, higher quantities and other metals (e.g., Cd, Hg, Pb, Sn, Sb) or metalloids (As, Po) can be highly toxic. Some examples are methylmercury and cadmium found in fish and shellfish [3,6–8], arsenic in rice and water [9,10], lead in water [11], and tin in canned foods [12]. Toxicity usually varies between chemical forms of metals (elemental–inorganic–organic) and between their oxidation states. For example, methyl mercury is much more toxic than elemental mercury, inorganic arsenic is much more toxic than organic arsenic and Cr (VI) is much more toxic than Cr (III), respectively [13–15].

1.2.3 Health Problems by Toxic Elements

Both, lack of food and bad quality of food have created throughout the centuries serious problems for people. Nowadays, it is calculated that over 3 billion people worldwide suffer from either deficiency or toxicity of some trace elements. Toxic elements can bio accumulate in the body and cause organ damage particularly to susceptible groups such as fetuses and young children. Overtime toxic metals' accumulation, in human body, can cause serious health problems: damaged or reduced mental and central nervous function, lower energy levels and problems in blood composition, lungs, kidneys, liver, and other vital organs. Also long-term exposure may result in slowly progressing physical, muscular, and neurological degenerative processes that mimic Alzheimer's disease, Parkinson's disease, muscular dystrophy, multiple sclerosis, and cancer. Furthermore, allergies are not uncommon. Many of these diseases are caused because toxic elements imitate the action of an essential element in the body, interfering with the metabolic process to cause illness. For example As, in the form of arsenite has the apparent ability to act as a functional mimic of estrogen (estradiol) at the site of the estrogen receptor and also radium imitates calcium to the point of being incorporated into human bone. Therefore, toxic elements are of particular concern in food because of their toxicity, especially for long-term (chronic) intake [3–4,16–22].

1.2.4 Legislation on Toxic Elements

In order to reduce risks regarding human health, European Commission has established maximum levels of heavy metals such as lead, mercury, and cadmium in different foods, by Commission Regulation (EC) No 1881/2006 [23] and its amendment Commission Regulation (EC) No 629/2008 [24] and also in animal feed by Commission Directive (EC) No 2002/32 [25]. Furthermore, European Commission laid down migration limits for lead and cadmium, which may migrate from ceramic articles to the foodstuff, and a declaration of compliance and performance criteria of the analytical method, by Council Directive (EEC) 84/500 [26] and its amendment Commission Directive (EC) 2005/31 [27]. Additionally, European Commission established migration limits for barium, cobalt, copper, iron, lithium, manganese, and zinc concerning plastic materials that intended to come into contact with foodstuffs and also limits for zinc, copper, lead, arsenic, chromium concerning the composition of copolymer powder raw material from which plastic will be made, by Commission Regulation (EU) No 10/2011 [28]. European Commission has laid down the methods of sampling and analysis for the official control of the levels of heavy metals and other contaminants by Commission Regulation (EC) No 333/2007 [29]. As regards water for human consumption, European Commission has also defined the maximum levels by Council Directive (EC) 98/83 [30]. Globally, there are also limits concerning heavy metals in foodstuffs and water, some of them are depicted beneath:

Internationally: WHO, FAO, Codex Alimentarius Commission (CODEX) General standard for contaminants and toxins in food and feed CODEX STAN 193-1995 with its amendment 2010 and in water [31,32].

Nationally: USA, Food and Drug Administration (FDA), EPA National Primary Drinking Water Regulations [33].

Australia and New Zealand, FSANZ: Standard 1.4.1—Contaminants and Natural Toxicants [34].

China, National Food Safety Standard-Maximum Levels of Contaminants in Food, Standard 2762 [35].

1.3 Nutritive Elements

1.3.1 Biological Function, Deficiency Problems and Bioavailability

Concerning food quality, as previously mentioned, trace amounts of these elements: Fe, Mn, Mo, Zn, Co, Ni, Cr, Se, Cu, Si, I, and F are necessary for proper human health, apart from H, C, N, O, Na, K, S, Cl, Mg, Ca, and P which are required in relatively large quantities in a diet. There is also a group of elements called ultra trace minerals, including V, Sn, Ni, As, and B, that are being investigated for possible biological function but currently do not have clearly defined biochemical roles [36].

Iron is one of the most important elements due to its chemical properties and physiological role. Iron contributes to normal cognitive function, to normal energy-yielding metabolism, to formation of red blood cells and hemoglobin, to oxygen transport in the body, in the immune system, to the reduction of tiredness and fatigue, and a significant role in the process of cell division. Iron deficiency, among others, causes anemia and cognitive performance problems [36–39].

Manganese is associated with the formation of connective and bony tissue, with growth and reproductive functions, and finally with carbohydrate and lipid metabolism. It is also involved in many enzymatic processes as a metalloenzyme or as a cofactor. Consequently, it contributes to normal energy-yielding metabolism, the maintenance of bones, the formation of connective tissue, and the protection of cells from oxidative stress. Deficiency of manganese results in a variety of structural and physiological defects, including reduced growth rate and skeletal abnormalities [37–39].

The essentiality of molybdenum is interpreted by the importance of the three molybdenum enzymes: xanthine oxidase, aldehyde oxidase, and sulfite oxidase. This way molybdenum contributes to normal sulfur amino acid metabolism. As a result, its deficiency causes syndromes related to the aforementioned enzymes, such as hypermethioninemia, hypouricemia, and hyperoxypurinemia. Zinc is involved in a plethora of metabolic functions because of its presence in over 300 metalloenzymes, such as carboxypeptidase, liver alcohol dehydrogenase, and carbonic anhydrase. Also, zinc plays a major role in protein synthesis and has a role in gene expression. Therefore zinc, among others, contributes to normal acid–base, carbohydrate, fatty acids, and macronutrient (e.g., vitamin A) metabolism, to normal cognitive function, to the maintenance of normal vision and bones, to the normal function of the immune system, to normal fertility and reproduction, and finally to the protection of cells from oxidative stress. Deficiency of zinc can cause several problems such as acrodermatitis hyperammonemia, growth retardation, hypogonadism, poor appetite, and diarrhea [37,39–41].

Cobalt is a key constituent of cobalamin, also known as vitamin B_{12}. It is, therefore, required for biosynthesis of vitamin B_{12} family of coenzymes. Deficiency of cobalt leads to anemia. However, cobalt is essential only in minute amounts and higher concentration is hazardous and toxic. As cobalt, nickel is required in minimal quantities. Nickel presents in the active site of urease. Deficiency of nickel can induce some dysfunction of fat metabolism. Chromium is an essential micronutrient for the maintenance of normal blood glucose levels. It also contributes to the normal macronutrient (e.g., lipids, proteins) metabolism. The beneficial impact of Cr on cholesterol fractions and the ratio of LDL and HDL have been observed. Chromium deficiency results in increased cholesterol levels, high blood glucose levels, coronary dysfunction, and some abnormalities of nerve stimulation of the extremities [16,36–37,42].

Another essential element for human health is selenium. More than 100 Se proteins, enzymes, and other compounds have been identified in biological samples. Selenium and its compounds have been recently regarded as the most effective antioxidants in both prevention and cure of cancer. The beneficial effects of selenium on human health are strongly dependent on its chemical form and concentration. Selenomethionine (SeMet) is involved in oxidation processes as both promotor and inhibitor agent. Selenium contributes to the maintenance of normal hair and nails and also contributes to the normal function of the immune system and thyroid function. Selenium deficiency causes muscle inflammation and weakness of muscles, fragile red blood cells, degeneration of pancreas, abnormal skin coloration, Keshan disease (cardiomyopathy), and Kashin–Beck disease (osteoarthropahty) [37,43–46].

Copper is a component of several proteins and metalloenzymes. It is prerequisite in oxidation–reduction reactions and hemoglobin synthesis. Metallothionein is the main intracellular Cu-protein involved in Cu transport within the body. Copper deficiency may lead to several health problems such as: slow growth, anemia, fertility problems, hair and weight loss, disorders of central nervous system, cardiovascular problems, osteoporosis and several other metabolic dysfunctions. Copper contributes to the function of the nervous system, energy-yielding metabolism, hair and skin pigmentation, iron transport, function of the immune system, protection of cells from oxidative stress, and finally to maintenance of normal connective tissues. Silicon is not easily available in humans due to its low water solubility, however, it is an essential nutrient. It is required for bone, cartilage, and connective tissue formation and is also necessary for proper development of skin and hair. Thus, Si has been reported to be a major ion in osteogenic cells [37,43,47].

Iodine is one of the very essential micronutrients that human body could not do without this element. It is concentrated mainly in thyroxin and triothyronine. These hormones are involved in most biological processes, from bone growth to reproduction. Thus, an adequate level of iodine in the body is very crucial. Iodine deficiency causes fetus anomalies, including abortion and stillbirth, impaired mental function and retarded physical development, commonly associated with congenital anomalies of children and adolescents and finally hypothyroidism with various complications in adults. Iodine contributes to normal function of several body systems and processes [37,48].

Since recommended intakes for calcium, magnesium, and phosphorus are much higher, they are regarded as major minerals and not trace elements. Magnesium is an essential element that activates more than 300 enzymes. In addition to being an enzyme cofactor, major physiological roles of magnesium include bone and teeth mineralization, muscle contraction, nerve impulse transmission, and blood clotting. Also magnesium contributes to reduction of fatigue, to normal energy-yielding metabolism, to normal protein synthesis, to normal psychological function, and finally has a role in the process of cell division. As a result, deficiency of magnesium can cause dizziness, muscle cramps, muscle weakness, and fatigue and severe magnesium deficiency could lead to hypocalcemia, hypokalemia, retention of sodium, low circulating levels of parathyroid hormone (PTH), neurological and muscular symptoms [49,50].

Although >99% of body calcium is located in the skeleton, its physiological role as an essential nutrient goes much further than maintaining skeletal integrity. In addition to its structural role in bone and teeth in the form of hydroxyapatite, calcium is needed for nerve signal transmission and muscle contraction. Also calcium contributes to normal blood clotting, energy-yielding metabolism, function of digestive enzymes, and finally has a role in the process of cell division and specialization. Calcium deficiency can cause hypocalcaemia, osteoporosis, cardiovascular disease, high blood pressure, kidney stones, and weight loss [36,51].

So, several inorganic nutrients are essential due to their vital role in the control of body biochemistry. Furthermore the beneficial activity, the absorption, the mobility and the bioavailability of these elements, as also their toxic effect, depend on the chemical form of elements (elemental–inorganic–organic) and on their oxidation states. For example, selenomethionine has higher beneficial activity than inorganic selenium. Also the presence of vitamins in a meal enhances the absorption of minerals. For example, vitamin C improves iron absorption and vitamin D aids in the absorption of calcium, phosphorous, and magnesium. On the other hand, a large amount of zinc in a diet decreases the absorption of iron and copper [36,52].

1.3.2 Worldwide Situation, Guidelines—Reference Intakes

Worldwide, more than 2 billion people are at high risk for at least one trace element deficiency, especially in low-income countries. According to World Health Organization (WHO), the most common deficiencies are iron, iodine, and zinc followed by selenium and copper. Although minerals are found in foods, they are usually only present in limited amounts. To obtain daily mineral requirements, diets must contain a wide variety of foods. However, much of the mineral content in these foods is poorly absorbed by the body. People on low-calorie diets for prolonged periods are particularly at high risk for developing mineral deficiencies.

In order to prevent nutrient deficiencies, but also to reduce the risk of chronic diseases such as osteoporosis, cancer, and cardiovascular disease, Scientific Committee on Food (SCF) of EC has established since 1993 average requirements (AR), population reference intakes (PRI), lowest threshold intake (LTI), and maximum safe intake for each nutrient. Nowadays, the European Food Safety Authority (EFSA) having taken into account new scientific evidence and recent recommendations that were issued at national and international level updated all these limits and added the adequate intake (AI) and reference intake ranges for macronutrients (RI). Nutrient requirements differ according to age, sex, and physiological condition. Also, separate reference values are established for pregnant and lactating women. But most of the EU member states have produced their own quantitative dietary recommendations under a variety of names (e.g., Dietary Reference Values, UK), with values adapted to different population groups (children, adolescents, pregnant women, or older people). However, EFSA strives to establish uniform dietary reference values designed to ensure a diet that provides energy and nutrients for lifelong optimal growth, development, function, and health. It seeks to remove the existing confusion over competing classification systems, such as the European Dietary Reference Values (DRV) and the Recommended Dietary Allowances (RDA) used in the US Dietary reference values (DRVs) indicate the amount of an individual nutrient that people need for good health depending on their age and gender. Finally, with Commission Regulation (EU) No 1169/2011, EU established daily reference intakes for vitamins and minerals which may be declared on food labels [43,53,54].

Other types of nutrient values used are RDA and tolerable upper intake levels (UL). Recommended Dietary Allowances (RDA) is the daily dietary intake level of a nutrient considered sufficient in each life stage and sex group. Tolerable upper intake levels (UL) is the maximum level of total chronic daily intake of a nutrient (from all sources) judged to be unlikely to pose a risk of adverse health effects to humans. This is the highest level of daily consumption that current data have shown to cause no side effects in humans when used indefinitely without medical supervision.

WHO/FAO and USDA (United States Department of Agriculture) have also established RDA and UL values. The Department of Nutrition for Health and Development WHO, in collaboration with FAO, continually reviews new research and information from around the world on human nutrient requirements and recommended nutrient intakes. This is a vast and never-ending task, given the large number of essential human nutrients. These nutrients include protein, energy, carbohydrates, fats and lipids, a range of vitamins, and a host of trace elements [55,56].

2. THEORETICAL ASPECTS

A Elemental Mass Spectrometry

2.1 ICP-MS

2.1.1 Throwback—Historical Overview

The first work on Inductively Coupled Plasma Mass Spectrometry (ICP-MS) was in 1980 by G. Houk et al. ICP-MS is undoubtedly the fastest growing trace element technique today. Since its commercialization in 1983, by the Canadian company Sciex, many ICP-MS systems have been installed worldwide, carrying out many diverse applications. The most common applications of ICP-MS today include environmental, geological, semiconductor, food, biomedical and nuclear fields. There is no question that the major reason for its unparalleled growth is its ability to carry out rapid multielement determinations at ultra-trace level. During the last 10 years, the technique has spread outside research laboratories, used for control activities and the industry. A number of desirable features contributed to this success. These include high sensitivity, multielement capability, wide linear dynamic range, high sample throughput and ability to discriminate between isotopes. With modern instruments and in the absence of spectroscopic interferences, the detection limits (DLs) of most trace elements are in the low ng L^{-1} range when samples digestates are analyzed. Another advantage is the suitability of ICP-MS as a selective online detector in hyphenated methods for the determination of element species. These methods gain further popularity, as the importance of elemental speciation in biological sciences, including food toxicology, is recognized [57–59].

2.1.2 Main Features and Operation Principles

There is a number of different ICP-MS designs available today, which share many similar components such as nebulizer, spray chamber, plasma torch, and detector, but can differ quite significantly in the design of the interface, ion focusing system, mass separation device, and vacuum chamber. A schematic diagram of an ICP-MS instrument is shown in Figure 1.

2.1.3 Sample Introduction

The sample, which usually must be in a liquid form, is pumped at around 1 mL/min, usually with a peristaltic pump into a nebulizer, where it is converted into a fine aerosol with argon gas at about 1 L/min. From the sample solution to be analyzed, small droplets are formed by the nebulization of the solution using an appropriate concentric or cross-flow pneumatic nebulizer/spray chamber system. Then, the fine aerosol emerges from the exit tube of the spray chamber

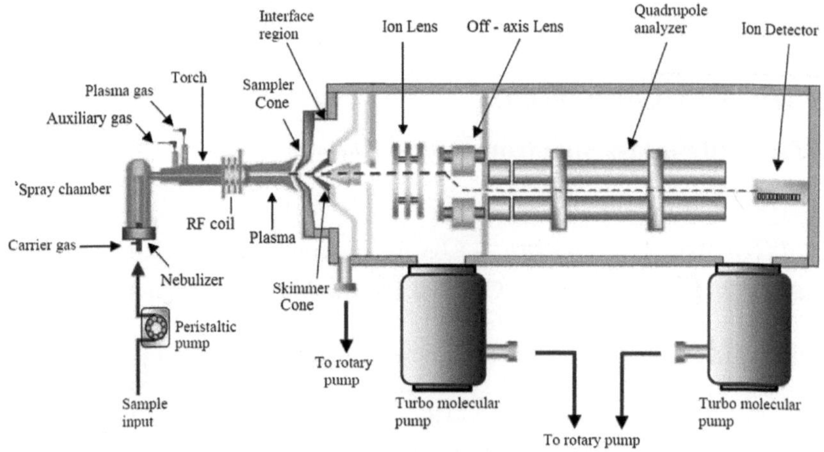

FIGURE 1 The basic components of an ICP-MS system.

and is transported into the plasma torch via a sample injector. Quite different solution introduction systems have been created for the appropriate generation of an aerosol from a liquid sample and for separation of large size droplets. Such an arrangement provides an efficiency of the analyte introduction in the plasma of 1–3% only.

Various efficient devices have been utilized for sample introduction into an inductive plasma source, for example, the application of several nebulizers, hyphenated techniques, hydride generation, laser ablation, and electrothermal vaporization. The role of the solution introduction system in an inductively coupled plasma source is to convert the liquid sample into a suitable form (e.g., using a nebulizer/spray chamber arrangement) that can be effectively vaporized into free atoms in order to generate ions. Gas chromatography has also been coupled to ICP-MS for selective analysis of gas mixtures. Several tools for sample introduction in an inductively coupled plasma source, including different nebulizers for solution introduction (such as a pneumatic nebulizer together with a spray chamber, ultrasonic nebulizer or microconcentric nebulizer with a desolvator, high-efficiency nebulizer, direct injection nebulizer, the application of hydride generation) into an inductively coupled plasma, hyphenated techniques for speciation analysis, the slurry technique, spark and laser ablation, and electrothermal evaporation for the analysis of solid samples [58–60].

2.1.4 Ion Source

The ICP has been described as an ideal ion source for inorganic mass spectrometry. Compared to established gaseous and solid state mass spectrometric techniques, the combination of an ICP ion source with a mass spectrometer is a relatively young analytical technique. The development of ICP ion sources was combined with fundamental studies of plasma characteristics with respect to plasma gas, electron number density, ion distribution of positive singly and doubly charged ions, and also negatively charged ions. ICP ion source is formed in a nearly chemically inert environment in a stream of a noble gas. A schematic of an ICP ion source including the quartz plasma torch and induction load coil together with sampler and skimmer cone as part of the interface region of a mass spectrometer is shown in Figure 2 [60].

2.1.5 Ion Formation

The roll of plasma torch is crucial. ICP plasma source dissociates the sample into its constituent atoms or ions. Plasma is formed by the interaction of an intense magnetical field, produced by radiofrequency (RF) passing through a copper coil, on a tangential flow of gas (for most applications only argon of the highest purity is usually employed as plasma gas), at about 15 L/min flowing through concentrically structured quartz tube (torch). This has the effect of ionizing the gas and, when seeded with a source of electrons from a high-voltage spark, forms a very high temperature plasma discharge (~10,000 K) at

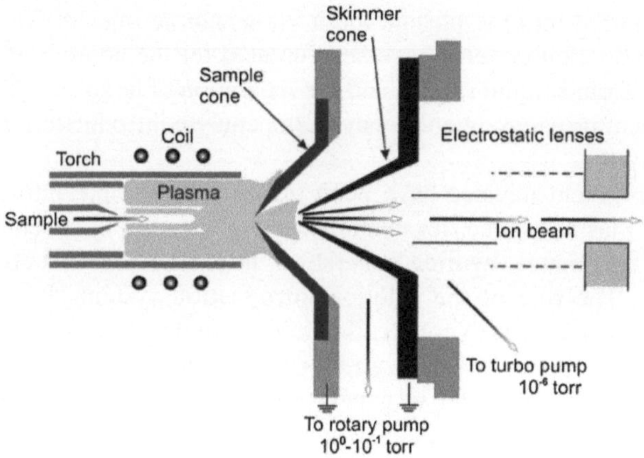

FIGURE 2 ICP ion source.

the open end of the tube. The plasma temperature and electron number densities are a function of the experimental parameters applied (RF power, nebulizer gas flow rate, solution uptake rate, torch design and others). This high temperature ensures almost complete decomposition of the sample into its constituent atoms, and the ionization conditions within the ICP result in highly efficient ionization of most elements in the periodic table and, importantly, these ions are almost exclusively singly charged. In most applications, ICP-MS operates at an RF power of the ICP of about 1200–1300 W. The ionization efficiency of an ICP source depends on the ionization energy, Ei, of the element to be analyzed. Elements with an ionization energy of less than about 8 eV are ionized with nearly 100% efficiency. With increasing first ionization energy, the ionization efficiency decreases [59,60].

The plasma torch is positioned horizontally and is used to generate positively charged ions and not photons. In fact, every attempt is made to stop the photons from reaching the detector because they have the potential to increase signal noise. The high number of ions produced, combined with very low backgrounds, provides the best detection limits available for most elements, normally in the parts per trillion (ppt) range. As we can notice in Figure 3, more than 85% of the elements that can be determined by a contemporary commercial ICP-MS have detection limits less than 1 ppt [58,59,61].

After ion generation in the ICP ion source, the positively charged ions are extracted from the argon plasma via the differentially pumped interface, between the sampler and skimmer cones, into the high vacuum of the mass analyzers. The two-stage differentially pumped interface is employed in each ICP mass spectrometer. A mechanical roughing pump maintains a vacuum of 1–2 Torr in the interface region. After several attempts, the problem of the extraction of ions formed in an atmospheric pressure ion source into the vacuum of a mass spectrometer was solved. This was made possible by the insertion of

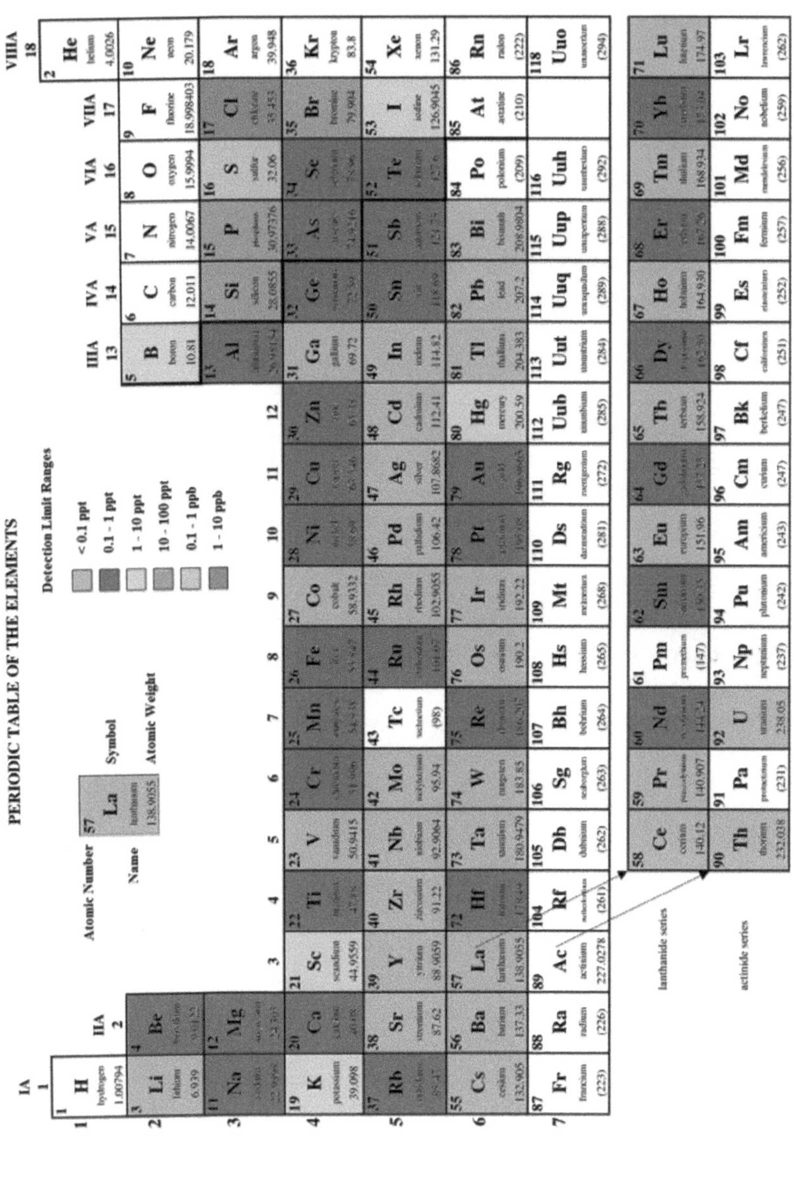

FIGURE 3 Detection limits of a contemporary ICP-MS.

an ion extraction interface with sampling cone and skimmer cone as the boundary to the atmospheric ion source, on the one side, and to ion optics as a part of the high-vacuum mass analyzer on the other side, respectively. The plasma, which expands through the sampling orifice (diameter of orifice 0.8–1.2 mm), produces a free jet, whereas the centerline flow of the jet passes through the skimmer orifice (diameter of orifice 0.4–1 mm) in the ion lens system. Smaller sampling orifices (<0.8 mm) have been applied, but in the case of a high salt content of the analyzed solution, clogging is a possibility. By differential pumping of the interface, the atmospheric pressure in the inductively coupled plasma ion source and the high vacuum of the mass analyzer can be matched. This interface region is one of the most critical areas of an ICP mass spectrometer because the ions must be transported efficiently and with electrical integrity from the plasma, which is at atmospheric pressure (760 Torr) to the mass spectrometer analyzer region at approximately 10^{-6} Torr [60].

Another problem is the capacitive coupling between the RF coil and the plasma, which produces a potential difference of a few hundred volts. If this was not eliminated, it would have resulted in an electrical discharge (called a secondary discharge or pinch effect) between the plasma and the sampler cone. This discharge increases the formation of interfering species and also dramatically affects the kinetic energy of the ions entering the mass spectrometer, making optimization of the ion optics very erratic and unpredictable. For this reason, it is absolutely critical that the secondary charge is eliminated by grounding the RF coil. There have been different approaches used over the years to achieve this, including a grounding strap between the coil and the interface, balancing the oscillator inside the RF generator circuitry, a grounded shield or plate between the coil and the plasma torch, or the use of a double interlaced coil where RF fields go in opposing directions. They all work differently, but basically achieve a similar result, which is to reduce or to eliminate the secondary discharge [58].

2.1.6 Ion Optics

After the ions have been successfully extracted from the interface region, they are directed into the main vacuum chamber by a series of electrostatic lens, called ion optics. The operating vacuum in this region is maintained at about 10^{-3} Torr with a turbomolecular pump Figure 1. There is a variety the ion optical region designs, but they serve the same function, which is to electrostatically focus the ion beam toward the mass separation device, while stopping photons, particulates, and neutral species from reaching the detector. The ion beam containing all the analytes and matrix ions exits the ion optics and passes into the heart of the mass spectrometer, where the ions are separated based on their atomic mass to charge ratio (*m/z*). The mass separation device is kept at an operating vacuum of approximately 10^{-6} Torr with a second turbomolecular pump. Many designs now exist for ion focusing optics specifically tailored for ICP-MS applications. The purpose of the ion optics is to guide analyte ions to the ion separation devices where mass analysis takes place. It should be recognized that the ion optics act as an ion energy filter and as such have to be compatible with the range of ion kinetic energies emerging from the plasma sampling interface [62].

Ion lenses generally take the form of metal discs with a central hole, tubes, or plates. As the ion beam effectively starts as a point source from the back of the skimmer cone, the ion beam shape is effectively circular, so most ion lenses are cylindrically symmetrical in quadrupole-based instruments. The exception only appears where the ion beam is deflected off-axis for reasons related to background reduction (stopping photons, particulates, and neutral species from reaching the detector). In magnetic sector ICP-MS instruments, the ion beam shape is deliberately modified from circular to rectangular by compressing the ion beam on one axis. This change is made so that the ions can be focused through slits. On a magnetic sector instrument, slit width controls mass resolution. All ion lenses must be electrically isolated from one another to operate correctly [62].

2.1.7 Mass Analyzer

As an essential part of a mass spectrometer, the ion separation system has the task of separating the fast flying ions with respect to their different *m/z*. The separated ion beams are then supplied to the ion detection system for spatial or time-resolved ion detection and registration. The mass spectrum is then the 2D representation of ion intensity as a function of the *m/z* ratio. So functionally, all mass analyzers perform two basic tasks: (1) separate ions according to their *m/z* and (2) measure the relative abundance of ions at each mass. These processes are achieved in a number of different ways, depending on the mass separation device, but they all have one common goal and that is to separate the ions of interest from all other nonanalyte, matrix, solvent, and argon-based ions [60,63].

Although ICP-MS was commercialized in 1983, the first 10 years of its development was based on traditional quadrupole mass filter technology to separate the ions of interest. This category of ICP-MS worked exceptionally well for most applications but proved to have limitations when determining specific elements or dealing with more complex sample matrices. This led to the development of alternative mass separation devices, which allowed ICP-MS to be used for applications which required higher resolution, faster data capture, and/or a reduction in polyatomic spectral interferences [58].

There are many different mass separation devices that can be attached to the ICP ion source. Commercially available instrumentation includes quadrupole-based, sector-based, and time-of-flight (TOF)-based instruments. They basically serve the same purpose, which is to allow analyte ions of a particular *m/z* through the detector and to filter out all the nonanalyte, interfering, and matrix ions. Depending on the design of the mass spectrometer, this is either a scanning process, where the ions arrive at the detector in a sequentially manner, or a simultaneous process, where the ions are either sampled or detected at the same time. Quadrupole mass spectrometers only allow ions of a specific *m/z* value to reach a detector, ejecting all other ions. Sector instruments use magnetic and electrical fields to focus a beam composed of spatially dispersed ions. Focusing permits ions of a particular *m/z* to reach the detector through a slit. In TOF instruments, ions travel through a tube at different velocities, depending on their *m/z*. All ions are detected, with lighter ions reaching the end of the tube before heavier ions. The significant reduction in cost of nonquadrupole-based ICP-MS in recent years, has led to a significant increase in the number of these instruments used compared to quadrupole [62].

2.1.8 Mass Resolution and Abundance Sensitivity

As slits used in mass spectrometer are between 0.1 and 1 mm width are used in mass spectrometers, the separated ion beams have a defined breath (b) close to the slits. The capability of the mass spectrometer for separating ion beams with different masses m and m + Δm (e.g., isotopes) is characterized by the mass resolution, R, which can be expressed as: $R = m/\Delta m$.

The mass resolution, R, of a mass spectrometer (or part of the mass spectrometer, e.g., of a sector field) gives information on the mass difference of two ion beams with masses m and m + Δm in order to separate and detect both ion beams clearly, Figure 4. That means the peak tailing increases the analyte intensity at mass m + Δm. A common definition of mass resolution is taken at 50% of peak height and is called full width at half maximum (FWHM). The required mass resolution for mass spectrometric separation of two neighboring peaks (e.g., of two atomic ions or of an atomic and a molecular ion) can be estimated using the known atomic masses from the IUPAC table value for atomic masses for m and Δm whereby the mass of the polyatomic or cluster ions is calculated

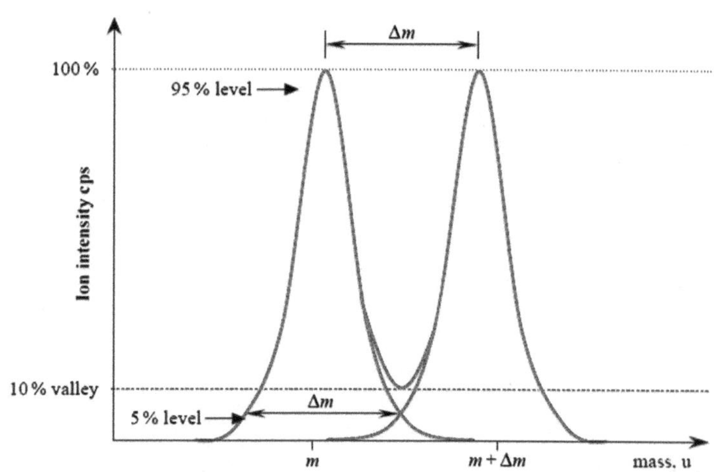

FIGURE 4 Demonstration of mass resolution definition.

as the sum of the masses of their atomic constituents. The theoretical mass resolution can be calculated from the geometric parameter for each ion separation system. In order to improve the mass resolution of a mass spectrometric system for solving interference problems, the slit width in the mass spectrometer is minimized. However, an increase in mass resolution results in a loss of ion intensity and sensitivity.

Another important property of mass spectrometric separation systems is the abundance sensitivity. Even under sufficient vacuum conditions there is a scattering of ions in the beam by the residual gas. Furthermore, by scattering of particles by the wall of the analyzer tube or by electrostatic repulsion in the ion beam itself or charging effects, so-called "peak tails" in mass spectra are observed, which result in an increasing energy spread of ions and consequently in a deterioration in the abundance sensitivity. The abundance sensitivity of a mass spectrometer is defined as:

Abundance sensitivity = Ion intensity at mass m + 1/Ion intensity at mass m [60].

2.1.8.1 Quadrupole

Quadrupoles are the cheapest choice of the above-mentioned mass spectrometry devices. They are versatile and robust machines, and account for the majority of the ICP-MS instrument sales. Laboratories that routinely analyze samples use quadrupole instruments for their high-throughput capabilities. Other machines, such as those based on ion traps or Fourier transform ion cyclotron resonance, are being developed for use in research laboratories.

Figure 1 represents schematically a generic quadrupole-based instrument. As can be seen, the mass analyzer is positioned between the ion optics and the detector and is maintained at a vacuum of approximately 10^{-6} Torr with an additional turbomolecular pump to the one that is used for the lens chamber. Assuming the ions are emerging from the ion optics at the optimum kinetic energy, they are ready to be separated according to their *m/z* by the mass analyzer. These mass spectrometers employ a quadrupole mass filter and have a resolution of something less than 1 amu (atomic mass unit). Thus, they are typically referred to as low-resolution instruments. Quadrupoles of commercial ICP-MS instruments can be operated in different modes. The entire mass range or selected ranges may be scanned in the scanning mode, or a selected number of masses can be measured in the peak hopping mode. Rapid semiquantitative scans on the entire mass range are typically performed in a few minutes [64–65].

A quadrupole consists of four cylindrical or hyperbolic metallic rods of the same length and diameter. They are typically made of stainless steel or molybdenum and sometimes coated with a ceramic coating for corrosion resistance. Quadrupoles used in ICP-MS are typically 15–20 cm in length, about 1 cm in diameter, and operate at a frequency of 2–3 MHz.

A quadrupole operates by placing both a direct current (DC) field and a time-dependent alternating current (AC) of radio frequency on opposite pairs of the four rods. By selecting the optimum AC/DC ratio on each pair of rods, ions of a selected mass are then allowed to pass through the rods to the detector, while the others are unstable and ejected from the quadrupole. Quadrupole scan rates are typically in the order of 2500 amu per second and can cover the entire mass range of 0–300 amu in about one-tenth of a second. However, real-world analysis speeds are much slower than this, and in practice, 25 elements can be determined in duplicate with good precision in 1–2 min, depending on the analytical requirements [58].

Quadrupole analyzers are frequently used as universal mass analyzers due to their high speed, high dynamic range, MS/MS capability, simplicity, easy sample introduction, and relatively low cost. A single quadrupole analyzer is typically used for elemental and isotope analysis in inorganic mass spectrometry and is sufficient for a multitude of applications in organic mass spectrometry. Triple quadrupole mass spectrometers are not used in inorganic mass spectrometry [60].

2.1.8.1.1 Collision/Reaction Cell Technology Due to limited resolution, spectroscopic interferences are a major problem in quadrupole ICP-MS. Among them, polyatomic interferences, caused by two or more atoms forming an ion detected at the same *m/z* as the analyte of interest, are particularly deleterious. To address this issue, collision and reaction cell technology has been developed. Multipole collision/reaction cells were originally designed for the study of organic molecules using tandem mass spectrometry. The more collision-induced fragment species generated, the better the chance of identifying the structure of the molecule. This was a very desirable capability in electrospray MS/MS or LC-MS/MS studies, but was a hindrance in inorganic mass spectrometry, where undesirable side reactions must be avoided. Dynamic reaction cell (DRC) technology was designed to address this problem.

A collision/reaction cell is a linear RF-driven multipole ion guide in an enclosed reaction volume that may be held under a constant pressure of a high purity reaction and/or collision gas. The cells are used to promote reactive and nonreactive collisions, with resultant benefits in interference reduction. Several ion–molecule chemistry schemes can be employed, using a variety of reaction gas reagents selected on the basis of thermodynamic and kinetic principles and data.

The dynamic reaction cell is similar to the collision cell since it is a pressurized multipole positioned between the ion optics and the analyzer quadrupole. In DRC technology, a quadrupole is placed inside an enclosed reaction chamber. DRC eliminates polyatomic interferences caused by the combination of plasma gases and sample-matrix constituents before they can enter the analyzing quadrupole. A reactive gas such as ammonia, methane, helium, or oxygen is bled into the cell, serving as a catalyst for ion–molecule chemistry to take place. By a number of different mechanisms, which are predominantly reactions rather than collisions, the gaseous molecules react with the interfering ions to convert them either into an innocuous species dissimilar to the analyte mass or a harmless neutral species. The analyte mass then emerges from the dynamic reaction cell free of its interference and is steered into the analyzer quadrupole for conventional mass separation. By careful optimization of the quadrupole electrical fields, unwanted side reaction by-products, which could potentially lead to new interferences, are prevented. This means that every time an analyte and interfering ions enter the dynamic reaction cell, the bandpass of the quadrupole can be optimized for that specific problem and then changed on-the-fly for the next one. This is seen schematically in Figure 5. This figure shows an analyte ion and an isobaric interference entering the dynamic reaction cell. The reaction gas reacts with the interfering species to form a positive gaseous ion and some harmless neutral species. The electrical field of the reaction cell quadrupole is then set to allow the transmission of the analyte ion to the analyzer quadrupole, free of the problematic isobaric interference. Based on current evidence, the DRC approach appears to offer improved interference reduction and lower detection limits for many elements prone to isobaric interferences.

In collision/reaction cells, postcell kinetic energy discrimination (KED) is used to control chemistry. A potential barrier between the cell and mass analyzer is created, typically by operating the cell at a DC offset potential somewhat lower than that of the analyzing quadrupole. Beam ions from the ICP retain a significant portion of their original kinetic energy and trajectory, whereas ions formed inside the cell are typically formed at lower kinetic energy and/or with off-axis trajectories. The lower energy ions formed in the cell are unable to surmount the potential barrier and are thus prevented from passing into the analyzing quadrupole and onto the ion detector [65].

2.1.8.2 Sector Field

Sector field (SF) is more expensive than the other types of ICP-MS instruments and require more expertise on the part of the operator. Although quadrupole mass analyzers represent over than 85% of all ICP-MS systems installed

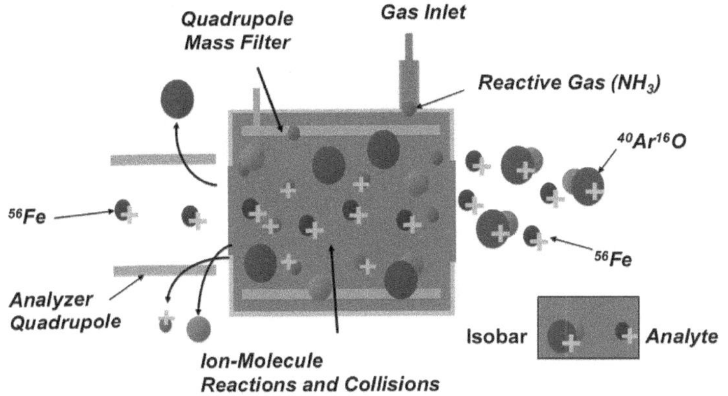

FIGURE 5 Schematic illustration of a DRC system exemplified by the "treatment" of ArO^+ interference which prevents the determination of Fe. *Courtesy of Perkin Elmer.*

worldwide, limitations in their resolving power has led to the development of high-resolution spectrometers based on the double-focusing magnetic sector design. As previously mentioned, a quadrupole-based ICP-MS system typically offers a resolution of 0.7–1.0 amu. This is quite adequate for most routine applications, but has proved to be inadequate for many elements that are prone to argon-, solvent-, and/or sample-based spectral interferences. These limitations drove research in the direction of traditional high-resolution, magnetic sector technology to improve quantitation by resolving the analyte mass away from the spectral interference. The resolution of a magnetic-sector instrument is independent of ion signals, particularly at low mass. These ICP-MS instruments, which were first commercialized in the late 1980s, offered resolving power of up to 10,000, compared to a quadrupole, which was on the order of around 300. This dramatic improvement in resolving power allowed elements prone to isobaric interferences such as Fe, K, As, V, and Cr to be determined with relative ease, even in complex sample matrices [58,64].

Nowadays instrumentation has typically been based on two different approaches—the "standard" and "reverse" Nier–Johnson geometry. Both these designs, which use the same basic principles, consist of two analyzers—a traditional electromagnet and an electrostatic analyzer (ESA). In the standard (sometimes called forward) design, the ESA is positioned before the magnet, and in the reverse design it is positioned after the magnet. A schematic of the reverse Nier–Johnson spectrometer is shown in Figure 6 [58].

The lowest practical resolution achievable with a double-focusing magnetic-sector instrument, using the widest entrance and exit slits, is approximately 300–400, whereas the highest practical resolution, using the narrowest entrance and exit slits, is approximately 10,000. Most commercial systems operate at

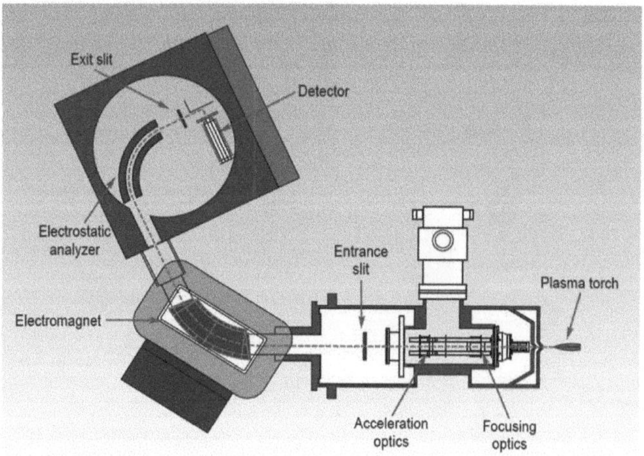

FIGURE 6 Schematic of a reverse Nier–Johnson double-focusing magnetic-sector mass spectrometer. *Courtesy of Thermo.*

fixed resolution settings—for example, low is typically 300–400; medium is typically 3000–4000, and high is typically 8000–10,000, the choice of settings will vary depending on the instrumentation. However, it should be emphasized that, as the resolution is increased, the sensitivity decreases. So even though extremely high resolution is available, detection limits will be compromised under these conditions. This loss in sensitivity could be an issue if low detection limits are required [58].

Besides high resolving power, another attractive feature of magnetic-sector instruments is their very high sensitivity combined with extremely low background levels. Also except for good detection capability, a clear benefit of the magnetic-sector approach is its ability to quantitate with excellent precision. Double-focusing magnetic-sector ICP-MS systems are valuable addition to the trace element toolkit, particularly for challenging applications that require good detection capability, exceptional resolving power, and very high precision [58].

When high-throughput trace analysis is required, sector field double-focusing ICP-MS in the low-mass resolution mode provides the lowest detection limits. It may be used with a simple dilution of the sample and an external calibration for the highest throughput. With a double-focusing sector field instrument, precision depends on the mass resolution, the best precision being achieved with flat-top peaks, which are obtained in low-mass resolution. However, manufacturers of multicollector sector field ICP-MS instruments have succeeded in preserving the flat-top peak shape at higher mass resolution, hence providing the best precision for isotope ratios while allowing the resolution of various spectroscopic interferences. Isotope ratio precision down to 0.002% RSD can be achieved, therefore allowing investigation of more subtle isotopic variations. On the other hand, multicollector instruments are significantly more expensive than the other types and still require appropriate corrections of mass bias, like all other ICP-MS instrument types [66].

2.1.8.3 Time-of-Flight Analyzer

The recent growth in ICP-TOF-MS, sales has come about because of its unique ability to sample all ions generated in the plasma at exactly the same time, which is ideally suited for multielement determinations of rapid transient signals, high-precision ratio analysis, and rapid data acquisition. The idea of a time-of-flight (TOF) analyzer for mass separation was first proposed by Stephens in 1946. TOF mass analyzers have a quite different mode of operation compared to other analyzers. In conventional analyzers, the ion signal is a continuous beam. In TOF mass spectrometry, the ion beam is pulsed so that the ions are either formed or introduced to the analyzer in "packets." These ion packets are introduced to the field-free region of a flight tube 30–100 cm long. The principle behind TOF analysis is that, if all ions are accelerated to the same kinetic energy, each ion will acquire a characteristic velocity dependent on its m/z ratio. Ions of different mass travel down the flight tube at different speeds, thereby separating spatially along the flight tube with lighter, faster ions reaching the detector before the heavier ions [58,63].

The TOF analyzer has a number of advantages over other analyzers used for ICP-MS. The fact that all masses in the spectrum can be monitored simultaneously is the main attraction for many ICP-MS users. Even for simple multielement analysis, scanning analyzers suffer from a decrease in sensitivity as the number of ions to be monitored is increased. The fact that over 30,000 simultaneous mass spectra per second can be obtained by the TOF analyzer, has many advantages for monitoring transient signals, such as those produced by laser ablation and chromatographic, capillary electrophoresis, and electrothermal vaporization sample introduction techniques. In addition, improved isotope ratio measurements have been reported because of the elimination of noise due to temporal fluctuations in the plasma and sample introduction systems [60,63].

However, the resolving power of commercial ICP-TOF-MS systems (Figure 7) is typically in the order of 500–2000, depending on the mass region, which makes them inadequate to resolve interferences from polyatomic species encountered in ICP-MS. In comparison, commercial high-resolution systems based on the double-focusing magnetic sector design offer resolving power up to 10,000, while commercial quadrupoles achieve 300–400. Also, as only <20% of ions are accelerated to the TOF tube, the detection limits are about one order of magnitude worse than those obtained by quadrupole-based instruments. As a result, TOF instruments are likely to find application in specific areas but are not expected to reach the popularity of quadrupole ICP-MS in the near future [58,64].

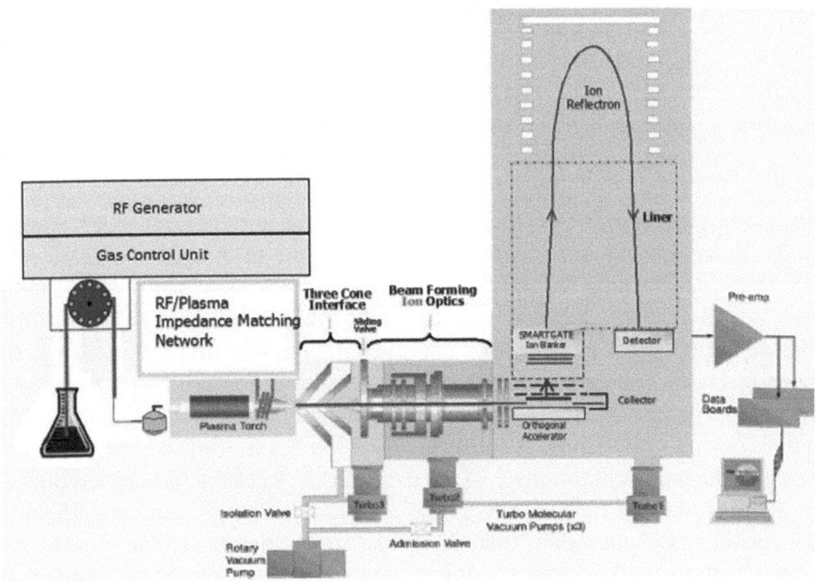

FIGURE 7 Schematic diagram of an ICP-TOF-MS. *Courtesy of GBC Scientific Equipment.*

2.1.8.4 Ion Trap Mass Analyzer

The ion trap possesses two end cap electrodes with hyperbolic surfaces and a ring electrode. The end cap electrodes are grounded whereas there is an RF voltage (1 MHz) on the ring electrode as shown in Figure 8. Ions are stored together in the ion trap and are detected by changing the experimental parameters. A commercial ion trap (IT) mass analyzer operates with a noble gas, e.g., helium, at a pressure of ~0.1 Pa. Ions which enter through a hole in the end cap oscillate in the ion trap, whereby the stability of the oscillating ions is determined by the *m/z* ratio of ions, the RF (radio frequency) and voltage supplied to the ring electrode. By changing the frequency of the RF generator, which excites the ion oscillation in the ion trap, ions with different masses are destabilized step by step. They then leave the ion trap and are detected. The mass spectrum of the ions ejected from the ion trap and separated according to their *m/z* ratio is then obtained by scanning the RF voltage.

This process has been further developed and ion traps are now successfully used for MS–MS and MS^n experiments. However, in spite of their high mass resolution, ion trap mass analyzers have not been widely applied in inorganic mass spectrometry and no commercial ICP-IT-MS is available on the analytical market. One reason could be the occurrence of space charge effects of the most abundant argon plasma ions in the ion trap analyzer, which results in in-trap collisions and scattering of the analyte ions of interest. As a result, a decrease in the sensitivity of analyte ions has been observed in comparison to ICP-MS with double-focusing sector field geometry. Ion traps have been applied in inorganic mass spectrometry for fundamental studies of ion—molecule reactions, especially for reducing disturbing isobaric interferences (e.g., of $^{90}Sr^+$, $^{90}Y^+$

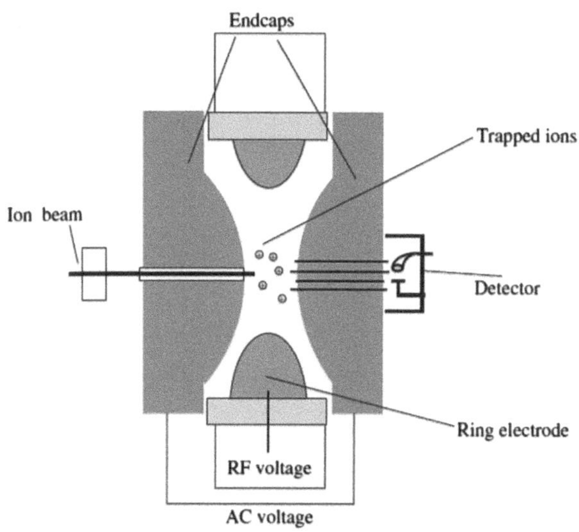

FIGURE 8 Schematic diagram of an ion-trap mass analyzer [63].

and $^{90}Zr^+$ for $^{90}Sr^+$ determination for isotope ratio measurements) by collision-induced dissociation and chemical reaction of ions with collision gas atoms or molecules [63,64,67].

The combination of the ICP-MS and ion trap techniques has remarkable potential for elemental and isotopic analysis. The ICP source offers sample tolerance and higher, more uniform ion yields than any other inorganic ion source. The ion trap technique, equally applicable to the lower mass range relevant to elemental analysis, offers enhanced atomic detection capability (necessary for certain applications) and reduction of matrix and plasma-related polyatomic interferences [67].

2.1.9 Ion Detection Systems

The final process is to convert the ions into an electrical signal with an ion detector. When the separated ion beams leave the mass analyzer system the ions are collected and detected using an appropriate ion detection system inserted in the ultrahigh vacuum of the mass spectrometer. Ion currents at the exit of the mass analyzer are in the range of 10^{-8}–10^{-19} A. The registration of both high and very small ion currents requires special fast ion detection systems.

The most common design used today is called a discrete dynode detector, which contain a series of metal dynodes along the length of the detector. In this design, when the ions emerge from the mass filter, they impinge on the first dynode and are converted into electrons. As the electrons are attracted to the next dynode, electron multiplication takes place, which results in a very high steam of electrons emerging from the final dynode. This electronic signal is then processed by the data-handling system in the conventional way and then converted into analyte concentration using ICP-MS calibration standards.

Other commonly used ion detection systems are the faraday cup, the secondary electron multiplier, combination of faraday cup and secondary electron multiplier, the channel electron multiplier and microchannel plates, the daly detector, the multiple ion collection system (MC-ICP-MS), the fluorescence screen, and the photographic ion detection. In ICP-MS or in TIMS mostly a system of several Faraday cups (up to 16) and/or ion counters (electron multipliers) is applied. Most detection systems used can handle up to eight orders of dynamic range, which means that they can be used to analyze samples from ppt levels, up to a few hundred parts per million (ppm) [58,60].

2.1.10 Interferences

Interferences in ICP-MS are caused when ions generated from the plasma, the sample, or a combination of the two carry a mass-to-charge ratio that is identical to that of the analyte ion. Some common interferences and the ions impacted are shown in Table 1. Interferences in ICP-MS are generally classified into two major groups: spectroscopic (spectral) and nonspectroscopic (matrix- and physical-based) interferences. Spectroscopic and nonspectroscopic interferences have been considered to be a severe impediment since the first days

TABLE 1 Common ICP-MS Interferences

Interferences	Interfered Analyte
$^{12}C^{15}N^+$, $^{1}H^{12}C^{14}N^+$, $^{13}C^{14}N^+$, $^{14}N_2$ spread	^{27}Al
$^{38}Ar^{1}H^+$	^{39}K
$^{40}Ar^+$	^{40}Ca
$^{35}Cl^{16}O^+$, $^{34}S^{16}O^{1}H^+$, $^{35}Cl^{16}O^+$, $^{38}Ar^{13}C^+$, $^{36}Ar^{15}N^+$, $^{36}Ar^{14}N^{1}H^+$, $^{37}Cl^{14}N^+$, $^{36}S^{15}N^+$, $^{33}S^{18}O^+$, $^{34}S^{17}O^+$	^{51}V
$^{35}Cl^{16}O^{1}H^+$, $^{40}Ar^{12}C^+$, $^{36}Ar^{16}O^+$, $^{37}Cl^{15}N^+$, $^{34}S^{18}O^+$, $^{36}S^{16}O^+$, $^{38}Ar^{14}N^+$, $^{36}Ar^{15}N^{1}H^+$, $^{35}Cl^{17}O^+$	^{52}Cr
$^{38}Ar^{16}O^{1}H^+$, $^{40}Ar^{14}N^{1}H^+$, $^{39}K^{16}O^+$, $^{37}Cl^{18}O^+$, $^{40}Ar^{15}N^+$, $^{38}Ar^{17}O^+$, $^{36}Ar^{18}O^{1}H^+$, $^{37}Cl^{17}O^{1}H^+$, $^{23}Na^{32}S^+$, $^{36}Ar^{19}F^+$	^{55}Mn
$^{40}Ar^{16}O^+$, $^{40}Ca^{16}O^+$, $^{40}Ar^{15}N^{1}H^+$, $^{38}Ar^{18}O^+$, $^{38}Ar^{17}O^{1}H^+$, $^{37}Cl^{18}O^{1}H^+$	^{56}Fe
$^{40}Ar^{16}O^{1}H^+$, $^{40}Ca^{16}O^{1}H^+$, $^{40}Ar^{17}O^+$, $^{38}Ar^{18}O^{1}H^+$, $^{38}Ar^{19}F^+$	^{57}Fe
$^{40}Ar^{35}Cl^+$, $^{59}Co^{16}O^+$, $^{36}Ar^{38}Ar^{1}H^+$, $^{38}Ar^{37}Cl^+$, $^{36}Ar^{39}K$, $^{43}Ca^{16}O_2$, $^{23}Na^{12}C^{40}Ar$, $^{12}C^{31}P^{16}O_2^+$	^{75}As
$^{40}Ar_2^+$, $^{32}S^{16}O_3^+$	^{80}Se

of the ICP-MS technique, limiting the maximum achievable analytical figures of merit. Each of them has the potential to be problematic in its own right, but modern instrumentation and good software combined with optimized analytical methodologies have minimized their negative impact on elemental determinations by ICP-MS.

Although interferences are reasonably well understood in ICP-MS, it can often be difficult and time-consuming to compensate for them, particularly in complex sample matrices. Having prior knowledge of the interferences associated with a particular set of samples will often dictate the sample preparation steps and the instrumental methodology used to analyze them [58,62].

2.1.10.1 Spectroscopic Interferences

Spectroscopic interferences are probably the largest class of interferences in ICP-MS. These interferences are a well-known consequence of the limited mass resolution affordable with ICP-MS. Spectroscopic interferences can be subdivided into isobaric atomic ions, multiply charged ions, polyatomic ions (such as oxides (MO^+), hydroxides (MOH^+), argides (ArM^+), having the same nominal mass-to-charge ratio (m/z) as the nuclide of interest) or intense adjacent signals, which overlap with the analytical signal of the isotope of interest. Atomic (isobaric) interferences are caused by overlapping isotopes of different elements with the same nominal mass. Isobaric overlaps are easy to predict and to overcome by selection of alternative isotopes or by elemental correction equations based on relative natural abundances. Multiply charged ions are found at fractions of their nominal mass, such that the observed mass = the nominal mass/N, where N is the charge on the ion (for example, $^{88}Sr_2^+$ on $^{44}Ca^+$). These ions have a smaller formation probability, because the second and higher ionization potentials are larger than the first and are not generally a significant problem unless the source of the multiply charged ion is a relatively easily ionized matrix element of very high concentration. Multiply charged ions are mainly found in the low mass range and require in general mass resolutions between 2000 and 10,000, easily within the reach of SF-ICP-MS instruments.

Molecular or polyatomic ion interferences are formed by the matrix elements, the atoms of the solvent and/or the atoms of the plasma gas. The latter two sources contribute a significant amount since the high number of atoms present increases the quantity of interference generated, even for interferences of low formation probability. Formation of polyatomic species is believed to result from processes including condensation reactions in the expansion region, collisional reactions in the boundary layer around the outside of the sampler cone and survival of the species through the plasma (e.g., refractory metal oxide ions). The formation of polyatomic ions is difficult to predict and depend on the matrix composition and concentration, as well as on nebulization and plasma conditions, resulting in significantly increased uncertainty on the measured analytical signal.

In the case of very high signal intensities, the adjacent lower abundant peak can be significantly interfered by the tailing of the major signal (e.g., measurement of ^{206}Pb in a ^{205}Tl matrix or tailing of ^{232}Th into the lower abundant Th isotopes). As previously discussed, the parameter that quantifies the ability of a mass spectrometer to measure low signals at adjacent masses to very high signals is called abundance sensitivity. Since this parameter is defined by the ratio of the larger signal at mass M to the smaller signals at masses M ± 1, application of high mass resolution can offer some apparent improvement in abundance sensitivity in cases where increasing the resolution has a larger relative effect on reducing the contribution of the peak tail to an adjacent mass than it does on signal reduction of the more intense peak [62].

Current ICP-MS instrumental software corrects all known atomic "isobaric" interferences, or those caused by overlapping isotopes of different elements, but does not correct for most polyatomic interferences. Such interferences are caused by polyatomic ions that are formed from precursors having numerous sources, such as the sample matrix, reagents used for preparation, plasma gases, and entrained atmospheric gases. A prior knowledge of polyatomic interferences cited in the literature for a particular analyte mass may be helpful to the analyst for selecting reagents and conditions that would preclude or at least reduce the possibility of their formation.

Polyatomic ions may originate from the plasma, entrained air, the solvent, or the matrix. In the first case, interferences can be generally corrected by blank subtraction. In the last case, the solution of interference problems, which mainly occur at $m/z \leq 80$ and in the presence of complex matrices, is more challenging, especially with quadrupole instruments.

In multielement analysis of food, it takes considerable effort to develop methods which account for all the possible matrices and any associated interference. The great variability of analytes and the constituent levels among different food commodities often cause a potential interference to be negligible in one matrix and not in another.

A number of analytical solutions have been proposed to overcome spectral interferences caused by matrix-induced polyatomic species in food analysis. For polyisotopic elements, the most obvious is the selection of an alternative isotope, provided it has a sufficient abundance for analyte detection with acceptable sensitivity. Suited sample preparation protocols have been developed by several authors to cope with interferences affecting one or few "difficult" elements, such as K, Ca, V, Cr, Fe, As, and Se. Furthermore, sample digestion procedures, which minimize C and Cl content of digestates have been described. In addition, separation of the analyte from polyatomic interferents is easily achieved in most analytical protocols used for speciation purposes. The best and probably most efficient way to remove spectral overlaps is to resolve them away using a high-resolution mass spectrometer such as double focusing magnetic sector mass analyzers. Use of mixed-gas plasma has also been used to reduce interferences of food matrices. Generally, the addition of N_2 to both the outer and the carrier Ar gas flow appears to be the most successful in the reduction of oxides and other polyatomic species such as ArO, ClO, and ArCl. In order to achieve signal deconvolution, correction equations are also used. Finally, another efficient way to cope with the problem of spectral interferences is through collision cell and dynamic reaction cell (DRC) systems in which interaction of the ion beam with a gas, or a mixture of gases, enables the chemical removal of polyatomic interferents through collisional dissociation and selective ion–molecule reactions [58,59].

2.1.10.2 Nonspectroscopic Interferences

Every experienced ICP-MS user is well aware of the impressive effect that nebulization of an organic solvent, a salt buildup on the sampler and skimmer cones, or an excessive concentration of matrix elements in a digestate, may have on measurements. Nonspectroscopic interferences have been defined as variations in the analytical signal caused by matrix-induced changes in sample transport efficiency, ionization in the plasma, or extraction and transfer of ions in the mass analyzer region. On account of the variety of different mechanisms involved, matrix effects may result in either, signal enhancement or suppression. Their severity depends on the analyte, the sample's physical properties (e.g., viscosity), matrix composition, instrumental setup, and operating conditions.

Regarding the instrument setup, both the sample introduction system and the interface/lens configuration and settings are important in determining the extent to which specific matrix effects are likely to occur.

Basically, there are three types of matrix-induced interferences. The first and simplest to overcome is often called a sample transport effect and is a physical suppression of the analyte signal, brought on by the level of dissolved solids or acid concentration in the sample. It is caused by the sample's impact on droplet formation in the nebulizer or droplet size selection in the spray chamber. In the case of organic matrices, it is usually caused by variations in the pumping rate of solvents with different viscosities. The second type of matrix suppression is caused when the sample affects the ionization conditions of the plasma discharge. This results in the signal being suppressed by varying amounts, depending on the concentration of the matrix components. The ionization conditions in the plasma are so fragile that higher concentrations of acid result in severe suppression of the analyte signal.

The common way to deal with physical interferences is to use internal standards. With this method of correction, a small group of elements are spiked into the samples, calibration standards, and blank. As the intensity of the internal standards changes, the element responses are updated, every time a sample is analyzed. The selection of the internal standards is performed, classically, by the following criteria: (1) they are not present in the sample; (2) the sample matrix or analyte elements do not spectrally interfere with them; (3) they do not spectrally interfere with the analyte masses; (4) they should not be elements that are considered environmental contaminants; (5) they are usually grouped with analyte elements of a similar mass range; for example, a low-mass internal standard is grouped with the low-mass analyte elements and so on up the mass range; (6) they should be of a similar ionization potential to the groups of analyte elements so they behave in a similar manner in the plasma; (7) some of the most common elements/masses reported to be good candidates for internal standards include ^{9}Be, ^{45}Sc, ^{59}Co, ^{74}Ge, ^{89}Y, ^{103}Rh, ^{115}In, ^{169}Tm, ^{175}Lu, ^{187}Re, and ^{232}Th.

Remarkable, also, is that internal standardization is used to compensate for long-term, signal drift produced by matrix components slowly blocking the sampler and skimmer cone orifices. Although total dissolved solids are usually kept below 0.2% in ICP-MS, this can still produce instability of the analyte signal over time with some sample matrices. It should also be noticed that the difference in intensities of the internal standard elements across the mass range will indicate the flatness of the mass response curve. The flatter the mass response curve (e.g., less mass discrimination), the easier it is to compensate for matrix-based suppression effects using internal standardization.

As it is reported, the magnitude of signal suppression in ICP-MS is increased with decreasing atomic mass of the analyte ion. It is believed that the major cause of this kind of suppression is the result of poor transmission of ions through the ion optics due to matrix-induced space-charge effects. This has the effect of defocusing the ion beam, which leads to poor sensitivity and detection limits, especially when trace levels of low-mass elements are being determined in the presence of large concentrations of high-mass matrices. Unless any compensation is made, the high-mass matrix element will dominate the ion beam, pushing the lighter elements out of the way [58,59].

2.1.11 Sample Preparation—Pretreatment

Fresh foods are commonly prepared as for the table and using stainless steel knives and clean plastic chopping boards that minimize contamination problems. Domestic blenders, coffee grinders, and food processors fitted with stainless steel, Ti, or ceramic blades can be extremely effective at homogenizing food samples. If the elements analyzed do not include volatile elements, for example Hg, samples can be freeze dried before homogenization. Freeze drying is unlikely to be effective for foods such as raw cereals or root vegetables because the resulting dried material is relatively hard to crush. Ensuring that the test material has been fully homogenized is difficult even through performing measurements in duplicate can help with this [63].

ICP-MS was originally developed for the analysis of liquid samples. If the sample is not a liquid, some kind of sample preparation has to be carried out to get it into solution. There is no question that collecting a solid sample, preparing it, and getting it into solution probably represents the most crucial steps in the overall ICP-MS analytical methodology, because of the potential sources of contamination from grinding, sieving, weighing, dissolving, and diluting the sample.

For solid foods, total element determination using conventional, e.g., liquid, sample introduction requires a dissolution step for analyte presentation to the instrument. The choice of the digestion method is made on the basis of several considerations, including efficiency in sample solubilization, potential for loss of volatile elements and contamination of samples, reagent-related interferences and matrix effects, sample throughput, detection limits requirements and safety aspects. Microwave (MW)-assisted digestion with HNO_3, generally in combination with H_2O_2, is a widely used method for the dissolution of food samples. In particular, microwave digestion with closed vessel systems has gained popularity, being a simple and fast dissolution technique that minimizes acid consumption, the risk of sample contamination, and the loss of volatile elements. HNO_3 is the acid of choice for ICP-MS analysis as it causes minimal interferences. Small amounts of other acids can be added for specific analytical purposes, e.g., HF for the solubilization of Si-bound elements (Al, Co, Cr, Ni, Th, U, V) in vegetable matrices containing substantial amounts of silicon. As previously stated, the use of HCl, $HClO_4$, H_2SO_4, is generally excluded to avoid resulting Cl, S interferences or cone blockage by salt formation (H_2SO_4). However, mixed acids can be successfully used for the heavier elements, where polyatomic interferences are less problematic. Alternative solutions to digestion with concentrated acids have also been reported for a rapid, multielement screening of biological and botanical materials, e.g., extraction with water-soluble tertiary amine solutions containing EDTA [58,64,68].

Solid foods in powder form can be analyzed directly by means of LA-ICP-MS or ETV-ICP-MS to eliminate time-consuming sample dissolution procedures. However, this requires the preparation of homogeneous powdered samples and the subsequent analytical determination is not as straightforward as the one based on liquid sample introduction, and requires matrix-matched solid samples.

Concerning liquid samples, these can be in many cases directly aspirated into the plasma following acidification or, alternatively, dilution and addition of suitable reagents. The traditional techniques for sample pretreatment are time-consuming and require large amounts of reagents, which are expensive, generate hazardous waste, and might contaminate the sample with the analytes. Advances in sample pretreatment over the last few decades have been propelled by the advance of microwave-assisted acid digestion, ultrasound-assisted, extraction and slurry preparation and direct solid sampling analysis [69].

Wet digestion methods are carried out in open vessels, in tubes, on a hot plate or in an aluminum heating block or in closed vessels at elevated pressure (digestion bombs) with thermal or microwave heating. Microwave-assisted digestion is an attractive method, especially for small samples. Extreme care should be exercised in using sealed pressure vessels since there is much anecdotal evidence of these vessels rupturing occasionally during conventional or microwave-assisted digestion of organic materials. The applicability of this technique is strictly dependent on the type of food: carbohydrates are easily distracted with nitric acid at 180°C, while fats, proteins, and amino acids cause incomplete digestion due to the relatively low oxidation potential of nitric acid at 200°C, these samples require the addition of sulfuric and/or perchloric acid with all the problems related to their use at high temperature and pressure. An alternative procedure is to decompose most of the organic matter using H_2O_2 followed by addition of nitric and/or other acids. H_2O_2 has the advantage of being eliminated, not increasing the ion content of the sample digestate.

The type of acid used in the preparation procedure can have important consequences in the measurement step. The most commonly used reagent in all atomic spectrometric techniques is nitric acid. In spite of occasionally observed signal suppression in its presence, no severe analytical problems are encountered in practice with nitric acid at concentrations up to 10%, sometimes higher, as long as its concentration is similar in calibration and sample solutions. Hydrogen peroxide and hydrochloric acid, added in mineralization procedures, are quite well accepted. Because of its high viscosity, utilization of sulfuric acid is usually avoided. Mineralization for ICP-MS increases the cost due to the requirement for suprapur materials. These materials are not only a requirement in ultratrace analysis but also for the MS system that must be protected from contamination.

2.1.12 Microwave-Assisted Digestion

Microwave (MW)-assisted digestion procedures based on the use of concentrated acids are an efficient and safe way to decompose many kinds of samples for the determination of the elemental fingerprint. Microwave heating has several advantages over conventional heating on a hot plate, etc., as the energy is generated in the digestion mixture and not transferred by conduction. Among the key advantages of this procedure are the much shorter digestion times and the reduced need for aggressive reagents to obtain complete digestion. There are two different systems available for MW-assisted digestion, pressurized closed-vessel systems, and open focused-MW systems that work under atmospheric pressure. Microwave-assisted digestion in closed vessels under pressure has gained popularity as a simple and fast dissolution technique that minimizes acid consumption, the risk of sample contamination, and loss of sample and volatile elements. One of the limitations is the time required for cooling before the vessels can be opened, which may take hours, depending on the type of equipment used. The main advantages are safety, versatility, control of microwave energy released to the sample, and the possibility for programed addition of solutions during the digestion.

2.1.12.1 Ultrasonic Extraction

Ultrasound has been employed for sample preparation in order to improve analytical throughput. Also chemical information of samples submitted to ultrasonic irradiation can be severely compromised since the collapse of cavitation bubble results in a strong local temperature increase and free-radical production, which could provoke analyte loss and gross analytical errors. Analyte losses were also observed, contrasting the results obtained for various metals in food samples pretreated with ultrasound devices with other sample preparation techniques and certified materials [69,70].

2.1.13 Direct Solid Sampling Analysis

Among the techniques that can be used for direct solid samples analysis in combination with are laser ablation (LA) and electrothermal vaporization (ETV). Direct solid sampling analysis offers a number of advantages, such as the reduced sample preparation time and hence a faster analysis; higher accuracy, as errors due to analyte loss and/or contamination are dramatically reduced; higher sensitivity due to the absence of any dilution; and the absence of any corrosive or toxic waste. The requirement of matrix-matched standards construction minimizes the application of LA-ICP-MS for food analysis. However, this mode of ICP-MS could prove useful in the future taking advantage of the ability to provide microscale information that could be extremely interesting for specific foods [60,68,71].

Laser is an acronym for Light Amplification by Stimulated Emission of Radiation. As a consequence of its light-amplifying property, a laser produces spatially narrow and extremely intense beams of radiation having identical frequency, phase, direction, and polarization properties. The cascade of photons is focused on a few millimeter area of the sample. Interaction between the laser beam allows the conversion of photon energy into thermal energy, which is responsible for the vaporization of most of the exposed solid surface. The material ablated is swept away with an argon stream to an ICP-MS and analyzed [64].

Specific food commodities:

1. Vegetables and Mushrooms

Analyses in support of dietary intake studies may require that root vegetables that are customarily eaten whole should not be peeled but are scrubbed before pretreatment and outer leaves which are not normally consumed should be discarded from leafy vegetables. However, peel and outer leaves may have to be analyzed in plant uptake studies and removal of surface contamination should be undertaken using cleaning methods which approximate treatment prior to consumption as closely as possible. Dicing the prepared vegetables, either using a stainless steel knife or an automated dicer/chopper, can be helpful but further blending may be needed if smaller sample sizes are taken for analysis. Mushrooms should be brushed clean before analysis, bearing in mind that local custom can dictate whether or not the outer skin is removed prior to consumption and that fungi can absorb significant amounts of water during washing.

2. Fruits

The outer skin or peel of fruit can be a useful indicator of surface contamination and may have to be analyzed separately from the remaining edible portions of hard fruits such as apples and pears. Freezing the pulp, allowing it to partially thaw and mixing the iced slurry well can also help produce a representative sample of flesh and juice.

Fruit juices, which have a complex organic matrix with a significant carbohydrate content, can be analyzed either following dilution, with subsequent centrifugation/filtration or after full sample digestion.

3. Cereals

Edible grains need to be separated from the chaff and analyses performed either on a subsample from a well-mixed bulk of individual grains or on a rough flour produced using a coffee grinder or equivalent.

4. Wines and beverages

Many wines and most other beverages only need to be shaken well before subsamples are taken. Sparkling wines, beers, etc., should be degassed in an ultrasonic bath before digestion and those wines which form a precipitate, for example, crusted port, and some red wines have to be decanted into another container before subsampling. If Pb is to be analyzed in wines, then care may be needed to ensure that no contamination arises where Pb foil caps have been used, as these can leave deposits on the lip of the bottle.

Dilution is usually applied to reduce the ethanol content of wine before analysis. Internal standard addition, isotope dilution analysis, matrix-matching, or ethanol removal by heating have also been reported as viable solutions to deal with matrix effects in the analysis of alcoholic beverages. However, depending on the nebulizer used, not only ethanol but also other organic and inorganic components can produce significant matrix effects that need to be taken care of, especially when red wines are analyzed.

5. Water

Although water is assumed as an "easy" matrix regarding the sample pretreatment samples, its analysis can be challenging, especially in samples with a high mineral or bicarbonate content. In such cases, as well as with more complex water matrices, the addition of an appropriate internal standard is recommended.

6. Dairy products

Dairy products should be transferred to the laboratory in an ice box, properly labeled and stored in refrigerator. The sample preparation should be done before the expiry date. As the other liquids, milk needs to be mixed by pouring the contents very well. Fresh milk dilution in alkaline media and multielement quantification with the standard addition method has been successfully even though in most studies milk and infant formulae are analyzed following full digestion. To improve LODs, milk can be freeze dried and this can also help with longer term storage.

Regarding cheese and yogurt samples, cheeses are cut into small cubes, mixed in plastic bags, and homogenized in a homogenizer. All containers and equipments have to be made from either plastic or stainless steel. Containers and equipments usually are washed by a solution of 2% sodium EDTA and 2% sodium citrate in order to remove ions. Finally, all surfaces should be washed by Milli-Q water before contact with samples.

7. Fats and oils

Oils need to be shaken well before subsampling but fats can probably be dealt with most effectively by melting and subsampling from the stirred molten bulk, although some fats may separate on melting, which can cause difficulties. Unless several different sources of fat are to be analyzed as a homogenate, most fats will have been effectively mixed during preparation and a representative slice taken through the bulk will probably be adequate.

8. Meat

Inedible material, e.g., skin, hair, etc., should be removed and discarded before the edible flesh is sampled. Meat samples were homogenized using a stainless steel rotating knife homogenizer blender. It is imperative that the blender is cleaned well between samples using dilute detergent followed by tap water and then Milli-Q water.

9. Seafood

The fish should be gutted, washed with pure water and portions of edible flesh taken for analysis avoiding the head, fins, and larger bones. Whether to include the skin or not may depend on the type of fish and local dietary custom. Fish species that are normally eaten whole should be blended without deboning. Removal of edible flesh from shellfish can be awkward and time-consuming but wherever possible fresh samples should be analyzed, which have been washed free of sand, etc. If practicable, viscera should be removed. It should be noted that freezing some crustaceans can cause undesirable liquefaction of the flesh on thawing. This can cause problems, for example, if brown and white crab meat needs to be analyzed separately. In all cases a food blender can be used to produce a satisfactory homogenate.

10. Honey

Concerning honey samples, acid mineralization in block digester, and microwave-assisted acid digestion have been reported. The methods were optimized in order to minimize the final acidity and the residual carbon content after the sample preparation [63,72–74].

2.1.14 Hyphenated Techniques

Hyphenated techniques involving ICP-MS are among the fastest growing research and application areas in atomic spectroscopy. This is because, by itself, ICP-MS does not give information on the chemical or structural form of the analytes present since all forms of the analytes are converted to positively charged atomic ions in the plasma. However, as an excellent elemental analyzer, it also performs as a superb detector for chromatography. Hyphenated ICP-MS is achieved through the coupling of the ICP-MS to a separation technique—normally a chromatographic separation. In this way, target analytes are separated into their constituent chemical forms or oxidation states before elemental analysis. The most common separation techniques are gas chromatography (GC) and liquid chromatography (LC), which includes ion chromatography (IC), but, other separation techniques, such as capillary electrophoresis (CE) and field flow fractionation (FFF) are also used. Because of ICP-MS ability to accurately distinguish isotopes of the same element, particularly now that collision/reaction cell technology has eliminated almost all interferences, ICP-MS is also capable of isotope dilution (ID) quantification.

In addition to the more conventional liquid phase separations, LC and IC for example, ICP-MS is also an excellent detector for separations carried out by GC. While other element-specific detectors exist for GC, none possesses the elemental coverage, sensitivity, or specificity of ICP-MS. All hyphenated ICP-MS systems require that a few simple conditions are met:

- The connecting interface must transmit the fractionated sample quantitatively from the separation system to the plasma of the ICP-MS in a form that the plasma can tolerate.
- The temporal resolution of the sample components must not be unacceptably degraded during transfer to the plasma.
- The chromatograph should communicate with the ICP-MS to allow synchronous separation and detection.
- The ICP-MS must be capable of transient signal acquisition at sufficient sampling frequency and over sufficient dynamic range to accommodate the resolution of the chromatograph and the required number of elements or isotopes per peak over their ranges of concentrations.

It must be possible to tune the ICP-MS under plasma conditions similar to those encountered during the chromatographic run. Generally, this entails introducing the tuning element(s) via the chromatographic interface. In general, using an ICP-MS as a detector for chromatography is a simple matter of connecting the outlet of the chromatographic column to the sample introduction system of the ICP-MS. If the sample is gaseous, as in GC, the transfer line should be passivated and heated to eliminate sample degradation and condensation and should terminate directly into the ICP torch. If the sample is a liquid, the transfer line will likely terminate in a nebulizer in order to generate an aerosol compatible with the plasma. This may require either a split flow or make-up flow in order to match the chromatographic flow with the nebulizer and plasma requirements. Depending on the total sample flow and choice of nebulizers, the use of a spray chamber may or may not be necessary.

2.1.14.1 LC-ICP-MS

Liquid chromatography coupled to inductively coupled plasma mass spectrometry (LC–ICP–MS) (Figure 9) has become the most popular technique for elemental speciation studies. For speciation analysis of samples that are not already in liquid form, three steps are generally required. The first step comprises the extraction of the element species from the sample, because so far, no techniques are available to do in situ speciation analysis at environmental concentrations. During this critical extraction step, the compounds should be quantitatively extracted and must not be changed. Combinations of water and/or organic solvents or weak acids are often employed for this purpose. In a second step, the compounds are then separated by liquid chromatography. After the separation of the compounds, a reliable detection step is necessary. ICP-MS is typically the detector of choice because of its excellent detection limits, wide dynamic range, multi-element capabilities, and the possibility to determine isotopes. The robustness of the ICP is certainly one of the reasons that make the hard ionization in the ICP superior to molecule-selective detection

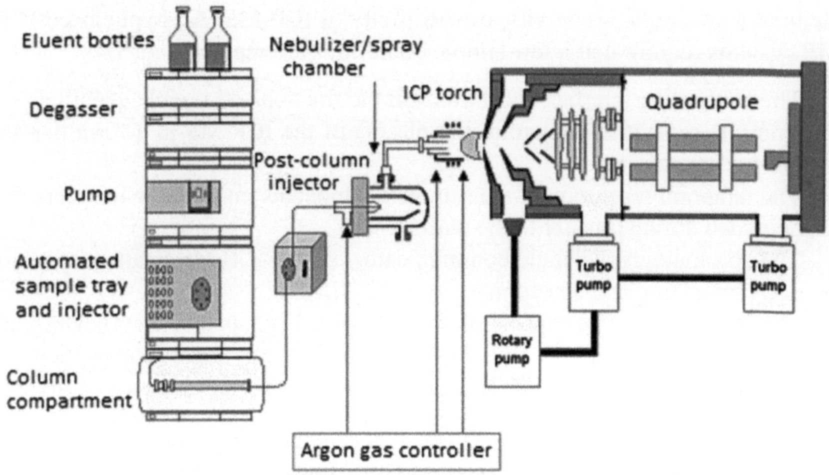

FIGURE 9 Schematic presentation of an LC-ICP-MS. *Courtesy of Agilent.*

using electrospray ionization. Compound independent quantification should be mentioned as another benefit of ICP-MS.

The feasibility of coupling LC-ICP-MS is mainly affected by the composition and flow of the mobile phases used to perform the chromatographic separation. The mobile phase flow rates of LC columns match the typical liquid flow rate of common nebulizers making the coupling of the two components simple. The connection between the outlet of the chromatographic column and the nebulizer is an inert plastic tubing, whose inner diameter and length are small enough to minimize peak broadening. The application range is wide on account of the different chromatographic principles exploited, i.e., size-exclusion, reversed-phase, ion-exchange. The use of ion-pairing reagents both in reversed-phase chromatography (RPC), in order to resolve ionic analytes (reversed-phase ion-pairing chromatography, RPIPC), and in anion-exchange chromatography, with the aim to achieve simultaneous separation of anionic, cationic, and neutral species, further extends the analytical capabilities of LC. Characteristic examples of Ion Chromatography Inductively Coupled Plasma Mass Spectrometry (IC-ICP-MS) uses are for the determination of Cr (VI), bromate, iodoacetic acids, and bromoacetic acids in drinking water.

A primary problem in the coupling of the two techniques is the low tolerance of ICP-MS for organic modifiers and salts (buffers, ion-pairing reagents) added to LC mobile phases to enhance separation. A compromise between chromatographic separation and detection must be adopted, and often additional measures to reduce the deleterious effects of organic solvents on the plasma and the interface—e.g., cooling of the spray chamber, desolvation, oxygen addition to the nebulizer gas and use of Pt cones—are required. Most LC-ICP-MS hyphenation is performed by using isocratic elution. Gradient elution may be used to improve retention times and partition functions but often difficulties arise when gradients are introduced into the ICP. Identifying optimal operating conditions for different mobile phase concentrations can be difficult. In addition, the changing load on the plasma due to variations in mobile phase composition can cause plasma instability. Time-resolved analysis is needed for acquisition of chromatographic signal. To this end, ICP-MS instruments are equipped with software capable of monitoring signal versus time at different *m/z* values.

With the continuing improvements in the robustness of ICP-MS instruments, chromatographic separations using organic solvents are feasible. The addition of oxygen to the plasma makes an isocratic separation with an organic solvent "routine." Challenges remain when gradient elution is required, and quantification becomes a bit more difficult, but possible solutions such as counter gradients are already published [64,75–78].

In some cases, it is the presence of the target element that is important, for example, Cr(III) or Cr(VI). In other cases, the element or elements are a simple way to identify and quantify a molecule present in a complex mixture, for example, using P as a means of quantifying organophosphorus compounds.

A clear trend in speciation analysis is toward higher accuracy of the results. Published semiquantitative results are nowadays the exception. Determination of the column recovery (ratio of the amount of an element injected onto the column/the sum of the species) is obligatory. The use of reference materials, when available, is compulsory although there is still a big need for more certified reference materials with certified species concentrations. Proper species standards are still not commercially available. Intercomparison studies of expert laboratories already show good agreement.

2.1.14.2 GC-ICP-MS

GC-ICP-MS is used for the analysis of volatile organic or organometallic compounds when no other GC detector can provide the required elemental or isotopic specificity or sensitivity. The hyphenation of ICP-MS with GC is more complicated than with LC but offers several analytical advantages. These include higher resolving power, 100% sample introduction efficiency, obtainment of a more stable plasma, fewer spectral interferences, reduced cone wear. Furthermore, because of the generally higher resolution of GC compared with LC, it is sometimes advantageous to create volatile derivatives of otherwise nonvolatile compounds for analysis by GC. When used as a detector for GC, ICP-MS provides several other advantages over alternative elemental detectors:

- ICP-MS is almost universal. Only hydrogen, helium, argon, fluorine, and neon cannot be measured directly due to ionization potentials (IPs) that are higher than argon. A few other elements such as nitrogen and short-lived radionuclides are not practical in most cases.
- ICP-MS can tolerate a wide range of GC carrier gases and flows.

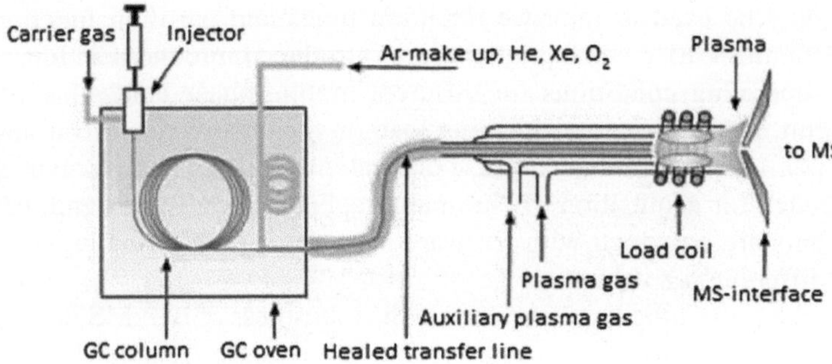

FIGURE 10 Schematic presentation of a GC-ICP-MS.

- ICP-MS does not typically suffer from suppression of analyte response due to coeluting compounds.

The GC-ICP-MS interface consists of a heated, passivated transfer line, and a special torch with a heated injector tube. Argon make-up gas supplied by the ICP-MS is preheated via a stainless steel heat exchanger inside the GC oven before being added to the GC column effluent at the head of the transfer line. In this way, the sample is maintained at constant high temperature from the end of the chromatographic column in the GC oven to the tip of the ICP injector (Figure 10) [79,80].

2.1.14.3 Multi-MS

While ICP-MS is an excellent quantitative elemental detector for liquid chromatography, it cannot, by itself, provide information about molecular weight or structure. Conversely, traditional molecular mass spectrometry (MS) techniques commonly used with LC can provide molecular weight and structural information, but lack the element specificity and sensitivity of ICP-MS. When used in parallel LC-ICP-MS/ESI-MS, elemental and molecular MS can work together to provide analytical information previously unavailable in a single analysis.

As a result, the combined application of elemental and molecular MS with LC has been increasingly exploited in bioinorganic analytical chemistry for the past 15 years. ICP-MS, the most popular element-specific detector has played a crucial role in this marriage because of its detection power, which can measure low parts per billion and parts per trillion (ppt) levels of elemental species. Additionally, its multielement and multiisotope detection capabilities have been invaluable for monitoring the degree of species transformation and for species quantification in speciation applications. The elemental specificity of ICP-MS, improved by continued developments in interference-reducing technology, has become a key to successful compound identification by combined elemental and molecular MS. LC-ICP-MS retention times, although insufficient on their own for this purpose, are invaluable in helping to locate target elemental species in very complex organic mass spectra by data mining, enabling their subsequent structural characterization. The capability to characterize elemental species at subnanomolar levels in complex samples by electrospray ionization MS/MS (ESI-MS/MS) has been further improved by significantly improved signal-to-noise ratio as obtained by using multistep sample preparation and preconcentration procedures. Additionally, efforts have been made to develop high-resolution separation systems compatible with both ICP and ESI. This means introduction of higher organic content and lower flow rates (down to nL/min) into the ICP-MS, which have driven the development of optimized ICP-MS instrumentation and chromatographic interfaces. For elements of high molecular diversity, successful strategies have made use of the power of accurate mass measurement. Due to the high probability of assigning different chemically possible formulas to the same mass, the need for combining a highly selective separation technique with element-specific detection, for example, by ICP-MS, and accurate mass measurement of the parent and fragment ions has been demonstrated by a number of speciation scientists.

In summary, the combination of LC with element (ICP-MS, for screening and quantification) and molecule-specific (ESI-MS/MS or MALDI-MS for identification) detection is considered a "must" in speciation research nowadays. Future developments of this multi-MS approach will be driven by the introduction of new concepts, technologies, emerging regulations, and needs in foodomics, for example, protein analysis.

2.1.14.4 FFF-ICP-MS

Field-flow fractionation (FFF) is a developing online fractionation technique that enables the separation of macromolecules, colloids, nano- and microparticles according to size, chemical composition, or density with excellent resolution over a size range from a few nanometers up to several microns. Many of the limitations of other separation techniques, including degradation, filtration, decreased resolution, or unwanted adsorption, can be overcome by FFF.

In FFF, separation is performed in an empty channel by application of an external separation field perpendicular to the solvent flow, which leads to an arrangement of the molecules or particles in different flow layers. The parabolic flow profile of the laminar solvent stream has different flow velocities in these solvent layers and, as a result, the analyte molecules or particles are separated.

The FFF technique allows the characterization of a sample without using a stationary phase, which can be a source of bias when using other separation techniques. Size information of the different elution volumes can be obtained by the use of mass or size-sensitive detectors, such as light scattering, by calibration of the FFF-channel or by calculation directly from FFF theory.

Since FFF enables the analysis of molar mass or particle size for a wide variety of samples, FFF-ICP-MS coupling is developing toward a key technique that connects both size and structural information in the nanometer and micrometer scale together with elemental composition and concentration data.

Since FFF works with a constant outlet flow between 0.1 and 2 mL/min, it can easily be coupled to an ICP-MS system via a simple capillary. The capillary tube should be as small as possible to prevent peak broadening, which could decrease the signal intensity. A split of the flow is not necessary due to the low flow rates in FFF, but it could be used for the introduction of an internal standard or acid for quantification. Standard concentric nebulizers are compatible with FFF-techniques.

One challenge in coupling ICP-MS with online separation techniques is the dilution of the narrow dispersed fractions. In addition, many samples often contain very low amounts of the analyte. Often preconcentration of the sample prior to injection into the separation system is necessary.

Examples of the practical use of the method can be found in sectors like agricultural and food science, biological, and pharmaceutical research. Some additional examples of the successful use of FFF-ICP-MS include bioanalysis (e.g., analysis of protein–metal complexes). Although FFF-ICP-MS has been used in a wide range of applications to date, it remains a relatively unknown and under used hyphenation technique. It is however a powerful technique that can provide an excellent tool for research activities in the future [81].

2.1.14.5 CE-ICP-MS and GE-ICP-MS

Electrophoretic techniques result in the separation of the species of interest in space rather than time. Gel electrophoresis (GE) is a common technique used for the separation of large molecules (e.g., biomacromolecules including peptides, proteins, and nucleotides) using gels made from agarose or polyacrylamide. In gel electrophoresis, a solution containing a mixture of analytes is applied to one end of a gel, either in a rectangular slab as in slab gel electrophoresis, or a gel filled capillary as in capillary gel electrophoresis. Its online coupling with ICP-MS detection has opened the doors to huge opportunities to get more detailed information about biomolecules [82].

The coupling of capillary electrophoresis (CE) and ICP-MS has been the latest to be realized due to the complex interface needed. However, in recent years CE-ICP-MS has increasingly received attention as a technique for speciation of metal and metalloid species. Capillary electrophoresis offers an alternative separation approach to chromatographies. The main favorable features of CE include:

1. high separation efficiency;
2. low reagent- and sample-volume consumption, which is particularly important when handling low sample amounts—volumes; and

3. comparatively mild separation conditions, which would help to preserve the original trace-element speciation in samples.

Among the different modes of CE, capillary zone electrophoresis (CZE) is the most commonly used with ICP-MS detection to date. A major problem of CE-ICP-MS is that it still suffers from relatively high DLs. However, recent developments delivered a number of improvements, making this hyphenated technique more promising.

CE is a technique in which the capillary is filled with conductive liquid buffer rather than a gel. After application of the sample, subsequent application of a high voltage across the length of the gel or capillary causes migration of the analytes along the length parallel to the applied voltage. Their rate (and direction) of migration depends on a variety of factors such as their overall charge and three-dimensional structure as well as the pH of the buffer and characteristics of the gel. Analytes are detected as they elute from the end of the capillary, much like chromatographic techniques.

More recently, gels can be scanned by a laser ablation device connected to an ICP-MS. As the laser moves across the gel, the components are vaporized and transported in a flowing gas stream to the ICP-MS for detection. Like any separation technique, care must be taken to ensure that the separation technique does not change the nature of the analytes. For example, many metalloproteins contain metals that are solvent-exposed and are thus easily stripped away by the high voltage applied across the length of the gel [64,82,83].

2.1.15 Conclusion, Strengths and Limitations

In conclusion, ICP-MS is the most versatile trace, ultratrace elemental and isotope analysis technique available today, that uses an ICP plasma source to dissociate the sample into its constituent ions. The ions are extracted from the plasma and passed into the mass spectrometer, where they are separated based on their atomic mass-to-charge ratio by a quadrupole or magnetic sector analyzer. The high number of ions produced, combined with very low backgrounds, provides the best detection limits available for most elements, normally in parts per trillion range. It is important to remember that detection limits can be no better than lab cleanliness allows. Quadrupole mass spectrometers are most common in ICP-MS, yet magnetic sector instrumentation fulfills exacting requirements that demand the ultimate detectability and isotope ratio measurement.

Strengths

- Ultra low detection limits (<ppt)
- Wide elemental coverage (73 elements)
- Fast analysis times (73 elements in 3 min)
- High-throughput analyses (160 samples × 73 elements = 11,680 analyses in 8 h)

- Well suited for high number of samples and/or elements
- Wide dynamic range (up to nine orders of magnitude)
- Ability to provide isotopic data (e.g., MC-ICP-MS)
- Fast semiquantitative screening
- Hybrid techniques
  LA-ICP-MS for solids
  LC-ICP-MS, GC-ICP-MS, etc., for speciation
- Easily interpreted spectra
- Rare Earth Elements determination

Limitations

- Initial capital cost, ~180,000 € for a quadrupole-based ICP-MS
- Operation cost, ~250 €/10 h
- Although there are some spectral interferences they are well defined
- Liquids analyzed should contain less than <0.2% dissolved solids
- Requirement for clean room conditions
- Need for suprapur acids to digestate

2.1.16 ICP-MS versus Other Techniques

ICP-MS not only offers extremely low detection limits in the sub parts per trillion range, but also enables quantitation up to the parts per million level. This unique capability makes the technique very attractive compared to other atomic spectrometric techniques: Atomic Absorption Spectrometry (AAS), including ElectroThermal (ETAAS), Graphite Furnace (GF-AAS), Hydride Generation (HG-AAS) and Cold Vapor (CV-AAS), and Inductively Coupled Plasma Atomic Emission Spectroscopy (ICP-AES), also referred to as Inductively Coupled Plasma Optical Emission Spectrometry (ICP-OES). Besides, the above ICP-MS has clear advantages in its multielement characteristics, speed of analysis, detection limits, isotopic capability, and high versatility (e.g., hybrid techniques, LC-ICP-MS). Compared to ICP-AES, which is also a fast multielement technique ICP-MS provides also isotopic information, much lower detection limits, and capability for rare earth elements.

The main disadvantage of ICP-MS in comparison to AAS techniques is the cost for instrument purchase and the running cost in terms of consumables, i.e., argon and suprapur acids. Furthermore, trained personnel are obligatory to carry out analytical work. When complex analyses are performed, large amounts of data can be generated which require careful processing and such procedures can be very time-consuming. Purpose-written software, e.g., spreadsheet macros, can be very efficient to handle multielement and multiisotope data, although visual inspection of analytical results by an experienced specialist is essential. A brief comparison with other techniques is shown in Table 2 [58,64,84].

TABLE 2 ICP-MS Comparison with Other Spectroscopic Techniques

Technique Characteristic	Flame AAS	GF-AAS	ICP-AES	ICP-MS
Detection limits	Very good for some elements	Excellent for some elements	Very good for most elements	Excellent for most elements
Sample throughput	10–15 s/element	3–4 min/element	1–60 elements in <3 min	73 elements in <3 min
Dynamic range	10^3	10^2	10^6	10^8
Precision				
Short term	0.1–1.0%	0.5–5%	0.1–2%	0.5–2%
Long term	1–2%	1–10%	1–5%	2–4%
Interferences				
Spectral	Few	Few	Many	Few
Chemical matrix	Many	Many	Very few	Some
Physical matrix	Some	Very few	Very few	Some
Dissolved solids	0.5–5%	>20%	0–20%	0.1–0.4%
Sample volumes required	Large	Very small	Medium	Very small to medium
Semiquantitative analysis	No	No	Yes	Yes
Isotopic analysis	No	No	No	Yes
Ease of use	Very easy	Moderately easy	Easy	Moderately easy
Method development	Easy	Difficult	Moderately easy	Difficult
Unattended operation	No	Yes	Yes	Yes
Capital costs	Low	Medium	High	Very high
Running costs	Low	Medium	High	High
Cost per analysis				
Few elements/high throughput	Low	High	Medium	Medium
Many elements/high throughput	Medium	High	Low-medium	Low-medium

B Isotopic Mass Spectrometry

For many decades, the accurate and precise determination of isotope ratios has remained a very strong interest to many researchers due to its important applications in earth, biological, food, medical, environmental, and archeological sciences. An important requirement for a technique used for studying, sometimes extremely small, variations in the isotopic composition of an element is the ability to perform analyses with a very high precision and accuracy. IRMS, MC-ICP-MS, and TIMS meet these features and so they are used for isotope analysis of foodstuffs.

2.2 IRMS

2.2.1 Isotopes

Isotopes are atoms with the same number of protons and electrons, but different numbers of neutrons N in their nuclei. This gives them different atomic weights for the same atomic number. Almost all of the 92 naturally occurring chemical elements are found in nature in more than one isotopic form, with the exception of 21 elements such as fluorine and phosphorous, which are monoisotopic. The vast majority of these isotopes are stable isotopes. Most elements consist of more than one stable isotope. Stable isotopes indicate those isotopes of an element which are stable and that do not decay through radioactive processes over time. Radioactive isotopes (e.g., ^{14}C) decay over time, whereas stable isotopes (e.g.,^{12}C, ^{13}C) do not. Carbon occurs in three isotopic forms: ^{12}C, ^{13}C, and ^{14}C, where the superscript number is the sum of the protons (6) and neutrons (6, 7 and, 8) in the carbon nucleus. Radioisotopes by contrast are not stable and hence, undergo radioactive decay during which the parent element is transformed into a lighter daughter element of a lower atomic number. The time required for this decay may vary widely ranging from nanoseconds to thousands of years. For example, the half-life of ^{14}C is 5730 ± 40 years [85,86].

The feature that characterizes a chemical element, defines its chemical nature and makes, e.g., carbon behaves differently from sulfur, is the number of protons in its nucleus since this is matched by the number of electrons surrounding the nucleus. The number of electrons and their quantum mechanical status define the nature and number of bonds a chemical element can form. Isotopes of a given element contain the same number of protons (and electrons) and hence share the same chemical characteristics but they contain different numbers of neutrons and are therefore of different atomic mass [85].

A brief listing of the stable isotopes and their abundances for the elements most commonly used in food research can be seen in Table 3.

2.2.2 Expression of Stable Isotope Abundances and δ-Notation

Isotope abundance is measured by instruments with dedicated channels per isotope (multicollector MS), expressing natural abundances of stable isotopes in a given material as the ratio of the minor (usually heavier) over the major

TABLE 3 Stable Isotopes and Their Abundances for the Elements Most Commonly Used in Food Studies

Chemical Element	Major Abundant Isotope (at %)	First Minor Abundant Isotope (at %)	Second Minor Abundant Isotope (at %)
Hydrogen	^{1}H 99.985	^{2}H 0.015	
Carbon	^{12}C 98.89	^{13}C 1.11	
Nitrogen	^{14}N 99.63	^{15}N 0.37	
Oxygen	^{16}O 99.76	^{18}O 0.20	^{17}O 0.04
Sulfur	^{32}S 95.02	^{34}S 4.22	^{33}S 0.76
Strontium	^{88}Sr 82.56	^{86}Sr 9.86	^{87}Sr 7.02
Lead	^{208}Pb 52.4	^{206}Pb 24.1	^{207}Pb 22.1

abundant (usually lighter) isotope of a given element (e.g., $^{2}H/^{1}H$ or $^{13}C/^{12}C$) has been adopted as the preferred notation. Since changes in isotope composition of light elements at natural abundance level are relatively minute and isotope abundances of the minor isotopes are very small, significant changes when given as percent happen typically in the second decimal place. Similarly, changes in the ratio of heavier–lighter isotope are usually confined to the numerals in the third and fourth decimal places. Owing to the minute nature of these changes in isotope abundance at natural abundance level, measured isotope ratios of a given material are expressed relative to a contemporaneously measured isotope ratio of a standard of known isotopic composition. To make the resulting figures more manageable, the "delta notation" (δ) was adopted. The abundance of stable isotopes is typically presented in delta notation (δ), in which the stable isotope abundance is expressed relative to a standard (Eqn (1)):

$$\delta = (\text{Rsample}/\text{Rstandard} - 1) \times 1000‰ \quad (1)$$

where R is the molar ratio of the heavy to light isotopes, e.g.,:

$$R = {}^{13}C/{}^{12}C \text{ or } {}^{2}H/{}^{1}H \text{ or } {}^{18}O/{}^{16}O \quad (2)$$

and Rstandard is the measured isotope ratio for the corresponding international reference material (RM) (e.g., for ^{13}C it is the Vienna Pee Dee Belemnite, VPDB). For the most commonly used light isotopes, the internationally recognized standards are shown in Table 4.

The results of this equation are referred to as per mil values (‰). A positive δ value is interpreted to mean the sample has a higher abundance of the heavier isotope than the international reference standard that defines the scale for a particular isotope, while a negative δ value is interpreted to mean the sample has a lower abundance of the heavier isotope than the international reference standard [85].

2.2.3 Causes of Stable Isotopes Variations

Variations in the abundances of stable isotopes among different compounds arise because the chemical bonding is stronger in molecules containing heavier isotopic forms, making it more difficult to break up the molecule in a chemical reaction (often termed *kinetic fractionation*), or because of differences in the physical properties of molecules containing heavier isotopic forms (often termed *diffusive and equilibrium fractionation*). With kinetic fractionation, the rate of an enzymatic reaction is faster with substrates that contain the lighter isotopic form than in reactions involving the heavier isotopic form. As a consequence, there will be differences in the abundances of the stable isotopes between substrate and product. Such differences will occur unless, of course, all of the substrate were consumed, in which case there would be no difference in the isotopic composition of substrate and product. Expression of a significant kinetic fractionation in most biological reactions involves substrates at branch points in metabolism, such as the initial fixation of carbon dioxide in photosynthesis. Equilibrium fractionation events reflect the observation that during equilibrium reactions, such as the equilibration of liquid and gaseous water, molecules with the heavier isotopic species are typically more abundant in the lower energy state phase. Diffusive fractionation events reflect the observation that heavier isotopic forms diffuse more slowly than lighter isotopic forms [86].

2.2.4 Isotopic Variations in Nature

The natural variations in isotopic abundance can be large. Some atmospheric gases, such as CO_2, N_2, and O_2, exhibit limited variation, while N_2O and CH_4 exhibit wide isotopic variation. The larger isotopic ranges in the latter two gases reflect both significant isotopic fractionation by microbes as well as different biological substrates which are used to produce these gases. Oceans, the largest volume of water on earth, exhibit only small variations in isotopic abundance and most of this is associated with changes in salinity; for this reason ocean water is used as a standard (Table 4). However, once water evaporates from the oceans and then

TABLE 4 Internationally Recognized Standards for the Most Commonly Used Light Isotopes

H	Standard mean ocean water (SMOW)	R = 0.0001558
C	Vienna pee dee Belemnite (VPDB)	R = 0.0112372
N	Atmospheric air (air)	R = 0.0036765
O	Standard mean ocean water (SMOW)	R = 0.0020052
S	Canyon diablo triolite (CDT)	R = 0.0450045

recondenses as precipitation, there are large isotopic variations that are dependent on both cloud temperature and the amount of residual moisture remaining within the cloud mass. Lakes and rivers reflect precipitation input values, but are often further enriched by evaporative processes, which favor the movement of lighter isotopic forms of water into the vapor phase. There can be significant biological fractionation against carbon dioxide during photosynthesis, which results in plants being isotopically depleted relative to the carbon dioxide substrate.

There is very limited isotopic variation in diatomic nitrogen. The microbial process of nitrogen fixation ($N_2 \rightarrow NH_4^+$) exhibits little isotopic fractionation and therefore these products have a similar nitrogen isotope ratio as the atmosphere. Yet subsequent nitrogen transformation reactions exhibit strong isotopic fractionations. In general, the isotopic composition of animals within a trophic level is enriched by about 3‰ relative to their food substrate. The wide variations in nitrogen dioxide reflect the large differences associated with aerobic versus anaerobic processes and industrial versus stratospheric processes. In addition, climate and soil conditions can influence $^{15}N/^{14}N$ in soil, and therefore the type of plants growing on it. Nitrogen-fixing plants show lower $\delta^{15}N$ than nonnitrogen-fixing plants. The $^{34}S/^{32}S$ ratio is more difficult to interpret due to the numerous factors influencing its nature of soil, industrial emissions, and sulfur-containing fertilizers.

The $^{87}Sr/^{86}Sr$ ratio depends only on the types of rocks and soils, and not on human activity, climate or season of production. ^{87}Sr is produced from ^{87}Rb by radioactive b-decay, whereas the abundance of primordial ^{86}Sr remains virtually constant in a given rock. Old acidic rocks such as granite show the highest ratios, mafic, and carbonate-rich rocks the lowest [87–90].

2.2.5 Isotopic Variations in Plants and Animals

Animals are a very complicated substrate because the isotopic abundances of their tissues and products are the summation of feeds ingested throughout all their life, plus the kinetic fractionations occurring in animal metabolism. The $^{13}C/^{12}C$ ratio for both milk fat and cheese protein gives information on the type of forage fed to the cows. This is because the $^{13}C/^{12}C$ ratio depends almost exclusively on the photosynthetic mechanism used by the plants for CO_2 fixation: the Hatch and Slack cycle or C4 (monocotyledonous plants such as sugar cane, corn, tropical grasses, desert plants, and marine plants) with $\delta^{13}C$ mean values of −12/−14‰, and the Calvin cycle or C3 (dicotyledons plants such as flowering plants, wheat, rice, rye and cotton) with $\delta^{13}C$ mean values of −26/−28‰. Also, crassulacean acid metabolism (CAM) plants, such as pineapple, cactus, and orchids, can utilize either the C3 or C4 metabolic systems, depending on sunlight, and therefore have $\delta^{13}C$ mean values of −10/−34‰.

Therefore the amount of C4 plants (mainly maize) in animals' diet can influence the $^{13}C/^{12}C$ ratio of products of animal origin. Differences in the $^{15}N/^{14}N$ ratio also result essentially from forage. Organic fertilizers and intensive farming methods increase the level of ^{15}N in the soil and consequently in the plants, milk, and cheese.

The $^{18}O/^{16}O$ ratio of milk depends on the water ingested and the proportion of fresh and dry fodder. Isotope ratios of precipitation and groundwater depend largely on temperature, latitude, altitude, and distance from the sea. In grass proper enrichment in ^{18}O occurs due to fractional evaporation of water. Therefore, during summer, when cows feed almost exclusively on fresh grass, a higher $^{18}O/^{16}O$ ratio is observed. Similarly, D/H gives the same climate and weather information as does $\delta^{18}O$. In summary, in animal tissues, the differences in $^{2}H/^{1}H$ and $^{18}O/^{16}O$ reflect climatic differences while $^{13}C/^{12}C$ reflect mainly differences in food resources [89,91].

2.2.6 Instrumentation

Isotope ratio mass spectrometers (IRMS) are specialized mass spectrometers that produce precise and accurate measurements of variations in the natural isotopic abundance of light stable isotopes. IRMS instruments are different from conventional organic mass spectrometers, in that they do not scan a mass range for characteristic fragment ions in order to provide structural information on the sample being analyzed. The breakthrough in classical isotope ratio mass spectrometry was the introduction of the dual inlet mass spectrometer by Urey in 1948. Since then, the instrument has been further developed and automated, leading to the systems currently available commercially.

The isotopes of the following elements are routinely measured by IRMS: carbon (isotopes: ^{13}C and ^{12}C, not ^{14}C), oxygen (isotopes: ^{16}O, ^{17}O, and ^{18}O), hydrogen (isotopes: ^{1}H, ^{2}H, but not ^{3}H), nitrogen (isotopes: ^{14}N and ^{15}N) and sulfur (isotopes: ^{32}S, ^{33}S, ^{34}S, and ^{36}S). Chlorine (Cl), Silicon (Si), and Selenium (Se) isotopes are less frequently analyzed [95].

A stable isotope ratio mass spectrometer (IRMS) consists of an inlet system, an electron ionization source, a magnetic sector analyzer for ion separation, a Faraday-collector detector array for ion registration and a computer-controlled data acquisition system. Figure 11 demonstrates how each of these sample introduction systems can be coupled to the same mass spectrometer.

2.2.7 Inlet System

The inlet system is designed to handle pure gases, principally CO_2, N_2, H_2, and SO_2 but also others such as O_2, N_2O, CO, CH_3Cl, SF_6, CF_4, and SiF_4. Neutral molecules from the inlet system are introduced into the ion source, where they are ionized via electron impact and accelerated to several kilovolts, and then separated by a magnetic field and detected by Faraday cups positioned along the image plane of the mass spectrometer [91,92].

Inlet systems for isotope ratio mass spectrometers are rather simple and clean devices consisting of valves, pipes, capillaries, connectors, and gauges. Homemade-inlet systems are often made of glass, but commercially available inlet systems are mostly designed from stainless steel components that have no cavities. All components and surfaces are carefully selected for maximum inertness toward the gases to be analyzed. The materials used as components of the

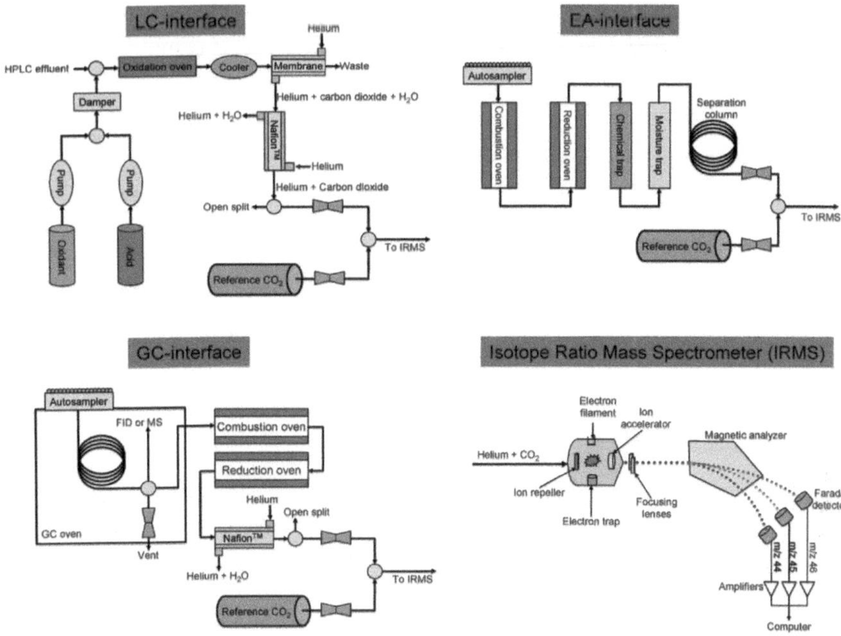

FIGURE 11 Schematic illustration of the three most common sample introduction systems/interfaces for carbon isotope measurements (as CO_2) and an isotope ratio mass spectrometer (IRMS). LC: liquid chromatography, EA: elemental analyzer, GC: gas chromatography [91].

valves deserve special attention. The highest quality valves are of "all-metal" design, with all wetted surfaces made either from stainless steel (the body and membranes) or from gold (the gaskets or seals and the valve seat) [92].

Common inlet systems for IRMS analysis are the Dual Inlet IRMS (DI-IRMS) and the Continuous Flow IRMS (CF-IRMS). Both of these techniques require solid, liquid, and gaseous samples to be converted into pure gases.

2.2.7.1 Dual-Inlet IRMS (DI-IRMS)

With a dual-inlet system, the samples for analysis are prepared (i.e., converted into simple gases) offline. The off-line sample preparation procedure utilizes a specially designed apparatus involving vacuum lines, compression pumps, concentrators, reaction furnaces, and microdistillation equipment. This technique is time-consuming, usually requires larger samples, and contamination and isotopic fractionation can occur at each of the steps. Once prepared, the pure gas is admitted into the IRMS via a variable volume gas reservoir, referred to as a bellow. The reference gas is also admitted into the mass spectrometer via a bellows system. The bellows system allows sample and reference comparisons to be made under generally identical circumstances. The bellows are connected to the ion source of the mass spectrometer through a capillary inlet line. A valve system, known as the changeover valve, between the capillaries and the MS, is used to alternate the capillary effluents between the source of the MS and a waste line. This ensures that a constant flow through the capillaries is maintained [93].

2.2.7.2 CF-IRMS

The CF-IRMS technique consists of a helium carrier gas that carries the analyte gas into the ion source of the IRMS. This technique is used to connect an IRMS to a range of automated sample preparation devices. Two of them are the Bulk Stable Isotope Analysis (BSIA) and the Compound Specific Isotope Analysis (CSIA).

While dual inlet is generally the most precise method for stable isotope ratio measurements, continuous flow mass spectrometry offers online sample preparation, smaller sample size, faster and simpler analysis, increased cost effectiveness, and the possibility of interfacing with other preparation techniques, including elemental analysis (EA), gas chromatography (GC), and liquid chromatography (LC) [93].

2.2.7.2.1 Bulk Stable Isotope Analysis (BSIA) BSIA is based on the use of an elemental analyzer, which is an automated sample preparation instrument for the conversion of the sample of interest into simple gases for IRMS analysis. The isotopic values obtained from these analyses represent the isotopic composition of all the components in the mixture as a whole. The instrument for BSIA consists of an elemental analyzer coupled with an isotope ratio mass spectrometer (EA-IRMS). CF-IRMS was originally used for the measurement of nitrogen isotopes, but has been extended to include carbon and sulfur, and more recently oxygen and hydrogen isotope ratios [93].

2.2.7.2.1.1 EA-IRMS. As previously mentioned, EA-IRMS is a bulk measurement technique, which provides representative data for the average isotopic signal of the entire sample. Without significant sample preparation, this method cannot elucidate how each constituent of the sample contributes to the total average value. In order to measure the average isotope ratios for nonvolatile liquids or solids, the bulk sample can simply be weighed and placed in a tin or silver capsule. The sample capsule is lowered into a combustion furnace through an autosampler carousel, at which time the sample is combusted at elevated temperatures under a flow of oxygen into NO_x, CO_2, SO_2, or H_2O. Depending on the isotopes of interest, the combustion products may need to be specifically treated to reduce interferences. In carbon isotope ratio analysis—by far the most common application—the combusted sample is carried by a helium gas stream into a reduction chamber where nitrous oxides are converted into N_2 and excess O_2 is removed. The analyte is next carried through a chemical trap to remove water that was produced from combustion, and then into the gas chromatograph where separation of CO_2 and N_2 is performed. Effluent from the elemental analyzer is then sent to the IRMS. Because the isotope ratios for questioned samples are reported relative to a reference gas standard, best results are obtained when the signal intensities for the two samples are of similar magnitude and are analyzed as closely together in time as possible.

2.2.7.2.2 Compound-Specific Isotope Analysis (CSIA) In compound-specific isotope analysis (CSIA), the isotopic compositions of individual compounds within the sample are measured.

2.2.7.2.2.1 GC-IRMS. By performing a separation prior to isotope ratio analysis, hyphenated techniques such as GC-IRMS and LC-IRMS can provide isotopic analysis of a complex mixture, thereby providing additional information and higher discriminatory power. IRMS instruments require a somewhat steady stream of a fixed gas (such as CO_2) for precise analysis. The sample first elutes from the GC column into an oxidation chamber, usually housed at the outlet of the GC oven. The oxidation chamber is normally a nonporous alumina tube that contains three separate twisted wires made of copper, nickel, and platinum. The samples are combusted at elevated temperatures into a combination of gases such as CO_2, NO_x, and H_2O. For $\delta^{13}C$ measurements, the combusted sample is then carried into a reduction chamber where nitrous oxides are converted into N_2 and any excess O_2 is removed. Since CO_2, NO_x, and H_2O will not condense at room temperature, the transfer line from the oxidation chamber to the reduction chamber does not need to be heated. The reduction chamber and subsequent valves, splitters, and pneumatic actuators, etc., are contained in a stand-alone interface system. To avoid H_2O from protonating CO_2 in the MS source—and causing deleterious isobaric interference of $^{12}CO_2H^+$ with the $^{13}CO_2{}^+$ peak at *m/z* 45—the analyte stream is passed through a semipermeable membrane such as Nafion. Here, a dry helium counterflow is used to remove the H_2O. The flow rate of the subsequent sample stream is carefully controlled to provide a stable flow rate to the IRMS ion source of approximately 0.5 mL/min. Deactivated fused silica capillaries are used throughout the interface systems to restrict the analyte flow to the required flow rates. The interface system also uses electronically controlled pneumatic actuators to toggle the flow of the effluent stream between that of the analyte and that of a reference gas, such as a cylinder of CO_2.

In conclusion the advantages of GC-IRMS are that gases can be purified easily and efficiently, analysis is extremely rapid, and sample size is reduced by several orders of magnitude. There is, however, a slight loss of precision relative to the traditional dual-inlet system, although it appears that modern GC-IRMS instruments are now mature for high precision analyses [91,94].

2.2.7.2.2.2 LC-IRMS. LC-IRMS applications are typically dedicated to carbon isotope ratio analyses. When the solution elutes from a liquid chromatograph (LC) column, the solution is directly injected into one of two interfaces. These two interfaces are (1) a moving wire interface, and (2) a wet-chemical oxidation interface. The wet-chemical oxidation method converts organic compounds present in the effluent of the LC column into CO_2 gas directly in the mobile phase. To reduce interferences, the HPLC mobile phase must be void of any organic or oxidizable components that could interfere with the results. It should be noted that because most LC separations are greatly enhanced with organic solvents or modifiers, this requirement poses significant restrictions on the potential application of LC-IRMS. The effluent from the LC column is then mixed with a stream of an oxidizing agent such as ammonium peroxodisulfate, and a catalyst such as phosphoric acid/ and silver nitrate. The mobile phase and the combined reagents pass through a capillary oxidation reactor where the organic compounds are converted into CO_2. A membrane exchanger separates CO_2 gases from the other gases (water vapor, oxygen, argon, etc.) that originate from the liquid phase. The CO_2 is then transferred through a gas-permeable membrane into a counterflow of helium. The CO_2 in the helium stream is then dried in an online gas drying semipermeable membrane and admitted to the isotope ratio mass spectrometer via an open split. The wet-chemical oxidation interface allows for the $^{13}C/^{12}C$ determination of organic compounds with a completely automated online high precision method [91,95].

2.2.8 The Ion Source: Electron Impact Ion Production

The major objectives in ion source system of IRMS instruments include:

- a linear relation between the ion current intensities and the measured ratios.
- no memory between subsequent introductions of sample and reference gases in the mass spectrometer. Memory sources include gas adsorption on welds, copper gaskets (SO_2), and polymer gaskets (CO_2, H_2).
- chemical inertness of the hot filament [92].

2.2.9 Separation and Detection of Ions in the Mass Spectrometer

2.2.9.1 Magnetic Sectors

High-precision isotope ratio mass spectrometers employ magnetic sectors for the separation of ions almost exclusively. Both permanent magnets and electromagnets are used in commercial instrumentation. The ions normally possess a kinetic energy between 2.5 and 10 keV, varying from instrument to instrument. Smaller instruments have lower acceleration potentials and smaller magnets. Larger instruments with higher accelerating voltage have higher sensitivities, higher resolutions and better peak shapes than instruments with lower accelerating voltages, all other things being equal. Ions entering a homogeneous magnetic field are deflected perpendicular to their flight direction and perpendicular to the magnetic field according to the Lorentzian rule. The result is a circular path with the radius:

$$r = 1/B \times \sqrt{(2mU/ze)}$$

with B being the magnetic field, e is the elementary charge ($=1.6 \times 10^{-19}$ C), z is the number of charges, U is the accelerating potential, and m is the mass of the ion. As an example, a singly charged ion of 44 amu (=atomic mass units or Daltons) that has been accelerated to 5 keV will describe a radius of 13.5 cm when traveling through a homogeneous magnetic field of 0.5 T. It should be noted that mass and translational energy are equivalent in the above equation. Any inhomogeneity of the kinetic energy of the ion beam will lead to a broadening of the ion image at the detector. This is especially important in light of the discussion above about the necessity to suppress ion–molecule reactions by applying a high draw out field to the ionization chamber. The effect is more severe for smaller mass spectrometers since the relative energy spread ΔE/U is larger for lower acceleration potentials U. For high-precision molecular mass determination in organic mass spectrometry, double focusing arrangements that reverse the dispersion due to the energy spread are commonly used.

A magnetic sector can be treated like an optical prism in geometrical optics: Monoenergetic ions of different mass are dispersed through the magnetic field. Light ions follow a path with a small radius whereas heavier ions describe circular paths with larger radii. Ions with identical mass coming from a focal point with a certain lateral spread α will be focused again after exiting the magnet. The focal points of different masses lie on a focal plane. When entrance and exit drift lengths are identical, the image at the detector plane will be the same size as the original beam at the entrance slit.

The choice between a permanent magnet and an electromagnet needs to be discussed in terms of the limitations that a particular choice imposes on the analytical performance. From a theoretical (i.e., ion optical) point of view, both types of magnets are identical. However, permanent magnets often do not reach the same homogeneity as electromagnets. In practice, the difference between the two types of magnets shows up in the way a particular mass is selected. The field of an electromagnet may be changed by applying a variable current to the coils thus allowing a scan of the complete mass spectrum. Because the field of a permanent magnet cannot be changed, mass selection is made by altering the accelerating voltage.

2.2.9.2 Multiple Faraday Cup Detectors

The use of multiple detectors to simultaneously monitor and integrate the ion currents of interest was introduced by A.O. Nier in 1947. Two ion currents, e.g., masses 44 and 45 from CO_2, hit a collector plate mounted behind a grounded slit and a pair of secondary electron suppressor shields. The advantage of the simultaneous measurement with two separate amplifiers is that fluctuations of the ion currents due to temperature changes, electron beam instability etc., cancel completely. The magnet and the accelerating voltage remain constant. No peak jumping is required, which eliminates the corresponding settling times. Moreover, each detector channel can be fitted with a high ohmic resistor appropriate for the mean natural abundance of the isotope ion current of interest. This principle of static multicollection is still in use but the collector plate has been replaced by deep Faraday cups, in order to minimize false detector currents generated from secondary electrons [92].

The Faraday cups (FC) are positioned so that the major ion currents simultaneously strike the middle of the entrance slit of the respective cups. Each incoming ion contributes one charge. No stray ions or electrons can enter the cup, and no secondary particles formed from the impact with the inner walls of the cups exit the cup. The ion currents are continuously monitored, then amplified, and finally transferred to a computer. The computer integrates the peak area for each isotopomer and calculates the corresponding ratios. For example, when analyzing CO_2, the data consist of three ion traces for the different isotopomers: $^{12}C^{16}O_2$, $^{13}C^{16}O_2$, and $^{12}C^{18}O^{16}O$, with their corresponding masses at *m/z* 44, 45, and 46 [93].

2.2.10 Isobaric Interferences

Isobaric interferences are ion currents in Faraday cups that belong to ionic species other than those being examined. The most prominent examples are the aforementioned correction for the ^{17}O moiety at mass 45 of CO_2 and the correction for the H_3^+ contribution to the mass 3 ion current. Other isobaric interferences and contributions include:

- ^{17}O and ^{13}C on mass 46 of CO_2
- CO^+ interference on mass 28 when measuring nitrogen
- N_2 interference on mass 28 when measuring ^{18}O on line on the CO masses 30 and 28
- N_2O interference on the masses 44, 45 and 46, and
- ^{18}O and ^{17}O contributions on masses 66 and 65 of SO_2.

2.2.11 Conclusions

Variations of stable isotope ratios in nature are mostly small. They are, however, important tracers that can reveal a wealth of information about processes that are happening or that have happened in the past. In order to read this information, the underlying principles need to be understood, and, most importantly, the measurements need to be made with the appropriate high precision. This has become possible with the development of the special instrumentation now common in the laboratories that specialize on isotope ratio analysis. All instrument designs have their merits and pitfalls which must be weighed in order to produce a consistent set of isotope ratio values. The future in isotope ratio mass spectrometry clearly belongs to the chromatographic techniques [92].

2.3 TIMS

Another isotope technique which is used for food analysis is Thermal Ionization Mass Spectrometry (TIMS). Thermal ionization mass spectrometry began soon after the discovery of the atom and understanding of the behavior of charged particles in electric and magnetic fields. TIMS enables measurement of isotope ratios with the highest precision, accuracy, and sensitivity.

2.3.1 Instrumentation

Modern instruments are composed of three primary components: (1) ion source, the region in which ions are produced, accelerated, and focused; (2) analyzer, the region in which the beam is separated based on mass/charge ratios; and (3) collector, a region in which the ion beams are measured either sequentially (single collector) or simultaneously (multicollector). The electronics of these instruments must operate to very close tolerances in order to produce isotope ratios that are precise to 0.01–0.001%.

TIMS is based on a magnetic sector mass spectrometer (Figure 12) that is capable of making very precise measurements of isotope ratios of elements that can be ionized thermally, usually by passing a current through a

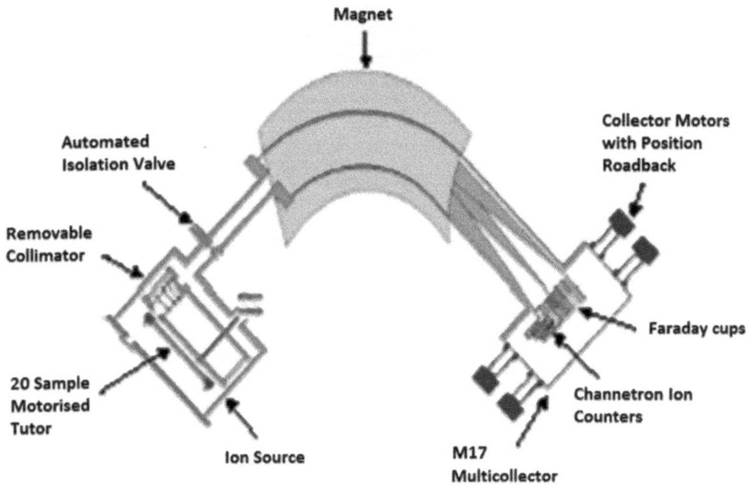

FIGURE 12 Schematic illustration of a commercial TIMS instrument.

thin metal ribbon or ribbons under vacuum. Samples are deposited on specially treated filaments (usually rhenium or tantalum), then carefully dried. A filament is put into the mass spectrometer under vacuum and a potential of 8–10 kV is applied to the filament. A current is passed through the filament causing the thin ribbon to heat, and as the temperature of the filament surpasses the vaporization temperature of the element/salt, neutral species and ions of the element are emitted from the hot surface of the filament. For single filament loads, the filament acts as both the vaporization and ionization filament. In a double or triple filament source, one filament acts as an ionization filament and the other filaments are loaded with sample and act as vaporization filaments. The neutral metal atoms emitted by the ionization filament are absorbed on the surface of the vaporization filament where ionization of the analyte takes place. The separation of the ionization and vaporization filaments allows for a finer control of evaporation rate as well as allowing the ionization filament to be at an elevated temperature, which typically results in a higher ionization ratio for most elements. For example, if the element to analyze is volatile or if its ionization energy is too large, the analyte forms a cloud of neutral atoms above the filament ribbon.

The ions created on the ribbon(s) are accelerated across an electrical potential gradient and focused into a beam via a series of slits and electrostatically charged plates. This ion beam then passes through a magnetic field and the original ion beam is dispersed into separate beams on the basis of their *m/z*. These mass-resolved beams are then directed into collectors where the ion beam is converted into voltage. The ratio of 1 F cup to another is used to calculate isotope ratios. Comparison of voltages corresponding to individual ion beams yield precise isotope ratios. Usually TIMS is used between else for the following elements Pb, Sr, U, Fe, Cu, Mg, Os, B, and Li isotope analysis [96–98].

In addition to the need to develop new methods that make sample preparation more rapid, and thus increase sample throughput, the potential of TIMS lies in further improvement of the precision attainable in isotope analysis, especially for inorganic nutrients. In TIMS, the precision is primarily limited by mass-dependent fractionation effects that occur during sample evaporation in the ion source. The lighter isotopes are commonly evaporated preferentially, resulting in a continuous change in the isotope ratios within and between measurements, with a consequent direct impact on precision. This effect depends not on absolute mass differences, but on relative mass differences of the evaporating species. This makes isotope analysis of some inorganic nutrients, for example Ca, Mg, or Fe, a particular challenge [97].

2.3.2 Strengths

The advantage of TIMS compared to other isotope ratio techniques include:

- the chemical and physical stability of the measurement environment, which lead to highly precise measurements,
- lower and more consistent average mass fractionation,
- relatively low production of molecular backgrounds and multiply charged ions,
- the use of single element solutions to eliminate isobaric interferences,
- production of ions with a restricted range of energies, eliminates need for energy filter,
- easily automated operation, and
- near 100% transmission of ions from source to collector.

2.3.3 Limitations

The disadvantages include:

- not all elements are easily ionized, which restricts applications to elements with low ionization potentials,
- ionization is not equally efficient for all elements, and is generally less than 1%,
- mass fractionation continually changes during analysis,
- extensive sample preparation, and
- accurate mass fractionation correction is limited to elements with three or more isotopes of which at least two are stable [96,97,99].

2.3.4 Comparison with MC-ICP-MS

Instrumental developments made MC-ICP-MS competitive to TIMS as in isotope analysis field. MC-ICP-MS allows isotope ratio measurements for various elements at precisions below 50 ppm, which was previously only possible by using TIMS. At the same time, isotopic analysis is less time-consuming. For TIMS analysis, the sample has to be separated from the matrix, presented in a solid state to the mass spectrometer and heated carefully to obtain a stable ion beam. On the other hand, by using a plasma ion source, the sample can be presented directly in solution or as an aerosol to the instrument, which can make sample digestion superfluous (e.g., when using a laser beam for direct sampling). In addition, ionization efficiencies are usually higher when using ICP-MS which makes it even more sensitive. Considering the obvious advantages of MC-ICP-MS over TIMS it seems as if TIMS has become probably superfluous.

However TIMS is less susceptible to isotopic fractionation than MC-ICP-MS. Fractionation effects in TIMS are usually in the order of parts per thousands, relatively stable and systematic, while isotopic fractionation in MC-ICP-MS, commonly referred to as mass bias, can be easily in the percent range and depend on instrument settings and time. Furthermore, chemical separation of the element from the matrix and the need for various quality control tests make measurements by MC-ICP-MS nearly as time-consuming as by TIMS. Thus, other factors become decisive such as robustness, running costs, and element-specific advantages [100].

2.4 MC-ICP-MS

Traditionally, thermal ionization mass spectrometry (TIMS) has been the technique of choice for achieving the highest accuracy and precision in isotope analysis. However, quite recent developments in multicollector inductively coupled plasma mass spectrometry (MC-ICP-MS) have brought a new dimension to this field, because it permits the precise measurement of isotope compositions for a wide range of elements. In addition to its simple and robust sample introduction, high sample throughput, and high mass resolution, the flat-topped peaks generated by this technique provide for accurate and precise determination of isotope ratios with precision reaching 0.001%, comparable to that achieved with TIMS. These features, in combination with the ability of the ICP source to ionize nearly all elements in the periodic table, have resulted in an increased use of MC-ICP-MS for such measurements in various sample matrices. To determine accurate and precise isotope ratios with MC-ICP-MS, utmost care must be exercised during sample preparation, optimization of the instrument, and mass bias corrections [101–103].

2.4.1 Instrumentation

An MC-ICP-MS Figure 13 is a hybrid mass spectrometer that combines the ionization efficiency of the ICP ion source with a magnetic sector mass spectrometer equipped with multiple Faraday cups, for the measurement of ions. The rationale behind this approach was to build an instrument that would readily permit the precise determination of the isotopic composition of elements with high ionization potentials that are difficult to analyze by thermal ionization mass spectrometry (TIMS). Also the ICP source allows flexibility in how

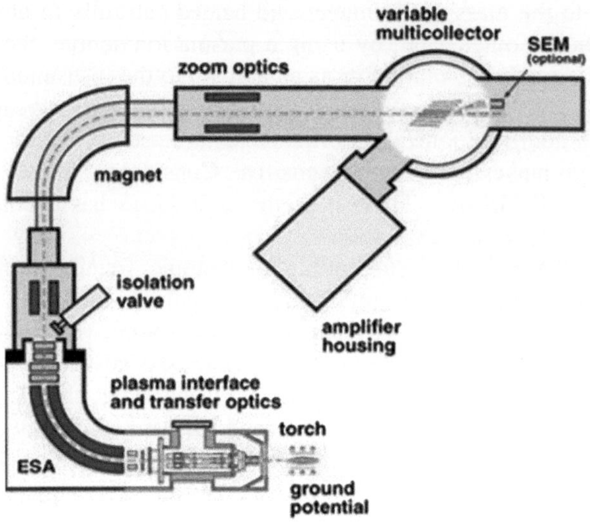

FIGURE 13 Schematic presentation of an MC-ICP-MS. *Courtesy of Thermo.*

samples are introduced to the mass spectrometer and allows the analysis of samples introduced either as an aspirated solution or as an aerosol produced by laser ablation.

The application of quadrupole ICP-MS instruments to isotope ratio measurements is limited by the comparatively poor precision of the analytical data. Even under ideal conditions, the precision of isotope ratio measurements is no better than ~2‰ and for many applications is less than 5‰. Similar or somewhat better results were achieved by ICP-TOF-MS and, in particular, with single-collector double focusing SF-ICP-MS instruments. Nonetheless, these techniques are still unable to rival TIMS in providing the most precise and accurate isotope ratio measurement capabilities for demanding applications. The precision of isotope ratio measurements obtained with such "conventional" ICP-MS instruments is limited by the single-collector ion detection systems and/or the poor peak shapes produced by the mass analyzer. Elements with high ionization potentials, however, are difficult to analyze by TIMS at high precision. Furthermore, the application of TIMS necessitates careful attention to sample preparation and is associated with low sample throughput.

MC-ICP-MS instruments were designed specifically to overcome the limitations of other mass spectrometric techniques. To achieve this goal, they combine the Ar ICP-source of conventional ICP-MS instruments with the magnetic sector analyzer and multiple-Faraday cup array of TIMS. At the high temperatures that are attained in the Ar plasma (~6000–8000 K), elements with first ionization potentials of <10 eV are ionized to an extent of 75% or more. Thus, with a plasma source, virtually all elements of the periodic table are accessible to isotopic analysis.

The magnetic sector mass analyzers of MC-ICP-MS instruments are similar to those used in TIMS, with the ultimate aim of achieving the flat-topped peaks necessary for high-precision isotope ratio measurements. To this end, all present instruments operate at a low mass resolution of approximately 400, but at minimal loss of transmission, by the use of wide source and collector slits. At such a mass resolution, however, ions of the same nominal *m/z* are not separated from one another by the mass analyzer (e.g., the separation of $^{56}Fe^{+}$ from $^{40}Ar^{16}O^{+}$ would require a resolution of about 2500).

The use of a detector array with multiple Faraday cups permits the simultaneous collection of the separated isotopes and this cancels out the effect of a "noisy" signal on the isotope ratio measurement. This technique of static multiple collection is particularly critical in ICP source mass spectrometry, because the ion beam that is produced in the plasma is significantly more unstable than the ion beam of TIMS, mainly due to short-term intensity fluctuation ("plasma flicker"). The Faraday cups are independently adjustable, such that the collector "coincidences" can be adjusted to permit isotopic analyses of a wide range of elements having isotopes that display different mass dispersions. The multiple collector can be configured with as many as nine Faraday cups and eight miniature ion counters. Multiple ion collectors—as a combination of Faraday cups and ion counters—are being increasingly applied in mass spectrometry where precise and accurate isotope ratio measurements of very low-abundance isotopes. In addition, the dynamic range for isotope ratio measurement can be significantly increased. Measured isotope ratios must be properly corrected for all instrumental biases, including mass fractionation. Once corrected, these ratios are suitable for plotting in any diagrams requiring atomic ratios [60,103].

In contrast to the single-collector instruments, with which the analyte signals are monitored sequentially, MC-ICP-MS instruments are able to monitor the intensity of several ion beams simultaneously, as a result of which short-term variations in signal intensity are affecting all isotopes to the same extent, such that these cause, practically, no detrimental effect on the isotope ratio precision. In this way, with the most recent MC instruments, isotope ratio precisions down to 0.002% RSD can be obtained, which makes MC-ICP-MS a very strong competitor for TIMS studies [102].

Multicollector inductively coupled plasma mass spectrometry (MC-ICP-MS) instruments are very complex, and proper optimization of the instrument prior to intensity measurements is important to ensure accuracy and precision. Optimization of the instrument should be aimed at achieving the following important conditions: (1) the peaks for all isotopes should be as flat as possible, (2) the most stable and highest sensitivity for the analyte and lowest blank, (3) use of Faraday detectors for all isotopes, if possible, (4) daily calibration of efficiencies or gain factors for Faraday cups, (5) although dynamic runs can provide accurate and precise data, static runs are preferred for ultimate results, (6) proper optimization of abundance sensitivity (the contribution of a neighboring peak to the intensity of a particular analyte isotope peak) for applications in which high abundance sensitivity is needed, for example, for the U–Th system, (7) optimized measurement time for each run so as to achieve the most precise results, and (8) use of an appropriate wash solution and wash time between samples to reduce any possible memory effects [101].

Of course, several "confounding factors" have to be taken into account when isotopic data with the highest precision and accuracy as possible are aimed at. First of all, it is important to find a suitable sample preparation method, most often including a separation procedure to isolate the element of interest from its concomitant matrix. This in order to avoid both spectral and nonspectral interferences, while during the measurement itself, the problems of possible spectral interference and the occurrence of mass discrimination effects, should definitely be tackled. The problem of spectral interferences has been the subject of continuous research, and various approaches have been developed to avoid or reduce the interferences. The most common strategies to do this with state-of-art single-collector instruments are the use of a higher mass resolution (sector-field instruments) or the use of chemical resolution in a collision/reaction cell (quadrupole-based instruments). Also in the field of multicollector ICP-MS, these strategies can be applied, so that nowadays, it is possible to perform isotopic analysis with an excellent isotope ratio precision under interference-free conditions by means of ICP-MS [102].

2.4.2 Fundamentals of MC-ICP-MS

The ions are produced by introducing the sample into an ICP which strips off electrons thereby creating positively charged ions. These ions are accelerated across an electrical potential gradient (up to 10 KV) and focused into a beam via a series of slits and electrostatically charged plates. This ion beam then passes through an energy filter, which results in a consistent energy spectrum in the ion beam and then through a magnetic field where the ions are separated on the basis of their *m/z*. These mass-resolved beams are then directed into collectors where the ions reaching the collectors are converted into voltage. Isotope ratios are calculated by comparing voltages from the different collectors.

2.4.3 Strengths

The advantages of MC-ICP-MS compared to other isotope ratio techniques include:

- the ionization efficiency is very high (near 100%) for most elements which enables analysis of most of the elements of the periodic table, including those with high ionization potential that are difficult to analyze by TIMS
- the MC-ICP-MS operates essentially as a steady state system during the analysis resulting in time invariant mass fractionation
- there is consistent mass bias variation across the mass range which allows the use of an adjacent element to calculate mass bias for those elements without more than two stable isotopes

- the MC-ICP-MS permits flexibility in sample introduction systems. Solutions can be introduced at atmospheric pressure, which allow ease in handling. Laser ablation systems can also be coupled with the MC-ICP-MS, which allows in situ isotopic measurements in solid materials.

2.4.4 Limitations

The disadvantages of MC-ICP-MS include:

- essentially all elements introduced into the plasma are ionized including doubly charged species, oxides, and argides; in order to achieve the highest precision and accuracy samples, need to be chemically purified.
- plasma instability can limit precision
- even though ionization efficiency of the plasma is near 100%, transmission of ions is lower than with TIMS because the plasma-generated ion must be transferred from atmospheric pressure to the high vacuum of the mass spectrometer. Many ions are lost during this difficult transfer.
- although similar elements can be used to determine mass bias corrections, in systems with only two isotopes, e.g., Yb for Lu; Tl for Pb, the mass bias response between the two elements is not identical and must be accounted for.

2.4.5 Isotope Dilution

Isotope dilution (ID) analysis is one of the most powerful and accurate analytical methods for determining quantitative data, concentrations of elements or element species, in the sample investigated. Isotope dilution techniques were developed more than 50 years ago and applied for major and trace element determination in TIMS. Today ID analysis can be used with every kind of mass spectrometer in different analytical fields where reliable results are required for the determination of element concentrations in bulk, on the surface or in speciation analysis. ID analysis is regarded as a definitive technique because the precision and accuracy obtainable are unsurpassed by alternative analytical methods. Isotope dilution analysis has the advantage that it can overcome problems associated with instrumental drift and matrix effects during mass spectrometric detection. Furthermore, isotope dilution analysis relies on the measurement of isotope ratios and not external calibration, so measurements meet the highest metrological standards [60,104].

The ID procedure involves the alteration of the natural isotopic abundance of an analyte in a sample by spiking with a standard of modified isotopic composition. A prerequisite for ID is that the target analyte should have more than one stable isotope, so, because this is the case for the majority of the elements, most of them can be investigated using this method. Two stable isotopes of the target analyte are chosen, which should ideally have a large difference in natural abundance. For best practice, the isotopically enriched analogue, the spike, should have the isotope of lowest natural abundance enriched to as high an abundance as possible, the spike isotope, with the lower abundance isotope heavily depleted, the reference isotope. The isotope amount ratio in the sample is measured, after spiking and equilibration, and entered into the isotope dilution equation, showed below, along with other parameters. The concentration of the analyte in the sample can then be calculated.

$$C_X = (C_S W_S M_X / W_X M_S) \times ((A_S - RB_S) / (RB_X - A_X))$$

C_X: concentration of the analyte in the sample
C_S: concentration of the analyte in the spike solution
W_S: mass of spike
W_X: mass of sample
M_X: molar mass of element in the sample
M_S: molar mass of element in the spike
A_S: abundance of reference isotope in the spike
B_S: abundance of spike isotope in the spike
A_X: abundance of reference isotope in the sample
B_X: abundance of spike isotope in the sample
R: reference and spike isotope amount ratio in the sample after spiking

In order to achieve the best accuracy and precision, a number of factors must be taken into account. The critical stages are following:

- Sample preparation
- Selection of the most appropriate isotopic internal standard
- Characterization of the isotopically enriched analogue
- Addition of the isotopically enriched analogue
- Blank correction
- Instrumental analysis

with the important sources of error given as:

- Less than complete equilibration between the sample and spike will lead to significant systematic errors;
- Isobaric and polyatomic ion interferences;
- Isotopic discrimination e.g., isotopic fractionation, detector deadtime, and mass bias during instrumental analysis [104];

The great accuracy and precision arising from the use of ID-ICP-MS, motivate an increasing number of ICP-MS users to implement it for the determination of trace elements down to μg/L [66].

The advantage of ID in comparison to other quantification strategies, such as external calibration or standard addition, is that analyte recovery does not have to be quantitative, provided that isotopic equilibrium has been achieved between all of the analyte and the added spike material. High accuracy analysis by ID is now well established [60].

3. APPLICATIONS

Nowadays ICP-MS and IRMS are the major techniques, concerning elemental and isotope analysis of foods. The instrumental progress and the technological improvement render ICP-MS and IRMS almost indispensable. They are popular not only in research but also routine laboratories. As we can see in Figure 14, publications of ICP-MS and IRMS are continuously increasing. There has been a rapid increase of ICP-MS publications, in the last 2 years to 283 from 179 in 2010–2011. On the other hand TIMS publications remain constant over the years.

Figure 15 presents the percentage of publications per technique. ICP-MS occupies the largest percentage of these publications: 1207 for ICP-MS, 321 for IRMS, and 37 for TIMS due to the wide field of studies which can be carried

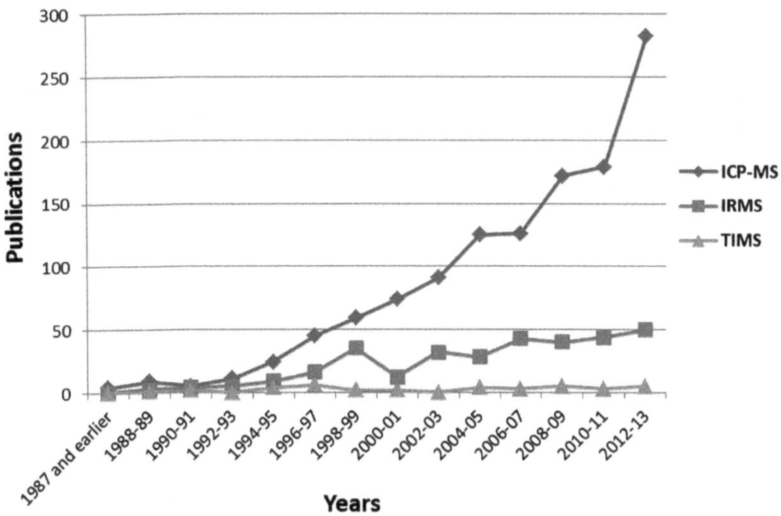

FIGURE 14 Time evolution of ICP-MS, IRMS, and TIMS publications concerning food, waters, wines, and beverages. *Data accessed from scopus, July 2014.*

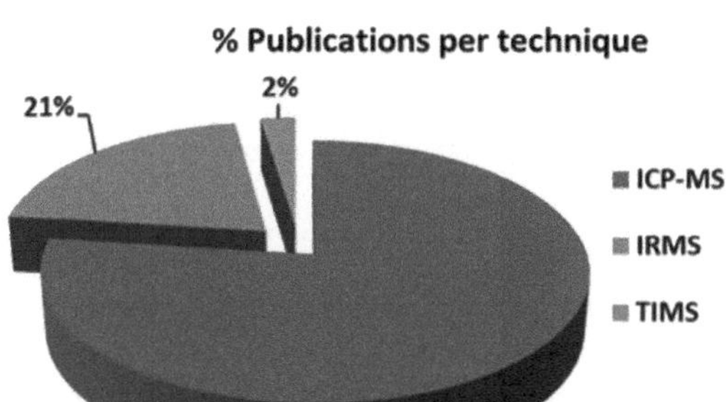

FIGURE 15 Percentage of publications per technique concerning food, waters, wines, and beverages.

out by this technique. These concern elemental analysis, speciation analysis by hyphenated ICP-MS, and isotope determinations by MC-ICP-MS. It should be noted that one-fifth of articles on ICP-MS, 256 articles, concern LC-ICP-MS for speciation studies, just 30 articles are with GC-ICP-MS, and only 24 articles with MC-ICP-MS, mainly for food authentication studies.

Combination of two or more techniques increases discrimination between samples providing good prospects for food authentication.

3.1 ICP-MS

3.1.1 Trends

Major use of ICP-MS in food is for determination of toxic and nutrient elements and food authentication (Figure 16). Major part of the work, ~1000 publications, concerns toxic and nutrient elements. Of these, ~460 articles are about toxic elements such as heavy metals (e.g., Cd, Hg, Pb, Sn, Sb), metalloids (e.g., As, Po), and radioactive elements (e.g., ^{226}Ra, ^{226}Th, ^{235}U). More than 200 are about nutrient elements such as Fe, Mg, Mn, Mo, Zn, and Co, while approximately 300 deal with both toxic and nutrient elements. Food authentication is also of major concern resulting in more than 130 articles. Emerging trend in application of nanomaterials in food is reflected in 30 publications while migration studies are also of increased interest.

Time evolution of publications per category is depicted in Figure 17 Studies concerning toxic and nutrient element content determination and food authentication are continuously increased especially from 1994 and onward. Food authentication output has doubled during the last 4 years period, from 30 to 62 publications. There is also a growth-concerning food nanomaterial determinations and migration studies during the last 4 years period.

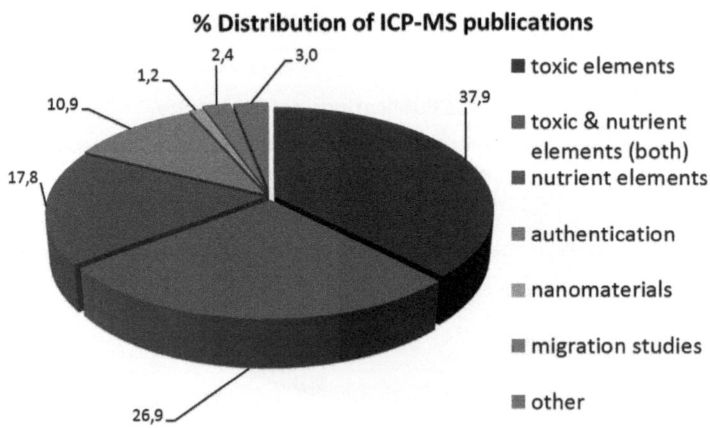

FIGURE 16 Percentage Distribution of ICP-MS publications according application field in food, waters, wines, and beverages.

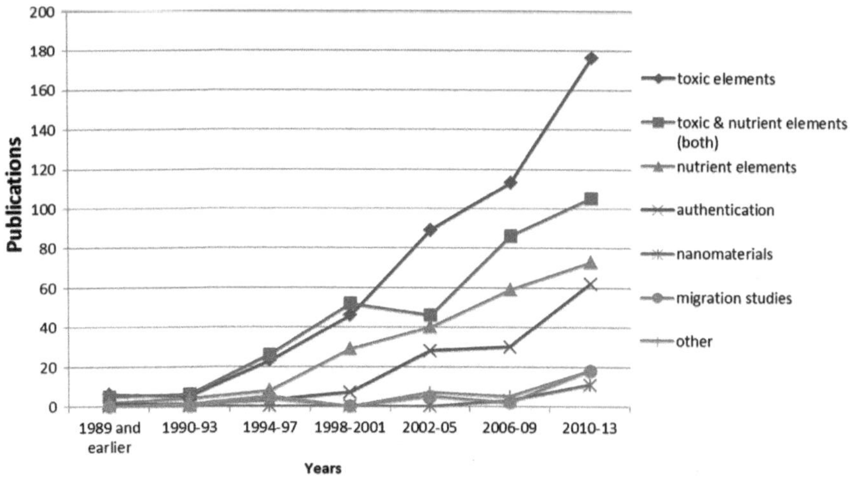

FIGURE 17 Time evolution of ICP-MS research.

3.1.2 Toxic and Nutritive Elements

Elemental food content is under thorough discussion as a quality parameter for their nutritional and toxicological relevance. The majority of ICP-MS applications concern the rather trivial field of toxic and nutritive elements (Figure 16) where the MS provides unsurpassed sensitivity and ability for rare earth elements. Regarding unsurpassed sensitivity, it is almost a requirement for determination of toxic and trace nutritive elements in foodstuff for infants and other vulnerable categories [105]. Due to the aforementioned the technique is dominant in the field displacing all others that are still used mostly in cases of limited budget for instruments. On detection and identification of essential elements, literature is focused on trace essential elements, such as iron, copper, and zinc, as well as ultratrace elements such as cobalt, manganese, and selenium where low detection limits by the MS are beneficial.

The intake of nutritive elements that play an important role in human biology could be inadequate due to low food content. On the other hand, toxic element content could render certain foods toxic, of special concern to infants and other vulnerable people. These cases the low detection limits provided by the MS could address the concerns. Food is the primary source of essential elements for humans and it is an important source of exposure to toxic elements. In this context, levels of essential and toxic elements must be determined routinely in consumed food products. Some specific examples are reported below:

Concentrations of arsenic, cadmium, mercury, and lead in common foods in children's, adolescents', adults', and seniors' daily intake were examined. Food samples were randomly acquired in seven cities of Catalonia, Spain. The dietary intake of these elements was determined by a total diet study. Calculations were carried out on the basis of recent data on the consumption of the selected food items. This work revealed that dietary intakes of As, Cd, Hg, and Pb by Catalonians are currently well below the respective provisional tolerable weekly intake (PTWI) [106]. In another article the composition of Icelandic dairy products (whole milk, fresh cheese (skyr), firm cheese (Gouda) skimmed milk, cream and whey) and meat (lamb meat and minced beef) regarding minerals and trace elements was investigated. The characteristics of their elemental seasonal variation were also studied [72]. Another interesting study examined the influence of organic selenium supplementation on the accumulation of toxic and essential trace elements involved in the antioxidant system of chicken. Chicks were used to investigate the effect of Se yeast supplementation on the accumulation of Cd, Cu, Fe, and Zn. The balance between Se, Cu, Fe, and Zn is important for proper antioxidant defense since they are an integral part of various antioxidant enzymes. Supplementation with Se-yeast, not only increased Se concentration but also reduced Cd concentration in the tissues. Selenium was negatively correlated with Cd and positively correlated with Zn, Cu, and Fe. Cadmium was negatively correlated with Zn and Cu while zinc correlated positively with Cu. Iron correlated negatively with Cu while it did not correlate with Zn and Cd [107].

Dietary exposure estimates of 18 elements from the 1st French Total Diet Study are shown in [108]. Samples were food typically consumed by the French population and were purchased, prepared, and cooked. A total of ~1000 individual food composites were collected and analyzed for arsenic, lead, cadmium, aluminum, mercury, antimony, chrome, calcium, manganese, magnesium, nickel, copper, zinc, lithium, sodium, molybdenum, cobalt, and selenium. Intakes were calculated from different food consumption patterns found in France for average and high consumers among adults and children. Dietary exposures of those consumers estimated and compared with existing nutritional reference values (Lowest Threshold Intake, LTI) or toxicological reference values (Provisional Tolerable Weekly Intake, PTWI or Upper Level, UL) of the respective element and from previous French studies. This study confirms for the populations concerned, the low probability of nutritional or health risks due to food consumption [108]. Additionally, the rapidly developing industry of food supplements requires a complete account of all the forms of the supplemented elements present as only some of them have beneficial value while others may even be harmful. So in another publication, 95 dietary supplement products were examined for arsenic, cadmium, mercury, and lead using microwave digestion and ICP-MS. The concentration (μg/kg) ranges were as follow: As, 5–3770; Cd, 10–368; Hg, 80–16,800; and Pb, 20–48,600. An assessment of estimated exposures/intakes for these elements is also presented [109].

Another article deals with essential and potentially toxic trace elements in honey. Dissolution of the samples for suitable presentation to the analytical systems was achieved by gentle heating at 50°C, sonication and addition of high-purity water. The ranges ascertained are as follows, μg/kg: As, 0.50–0.70; Cd, 0.50–0.74; Cr, 1.03–3.93; Cu, 144–216; Fe, 191–651; Mn, 223–580; Ni, 17–49;

Pb, 3.20–186; Pt, 0.50; Sn, 4–27; V, 1.22–1.94; and Zn 565–1144. As a rule, concentrations of elements in honey from different beehives were similar. A few exceptions were noted for As, Cu, Fe, Ni, and Zn [110].

Concentrations of eight essential trace elements Co, Cr, Cu, Fe, Mg, Ni, Se, and V and of seven nonessential—toxic elements Ag, Al, As, Au, Pt, Sc, and Ti in infant formulas and human milk were determined by SF-ICP-MS in another publication. As previously mentioned, this advanced instrumentation can separate spectral overlaps from the analyte signal, providing an advantage to conventional ICP-MS. Further, superior detection limits in the picogram per liter range can be obtained. Concentrations of Ag and Au showed large variations in human milk that might be associated with dental fillings and jewelry. Pt concentrations were very low with most of the samples below the method detection limit of 0.01 μg/L. Human milk concentrations of Co, Fe, Mn, Ni, and Se were at the low end of the corresponding reference ranges. Concentrations of Cr in human milk were five times higher than the high end of the reference range. For Al, As, and V most of the samples had concentrations well within the reference ranges. All elemental concentrations in infant formulas (except for Cr) were approximately one order of magnitude higher than in human milk [111].

Another group investigated the level of 18 trace elements in freeze-dried samples of blueberry and strawberry. Minor concentrations of nutritional elements were found in each freeze-dried berry. Toxic trace elements were found down the safe limit for human consumption. The overall quality of the blueberry surpassed that of the strawberry. The results certify that the two freeze-dried berries have potential for human consumption in value-added products [112]. Furthermore, another recent article examines the minor and trace elements in aromatic spices by microwave-assisted digestion and ICP-MS. This study aimed at analyzing the concentrations of 23 minor and trace elements in aromatic spices, after wet digestion by microwave system. The analytical method was validated by linearity, detection limits, precision, accuracy and recovery experiments, obtaining satisfactory values in all cases. Results indicated the presence of variable amounts of both minor and trace elements in the selected aromatic spices. Manganese was high in cinnamon (879.8 μg/g) followed by cardamom (758.1 μg/g) and clove (649.9 μg/g), strontium and zinc were high in ajwain (489.9 μg/g and 84.95 μg/g, respectively), while copper was high in mango powder (77.68 μg/g). On the whole, some of the minor and essential trace elements were found to have good nutritional contribution in accordance to RDA. The levels of toxic trace elements, including As, Cd, and Pb were very low and did not found to pose any threat to consumers [113].

The essential role of many trace elements, their implication in prevention of a number of diseases and the low levels of some of these elements in the diet in many countries, caused the increasing interest in the supplementation of food and feed with mineral elements. The most popular element of concern has been selenium, supplied in the form of yeast but also supplements for other metals, e.g., Zn, Fe, Mn, and Cr, are proliferated on the market [114]. Particular attention is paid in the development of Se-enriched food and nutritional products. The discovery that selenium may help prevent certain forms of cancer has prompted a pronounced interest in foods that are rich in Se. Thus, determination of nutritional value and safety of selenium-enriched watermelon by ICP-MS was studied in another article [115]. The changes of trace elements and heavy metals contents in watermelon were determined by ICP-MS after the application of selenium fertilizer. The results showed that selenium content in watermelon increased significantly after spraying selenium-enriched foliage fertilizer, which reached the requirement of selenium-enriched products. Especially contents of Li, Mg, Sr, Mn, Mo increased significantly. Content of heavy metals Cu, As, Cd, Pb, Cr in watermelon decreased significantly after spraying selenium-enriched foliage fertilizer compared to control. Concluding, application of selenium fertilizer can improve the nutrition of beneficial trace elements and the edible safety of watermelon in heavy metals [115,116].

In another study, the mineral content and levels of trace elements in the main exotic food supplements, colloquially called superfoods, have been here determined using ICP-MS after microwave digestion. The selected products were goji berries, goji juices, goji capsules, pomegranate juices, pomegranate capsules, chia seeds, acaí juices, mangosteen juices, and mixtures of berries [117]. The inorganic content of these products has scarcely, or not at all, been described in scientific literature and, taking into account the increase in consumers' interest for these supplements, provides valuable information for human health. A cranberry-certified reference material (SRM 3283) and recovery experiments over different samples were performed for method validation. The obtained results were discussed using the recommended daily allowance for minerals provided by the European Commission. Also a comparison between the different supplements was carried out [117]. In addition, another interesting study compares the nutritional value and food safety aspects of organically and conventionally produced wheat flours. This study has been prompted due to the growing interest in organic agriculture. Obtained results showed that organic samples had significantly lower protein content and lower levels of Ca, Mn, and Fe compared to conventional samples. Protein digestibility and levels of K, Zn, and Mo were significantly higher in organic than in conventional wheat flours. Regarding undesirable metals, significantly higher levels of As and Cd were found in conventional compared to organic wheat flours. This study revealed that organic agriculture has the potential to yield products with some relevant improvements in terms of high-quality proteins and microelements contents, while the reduction in contamination with toxic elements and mycotoxins may be accomplished [118].

As previously mentioned, ICP-MS is used also for radioactive elements determination in foodstuffs. For example, a rapid fusion method was developed in order to determine very low levels of plutonium isotopes in large rice samples. The recent accident at Fukushima Nuclear Power Plant in March 2011 reinforces the need to have rapid, reliable radiochemical analyses for radionuclides in environmental and food samples. Public concern regarding foods, particularly foods such as rice in Japan, highlights the need for analytical techniques that will allow very large sample aliquots of rice to be used for analysis so that very low levels of plutonium isotopes may be detected. This new method utilizes a furnace ashing step, a rapid sodium hydroxide fusion method, a lanthanum fluoride matrix removal step and a column separation process with TEVA Resin™ cartridges. The method can be applied to rice sample aliquots as large as 5 kg. Plutonium isotopes can be determined using ICP-MS. The method showed high chemical recoveries and effective removal of interferences. The rapid fusion technique is a rugged sample digestion method that ensures that any refractory plutonium particles are effectively digested. This method can easily allow detection of plutonium isotopic ratios [119]. Moreover, another interesting article concerns Uranium in Kosovo's drinking water. The results of this publication are an initiation to capture the drinking water and/or groundwater elemental situation as it is distributed to the population of Kosovo. Around 950 drinking water samples were analyzed by ICP-MS. The results are the first countrywide interpretation of the uranium concentration in drinking water and/or groundwater, directly following the Kosovo war of 1999. More than 98% of the samples had uranium concentrations above 0.01 μg/L, which was also the limit of quantification. Concentrations up to 166 μg/L were found with a mean of 5 μg/L and median 1.6 μg/L. Around 3% of the analyzed samples exceeded the WHO maximum acceptable concentration of 30 μg/L, and about 45% of the samples exceeded the 2 μg/L German maximum acceptable concentrations recommended for infant food preparations [120].

3.1.2.1 Speciation

Over the last two decades, the development of speciation methodology has been driven forward to a great extent by hyphenated techniques using ICP-MS as a detection system. The main reason for this success story is the ease of interfacing the different separation techniques to the ICP-MS, the wide range of accessible elements and the detection power provided by ICP-MS. While today speciation analysis is well established within the area of research, its routine application in the general field of testing and analysis is still in development.

Applications of hyphenated ICP-MS fall into the general category termed speciation analysis. In all cases, the fractionation device (chromatograph or other) is used to separate the species from each other and the matrix, and the ICP-MS is used to detect the species of interest. The analyte species may be as simple as elemental ions of various oxidation states in solution, or as complex as mixtures of pesticides or biomolecules [121–123]. In all cases though, the ICP-MS is simply acting as an elemental detector. The fractionation device serves to separate the various components in the sample before detection as well as to provide additional information in the form of retention time. Often this combination is sufficient in order to identify and quantify the target analytes. However, analysis of standards or the use of additional mass spectrometric techniques can provide further confirmation of identification.

The total amount of an element present in food tells us very little about its possible absorption and fate inside the body. So, the chemical form of the element must be assessed as this form critically influences its bioavailability, essentiality or toxicity [59,123]. To gain an understanding of the function, toxicity and distribution of elements, it is necessary to determine not only the presence and concentration of the elements of interest, but also their speciation, by identifying and characterizing the compounds within which each is present.

A common example would be the measurement of Cr(VI) (toxic) and Cr(III) (essential nutrient) as opposed to total Cr. Similar examples of elemental speciation include As(III)/As(V), Se(IV)/Se(VI), and other elements that can exist in different stable oxidation states. Furthermore, arsenic and selenium in particular, commonly exist in various organic forms. In the case of more complex molecules such as pesticides or biomolecules, the ICP-MS is able to identify and quantify the presence of a particular element or elements in molecular chromatographic peaks. When used in conjunction with organic MS techniques, this technique can permit quick screening for molecules (peaks) containing specific elements in a complex mixture, prior to analysis by organic MS. With modern, integrated systems and software, simultaneous analysis by ICP-MS and organic (for example, electrospray ionization, ESI) MS is also possible, using a split flow from a single chromatographic device.

Elemental speciation is important in many application areas and is becoming particularly important in the food, and in the foodomics (a new approach to food and nutrition that studies the food along nutrition to reach the main objective: the optimization of human health and well-being, connecting food components with health) [124]. This is because, for many elements, properties such as those listed below depend on the species or chemical form of the element present in the sample:

A great deal of research has been performed to identify compounds containing metals and metalloids that are toxic to humans or other organisms. Toxic elements studied include arsenic, antimony, tin, mercury, lead, aluminum, chromium, cadmium and mercury, and studies have focused both on environmental species that might be ingested, as well as species found in tissue and bodily fluids. Speciation analysis of tin, lead, mercury, and antimony compounds has focused mainly on the detection and identification of small organic derivatives of these metals, as well as detection of the free metal ion. Speciation of aluminum, chromium, and cadmium has been focused on biologically synthesized proteins and small molecules that act as ligands to bind and deactivate the toxic metals. Such ligands include citrate, aconitate, and malate ions, transferrin, albumin, ceruloplasmin, metallothionein, and phytochelatins. The latter two of this list have been studied in detail due to their ability to bind and sequester a wide variety of toxic metals. Oxidation state has been of great interest as well, since the toxicity of elements such as antimony, arsenic, and chromium are highly dependent upon oxidation state [116].

Examples of ICP-MS in speciation are many and cover a broad variety of food applications:

- Organotin species, e.g., dimethyl-, tributyl- and triphenyltinchloride, species in food products and drinking water, were determined by LC-ICP-MS, GC-ICP-MS, and CE-ICP-MS [125–128].
- Mercury species in foods and especially in fish were determined by LC-ICP-MS and GC-ICP-MS

Common forms include inorganic mercuric (II), methyl mercury cation (MeHgX), and dimethyl mercury (Me_2Hg). Humic matter can methylate mercury, reduce Hg(II) to Hg^0 and form complexes that are considered to be of greater toxicity than the inorganic starting form. Approximately 50–90% of total mercury in coastal waters and estuaries is bound to humic matter. Because methyl mercury appears to bioaccumulate in organisms, it is of greater concern than other forms of this metal. Methylmercury is especially known to bioaccumulate and biomagnify up the marine food chain. Fish from high levels of the marine food chain may contain relatively high concentrations of mercury, and most (>70%) of the mercury found in muscle is methylmercury. In aquaculture, marine protein is the dominant source of methylmercury, and this raises some concern with regards to fish welfare and consumer safety. Also it is reported that methylmercury is accumulated in fish muscle, where it is incorporated into larger peptides or proteins [129–133].

- Arsenic species such as arsenite, arsenate along with organic arsenic compounds: monomethylarsonic acid, dimethylarsinic acid (DMAA), arsenobetaine, arsenocholine, trimethylarsine oxide, tetramethylarsonium ion, and several arsenosugars and arsenolipids in food products and drinking water.

Arsenic exists in either its elemental form or the As(III) or As(V) valence states. The (III) and (V) forms can be either inorganic (e.g., As_2O_3) or contain organic groups (e.g., methyl arsine). Commonly reported organic forms include monomethyl arsonic acid (MMAA) and dimethylarsinic acid (DMAA). Monomethyl arsonous acid (MMAA-As(III)) and dimethyl arsonous acid (DMAA-As(III)), formed by the action of hydrogen sulfide on MMAA and DMAA. Bioincorporated forms include arsenobetaine (AB, present in marine organisms and used as an indicator of arsenic uptake) and arsenocholine (AC). Certain researchers have concluded in that the toxicity of arsenic is dependent on its species. The inorganic arsenics are more toxic than the organoarsenicals, and the trivalent forms more toxic than the pentavalents. Therefore much effort has been devoted to methods that will identify the particular molecule in terms of its bonded groups and oxidation state [9,133–137].

- Lead in wine and foodstuffs

The most important organic forms are the tetraalkyl leads (TALs). Researchers have found that these compounds can be readily absorbed by the lungs and also penetrate the skin and biological membranes. They therefore are considered more harmful than inorganic lead. TALs are degraded by sunlight and ozone to trialkyl lead compounds, which are further degraded to inorganic leads via a dialkyl lead intermediate [133,138,139].

- Chromium in drinking water, mainly by IC-ICP-MS, and foodstuffs

Chromium can act either as an essential micronutrient or a chemical carcinogen, depending on its oxidation state. Of the two most stable chromium oxidation states, Cr(VI) is classified as a human carcinogen based on studies showing an increase in lung cancer. The Cr(III) is considered to be essential for the maintenance of lipid, protein, and glucose metabolism, but Cr(VI) is reported to be toxic due to its facile penetration of biological membranes and its oxidizing potential [133,137,140–143].

- Selenium in foodstuffs

The vast majority of studies in the area of elemental speciation in nutritional samples have been devoted to selenium speciation. Detailed speciation of selenium is complicated by its four possible oxidation states, -II, selenide; 0, elemental; +IV, selenite and +VI, selenite; and their selective complexation and/or bonding. Volatile species include dimethyl selenide (DMSe) and dimethyl diselenide (DMDSe). In addition to oxidation state, selenium in water samples is divided into two classes: dissolved Se that passes through a 0.45 mM filter and particulate Se (>0.45 mM). Particulate selenium is associated with sediments and other suspended solids. In sediments, selenium may be associated with organic material, iron and manganese oxides, carbonates, or other mineral phases, either adsorbed to or coprecipitated with these phases. Se(II) can be covalently bound to the organic portion of the sediment and other materials to give organoselenides [116,133].

In addition, selenium is an essential element for human health. It has been recognized as an antioxidant and chemopreventive agent in cancer. Moreover, selenium is known to develop its biological activity via selenocysteine residue in the catalytically active center of selenoproteins. The main source of selenium in humans is the diet. However, in several regions of the world the content of selenium in diet has been estimated insufficient for a correct expression of the proteins. The beneficial effects of selenium on human health are strongly dependent on its chemical form and concentration. Particular attention is paid also how the application of hyphenated techniques above is mandatory to get reliable results on selenium metabolism [144]. Moreover, as an example an article deals with speciation in selenium-enriched shiitake mushroom, *Lentinula edodes*. The major selenium compound in an aqueous extract of this popular mushroom in eastern Asian, fortified with selenium (Se) was identified by LC-ICP-MS. Sixty-eight per cent of the total Se in the selenized shiitake was extracted with water, and 49.8% of the Se in the water extract was eluted in the high molecular mass fraction (>40,000 kDa) before incubation at 37°C. After incubation,

~40% of the Se in the water extract was eluted in a lower molecular mass fraction and the Se eluted in the high molecular mass fraction had decreased to 14.0%, suggesting that the major selenium compound in the water extract was initially in a form bound to macromolecule(s) and was then enzymatically liberated from the macromolecule(s). The retention time of the liberated selenium compound in LC-ICP-MS matched that of selenomethionine (SeMet), and the masses of molecular and fragment ions detected by LC-MS also suggested that the selenium compound was SeMet. The selenized shiitake accumulated Se as SeMet, and SeMet might be bound to the water extractable high molecular mass protein(s) [145].

Furthermore, the discovery that individuals in many countries have a diet deficient in selenium has spurred the introduction of a wide variety of selenium supplements. A number of research groups have analyzed these supplements as well as natural sources of selenium to determine the species of selenium present, and others have studied to elucidate the mechanisms by which selenium is metabolized and excreted [146,147]. In addition to studies of the utilization of selenium, pathways of toxic action are of interest as well, since the margin between the nutritionally required and toxic amounts is very narrow relative to those observed for other elements. Work performed thus far has identified the amino acids selenomethionine, selenoethionine, and selenocysteine as likely cancer-preventative species, although a number of selenoamino acid derivatives, selenium-containing glutathione derivatives, and smaller organoselenium species have been detected and identified in selenium-enriched yeast [116]. Moreover, quantification of Se-Methylselenocysteine and its γ-glutamyl derivative from naturally Se-enriched green bean (*Phaseolus vulgaris vulgaris*) was investigated. LC-TOF-MS, Orbitrap, and ICP-MS methods were addressed to identify and quantify selenium species from a naturally Se-enriched green bean sample after proteolytic digestion. While selenomethionine (10.1 mg/kg as Se) and selenate (9.5 mg/kg as Se) could be quantified in a straightforward way by anion exchange LC-ICP-MS technique, a multistep purification protocol was required to identify Se-methylselenocysteine and γ-glutamyl-Se-methylselenocysteine in an unambiguous way prior to quantification by using either in-source fragmentation (LC-TOF-MS) or collision-induced dissociation (LC-Orbitrap MS). Finally, Se-methylselenocysteine (2.6 mg/kg as Se) and γ-glutamyl-Se-methylselenocysteine (1.2 mg/kg as Se) could contribute to the overall selenium recovery of 72%. This sample is the first of the Faboideae subfamily and *Phaseolus* ssp. to be speciated to such an extent for selenium including γ-glutamyl-Se-methylselenocysteine, a highly potential selenium species, which makes this bean material an ideal candidate for functional food purposes [148].

Size exclusion chromatography (SEC) was used to separate selenium species present in the different yeast extracts and showed that about 75% of yeast selenium was bound to high molecular mass compounds (proteins, nucleic acids, or cell walls) or present in water-soluble proteins. There are also reported two interesting articles that concern Se-containing metabolomes and anionic Se species in Se-rich yeast by SEC-ICP-MS [149,150]. Fish accumulate important amounts of selenium and constitute an important dietary source of the element. Size exclusion chromatography coupled to ICP-MS was used to study the soluble selenocompounds in fish, showing large differences among selenocompounds present in different fish species [151]. Nuts could be used as Se supplements because high levels of Se have been found in nuts. The elemental distribution among different fractions (liquid extract, low molecular weight, and protein fraction) and Se speciation has been reported in different types of nuts. For example, Brazilian nuts have been classified as the foodstuffs that contain the highest level of selenium. In this work, various sample preparation approaches, including microwave extractions and enzymatic treatments, are examined with the goal of species preservation and subsequent selenium speciation. Most effective of these approaches has been proved an enzymatic treatment with Proteinase K. Extracts were evaluated against available standards for the commercially obtainable selenoamino acids, selenomethionine (SeMet), selenoethionine (SeEt), and selenocystine (SeCys). Selenomethionine was demonstrated to be the most abundant of these selenoamino acids [152].

- Pesticides and herbicides

For example, SPME-enantioselective gas chromatography with ICP-MS detection was investigated for the chiral speciation of the pesticide ruelene in river water, red wine, orange, and tomato juices. Also analysis of phosphorus herbicides was studied by ion-pairing reversed-phase liquid chromatography coupled to ICP-MS with octapole reaction cell [121,153].

Another interesting publication reported the determination of iodine and bromine compounds in foodstuffs by CE-ICP-MS. Samples containing ionic iodine (I^- and IO_3^-) and bromine (Br^- and BrO_3^-) species were subjected to electrophoretic separation before injection into the microconcentric nebulizer. The separation has been achieved in a 50 cm length × 75 μm id fused-silica capillary. This method has been applied to determine iodine and bromine species in NIST SRM 1573a tomato leaves reference material and a salt and seaweed samples obtained locally. A microwave-assisted extraction method was used for the extraction of these compounds. Over 87% of the total iodine and 83% of the total bromine are extracted using a 10% m/v tetramethylammonium hydroxide (TMAH) solution in a focused microwave field within a period of 10 min. The spike recoveries were in the range of 94–105% for all the determinations. The major species of iodine and bromine in tomato leaves, salt, and seaweed are BrO_3^-, IO_3^-, I^-, and Br^-, respectively [154].

3.1.2.2 Metallomics

Metallomics is a field that receives great attention as a new frontier in the investigation of trace elements in biology and is expected to develop as an interdisciplinary science complementary to genomics and proteomics. It covers the research fields related to biometals and their role in biological, environmental, and clinical systems. So the emerging field of metallomics refers to the entirety of research activities aimed at the understanding of the molecular mechanisms of metal-dependent life processes. The metallomics approach is very useful in metal binding characterization in foods and thus to the relation with human diet. Size exclusion chromatography coupled to ICP-MS (SEC-ICP-MS) has been widely used for the speciation of polysaccharides binding to cations through negatively charged oxygen functions for metal speciation in fresh fruit and vegetables [155]. SEC and cation-exchange separation, based on solid-phase extraction cartridges coupled to ICP-MS, have been used to obtain complementary information about the metal species present in tea infusion. These techniques confirm that metal-binding organic ligands in the tea infusion are large polyphenolic compounds, which are probably the Al-binding ligands in the tea infusion [156,157].

Concerning food supplements, the relevant metallomics studies focus on two fields [114]:

1. The characterization of the chemical forms and their bioavailability in the supplements. The knowledge of virtually all forms present at levels exceeding 0.1% is required to comply with regulations of governmental agencies issuing sales authorizations. The elemental speciation pattern is also a precious tracer of the origin of the product and its control allows a better optimization of the biotechnological production process.
2. The transport and fate in the target organisms, such as animals (meat, milk, and eggs from Se-supplemented animals become functional food themselves) and humans.

3.1.3 Authentication

Food authentication is the process by which a food is verified as complying with its label description, for example, for its origin, geographical and genetic, production method, e.g., organic farming, traditional procedures, processing technologies, e.g., irradiation, freezing, microwave heating. Of special concern is the declaration of specific quality attributes in high-value products. Proof of provenance has become an important topic in the context of food safety, food quality, and consumer protection in accordance with national legislation and international standards and guidelines. It is of outmost importance to identify and ensure the origin of a commodity and thereby the region where it was produced. Due to the globalization of food markets and the increased import of products from other countries, customers over the years seem to be more and more interested in the geographical origin and the concomitant assumed quality of the products they eat and drink. Also from an economical and legal point of view, the quality of food and drinks and the methods which are used to control and assure it are of great concern.

Authenticity has probably always been a major concern of many consumers, and it is still gaining more and more importance. Therefore, since the beginning of the twentieth century, organizations that set standards for and control the origin of ingredients and the production process, have been established all over the world e.g., the French "Institut National des Appellations d'Origine" (INAO), Italy's "Denominazione di Origine Controllata," Spain's "Denominación de Origen," South Africa's "Wine of Origin" or the United States' "American Viticultural Areas." As the production of consumer goods according to these standardized procedures can often be rewarded with higher purchase prices, unfortunately, nowadays the production of counterfeit food and illegal food trades is a common fact. Therefore, there is a growing need for control organs to have access to reliable analytical methods that can give a decisive answer about the authenticity of foodstuffs [102].

In Europe, origin is one of the main authenticity issues dealing with food. European Union legislation reserving specific names for foods and beverages of a particular quality or reputation has been abundant since the dawn of the European integration process. Eventually culminating in the introduction of a regulatory framework for wines and spirits and quality schemes for food products: protected denominations of origin (PDO), protected geographical indication (PGI), traditional specialities guaranteed (TSG), and recently optional quality terms (OQT, "mountain product" and "product of island farming") (1151/2012 EU regulation) [158,159]. The purpose of these EU schemes is to protect the reputation of the regional foods, promote rural, and agricultural activity, help producers obtain a premium price for authentic products, and eliminate the unfair competition and misleading of consumers by nongenuine products, which may be of inferior quality or different flavor.

Trace element availability to plants depends on several factors such as soil pH, humidity, porosity, clay, and humic complexes, etc. Consequently, the range of soils present and bioavailability mean that elemental composition may provide unique markers in food characteristic of the geographical origin. Alkaline metals especially Rubidium (Rb) and Cesium (Cs) being easily mobilized in the soil and easily transported into plants, are good indicators of geographical identity. Although fertilization conditions, crop year, variety, soil type, and production year cause concentration variations of several elements, these variations are smaller than those observed between production places when appropriate elements are selected. Rare-earth elements can also be a reliable indicator for the determination of geographical origin. Rare-earth fingerprint is directly linked to the geology of the area and could be minimally influenced by different agricultural practices and harvest year. The multielement composition of animal tissues reflects, to some extent, that of the vegetation that they eat. The vegetation is the compositional reflection of the bioavailable and mobilized nutrients present in the underlying soils from which they were cultivated. Chemometric techniques are used for statistical analyses of the obtained ICP-MS data [160–162].

ICP-MS is one of the best choices for authentication studies, especially for accurate geographical origin verification. The relative abundance of natural strontium (Sr) isotopes is related to local geological conditions and may therefore provide information on the origin of raw cheese and other products. Values of Sr isotope abundance ratios in terrestrial vegetation are linked with the Sr isotopic composition of the soil, which is influenced by bedrock, soil/water properties, and atmospheric inputs. Other isotope systems that are used are $^{20X}Pb/^{20X}Pb$ and $^{187}Os/^{188}Os$ and $^{11}B/^{10}B$ [102,160].

ICP-MS can screen the geographical origin of food products by the analysis of numerous inorganic elements and obtaining fingerprints of the element pattern. It has been proven to be a reliable technique for provenance determination in all types of foods, selected examples are following:

- Wines: Determination of regional origin of 112 Spanish and English wines [163]. Elemental profile of Okanagan Valley wines in Canada correlate strongly with vineyard of origin [164]. Isotope ratios and concentrations for lead measured by ICP-TOF-MS differentiate 20 wines from the five continents [165]. $^{87}Sr/^{86}Sr$ ratio is also another good choice for regional origin determination of wines by SF-ICP-MS [166]. Recently, intraregional classification of wine was achieved using the elemental profile with B, Ba, Cs, Cu, Mg, Rb, Sr, Tl, and Zn as suitable indicators. Results for the Stellenbosch wine district in the Western Cape Wine Region, South Africa, comprising an area of less than 1000 km^2, suggest that classification of wines from different estates, 120 wines from 23 estates, is indeed possible [167].
- Beverages—spirits: Elemental concentrations and lead isotope ratio measured by SF-ICP-MS distinguish alcoholic beverages origin [168]. Yerba mate (Ilex paraguariensis), is a widely consumed beverage in South America that nowadays expands worldwide. Classification to the country of origin is based on element concentrations using various chemometric tools such as Principal Component Analysis (PCA), KNN, SIMCA, PLS-DA, and support vector machine discriminant analysis (SVM-DA) [169].
- Vegetables: Elemental profile assisted distinction of Japanese and Chinese welsh onions from different regions [170]. MC-ICP-MS after optimized Rb/Sr separation differentiated Marchfeld asparagus (PGI), Austria from other asparagus samples from Hungary, Slovakia, Peru, the Netherlands, and Germany using Sr isotope ratio measurements [171]. Differentiation according their geographical origin was achieved for tomatoes and triple concentrated tomato pastes from different Italian regions and China, Greece, and California. The origin of tomato fruits and the areas of production as "Italy" and "non-Italy" of the triple concentrated pastes were evaluated by three supervised pattern recognition procedures, LDA, SIMCA, and K-nearest neighbors (KNN) [172]. Determination of Tropea red onion (*Allium cepa* L. var. Tropea, PGI) origin by its elemental profile was achieved using LDA, SIMCA, and back-propagation artificial neural network BP-ANN. Tropea red onion is among the most highly appreciated Italian products [173]. Identification of species of the *Euterpe* genus by rare earth elements and linear discriminant analysis was successful for açaí (*Euterpe oleracea* Mart.) and juçara (*Euterpe edulis*

Mart.). These fruits are almost identical, rich in energy, minerals, vitamins, and natural compounds with antioxidant and anti-inflammatory properties. REE content, in particular Sm, Th, La, Pr, Gd, and especially Ce and Nd were the most important indices [174].

- Vegetable oils: Discrimination of olive oils produced in four Greek regions according to their geographical origin was achieved using the rare earth elements content and supervised chemometric techniques. Different ANNs models, Multi-layer Perceptions (MLPs) and Radial Basis Function (RBF), were developed and evaluated. Two additional supervised techniques, discriminant analysis (DA) and classification trees (CTs), were also applied for data pretreatment and comparison purposes [175]. Multielement analysis of virgin olive oils from different Italian regions by ICP-MS and the data procession by linear discriminant analysis (LDA) facilitated classification of unknown samples [176]. Geographical origin identification of Styrian pumpkin seed oil, a high priced local product which is protected by the European Union, was achieved through the rare earth elements profile and discriminant analysis [177].
- Fruit juices: Discrimination of regional origin of orange juices and peel extracts by elemental fingerprinting. These results could be related to differences in soil and rootstock [178].
- Cereal—pulses: Discrimination between "Fava Santorinis" (PDO) and other yellow split peas using four classification methods, orthogonal projection analysis (OPA), Mahalanobis distance (MD), partial least squares discriminant analysis (PLS-DA), and k nearest neighbors (KNN), was achieved fusing rare earth with and trace element data [179]. Boron, holmium, gadolinium, magnesium, rubidium, selenium, and tungsten combined with $^{13}C/^{12}C$ and $^{18}O/^{16}O$ ratios differentiate rice samples cultivated in the United States, Europe, and Basmati regions [180]. Discrimination of Piedmont hazelnuts, Italy (PGI) from others was achieved through lanthanides fingerprint. The "Tonda Gentile delle Langhe" (TGL) variety is acknowledged all over the world as the best one, and it is particularly appreciated when used to provide flavor in chocolate products. Authentication is prerequisite to safeguard this variety against fraud, which can occur when the product is partially or totally substituted with hazelnuts of lower quality [181]. Classification of wheat from four major Chinese wheat-producing regions was through their elemental profile [182].
- Dairy products: Multielement profile (P, S, K, Ca, V, Cr, Mn, Fe, Co, Zn, Ga, Rb, Sr, Mo, Cs, and Ba) was instrumental for cow and buffalo milk differentiation. The sources of food and water available for consumption by the animals were also studied and results showed no correlation between the elemental composition of the dietary components and milk [183].
- Meat: Determination of poultry and dried beef geographic origin by elemental concentrations was achieved through SF-ICP-MS. Twenty five poultry breast filets samples originated from Switzerland, France, Germany, Hungary, Brazil, and Thailand, and the 23 dried beef samples, made from M. biceps femoris and M. semitendinosus, were produced in Switzerland, Austria, Australia,

United States, and Canada out of raw meat originating either from these or from other countries. For both poultry meat and dried beef, a differentiation of the origins was possible using those elements, which were significantly different across countries: As, Na, Rb, and Tl in poultry; B, Ca, Cd, Cu, Dy, Eu, Ga, Li, Ni, Pd, Rb, Sr, Te, Tl, Tm, V, Yb, and Zn in beef [184]. Determining the geographical origin of mutton from different regions of China was achieved by the elemental profile [185].

- Fish and fish products: Distinction of commercial marine species from the East China Sea according to their geographical origin by their elemental fingerprint using multivariate statistical analysis. PLS-DA and probabilistic neural network (PNN) were proven that they both precisely predict the origin of the marine species [186]. Origin distinction between vendace and whitefish caviars from brackish and freshwaters by differences in elemental concentrations or sample-specific isotopic composition (Sr and Os) variations using SF-ICP-MS and MC-ICP-MS [187].
- Honeys: Distinction between honeydew, buckwheat, and rape honey and also between different areas in Poland was achieved through the mineral profile with the aid of CA and PCA. CA showed three clusters corresponding to the three botanical origins of honey. PCA permitted the reduction of 13 variables to four principal components explaining 77.19% of the total variance. The first most important principal component was strongly associated with the value of K, Al, Ni, and Cd [188].
- Coffees and tea: Tea samples were classified according to geographical origin by elemental fingerprinting with principal component analysis (PCA) and cluster analysis (CA), as exploratory techniques, and linear discriminant analysis (LDA) and soft independent modeling of class analogy (SIMCA) [189]. Distinction between 20 green coffee beans from different geographical origins was achieved through Sr and O isotope ratios. The final results allowed discrimination of local provenances investigated in this study by principal component analysis (PCA) and exhibit the potential to proof authenticity of world coffees [190].
- Food ingredients: Determination of Korean and Chinese ginsengs origin according to their $^{87}Sr/^{86}Sr$ ratios [191]. Elemental and Sr isotope ratios fingerprinting of Szegedi paprika (PDO) by MC-ICP-MS and classification of authentic and purchased paprika from different known, declared and unknown geographical origin using multivariate statistical tools [192]. Identification of 27 Saffron spices produced in three Italian regions, Abruzzo (L'Aquila), Umbria by their mineral composition [193].
- Organic foods: Concerning organic foods, the elemental uptake of a specific crop and thus its chemical composition are influenced by many factors, including plant species, cultivar, physiological age, soil fertility, climate, crop rotation, and fertilization strategy as well as pest and weed control management. Organic and conventional farming differ strongly in the last two factors. In particular, organic farming severely restricts the use of artificial fertilizers;

instead, use is made of organic fertilizers and nitrogen-fixing plant species. The extent and rate of release of nutrients are thus different in the two farming practices, and this has an impact on the overall crop chemical composition [194–195]. Organic wheat, barley, fava bean, and potato were characterized on their elemental fingerprinting. Crops were cultivated in 2 years at three different locations, each accommodating one conventional and two organic cropping systems [161]. Only eight of the minerals investigated, namely Zn, Se, Ba, U, Dy, Tl, Th, and Mo were found important for the authenticity of organic coffee by elemental profile and the use of data mining techniques while classification was achieved by Multilayer Perceptron (MLP), Support Vector Machine (SVM), and Naïve Bayes (NB) [196]. Authenticity of free range pastured hens' eggs by 15 elements (As, Ba, Cd, Co, Cs, Cu, Fe, Mg, Mn, Mo, Pb, Se, Sr, V, and Zn) database. The egg classification, free-range eggs compared with battery eggs, was carried out using two machine learning algorithms, decision trees, and naïve Bayes (NB) [197].

3.1.4 Migration Studies— Food Nanomaterials— Others

Regarding the uses of ICP-MS in food quality and safety, there are two emerging fields that should be also reported. These are migration and food nanomaterials. In a very recent study the migration and the characterization of nanosilver from food containers was investigated by AF^4-ICP-MS. In this work, silver migration from commercial food containers was evaluated according to European Regulation 10/2011. Several experimental parameters affected silver release: food simulant, temperature, exposition time, and sampled bag area. Results demonstrated a significant silver nanoparticle (AgNP) migration into aqueous and acidic simulants. The amount of silver migrated increased with storage time and temperature although, in general, silver showed a low tendency to migrate into food simulants. However, the food simulant did not seem to be a really outstanding variable for long-term storage. ICP-MS was used to confirm the presence of AgNPs in the simulants. The low limit of detection achieved (0.4 μg/L) allowed the identification of AgNPs and their size characterization (40–60 nm) [198]. In a similar work, the migration potential of nano silver particles from food contact polyolefins was investigated [199]. Migration and exposure assessment of silver from a PVC nanocomposite was also investigated [200]. In another recent work, the influence of storage time and temperature on Sb migration from PET bottles into mineral water was studied in short-term tests (up to 15 days) and long-term studies (up to 220 days). Samples were stored in three different colored bottles. Sb migration was assayed by HPLC-ICP-MS for speciation analysis. Migration studies showed that waters stored at 4 and 20°C were not subjected to Sb migration. At 40°C there was a significant increase in Sb concentration, although the maximum limit established by the EU was not exceeded, whereas at 60°C samples were subjected to considerable Sb migration after 30 days of storage. In this case, the maximum limit established by the European Union was exceeded and both Sb (V) and Sb (III) were detected [201]. Another interesting article describes that the stainless steel leaches nickel and chromium into foods during cooking. This study examined stainless steel grades, cooking time, repetitive cooking cycles, and multiple types of tomato sauces for their effects on nickel and chromium leaching [202].

Another study aimed to provide a protocol for sampling and analysis of metal elements migrating from carbon steel cylinders, used for gas storage and distribution, to food gases, i.e., those gases, such as CO_2, N_2, and O_2, employed by food and beverage industries. The concentrations of 23 selected elements were analyzed by ICP-MS. This work concluded in that the effects of migration of contaminants from carbon steel cylinders do not have a significant influence on the quality of food gases, independently of the type of food gas and carbon steel composition [203]. In another publication, a method for titanium in Nano-Titanium(IV) oxide composite food packaging determination with an ICP-MS was developed. Microwave digestion was optimized using different acid combinations. The method showed good reproducibility, repeatability, and recovery [204]. Furthermore an analytical method based on ICP-MS was developed for the determination of Ti in food simulants (3% (w/v) aqueous acetic acid and 50% (v/v) aqueous ethanol). The method was used to determine the migration of Ti from nano-TiO_2-PE films used for food packaging into food simulants under different temperature and migration time conditions. It concluded in that increasing the additive content in the film promotes the migration of nanoparticles and also the migration of nanoparticles might occur via dissolution from the surface and cut edges of the solid phase (film) into the liquid phase [205]. In a similar research, silicon migration from high barrier-coated multilayer polymeric films to selected food simulants after microwave processing treatments was investigated [206]. A method was also developed for simultaneous determination of 19 chemical elements (Al, V, Cr, Mn, Co, Cu, Ga, As, Rb, Y, Zr, Cd, Te, Ce, Pr, Nd, Sm, Er, and Pb) in simulated foods (4%, 6%, and 8% (w/v) aqueous acetic acid) by microwave-assisted digestion and ICP-MS. The migration of these chemical elements from ceramic packaging into the simulated foods and into two types of mature vinegar was examined. The simulated foods showed matrix effects, compared with stock solution solvents (5% ultra-pure nitric acid). The migration of chemical elements from ceramic packaging into simulated foods was affected by migration time, temperature, and acidity of the simulated food. The chemical elements showed different migration behaviors in black and white mature vinegar, and the amount of migration also varied depending on the inner surface of the ceramic packaging [207]. An interesting work investigated the quantification of metal release from stainless steel electrodes during conventional and pulsed ohmic heating [208]. Another recent study investigates the migration of silver from commercial plastic food containers, with declared content of “nano-” or “microsilver,” and implications for consumer exposure assessment [209].

The nanoparticle release from nanosilver antimicrobial food containers was also investigated. Polymer nanocomposites incorporating metal or metal oxide nanoparticles have been developed to improve their characteristics (flexibility, gas barrier properties, antimicrobial or antioxidant properties, etc.). Among them silver nanoparticles are used because of their antimicrobial effect in many daily life materials, i.e., food packaging. In this paper, the results of migration studies (with different simulant solutions and times) in three commercial nanosilver plastic food containers are presented. Migration solutions were evaluated by ICP-MS and SEM-EDX analysis. Silver in dissolved form and silver as nanoparticles were also analyzed as a key aspect for the toxicity. Size and morphology of the silver nanoparticles change for the different samples (ranging between 10 and 60 nm) and migration of other nanosized materials was also confirmed [210]. In another publication, aluminum migration to orange juice in laminated paperboard packages was investigated. The analytical procedure includes MW-assisted digestion of orange juice with concentrated nitric acid followed by determination of Al concentration by double-focusing sector field inductively coupled plasma mass spectrometry (SF-ICP-MS) [211]. Finally, migration from paper and board food packaging materials was also studied [212].

Nanoparticles are increasingly applied on a broad range of applications and may also play a vital role as food additives. The application of nanomaterials is leading to innovative developments in industry, agriculture, consumer products, and food or related sectors. However, due to the special properties of these materials, there are concerns about their safety, especially because of our limited knowledge of human health effects and the fact that constantly new nanomaterials and applications thereof are being produced. The development of analytical techniques is a key element to understand the benefits as well as the risks of the application of such materials. Therefore, two very recent publications are dealing with detection and characterization of silver nanoparticles in chicken meat [213,214]. In the first, a method of analysis of nano-silver (AgNPs) in chicken meat was developed by asymmetric flow field flow fractionation with detection by inductively coupled plasma-mass spectrometry, AF^4-ICP-MS. In another four very recent publications, characterization of titanium dioxide nanoparticles in food products was studied [215–218]. Titanium dioxide (TiO_2) is a common food additive used to enhance the white color, brightness, and sometimes flavor of a variety of food products. Three principally different methods have been used to determine the size distribution of TiO_2 particles: electron microscopy, asymmetric flow field-flow fractionation combined with inductively coupled mass spectrometry, and single-particle inductively coupled mass spectrometry. Furthermore, other publications are dealing with Ag nanoparticles detection by single-particle inductively coupled plasma-mass spectrometry, sp-ICP-MS [219,220].

Other studies concern mainly dietary intake studies. For example, a recent publication investigates the iron stable isotope fingerprint of the human diet. As it is known, the stable isotopes of iron disclose the metabolic pathways of iron within the human food chain. The iron concentrations and the stable isotope composition of 60 food products that are representative of the average German diet have been measured with MC-ICP-MS. It was found that vegetables fall within the range typical of strategy I plants (–0.1 to –1.4‰ in $\delta^{56}Fe$), crop products and processed crop foods into the range typical of strategy II plants (–0.6 to +0.4‰), and animal products into the ^{54}Fe-enriched range known for animal tissue and blood (–1.1 to –2.7‰). Weighting these isotope compositions by the average iron dietary sources, a representative composition of European vegetarian diet of –0.45‰ was found, whereas that of omnivores is –0.82‰. For human blood, known to be enriched in light iron isotopes, fractionation factors for iron absorption of –2.0 and –2.3‰ for vegetarians (female and male, respectively) and –1.3 and –1.5‰ for omnivores (female and male, respectively) was found. Knowing these fractionation factors is a prerequisite for using stable iron isotope ratios in blood as monitors of intestinal iron uptake regulation [221].

3.2 IRMS

3.2.1 Trends

There are two major applications of IRMS concerning food, waters, and beverages are authentication and nutrient studies. Authentication studies can be subdivided in adulteration and determination of origin. IRMS detects small differences in stable isotopes that can be correlated to different origins or even to adulterations of products. Figure 18 depicts IRMS publications distribution, about half (177 publications) concern adulteration studies while determination of origin about one-third (90 publications). The rest 50 publications and nutrient studies. There are also some other publications concerning mainly food safety. All these IRMS applications are based on the measurement of stable isotope ratios of a product or of a specific component such as a fraction, an ingredient or a target molecule (or group of molecules) of the product.

Time evolution of IRMS research is shown in Figure 19. IRMS research on adulteration and determination of origin are continuously increased especially from 1997 and onward. On the other hand, nutrient studies publications are constant from 2001. An overview of the main IRMS applications currently in the literature is presented below.

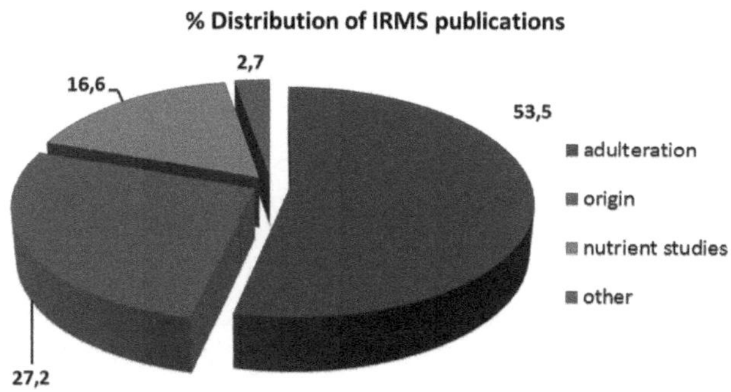

FIGURE 18 Percentage distribution of IRMS publications according application field in food, waters, wines, and beverages.

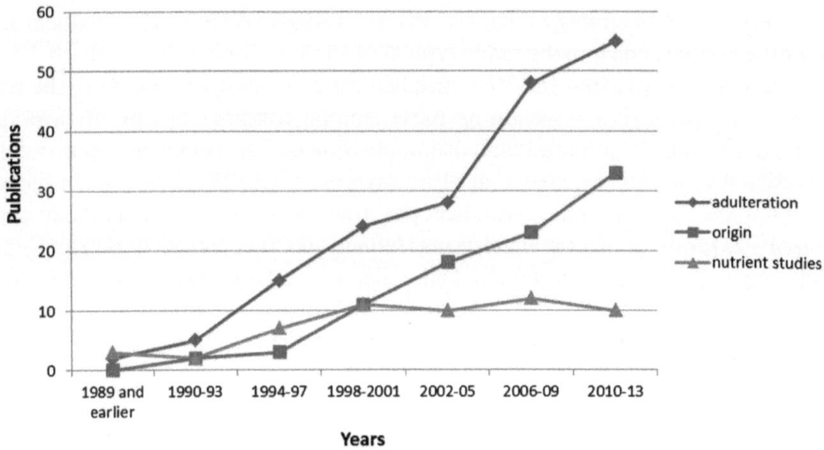

FIGURE 19 Time evolution of IRMS research in 4 years period.

3.2.2 Authentication

3.2.2.1 Adulteration

IRMS has been proven to be a very efficient technique for uncovering fraudulent practices since its early development in the 1980s. Characteristic examples are the detection of undeclared addition of cane or corn sugars to apple juice [222] and to orange juice [223] by the variability of the $^{13}C/^{12}C$ ratio. A list of the food categories in which IRMS was successfully applied in, regarding adulteration detection, is following:

- Wines-beverages-spirits; Detection of ethanol enrichment by the variability of the $^{13}C/^{12}C$ ratio in wines [224], detection of added (exogenous) CO_2 in sparkling drinks by the variability of the $^{13}C/^{12}C$ ratio with GC-IRMS [225–227], detection of wine enrichment by bot sucrose addition and must concentration via δ^2H and $\delta^{18}O$ isotope ratios of wine water [228], investigation of Scotch whisky brand authenticity trough congeners using GC-IRMS by the variability of the $^{13}C/^{12}C$ ratio [229]. In another interesting work Tequila authenticity was assessed by GC-IRMS analysis of $^{13}C/^{12}C$ and $^{18}O/^{16}O$ ratios of ethanol [230]. Authenticity of carbon dioxide bubbles of French ciders was achieved through $^{13}C/^{12}C$ ratios [231].
- Vegetable oils: Detection of maize oil as an adulterant is based on the differences in $\delta^{13}C$ value between plants. This results from the different biosynthetic paths by different plants, C3 or C4 plants that use different modes of carbon dioxide fixation resulting in different $^{13}C/^{12}C$ ratios assessed through GC-IRMS [232]. Detection of olive oil adulteration with pomace oil was also based on differences in $\delta^{13}C$ values of aliphatic alcoholic fraction by GC-IRMS [233]. Another example is the detection of corn oil in adulterated sesame oil by GC-IRMS, based on $\delta^{13}C$ variability [234].

- Juices; Detection of added water in orange juice by determination of $^{18}O/^{16}O$ ratios of water [235]. Detection of exogenous citric acid in fruit juices by the variability of the $^{2}H/^{1}H$ and $^{13}C/^{12}C$ ratios [228,236].
- Syrups: There are several publications concerning IRMS use for uncovering syrups fraudulent practices: detection of maple syrup adulteration by exogenous sugar addition (beet and cane sugar) was via $\delta^{13}C$ variability [237]. Detection of adulteration in mulberry pekmez with various sugar syrups based on different $\delta^{13}C$ values [238]. In another publication, a collaborative study of $^{13}C/^{12}C$ ratio of fermentation ethanol for detecting some sugar additions in maple syrup is reported. The maple syrups were completely fermented with yeast, and the alcohol was distilled with a quantitative yield (>96%). The $\delta^{13}C$ of ethanol becomes less negative when exogenous sugar derived from plants exhibiting a C4 metabolism (e.g., corn or cane) is added to a maple syrup obtained from plants exhibiting a C3 metabolism (most common fruits except pineapple). Conversely, the $\delta^{13}C$ of ethanol becomes more negative when exogenous sugar derived from C3 plants (e.g., beet, wheat, rice) [239].
- Milk: Distinction between raw and reconstituted milk based on differences in $\delta^{2}H$ and $\delta^{18}O$ values [240].
- Honey and royal jelly: honey adulteration assessment is mostly through IRMS. Adulteration practices uncovered were addition of high fructose corn syrup as a sugar source and biotechnologically produced yeast powder as protein source in royal jelly by the variability of the $^{13}C/^{12}C$ and $^{15}N/^{14}N$ ratios [241]. Detection of honey adulteration with various sugar syrups were uncovered by $\delta^{13}C$ variability [242]. Another work reports the investigation of commercial honey adulteration based on $^{13}C/^{12}C$ by EA-IRMS [243].
- Spices: Another successful application of IRMS described in several publications concerns the distinction between natural and synthetic vanillin, based on different $\delta^{13}C$ and $\delta^{2}H$ values. Vanilla remains one of the most important and widely used flavors in food industry. Natural vanilla flavor, extracted from the pods of the tropic orchid vanilla, is considerably more expensive than synthetic vanillin. The disparity of prices between natural vanillin and that derived from other sources has given rise to many cases of fraudulent adulteration, and for more than 30 years, strenuous efforts have been made to authenticate sources of vanillin [244–246]. Distinction between natural and synthetic allyl isothiocyanate in mustard oils is based on differences in $\delta^{13}C$, $\delta^{15}N$, and $\delta^{34}S$. This molecule can be synthesized at a much lower price than its cost when extracted from mustard seeds [247].
- Essential oils: Essential oils can be adulterated, by addition of synthetic compounds related to the composition of the oil or of natural compounds or oil fractions from other, cheaper, essential oils, to reduce the cost of the essential oil. For essential oils containing citronellal and citral adulteration detection was by GC-IRMS based on different $\delta^{13}C$ and $\delta^{2}H$ values [248]. Cold-pressed mandarin essential oils are products of great economic importance in many parts of the world and are used in perfumery, as well as in food products.

Detection of adulterated mandarin essential oils by GC-IRMS was based on different $\delta^{13}C$ values [249].

- Food ingredients: within the European Community only L-tartaric acid extracted from grapes (therefore natural) is allowed. Detection of L-tartaric acid "naturalness" was based on $\delta^{13}C$ and $\delta^{18}O$ differences [250]. Discrimination of natural and synthetic caffeine contained in a big variety of drinks via high-temperature reversed-phase LC-IRMS by $\delta^{13}C$-values [251] and also via $\delta^{13}C$ and $\delta^{18}O$ values by EA-IRMS [252]. Detection of synthetic acetic acid in vinegar by the variability of $\delta^{13}C$ and $\delta^{18}O$ [253].

3.2.2.2 Determination of Origin

Determination of origin attracts the interest of the scientific community as shown in Figure 19. Most of the work (88%) concerns geographical origin, while a fraction (6%) is devoted to the botanical origin and (6%) to organic food.

IRMS is a technique that can distinguish chemically identical compounds based on their isotope content. The ratio of the stable isotopes of the elements that constitute almost all biological material, $^{13}C/^{12}C$, $^{15}N/^{14}N$, $^{18}O/^{16}O$, and $^{2}H/^{1}H$ can be determined. In addition other elements like $^{34}S/^{32}S$ and $^{37}Cl/^{35}Cl$ can be included to improve the discriminative power. The isotopic composition of proteins, carbohydrates, fats, and minerals of agricultural products depends on various factors such as latitude, temperature, and local agricultural practices. Some of those factors can be expected to be indicative for the geographical origin, others are more related to the production factors. These factors include the use of fertilizers, certain feedings stuffs in the diet of farm animals, seasonal variations, and geological factors (e.g., soil composition, altitude, etc.). Factors that affect the stable isotope ratio (e.g., isotopic fractionation) can be used in assigning the regional origin of agricultural products. For example, the $^{18}O/^{16}O$ ratio is highly dependent on the distance to the ocean and on the altitude above sea level of the production site. IRMS data are collected for several elements and interpreted using chemometric methods. Furthermore, IRMS combined with or without other techniques (e.g., elemental analysis, NMR, GC) and/or chemometric methods has been applied to determine the origin of a variety of food products [102,254–256].

IRMS has been accepted as an effective tool for origin determination on different types of foods:

- Wines: At the early beginning, IRMS was mostly used for wines' geographical origin; $^{13}C/^{12}C$ of ethanol and $^{18}O/^{16}O$ of water were used for investigation of exact origin among different wine-producing Italian regions [257]. Reasonable differentiation results are achieved at a microregional scale of red wines from Valencia (Spain) in terms of geographic provenance and even grapevine genotypic features by $\delta^{13}C$ values of fermentative ethanol [258]. $\delta^{13}C$ values were also used for characterization of the geographical origin of Bordeaux wines

[259]. In another work, $\delta^{18}O$ was proven to be a reliable marker for EU wines origin discrimination [260]. Also, regarding wine's ethanol characterization, $\delta^{13}C$ values can be used as measured by LC-IRMS. This way the measurement of $\delta^{13}C$ was easier than with GC-IRMS that requires previous ethanol isolation [261]. Differences in $^{18}O/^{16}O$ ratios reflect the specific climatic and harvesting condition of each vineyard of Romanian wines [262].

- Beverages-spirits: Investigation of Scotch whisky geographical origin was achieved using GC-IRMS by $\delta^{15}N$ ratio [229]. The reputation of particular countries and plantations within those countries, for the production of high quality tea means that specific producers can ask a significantly higher price for their product. This leads to a temptation for unscrupulous producers to fraudulently label their product as coming from one of these areas to take advantage of this higher price. Provenance of tea (*Camellia sinensis*) was revealed through determination of $\delta^{2}H$ and $\delta^{13}C$ values [263].
- Vegetables—fruits and juices: Characterization of the geographical origin of blueberries (*Vaccinium corymbosum* L.) was achieved through $^{18}O/^{16}O$ ratios of juice and $^{13}C/^{12}C$ and $^{15}N/^{14}N$ of pulp [264]. Characterization of commercial Slovenian and Cypriot fruit juices was based on $\delta^{13}C$ values in the pulp, sugars, and ethanol and measurements of $\delta^{18}O$ in water [265]. Identification of Chinese Schisandra fruits' geographical origins based on $\delta^{13}C$ variability by EA-IRMS [266]. Differentiation between organically and conventionally grown vegetables was by $^{15}N/^{14}N$ ratios [267].
- Vegetable oils: Characterization of olive oil according to their geographical origin via the variability of the $^{13}C/^{12}C$ ratio of individual fatty acids was by GC-IRMS [268]. Verification of geographical origin and distinction between raw and refined camelina or other less-expensive oils was achieved by $\delta^{13}C$ values of fatty acids. The use of $\delta^{13}C$-18:2ω6 versus $\delta^{13}C$-18:3ω3 correlation can clarify cases where impurity or adulteration is suspected [269]. Characterization of authentic Italian extra-virgin olive oils is described by $^{18}O/^{16}O$, $^{13}C/^{12}C$, and $^{2}H/^{1}H$ ratios [270].
- Cereal—wheat: IRMS was able to indicate wheat's geographical origin (e.g., Italian durum wheat semolina) [271]. Geographical origin determination of distillers dried grains and solubles was $^{2}H/^{1}H$, $^{13}C/^{12}C$, $^{15}N/^{14}N$, $^{18}O/^{16}O$173, and $^{34}S/^{32}S$ isotope ratios [272].
- Milk and dairy products: Characterization of cow milk according to geographic origin [273,274]. Characterization of the geographical origin of buffalo milk and mozzarella cheese was achieved through $^{13}C/^{12}C$ and $^{15}N/^{14}N$ isotopic ratios [275]. Parmigiano Reggiano (PDO hard cheese) authentication was achieved through $\delta^{13}C$, $\delta^{2}H$, $\delta^{15}N$, $\delta^{34}S$ variability [256]. A very recent work describes the discrimination between organic and conventional milk using $\delta^{13}C$ and $\delta^{15}N$ [276].
- Meat: lamb meat distinction according to their country of origin and feeding regime was achieved by analysis of $^{13}C/^{12}C$ and $^{15}N/^{14}N$ ratios [277]. Combined C, N, and S stable isotope ratio analysis leads to verification of the

geographical origin, feeding history, and separation of organically and conventionally produced beef cattle [278]. In a similar work, beef meat geographical origin discrimination was achieved by $^{18}O/^{16}O$ ratios [279]. Corn-fed poultry authentication has been achieved through $\delta^{13}C$ values [280]. Regional geographical traceability of cattle in China was achieved through $\delta^{13}C$ and $\delta^{15}N$ variability [281]. Origin authentication of pork was through differences in $\delta^{13}C$ and $\delta^{15}N$ values [282]. Muscle $^{13}C/^{12}C$ and $^{34}S/^{32}S$ and plasma $^{13}C/^{12}C$ and $^{18}O/^{16}O$ ratios reveal the feeding changes of lambs [283]. Authentication of PDO dried-cured hams according to their origin, breeding and processing conditions was by $^{2}H/^{1}H$, $^{13}C/^{12}C$, $^{15}N/^{14}N$, $^{18}O/^{16}O$, and $^{34}S/^{32}S$ ratios [284].

- Fish: Consumers are interested in being assured that fish are wild grown as claimed and not farmed. Discrimination between wild and farmed gilt-head sea bream (*Sparus aurata*) was achieved through $\delta^{13}C$ and $\delta^{15}N$ values. These are the most informative parameters of fish diet. $\delta^{13}C$ gives a tool to distinguish between wild and farmed gilt-head sea bream and $\delta^{15}N$ being more informative on the geographical origin of fish. This fact could be related more to differences in feed mixtures given to farmed fish than to geographic aspects [285]. A similar work also concerning wild gilt-head sea bream authentication, proposes the $\delta^{15}N$ of phospholipid choline nitrogen and $\delta^{13}C$ of fatty acids as markers [286].
- Honey and similar products: Determination of the geographical origin of black locust, lime, and chestnut honey by $\delta^{13}C$ and $\delta^{15}N$ values [243,287]. Geographical and botanical origin discrimination of Romanian honey was based on $\delta^{13}C$, $\delta^{18}O$, and $\delta^{2}H$ variability [288].
- Spices: Confirmation of vanillins' origin in ice cream and yoghurt was based on $^{13}C/^{12}C$ ratios by EA-IRMS [289].
- Essential oils: discrimination of bergamot oils according to their geographic provenance by GC-IRMS [290].
- Food ingredients: Origin and naturalness assessment of decanal, linalool, and linalyl acetate based on $^{2}H/^{1}H$ ratio measurements by GC-IRMS [291]. Discrimination between 68 green coffee bean samples from 20 different geographic origins distributed over Central America, Pacific, South America, Africa, Asia, and Oceania was by their isotopic composition ($\delta^{13}C$, $\delta^{15}N$, and $\delta^{18}O$) [292]. Characterization of the vinegar's acetic acid botanical origin by $\delta^{13}C$ and $\delta^{18}O$ values [253]. Another article investigates whether the analysis of stable isotope ratios D/H and $^{13}C/^{12}C$ in ethanol and acetic acid and of $^{18}O/^{16}O$ in water can be applied to the ingredients of "aceto balsamico di Modena IGP" (ABM) and wine vinegar in order to evaluate their authenticity [293].

3.2.3 Nutrient Intake and Bioavailability

IRMS is useful in bioavailability and nutrient intake studies. Consumption of anthocyanin-rich foods benefits the cardiovascular system. A recent study uses LC-IRMS to reveal the absorption, distribution, metabolism, and elimination of anthocyanin-rich foods through a $^{13}C_5$-labeled anthocyanin [294]. Carbon isotope ratio of alanine in red blood cells was proposed as a new biomarker for the clarification of sugar-sweetened beverage intake contribution to obesity and chronic disease risk [295].

In another nutrient study, serum carbon and nitrogen stable isotopes as potential biomarkers of dietary intake and their relation with incident type 2 diabetes (the EPIC-Norfolk study) were investigated. Stable-isotope ratios of carbon and nitrogen have been proposed as potential nutritional biomarkers to distinguish between meat, fish, and plant-based foods consumption. The isotope ratios $\delta^{13}C$ and $\delta^{15}N$ may both serve as potential biomarkers of fish protein intake, whereas only $\delta^{15}N$ may reflect broader animal-source protein intake in a European population. The inverse association of $\delta^{13}C$ but a positive association of $\delta^{15}N$ with incident diabetes should be interpreted in the light of knowledge of dietary intake and may assist in identifying dietary components that are associated with health risks and benefits [296].

Another interesting work concluded that ^{13}C amino acids fingerprints could provide a powerful in situ assay of the biosynthetic sources of amino acids (plant, fungal, and bacterial origins) and a potential new tool for understanding nutritional linkages in food webs [297]. In case of muscle-based foods, the incorporation of dietary isotopic signals into muscle tissue is a dynamic process and it is not known whether all muscles, or locations within a muscle, have the same isotopic composition. Therefore, an experiment was conducted in which 28 lambs were switched from a control diet to an isotopically distinct experimental diet offered at two different energy allowances. Small, albeit significant, differences were detected in tissue carbon turnover within the muscle Longissimus dorsi. Intermuscular comparison showed similar carbon half-lives for four of the five analyzed muscles. The results also clearly demonstrated that the energy allowances had a significant impact on intra- and intermuscular carbon turnover. The findings of similar tissue carbon turnover of several muscles sold as meats and the comparable tissue-diet fractionation of the analyzed muscles, both within the same EA, will enable scientists to analyze different meat samples according to availability without introducing large biases. However, EA must be considered a factor of uncertainty. Significant differences in tissue turnover can exist within one muscle. Even though these differences were small, biases can be avoided by collecting samples from the same location throughout an experiment where possible. Furthermore, calculated differences in diet-tissue fractionation and half-lives of different muscles sampled from animals receiving different EAs may affect results significantly. However, the differences between muscles collected from animals on the same EA were rather small, while the diet-tissue fractionation and half-lives of four muscles commonly consumed as meats were very similar. Calculation of the animal growth factors revealed a strong link between EA and tissue turnover, as the contribution of growth to tissue turnover was highest in animals receiving a high EA. The finding that EA can affect muscle tissue growth, diet-tissue fractionation, tissue half-lives, or growth contribution toward turnover must be considered as a source of uncertainty when IRMS is applied in meat authentication because the EA is usually unknown in surveys of meats and should therefore not be underestimated [298].

IRMS investigates the possible benefits and risks of protein consumption. Leucine can be a potential marker of protein intake. As it is known, high protein or meat intake might be a risk factor for metabolic disorders [299]. A similar study resulted in that the choice of dietary protein of vegetarians and omnivores is reflected in their hair protein ^{13}C and ^{15}N abundance, using EA-IRMS and GC-IRMS [300]. Another study revealed that carbon and nitrogen-stable isotopic composition of hair protein and amino acids can be used as biomarkers for animal-derived dietary protein intake in humans [301].

Examples of the use of IRMS in bioavaibility studies concern iron [302] and the investigation of the natural abundance of the minor isotope of carbon, ^{13}C, in common foodstuffs in the British diet in order to highlight the difference in isotopic abundance between northern European foodstuffs and North American foodstuffs [303]. IRMS is also used for investigation of specific compound biosynthesis e.g., human milk oligosaccharides (HMO) from ^{13}C-enriched food. For example, incorporation of orally applied ^{13}C-galactose into milk lactose and oligosaccharides [304].

3.2.4 Other

There are also some other applications of IRMS regarding food. A very recent interesting work is the use of GC-IRMS in quantifying the contribution of grape hexoses to wine volatiles by high-precision [U^{13}C]-glucose tracer studies [305]. IRMS is also implemented for food safety purposes. A characteristic example is the detection of the abuse of 17β-estradiol in cattle by the distinction between endogenous steroids and their synthetic homologues on the basis of their $^{13}C/^{12}C$ isotopic ratio by GC-IRMS [306–313]. The use of these steroids as growth-promoting agents in food production is banned under European Union legislation.

3.3 TIMS

Major applications of TIMS concerning food, waters, and beverages are authentication, safety and bioavailability and nutrient studies:

- Authentication: The radiogenic isotopic compositions of inorganic heavy elements such as Sr, Nd, and Pb of the food chain usually constitute a reliable geographic fingerprint. Their isotopic ratios are inherited by the geological substratum of the territory of production. The Sr isotope composition of geomaterials (i.e., rocks and soils) is largely variable, and it depends upon the age of the rocks and their nature (e.g., genesis, composition). For example, distinction of wines geographic provenance is achieved by $^{87}Sr/^{86}Sr$ ratios [314]. In the same way, determination of tomatoes (berries, "passata," tinned tomatoes, sauce, double and triple concentrate) geographical origin is made by virtue of different $^{87}Sr/^{86}Sr$ ratios and δ‰ values [315]. Another recent work describes authentication of PDO Nîmes (France) olive oil by $^{87}Sr/^{86}Sr$ ratios and methodology development for Sr extraction from olive oils [316]. Furthermore, differentiation of cattle's beef geographical origin, located in 12 different European regions within France, Germany, Greece, Ireland, Italy, Spain, and the UK has been managed by strontium isotope ratio $^{87}Sr/^{86}Sr$ variations [317].

- Safety: Uranium isotope ratio measurement in water samples collected around the 30 km Chernobyl exclusion zone. ^{234}U, ^{235}U, and ^{238}U are naturally occurring alpha emitting long-lived radionuclides, which are taken up daily at low levels with food and drinking. IUPAC has established natural isotopic composition of $^{235}U/^{238}U$ to be 0.00725. Therefore, isotope ratio measurements are important to provide information on the origin of uranium [318]. Finally, TIMS has been used in a study for elevated lead levels in drinking water [319].
- Nutrient Intake—Bioavailability: A number of minerals contained in foods are essential nutrients for humans and animals. While most vitamins are very well absorbed, most essential minerals are not. Usual absorption of minerals ranges from less than 1% to over 90%. The bioavailability of dietary minerals must be considered when determining whether the diet contains enough, too little, or too much. By using stable isotope tracers as labels, the metabolic fate of minerals in a specific day's diet, a specific meal, or a food can be distinguished from minerals from other sources. TIMS is routinely used to study bioavailability of Zn, Cu, and Fe [320]. Quantification of ferritin-bound iron in plant samples using isotope tagging and species-specific isotope dilution mass spectrometry is achieved by means of a biosynthetically produced ^{57}Fe-labeled ferritin spike and negative thermal ionization mass spectrometry. Ferritin is nature's predominant iron storage protein. This protein has been identified as a target molecule for increasing iron content in plant staple foods in order to combat dietary iron deficiency, a major public health problem in developing countries [321]. The bioavailability of iron in different weaning foods and the enhancing effect of a fruit drink containing ascorbic acid was also managed by TIMS [322]. Moreover, the relationship assessment between both serum hepcidin and serum prohepcidin with nonheme-iron absorption in the presence and absence of food has been also studied with the use of TIMS. In addition, serum hepcidin is significantly associated with iron absorption from food and supplemental sources in healthy young women. Hepcidin is a key regulator of iron homeostasis [323]. Iron and zinc bioavailability in children, from Chinese traditional diets consumed commonly in Shandong rural region on the basis of the contents of nutritional factors known to promote or inhibit food iron and zinc absorption is done by TIMS [324,325]. A final example of TIMS applications is iron absorption in infants with the use of ^{57}Fe and ^{58}Fe as labels [326].

4. CONCLUSION AND FUTURE OUTLOOK

Nowadays, mass spectrometry techniques are the basic choice for food safety and quality assessment. Regarding elemental and isotope food analysis, ICP-MS and IRMS techniques are the most reliable and robust analytical tools.

Globalization and the growing complexity of the food chain, combined with recent food scares such as the horse meat scandal, the melamine adulterated milk and numerous others, have raised consumer awareness regarding the quality and authenticity of the food they consume. Authenticity has probably always been a major concern of many consumers, and it is still gaining more and more importance. In Europe, origin is one of the main authenticity issues dealing with food. The recent trend in European legislation is the protection of "mountain products" and "products of island farming." These two labels are the very recent companions to PDO and PGI labels that will be implemented in EU legislation in the near future. The information includes, beyond the others, a characterization on the morphology of the geographic region. These steams from the consumer perception of special quality attributed to mountain and island products. In the case of cultivated species, EU indicates that a reference should be made to the country in which the product undergoes the final development stage.

High quality products with geographical indications and designations of origin are generally high-priced and bring in a higher benefit to the producers than other similar products. So there is a need to protect such products by detecting possible commercial frauds. These products are defined by geographical origin, know-how, certain farming methods (e.g., organic foods) and in some cases by feeding diet and animal breed. At present, knowledge of the isotopic and elemental food fingerprint looks as a plausible way to establish the geographical origin of food products. The combination of elemental concentrations and isotopic variation data by chemometric tools is one of the most used approaches nowadays [327].

Inductively coupled plasma mass spectrometry is a powerful technique allowing multielemental ultratrace analysis for a wide variety of samples that is, nowadays, a routine in food analysis laboratories. ICP-MS has clear advantages in its multielement capability and throughput comparable only by ICP-OES that is however inferior in terms of detection limits, isotopic capability and ability for rare earths. It should be noted that the lately increased interest in rare earths will grow stronger in the near future. Although laser ablation provides capabilities for direct analysis of solids, it carries the penalty concerning calibration samples in the solid form. However, LA-ICP-MS could provide solutions when spatial analysis of food samples is important. Method detection limits reach the ppq (pg/L) range for elements with high *m/z* ratios which makes the technique especially useful for determination of such low-abundant elements as REE, actinide elements, etc.

Today, ICP-MS technique has become popular for the elemental analysis of food, notwithstanding the expensiveness of the instruments and the high running cost in terms of consumables and maintenance. Many governmental institutions have turned to ICP-MS for their institutional and control activities. On the other hand, the technique is being increasingly used for quality control and compliance with food regulations by major food industries or independent analytical laboratories. As a result, atomic absorption spectroscopy (AAS) is no longer being prescribed as the exclusive detection technique by legislation laying down the analytical methods for the official control of trace elements in food. Rather, either performance criteria are established and any validated method that fulfils these requirements is accepted, or ICP-MS is explicitly placed in the list of recommended techniques.

The determination of essential and toxic trace elements in nutrition has acquired great importance. However, elements may occur in food in many chemical forms, i.e., as building blocks of essential macronutrients (e.g., proteins, carbohydrates, lipids, etc.) and micronutrients (e.g., vitamins, enzymes) and as minerals and trace elements. Today it is widely recognized that the nature and amount of a given element chemical species in a food matrix, rather than the corresponding element total analytical concentration, will determine its bioavailability (e.g., in human milk), metabolism, transport/storage in the body, and eventually its biochemical essential role and so its nutritional value. In other words, chemical element species information (apart from total elemental concentration) is needed for a sound assessment of how the considered element is absorbed, retained, metabolized, etc., and to decide whether its effects are beneficial (essential elements), toxic, or rather the element has no adverse impact at a specific concentration, but it can become therapeutic at higher levels of concentration. In brief, the consideration of the essentiality (or toxicity) of a given trace element is deeply related to its chemical form in the food. Therefore, aiming at assessing the role of trace elements in food and nutrition, problem-related chemical speciation is becoming a most important analytical strategy. Study areas have included the identification of metalloproteins and metalloenzymes, with subsequent analysis of their structure and function, the detection of biological metal-complexing ligands, and the differentiation between oxidation states of metals such as chromium, for which the Cr(III) oxidation state is essential and the Cr (VI) oxidation state is toxic. The best and almost the only way for all these studies are by hyphenated ICP-MS techniques.

The emerging research fields of metallomics and foodomics require advanced mass spectrometry technologies, such as ICP-MS and IRMS. Advantages of mass spectrometry include the abilities to perform speciation studies when coupled with separation techniques, to measure isotopic composition in nutritional studies and to identify sources of environmental exposure.

There are also certain concerns on the safety of nanomaterials in food products and migration from packaging materials. ICP-MS has been used during the very recent years to elucidate potential hazards resulting from metallic particles introduced to food products by nanomaterials used.

Furthermore, one of the major advantages of isotope techniques is that multiple isotopes of the same element can be used simultaneously and multiple elements can be studied simultaneously. The use of stable isotopes for studies of bioavailability of elements in foods has gained widespread interest in recent years. The approach is expected to be applied to an increasing number of food science and nutrition problems in the future.

IRMS has been shown to have wide versatility when coupling with several different interfaces. In determining which interface would be best suited for coupling to the IRMS, the sample itself is the most important determining factor. Nonvolatile substances such as amino acids and fatty acids can be most easily measured with EA-IRMS. Even though this technique only provides an average isotope ratio value for the entire sample. Analysis can typically be performed on samples as small as 0.5 mg and often avoids the complex sample preparation procedures that are usually needed for GC- or LC-IRMS analysis. With that said, it is important to note that GC-IRMS can be used for most volatile organic substances without sample preparation. Also, over the years sample introduction method of LC-IRMS is developed and is becoming more mature for routine labs.

Regardless the sample introduction method, a noticeable gap in the market exists for IRMS standards, although there are increasing several suppliers of bulk isotope ratio standards such as polyethylene, sugar, and flour. Another major issue is the rendering of compatible isotopic measurement referencing strategies. IRMS offers the potential of unlimited applications for nonvolatile and volatile compounds while achieving higher accuracy and precision via increased automation. Through continued progress in fundamental understanding and application development, IRMS is transiting to a more routine technique. Future instrumentation goals could then focus on shrinking the footprint and cost of the instrumentation, reducing analysis times, obtaining higher resolution data, and perhaps even looking toward miniaturization or portable instruments.

REFERENCES

[1] X. Wanga, S. Wangb, Z. Cai, The latest developments and applications of mass spectrometry in food-safety and quality analysis, TrAC 52 (2013) 170–185.

[2] D. Aiello, D. De Luca, E. Gionfriddo, A. Naccarato, A. Napoli, E. Romano, A. Russo, G. Sindona, A. Tagarelli, Review: multistage mass spectrometry in quality, safety and origin of foods, Eur. J. Mass. Spectrom. 17 (1) (2011) 1–31.

[3] L. Järup, Hazards of heavy metal contamination, Br. Med. Bull. 68 (2003) 167–182.

[4] J. Zukowska, M. Biziuk, Methodological evaluation of method for dietary heavy metal intake, J. Food Sci. 73 (2) (2008) R21–R29.

[5] M. Demont, K. Boutakhrit, V. Fekete, F. Bolle, J. Van Loco, Food Chem. Toxicol. 50 (3–4) (2012) 734–743.

[6] K.R. Mahaffey, Fish and shellfish as dietary sources of methylmercury and the o-3 fatty acids, eicosahexaenoic acid and docosahexaenoic acid: risks and benefits, Environ. Res. 95 (2004) 414–428.

[7] P. Sivapcrumal, T.V. Sankar, P.G. Viswanathan Nair, Heavy metal concentrations in fish, shellfish and fish products from internal markets of India vis-a-vis international standards, Food Chem. 102 (2007) 612–620.

[8] X. Wang, T. Sato, B. Xing, S. Taob, Health risks of heavy metals to the general public in Tianjin, China via consumption of vegetables and fish, Sci. Total Environ. 350 (2005) 28–37.

[9] J. Abedin, M.S. Cresser, A.A. Meharg, J. Feldmann, J. Cotter-Howells, Arsenic accumulation and metabolism in rice (Oryza sativa L.), Environ. Sci. Technol. 36 (2002) 962–968.

[10] T. Uchinoa, T. Roychowdhurya, M. Andob, H. Tokunagaa, Intake of arsenic from water, food composites and excretion through urine, hair from a studied population in West Bengal, India, Food Chem. Toxicol. 44 (4) (2006) 455–461.

[11] J. Hu, Y. Ma, L. Zhang, F. Gan, Y.-S. Ho, A historical review and bibliometric analysis of research on lead in drinking water field from 1991 to 2007, Sci. Total Environ. 408 (7) (2010) 1738–1744.

[12] S. Blunden, T. Wallace, Tin in canned food: a review and understanding of occurrence and effect, Food Chem. Toxicol. 41 (2003) 1651–1662.

[13] EPA, U.S. Environmental Protection Agency—Hazard summary, Mercury compounds. Available at: http://www.epa.gov/ttn/atw/hlthef/mercury.html (Last accessed: 6/2014).

[14] EPA, U.S. Environmental Protection Agency—Hazard summary, Arsenic compounds. Available at: http://www.epa.gov/ttn/atw/hlthef/arsenic.html (Last accessed: 6/2014).

[15] EPA, U.S. Environmental Protection Agency—Hazard summary, Chromium compounds. Available at: http://www.epa.gov/ttn/atw/hlthef/chromium.html (Last accessed: 6/2014).

[16] A. Kabata-Pendias, A.B. Mukherjee (Eds.), Trace Elements from Soil to Human, Springer, Berlin, 2007.

[17] M.A. Lynes, Y.J. Kang, S.L. Sensi, G.A. Perdrizet, L.E. Hightower, Heavy metal ions in normal physiology, toxic stress, and cytoprotection, Ann. N.Y. Acad. Sci. 1113 (2007) 159–172.

[18] V. Mudgal, N. Madaan, A. Mudgal, R.B. Singh, S. Mishra, Effect of toxic metals on human health, Open Nutraceuticals J. 3 (2010) 94–99.

[19] P. Koedrith, Y.R. Seo, Advances in carcinogenic metal toxicity and potential molecular markers, Int. J. Mol. Sci. 12 (12) (2011) 9576–9595.

[20] C.C. Bridges, R.K. Zalups, Molecular and ionic mimicry and the transport of toxic metals, Toxicol. Appl. Pharmacol. 204 (2005) 274–308.

[21] EPA, U.S. Environmental Protection Agency—Radiation protection, Radium. Available at: http://www.epa.gov/radiation/radionuclides/radium.html#affecthealth (Last accessed: 6/2014).

[22] P. Hajeb, J.J. Sloth, Sh. Shakibazadeh, N.A. Mahyudin, L. Afsah-Hejri, Toxic elements in food: occurrence, binding, and reduction approaches, Compr. Rev. Food Sci. Food Saf. 13 (4) (2014) 457–472.

[23] Commission Regulation (EC) No 1881/2006 of 19 December 2006 setting maximum levels for certain contaminants in foodstuffs. Available at: http://eur-lex.europa.eu/LexUriServ/LexUriServ.do?uri=CONSLEG:2006R1881:20100701:EN:PDF (Last accessed: 6/2014).

[24] Commission Regulation (EC) No 629/2008 of 2 July 2008 amending Regulation (EC) No 1881/2006 setting maximum levels for certain contaminants in foodstuffs. Available at: http://eur-lex.europa.eu/LexUriServ/LexUriServ.do?uri=OJ: L:2008:173:0006:0009:EN:PDF (Last accessed: 6/2014).

[25] Directive 2002/32/EC of the European Parliament and of the Council of 7 May 2002 on undesirable substances in animal feed. Available at: http://eur-lex.europa.eu/LexUriServ/LexUriServ.do?uri=CONSLEG:2002L0032:20061020:EN:PDF (Last accessed: 6/2014).

[26] Council Directive (EEC) No 500/84 of 15 October 1984 on the approximation of the laws of the Member States relating to ceramic articles intended to come into contact with foodstuffs. Available at: http://eur-lex.europa.eu/LexUriServ/LexUriServ.do?uri=CONSLEG:1984L0500:20050520:EN:PDF (Last accessed: 6/2014).
[27] Commission Directive 2005/31/EC of 29 April 2005 amending Council Directive 84/500/EEC as regards a declaration of compliance and performance criteria of the analytical method for ceramic articles intended to come into contact with foodstuffs. Available at: http://eur-lex.europa.eu/LexUriServ/LexUriServ.do?uri=OJ: L:2005:110:0036:0039:en:PDF (Last accessed: 6/2014).
[28] Commission Regulation (EU) No 10/2011 of 14 January 2011 on plastic materials and articles intended to come into contact with food. Available at: http://eur-lex.europa.eu/LexUriServ/LexUriServ.do?uri=CONSLEG:2011R0010:20111230:EN: PDF (Last accessed: 6/2014).
[29] Regulation (EC) No 396/2005 of the European Parliament and of the Council of 23 February 2005 on maximum residue levels of pesticides in or on food and feed of plant and animal origin and amending Council Directive 91/414/EEC. Available at: http://eur-lex.europa.eu/LexUriServ/LexUriServ.do?uri=CONSLEG:2005R0396:20110101:EN: PDF (Last accessed: 6/2014).
[30] Council Directive 98/83/EC of 3 November 1998 on the quality of water intended for human consumption. Available at: http://eur-lex.europa.eu/LexUriServ/LexUriServ.do?uri=OJ: L:1998:330:0032:0054:EN: PDF (Last accessed: 6/2014).
[31] FAO, Codex General Standard for Contaminants and Toxins in Food and Feed, Codex Stan 193–1995. Available at: http://www.fao.org/fileadmin/user_upload/agns/pdf/CXS_193e.pdf (Last accessed: 6/2014).
[32] WHO, Guidelines for Drinking-water Quality, Third Edition, Incorporating the First and Second, Addenda. Available at: http://www.who.int/water_sanitation_health/dwq/fulltext.pdf (Last accessed: 6/2014).
[33] EPA, U.S. Environmental Protection Agency, Drinking Water Contaminants, National Primary Drinking Water Regulations. Available at: http://water.epa.gov/drink/contaminants/index.cfm#Inorganic (Last accessed: 6/2014).
[34] Australia New Zealand Food Standards Code - Standard 1.4.1-Contaminants and Natural Toxicants - F2011C00542. Available at: http://www.comlaw.gov.au/Details/F2011C00542/Download (Last accessed: 6/2014).
[35] Peoples Republic of China–Administrative Measures for National Food Safety Standards. Available at: http://gain.fas.usda.gov/Recent%20GAIN%20Publications/Administrative%20Measures%20for%20National%20Food%20Safety%20Standards_Beijing_China%20-%20Peoples%20Republic%20of_3-11-2013.pdf (Last accessed: 6/2014).
[36] J. Bogden, L.M. Klevay (Eds.), Clinical Nutrition of the Essential Trace Elements and Minerals: The Guide for Health Professionals, Humana Press, Totowa, New Jersey, 2000.
[37] S.B. Goldhaber, Trace element risk assessment: essentiality vs toxicity, Regul. Toxicol. Pharmacol. 38 (2003) 232–242.
[38] M. Farina, D. Silva Avila, J. Batista Teixeira da Rocha, M. Aschner, Metals, oxidative stress and neurodegeneration: a focus on iron, manganese and mercury, Neurochem. Int. 62 (2013) 575–594.
[39] C.G. Fraga, Relevance, essentiality and toxicity of trace elements in human health, Mol. Aspects Med. 26 (2005) 235–244.
[40] M. Stefanidou, C. Maravelias, A. Dona, C. Spiliopoulou, Zinc: a multipurpose trace element, Arch. Toxicol. 80 (2006) 1–9.
[41] K. Kaur, R. Gupta, S.A. Saraf, S.K. Saraf, Zinc: the metal of life, Compr. Rev. Food Sci. Food Saf. 13 (4) (2014) 358–376.

[42] D. Volkmer, B. Hommerich, K. Griesar, W. Haase, B. Krebs, Dinuclear nickel(II) complexes as models for the active site of urease, Inorg. Chem. 35 (13) (1996) 3792–3803.
[43] Regulation (EU) No 1169/2011 of the European Parliament and of the Council of 25 October 2011 on the provision of food information to consumers, amending, Regulations (EC) No 1924/2006 and (EC) No 1925/2006 of the European Parliament and of the Council, and repealing Commission Directive 87/250/EEC, Council Directive 90/496/EEC, Commission Directive 1999/10/EC, Directive 2000/13/EC of the European Parliament and of the Council, Commission Directives 2002/67/EC and 2008/5/EC and Commission Regulation (EC) No 608/2004. Available at: http://eur-lex.europa.eu/LexUriServ/LexUriServ.do?uri=OJ:L:2011:304:0018:0063:EN: PDF (Last accessed: 6/2014).
[44] K.M. Brown, J.R. Arthur, Selenium, selenoproteins and human health: a review, Public Health Nutr. 4 (2B) (2001) 593–599.
[45] S.J. Fairweather-Tait, Y. Bao, M.R. Broadley, R. Collings, D. Ford, J.E. Hesketh, R. Hurst, Selenium in human health and disease, Antioxid. Redox Signaling 14 (7) (2011) 1337–1383.
[46] L.V. Papp, J. Lu, A. Holmgren, K.K. Khanna, From selenium to selenoproteins: synthesis, identity, and their role in human health, Antioxid. Redox Signaling 9 (7) (2007) 775–806.
[47] S. Chan, B. Gerson, S. Subramaniam, The role of copper, molybdenum, selenium, and zinc in nutrition and health, Clin. Lab Med. 18 (4) (1998) 673–685.
[48] F. Delange, The disorders induced by iodine deficiency, Thyroid 4 (1) (1994) 107–128.
[49] J.A. Cowan, Structural and catalytic chemistry of magnesium-dependent enzymes, BioMetals 15 (2002) 225–235.
[50] M.J. Laires, C.P. Monteiro, M. Bicho, Role of cellular magnesium in health and human disease, Front. Biosci. 9 (2004) 262–276.
[51] H. Theobald, Dietary calcium and health, BNF Nutr. Bull. 30 (3) (2005) 237–277.
[52] S. Fairweather-Tait, R.F. Hurrell, Bioavailability of minerals and trace elements, Nutr. Res. Rev. 9 (1996) 295–324.
[53] WHO—WFR—UNICEF Statement, Preventing and controlling micronutrient deficiencies in populations affected by an emergency. Available at: http://www.who.int/nutrition/publications/micronutrients/WHO_WFP_UNICEFstatement.pdf (Last accessed: 6/2014).
[54] EFSA, Scientific Opinion on principles for deriving and applying Dietary Reference Values, EFSA, Panel on Dietetic Products, Nutrition and Allergies (NDA). Available at: http://www.efsa.europa.eu/en/efsajournal/doc/1458.pdf (Last accessed: 6/2014).
[55] WHO, Dietary recommendations/Nutritional requirements. Available at: http://www.who.int/nutrition/topics/nutrecomm/en/ (Last accessed: 6/2014).
[56] USDA, National Agricultural Library. Available at: http://fnic.nal.usda.gov/dietary-guidance/dietary-reference-intakes/dri-reports (Last accessed: 6/2014).
[57] R.S. Houk, V.A. Fassel, G.D. Flesch, H.J. Svec, A.L. Gray, C.E. Taylor, Inductively coupled argon plasma as an ion source for mass spectrometric determination of trace elements, Anal. Chem. 52 (1980) 2283.
[58] R. Thomas, Practical Guide to ICP-MS, Marcel Dekker, New York, 2004.
[59] F. Cubadda, Inductively coupled plasma-mass spectrometry for the determination of elements and elemental species in food: a review, J. AOAC Int. 87 (1) (2004) 173–204.
[60] J.S. Becker, Inorganic Mass Spectrometry Principles and Applications, John Wiley & Sons, Chichester, 2007.
[61] Perkin Elmer http://www.perkinelmer.com/ (Last accessed: 7/2014).
[62] S.M. Nelms (Ed.), Inductively Coupled Plasma Mass Spectrometry Handbook, Blackwell Publishing, CRC Press, Oxford, 2005.

[63] S.J. Hill, Inductively Coupled Plasma Spectrometry and its Applications, Blackwell Publishing, Oxford, 2007.

[64] Y. Picó (Ed.), Food Toxicants Analysis: Techniques, Strategies and Developments, F. Cubadda, Inductively Coupled Plasma Mass Spectrometry, Elsevier, Amsterdam, 2007.

[65] A.A. Ammann, Inductively coupled plasma mass spectrometry (ICP MS): a versatile tool, J. Mass Spectrom. 42 (2007) 419–427.

[66] D. Beauchemin, Inductively coupled plasma mass spectrometry, Anal. Chem. 78 (2006) 4111–4136.

[67] C.J. Barinaga, D.W. Koppenaal, Ion-trap mass spectrometry with an inductively coupled plasma source, Rapid Commun. Mass Spectrom. 8 (1994) 71–76.

[68] E.P. Nardi, F.S. Evangelista, L. Tormen, T.D. SaińtPierre, A.J. Curtius, S.S. Souza Jr., F. Barbosa, The use of inductively coupled plasma mass spectrometry (ICP-MS) for the determination of toxic and essential elements in different types of food samples, Food Chem. 112 (3) (2009) 727–732.

[69] M.G.A. Korn, E.S.B. Morte, D.C.M.B. Santos, J.T. Castro, J.T.P. Barbosa, A.P. Teixeira, A.P. Fernandes, B. Welz, W.P.C. Santos, E.B.G.N. Santos, M. Korn, Sample preparation for the determination of metals in food samples using spectroanalytical methods—a review, Appl. Spectrosc. Rev. 43 (2008) 67–92.

[70] E.M.M. Flores, Microwave-assisted Sample Preparation for Trace Element Determination, Elsevier, Amsterdam, 2014.

[71] Z. Pedrero, Y. Madrid, C. Cámara, E. Schram, J.B. Luten, I. Feldmann, L. Waentig, H. Hayen, N. Jakubowski, Screening of selenium containing proteins in the Tris-buffer soluble fraction of African catfish (Clarias gariepinus) fillets by laser ablation-ICP-MS after SDS-PAGE and electroblotting onto membranes, J. Anal. Spectrom. 24 (6) (2009) 775–784.

[72] O. Reykdal, S. Rabieh, L. Steingrimsdottir, H. Gunnlaugsdottir, Minerals and trace elements in Icelandic dairy products and meat, J. Food Compos. Anal. 24 (7) (2011) 980–986.

[73] N. Khan, I.S. Jeong, I.M. Hwang, J.S. Kim, S.H. Choi, E.Y. Nho, J.Y. Choi, K.S. Park, K.S. Kim, Analysis of minor and trace elements in milk and yogurts by inductively coupled plasma-mass spectrometry (ICP-MS), Food Chem. 147 (2014) 220–224.

[74] A.B.P. Leme, S.R. Bianchi, R.L. Carneiro, A.R.A. Nogueira, Optimization of sample preparation in the determination of minerals and trace elements in honey by ICP-MS, Food Anal. Method. 7 (5) (2014) 1009–1015.

[75] M. Grottia, A. Terola, J.L. Todolíb, Speciation analysis by small-bore HPLC coupled to ICP-MS, TrAC Trends Anal. Chem. 61 (2014) 92–106.

[76] S. Imanaka, H. Hayashi, Behavior of hexavalent chromium in the water supply system by IC-ICP-MS method, Water Sci. Technol. Water Supply 13 (1) (2013) 96–103.

[77] H. Shi, C. Adams, Rapid IC-ICP/MS method for simultaneous analysis of iodoacetic acids, bromoacetic acids, bromate, and other related halogenated compounds in water, Talanta 79 (2) (2009) 523–527.

[78] M. Montes-Bayón, K. DeNicola, J.A. Caruso, Liquid chromatography-inductively coupled plasma mass spectrometry, J. Chromatogr. A 1000 (1–2) (2003) 457–476.

[79] B. Bouyssiere, J. Szpunar, R. Lobinski, Gas chromatography with inductively coupled plasma mass spectrometric detection in speciation analysis, Spectrochim. Acta Part B 57 (5) (2002) 805–828.

[80] B. Bouyssiere, J. Szpunar, G. Lespes, R. Lobinski, Gas chromatography with inductively coupled plasma mass spectrometric detection (GC-ICP MS), Adv. Chromatogr. 42 (2003) 107–137.

[81] S. Dubascoux, I. Le Hécho, M. Hassellöv, F. Von Der Kammer, M. Potin Gautier, G. Lespes, Field-flow fractionation and inductively coupled plasma mass spectrometer coupling: history, development and applications, J. Anal. Spectrom. 25 (5) (2010) 613–623.
[82] S.R. Haider, B.L. Sharp, H.J. Reid, On-line coupling of gel electrophoresis and inductively coupled plasma-mass spectrometry, TrAC Trends Anal. Chem. 30 (11) (2011) 1793–1808.
[83] G. Alvarez-Llamas, M.R.F. de la Campa, A. Sanz-Medel, ICP-MS for specific detection in capillary electrophoresis, TrAC Trends Anal. Chem. 24 (1) (2005) 28–36.
[84] X. Hou, B.T. Jones, R.A. Meyers (Eds.), Inductively Coupled Plasma/Optical Emission Spectrometry, Encyclopedia of Analytical Chemistry, John Wiley & Sons, Chichester, 2000.
[85] W. Meier-Augenstein, H.F. Kemp, Stable Isotope Analysis: General Principles and Limitations, Encyclopedia of Forensic Science, Wiley, Chichester, 2012.
[86] H.A. Mooney, J.G. Canadell (Eds.), The Earth System: Biological and Ecological Dimensions of Global Environmental Change, Stable Isotopes, Encyclopedia of Global Environmental Change, vol. 2, John Wiley & Sons, Chichester, 2002.
[87] J. Hoefs (Ed.), Stable Isotope Geochemistry, sixth ed., Springer, Berlin, 2009.
[88] J.E. Staller, R.H. Tykot, B.F. Benz (Eds.), Histories of Maize, Academic Press, Amsterdam, 2006.
[89] S. Ghidini, A. Ianieri, E. Zanardi, M. Conter, T. Boschetti, P. Iacumin, P.G. Bracchi, Stable isotopes determination in food authentication: a review, Ann. Fac. Medic. Vet. di Parma 26 (2006) 193–204.
[90] R. Michener, K. Lajtha, Stable Isotopes in Ecology and Environmental Science, second ed., Blackwell, Oxford, 2007.
[91] Z. Muccio, G.P. Jackson, Isotope ratio mass spectrometry, Analyst 134 (2009) 213–222.
[92] P.A. de Groot (Ed.), Handbook of Stable Isotope Analytical Techniques, Volume-I, Elsevier B.V., Amsterdam, 2004.
[93] S. Benson, C. Lennard, P. Maynard, C. Roux, Forensic applications of isotope ratio mass spectrometry, Forensic Sci. Int. 157 (2006) 1–22.
[94] A.L. Sessions, Isotope-ratio detection for gas chromatography, J. Sep. Sci. 29 (2006) 1946–1961.
[95] M. Krummen, A.W. Hilkert, D. Juchelka, A. Duhr, H.-J. Schluter, R. Pesch, A new concept for isotope ratio monitoring liquid chromatography/mass spectrometry, Rapid Commun. Mass Spectrom. 18 (2004) 2260–2266.
[96] F. Albarède, B. Beard, Analytical methods for non-traditional isotopes, Rev. Mineral. Geochem. 55 (2004) 113–152.
[97] T. Walczyk, The potential of inorganic mass spectrometry in mineral and trace element nutrition research, Fresenius, J. Anal. Chem. 370 (2001) 444–453.
[98] M.S. Fantle, T.D. Bullen, Essentials of iron, chromium, and calcium isotope analysis of natural materials by thermal ionization mass spectrometry, Chem. Geol. 258 (2009) 50–64.
[99] K.K. Turekian, H.D. Holland, Treatise on Geochemistry, second ed., 15, Elsevier, 2013, pp. 337–354.
[100] T. Walczyk, TIMS versus multicollector-ICP-MS: coexistence or struggle for survival? Anal. Bioanal. Chem. 378 (2004) 229–231.
[101] L. Yang, Accurate and precise determination of isotopic ratios by mc-icp-ms: a review, Mass Spectrom. Rev. 28 (6) (2009) 990–1011.
[102] L. Balcaen, L. Moens, F. Vanhaecke, Determination of isotope ratios of metals (and metalloids) by means of inductively coupled plasma-mass spectrometry for provenancing purposes, Spectrochim. Acta Part B 65 (2010) 769–786.

[103] M. Rehkämper, M. Schönbächler, C.H. Stirling, Multiple collector ICP-MS: Introduction to instrumentation, measurement techniques and analytical capabilities, Geostand. Newsl. 25 (1) (2001) 23–40.

[104] R. Clough, J. Truscatt, S.T. Belt, E.H. Evans, B. Fairman, T. Catterick, Isotope dilution ICP-MS for speciation studies, Appl. Spectrosc. Rev. 38 (1) (2003) 101–132.

[105] E.-M. Rintala, P. Ekholm, P. Koivisto, K. Peltonen, E.-R. Venäläinen, The intake of inorganic arsenic from long grain rice and rice-based baby food in Finland–low safety margin warrants follow up, Food Chem. 150 (2014) 199–205.

[106] J.M. Llobet, G. Falcó, C. Casas, A. Teixidó, J.L. Domingo, Concentrations of arsenic, cadmium, mercury, and lead in common foods and estimated daily intake by children, adolescents, adults, and seniors of Catalonia, Spain, J. Agric. Food Chem. 51 (3) (2003) 838–842.

[107] A.C. Pappas, E. Zoidis, C.A. Georgiou, N. Demiris, P.F. Surai, K. Fegeros, Influence of organic selenium supplementation on the accumulation of toxic and essential trace elements involved in the antioxidant system of chicken, Food Addit. Contam. Part A 28 (4) (2011) 446–454.

[108] J.-C. Leblanc, T. Guérin, L. Noël, G. Calamassi-Tran, J.-L. Volatier, P. Verger, Dietary exposure estimates of 18 elements from the 1st french total diet study, Food Addit. Contam. 22 (7) (2005) 624–641.

[109] S.P. Dolan, D.A. Nortrup, P.M. Bolger, S.G. Capar, Analysis of dietary supplements for arsenic, cadmium, mercury, and lead using inductively coupled plasma mass spectrometry, J. Agric. Food Chem. 51 (5) (2003) 1307–1312.

[110] S. Caroli, G. Forte, A.L. Iamiceli, B. Galoppi, Determination of essential and potentially toxic trace elements in honey by inductively coupled plasma-based techniques, Talanta 50 (2) (1999) 327–336.

[111] M. Krachler, T. Prohaska, G. Koellensperger, E. Rossipal, G. Stingeder, Concentrations of selected trace elements in human milk and in infant formulas determined by magnetic sector field inductively coupled plasma-mass spectrometry, Biol. Trace Elem. Res. 76 (2) (2000) 97–112.

[112] H. Zhang, Z.-Y. Wang, X. Yang, H.-T. Zhao, Y.-C. Zhang, A.-J. Dong, J. Jing, J. Wang, Determination of free amino acids and 18 elements in freeze-dried strawberry and blueberry fruit using an amino acid analyzer and ICP-MS with micro-wave digestion, Food Chem. 147 (2014) 189–194.

[113] N. Khan, J.Y. Choi, E.Y. Nho, N. Jamila, G. Habte, J.H. Hong, I.M. Hwang, K.S. Kim, Determination of minor and trace elements in aromatic spices by micro-wave assisted digestion and inductively coupled plasma-mass spectrometry, Food Chem. 158 (2014) 200–206.

[114] S. Mounicou, J. Szpunar, R. Lobinski, Metallomics: the concept and methodology, Chem. Soc. Rev. 38 (2009) 1119–1138.

[115] P. Sun, Y. Rui, Determination of nutritional value and safety of selenium-enriched watermelon by ICP-MS, Asian J. Chem. 26 (7) (2014) 2185–2186.

[116] A.L. Rosen, G.M. Hieftje, Inductively coupled plasma mass spectrometry and electrospray mass spectrometry for speciation analysis: applications and instrumentation, Spectrochim. Acta Part B 59 (2004) 135–146.

[117] E.J. Llorent-Martínez, M.L. Fernández-de Córdova, P. Ortega-Barrales, A. Ruiz-Medina, Characterization and comparison of the chemical composition of exotic superfoods, Microchem. J. 110 (2013) 444–451.

[118] I.V. Vrček, D.V. Čepo, D. Rašić, M. Peraica, I. Žuntar, M. Bojić, G. Mendaš, M. Medić-Šarić, A comparison of the nutritional value and food safety of organically and conventionally produced wheat flours, Food Chem. 143 (2014) 522–529.

[119] S.L. Maxwell, B.K. Culligan, J.B. Hutchison, Rapid fusion method for determination of plutonium isotopes in large rice samples, J. Radioanal. Nucl. Chem. 298 (2) (2013) 1367–1374.

[120] F. Berisha, W. Goessler, Uranium in Kosovo's drinking water, Chemosphere 93 (9) (2013) 2165–2170.

[121] N. Fidalgo-Used, M. Montes-Bayón, E. Blanco-González, A. Sanz-Medel, SPME-enantioselective gas chromatography with ECD and ICP-MS detection for the chiral speciation of the pesticide ruelene in environmental samples, J. Anal. Spectrom. 21 (9) (2006) 876–883.

[122] V. Gómez-Jacinto, T. García-Barrera, I. Garbayo, C. Vílchez, J.L. Gómez-Ariza, Metallomic study of selenium biomolecules metabolized by the microalgae *Chlorella sorkiniana* in the biotechnological production of functional foods enriched in selenium, Pure Appl. Chem. 84 (2) (2012) 269–280.

[123] J. Szpunar, Bio-inorganic speciation analysis by hyphenated techniques, Analyst 125 (2000) 963–988.

[124] F. Capozzi, A. Bordoni, Foodomics: a new comprehensive approach to food and nutrition, Genes Nutr. 8 (1) (2013) 1–4.

[125] Y.-X. Liu, Y.-Q. Wan, L. Guo, Determination of organotin compounds in wine by microwave -assisted extraction and high performance liquid chromatography-inductively coupled plasma mass spectrometry, Anal. Lett. 47 (2) (2014) 343–355.

[126] U.T. Kumar, J.G. Dorsey, J.A. Caruso, E.H. Evans, Speciation of inorganic and organotin compounds in biological samples by liquid chromatography with inductively coupled plasma mass spectrometric detection, J. Chromatogr. A 654 (2) (1993) 261–268.

[127] N. Poperechna, K.G. Heumann, Simultaneous multi-species determination of trimethyllead, monomethylmercury and three butyltin compounds by species-specific isotope dilution GC-ICP-MS in biological samples, Anal. Bioanal. Chem. 383 (2) (2005) 153–159.

[128] J. Sun, B. He, Y. Yin, L. Li, G. Jiang, Speciation of organotin compounds in environmental samples with semi-permanent coated capillaries by capillary electrophoresis coupled with inductively coupled plasma mass spectrometry, Anal. Methods 2 (12) (2010) 2025–2031.

[129] S.C. Hight, J. Cheng, Determination of methylmercury and estimation of total mercury in seafood using high performance liquid chromatography (HPLC) and inductively coupled plasma-mass spectrometry (ICP-MS): method development and validation, Anal. Chim. Acta 567 (2) (2006) 160–172.

[130] A.I. Cabañero, C. Carvalho, Y. Madrid, C. Batoréu, C. Cámara, Quantification and speciation of mercury and selenium in fish samples of high consumption in Spain and Portugal, Biol. Trace Elem. Res. 103 (1) (2005) 17–35.

[131] H. Amlund, A.-K. Lundebye, M.H.G. Berntssen, Accumulation and elimination of methylmercury in Atlantic cod (Gadus morhua L.) following dietary exposure, Aquat. Toxicol. 83 (4) (2007) 323–330.

[132] B.L. Batista, J.L. Rodrigues, S.S. De Souza, V.C. Oliveira Souza Jr., F. Barbosa, Mercury speciation in seafood samples by LC-ICP-MS with a rapid ultrasound-assisted extraction procedure: application to the determination of mercury in Brazilian seafood samples, Food Chem. 126 (4) (2011) 2000–2004.

[133] M.H. Mach, I.,B. Nott, J.W. Scott, R.F. Maddalone, N.T. Whiddon, Metal speciation: survey of environmental methods of analysis, Water Air Soil Pollut. 90 (1996) 269–279.

[134] S. Lischka, U. Arroyo-Abad, J. Mattusch, A. Kühn, Ch. Piechotta, The high diversity of arsenolipids in herring fillet (Clupea harengus), Talanta 110 (2013).

[135] U. Kohlmeyer, J. Kuballa, E. Jantzen, Simultaneous separation of 17 inorganic and organic arsenic compounds in marine biota by means of high-performance liquid chromatography/inductively coupled plasma mass spectrometry, Rapid Commun. Mass Spectrom. 16 (10) (2002) 965–974.

[136] A.A. Carbonell-Barrachina, X. Wu, A. Ramírez-Gandolfo, G.J. Norton, F. Burló, C. Deacon, A.A. Meharg, Inorganic arsenic contents in rice-based infant foods from Spain, UK, China and USA, Environ. Pollut. 163 (2012) 77–83.

[137] M. Jabłońska-Czapla, S. Szopa, K. Grygoyć, A. Łyko, R. Michalski, Development and validation of HPLC-ICP-MS method for the determination inorganic Cr, As and Sb speciation forms and its application for Pławniowice reservoir (Poland) water and bottom sediments variability study, Talanta 120 (2014) 475–483.

[138] J. Szpunar, P. Pellerin, A. Makarov, T. Doco, P. Williams, B. Medina, R. Łobiński, Speciation analysis for biomolecular complexes of lead in wine by size-exclusion high-performance liquid chromatography-inductively coupled plasma mass spectrometry, J. Anal. Spectrom. 13 (8) (1998) 749–754.

[139] L.-F. Chang, S.-J. Jiang, A.C. Sahayam, Speciation analysis of mercury and lead in fish samples using liquid chromatography-inductively coupled plasma mass spectrometry, J. Chromatogr. A 1176, (1–2) (2007) 143–148.

[140] F. Sacher, B. Raue, J. Klinger, H.-J. Brauch, Simultaneous determination of Cr(III) and Cr(VI) in ground and drinking waters by IC-ICP-MS, Int. J. Environ. Anal. Chem. 74 (1–4) (1999) 191–201.

[141] Y.-L. Chang, S.-J. Jiang, Determination of chromium species in water samples by liquid chromatography-inductively coupled plasma-dynamic reaction cell-mass spectrometry, J. Anal. Spectrom. 16 (8) (2001) 858–862.

[142] A.A. Ambushe, R.I. McCrindle, C.M.E. McCrindle, Speciation of chromium in cow's milk by solid-phase extraction/dynamic reaction cell inductively coupled plasma mass spectrometry (DRC-ICP-MS), J. Anal. Spectrom. 24 (4) (2009) 502–507.

[143] B. Novotnik, T. Zuliani, J. Ščančar, R. Milačič, Chromate in food samples: an artefact of wrongly applied analytical methodology? J. Anal. Spectrom. 28 (4) (2013) 558–566.

[144] Z. Pedrero, Y. Madrid, Novel approaches for selenium speciation in foodstuffs and biological specimens: a review, Anal. Chim. Acta 634 (2) (2009) 135–152.

[145] Y. Ogra, K. Ishiwata, J.R. Encinar, R. Łobiński, K.T. Suzuki, Speciation of selenium in selenium-enriched shiitake mushroom, *Lentinula edodes*, Anal. Bioanal. Chem. 379 (5–6) (2004) 861–866.

[146] P.O. Amoako, P.C. Uden, J.F. Tyson, Speciation of selenium dietary supplements; formation of S (methylseleno)cysteine and other selenium compounds, Anal. Chim. Acta 652 (1–2) (2009) 315–323.

[147] J. Gromadzińska, E. Reszka, K. Bruzelius, W. Wąsowicz, B. Åkesson, Selenium and cancer: biomarkers of selenium status and molecular action of selenium supplements, Eur. J. Nutr. 47 (2) (2008) 29–50.

[148] S. Shao, X. Mi, L. Ouerdane, R. Lobinski, J.F. García-Reyes, A. Molina-Díaz, A. Vass, M. Dernovics, Quantification of Se-methylselenocysteine and its γ-glutamyl derivative from naturally Se-enriched green bean (*Phaseolus vulgaris vulgaris*) after HPLC-ESI-TOF-MS and orbitrap MSn-Based identification, Food Anal. Method. 7 (5) (2014) 1147–1157.

[149] S.G. Casal, J. Far, K. Bierla, L. Ouerdane, J. Szpunar, Study of the Se-containing metabolomes in Se-rich yeast by size-exclusion–cation-exchange HPLC with the parallel ICP MS and electrospray orbital ion trap detection, Metallomics 2 (8) (2010) 535–548.

[150] M. Dernovics, J. Far, R. Lobinski, Identification of anionic selenium species in Se-rich yeast by electrospray QTOF MS/MS and hybrid linear ion trap/orbitrap MSn, Metallomics 1 (4) (2009) 317–329.

[151] S. Yoshida, M. Haratake, T. Fuchigami, M. Nakayama, Selenium in seafood materials, J. Health Sci. 57 (3) (2011) 215–224.

[152] A.P. Vonderhcide, K. Wrobel, S.S. Kannamkumarath, C. B'Hymer, M. Montes-Bayón, C.P. De León, J.A. Caruso, Characterization of selenium species in Brazil nuts by HPLC-ICP-MS and ES-MS, J. Agric. Food Chem. 50 (20) (2002) 5722–5728.
[153] B.B.M. Sadi, A.P. Vonderheide, J.A. Caruso, Analysis of phosphorus herbicides by ion-pairing reversed-phase liquid chromatography coupled to inductively coupled plasma mass spectrometry with octapole reaction cell, J. Chromatogr. A 1050 (2004) 95–101.
[154] J.-H. Chen, K.-E. Wang, S.-J. Jiang, Determination of iodine and bromine compounds in foodstuffs by CE-inductively coupled plasma MS, Electrophoresis 28 (22) (2007) 4227–4232.
[155] J. Szpunar, P. Pellerin, A. Makarov, T. Doco, P. Williams, R. Lobinski, Speciation of metal-carbohydrate complexes in fruit and vegetable samples by size-exclusion HPLC-ICP-MS, J. Anal. Spectrom. 14 (4) (1999) 639–644.
[156] J.L. Gómez-Ariza, T. Garcia-Barrera, F. Lorenzo, V. Bernal, M.J. Villegas, V. Oliveira, Use of mass spectrometry techniques for the characterization of metal bound to proteins (metallomics) in biological systems, Anal. Chim. Acta 524 (2004) 15–22.
[157] K.E. Odegard, W. Lund, Multi-element speciation of tea infusion using cation-exchange separation and size-exclusion chromatography in combination with inductively coupled plasma mass spectrometry, J. Anal. Spectrom. 12 (4) (1997) 403–408.
[158] Council Regulation (EEC) No 2081/92 of 14 July 1992 on the protection of geographical indications and designations of origin for agricultural products and foodstuffs. Available at: http://eur-lex.europa.eu/LexUriServ/LexUriServ.do?uri=CELEX:31992R2081:EN: HTML (Last accessed: 7/2014).
[159] Regulation (EU) No 1151/2012 of the European Parliament and of the Council of 21 November 2012 on quality schemes for agricultural products and foodstuffs. Available at: http://eur-lex.europa.eu/LexUriServ/LexUriServ.do?uri=OJ: L:2012:343:0001:0029:en:PDF (Last accessed: 7/2014).
[160] S. Kelly, K. Heaton, J. Hoogewerff, Tracing the geographical origin of food: the application of multi-element and multi-isotope analysis, Trends Food Sci. Technol. 16 (2005) 555–567.
[161] K.H. Laursen, J.K. Schjoerring, J.E. Olesen, U. Halekoh, M. Askegaard, S. Husted, Multi-elemental fingerprinting as a tool for authentication of organic wheat, barley, faba bean, and potato, J. Agric. Food Chem. 59 (9) (2011) 4385–4396.
[162] S.A. Drivelos, C.A. Georgiou, Multi-element and multi-isotope-ratio analysis to determine the geographical origin of foods in the European Union, TrAC Trends Anal. Chem. 40 (2012) 38–51.
[163] M.J. Baxter, H.M. Crews, M.J. Dennis, I. Goodall, D. Anderson, The determination of the authenticity of wine from its trace element composition, Food Chem. 60 (3) (1997) 443–450.
[164] J.D. Greenough, H.P. Longerich, S.E. Jackson, Element fingerprinting of Okanagan Valley wines using ICP-MS: Relationships between wine composition, vineyard and wine colour, Aust. J. Grape Wine Res. 3 (2) (1997) 75–83.
[165] X. Tian, H. Emteborg, M. Barbaste, F.C. Adams, Accuracy and precision of lead isotope ratios in wines measured by axial inductively coupled plasma time of flight mass spectrometry, J. Anal. Spectrom. 15 (7) (2000) 829–835.
[166] C.M. Almeida, M.T.S.D. Vasconcelos, ICP-MS determination of strontium isotope ratio in wine in order to be used as a fingerprint of its regional origin, J. Anal. Spectrom. 16 (6) (2001) 607–611.
[167] P.P. Coetzee, F.P. Van Jaarsveld, F. Vanhaecke, Intraregional classification of wine via ICP-MS elemental fingerprinting, Food Chem. 164 (2014) 485–492.
[168] I. Rodushkin, F. Ödman, P.K. Appelblad, Multielement determination and lead isotope ratio measurement in alcoholic beverages by high-resolution inductively coupled plasma mass spectrometry, J. Food Compos. Anal. 12 (4) (1999) 243–257.

[169] M.C.A. Marcelo, C.A. Martins, D. Pozebon, V.L. Dressler, M.F. Ferrão, Classification of yerba mate (Ilex paraguariensis) according to the country of origin based on element concentrations, Microchem. J. 117 (2014) 164–171.

[170] K. Ariyama, H. Horita, A. Yasui, Application of inorganic element ratios to chemometrics for determination of the geographic origin of Welsh onions, J. Agric. Food Chem. 52 (19) (2004) 5803–5809.

[171] S. Swoboda, M. Brunner, S.F. Boulyga, P. Galler, M. Horacek, T. Prohaska, Identification of Marchfeld asparagus using Sr isotope ratio measurements by MC-ICP-MS, Anal. Bioanal. Chem. 390 (2) (2008) 487–494.

[172] G. Lo Feudo, A. Naccarato, G. Sindona, A. Tagarelli, Investigating the origin of tomatoes and triple concentrated tomato pastes through multielement determination by inductively coupled plasma mass spectrometry and statistical analysis, J. Agric. Food Chem. 58 (6) (2010) 3801–3807.

[173] E. Furia, A. Naccarato, G. Sindona, G. Stabile, A. Tagarelli, Multielement fingerprinting as a tool in origin authentication of pgi food products: Tropea red onion, J. Agric. Food Chem. 59 (15) (2011) 8450–8457.

[174] V.S. Santos, V. Nardini, L.C. Cunha Jr., F. Barbosa Jr., G.H. De Almeida Teixeira, Identification of species of the *Euterpe* genus by rare earth elements using inductively coupled plasma mass spectrometry and linear discriminant analysis, Food Chem. 153 (2014) 334–339.

[175] E.G. Farmaki, N.S. Thomaidis, K.S. Minioti, E. Ioannou, C.A. Georgiou, C.E. Efstathiou, Geographical characterization of greek olive oils using rare earth elements content and supervised chemometric techniques, Anal. Lett. 45 (8) (2012) 920–932.

[176] C. Benincasa, J. Lewis, E. Perri, G. Sindona, A. Tagarelli, Determination of trace element in Italian virgin olive oils and their characterization according to geographical origin by statistical analysis, Anal. Chim. Acta 585 (2) (2007) 366–370.

[177] D. Joebstl, D. Bandoniene, T. Meisel, S. Chatzistathis, Identification of the geographical origin of pumpkin seed oil by the use of rare earth elements and discriminant analysis, Food Chem. 123 (4) (2010) 1303–1309.

[178] W.A. Simpkins, H. Louie, M. Wu, M. Harrison, D. Goldberg, Trace elements in Australian orange juice and other products, Food Chem. 71 (4) (2000) 423–433.

[179] S.A. Drivelos, K. Higgins, J.H. Kalivas, S.A. Haroutounian, C.A. Georgiou, Data fusion for food authentication. Combining rare earth elements and trace metals to discriminate "fava Santorinis" from other yellow split peas using chemometric tools, Food Chem. 165 (2014) 316–322.

[180] S. Kelly, M. Baxter, S. Chapman, C. Rhodes, J. Dennis, P. Brereton, The application of isotopic and elemental analysis to determine the geographical origin of premium long grain rice, Eur. Food Res. Technol. 214 (1) (2002) 72–78.

[181] M. Oddone, M. Aceto, M. Baldizzone, D. Musso, D. Osella, Authentication and traceability study of hazelnuts from piedmont, Italy, J. Agric. Food Chem. 57 (9) (2009) 3404–3408.

[182] H. Zhao, B. Guo, Y. Wei, B. Zhang, S. Sun, L. Zhang, J. Yan, Determining the geographic origin of wheat using multielement analysis and multivariate statistics, J. Agric. Food Chem. 59 (9) (2011) 4397–4402.

[183] C. Benincasa, J. Lewis, G. Sindona, A. Tagarelli, The use of multi element profiling to differentiate between cow and buffalo milk, Food Chem. 110 (1) (2008) 257–262.

[184] B.M. Franke, M. Haldimann, J. Reimann, B. Baumer, G. Gremaud, R. Hadorn, J.-O. Bosset, M. Kreuzer, Indications for the applicability of element signature analysis for the determination of the geographic origin of dried beef and poultry meat, Eur. Food Res. Technol. 225 (3–4) (2007) 501–509.

[185] S. Sun, B. Guo, Y. Wei, M. Fan, Multi-element analysis for determining the geographical origin of mutton from different regions of China, Food Chem. 124 (3) (2011) 1151–1156.

[186] L. Guo, L. Gong, Y. Yu, H. Zhang, Multi-element fingerprinting as a tool in origin authentication of four east china marine species, J. Food Sci. 78 (12) (2013) C1852–C1857.

[187] I. Rodushkin, T. Bergman, G. Douglas, E. Engström, D. Sörlin, D.C. Baxter, Authentication of Kalix (N.E. Sweden) vendace caviar using inductively coupled plasma-based analytical techniques: evaluation of different approaches, Anal. Chim. Acta 583 (2) (2007) 310–318.

[188] M. Chudzinska, D. Baralkiewicz, Estimation of honey authenticity by multielements characteristics using inductively coupled plasma-mass spectrometry (ICP-MS) combined with chemometrics, Food Chem. Toxicol. 48 (1) (2010) 284–290.

[189] A. Moreda-Piñeiro, A. Fisher, S.J. Hill, The classification of tea according to region of origin using pattern recognition techniques and trace metal data, J. Food Compos. Anal. 16 (2) (2003) 195–211.

[190] C. Rodrigues, C. Máguas, T. Prohaska, Strontium and oxygen isotope fingerprinting of green coffee beans and its potential to proof authenticity of coffee, Eur. Food Res. Technol. 232 (2) (2011) 361–373.

[191] S.-M. Choi, H.-S. Lee, G.-H. Lee, J.-K. Han, Determination of the strontium isotope ratio by ICP-MS ginseng as a tracer of regional origin, Food Chem. 108 (3) (2008) 1149–1154.

[192] M. Brunner, R. Katona, Z. Stefánka, T. Prohaska, Determination of the geographical origin of processed spice using multielement and isotopic pattern on the example of Szegedi paprika, Eur. Food Res. Technol. 231 (4) (2010) 623–634.

[193] A.A. D'Archivio, A. Giannitto, A. Incani, S. Nisi, Analysis of the mineral composition of Italian saffron by ICP-MS and classification of geographical origin, Food Chem. 157 (2014) 485–489.

[194] E. Capuano, R. Boerrigter-Eenling, G. van der Veer, S.M. van Ruth, Analytical authentication of organic products: an overview of markers, J. Sci. Food Agric. 93 (1) (2013) 12–28.

[195] K.H. Laursen, J.K. Schjoerring, S.D. Kelly, S. Husted, Authentication of organically grown plants—advantages and limitations of atomic spectroscopy for multi-element and stable isotope analysis, TrAC Trends Anal. Chem. 59 (2014) 73–82.

[196] R.M. Barbosa, B.L. Batista, R.M. Varrique, V.A. Coelho, A.D. Campiglia, F. Barbosa, The use of advanced chemometric techniques and trace element levels for controlling the authenticity of organic coffee, Food Res. Int. 61 (2014) 246–251.

[197] R.M. Barbosa, L.R. Nacano, R. Freitas, B.L. Batista, F. Barbosa, The use of decision trees and naïve bayes algorithms and trace element patterns for controlling the authenticity of free-range-pastured hens' eggs, J. Food Sci. 79 (9) (2014) 1672–1677.

[198] G. Artiaga, K. Ramos, L. Ramos, C. Cámara, M. Gómez-Gómez, Migration and characterisation of nanosilver from food containers by AF4-ICP-MS, Food Chem. 166 (2015) 76–85.

[199] J. Bott, A. Störmer, R. Franz, A comprehensive study into the migration potential of nano silver particles from food contact polyolefins, ACS Symp. Ser. 1159 (2014) 51–70.

[200] M. Cushen, J. Kerry, M. Morris, M. Cruz-Romero, E. Cummins, Migration and exposure assessment of silver from a PVC nanocomposite, Food Chem. 139 (1–4) (2013) 389–397.

[201] S. Carneado, E. Hernández-Nataren, J.F. López-Sánchez, A. Sahuquillo, Migration of antimony from polyethylene terephthalate used in mineral water bottles, Food Chem. 166 (2015) 544–550.

[202] K.L. Kamerud, K.A. Hobbie, K.A. Anderson, Stainless steel leaches nickel and chromium into foods during cooking, J. Agric. Food Chem. 61 (39) (2013) 9495–9501.

[203] F. Tassi, F. Capecchiacci, O. Vaselli, Migration processes of metal elements from carbon steel cylinders to food gases, Packag. Technol. Sci., in press.

[204] Q.-B. Lin, H. Li, H.-N. Zhong, Q. Zhao, D.-H. Xiao, Z.-W. Wang, Determination of titanium in nano-titanium (IV) oxide composite food packaging by microwave digestion and inductively coupled plasma atomic emission spectrometry and inductively coupled plasma mass spectrometry, Anal. Lett. 47 (12) (2014) 2095–2103.

[205] Q.-B. Lin, H. Li, H.-N. Zhong, Q. Zhao, D.-H. Xiao, Z.-W. Wang, Migration of Ti from nano-TiO_2-polyethylene composite packaging into food simulants, Food Addit. Contam. Part A 31 (7) (2014) 1284–1290.

[206] S. Dhawan, S.S. Sablani, J. Tang, G.V. Barbosa-Cànovas, J.L. Ullman, K. Bhunia, Silicon migration from high-barrier coated multilayer polymeric films to selected food simulants after microwave processing treatments, Packag. Technol. Sci. 27 (8) (2014) 625–638.

[207] Q.-B. Lin, Y. Chen, H. Song, H.-J. Wu, X.-Y. Wang, Kinetic migration of chemical elements from ceramic packaging into simulated foods and mature vinegar, Packag. Technol. Sci. 27 (1) (2014) 59–67.

[208] G. Pataro, G.M.J. Barca, R.N. Pereira, A.A. Vicente, J.A. Teixeira, G. Ferrari, Quantification of metal release from stainless steel electrodes during conventional and pulsed ohmic heating, Innovative Food Sci. Emerging Technol. 21 (2014) 66–73.

[209] N. von Goetz, L. Fabricius, R. Glaus, V. Weitbrecht, D. Günther, K. Hungerbühler, Migration of silver from commercial plastic food containers and implications for consumer exposure assessment, Food Addit. Contam. Part A 30 (3) (2013) 612–620.

[210] Y. Echegoyen, C. Nerín, Nanoparticle release from nano-silver antimicrobial food containers, Food Chem. Toxicol. 62 (2013) 16–22.

[211] I. Rodushkin, A. Magnusson, Aluminium migration to orange juice in laminated paperboard packages, J. Food Compos. Anal. 18 (5) (2005) 365–374.

[212] L. Castle, C.P. Offen, M.J. Baxter, J. Gilbert, Migration studies from paper and board food packaging materials. 1. Compositional analysis, Food Addit. Contam. 14 (1) (1997) 35–44.

[213] K. Loeschner, J. Navratilova, C. Købler, K. Mølhave, S. Wagner, F. von Der Kammer, E.H. Larsen, Detection and characterization of silver nanoparticles in chicken meat by asymmetric flow field flow fractionation with detection by conventional or single particle ICP-MS, Anal. Bioanal. Chem. 405 (25) (2013) 8185–8195.

[214] R.J.B. Peters, Z.H. Rivera, G. Van Bemmel, H.J.P. Marvin, S. Weigel, H. Bouwmeester, Development and validation of single particle ICP-MS for sizing and quantitative determination of nano-silver in chicken meat characterisation of nanomaterials in biological samples, Anal. Bioanal. Chem. 406 (16) (2014) 3875–3885.

[215] R.J.B. Peters, G. Van Bemmel, Z. Herrera-Rivera, H.P.F.G. Helsper, H.J.P. Marvin, S. Weigel, P.C. Tromp, A.G. Oomen, A.G. Rietveld, H. Bouwmeester, Characterization of titanium dioxide nanoparticles in food products: analytical methods to define nanoparticles, J. Agric. Food Chem. 62 (27) (2014) 6285–6293.

[216] Y. Yang, K. Doudrick, X. Bi, K. Hristovski, P. Herckes, P. Westerhoff, R. Kaegi, Characterization of food-grade titanium dioxide: the presence of nanosized particles, Environ. Sci. Technol. 48 (11) (2014) 6391–6400.

[217] P. Krystek, J. Tentschert, Y. Nia, B. Trouiller, L. Noël, M.E. Goetz, A. Papin, A. Luch, T. Guérin, W.H. De Jong, Method development and inter-laboratory comparison about the determination of titanium from titanium dioxide nanoparticles in tissues by inductively coupled plasma mass spectrometry characterisation of nanomaterials in biological samples, Anal. Bioanal. Chem. 406 (16) (2014) 3853–3861.

[218] T. Potouridis, J. Völker, H. Alsenz, M. Oetken, W. Püttmann, Using ICP-qMS to trace the uptake of nanoscale titanium dioxide by microalgae-potential disadvantages of vegetable reference material, Anal. Bioanal. Chem. 406 (11) (2014) 2495–2502.

[219] D.M. Mitrano, E.K. Lesher, A. Bednar, J. Monserud, C.P. Higgins, J.F. Ranville, Detecting nanoparticulate silver using single-particle inductively coupled plasma-mass spectrometry, Environ. Toxicol. Chem. 31 (1) (2012) 115–121.

[220] T.P.J. Linsinger, R. Peters, S. Weigel, International interlaboratory study for sizing and quantification of Ag nanoparticles in food simulants by single-particle ICPMS characterisation of nanomaterials in biological samples, Anal. Bioanal. Chem. 406 (16) (2014) 3835–3843.

[221] F. Von Blanckenburg, J. Noordmann, M. Guelke-Stelling, The iron stable isotope fingerprint of the human diet, J. Agric. Food Chem. 61 (49) (2013) 11893–11899.

[222] L.W. Doner, H.W. Krueger, R.H. Reesman, Isotopic composition of carbon in apple juice, J. Agric. Food Chem. 28 (2) (1980) 362–364.

[223] L.W. Doner, D.D. Bills, Stable carbon isotope ratios in orange juice, J. Agric. Food Chem. 29 (4) (1981) 803–804.

[224] A. Rossmann, H.-L. Schmidt, F. Reniero, G. Versini, I. Moussa, M.H. Merle, Stable carbon isotope content in ethanol of EC data bank wines from Italy, France and Germany, Eur. Food Res. Technol. 203 (3) (1996) 293–301.

[225] A.I. Cabañero, T. San-Hipólito, M. Rupérez, GasBench/isotope ratio mass spectrometry: a carbon isotope approach to detect exogenous CO_2 in sparkling drinks, Rapid Commun. Mass Spectrom. 21 (20) (2007) 3323–3328.

[226] G. Calderone, C. Guillou, F. Reniero, N. Naulet, Helping to authenticate sparkling drinks with $^{13}C/^{12}C$ of CO_2 by gas chromatography-isotope ratio mass spectrometry, Food Res. Int. 40 (3) (2007) 324–331.

[227] G. Calderone, N. Naulet, C. Guillou, F. Reniero, A.I.B. Cortes, Analysis of the ^{13}C natural abundance of CO_2 gas from sparkling drinks by gas chromatography/combustion/isotope ratio mass spectrometry, Rapid Commun. Mass Spectrom. 19 (5) (2005) 701–705.

[228] B.-L. Yunianta Zhang, M. Lees, G.J. Martin, Stable isotope fractionation in fruit juice concentrates: application to the authentication of grape and orange products, J. Agric. Food Chem. 43 (9) (1995) 2411–2417.

[229] I.G. Parker, S.D. Kelly, M. Sharman, M.J. Dennis, D. Howie, Investigation into the use of carbon isotope ratios ($^{13}C/^{12}C$) of Scotch whisky congeners to establish brand authenticity using gas chromatography-combustion-isotope ratio mass spectrometry, Food Chem. 63 (3) (1998) 423–428.

[230] B.O. Aguilar-Cisneros, M.G. López, E. Richling, F. Heckel, P. Schreier, Tequila authenticity assessment by headspace SPME-HRGC-IRMS analysis of $^{13}C/^{12}C$ and $^{18}O/^{16}O$ ratios of ethanol, J. Agric. Food Chem. 50 (26) (2002) 7520–7523.

[231] L. Gaillard, F. Guyon, M.-H. Salagoity, B. Medina, Authenticity of carbon dioxide bubbles in French ciders through multiflow-isotope ratio mass spectrometry measurements, Food Chem. 141 (3) (2013) 2103–2107.

[232] S.E. Woodbury, R.P. Evershed, J.B. Rossell, R.E. Griffith, P. Farnell, Detection of vegetable oil adulteration using gas chromatography combustion/isotope ratio mass spectrometry, Anal. Chem. 67 (15) (1995) 2685–2690.

[233] F. Angerosa, L. Camera, S. Cumitini, G. Gleixner, F. Reniero, Carbon stable isotopes and olive oil adulteration with pomace oil, J. Agric. Food Chem. 45 (8) (1997) 3044–3048.

[234] H.-Y. Seo, J. Ha, D.-B. Shin, S.-L. Shim, K.-M. No, K.-S. Kim, K.-B. Lee, S.-B. Han, Detection of corn oil in adulterated sesame oil by chromatography and carbon isotope analysis, J. Am. Oil Chem. Soc. 87 (6) (2010) 621–626.

[235] E. Jamin, R. Guérin, M. Rétif, M. Lees, G.J. Martin, Improved detection of added water in orange juice by simultaneous determination of the oxygen-18/oxygen-16 isotope ratios of water and ethanol derived from sugars, J. Agric. Food Chem. 51 (18) (2003) 5202–5206.

[236] E. Jamin, F. Martin, R. Santamaria-Fernandez, M. Lees, Detection of exogenous citric acid in fruit juices by stable isotope ratio analysis, J. Agric. Food Chem. 53 (13) (2005) 5130–5133.

[237] P. Tremblay, R. Paquin, Improved detection of sugar addition to maple syrup using malic acid as internal standard and in ^{13}C isotope ratio mass spectrometry (IRMS), J. Agric. Food Chem. 55 (2) (2007) 197–203.

[238] M. Tosun, Detection of adulteration in mulberry pekmez samples added various sugar syrups with $^{13}C/^{12}C$ isotope ratio analysis method, Food Chem. 165 (2014) 555–559.

[239] E. Jamin, F. Martin, G.G. Martin, Determination of the $^{13}C/^{12}C$ ratio of ethanol derived from fruit juices and maple syrup by isotope ratio mass spectrometry: collaborative study, J. AOAC Int. 87 (3) (2004) 621–631.

[240] G.P. Lin, Y.H. Rau, Y.F. Chen, C.C. Chou, W.G. Fu, Measurements of δD and $\delta^{18}O$stable isotope ratios in milk, J. Food Sci. 68, (7) (2003) 2192–2195.

[241] A. Stocker, A. Rossmann, A. Kettrup, E. Bengsch, Detection of royal jelly adulteration using carbon and nitrogen stable isotope ratio analysis, Rapid Commun. Mass Spectrom. 20 (2) (2006) 181–184.

[242] M. Tosun, Detection of adulteration in honey samples added various sugar syrups with $^{13}C/^{12}C$ isotope ratio analysis method, Food Chem. 138, (2–3) (2013) 1629–1632.

[243] A. Simsek, M. Bilsel, A.C. Goren, $^{13}C/^{12}C$ pattern of honey from Turkey and determination of adulteration in commercially available honey samples using EA-IRMS, Food Chem. 130 (4) (2012) 1115–1121.

[244] M. Greule, L.D. Tumino, T. Kronewald, U. Hener, J. Schleucher, A. Mosandl, F. Keppler, Improved rapid authentication of vanillin using $\delta^{13}C$ and $\delta^{2}H$ values, Eur. Food Res. Technol. 231 (6) (2010) 933–941.

[245] G.S. Remaud, Y.-L. Martin, G.G. Martin, G.J. Martin, Detection of sophisticated adulterations of natural vanilla flavors and extracts: application of the SNIF-NMR method to vanillin and p-hydroxybenzaldehyde, J. Agric. Food Chem. 45 (3) (1997) 859–866.

[246] F.F. Bensaid, K. Wietzerbin, G.J. Martin, Authentication of natural vanilla flavorings: isotopic characterization using degradation of vanillin into guaiacol, J. Agric. Food Chem. 50 (22) (2002) 6271–6275.

[247] G.S. Remaud, Y.-L. Martin, G.G. Martin, N. Naulet, G.J. Martin, Authentication of mustard oils by combined stable isotope analysis (SNIF-NMR and IRMS), J. Agric. Food Chem. 45 (5) (1997) 1844–1848.

[248] T. T-Nhu-Trang, H. Casabianca, M.-F. Grenier-Loustalot, Authenticity control of essential oils containing citronellal and citral by chiral and stable-isotope gas-chromatographic analysis, Anal. Bioanal. Chem. 386 (2006) 2141–2152.

[249] L. Schipilliti, P.Q. Tranchida, D. Sciarrone, M. Russo, P. Dugo, G. Dugo, L. Mondello, Genuineness assessment of mandarin essential oils employing gas chromatography-combustion-isotope ratio MS (GC-C-IRMS), J. Sep. Sci. 33 (4–5) (2010) 617–625.

[250] F. Serra, F. Reniero, C.G. Guillou, J.M. Moreno, J.M. Marinas, F. Vanhaecke, ^{13}C and ^{18}O isotopic analysis to determine the origin of L-tartaric acid, Rapid Commun. Mass Spectrom. 19 (10) (2005) 1227–1230.

[251] L. Zhang, D.M. Kujawinski, E. Federherr, T.C. Schmidt, M.A. Jochmann, Caffeine in your drink: natural or synthetic? Anal. Chem. 84 (6) (2012) 2805–2810.

[252] E. Richling, C. Höhn, B. Weckerle, F. Heckel, P. Schreier, Authentication analysis of caffeine-containing foods via elemental analysis combustion/pyrolysis isotope ratio mass spectrometry (EA-C/P-IRMS), Eur. Food Res. Technol. 216 (6) (2003) 544–548.

[253] F. Thomas, E. Jamin, ^{2}H NMR and ^{13}C-IRMS analyses of acetic acid from vinegar, ^{18}O-IRMS analysis of water in vinegar: International collaborative study report, Anal. Chim. Acta 649 (1) (2009) 98–105.

[254] D.M.A.M. Luykx, S.M. van Ruth, An overview of analytical methods for determining the geographical origin of food products, Food Chem. 107 (2008) 897–911.

[255] I.J. Košir, M. Kocjančič, N. Ogrinc, J. Kidrič, Use of SNIF-NMR and IRMS in combination with chemometric methods for the determination of chaptalisation and geographical origin of wines (the example of Slovenian wines), Anal. Chim. Acta 429 (2) (2001) 195–206.

[256] F. Camin, R. Wehrens, D. Bertoldi, L. Bontempo, L. Ziller, M. Perini, G. Nicolini, M. Nocetti, R. Larcher, H, C, N and S stable isotopes and mineral profiles to objectively guarantee the authenticity of grated hard cheeses, Anal. Chim. Acta 711 (2012) 54–59.

[257] A. Monetti, G. Versini, F. Reniero, Classification of Italian wines on a regional scale by means of a multi-isotopic analysis, Dev. Food Sci. 37 (C) (1995) 1723–1730.

[258] J.E. Giménez-Miralles, D.M. Salazar, I. Solana, Regional origin assignment of red wines from Valencia (Spain) by ^{2}H NMR and ^{13}C IRMS stable isotope analysis of fermentative ethanol, J. Agric. Food Chem. 47 (7) (1999) 2645–2652.

[259] G.J. Martin, M. Mazure, C. Jouitteau, Y.-L. Martin, L. Aguile, P. Allain, Characterization of the geographic origin of Bordeaux wines by a combined use of isotopic and trace element measurements, Am. J. Enol. Vitic. 50 (4) (1999) 409–417.

[260] A. Rossmann, F. Reniero, I. Moussa, H.-L. Schmidt, G. Versini, M.H. Merle, Stable oxygen isotope content of water of EU data-bank wines from Italy, France and Germany, Eur. Food Res. Technol. 208 (5–6) (1999) 400–407.

[261] A.I. Cabañero, J.L. Recio, M. Rupérez, Isotope ratio mass spectrometry coupled to liquid and gas chromatography for wine ethanol characterization, Rapid Commun. Mass Spectrom. 22 (20) (2008) 3111–3118.

[262] V. Avram, D.A. Magdas, C. Voica, G. Cristea, C. Cimpoiu, A. Hosu, C. Marutoiu, Isotopic oxygen ratios and trace metal determination in some romanian commercial wines, Anal. Lett. 47 (4) (2014) 641–653.

[263] T.S. Pilgrim, R.J. Watling, K. Grice, Application of trace element and stable isotope signatures to determine the provenance of tea (*Camellia sinensis*) samples, Food Chem. 118 (4) (2010) 921–926.

[264] F. Camin, M. Perini, L. Bontempo, L. Giongo, Multi-element (H, C, N, O) stable isotope characterization of blueberries, Acta Hortic. 810 (2009) 697–704.

[265] N. Ogrinc, K. Bat, I.J. Kosir, T. Golo, R. Kokkinofta, Characterization of commercial Slovenian and cypriot fruit juices using stable isotopes, J. Agric. Food Chem. 57 (15) (2009) 6764–6769.

[266] G. Li, Z. Wu, Y. Wang, X. Dong, B. Li, W. He, S. Wang, J. Cui, Identification of geographical origins of Schisandra fruits in China based on stable carbon isotope ratio analysis, Eur. Food Res. Technol. 232 (5) (2011) 797–802.

[267] M. Šturm, S. Lojen, Nitrogen isotopic signature of vegetables from the Slovenian market and its suitability as an indicator of organic production, Isot. Environ. Health Stud. 47 (2) (2011) 214–220.

[268] J.E. Spangenberg, S.A. Macko, J. Hunziker, Characterization of olive oil by carbon isotope analysis of individual fatty acids: implications for authentication, J. Agric. Food Chem. 46 (10) (1998) 4179–4184.

[269] R. Hrastar, M.G. Petrišič, N. Ogrinc, I.J. Košir, Fatty acid and stable carbon isotope characterization of *Camelina sativa* oil: implications for authentication, J. Agric. Food Chem. 57 (2) (2009) 579–585.

[270] F. Camin, R. Larcher, M. Perini, L. Bontempo, D. Bertoldi, G. Gagliano, G. Nicolini, G. Versini, Characterisation of authentic Italian extra-virgin olive oils by stable isotope ratios of C, O and H and mineral composition, Food Chem. 118 (4) (2010) 901–909.

[271] M.A. Brescia, G. Di Martino, C. Fares, N. Di Fonzo, C. Platani, S. Ghelli, F. Reniero, A. Sacco, Characterization of Italian durum wheat semolina by means of chemical analytical and spectroscopic determinations, Cereal Chem. 79 (2) (2002) 238–242.
[272] T. Nietner, S.A. Haughey, N. Ogle, C. Fauhl-Hassek, C.T. Elliott, Determination of geographical origin of distillers dried grains and solubles using isotope ratio mass spectrometry, Food Res. Int. 60 (2014) 146–153.
[273] J.-P. Renou, C. Deponge, P. Gachon, J.-C. Bonnefoy, J.-B. Coulon, J.-P. Garel, R. Vérité, P. Ritz, Characterization of animal products according to geographic origin and feeding diet using nuclear magnetic resonance and isotope ratio mass spectrometry: cow milk, Food Chem. 85 (1) (2004) 63–66.
[274] P. Ritz, P. Gachon, J.-P. Garel, J.-C. Bonnefoy, J.-B. Coulon, J.-P. Renou, Milk characterization: effect of the breed, Food Chem. 91 (3) (2005) 521–523.
[275] M.A. Brescia, M. Monfreda, A. Buccolieri, C. Carrino, Characterisation of the geographical origin of buffalo milk and mozzarella cheese by means of analytical and spectroscopic determinations, Food Chem. 89 (1) (2005) 139–147.
[276] I.-M. Chung, I. Park, J.-Y. Yoon, Y.-S. Yang, S.-H. Kim, Determination of organic milk authenticity using carbon and nitrogen natural isotopes, Food Chem. 160 (2014) 214–218.
[277] E. Piasentier, R. Valusso, F. Camin, G. Versini, Stable isotope ratio analysis for authentication of lamb meat, Meat Sci. 64 (3) (2003) 239–247.
[278] O. Schmidt, J.M. Quilter, B. Bahar, A.P. Moloney, C.M. Scrimgeour, I.S. Begley, F.J. Monahan, Inferring the origin and dietary history of beef from C, N and S stable isotope ratio analysis, Food Chem. 91 (3) (2005) 545–549.
[279] J.-P. Renou, G. Bielicki, C. Deponge, P. Gachon, D. Micol, P. Ritz, Characterization of animal products according to geographic origin and feeding diet using nuclear magnetic resonance and isotope ratio mass spectrometry. Part II: beef meat, Food Chem. 86 (2) (2004) 251–256.
[280] C.N. Rhodes, J.H. Lofthouse, S. Hird, P. Rose, P. Reece, J. Christy, R. Macarthur, P.A. Brereton, The use of stable carbon isotopes to authenticate claims that poultry have been corn-fed, Food Chem. 118 (4) (2010) 927–932.
[281] B.L. Guo, Y.M. Wei, J.R. Pan, Y. Li, Stable C and N isotope ratio analysis for regional geographical traceability of cattle in China, Food Chem. 118 (4) (2010) 915–920.
[282] K.S. Kim, J.S. Kim, I.M. Hwang, I.S. Jeong, N. Khan, S.I. Lee, D.B. Jeon, Y.H. Song, Application of stable isotope ratio analysis for origin authentication of pork, Korean J. Food Sci. Anim. Resour. 33 (1) (2013) 39–44.
[283] L. Biondi, M.G. D'Urso, V. Vasta, G. Luciano, M. Scerra, A. Priolo, L. Ziller, L. Bontempo, P. Caparra, F. Camin, Stable isotope ratios of blood components and muscle to trace dietary changes in lambs, Animal 7 (9) (2013) 1559–1566.
[284] M. Perini, F. Camin, J. Sánchez Del Pulgar, E. Piasentier, Effect of origin, breeding and processing conditions on the isotope ratios of bioelements in dry-cured ham, Food Chem. 136 (3–4) (2013) 1543–1550.
[285] J.M. Moreno Rojas, F. Serra, I. Giani, V.M. Moretti, F. Reniero, C. Guillou, the use of stable isotope ratio analyses to discriminate wild and farmed gilt-head sea bream (*Sparus aurata*), Rapid Commun. Mass Spectrom. 21 (2) (2007) 207–211.
[286] D.J. Morrison, T. Preston, J.E. Bron, R.J. Hemderson, K. Cooper, F. Strachan, J.G. Bell, Authenticating production origin of gilt-head sea bream (*Sparus aurata*) by chemical and isotopic fingerprinting, Lipids 42 (6) (2007) 537–545.
[287] U. Kropf, M. Korošec, J. Bertoncelj, N. Ogrinc, M. Nečemer, P. Kump, T. Golo, Determination of the geographical origin of Slovenian black locust, lime and chestnut honey, Food Chem. 121 (3) (2010) 839–846.

[288] O.-R. Dinca, R.E. Ionete, R. Popescu, D. Costinel, G.-L. Radu, Geographical and botanical origin discrimination of romanian honey using complex stable isotope data and chemometrics, Food Anal. Method. 8 (2) (2014) 401–412.

[289] G. Lamprecht, K. Blochberger, Protocol for isolation of vanillin from ice cream and yoghurt to confirm the vanilla beans origin by ^{13}C-EA-IRMS, Food Chem. 114 (3) (2009) 1130–1134.

[290] L. Schipilliti, G. Dugo, L. Santi, P. Dugo, L. Mondello, Authentication of bergamot essential oil by gas chromatography-combustion- isotope ratio mass spectrometer (GC-C-IRMS), J. Essent. Oil Res. 23 (2) (2011) 60–71.

[291] K. Hör, C. Ruff, B. Weckerle, T. König, P. Schreier, Flavor authenticity studies by ^{2}H/^{1}H ratio determination using on-line gas chromatography pyrolysis isotope ratio mass spectrometry, J. Agric. Food Chem. 49 (1) (2001) 21–25.

[292] C.I. Rodrigues, R. Maia, M. Miranda, M. Ribeirinho, J.M.F. Nogueira, C. Máguas, Stable isotope analysis for green coffee bean: a possible method for geographic origin discrimination, J. Food Compos. Anal. 22 (5) (2009) 463–471.

[293] M. Perini, M. Paolini, M. Simoni, L. Bontempo, U. Vrhovsek, M. Sacco, F. Thomas, E. Jamin, A. Hermann, F. Camin, Stable isotope ratio analysis for verifying the authenticity of balsamic and wine vinegar, J. Agric. Food Chem. 62 (32) (2014) 8197–8203.

[294] C. Czank, A. Cassidy, Q. Zhang, D.J. Morrison, T. Preston, P.A. Kroon, N.P. Botting, C.D. Kay, Human metabolism and elimination of the anthocyanin, cyanidin-3-glucoside: a ^{13}C-tracer study, Am. J. Clin. Nutr. 97 (5) (2013) 995–1003.

[295] K. Choy, S.H. Nash, A.R. Kristal, S. Hopkins, B.B. Boyer, D.M. O'Brien, The carbon isotope ratio of alanine in red blood cells is a new candidate biomarker of sugar-sweetened beverage intake, J. Nutr. 143 (6) (2013) 878–884.

[296] P.S. Patel, A.J.M. Cooper, T.C. O'Connell, G.G.C. Kuhnle, C.K. Kneale, A.M. Mulligan, R.N. Luben, S. Brage, K.-T. Khaw, N.J. Wareham, N.G. Forouhi, Serum carbon and nitrogen stable isotopes as potential biomarkers of dietary intake and their relation with incident type 2 diabetes: the EPIC-Norfolk study, Am. J. Clin. Nutr. 100 (2) (2014) 708–718.

[297] T. Larsen, D. Lee Taylor, M.B. Leigh, D.M. O'Brien, Stable isotope fingerprinting: a novel method for identifying plant, fungal, or bacterial origins of amino acids, Ecology 90 (12) (2009) 3526–3535.

[298] S.M. Harrison, F.J. Monahan, A.P. Moloney, S.D. Kelly, F. Cuffe, J. Hoogewerff, O. Schmidt, Intra-muscular and inter-muscular variation in carbon turnover of ovine muscles as recorded by stable isotope ratios, Food Chem. 123 (2) (2010) 203–209.

[299] K.J. Petzke, S. Lemke, Hair protein and amino acid ^{13}C and ^{15}N abundances take more than 4 weeks to clearly prove influences of animal protein intake in young women with a habitual daily protein consumption of more than 1 g per kg body weight, Rapid Commun. Mass Spectrom. 23 (16) (2009) 2411–2420.

[300] K.J. Petzke, H. Boeing, C.C. Metges, Choice of dietary protein of vegetarians and omnivores is reflected in their hair protein ^{13}C and ^{15}N abundance, Rapid Commun. Mass Spectrom. 19 (11) (2005) 1392–1400.

[301] K.J. Petzke, H. Boeing, S. Klaus, C.C. Metges, Carbon and nitrogen stable isotopic composition of hair protein and amino acids can be used as biomarkers for animal-derived dietary protein intake in humans, J. Nutr. 135 (6) (2005) 1515–1520.

[302] J. Eagles, Stable isotope ratio mass spectrometry for iron bioavailability studies, Anal. Chem. 57 (2) (1985) 469–471.

[303] D.J. Morrison, B. Dodson, C. Slater, T. Preston, (13)C natural abundance in the British diet: Implications for (13)C breath tests, Rapid Commun. Mass Spectrom. 14 (15) (2000) 1321–1324.

[304] S. Rudloff, S. Obermeier, C. Borsch, G. Pohlentz, R. Hartmann, H. Brösicke, M.J. Lentze, C. Kunz, Incorporation of orally applied ^{13}C-galactose into milk lactose and oligosaccharides, Glycobiology 16 (6) (2006) 477–487.

[305] M.A. Nisbet, H.J. Tobias, J.T. Brenna, G.L. Sacks, A.K. Mansfield, Quantifying the contribution of grape hexoses to wine volatiles by high-precision [U^{13}C]-glucose tracer studies, J. Agric. Food Chem. 62 (28) (2014) 6820–6827.

[306] G. Janssens, S. Mangelinckx, D. Courtheyn, S. Prévost, G. De Poorter, N. De Kimpe, B. Le Bizec, Application of gas chromatography-mass spectrometry/combustion/isotope ratio mass spectrometry (GC-MS/C/IRMS) to detect the abuse of 17β-estradiol in cattle, J. Agric. Food Chem. 61 (30) (2013) 7242–7249.

[307] J.P. Scarth, A. Clarke, P. Teale, A. Mill, R. Macarthur, J. Kay, Detection of endogenous steroid abuse in cattle: results from population studies in the UK, Food Addit. Contam. Part A 28 (1) (2011) 44–61.

[308] M. Doué, E. Bichon, G. Dervilly-Pinel, V. Pichon, F. Chapuis-Hugon, E. Lesellier, C. West, F. Monteau, B. Le Bizec, Molecularly imprinted polymer applied to the selective isolation of urinary steroid hormones: an efficient tool in the control of natural steroid hormones abuse in cattle, J. Chromatogr. A 1270 (2012) 51–61.

[309] B. Le Bizec, G. Pinel, J.-P. Antignac, Options for veterinary drug analysis using mass spectrometry, J. Chromatogr. A 1216 (46) (2009) 8016–8034.

[310] H. Noppe, B. Le Bizec, K. Verheyden, H.F. De Brabander, Novel analytical methods for the determination of steroid hormones in edible matrices, Anal. Chim. Acta 611 (1) (2008) 1–16.

[311] E. Bichon, F. Kieken, N. Cesbron, F. Monteau, S. Prévost, F. André, B. Le Bizec, Development and application of stable carbon isotope analysis to the detection of cortisol administration in cattle, Rapid Commun. Mass Spectrom. 21 (16) (2007) 2613–2620.

[312] M. Hebestreit, U. Flenker, C. Buisson, F. Andre, B. Le Bizec, H. Fry, M. Lang, A.P. Weigert, K. Heinrich, S. Hird, W. Schänzer, Application of stable carbon isotope analysis to the detection of testosterone administration to cattle, J. Agric. Food Chem. 54 (8) (2006) 2850–2858.

[313] C. Buisson, M. Hebestreit, A.P. Weigert, K. Heinrich, H. Fry, U. Flenker, S. Banneke, S. Prevost, F. Andre, W. Schaenzer, E. Houghton, B. Le Bizec, Application of stable carbon isotope analysis to the detection of 17β-estradiol administration to cattle, J. Chromatogr. A 1093 (1–2) (2005) 69–80.

[314] S. Marchionni, E. Braschi, S. Tommasini, A. Bollati, F. Cifelli, N. Mulinacci, M. Mattei, S. Conticelli, High-precision ^{87}Sr/^{86}Sr analyses in wines and their use as a geological fingerprint for tracing geographic provenance, J. Agric. Food Chem. 61 (28) (2013) 6822–6831.

[315] P.R. Trincherini, C. Baffi, P. Barbero, E. Pizzoglio, S. Spalla, Precise determination of strontium isotope ratios by TIMS to authenticate tomato geographical origin, Food Chem. 145 (2014) 349–355.

[316] S. Medini, M. Janin, P. Verdoux, I. Techer, Methodological development for ^{87}Sr/^{86}Sr measurement in olive oil and preliminary discussion of its use for geographical traceability of PDO Nîmes (France), Food Chem. 171 (2015) 78–83.

[317] S. Rummel, C.H. Dekant, S. Hölzl, S.D. Kelly, M. Baxter, N. Marigheto, C.R. Quetel, R. Larcher, G. Nicolini, H. Fröschl, H. Ueckermann, J. Hoogewerff, Sr isotope measurements in beef-analytical challenge and first results, Anal. Bioanal. Chem. 402 (9) (2012) 2837–2848.

[318] S.K. Sahoo, S. Yoshida, S. Tokonami, H. Yonehara, N.Y. Tsygankov, P.V. Zamostyan, Isotopic composition of uranium in soil and ground water samples collected around 30 km Chernobyl exclusion zone, Radioprotection 44 (5) (2009) 785–790.

[319] B.L. Gulson, M. James, A.M. Giblin, A. Sheehan, P. Mitchell, Maintenance of elevated lead levels in drinking water from occasional use and potential impact on blood leads in children, Sci. Total Environ. 205 (2–3) (1997) 271–275.

[320] J.R. Turnlund, Bioavailability of dietary minerals to humans: the stable isotope approach, Crit. Rev. Food Sci. Nutr. 30 (4) (1991) 387–396.
[321] M. Hoppler, C. Zeder, T. Walczyk, Quantification of ferritin-bound iron in plant samples by isotope tagging and species-specific isotope dilution mass spectrometry, Anal. Chem. 81 (17) (2009) 7368–7372.
[322] S. Fairweather-Tait, T. Fox, S.G. Wharf, J. Eagles, The bioavailability of iron in different weaning foods and the enhancing effect of a fruit drink containing ascorbic acid, Pediatr. Res. 37 (4) (1995) 389–394.
[323] M.F. Young, R.P. Glahn, M. Ariza-Nieto, J. Inglis, G. Olbina, M. Westerman, K.O. O'Brien, Serum hepcidin is significantly associated with iron absorption from food and supplemental sources in healthy young women, Am. J. Clin. Nutr. 89 (2) (2009) 533–538.
[324] X. Cui, J. Sun, X.Y. Ling, C.F. Zhao, Z.J. Liu, G.F. Xu, Iron bioavailability in 4- to 6-year-old children of rural region in China, Trace Elem. Electrolytes 24 (1) (2007) 28–33.
[325] X. Cui, J. Sun, X.Y. Ling, C.F. Zhao, Z.J. Liu, G.F. Xu, Study by stable isotope on zinc fractional absorption in Chinese children of Shandong rural region, Trace Elem. Electrolytes 22 (2) (2005) 99–104.
[326] P. Kastenmayer, L. Davidsson, P. Galan, F. Cherouvrier, S. Hercberg, R.F. Hurrell, A double stable isotope technique for measuring iron absorption in infants, Br. J. Nutr. 71 (3) (1994) 411–424.
[327] G. Danezis, C. Georgiou, Emerging Trends in Elemental and Isotopic Food Analysis, under review.

Chapter 4

Ambient Ionization Techniques

Marinella Farré[1,*] and Damià Barceló[1,2]
[1]*Institute of Environmental Assessment and Water Research (IDAEA-CSIC), Barcelona, Spain;*
[2]*Catalan Institute of Water Research (ICRA), Girona, Spain*
Corresponding author: E-mail: mfuqam@cid.csic.es

Chapter Outline

1. Introduction 245
2. Principles and Classes of Ambient Ionization MS Techniques 248
2.1 Momentum Desorption: Solid–Liquid Extraction Ambient Ionization Techniques 249
2.2 Techniques Based on the Thermal Desorption and Chemical Ionization 254
2.3 Techniques Based on Laser Ablation and Ionization 257
3. Application of Ambient Ionization MS Techniques in Food Analysis 259
4. Conclusions 267
Acknowledgments 268
References 268

1. INTRODUCTION

Contaminants may be present in food due to anthropogenic and natural sources, or as consequence of various stages of its production, including food processing, packaging, transport, and storage. Under the point of view of food safety, the main risks to be considered are:

- Risks associated to the accumulation or transfer of complex mixtures of contaminants that include anthropogenic pollutants (such as industrial residues, plasticizers, and pesticides) and natural contaminants (e.g., biotoxins).
- Contingencies associated to the globalization that involve differences in food standards in different geographical areas as well as those related to the transport and storage where different types of ingredients and materials sharing the same spaces.
- The increasing complexity of foodstuffs that contain multiple and processed ingredients, which are often sharing production lines.
- Food frauds.

Comprehensive Analytical Chemistry, Vol. 68. http://dx.doi.org/10.1016/B978-0-444-63340-8.00004-2

During the recent years, a series of improvements in food analytical chemistry have been achieved, which can be classified in [1]:

- Advances in sample pretreatments,
- Progresses in biological techniques,
- Developments of separation techniques and,
- Improvements of spectroscopic techniques.

Due to the complexity and variety of food matrices, a sample pretreatment process with different steps and sample manipulations is in general required. During the sample pretreatment is when the biggest percentage of variability of the whole analytical process is introduced. In addition, sample preparation is also the most polluting part of the process because of the use of organic solvents to selectively enrich the target compounds and to remove interfering substances. In the case of very complex samples, for example, with a high content of lipids, the elimination of interfering compounds requires a previous sample pretreatment such as alkaline saponification followed by solvent extraction or a further purification step. Conventional approaches such as Soxhlet extraction continues being used for the extraction of nonpolar analytes. However, at present, an important effort has been done to reduce the number of steps and the use of solvents during the sample pretreatment. Among the extraction methods in food analytical chemistry solid-phase extraction (SPE), headspace (HS), microextraction techniques (ME), pressurized liquid extraction (PLE), supercritical fluid extraction (SFE), microwaves-assisted extraction (MAE), ultrasonic-assisted extraction, (UAE) and purge & trap are currently the most commonly used. Thanks to the progress of advanced adsorbent materials, recent developments in sample preparation and purification processes have been achieved. Among them, restricted access materials (RAM), molecularly imprinted polymers (MIPs), techniques based on large-size particles and sorbents based on nanomaterials should be highlighted and successful examples of application have been recently reported [2–7]. However, sample preparation continues being an investment in time of analysis, sample manipulation, and analytical costs.

Biological approaches are a good alternative to the instrumental analysis for the fast screening of organic contaminants in food. Biosensors and the immunological techniques present as main advantages—rapid responses, simplified or not required sample pretreatment and a low or not required use of solvents, potential miniaturization, and automation. However, should be pointed out that both groups of techniques have as main limitation sometimes the lack of specificity due to potential cross-reactivity with other structurally related substances. So in the case of positive results, a confirmatory analysis using other techniques is required.

Among the state of the art of analytical techniques, chromatography or electrophoresis coupled to advanced mass spectrometry are the techniques of choice. The number of applications based on liquid chromatography (LC) and capillary electrophoresis (CE) progressively increased in food analysis during the last two decades. During the last few years, an important contribution has been produced, thanks to new materials employed for the ultra-performance liquid chromatography (UPLC), on-chip CE, monolithic columns, and the use of new separation mechanisms as hydrophilic interaction liquid chromatography (HILIC) allows to reduce the analysis time and to improve the resolution power and efficiency. The number of applications based on gas chromatography (GC) coupled to different analyzers, specially coupled to MS systems, continue to play a crucial role in the field of residue analysis in food, and in particular, for the determination of some groups of contaminants that are not so sensitive to atmospheric pressure ionization (API) or for certain food commodities (e.g., with high fat content) [8–12].

Trace detection of contaminants at sub-ppt concentrations is becoming routine for many analytes. Such progress is mostly due to development of hyphenated chromatography-mass spectrometry (MS) techniques, which are today the methods of choice for the determination of trace organic contaminants in food samples. Hyphenated separation techniques as heart-cutting multidimensional chromatography (e.g., LC–LC, GC–GC, LC–GC, and LC–CE) and comprehensive two-dimensional approaches (GCxGC and LCxLC) have emerged as an alternatives to analyze complex samples in a situation in which the technological improvements, such as new column technologies, seem to be close to their maximum level. The peak capacity enhancement, achievable by multidimensional chromatography is much higher than that obtained after improvement by any mean one-dimensional separations. Multidimensional chromatography allows combination of two or more steps, increasing significantly the separation capability of the corresponding one-dimensional techniques. The number of applications regarding the use of such comprehensive techniques increases every year, also in the food analysis [13,14].

During recent years, *Foodomics* has emerged as a new discipline in food analysis that studies the food and nutrition domains through the application and integration of advanced analytical techniques and the so-call *omic* technologies (transcriptomics, genomics, proteomics, and metabolomics) to improve consumer's health and knowledge [15,16]. *Foodomics* can help solve some of the new challenges that modern food safety, quality, and traceability have to face. These challenges encompass the multiple analyses of contaminants and allergens; the establishment of more powerful analytical methodologies to guarantee food origin, traceability, and quality; the discovery of biomarkers to detect unsafe products; the capability to detect food safety problems before they grow and affect more consumers.

Most of the advances in food analytical chemistry, including Foodomics are based on the use of the state of the art in mass spectrometry. In this sense, most common MS/MS instruments are triple quadrupole (QqQ) and quadrupole liner ion trap (QqLIT), but quadrupole time-of-flight (QqToF) and Orbitrap have gained popularity during recent years. In spite of the high number of analytical methods based on the use of QqQ or QqLIT, the quantification of some contaminants can present some drawbacks: (1) for a reliable quantification two selected reaction monitoring (SRMs) are required and some analytes present only one transition; (2) some transitions are unspecific; (3) SRM methods are typically limited to about 100 or 150 target analytes depending on the chromatographic separation resulting in a loss of sensitivity. High-resolution mass analyzers, such as QqToF and Orbitrap can overcome some of these drawbacks. QqToF-MS systems that allow MS^2 experiments to be performed to provide fragmentation patterns, together with accurate mass measurements of product ions (mass accuracy in the low ppm range) [17]. Fourier transform ion cyclotron resonance (FT-ICR) is one of the most sensitive ion detection methods with mass accuracy of <1 ppm [18], but the high cost of instrumentation currently restricts their application [19]. An alternative to these methods is the Orbitrap mass analyzer, which was launched in 2005. However, it suffers from a slower data acquisition rate compared to QqToF instruments, and therefore it is not appropriate for fast chromatography [20]. This system provides outstanding mass accuracy, mass resolution, and reliable high-sensitivity MS^n performance.

During the recent years, one of the more outstanding approaches for rapid and confident food analysis comes from the union between ambient ionization and mass spectrometry. This group of techniques simplified the analysis, because in general sample preparation is not required or can be very much simplify, offering at the same time the specificity of mass spectrometry. These techniques are in addition adequate for throughput analysis and can be applied in nondestructive approaches for food control [21,22].

In the last decade, ambient desorption ionization techniques for mass spectrometry were developed, for which the samples in either solid or liquid state can be directly ionized under ambient conditions without any sample preparation [23–25]. The first reported technique was desorption electrospray ionization (DESI) in 2004 [25] followed by direct analysis in real time (DART) [26]. Since then, more than 30 different ambient ionization techniques for MS have been presented [27]. These techniques open a new window for food analysis eliminating sample pretreatment and time of analysis.

This chapter presents the state of the art in ambient ionization techniques for the analysis of food contaminants, with special emphasis on emerging contaminants.

2. PRINCIPLES AND CLASSES OF AMBIENT IONIZATION MS TECHNIQUES

In most ambient ionization mass spectrometry systems, the atmospheric ionization techniques-electrospray ionization (ESI) and atmospheric pressure chemical ionization (APCI) are the predominant ionization processes. The analytes are in general desorbed by three main processes; solid–liquid extraction using charged droplets, or thermally desorbed and chemically ionized, or by laser ablation and ionization. Therefore, according to the desorption mechanism it can be classified as follows:

- Momentum desorption: Solid–liquid extraction ambient ionization techniques
  - Desorption electrospray ionization (DESI)
  - Reactive desorption electrospray ionization (reactive DESI)
  - Geometry-independent desorption electrospray ionization (GI-DESI)
  - Transmission mode desorption electrospray ionization (TM-DESI)
  - Electrode-assisted desorption electrospray ionization (EADESI)
  - Easy ambient sonic spray ionization (EASI)
  - Desorption ionization by charge exchange (DICE)
  - Liquid extraction surface analysis (LESA)
  - Nanospray desorption electrospray ionization (nano-DESI)
- Thermal desorption and chemical ionization
  - Direct analysis in real time (DART)
  - Atmospheric pressure glow discharge ionization (APGDI)
  - Atmospheric pressure thermal desorption/ionization (APTDI)
  - Atmospheric pressure solids analysis probe (ASAP)
  - Desorption atmospheric pressure chemical ionization (DAPCI)
  - Desorption atmospheric pressure photoionization (DAPPI)
  - Dielectric barrier discharge ionization (DBDI)
  - Desorption corona beam ionization (DCBI)
  - Helium atmospheric pressure glow discharge ionization (HAPGDI)
  - Plasma-assisted desorption ionization (PADI)
- Laser ablation and ionization
  - Laser Spray Ionization (LSI)
  - Electrospray laser desorption ionization (ELDI)
  - Matrix-assisted laser desorption electrospray ionization (MALDESI)
  - Laser ablation electrospray ionization (LAESI)
  - Laser-assisted desorption electrospray ionization (LADESI)
  - Infrared laser-assisted desorption electrospray ionization (IR-LADESI)
  - Infrared laser ablation metastable-induced chemical ionization (IR-LAMICI)
  - Laser electrospray mass spectrometry (LEMS)
  - Laser desorption spray postionization (LDSPI)
  - Laser desorption atmospheric pressure chemical ionization (LD-APCI)
  - Laser diode thermal desorption (LDTD)

2.1 Momentum Desorption: Solid–Liquid Extraction Ambient Ionization Techniques

This group of techniques is based on the use of a charged spray, to create a liquid film on the sample surface. Then, the surface molecules are extracted into the solvent. The primary droplets impacting the surface of the sample produces the secondary droplets that are the source of ions for the mass spectrometer.

In general, the ionization mechanism of this group of techniques is the electrospray ionization (ESI).

Desorption electrospray ionization (DESI) was developed by Professor Graham Cooks' group in 2004 at Purdue University [25]. This was the pioneering technique in ambient mass spectrometry consisting on an atmospheric pressure ion source that achieves the ionization under ambient air conditions. Table 1 summarizes main ambient air ionization techniques and their chronological publication.

DESI uses a spray of electrospray solvent composed of charged droplets as the ionizing agent. This technique is by far the most widely applied technique in open air mass spectrometry. In particular, it has been used for surface analysis, including industrial polymers, environmental monitoring, analysis of pharmaceuticals, and food analysis, among many others. In the Section 3, different applications in food analysis are summarized. In Figure 1 are shown the general scheme of DESI source, which can be related with typical ESI in terms of enabling the analysis of substances over a large mass range without significant fragmentation [27].

During the recent years, different analytical techniques based on modifications of DESI have been developed. Among them, **reactive DESI** is based on the use of specific chemical reagents to produce charged droplets with characteristic reactants that produce a series of selective reactions with the analytes. Reactive DESI provided significant improvements in ionization efficiency of apolar compounds in raw undiluted complex matrices as compared to conventional DESI. Another improvement was achieved with the **geometry-independent DESI (GI-DESI)**. In GI-DESI, the electrospray plume, the sample surface, and the MS-inlet capillary are all positioned within a pressure-tight enclosure [28,29]. This approach allows DESI to be performed without need for careful control of the geometry of the sprayer and MS-inlet capillary positions and angles relative to the sample surface. Another configuration was adopted by **transmission mode desorption electrospray ionization (TM-DESI)**, which passes the electrospray plume through a sample predeposited on a mesh substrate [29]. TM-DESI is an interface designed to simplify the sample preparation process for liquid samples and sample extracts. This technique offers the possibility of high-throughput analysis of target compounds [30]. However, it is not applicable to bulk materials. For the rapid, noncontact surface analysis of materials under ambient conditions, the **electrode-assisted desorption electrospray ionization (EADESI)** [31] is a good option. Under this configuration, the ions are generated by the impact of highly charged solvent droplets produced on a sharp-edged tip subjected to a high voltage. EADESI needs lower voltage for ionization and presents lower background and proper capabilities in obtaining spectra of high-molecular weight compounds. Another technique is the **easy ambient sonic spray ionization (EASI)**, which is based on the sonic spray ionization (SSI) principle developed by Hirabayashi et al. in 1994 [32]. In SSI,

TABLE 1 Chronological Publication of Main Ambient Air Ionization Techniques and Their ionization Principle

Technique	Acronym	Year of Publication	Ionization Principle
Desorption electrospray ionization	DESI	2004	ESI
Surface sampling probe	SSP	2004	
Direct analysis in real time	DART	2005	APCI
Ambient solid analysis probe	ASAP	2005	APCI
Electrospray laser desorption/ionization	ELDI	2005	ESI
Fused droplet electrospray ionization	FD-ESI	2005	ESI
Desorption atmospheric pressure chemical ionization	DAPCI	2005	APCI
MALDI-assisted electrospray ionization	MALDESI	2006	ESI
Extractive electrospray ionization	EESI	2006	ESI
Desorption sonic spray ionization	DeSSI	2006	ESI
Plasma-assisted Desorption/ ionization	PADI	2007	APCI
Dielectric discharge barrier ionization	DBDI	2007	APCI
Helium atmospheric pressure glow discharge ionization	HAPGDI	2007	APCI
Neutral desorption extractive electrospray ionization	ND-EESI	2007	ESI
Laser ablation electrospray ionization	LAESI	2007	ESI
Atmospheric pressure thermal desorption ionization	APTDI	2007	APCI
Desorption atmospheric Pressure Photo ionization	DAPPI	2007	APPI

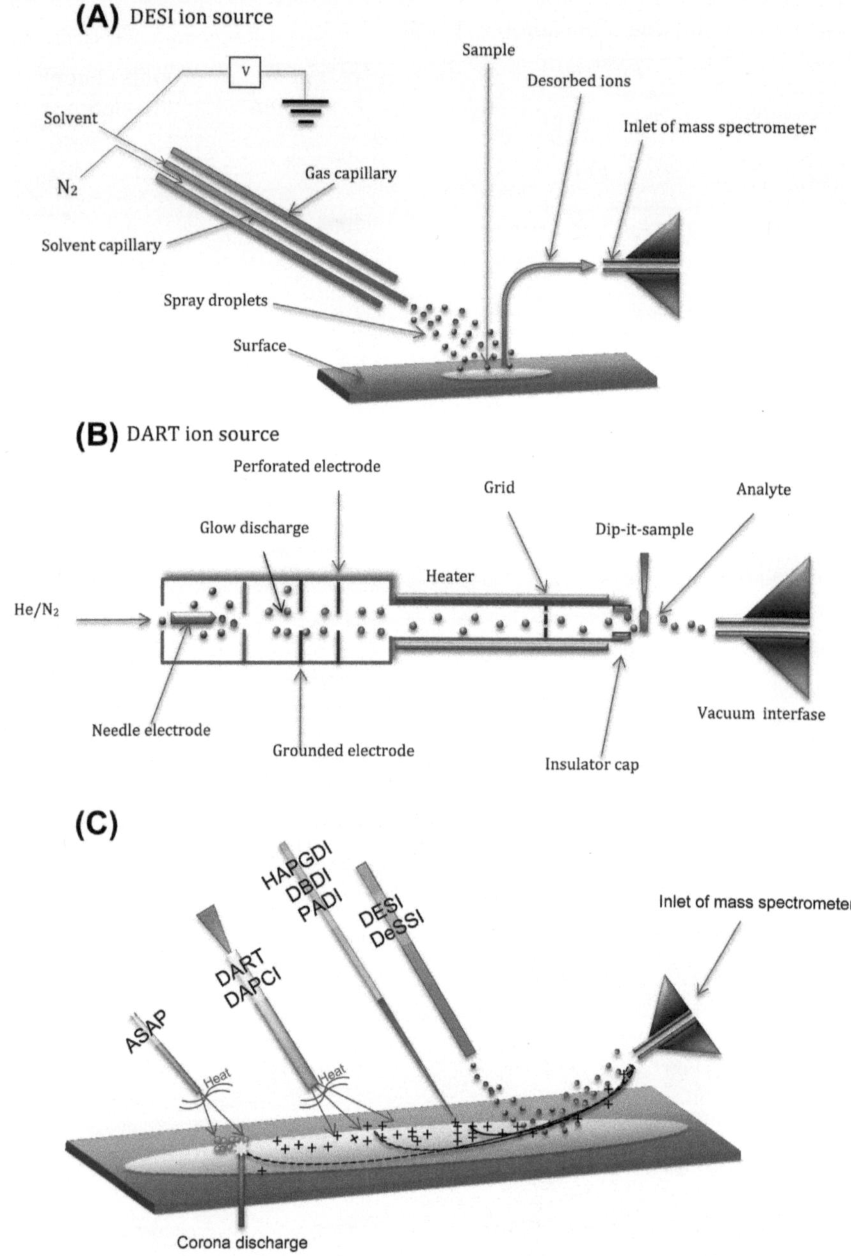

FIGURE 1 (A) and (B) schemes of DESI and DART, respectively. (C) Schematic summary of different ambient air ionization techniques.

the charged droplets are produced by spraying at sonic speed an acidified methanol solution of the analyte. The very minute droplets are charged with both negative and positive ions. However, the droplets have a limited charge capacity, which arises from a statistically imbalanced distribution of anions and cations. Then, the droplets desolvation is achieved at room temperature. By the assistance of nitrogen compressed, ions are formed via SSI. Haddad et al. in 2006 applied this principle to desorb and ionize analytes from surfaces under ambient air conditions [33,34]. The main advantages of EASI are the great simplicity, since just a compressed gas is necessary (N_2 or air), and the ability to produce positive- and negative-charged droplets, simultaneously. It does not require voltages, electrical discharges, UV or laser beams, or high temperature and can be easily implemented in any mass spectrometer equipped with an API source. Other advantages are the low charge concentrations [34,35], which rebounds in solvent noise reduction and an increasing the ration signal-to-noise, no thermal degradation. In addition in solid samples, the high velocity of EASI spray favors deep matrix penetration, providing quite homogeneous sampling and long-lasting ion signals. In addition, EASI thanks to the soft ionization favors the analysis of labile molecules and complex mixtures.

Among desorption techniques, the **desorption ionization by charge exchange (DICE)** [36] is another one for mass spectrometric analysis of substances present on solid surfaces was developed. In this case, a nebulized spray containing molecular ions of a solvent such as toluene can be generated by passing the solvent through a stainless steel capillary held at a high voltage. When the stream of charged droplets produced in this way is directed onto a solid surface, the analytes present on the surface are desorbed and ionized by a charge exchange process. This technique was shown to desorb and generate positively charged molecular ions from compounds that are not readily ionized by some other ambient methods, under positive-ion generation mode. **Liquid extraction surface analysis mass spectrometry (LESA-MS)** [37] is another novel surface profiling technique that combines microliquid extraction from a solid surface with nanoelectrospray mass spectrometry. A similar technique is the recent **nano-spray Desorption Electro-Spray Ionization technique nano-DESI [38]** that analyses tiny samples of atmospheric aerosols. This new technique requires no sample preparation and provides a reliable, simple way to introduce these samples into a mass spectrometer for detailed analysis of their molecular composition. Finally, a dual mode self-pumping technique, the **Venturi-EASI (V-EASI)** was derived for both solid and liquid sample analysis [39]. This technique adds a Venturi self-pumping of a proper SSI solvent solution to its precursor technique (EASI) that forms a stream of bipolar charged droplets that bombard the sample surface, causing desorption and ionization of the analyte molecules.

2.2 Techniques Based on the Thermal Desorption and Chemical Ionization

An electrical discharge is used to produce reactive species in a flowing gas prior to interaction with the sample. Heat is often used to assist in desorption of volatile species from the sample. Ions are formed by chemical ionization in the gas phase. In general, this group of techniques is based on atmospheric pressure chemical ionization (APCI) but some of them are based on atmospheric pressure photoionization (APPI).

The first technique developed within this group was the **direct analysis in real time (DART)** by Cody et al. [26] In the same manner that DESI can be related with ESI, DART can be more related to APCI [24], since it is based on the formation of a (distal) plasma discharge in a heated gas stream. The operation of DART involves the atmospheric pressure interactions of long-lived electronic excited state atoms or vibronic excited state molecules with the sample and atmospheric gases. In figure 2, a schematic diagram of the DART principle is presented. In this source, gas flow suffers an electrical glow discharge produced by an applied potential that generates ions, electrons, and excited state neutral species (atoms and molecules), referred to as metastables. Helium, nitrogen, and/or argon are the gases commonly used in DART. The chamber has perforated intermediate lenses or grids that eliminate most of the ions and electrons but leave the neutral gas atoms/molecules including metastable species. These gaseous species exiting the discharge chamber pass through an optional gas heater which adjusts the gas temperature (thermal analyte desorption). At the exit of the DART source, there is a grid that acts as an electrode promoting drifting of ions toward the inlet of the mass spectrometer's atmospheric interface and an insulation cap that protects the sample from any exposure to the grid. Ionization occurs when the DART gas makes contact with the sample. The DART ion source can operate in either positive or negative mode and ions formed by DART depend on the nature of the gas, ion polarity, and whether dopants are present.

In any case, both techniques (DESI and DART) produce ions of the intact analyte; therefore, their utility is dependent on the ability to obtain an elemental composition using accurate mass spectrometers (TOF or orbitrap), characteristic fragmentation patterns (QqQ or QqLIT), or both (QqTOF or different orbitrap combinations). However, most published data for DESI have been obtained in conjunction with MS/MS, whereas, in the case of DART systems, data are in general associated with accurate mass measurement coupling the DART source to a TOF-MS or and Orbitrap-MS instrument.

Comparing these two techniques one advantage of DART over DESI is that the adduct ions with metals and doubly charged ions are not formed. In addition, using DART the introduction of solvents in the ion source is not required. On the other hand, an advantage of DESI over DART is the wider mass range of DESI. DESI applies to both small and large molecules.

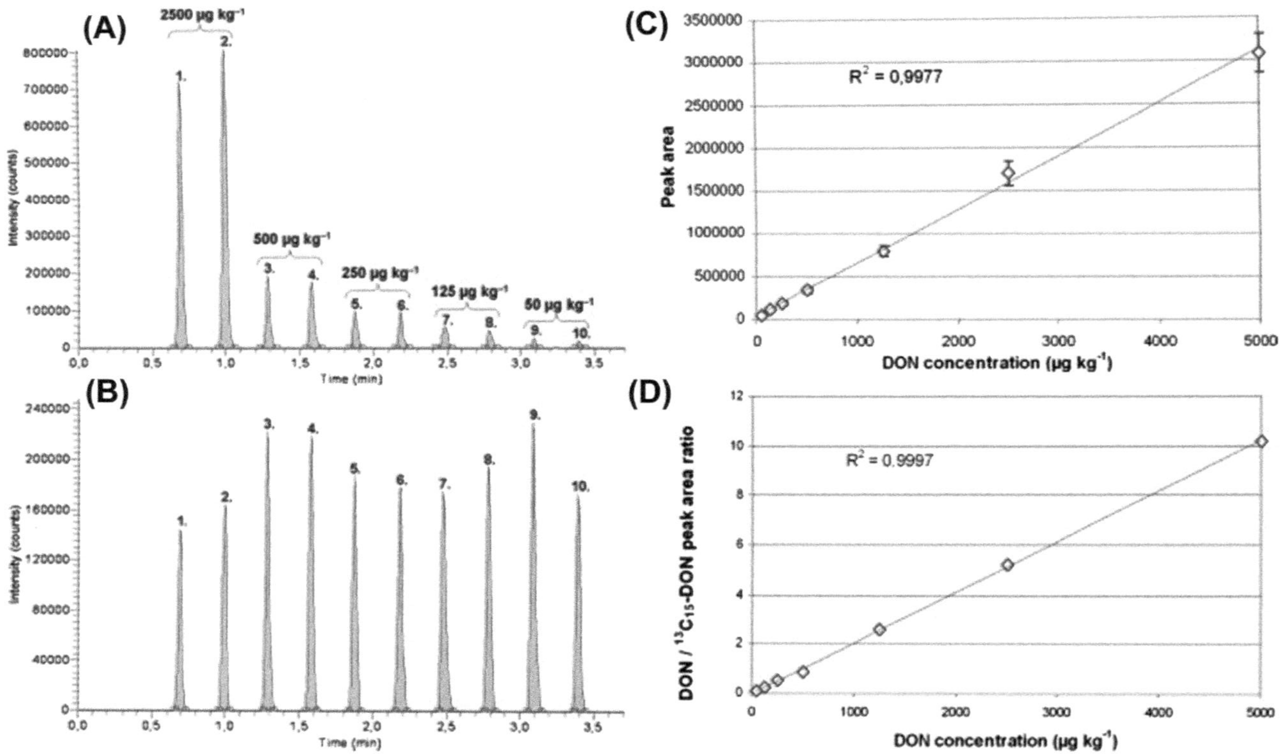

FIGURE 2 Calibrations employing matrix-matched standards and isotope dilution using DART-Orbitrap-MS method. (A) Extracted target ion record: DON (m/z 331.0943 ± 4 ppm), concentration in the range 50–2500 μg/kg; (B) extracted target ion record: 13C15-DON (m/z 346.1446 ± 4 ppm), concentration 500 μg/kg; (C) external calibration curve. (D) Isotope dilution calibration curve. Error bars are standard deviations calculated from three repeated injections. *Reproduced Ref. [60] with permission from Elsevier.*

DART is applicable to small molecules and does not ionize large molecules such as proteins. In summary, DART and DESI are different methods that complement each other.

Other relevant techniques based on the APCI and ionization techniques are **plasma-assisted desorption ionization (PADI), Helium atmospheric pressure glow discharge ionization (HAPGDI)**, atmospheric pressure solids analysis probe (ASAP), desorption atmospheric pressure chemical ionization (DAPCI) and Dielectric barrier discharge ionization (DBDI). In PADI [40], a nonthermal atmospheric plasma is directed onto the sample surface. This technique is characterized by a lower operating voltage, nonelimination of ions and electrons that also interact with the sample and higher current characteristics than corona discharges. Desorption occurs from the surface without heating of the sample and the subsequent ionization products are detected in real time. APGDI is based on apply a DC glow discharge to produce a plasma from an inert gas in ambient air conditions. The plasma species are produced into a region of reduced pressure to react with the volatile organic compounds [41]. The analytes are introduced, at reduced pressure into the glow discharge region for postionization. The **flowing afterglow/atmospheric pressure glow discharge ionization (FA-APGDI)** technique uses a glow discharge in the flowing afterglow mode to produce reacting ions and excited metastable species from He. The resulting species react with the analyte to produce the ionization.

Another technique that has been developed based on thermal desorption and chemical ionization is the **atmospheric pressure solids analysis probe (ASAP)** [42]. In this case, a hot nitrogen stream is used to impact the sample surface; the analytes are then transferred to the APCI needle tip and ionized by the corona discharge.

Desorption atmospheric pressure chemical ionization (DAPCI) is also an ambient air technique related to atmospheric pressure chemical ionization (APCI). In DAPCI, a corona discharge is generated and the reagent ions produced are directed pneumatically toward a surface using a high-velocity carrier gas (e.g., nitrogen). The analytes of the sample surface are desorbed and ionized by a two-step mechanism involving thermal desorption followed by gas-phase ionization. DAPCI is sensitive and efficient for the detection of low-polarity analytes. In fact, the first study using DAPCI appeared in a paper on the analysis of nitro-aromatic explosives by negative-ion DESI [43]. The main difference between DESI and DAPCI is the use of a hot solvent vapor and the application of a corona discharge instead of the charged solvent mist that is used in DESI. A similar technique to DAPCI is **desorption corona beam ionization (DCBI)** [44], which uses the reactive species that are generated by a corona discharge. In this case also, a high DC voltage about 1–5 KV is applied to induce the corona discharge from He. The reactive species formed from the He react with the analytes on the simple surface. This technique is adequate to a high number of low-molecular weight analytes.

In the case of DART, PADI, HAPGDI, APGDI, FA-APGDI, ASAP, and DAPCI and electrical discharge produce the neutral metastable species in a gas. These species react with water and air molecules to produce the reactive ionic species. The gas-phase ions used as reagents in chemical ionization can also be produced directly in the discharge by virtue of the high voltages involved.

On the other hand, based on the APPI as ionization mechanisms, **the desorption atmospheric pressure photoionization (DAPPI)** was developed with the aim to further improve the efficiency of ambient MS of low-polarity compounds. In DAPPI, the analytes from a sample surface are desorbed by a heated jet of solvent vapor and nebulizer gas. In this case, the use of a microchip nebulizer consisting on a glass for mixing gas and solvent flow (dopant) turned out to be advantageous for handling and for highly responsive temperature control of the gas. A krypton discharge lamp is used to irradiate the vapor phase immediately above the surface where ionization of the analyte occurs.

A very sensitive and selective method is the **atmospheric pressure thermal desorption ionization (APTDI)**, which consist in a technique based on chemical ionization. In this case, specific salts are heated and ions are directed toward the sample surface by a heated inert gas stream. In function of the analytes, the ionizing mixture can be very selective and sensitive as well.

2.3 Techniques Based on Laser Ablation and Ionization

Among the different laser-based techniques under ambient air conditions that have appeared during the last 10 years, **Laser Spray Ionization (LSI)** is the only one consisting on the direct desorption and ionization process by irradiation of the sample surface with protons.

When the reverse side of a glass slide with a predeposited dry sample is irradiated with a pulsed laser beam under ambient conditions, multiply charged ions are obtained. The ionization mechanism of LSI technique appears to occur through formation of matrix/analyte clusters under the action of laser irradiation, evaporation of the matrix from the clusters, and then formation of highly charged analyte ions through either charge condensation or ion ejection processes, as in ESI [45].

Several other hybrid ambient ionization techniques combining laser desorption/ablation with ESI have been reported, but based on a two-step process using charged ESI solvent species followed by the postionization. The main differences among these techniques are the presence or absence of matrix and the type of laser used (i.e., laser wavelength or pulse) [45].

Electrospray laser desorption ionization (ELDI) is a hybrid technique that combines the properties of both ESI and laser desorption allows the direct and sensitive analyses of a wide range of analytes on sample surfaces. ELDI-MS has been used for both small organic molecules and biomolecules (i.e., proteins). One of the main properties of this technique is the analysis of biomolecules without sample pretreatment. Chemical and biomolecules dissolved in solvents (e.g., water, methanol, methylene chloride, hexane, and toluene) can be characterized directly by ELDI-MS using fine carbon powders suspended in the solution, which absorb the laser energy [46]. ELDI-MS has been also used for molecular profiling and imaging of fungus and plant slices and for the surface analysis of TLC plates [47]. However, the main analytical limitation of this technique is that only polar compounds can be ionized in the ESI plume and detected by ELDI-MS.

Another technique based on laser ablation is **matrix-assisted laser desorption electrospray ionization (MALDESI)** [48,49]. In this technique, a pulsed laser is used to desorb or ablate particles, cluster of molecules from a simple surface. As when conventional matrix-assisted laser desorption/ionization (MALDI) is employed, the simple is typically mixed with a material that absorbs the energy from the laser and aids the desorption process. Different applications of MALDESI have been reported for proteins and peptides analysis [49] and proteomics [50].

A related approach is **laser ablation electrospray ionization (LAESI)**, which is based on the use of a mid-IR pulse laser for desorption [146]. This is also an electrospray-like ionization technique. The use of a mid-IR pulse laser (2.94 μm) for desorption means that the samples must be water-rich so that they undergo high absorption of laser energy. Main applications of LAESI have been in biological matrices and biomolecule analysis [50–53]. Another close related approach is the **infrared laser-assisted desorption electrospray ionization (IR-LADESI)** [54]. This technique employs an IR laser beam (2.94 μm) for sample desorption. IR-LADESI differs from LAESI only in terms of the energy (1.0 mJ) and incident angle (45°) of the laser beam. Other several approaches are the **laser electro-spray mass spectrometry (LEMS or fs-ELDI)** that uses a nonresonant, femtosecond-duration (70 fs) laser pulse for sample desorption, **the laser desorption spray post-ionization mass spectrometry (LDSPI-MS)**.

Among **laser techniques using APCI species for postionization**, the laser desorption atmospheric pressure chemical ionization (LD-APCI) and laser diode thermal desorption (LDTD) should be pointed out. The analytes produced by laser desorption/ablation can be ionized by reacting with charged solvent species generated through APCI. Because the ionization processes occur mainly through IMRs between polar analytes and singly charged solvent species, only singly charged analyte ions are produced. Therefore, this technique is only useful for small compounds analysis with a wide range of polarities [45]. **Laser desorption/atmospheric pressure chemical ionization (LD-APCI)** uses an inferred CO_2 pulse laser and the desorbed species are further ionized through reactions with charged ion species generated by APCI [55]. **Laser diode thermal desorption (LDTD)** combines ultrafast thermal desorption with APCI. The thermal desorption process is accomplished by irradiation of the rear side of a stainless-steel sheet by an inferred diode laser for a short period of time (5 s). The analytes of the sample applied on the other part of the sheet are desorbed and the ionization process is carried out by a corona discharge undergoing the APCI process [56].

3. APPLICATION OF AMBIENT IONIZATION MS TECHNIQUES IN FOOD ANALYSIS

During recent years, several applications of ambient ionization mass techniques have been reported for the analysis of chemical contaminants in food [57]. These techniques are very attractive for food analysis applications due to the elimination of the sample preparation step. However, reproducibility and matrix effects are two key challenges of ambient air techniques in food analysis because:

1. Ion suppression is a concern issue in complex samples subjected to mass spectrometry, in particular without any sample preparation.
2. The lack of reproducibility under some configurations could prevent them from becoming a real alternative to such a well-established quantitative method.

In spite of that, a great effort of development has been carried out and ambient air ionization approaches are constantly being introduced by the numerous advantages they present. Among the different techniques that have been presented in Section 2, DESI and DART are the best introduced techniques. Different examples of application are presented in Table 2.

TABLE 2 Examples of Ambient Air Ionization Techniques in Food Analysis

Analyte	Sample Matrix	Analytical Technique	References
Alternaria toxins, zearalenone, and sterigmatocystin	Cereals	DART-Orbitrap-MS	[60]
Ametryn, amitraz, azoxystrobin, bitertanol, buprofezin, imazalil, imazalil metabolite, isofenphos-methyl, malathion, nitenpyram, prochloraz, spinosad, terbuthylazine, thiabendazole, and thiacloprid	Fruits peels and foodstuff	DESI-LIT-MS	[87]
Atrazine	Cabbage leaf	DESI-LIT-MS	[62]
Authentication analysis of via TAG profiles	Brazil nut oil	EASI-MS	[78]
Clenbuterol	Urine	DESI-LIT-MS	[72]
Composition of phospholipids and triacylglycerols	Commercial lecithins	EASI-MS	[77]
Contaminants from packaging	Food	AMI-AMS	[74]
DEET, alachlor, atrazine	Vegetables	DESI-LIT-MS	[63]
Fluoroquinolone antibiotics	Food	DESI-MS^2	[71]

Continued

TABLE 2 Examples of Ambient Air Ionization Techniques in Food Analysis—cont'd

Analyte	Sample Matrix	Analytical Technique	References
Freshness degree	Lotus seeds	DAPCI-MS	[88]
Isopropylthioxanthone	Milk, yoghurt, fat	DART-TOF-MS	[89]
Lipid components	Lipids	DAPPI-MS DESI-MS	[90]
Maturation degree	Fruits and vegetables	EESI-QTOF-MS	[79]
Melamine	Pet food	DART-TOF-MS	[91]
Melamine	Milk	DART-TOF-MS	[92]
Melamine	Milk	EESI-QTOF-MS	[80]
Pesticides (alachlor, atrazine, carbendazim, carbofuran, dinoseb, isoproturon, metolachlor, metolcarb, propoxur, and simazine)	Fruit	APGDI-MS^2	[93]
Strobilurin fungicides	Wheat grain	DART-TOF	[65]
Strobilurin fungicides	Wheat grain	DESI-LIT-MS	[65]
Sugar monomers, oligomers, and citric acid	Fruits	AP-MALDI	[84]
The maturation degree through TAG profile	*Jatropha curcas* L. seeds	EASI-MS	[76]
Thiram	Fruits	DESI-MS^2	[64]
Thiram, ziram	Fruits	DART-TOF DART-Orbitrap-MS DESI-MS^2	[68]
Traceability	Dried sea cucumber products	DAPCI-MS	[94]
Unknown xenobiotics residues	Fruits	DART-Orbitrap-MS	[69]
Unsaturated fatty acyl hydroperoxides	Oils	EASI-MS	[75]
Vanilla	Flavored foods	ASAP-MS^2	[95]
α-tocoferol	Sage leaf	DAPPI-MS	[82]

Among the different applications of ambient ionization approaches in food analysis, the main developments have been performed in the fields of the analysis of contaminant residues and food quality control.

- Analysis of food contaminants
  - Natural origin such as biotoxins
  - Anthropogenic origin
    - Compounds used in food production, such as pesticides
    - Residues from packaging migration
- Food Quality Control
  - Autooxidation of oils and fats
  - Proteins and lipoproteins
  - Triacylglycerols and fats
  - Carbohydrates
  - Food frauds
    - Addition of foreign compounds
    - Authenticity

A high number of applications have been developed for the analysis of biotoxins, including mycotoxins, plant toxins, and marine toxins. For example, the use of DESI was described for plant alkaloids [58] and mycotoxins, such as fumonisin B1 [59]. Vaclavik et al. [60], demonstrated the capability of DART coupled to Orbitrap-MS to quantify zearalenone, Alternaria toxins, and sterigmatocystin in cereals. A QuEChERS extraction was previously performed and the method demonstrated its suitability for high-throughput control of maximum residue limits (MRLs) of deoxynivalenol and zearalenone, established in the EU regulation for unprocessed wheat or maize. Satisfactory accuracy of measurements was achieved by employing, both tested calibration approaches, matrix-matched calibration and isotope dilution (Figure 2). A high-throughput method for the control of 32 mycotoxins (Fusarium and Alternaria toxins, aflatoxins, ergot alkaloids, ochratoxins, and sterigmatocystin) in beer was proposed by Vaclavik et al. [60] In this method, due to the broad range of their physicochemical properties, the sample preparation step was simplified as much as possible to avoid analyte losses. The addition of acetonitrile to beer samples enabled precipitation of abundant matrix components. The clean-up efficiency was controlled by DART ion source. For determination of analytes, HPLC-orbitrap-MS was employed. This approach presents a new use of ambient air ionization as control technique in food analysis. Recently, imprints of potato sprout (*Solanum tuberosum* L.) were successfully imaged by DESI-MS on TLC plates through blotting assisted by heating and/or solvent extraction. Ion images showing the distribution of glycoalkaloid toxins in potato sprout [61]. DESI-MS images revealed little difference in signal intensity as well as image quality under different imprint conditions. Higher intensities at the sprout tip compared with the rest of the potato were observed for α-chaconine at m/z 852 and α-solanine at m/z 868. Also, higher intensities of α-chaconine compared with α-solanine was seen in all DESI-MS images regardless of the method of imprinting. The images obtained from imprints produced by regular blotting and thermally assisted blotting (Figure 3(A) and (B)) exhibit images with better overall quality compared with the images obtained from imprints produced by solvent extraction blotting with or without thermal assistance (Figure 3(B) and (D)). The use of solvents allows the diffusion of chemical compounds on the TLC plate surface that produces smeared images, but all four imprint conditions gave consistently good results.

Different applications have also been developed for the determination of pesticides in fruits and vegetables. Zhang et al. [62], assessed the use of DESI-MS^2 for the atrazine on cabbage leaf. In another study, Mulligan et al. [63], showed DESI-MS and DESI-MS^2 data for agrochemicals in vegetables. The analysis of thiram was studied in different fruits using DESI-MS^2 by

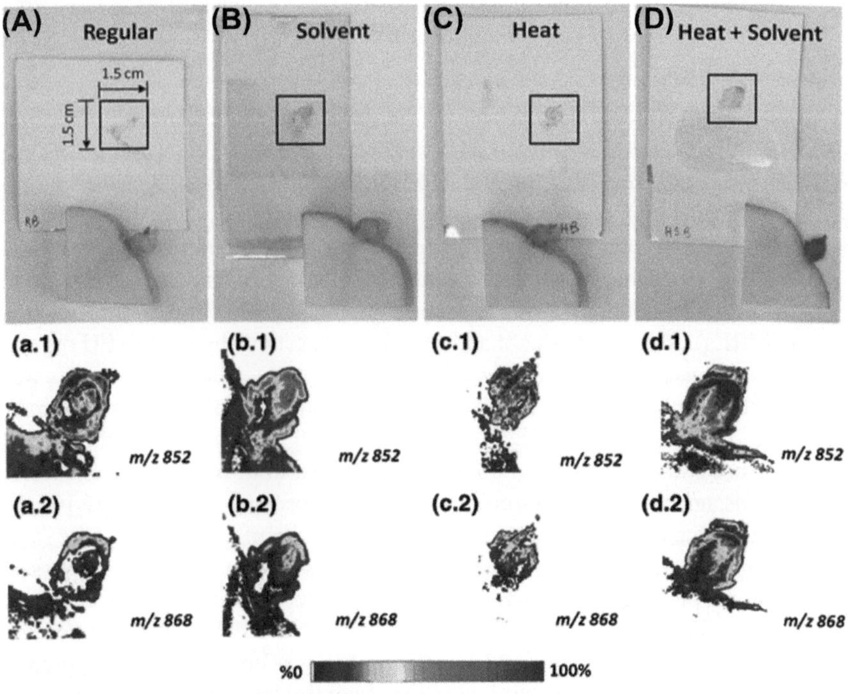

FIGURE 3 Imaging analysis of imprinted potato sprout. (A) Potato sprout imprinted directly onto a TLC plate without further manipulation. (B) Potato sprout imprinted directly onto the TLC plate after solvent deposition. (C) Potato sprout imprinted directly onto the TLC plate with heat applied. (D) Potato sprout imprinted directly onto the TLC plate with solvent mixture deposited and heat applied. (a.1–d.1) Ion images of different blotting samples showing the distribution of m/z 852 species, α-chaconine. (a.2–b.2) Ion images of different blotting samples showing the distribution of m/z 868 species, α-solanine. All ion images are the same size (1.5 × 1.5 cm) and are plotted on the same color scale, which is depicted below the panels to visualize the relative ion intensities from 0% (white) to 100% (red (dark gray in print versions)). All images were normalized by the same color scale (100% represents 2.38 × 103 counts).

Nielen et al. [64]. In this study, it was shown that an extraction with acetonitrile is required. Schurek et al. [65], used a DART-TOF-MS for the determination of strobilurin fungicides in ethyl acetate extracts prepared from wheat grains. Pesticide residues were detected using a positive-ionization mode, and the limit of quantification (LOQs) were in the range 0.005–0.03 mg/kg, which enabled control of MRLs in the range 0.05–0.3 mg/kg, for these compounds, as set by the EU. A rapid screening method for pesticides was developed by Edison et al. [66,67], to promote more efficient processing of produce entering the United States. Foam swabs were used to recover a multiclass mixture of 132 pesticides from the surfaces of grapes, apples, and oranges. The swabs were analyzed using DART ionization coupled with a high-resolution Exactive Orbitrap-MS. By using DART helium temperature gradient from 100 to 350°C over 3 min, a minimal separation of analytes based on volatility differences was achieved. This, combined with the Exactive's mass resolution of 100,000, allowed the chromatographic step, along with the typical compositing and extraction steps associated with conventional approaches by GC/MS or LC/MS, to be eliminated. Detection of 86% of the analytes present was consistently achieved at levels of 2 ng/g (per each apple or orange) and 10 ng/g (per grape). In addition, field samples that had undergone traditional sample preparation using QuEChERS (quick, easy, cheap, rugged, and safe) were analyzed using both LC/MS and DART-MS and the results from the two techniques were found to be comparable in terms of identification of the pesticides present. The use of swabs greatly increased sample throughput by reducing sample preparation and analysis time. The use of ambient mass spectrometry was explored by Cajka et al. [68], for the determination of two fungicides, thiram, and ziram in fruit. In this work, two complementary approaches were investigated. First, DART combined with medium-high resolution/accurate mass time-of-flight mass spectrometry (TOFMS) and high-resolution/accurate mass Orbitrap-MS. Second, DESI combined with tandem-in-time mass spectrometry (MS^2). With both techniques, thiram was detected. With DART, Ziram detection was also possible. In the case of thiram, the lowest calibration levels were 1 mg/kg (DART-TOFMS, DESI-MS^2) and 0.1 mg/kg (DART-Orbitrap-MS). For ziram, the achieved lowest calibration levels were 0.5 mg/kg (DART-TOFMS) and 1 mg/kg (DART-Orbitrap-MS). In all cases, the limits were enough to test samples against EU-MRLs for a number of fruit crops. In another application, Farré et al. [69] have studied the presence of xenobiotics in fruit peel by exposing it (without any pretreatment) to a DART coupled to a high-resolution Orbitrap-MS. In this study, the analysis of one sample was achieved in less than 1 min, and the benefits of both high-resolution and tandem mass spectrometry to elucidate nontarget or unknown compounds were obtained. The lowest imazalil concentration that was detected under this configuration was 1 ng (equivalent to a concentration of ca. 300 μg/kg), which is well below the maximum residue limit. A comparison of the results obtained using the direct peel screening DART-based method was made with those obtained by DART analysis of solvent extracts, as well as those obtained analyzing these extracts by UHPLC-Orbitrap, and good agreement was found. The application of ambient air techniques in pesticides residue analysis was reviewed by García-Reyes et al. [70] and Farré et al. [1].

Another field of application of novel ambient air techniques has been the determination of veterinary drug residues in food. However, just a small number of applications have been informed yet. Nielen et al. [71], used the fluoroquinolone antibiotic enrofloxacin as a model compound in a study of the performance of DESI-MS^2 analysis of immunoaffinity-captured antibiotics directly on a surface-plasmon resonance (SPR) biosensor chip. In another work, Lin et al. [72] developed a DESI-MS^2 screening method of clenbuterol in urine. The absolute limit of detection (LOD) using a liner ion trap (LIT) was 5 pg. Urine samples were extracted using a C18 SPE cartridge. The methanol eluate was analyzed by DESI-MS^2, and the LOD was 2 ng/mL. However, this approach has not fulfilled the requirements of current legislation.

The combined use of desorption by a continuous wave near-infrared diode laser and ionization by a DBDI-based probe (laser desorption dielectric barrier discharge ionization mass spectrometry (LD-DBDI- MS)) as an ambient ionization method for the mass spectrometric detection of nonvolatile chemicals on surfaces was recently presented by Gilbert-López et al. [73] Examples on the application of this approach for the analysis of nonvolatile chemicals were presented in both positive and negative ionization modes, including pesticides, pharmaceuticals, and explosives (reserpine, roxithromycin, propazine, prochloraz, spinosad, ampicillin, dicloxacillin, enrofloxacin, tetracycline, oxytetracycline, erythromycin, spinosad, cyclo-1,3,5,7-tetramethylene tetranitrate (HMX), and cyclo-1,3,5-trimethylene trinitramine (RDX)). A comparative evaluation revealed that the use of the laser is advantageous, compared to just heating the substrate surface.

Other recent applications have been related to the control of contaminants from packaging materials and inks. Setoff is the unintentional transfer of substances used in printing from the external printed surface of food packaging to the inner, food-contact surface. Bentayeb et al. [74] have presented ambient ionization–accurate mass spectrometry (AMI–AMS) to detect and identify setoff compounds. AMI mass spectra from inner and outer surfaces of printed and nonprinted food packaging were compared to detect and identify nonvisible setoff components. In this study, AMI–AMS was able to detect and identify seven different compounds. Setoff on the packaging samples was confirmed using GC–MS analysis of single-sided solvent extracts. N-Ethyl-2(and 4)-methylbenzenesulfonamide, 2,4-diphenyl-4-methyl-1(and 2)-pentene, and 2,4,7,9-tetramethyl-5-decyne-4,7-diol were identified on the food-contact layer at concentrations from 0.21 to 2.7 ± 1.6 $\mu g/dm^2$, corresponding to nearly milligram per kilogram concentrations in the packaged food.

Authentication and quality control is another emerging field of application of ambient mass spectrometric approaches. Several works have been recently reported using ESI-MS in oil quality control. For example, the oxidation is the major cause of fat and oil deterioration, which produces rancid flavors that drastically reduces their quality and nutritional value. The primary oxidation products are triacylglycerols (TAG) containing unsaturated fatty acyl hydroperoxides formed by a chain process occurring through free radical intermediates. The Rancimat method is currently the gold standard for measuring the oxidative stability of oils and fats. Simas et al. [75] reported the use of EASI-MS to monitor the most immediate TAG hydroperoxide products with comparable results to those obtained via the classical Rancimat test. In another work, the maturation of *Jatropha curcas* L. seeds was carried out via the monitoring TAG profile of the oil [76] using as well EASI-MS. Gabriel et al. used the same technique to characterize commercial lecithins [77], which are composed mainly of phospholipids and triacylglycerols. The analysis of the commercial lecithins, including their fraction of phospholipids, normally involves laborious and expensive protocols, whereas EASI-MS is shown to be an efficient technique for the analysis of lipids. Characteristic profiles of phosphatidylcholines and triacylglycerols are detected by EASI (+)-MS, whereas EASI (−)-MS provided phosphatidylethanolamines, glycophospholipids and free fatty acids profiles. Figure 4 shows representative EASI(−)-MS profiles from different commercial lecithins. Note that several free fatty acids (FFA), phosphatidylethanolamine (PE), lysophosphatidylethanolamine (LPE), acetylated phosphatidylethanolamine (PEAcet) and glycophospholipids (GPL) are now detected. The main ions were those of m/z 279 (FFA18:2), m/z 714 (PE34:2), m/z 738 (PE36:4), and m/z 833 (GPL). Acetylated lecithins also displayed characteristic acetylated derivatives. EASI-MS data were also compared to MALDI-MS, and found to display

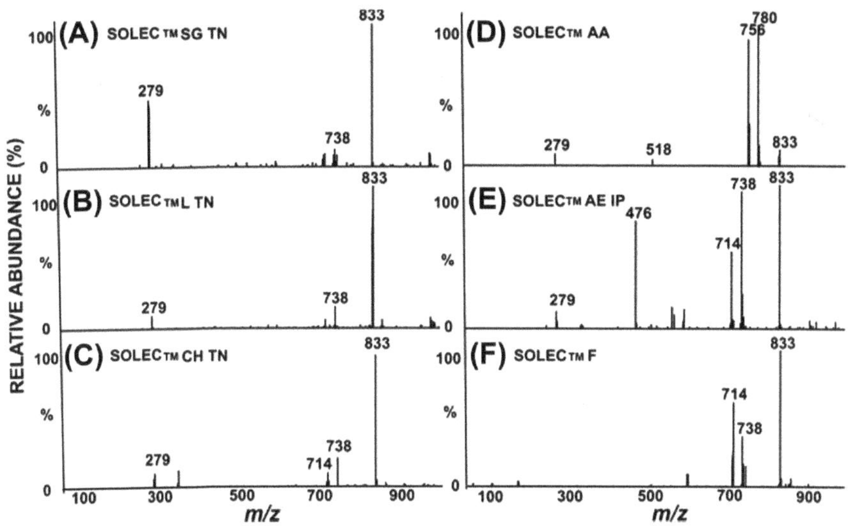

FIGURE 4 EASI(−)-MS for the commercial lecithins. Main ions assignment: m/z 279 (Linoleic Acid); m/z 476 (LPE 18:2); m/z 518 (LPEAcet 18:2); m/z 714 (PE 16:0/18:2); m/z 738 (PE 18:2/18:2 and/or 18:1/18:3); m/z 756 (PEAcet 16:0/18:2); m/z 780 (PEAcet 18:2/18:2 and/or 18:1/18:3); m/z 833 (Glycophospholipid).

richer compositional information. The industrial process applied to lecithin fabrication was also characterized via typical EASI-MS profiles [77]. Another related application was presented by Funasaki et al. [78] In this work, EASI-MS was used to evaluate and the authentication analysis of Brazil nut oil via TAG profiles.

Maturity is an essential factor that determines storage life and final quality of most fruits and vegetables. For the purpose of maturity differentiation, EESI-QTOF-MS was applied by Chen et al. [79] for rapid fingerprinting of compounds released from various fruits. Ripening stages of bananas, grapes, and strawberries are successfully differentiated by performing principal component analysis (PCA) of the mass spectral fingerprints of the fruits. Figure 5 illustrated the results obtained for strawberries. The mass spectra were quite

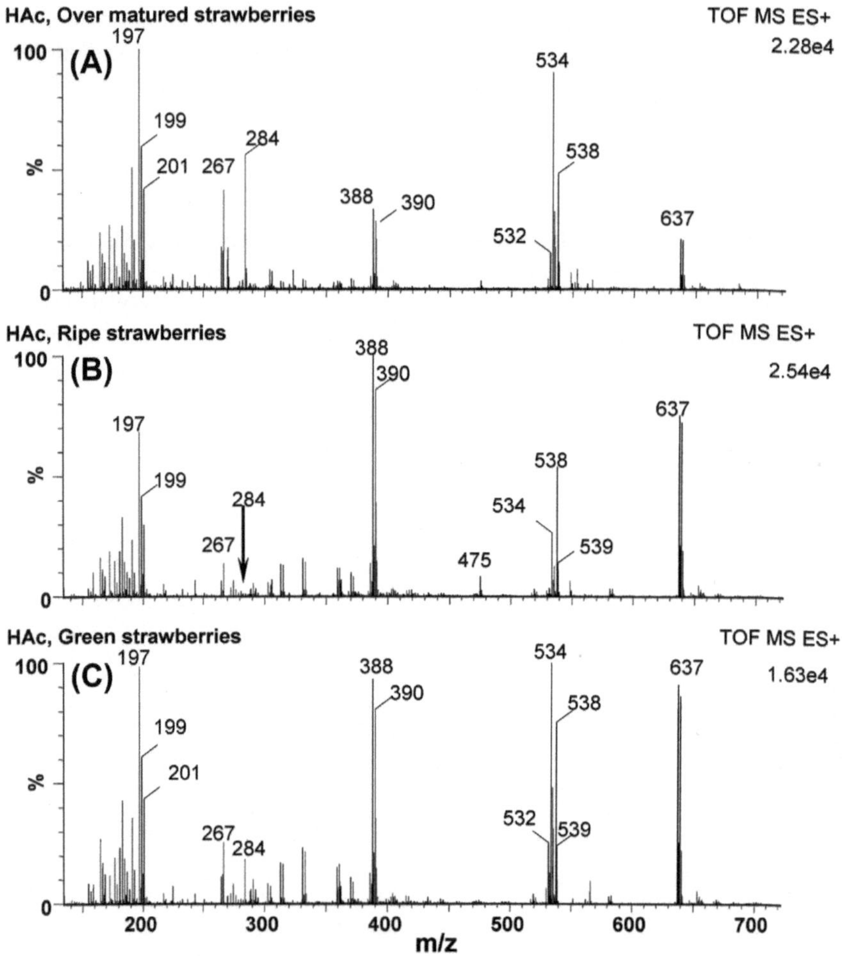

FIGURE 5 EESI-QTOF mass spectra of strawberries: (A) overripe strawberries; (B) normally ripe strawberries; and (C) green strawberries.

similar when these strawberries were inspected by extractive electrospray ionization (EESI). However, a close look revealed that the difference between these mass spectral fingerprints is still noticeable (shown in Figure 5(A)–(C)). For example, the intensities of peaks at m/z 267, 284, 388, 390, 534, and 673 varied in the mass spectra of strawberries of different qualities. The data indicated that EESI-QTOF-MS is a useful noninvasive tool for rapid investigation and differentiation of maturity and quality of fruits without sample preparation.

Another application of ambient ionization mass spectrometry is the rapid screening of food frauds. For example, EESI has been applied to the detection of melamine in untreated milk and wheat gluten [80].

PGDI has been employed to differentiate quality characteristics of Chinese liquors [81]. In this work, a homemade glow discharge gun was used to analyze volatile components from liquor samples. The fingerprint mass spectra of several Chinese liquors were obtained and combined with principal component analysis; this new method was successfully applied to differentiate different brands of Chinese liquors without any sample pretreatment. Compared with conventional methods, this novel approach has the advantages of easy operation, high speed, and high efficiency, which make it a potential tool in the fields of food safety, atmosphere chemistry.

When foods are exposed to ambient environments, they are prone to rapid spoilage and changes. For this reason, high speed of imaging is of particular importance in food analysis. Therefore, currently there is a demand for rapid imaging of food products. Mass spectrometry is a very attractive approach for chemical profiling because their high molecular specificity. Mass spectrometry for food imaging has increased during the last decade, and nowadays the last advance is performed in terms of ambient ionization techniques that allow direct and rapid analysis in ambient environment. Different techniques such as DAPPI, ELDI, DAPCI, and prove electrospray ionization (PESI) have been applied recently in food imaging. Among them different application for food nutrients assessment have been presented such as the content of α-tocoferol in sage leaf by DAPPI [82] and α-solanine in potatoes using ELDI or PESI [83]. AP-MALDI was used to imaging sugar monomers, oligomers and citric acid in different fruits [84]. DESI has been applied to determine the alkaloids in nutmeg. Also DESI was used to determine diterpenes in Stevia [85]. Finally, some application in terms of food safety has been also presented such as the determination of melamine content in milk of eggs [86].

4. CONCLUSIONS

Ambient ionization mass spectrometric techniques in food analysis have brought the promise of simple, high-throughput qualitative and quantitative analyses of both major and trace components. However, some limitations should be resolved. Firstly, MRLs are typically expressed in mg/kg of homogenized food product. In this case, data usually are related to pg/mm^2 obtained after direct detection of a contaminant from the surface of food. Moreover, contaminants are often not evenly distributed on the surface of the intact product. Secondly, the lack of separation is often resulting in matrix interferences. In this sense, advanced MS instruments combining the features of both high resolution and MS^n mass analyzers may provide solutions to obtain higher levels of confidence in the process of identification and quantification. Lastly, but not least, the unique potential for fingerprint-based authentication might provide a challenging option in food quality and safety.

ACKNOWLEDGMENTS

This work was supported by the Spanish Ministry of Science and Innovation, through the projects SCARCE Consolider-Ingenio 2010 CSD2009-00065 and by King Saud University grant number (KSU-VPP-105).

REFERENCES

[1] M. Farré, D. Barceló, Analysis of emerging contaminants in food, TrAC – Trends Anal. Chem. 43 (2013) 240–253.

[2] J. Chico, S. Meca, R. Companyó, M.D. Prat, M. Granados, Restricted access materials for sample clean-up in the analysis of trace levels of tetracyclines by liquid chromatography. Application to food and environmental analysis, J. Chromatogr. A 1181 (1–2) (2008) 1–8.

[3] O. Núñez, H. Gallart-Ayala, C.P.B. Martins, P. Lucci, New trends in fast liquid chromatography for food and environmental analysis, J. Chromatogr. A 1228 (2012) 298–323.

[4] M. Soylak, E. Yilmaz, M. Ghaedi, M. Montazerozohori, Solid phase extraction on multiwalled carbon nanotubes and flame atomic absorption spectrometry combination for determination of some metal ions in environmental and food samples, Toxicol. Environ. Chem. 93 (5) (2011) 873–885.

[5] V.F. Samanidou, E.G. Karageorgou, Carbon nanotubes in sample preparation, Current Organic Chem. 16 (14) (2012) 1645–1669.

[6] E.G. Karageorgou, V.F. Samanidou, Monolithic columns: the new era in the analysis of organic compounds by liquid chromatographic techniques, Current Organic Chem. 14 (19) (2010) 2310–2328.

[7] M.A. González-Curbelo, M. Asensio-Ramos, A.V. Herrera-Herrera, J. Hernández-Borges, Pesticide residue analysis in cereal-based baby foods using multi-walled carbon nanotubes dispersive solid-phase extraction, Anal. Bioanal. Chem. (2012) 1–14.

[8] P. Plaza-Bolaños, A.G. Frenich, J.L.M. Vidal, Polycyclic aromatic hydrocarbons in food and beverages. Analytical methods and trends, J. Chromatogr. A 1217 (41) (2010) 6303–6326.

[9] J. Falandysz, M. Rose, A.R. Fernandes, Mixed poly-brominated/chlorinated biphenyls (PXBs): widespread food and environmental contaminants, Environ. Int. 44 (1) (2012) 118–127.

[10] F. Hernández, T. Portolés, E. Pitarch, F.J. López, Gas chromatography coupled to high-resolution time-of-flight mass spectrometry to analyze trace-level organic compounds in the environment, food safety and toxicology, TrAC – Trends Anal. Chem. 30 (2) (2011) 388–400.

[11] D.A. Lambropoulou, I.K. Konstantinou, T.A. Albanis, Recent developments in headspace microextraction techniques for the analysis of environmental contaminants in different matrices, J. Chromatogr. A 1152 (1–2) (2007) 70–96.

[12] H. Sun, X. Ge, Y. Lv, A. Wang, Application of accelerated solvent extraction in the analysis of organic contaminants, bioactive and nutritional compounds in food and feed, J. Chromatogr. A 1237 (2012) 1–23.
[13] P.J. Marriott, S.T. Chin, B. Maikhunthod, H.G. Schmarr, S. Bieri, Multidimensional gas chromatography, TrAC – Trends Anal. Chem. 34 (2012) 1–20.
[14] M. Herrero, E. Ibáñez, A. Cifuentes, J. Bernal, Multidimensional chromatography in food analysis, J. Chromatogr. A 1216 (43) (2009) 7110–7129.
[15] A. Cifuentes, Foodomics: the necessary route to boost quality, safety and bioactivity of foods, Electrophoresis 35 (11) (2014) 1517–1518.
[16] C. Ibañez, V. García-Cañas, A. Valdés, C. Simó, Novel MS-based approaches and applications in food metabolomics, TrAC – Trends Anal. Chem. 52 (2013) 100–111.
[17] T. Reemtsma, LC-MS and strategies for trace-level analysis of polar organic pollutants, J. Chromatogr. A 1000 (2003) 477–501.
[18] S.C. Brown, G. Kurppa, J.L. Dasseux, Metabolomics applications of FT-ICR mass spectrometry, Mass Spectrom. Rev. 24 (2005) 223–231.
[19] A. Makarov, E. D, A. Kholomeev, W. Balschun, O. Lange, K. Strupat, S. Horning, Performance evaluation of a hybrid linear ion Trap/Orbitrap mass spectrometer, Anal. Chem. 78 (2006) 2113–2120.
[20] A. Tolonen, M. Turpeinen, O. Pelkonen, Liquid chromatography-mass spectrometry in in vitro drug metabolite screening, Drug Discov. Today 14 (3–4) (2009) 120–133.
[21] L.P. Li, B.S. Feng, J.W. Yang, C.L. Chang, Y. Bai, H.W. Liu, Applications of ambient mass spectrometry in high-throughput screening, Analyst 138 (11) (2013) 3097–3103.
[22] X. Wang, S. Wang, Z. Cai, The latest developments and applications of mass spectrometry in food-safety and quality analysis, TrAC – Trends Anal. Chem. 52 (2013) 170–185.
[23] R.G. Cooks, Z. Ouyang, Z. Takats, J.M. Wiseman, Ambient mass spectrometry, Science 311 (5767) (2006) 1566–1570.
[24] A. Venter, M. Nefliu, R. Graham Cooks, Ambient desorption ionization mass spectrometry, TrAC – Trends Anal. Chem. 27 (4) (2008) 284–290.
[25] Z. Takats, J.M. Wiseman, B. Gologan, R.G. Cooks, Mass spectrometry sampling under ambient conditions with desorption electrospray ionization, Science 306 (5695) (2004) 471–473.
[26] R.B. Cody, J.A. Laramée, H.D. Durst, Versatile new ion source for the analysis of materials in open air under ambient conditions, Anal. Chem. 77 (8) (2005) 2297–2302.
[27] D.J. Weston, Ambient ionization mass spectrometry: current understanding of mechanistic theory; analytical performance and application areas, Analyst 135 (4) (2010) 661–668.
[28] A. Venter, R.G. Cooks, Desorption electrospray ionization in a small pressure-tight enclosure, Anal. Chem. 79 (16) (2007) 6398–6403.
[29] J.E. Chipuk, J.S. Brodbelt, Transmission mode desorption electrospray ionization, J. Am. Soc. Mass Spectrom. 19 (11) (2008) 1612–1620.
[30] J.E. Chipuk, J.S. Brodbelt, The Influence of material and mesh characteristics on transmission mode desorption electrospray ionization, J. Am. Soc. Mass Spectrom. 20 (4) (2009) 584–592.
[31] A. Ozdemir, C.H. Chen, Electrode-assisted desorption electrospray ionization mass spectrometry, J. Mass Spectrom. 45 (10) (2010) 1203–1211.
[32] A. Hirabayashi, M. Sakairi, H. Koizumi, Sonic spray mass spectrometry, Anal. Chem. 67 (17) (1995) 2878–2882.
[33] R. Haddad, H.M.S. Milagre, R.R. Catharino, M.N. Eberlin, Easy ambient sonic-spray ionization mass spectrometry combined with thin-layer chromatography, Anal. Chem. 80 (8) (2008) 2744–2750.

[34] R. Haddad, R. Sparrapan, M.N. Eberlin, Desorption sonic spray ionization for (high) voltage-free ambient mass spectrometry, Rapid Commun. Mass Spectrom. 20 (19) (2006) 2901–2905.

[35] G.J. Van Berkel, S.P. Pasilis, O. Ovchinnikova, Established and emerging atmospheric pressure surface sampling/ionization techniques for mass spectrometry, J. Mass Spectrom. 43 (9) (2008) 1161–1180.

[36] C.-C. Chan, M. Bolgar, S. Miller, A. Attygalle, Desorption ionization by charge exchange (DICE) for sample analysis under ambient conditions by mass spectrometry, J. Am. Soc. Mass Spectrom. 21 (9) (2010) 1554–1560.

[37] D. Eikel, M. Vavrek, S. Smith, C. Bason, S. Yeh, W.A. Korfmacher, J.D. Henion, Liquid extraction surface analysis mass spectrometry (LESA-MS) as a novel profiling tool for drug distribution and metabolism analysis: the terfenadine example, Rapid Commun. Mass Spectrom. 25 (23) (2011) 3587–3596.

[38] P.J. Roach, J. Laskin, A. Laskin, Molecular characterization of organic aerosols using nanospray-desorption/electrospray ionization-mass spectrometry, Anal. Chem. 82 (19) (2010) 7979–7986.

[39] V.G. Santos, T. Regiani, F.F.G. Dias, W. Romao, J.L.P. Jara, C.F. Klitzke, F. Coelho, M.N. Eberlin, Venturi easy ambient Sonic-Spray ionization, Anal. Chem. 83 (4) (2011) 1375–1380.

[40] L.V. Ratcliffe, F.J.M. Rutten, D.A. Barrett, T. Whitmore, D. Seymour, C. Greenwood, Y. Aranda-Gonzalvo, S. Robinson, M. McCoustra, Surface analysis under ambient conditions using plasma-assisted desorption/ionization mass spectrometry, Anal. Chem. 79 (16) (2007) 6094–6101.

[41] S.A. McLuckey, G.L. Glish, K.G. Asano, B.C. Grant, Atmospheric sampling glow discharge ionization source for the determination of trace organic compounds in ambient air, Anal. Chem. 60 (20) (1988) 2220–2227.

[42] C.N. McEwen, R.G. McKay, B.S. Larsen, Analysis of solids, liquids, and biological tissues using solids probe introduction at atmospheric pressure on commercial LC/MS instruments, Anal. Chem. 77 (23) (2005) 7826–7831.

[43] Z. Takats, I. Cotte-Rodriguez, N. Talaty, H. Chen, R.G. Cooks, Direct, trace level detection of explosives on ambient surfaces by desorption electrospray ionization mass spectrometry, Chem. Commun. 15 (2005) 1950–1952.

[44] H. Wang, W. Sun, J. Zhang, X. Yang, T. Lin, L. Ding, Desorption corona beam ionization source for mass spectrometry, Analyst 135 (4) (2010) 688–695.

[45] M.Z. Huang, S.C. Cheng, Y.T. Cho, J. Shiea, Ambient ionization mass spectrometry: a tutorial, Analytica Chimica Acta 702 (1) (2011) 1–15.

[46] J. Shiea, C.H. Yuan, M.Z. Huang, S.C. Cheng, Y.L. Ma, W.L. Tseng, H.C. Chang, W.C. Hung, Detection of native protein ions in aqueous solution under ambient conditions by electrospray laser desorption/ionization mass spectrometry, Anal. Chem. 80 (13) (2008) 4845–4852.

[47] S.Y. Lin, M.Z. Huang, H.C. Chang, J. Shiea, Using electrospray-assisted laser desorption/ionization mass spectrometry to characterize organic compounds separated on thin-layer chromatography plates, Anal. Chem. 79 (22) (2007) 8789–8795.

[48] J.S. Sampson, A.M. Hawkridge, D.C. Muddiman, Construction of a Versatile high Precision ambient ionization source for direct analysis and imaging, J. Am. Soc. Mass Spectrom. 19 (10) (2008) 1527–1534.

[49] J.S. Sampson, A.M. Hawkridge, D.C. Muddiman, Generation and detection of multiply-charged peptides and proteins by matrix-assisted laser desorption electrospray ionization (MALDESI) Fourier transform ion cyclotron resonance mass spectrometry, J. Am. Soc. Mass Spectrom. 17 (12) (2006) 1712–1716.

[50] P. Nemes, A.A. Barton, Y. Li, A. Vertes, Ambient molecular imaging and depth profiling of live tissue by infrared laser ablation electrospray ionization mass spectrometry, Anal. Chem. 80 (12) (2008) 4575–4582.

[51] P. Nemes, A. Vertes, Laser ablation electrospray ionization for atmospheric pressure, in vivo, and imaging mass spectrometry, Anal. Chem. 79 (21) (2007) 8098–8106.

[52] P. Nemes, A.A. Barton, A. Vertes, Three-dimensional imaging of metabolites in tissues under ambient conditions by laser ablation electrospray ionization mass spectrometry, Anal. Chem. 81 (16) (2009) 6668–6675.

[53] P. Nemes, A. Vertes, Ambient mass spectrometry for in vivo local analysis and in situ molecular tissue imaging, TrAC – Trends Anal. Chem. 34 (2012) 22–33.

[54] Y.H. Rezenom, J. Dong, K.K. Murray, Infrared laser-assisted desorption electrospray ionization mass spectrometry, Analyst 133 (2) (2008) 226–232.

[55] J.J. Coon, W.W. Harrison, Laser desorption-atmospheric pressure chemical ionization mass spectrometry for the analysis of peptides from aqueous solutions, Anal. Chem. 74 (21) (2002) 5600–5605.

[56] J. Wu, C.S. Hughes, P. Picard, S. Letarte, M. Gaudreault, J.F. Lévesque, D.A. Nicoll-Griffith, K.P. Bateman, High-throughput cytochrome P450 inhibition assays using laser diode thermal desorption-atmospheric pressure chemical ionization-tandem mass spectrometry, Anal. Chem. 79 (12) (2007) 4657–4665.

[57] O. Lacina, P. Hradkova, J. Pulkrabova, J. Hajslova, Simple, high throughput ultra-high performance liquid chromatography/tandem mass spectrometry trace analysis of perfluorinated alkylated substances in food of animal origin: milk and fish, J. Chromatogr. A 1218 (28) (2011) 4312–4321.

[58] N. Talaty, Z. Takats, R.G. Cooks, Rapid in situ detection of alkaloids in plant tissue under ambient conditions using desorption electrospray ionization, Analyst 130 (12) (2005) 1624–1633.

[59] C.M. Maragos, M. Busman, Rapid and advanced tools for mycotoxin analysis: a review, Food Addit. Contam. – Part A Chem. Anal. Control Expo. Risk Assess. 27 (5) (2010) 688–700.

[60] L. Vaclavik, M. Zachariasova, V. Hrbek, J. Hajslova, Analysis of multiple mycotoxins in cereals under ambient conditions using direct analysis in real time (DART) ionization coupled to high resolution mass spectrometry, Talanta 82 (5) (2010) 1950–1957.

[61] E.C. Cabral, M.F. Mirabelli, C.J. Perez, D.R. Ifa, Blotting assisted by heating and solvent extraction for DESI-MS imaging, J. Am. Soc. Mass Spectrom. 24 (6) (2013) 956–965.

[62] X.Z. Zhang, C.J. Li, S.S. Chen, X.J. Li, H. Han, X.D. Ma, Direct determination of atrazine residue on Chinese cabbage leaf using desorption electrospray ionization-tandem mass spectrometry and its application for diagnosing atrazine drift phytotoxicity, J. AOAC Int. 92 (5) (2009) 1587–1592.

[63] C.C. Mulligan, N. Talaty, R.G. Cooks, Desorption electrospray ionization with a portable mass spectrometer: In situ analysis of ambient surfaces, Chem. Commun. 16 (2006) 1709–1711.

[64] M.W.F. Nielen, H. Hooijerink, G.R. Marchesini, W. Haasnoot, 2009.

[65] J. Schurek, L. Vaclavik, H. Hooijerink, O. Lacina, J. Poustka, M. Sharman, M. Caldow, M.W.F. Nielen, J. Hajslova, Control of strobilurin fungicides in wheat using direct analysis in real time accurate time-of-flight and desorption electrospray ionization linear ion trap mass spectrometry, Anal. Chem. 80 (24) (2008) 9567–9575.

[66] S.E. Edison, L.A. Lin, B.M. Gamble, J. Wong, K. Zhang, Surface swabbing technique for the rapid screening for pesticides using ambient pressure desorption ionization with high-resolution mass spectrometry, Rapid Commun. Mass Spectrom. 25 (1) (2011) 127–139.

[67] S.E. Edison, L.A. Lin, L. Parrales, Practical considerations for the rapid screening for pesticides using ambient pressure desorption ionisation with high-resolution mass spectrometry, Food Addit. Contam. – Part A Chem. Anal. Control Expo. Risk Assess. 28 (10) (2011) 1393–1404.

[68] T. Cajka, K. Riddellova, P. Zomer, H. Mol, J. Hajslova, Direct analysis of dithiocarbamate fungicides in fruit by ambient mass spectrometry, Food Addit. Contam. – Part A Chem. Anal. Control Expo. Risk Assess. 28 (10) (2011) 1372–1382.

[69] M. Farré, Y. Picó, D. Barceló, Direct peel monitoring of xenobiotics in fruit by direct analysis in real time coupled to a linear quadrupole ion trap-orbitrap mass spectrometer, Anal. Chem. 85 (5) (2013) 2638–2644.

[70] J.F. García-Reyes, B. Gilbert-López, A. Agüera, A.R. Fernández-Alba, A. Molina-Díaz, The potential of ambient desorption ionization methods combined with high-resolution mass spectrometry for pesticide testing in food, Comprehensive Anal. Chem. 58 (2012) 339–366.

[71] M.W.F. Nielen, H. Hooijerink, F.C. Claassen, M.C. van Engelen, T.A. van Beek, Desorption electrospray ionisation mass spectrometry: a rapid screening tool for veterinary drug preparations and forensic samples from hormone crime investigations, Analytica Chimica Acta 637 (1–2) (2009) 92–100.

[72] Z. Lin, S. Zhang, M. Zhao, C. Yang, D. Chen, X. Zhang, Rapid screening of clenbuterol in urine samples by desorption electrospray ionization tandem mass spectrometry, Rapid Commun. Mass Spectrom. 22 (12) (2008) 1882–1888.

[73] B. Gilbert-López, M. Schilling, N. Ahlmann, A. Michels, H. Hayen, A. Molina-Díaz, J.F. Garci-Reyes, J. Franzke, Ambient diode laser desorption dielectric barrier discharge ionization mass spectrometry of nonvolatile chemicals, Anal. Chem. 85 (6) (2013) 3174–3182.

[74] K. Bentayeb, L.K. Ackerman, T.H. Begley, Ambient Ionization–Accurate mass spectrometry (AMI-AMS) for the identification of nonvisible set-off in food-contact materials, J. Agric. Food Chem. 60 (8) (2012) 1914–1920.

[75] R.C. Simas, D. Barrera-Arellano, M.N. Eberlin, R.R. Catharino, V. Souza, R.M. Alberici, Triacylglycerols oxidation in oils and fats monitored by easy ambient sonic-spray ionization mass spectrometry, J. Am. Oil Chemists' Soc. 89 (7) (2012) 1193–1200.

[76] K.C. Cardoso, M.J. Da Silva, R. Grimaldi, M. Stahl, R.C. Simas, I.B.S. Cunha, M.N. Eberlin, Alberici RM: TAG profiles of *Jatropha curcas* L. Seed oil by easy ambient sonic-spray ionization mass spectrometry, J. Am. Oil Chemists' Soc. 89 (1) (2012) 67–71.

[77] D.F. Gabriel, M.A. Rosana, G.P. Gustavo, C.C. Elaine, N.E. Marcos, B.-A. Daniel, Direct characterization of commercial lecithins by easy ambient sonic-spray ionization mass spectrometry, Food Chem. 135 (3) (2012) 1855–1860.

[78] M. Funasaki, R.S. Oliveira, S.P. Zanotto, C.R.F. Carioca, R.C. Simas, M.N. Eberlin, R.M. Alberici, Brazil nut oil: quality control via triacylglycerol profiles provided by easy ambient sonic-spray ionization mass spectrometry, J. Agric. Food Chem. 60 (45) (2012) 11263–11267.

[79] H. Chen, Y. Sun, A. Wortmann, H. Gu, R. Zenobi, Differentiation of maturity and quality of fruit using noninvasive extractive electrospray ionization quadrupole time-of-flight mass spectrometry, Anal. Chem. 79 (4) (2007) 1447–1455.

[80] H. Chen, A. Venter, R.G. Cooks, Extractive electrospray ionization for direct analysis of undiluted urine, milk and other complex mixtures without sample preparation, Chem. Commun. 19 (2006) 2042–2044.

[81] C. Zhen, Y. Zhou, N. Zhang, J. Wang, C. Xiong, S. Chen, Z. Nie, Differentiation of Chinese liquors by using ambient glow discharge ionization mass spectrometry, Analyst 138 (13) (2013) 3830–3835.

[82] J. Pól, V. Vidová, G. Kruppa, V. Kobliha, P. Novák, K. Lemr, T. Kotiaho, R. Kostiainen, V. Havlíček, M. Volný, Automated ambient desorption–ionization platform for surface imaging integrated with a commercial fourier transform ion cyclotron resonance mass spectrometer, Anal. Chem. 81 (20) (2009) 8479–8487.

[83] E. Handberg, K. Chingin, N. Wang, X. Dai, H. Chen, Mass spectrometry imaging for visualizing organic analytes in food, Mass Spectrom. Rev. (2014).

[84] Y. Li, B. Shrestha, A. Vertes, Atmospheric pressure molecular imaging by infrared MALDI mass spectrometry, Anal. Chem. 79 (2) (2006) 523–532.

[85] A.U. Jackson, A. Tata, C. Wu, R.H. Perry, G. Haas, L. West, R.G. Cooks, Direct analysis of Stevia leaves for diterpene glycosides by desorption electrospray ionization mass spectrometry, Analyst 134 (5) (2009) 867–874.

[86] S. Yang, J. Ding, J. Zheng, B. Hu, J. Li, H. Chen, Z. Zhou, X. Qiao, Detection of melamine in milk products by surface desorption atmospheric pressure chemical ionization mass spectrometry, Anal. Chem. 81 (7) (2009) 2426–2436.

[87] J.F. García-Reyes, A.U. Jackson, A. Molina-Díaz, R.G. Cooks, Desorption electrospray ionization mass spectrometry for trace analysis of agrochemicals in food, Anal. Chem. 81 (2) (2009) 820–829.

[88] L. Luo, Z. Zhao, X. Dai, X. Zhang, Y. Liu, X. Zhang, W. Zhang, Y. Ouyang, Surface desorption atmospheric pressure chemical ionization mass spectrometry for identification of lotus seeds freshness based on PCA and BP-ANN, Nongye Gongcheng Xuebao/Trans. Chinese Soc. Agric. Engin. 29 (7) (2013) 261–266.

[89] G. Morlock, W. Schwack, Determination of isopropylthioxanthone (ITX) in milk, yoghurt and fat by HPTLC-FLD, HPTLC-ESI/MS and HPTLC-DART/MS, Anal. Bioanal. Chem. 385 (3) (2006) 586–595.

[90] N.M. Suni, H. Aalto, T.J. Kauppila, T. Kotiaho, R. Kostiainen, Analysis of lipids with desorption atmospheric pressure photoionization-mass spectrometry (DAPPI-MS) and desorption electrospray ionization-mass spectrometry (DESI-MS), J. Mass Spectrom. 47 (5) (2012) 611–619.

[91] T.M. Vail, P.R. Jones, O.D. Sparkman, J Anal. Toxicol. 3 (1) (2007).

[92] L. Vaclavik, J. Rosmus, B. Popping, J. Hajslova, Rapid determination of melamine and cyanuric acid in milk powder using direct analysis in real time-time-of-flight mass spectrometry, J. Chromatogr. A 1217 (25) (2010) 4204–4211.

[93] M.C. Jecklin, G. Gamez, D. Touboul, R. Zenobi, Atmospheric pressure glow discharge desorption mass spectrometry for rapid screening of pesticides in food, Rapid Commun. Mass Spectrom. 22 (18) (2008) 2791–2798.

[94] Z. Wu, H. Chen, W. Wang, B. Jia, T. Yang, Z. Zhao, J. Ding, X. Xiao, Differentiation of dried sea cucumber products from different geographical areas by surface desorption atmospheric pressure chemical ionization mass spectrometry, J. Agric. Food Chem. 57 (20) (2009) 9356–9364.

[95] P.J. Lee, A.M. Ruel, M.P. Balogh, P.B. Young, J.A. Burgess, Analysis of vanilla flavoured food products using atmospheric-pressure solids analysis probe, Agro Food Industry Hi-Tech 21 (3 Suppl.) (2010) 25–28.

Chapter 5

High-Performance Ion Mobility Spectrometry

Wenjie Liu[1,2] **and Herbert H. Hill Jr**[1,2,*]
[1]College of Life Science, Tarim University, Alar, Xinjiang, China; [2]Department of Chemistry, Washington State University, Pullman, WA, USA
**Corresponding author: E-mail: hhhill@wsu.edu*

Chapter Outline

1. Introduction 276
2. Ion Mobility Spectrometer 276
 2.1 Classification of Ion Mobility Spectrometers 276
 2.2 Sample Introduction and Ionization Method of Ion Mobility Spectrometer 277
 2.3 The Performance of IMS: Resolving Power and Peak Capacity 279
 2.4 Hyphenation of Ion Mobility Spectrometry and Mass Spectrometry 280
3. Analysis of Food Flavors and Composition with Ion Mobility Spectrometry 282
 3.1 Food Flavor Compounds 282
 3.2 Foodborne Microbial Detection with Ion Mobility Spectrometry 285
 3.3 Biogenic Amines 286
 3.4 Other Flavor-Related Compounds 288
4. Determination of Food Compositions and Food Additives with IMS 290
 4.1 Amino Acids 290
 4.2 Flavonoids 290
 4.3 Caffeine and Theophylline 290
 4.4 Food Additives and Nutritional Supplements 292
 4.5 Detection of Food Adulterants from Dietary Supplements and Other Food with IMS 294
5. Analysis of Food Contaminants and Residue with Ion Mobility Spectrometry 297
 5.1 The Detection of Pesticides 297
 5.2 Analysis of Veterinary Drugs 301
 5.3 Analysis of Nitrite and Nitrate in Food 302
 5.4 Analysis of Microbial Toxins in Food 302
6. Conclusions 303
References 303

Comprehensive Analytical Chemistry, Vol. 68. http://dx.doi.org/10.1016/B978-0-444-63340-8.00005-4

1. INTRODUCTION

Ion mobility spectrometry (IMS) is a high-throughput separation method that characterizes chemical substances based upon their gas-phase ion mobilities [1,2]. The principle behind IMS is the separation of ionized analytes that migrate through an electric field at characteristic velocities according to mass, charge, size, and shape [3]. The drift velocity, v_d, of a gas-phase ion swarm that drifts under the influence of an electric field, (E in V/cm), against a counter flow of a neutral buffer gas at atmosphere pressure. v_d can be determined by the travel time of ions (t_d) over the drift tube length (L).

$$v_d = L/t_d \tag{1}$$

To normalize the drift velocity in different electric fields, the mobility coefficient K (cm^2/V s) is defined as,

$$K = v_d/E \tag{2}$$

In order to calibrate experimental results under various temperature and air pressure, reduced mobility coefficient (K_o, cm^2/V s) is used to exclude the effects of temperature and pressure on the mobility coefficient, which is given by,

$$K_o = K\frac{273}{T}\frac{P}{760} \tag{3}$$

Ion mobility spectrometry (IMS) is a rapidly developing analytical method used to separate and identify ionized molecules in the gas phase based on their mobility in drift gas. IMS has demonstrated some unique features such as rapid separation, low cost, and high sensitivity in chemical-warfare agent detection [4,5], explosive detection [1,6,7], illicit drugs screening [8,9], adulterants screening in herbal supplement [10,11]. Most recently, there are more reports about the food contaminants analysis using stand-alone ion mobility spectrometers [12,13].

2. ION MOBILITY SPECTROMETER

2.1 Classification of Ion Mobility Spectrometers

Mainly, there are two different types of commercially available stand-alone ion mobility spectrometers widely used in food science. These two types of instruments are based on drift tube ion mobility spectrometry (DT-IMS) and differential mobility spectrometry (DMS). The later, also known as field asymmetric ion mobility spectrometry (FAIMS) is operated under strong oscillating electric fields [14]. FAIMS and DMS typically operate with oscillating electric fields of 10,000 V/cm that are applied orthogonally to the buffer gas flow while DT-IMS have static electric fields between 200 and 600 V/cm that are applied axially, opposing the buffer gas flow. Except for low molecular weight ions,

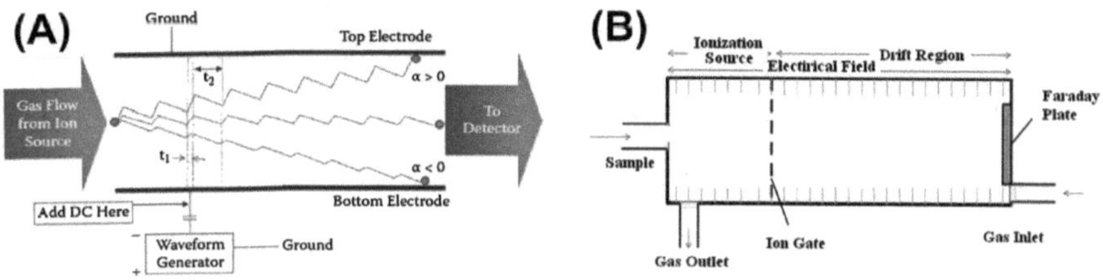

FIGURE 1 Schematic of differential mobility spectrometer (A) and drift time ion mobility spectrometer (B).

DT-IMS generally affords better ion separation than DMS. A schematic of a typical FAIMS instrument can been seen in Figure 1(A), while a typical DT-IMS including a drift tube, a detector, an ionization source, and an ion shutter can be seen in Figure 1(B). The ion shutter separates the drift tube into two regions: the reaction region and the drift region. A high voltage applies to the drift region and guide the ions fly through the drift tube and impact the detector to form mobility peaks. The drift gas is introduced in the low field region of the drift tube near the ion detector and flows counter to the ion flow, sweeping neutral interfering molecules out of the drift region of the spectrometer to avoid ion–neutral interactions in the drift tube [15].

2.2 Sample Introduction and Ionization Method of Ion Mobility Spectrometer

Sample introduction is recognized as one of the most important steps of an analytical procedure and is heavily dependent on the type of sample analyzed. Originally, ion mobility spectrometry was invented for the detection of gas species, and ion mobility spectrometry itself is a gaseous ion analytical technique, so the vapor sample introduction method was intensively studied. The most natural method to introduce a vapor sample into an IMS instrument is by directly conducting a carrier gas including the vapor analyte to the ionization source, especially for continuous monitoring. For an analyte with a relative high vapor pressure, ambient temperature carrier gas is usually sufficient for sampling because of the high sensitivity of IMS detection. This is also the most "clean" and selective method for sample introduction as most of interfering compounds are excluded. One common approach to avoid unwanted contaminants and particles from entering the IMS is to position a silicone membrane at the entrance of the ionization region. Selective membranes can be used to improve the selectivity of sampling and thus improved and S/N of spectrum.

For less-volatile samples, it is necessary to convert samples to vapor phase with heating or other techniques. Thermal desorption of solid samples produce analyte vapors that can be introduced to the ionization source of IMS with a carrier gas. Most commercial instruments use thermal desorption methods for introducing samples. An early thermal desorption device was simply a platinum wire, which liquid sample and sample solution could loaded onto. Basically it was a solid phase extraction (SPE) device without the coatings used by solid phase microextraction (SPME) methods today. The platinum wire with sample was then inserted into the heated injection port of IMS and the sample was vaporized and transferred to ionization source by carrier gas. For modern IMS instrument, "direct swabbing and analysis" is common. This approach uses a swab sampling which is wiped over the sample to pick up the sample, say from a surface, and then transferred to IMS for thermo desorption. Most recently, ambient pressure laser desorption and laser-induced acoustic desorption shows great potential when coupled with ion mobility spectrometry [16–18].

Though derivatization is a common method used in gas chromatography (GC) sample introduction for nonvolatile samples, this technique is seldom used in IMS. To introduce nonvolatile samples and thermally labile molecules to IMS, electrospray ionization (ESI) showed excellent performance as both sample introduction and ionization method [19,20]. An advantage of ESI is that both, inorganic and organic ions can be introduced into IMS through the electrospray process, opening up the possibility for the rapid separation of anions and cations in the gas phase rather that the slow processes that are involved using ion chromatography or capillary electrophoresis. In an early application, the anions, chloride, nitrite, formate, nitrate, and acetate, were all separated by IMS after ESI introduction. Another interesting attempt is the analysis of metal cations with ESI, the metal salt solution is sprayed directly into IMS. Aqueous phase metal cations can be detected by IMS, and most cations produce single, easily quantifiable responses. Average detection limits for the cations studied were in the low ng/μl range making it possible to develop a sensitive method for the direct field analysis of inorganic cations in water [21].

For volatile and semivolatile samples, radioactive ^{63}Ni source is the long-favored ionization source provided low-cost, stable, and reliable operation. The ionization process of ^{63}Ni is well understood for both positive and negative ions. The maximum energy of the electrons emitted from the ^{63}Ni source is 67 keV, with an average energy near 17 keV. Almost all the energy from this source is dissipated in air at an ambient pressure within 10–15 mm, from the surface of the radioactive material, which is the optimum diameter of the ionization source is a cylinder for transferring the carrier gas. The electrons emitted from the ^{63}Ni produce ions and secondary electrons and this process is repeated until the secondary electrons are no longer energetic enough to ionize the gas molecules of the supporting atmosphere. The formation of an ion pair requires about 35 eV, so each beta particle emitted from the source could ideally produce an average about 250 ions assuming that 50% of the electrons are directed into the foil. Negative ions may be produced also through electron attachment processes, which, in most cases, proceed efficiently when the electrons are at thermal energy, thus further increasing the ion yield.

Another alternative ionization source share similar characteristic with radioactive ^{63}Ni is the corona discharge (CD) ion source. CD sources with various designs and electronics control have been developed for or explored with IMS, and CDs are available in some production instruments [22]. To form a CD, a sharp needle or thin wire is placed 2–8 mm from a metal plate (point to face discharge) or discharge electrode (point to point discharge) with a voltage difference of 1–3 kV between the needle and electrode. An electric discharge develops in the gap between the needle or wire and the opposing conductor, and the ions formed in the gap can closely resemble those found in the ^{63}Ni ion source. These ions are then available for subsequent ion–molecule reactions with the sample to afford ions for separation.

To simplify the sample preparation process, any direct ionization methods from solid samples or sample surfaces are welcomed by food analysts. Generally, there are three major ionization sources for surface analysis of solid samples, including direct analysis in real time (DART) [23], desorption electrospray ionization (DESI) [24], and matrix-assisted laser desorption ionization (MALDI) [25]. All of above three techniques function both as the sample introduction method and the ionization method.

The DART ionization source consists of a discharge chamber containing a cathode and anode; this produces plasmas containing ions, electrons, and excited-state species. Neutral metastable helium atoms or nitrogen molecules are extracted from the plasma and directed to a sample surface, where Penning ionization makes product ions that can be analyzed by mass spectrometry (MS) or IMS [26]. DART sources have been used to ionize a number of classes of compounds, as solids are adsorbed directly onto a variety of surfaces. An integrated benchtop instrument equipped with DART and resistive glass drift tube was provided by Photonis is now available which provides DART-IMS analysis of solid, liquid, or gas analytes, and the sample preparation is easy, and identifiable results can be seen in less than 10 s.

2.3 The Performance of IMS: Resolving Power and Peak Capacity

The performance of ion mobility spectrometry is expressed as resolving power. Resolving power is a measure of the sharpness of an IMS peak. It is related to theoretical plates used in chromatography and the one-peak definition of resolution in MS. For ion mobility spectrometry, resolving power (R_p) is defined as the drift time (t_d) of the ion divided by the temporal peak width at half height ($w_{0.5}$) of the ion mobility peak [27]. This relationship is given by Eqn (4):

$$R_p = \frac{t_d}{w_{0.5}} \tag{4}$$

For a conventional gradient elution HPLC separation having a theoretical plate count of around 35,000, the equivalent resolving power would be

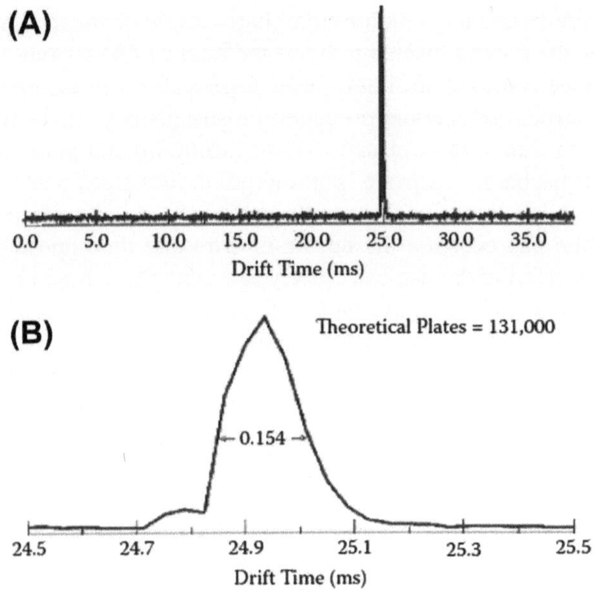

FIGURE 2 Measuring resolving power in IMS. (A) The mass-selected ion mobility spectrum of methamphetamine obtained with an atmospheric drift tube IMS interfaced to a quadrupole mass spectrometer. (B) Expanded scale of mobility peak for the calculation of resolving power. *Reproduced from Ref. [31], Figure 8.7, p. 172.*

about 80. Well-designed IMS instruments having a 20 cm drift tube can produce much higher resolving powers than HPLC. For the ion mobility spectrum shown in Figure 2, the ion mobility peak produced by methamphetamine with mass detector had a drift time of 24.94 ms and a peak width at half height of 0.154 ms. Thus, the measuring resolving power for this spectrum was 24.94/0.154 = 161 [28].

2.4 Hyphenation of Ion Mobility Spectrometry and Mass Spectrometry

Though stand-alone ion mobility spectrometers are successfully applied for explosive detection, chemical-warfare agent detection, and screening for illicit drugs, they are not capable of analyzing complex samples such as biological, food, and environmental samples. In many cases, identification of molecular species is required, which is beyond the ability of stand-alone ion mobility spectrometry, thus a variety of hyphenated ion mobility spectrometry and mass spectrometry (IMMS) instruments have been developed for application to complex samples.

When ion mobility is hyphenated to mass spectrometry there are a number of additional advantages. The primary advantage is that ion mobility provides a rapid preseparation step prior to MS. Separation based on ion size as well as mass provides a two-dimensional analysis of each ion, generating an increase in peak capacity over that possible by MS alone.

Generally, an IM-MS instrument must perform five basic functions: sample introduction, sample ionization, ion mobility separation, mass analysis, and ion detection. Because there are various types of ion mobility spectrometers and mass spectrometers, the combination of IM-MS produces a large number of instrument types. Table 1 lists some used components in IM-MS instruments construction [29].

Of them, ambient pressure ion mobility spectrometry, reduced-pressure ion mobility spectrometry and traveling-wave ion mobility spectrometry coupled with time-of flight mass spectrometers are available instruments provided by Tofwerk, Agilent, and Waters, respectively. IM-TOF provides extra separation for isomers and isobars that cannot be resolved by TOF alone thus reveals greater detail from complex samples. For example, the discrimination of pesticides Aldicarb-sulfone ($C_7H_{14}N_2O_4S$, $[M + Na]^+ = 245.056649$) and Acetamiprid ($C_{10}H_{11}ClN_4$, $[M + Na]^+ = 245.056445$), which have a 0.2 mDa mass difference. A mass resolution of about two million is required to separate these two compounds in an m/z domain. Due to the structural differences between these compounds, they can be easily separated in the drift time dimension.

Field asymmetric ion mobility spectrometry can be used as an ion filter to exclude chemical noise from a mass spectrometer. With FAIMS/MS better quantification precision and lower limits of detection can be achieved for the trace level determination of food contaminants. For the detection of

TABLE 1 Components of Ion Mobility Mass Spectrometers

Sample Introduction	Ionization Sources	Ion Mobility Spectrometers	Mass Spectrometers
Vapor introduction	Radioactive ionization	Ambient pressure ion mobility spectrometry	Time-of-flight
Thermal desorption	Corona discharge ionization	Reduced-pressure ion mobility spectrometry	Quadrupole
Gas chromatography	Photoionization	Aspiration ion mobility spectrometry	Ion trap
Direct infusion	Electrospray ionization	Field asymmetric ion mobility spectrometry	Fourier transform ion cyclotron
Liquid chromatography	Secondary electrospray ionization	Differential mobility analysis	Magnetic sector
Solid samples	Matrix-assisted laser desorption ionization	Traveling-wave ion mobility spectrometry	Tandem instruments

oxo-anions (XO^{3-}) of nitrogen, chlorine, bromine, and iodine in water, the combination of ESI with FAIMS and mass spectrometry (ESI-FAIMS-MS) was evaluated and the results showed that this system yielded mass spectra that are virtually free of common spectral interferences [30]. The unique separation and focusing properties of the FAIMS analyzer improved signal-to-background ratios (S/B) for all these analytes by 3–4 orders of magnitude compared with conventional ESI-MS and produced parts-per-trillion (ppt) detection limits for ClO^{3-}, BrO^{3-}, and IO^{3-} in methanolic solution.

Advantages of the addition of IMS to MS include the following [31]:

1. Rapid separation prior to mass analysis;
2. Increased sample throughput of complex samples;
3. Separation of isomers and isobars;
4. Reduction in chemical noise;
5. Measurement of ion size as collision cross-section Ω;
6. Correlation of mobility with mass in the form of trend bands or lines;
7. Separation of multiply charged ions into families of charge states;
8. Ion identification using Ω/m;
9. Alternate separation patterns with multiple buffer gases; and
10. Protection of MS from contamination from complex samples.

The hyphenation of ion mobility spectrometry and mass spectrometry is still undergoing rapid development and there are some hi-end IMMS instruments available from Exellims, Tofwerk, Waters, Agilent, and Thermo-Fisher. Though IM-MSs are still relative expensive instruments, the value-added information that comes from measuring mobility and mass simultaneously over that gained from their individual application makes them attractive in all applications. In near future, we can expect that almost all MSs in the instrument market will be fitted with some type of mobility-discriminating device.

3. ANALYSIS OF FOOD FLAVORS AND COMPOSITION WITH ION MOBILITY SPECTROMETRY

3.1 Food Flavor Compounds

Food flavor compounds are food aroma and taste-related molecules, which are the major factors influencing people perception about food. These sensations are caused by numerous volatile organic compounds which are complex. Odor-active compounds are usually found in very low concentrations, ranging from a few hundred parts per million (ppm) for strongly flavored food products to less than 10 ppm for weakly flavored foods. In addition, many of these flavor compounds are highly reactive and thermally labile, thus, some of the volatile components from food decay exponentially over time, which is a good indicator for freshness. These aspects make flavor analysis a challenging task. Various methods have been developed to cope with the complexity and to resolve practical problems related to food flavors. The first few steps usually involve the trap and concentration of volatiles and semi volatiles from their original food matrix. A number of different techniques, ranging from conventional solvent extraction and distillation to the newly developed direct thermal desorption (DTD) and SPME are available for isolating flavors from diverse food systems. Each of these techniques has its own applications with advantages and limitations. There is no single ideal method to extract flavor compounds from food matrices that is truly representative of the food flavor profile.

Coffee is an internationally popular drink that has a very aromatic mixture of complex organic compounds. The freshness of ground coffee can be related to the concentration of volatile compounds. The volatiles of freshly ground, opened coffee and artificially aged coffee were compared by analysis with a portable IMS with a ^{63}Ni ionization source [32]. Partial least squares (PLSs) were used to predict the actual coffee age from the IMS spectra (Figure 3). Coffee age can be determined using an IMS to monitor the volatile components from coffee grounds.

"Electronic nose" devices mimic the discrimination of the mammalian olfactory system for smells [33]. An electronic nose system based on IMS was already used for measuring the storage time as well as sensory quality changes during storage of a pork meat pizza topping product. This study concluded that the IMS-based electronic nose, MGD-1, showed clear ability to detect quality changes in the pizza topping product during storage. Prediction of storage time by the electronic nose IMS data was very good with a correlation, $r = 0.978$ ($RMSEP = 1.20$) [34]. The perceived sensory quality changes during storage as measured by sensory descriptive analysis showed a clear effect of storage, which could be divided into three main dimensions relating to oxidation, meat odor, and texture. The quality changes detected by the electronic nose were clearly related to changes in odors. Further study produced evidence that the electronic nose system would be a relevant device for at- or online implementation in quality control of pork meat pizza topping product. Sensory quality changes during storage were fairly well and significantly predicted for descriptors especially related to odor, flavor, and color characterizing the quality changes of the samples as detected by the electronic nose during storage.

The same ion mobility-based electronic nose also was used to determine the separation of various hard and extra-hard cheese samples via headspace volatile composition analysis [35]. Nine-month ripened Emmental cheeses were successfully discriminated from Emmental cheeses with different ripening time, geographical origin, and variety.

Another work was aimed at identifying the feeding regime used with the pigs from volatile compounds in subcutaneous fat samples with ion mobility spectrometry. A portable UV-IMS instrument was used for vapors detection [36]. The IMS data require chemometric processing of the raw IMS spectra. UV-IMS appeared to accurately detect when pigs were fed in a way that mimics the fatty acid profile of free-range pigs. IMS provides an effective new tool for

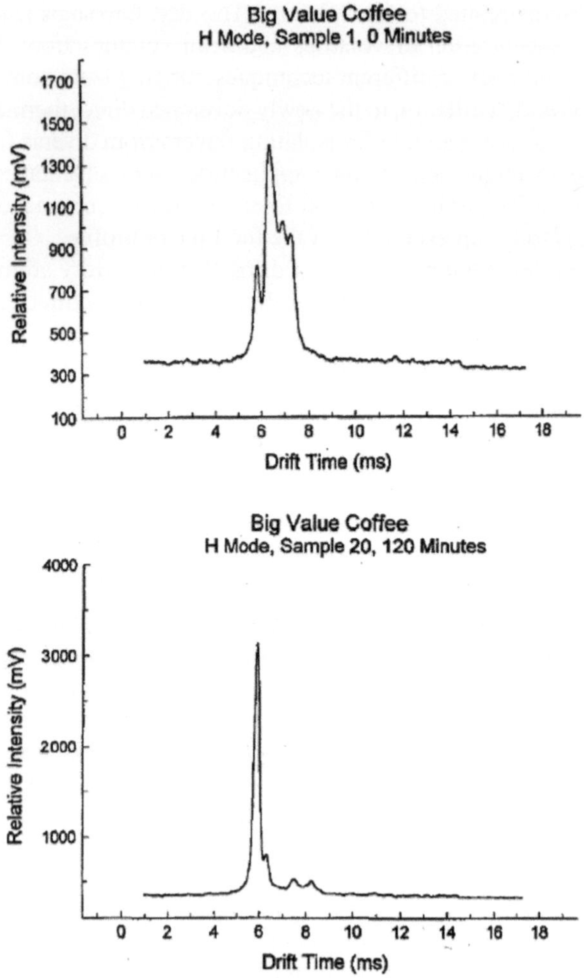

FIGURE 3 Ion mobility spectra for fresh and aged coffee. *Reproduced from Ref. [32], Figure 2, p. 84.*

detecting potentially fraudulent feeding regimes that cannot be accomplished by simply using GC.

Long chain omega-3 polyunsaturated fatty acids (PUFAs), eicosapentaenoic acid (EPA) and docosahexaenoic acid (DHA) in infant foods and dietary supplements have showed success in food markets for the health beneficial effects. However, omega-3 acids are chemically vulnerable compounds suffer from lipid oxidation, which lead to the formation of unpleasant fishy off-flavors. The main secondary products from lipid degradation, including hexanal, 2-butanone, acetone, and dimethyl sulfide were successfully identified and quantified from milk and linseed oil samples with online headspace-multicapillary column-IMS hyphenation, a commercial available instrument from Gesellschaft für Analytische Sensorsysteme (Dortmund, Germany), without any sample pretreatment [37]. In this combination, the multicapillary column is used as an interface between the headspace and the IMS, providing an efficient and rapid preseparation of the volatile compounds ahead of IMS separation. The limits of detection were in the interval 0.3 μg/L (for hexanal in milk) to 3.0 μg/L (for dimethyl sulfide in linseed oil) while the limits of quantification varied between1.1 μg/L (for hexanal in milk) and 9.6 μg/L (for dimethyl sulfide in linseed oil). The proposed protocol offers advantages such as portability, low cost, and maintenance with enough sensitivity compared to other alternative method such as GCMS when applied in food processing and preservation.

3.2 Foodborne Microbial Detection with Ion Mobility Spectrometry

The common approach to detect microbiological contamination is to inoculate a diluted sample of the homogenized food onto agar or into a broth of an appropriate selective medium and then incubated for 24–72 h. Incubation is followed by a visual count of the colonies on the agar or observing a colorimetric/fluorometric change or production of gas from the broth. These traditional methods are both slow to produce results and labor intensive to perform. Rapid methods of detection are needed to expedite monitoring of microbial contamination of foods. A ^{63}Ni ionization IMS equipped with an automated sampling system aimed to rapidly detect and enumerate microorganisms has been reported [38]. The headspace volatile components produced by bacterial growth was found to be related to the numbers of bacteria, and two different volatile markers were found for different microbes, including *Escherichia coli*, *Staphylococcus aureus*, *Yersinia enterocolitica*, *Listeria* spp., and *Aeromonas hydrophila*.

Ethanol, acetone, 2-Me-2-propanol and 2-pentanol, off-flavors emitted by mold, can be simultaneously monitored using a ^{63}Ni ionization IMS [39]. The direct analysis of these volatiles offers a simple, fast screening method for online monitoring of air quality during food processing. In a more detailed study using a mini IMS instrument, the volatile organic emissions of three molds, *Aspergillus niger*, *Penicillium chrysogenum*, and *Chaetomium globosum*, were characterized and compared for different growth stages. The assignment of IMS peaks to specific ionic species was achieved with GC-MS results [40]. A principal components analysis (PCA) was performed in order to illustrate the difference of the IMS spectra obtained from headspace of the investigated molds. It was observed that the formation of MVOCs (mold volatile organic compounds) in type and amount depends significantly on the specific mold strain and its cultivation status. IMS proved to be a very sensitive detection tool for MVOCs can be applied to indicate in-line active mold growth rapidly and can be used for monitoring of molds life cycle.

Gray mold (*Botrytis cinerea*) is a common phytopathogen of fruits, which causes serious postharvest loss. To indicate the health status of strawberry, a multicapillary column-ion mobility spectrometry (MCC-IMS) was successfully used to evaluate the development of strawberry aroma during shelf life and detect the degree of *B. cinerea* infection through changes in the volatile profile [41]. Results revealed that HS MCC-IMS could be used to follow the evolution of infection during times of high throughput (Figure 4). HS MCC-IMS also showed potential for predicting *B. cinerea* infection with sensitivities similar to those of ELISA. HS-SPME fast GC-MS predicted *B. cinerea* infection in strawberry in a comparable way although a longer analysis time was required.

3.3 Biogenic Amines

Biogenic amines are a class of small molecules that are produced and metabolized in vivo, which play an important role in human pathology and physiology. Biogenic amines have been found in a variety of foods, including seafood, fish, meat products, milk products, and some fermented products such as beer and wine. Some biogenic amines in foods, which are formed by microbial decarboxylation of amino acids, imply the deterioration of food quality [42,43]. The determination of biogenic amines is a time-consuming process that required sample extraction, purification of extracts, and derivatization. Derivatization is a process by which the original analyte is modified for chromatography or detection. For gas chromatography biogenic amines are derivatized to decrease their polarity and increase vapor pressure. For liquid chromatography, biogenic amines are derivatized to add a chromophore to the structure for ease of detection by UV-Vis adsorption or by florescence detection.

Free biogenic amines are easily ionized by gas-phase ionization methods used in IMS and thus can be detected sensitively by IMS. Headspace vapor monitoring of foods for volatile biogenic amines can disclose the extent of spoilage. Sample preparation is quite simple, just add a few drops of an alkaline solution to sample solution and then heat it up. In one study, the limit of detection for trimethylamine (TMA) was found to be 2 ng and the time required for the whole analysis process was less than 2 min, with a short- and long-term reproducibility of 15% and 25%, respectively. As expected, spoilage increased with storage time at room temperature and the IMS results were correlated with microorganism populations [44]. To obtain a reliable discriminate result for spoilage food from fresh food, spectra from similar type of samples should be collected and a spectral library developed. Fuzzy rule-building expert system (FuRES) provided better results than partial least square results from biogenic amine ion mobility spectra [45].

Several approaches using ^{63}Ni and corona discharge as ionization source have been developed for rapid detection of volatile biogenic amines, including putrescine, cadaverine, spermidine, spermine, and others [46]. The ion mobility spectra, calibration curves, and analytical parameters of the compounds

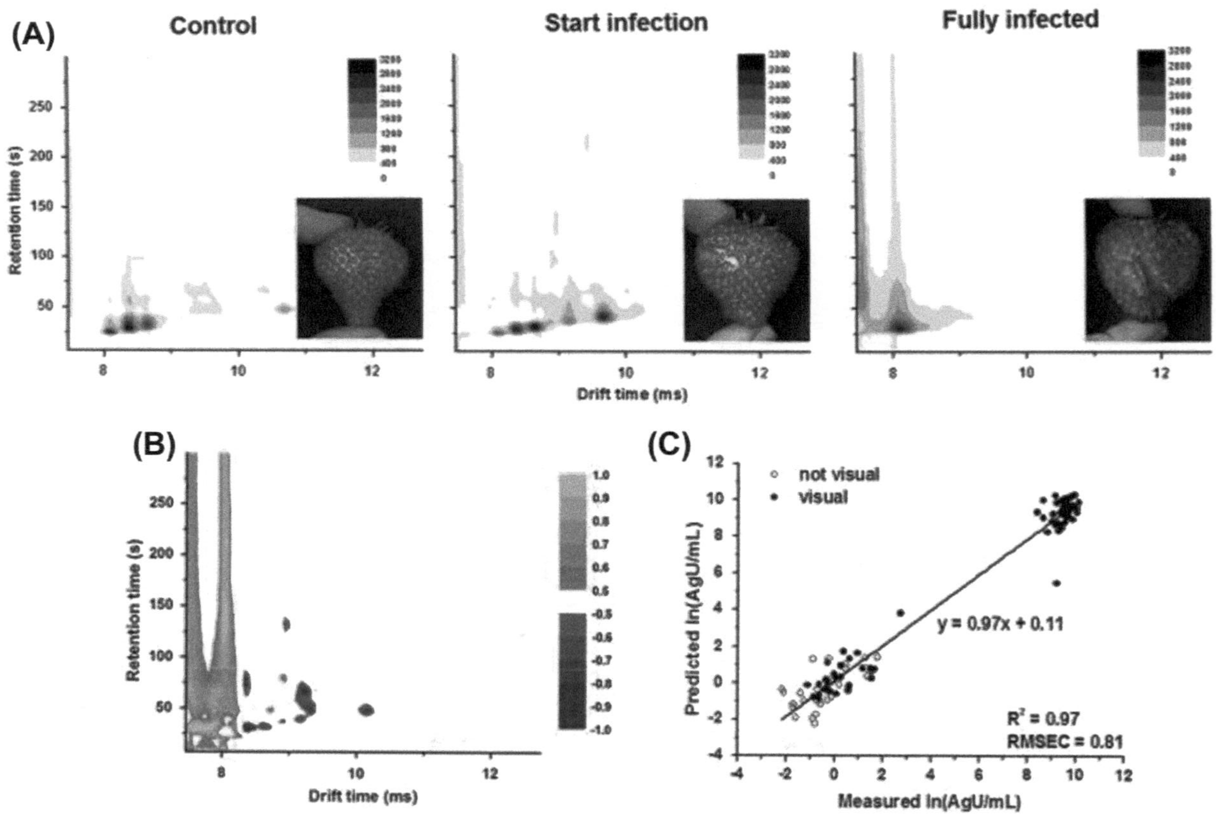

FIGURE 4 (A) HS MCC-IMS chromatograms for strawberries with increasing levels of *Botrytis cinerea* infection. The color indicates peak abundance. (B) Color map indicating the correlation between the HS MCC-IMS peaks and *B. cinerea* infection expressed in ln(AgU/mL). The color is based on the VID correlation and indicates a positive correlation or a negative correlation with *B. cinerea* infection. (C) Predicted versus measured level of *B. cinerea* infection expressed in ln(AgU/mL) based on a Partial Least Squares regression. *Reproduced from Ref. [41], Figure 3, p. 409.*

were obtained with and without using n-Nonylamine as the reagent gas dopant. These results indicate that it would be possible to analyze the biogenic amines without using n-Nonylamine as the reagent gas. The proposed method is a derivative, free, fast, and simple technique for the analysis of biogenic amines [47]. Another report demonstrated that ion mobility spectrometry is sensitive and accurate enough to detect and quantify levels of biogenic amine present in saliva, which is probably the most complex matrix of any biological fluid [48].

3.4 Other Flavor-Related Compounds

Hyphenated GC-differential ion mobility spectrometry (GC-DMS) is quite attractive for in situ detection of many kinds of volatile organic compounds (VOCs) appearing in the headspaces of selected foodstuff.

Organoleptic deviations in wines have been detected as early as 10 years ago in some wine-growing regions. Organoleptic compounds have been correlated to the presence of certain kinds of fungi on grapes, the presence of which increases under high humidity and on dense grapes. Organoleptic deviations are often let to off-flavors described as earthy-muddy, for which geosmin is the main responsible molecule. GC-DMS showed sufficiently selective and sensitive without any preliminary treatment of samples for rapid detection, identification, and quantification of geosmin in the headspace of grape must and wine (Figure 5). Other natural contaminants in wine including 2-MIB, 1-octen-3-one, 1-octen-3-ol, and pyrazines can be detected simultaneously with good separation and sensitivity [49].

Another compound 2,4,6-Trichloroanisole (2,4,6-TCA) can also cause unpleasant musty odors of wines. The analytical method for the determination of 2,4,6-TCA in wine by IMS consist of several simple steps [50]. An SPE and a headspace ionic liquid single-drop microextraction (SDME) are used to eliminate interferents and to extract the analyte from the wine sample. Compared with other analytical methodologies, IMS has lower LODs, requires less extraction time,

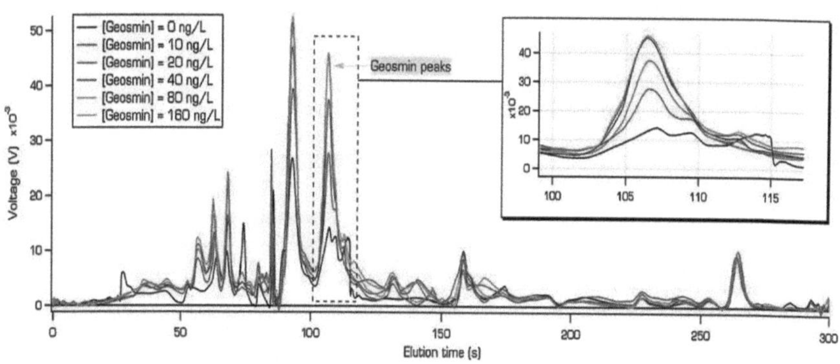

FIGURE 5 Geosmin in Elbling wine: GC/DMS smoothed chromatograms collected at dispersion field of 22 kV/cm (V_{RF} of 1100 V) and with compensation voltage of −1.72 V. *Reproduced from Ref. [49], Figure 2, p. 1039.*

and exhibits similar repeatability. The gas-phase ion chemistry of TCA in positive and negative modes were investigated by a stand-alone IMS to understand the ion behavior of TCA [51]. For the low level of TCA in wine (less than 10 ng/L), a preconcentration step and a sensitive analytical device such as IMS are both required.

Ion mobility spectrometry has also been applied to the discrimination of different grades of olive oil. Extra virgin olive oil (EVOO), olive oil (OO), and pomace olive oil (POO) were analyzed with two IMS instruments: one was an ion mobility spectrometry (IMS) device using an ultraviolet (UV) ionization source and the other was a gas chromatographic (GC) column coupled to an IM Spectrometer with a tritium source [52]. Experimental results indicate that both methods are capable of classifying three grades of olive oils, and GC-IMS provides a vanguard methodology for classifying olive oils according to commercial grade and avoiding fraud. Improved results were obtained with a commercial GC-IMS instrument equipped with a multicapillary column (MCC) [53].

To improve basic analytical properties such as sensitivity and selectivity, a preconcentrator with a Tenax TA trap was coupled to a UV-IMS for the analyzation of grade discrimination of olive oil [54]. The analytical performance of this combination was assessed for volatile aldehydes with 3 to 6 carbon atoms present in olive oil. The online coupling of Tenax TA trap to IMS enables qualitative and quantitative analysis of selected volatile aldehydes from olive oil in different grades and requires no sample preparation, which allows real-time analyses and produces no waste (Figure 6).

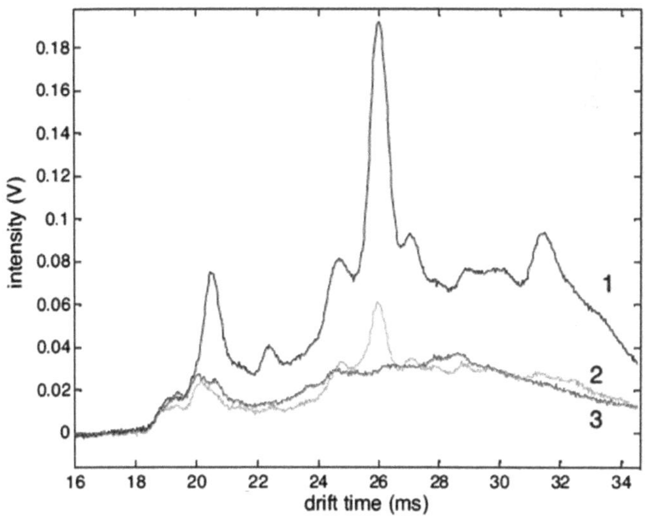

FIGURE 6 Ion mobility spectra for three different types of olive oil at 150 °C. (a) Extra virgin olive oil (blue (dark gray in print versions) line, (1)); (b) olive oil (green (light gray in print versions) line, (2)); and (c) pomace olive oil (red (gray in print versions) line, (3)). *Reproduced from Ref. [52], Figure 1, p. 110.*

4. DETERMINATION OF FOOD COMPOSITIONS AND FOOD ADDITIVES WITH IMS

4.1 Amino Acids

Amino acids are an important nutrient source for human. Some amino acids, such as histidine and isoleucine, are essential amino acids because they cannot be synthesized by humans and have to be provided through food. The determination of amino acid is generally accomplished by liquid chromatography, requiring tedious sample preparation and derivatization for detection. The first report of the separation of amino acids by ion mobility spectrometry was in 2000 when the mobilities of the 20 common amino acids were determined by electrospray ionization IMS (Table 2). With ESI-IMS, rapid separation and sensitive detection was demonstrated without the need for derivatization [55]. Further investigations, using different drift gases (N_2, Ar, CO_2), improved the separation selectivity of amino acids. The use of different drift gases alters the separation factor (α) in the IMS spectra and, thus, affects the separation selectivity [56]. It was found that each of the studied amino acids could be uniquely identified from the others, with the exception of alanine and glycine; their drift times were too close to resolve even with different drift gases [57].

4.2 Flavonoids

Because of their significant contribution to the human diet (1 g/day), antioxidant behavior and chemopreventive properties, there is a great importance in characterizing the chemistry of flavonoids in various plant-related products. By using a dual gate-ion mobility-quadrupole ion trap mass spectrometer, three sets of isobaric flavonoids were resolved from a mixture prior to mass analysis [58]. Silver has demonstrated the ability to form robust complexes with flavonoid diglycosides that may provide a wide variety of collision-assisted dissociation (CAD) pathways. Combined with silver adduction, ESI-IMS-ion trap MS offers a unique method to separate and conclusively identify isomeric flavonoid diglycosides in simple mixtures. For example, quercetin, the aglycone of rutin, displays a reduced mobility of 1.22 whereas hesperetin, the aglycone of hesperidin and neohesperidin, possesses a reduced mobility of 1.18. Protonated quercetin and hesperetin could be baseline resolved by ion mobility spectrometry (Figure 7).

4.3 Caffeine and Theophylline

Caffeine and theophylline are important naturally abundant xanthenes found in food products such as tea, coffee, coca beans, and some energy drinks. For its chemoprevention property, green tea has recently received much attention. Caffeine is one of the major parameters of different quality grade of green tea.

TABLE 2 List of Drift Times, Reduced Mobilities, and Limits of Detection for 20 Common Amino Acid Using ESI-IMS

Amino Acid[a]	Molecular Mass	Drift Tune (ms)	K_o	Absolute LOD[b] (pg)
Glycine	76	11.28	1.968	160 ± 30
Serine	106	12.20	1.819	130 ± 30
Alanine	90	12.30	1.805	130 ± 10
Proline	116	12.42	1.787	170 ± 20
Threonine	120	12.63	1.757	160 ± 10
Cysteine	122	12.70	1.748	230 ± 40
Valine	118	12.90	1.721	110 ± 20
Asparagine	133	12.95	1.714	210 ± 40
Aspartic acid	134	13.03	1.704	290 ± 50
Glutamic acid	148	13.50	1.644	240 ± 20
Glutamine	147	13.55	1.638	290 ± 90
Isoleucine	132	13.60	1.632	120 ± 10
Leucine	132	13.72	1.618	250 ± 30
Histidine	156	13.80	1.608	120 ± 10
Lysine	147	13.85	1.603	50 ± 2
Methionine	150	13.90	1.597	320 ± 80
Arginine	175	14.77	1.503	50 ± 7
Phenylalanine	166	14.85	1.495	96 ± 5
Tyrosine	182	15.42	1.440	160 ± 20
Tryptophan	205	16.45	1.349	700 ± 100

[a]*All ions were protonated molecules* $(M + H)^+$.
[b]*Limit of detections (LODs) were the amounts required to produce signals three times those of the noise levels.*
Reproduced from Ref. [54], Table 1, p. 436.

Standard methods of analysis for the simultaneous determination of caffeine and theophylline have been developed with UV-spectroscopy, MS and HPLC. With a molecular imprinted polymer-(MIP)-SPE method and ESI-IMS, the simultaneous analysis of these two compounds was achieved within 1 min, rather than 15–30 min required by HPLC [59]. With the MIP-SPE approach, only the caffeine peak appeared in the IMS spectrum, providing excellent selectivity and

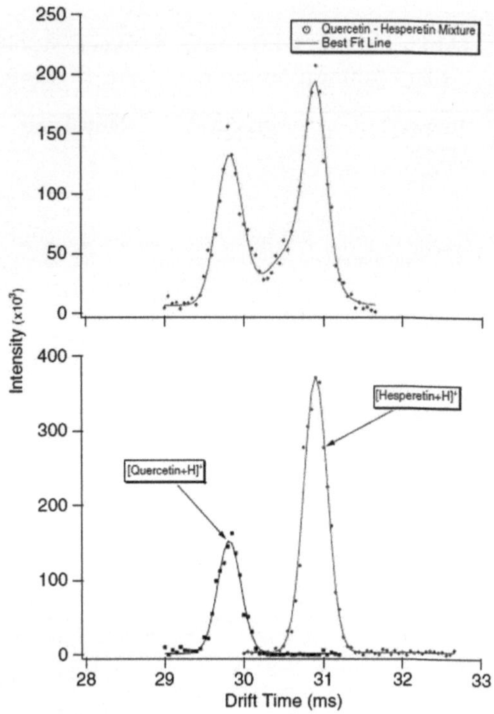

FIGURE 7 Separation of the protonated aglycone isomers, quercetin and hesperetin. *Reproduced from Ref. [58], Figure 4, p. 436.*

precision. Considering its portability, ease of operation, speed of analysis, sensitivity, selectivity and cost per analysis, IMS could be applied for the routine, simultaneous analysis of caffeine and theophylline from complex food matrices.

4.4 Food Additives and Nutritional Supplements

Food additives are substances added to food intentionally to preserve flavor or enhance the taste and appearance of food. Widely used in modern food industry, there are increasing concerns about the health risks of food additives (such as colorings, flavorings, and preservatives). Some authors claim that artificial sweeteners, such as saccharin, cyclamate, and aspartame, do not pose carcinogenic threats. Other studies, however, indicate that these sweeteners may impose human cancer risk. In addition, some studies present evidence that some food additives produce allergic responses and some mixtures of additives may increase the level of hyperactivity in children with behavior problems. There is a growing need in the food industries for rapid, low-cost, and sensitive analytical methods for quality assessment and control of food. These methods are required for the routine screening of raw materials, during production, and the final commercial products to avoid deterioration and changes in specifications. An ESI-IMS coupled to a quadrupole mass spectrometer was used to characterize the

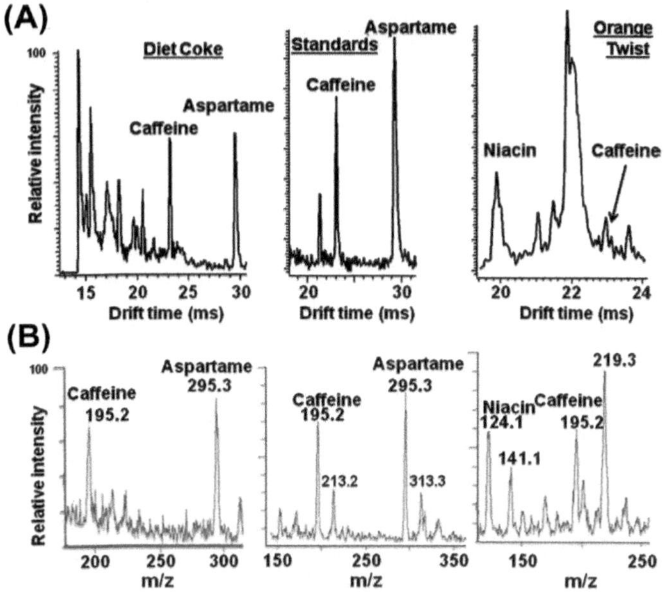

FIGURE 8 Mobility (A) and mass spectra (B) of energetic beverages and standard solutions. 5.1 Spectra of Diet coke showing the protonated peaks of caffeine at 23.2 ms and m/z 195.2 and aspartame at 29.5 ms and m/z 295.3. 5.2 Standard solution of caffeine and aspartame (500 μM) showing the protonated peaks of caffeine at 23.0 ms and m/z 195.2 and aspartame at 29.4 ms and m/z 295.3. The peaks at m/z 213.2 and 313.3 are most probably water clusters. 5.3 Spectra of an antioxidant water (Snapple®) showing the protonated peaks of niacin at 19.8 ms and 124.1 and caffeine at 23.0 ms and m/z 195.2. Mobility peaks were identified by SIM-IMS; mobility and mass spectra were averaged 250 and 500 times, respectively. *Reproduced from Ref. [60], Figure 2, p. 95.*

chemical compositions of beverages including diet coke, antioxidant water, and energy drinks [60]. Aspartame is an artificial, nonsaccharide sweetener, 180 times sweeter than sugar that is used in sugar-free beverages. The IMS protonated peaks of caffeine and aspartame in diet coke are shown in Figure 8; caffeine occurred at 23.2 ms and aspartame at 29.5 ms, as single peaks in the mobility spectrum. The peaks were assigned to operate the instrument in IMS-MS in selected ion monitoring (SIM) mode. The protonated peaks of niacin and caffeine from a sample of antioxidant water, identified by IMS-MS-SIM, occurred at 19.8 ms and 23.0 ms, respectively, in the mobility spectrum of the beverage. These spectra reveal the capabilities of IMS-MS for the rapid separation and identification of food additives in beverages.

Artificial coloring and sweetening reagents such as Brilliant Blue FCF, Allura Red AC, and sodium cyclamate are permitted in a certain range of food and beverages; however, the dosage and variety of food are strictly controlled. The Directspray ESI source has been used for these studies to show its capabilities for ionizing these compounds. Most of the dyes are salt compounds so they can be readily detected with ESI-high performance (HP) IMS [12].

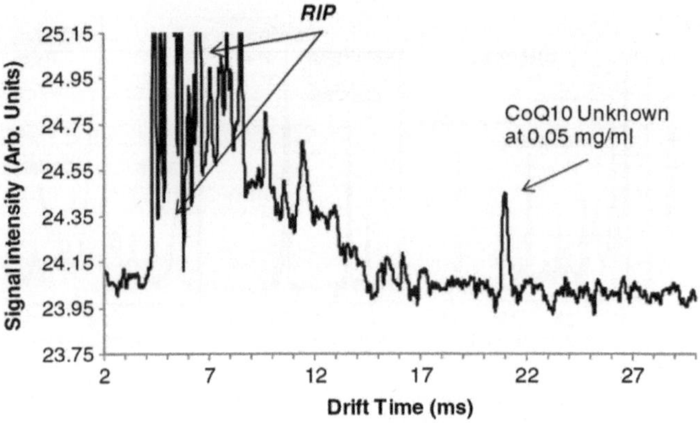

FIGURE 9 Ion mobility spectrum for coenzyme Q10 (CoQ10) sample at 0.05 mg/mL concentration using ESI-IMS in positive mode. The reactant ion peaks are labeled as RIP. Coenzyme Q10 appeared as a sole peak at 20.64 ms, with a reduced mobility of 0.52. *Reproduced from Ref. [61], Figure 10, p. 76.*

Malachite green and Rhodamine B can be detected as positive ions in the IMS while sodium cyclamate, sodium saccharin, Sudan I, sunset yellow FCF, allura red AC, tartrazine, amaranth, new coccine and brilliant blue FCF are separated and detected as negative ions.

Nutritional supplements have become worldwide popular products and are often welcomed by the health care consumer groups. The nutritional ingredients, however, span a wide range of volatilities and polarities, making them difficult to analyze by a single chromatographic method. In addition, for liquid chromatographic separations, large varieties of compounds do not contain chromophores and cannot be detected by UV detector unless they are derivatized. ESI-IMS is a promising method for the rapid screening of nutritional supplements compared to the traditional chromatographic methods used. Calibration curves have been established for selected compounds, such as L-Leucine, sucralose, epigallocatechin gallate (EGCG), guggulsterones, and coenzyme Q10 (Figure 9), with a reasonable linear response range and resolving powers in the range of 70–90 (Table 3). The ESI-IMS method can potentially be used for incoming material inspection, activity ingredient concentration verification, etc. Compared to established HPLC-based methods, ESI-HPIMS can be used for high-speed and low-cost analysis of polar-nonvolatile molecules [61].

4.5 Detection of Food Adulterants from Dietary Supplements and Other Food with IMS

Dietary supplement is intended to provide nutrients that may otherwise not be consumed in sufficient quantities from regular food. According to FDA's definition, dietary supplements are considered a food and not a drug, thus supplement manufacturers are not required to demonstrate product efficacy before

TABLE 3 Some Reduced Mobilities for Nutritional Supplements

Nutritional Supplement	Molecular Weight	Drill Time (ms)	K_0 (cm^2/V s)
L-Leucine	131	6.781	1.58
Sucralose	397	10.523	1.03
Epigallocatechin gallate	458	12.465	0.86
Guggulsterones E&Z	312	10.379	1.04
Coenzyme Q10	863	20.654	0.52

Reproduced from Ref. [61], Table 3, p. 77.

advertising and commercializing a product. Due to increasing health risks and even mortality related to dietary supplements, the regulation of products that claim bodybuilding effect, antioxidants, liver protection, and sexual enhancement tend to be stricter.

Portable ion mobility spectrometers with thermo desorption ports and ^{63}Ni ionization sources were evaluated for screening herbal dietary supplement extracts containing sibutramine, a commonly undisclosed weight loss drugs [10]. Sibutramine is associated with increased cardiovascular events and strokes and has been withdrawn from the market in many countries and regions. With only minimum steps of sample treatment, a portable IMS instrument using a thermo desorber can be applied to the analysis of low levels of sibutramine with an overall LOD of 2 ng. This sample preparation method followed by IMS detection is simple enough to be used to examine product at ports of entry to prevent adulterated materials from entering the marketplace.

Other adulterants in weight loss supplements can be screened with IMS including, bumetanide, desmethylsibutramine, didesmethylsibutramine, and other molecules. The LOD of these compounds varies from 1 to 1000 ng. Isopropanol was found to be the best extraction solvent for analytical speed and robustness. Both benchtop and portable IMS instruments are capable for routine screening tasks [62].

Type-5 phosphodiesterase (PDE-5) inhibitors including sildenafil, vardenafil and tadalafil, and avanafil have been approved by the FDA to treat erectile dysfunction (ED). However, the popularity of the PDE-5 inhibitor drugs has led to extensive counterfeiting in the herbal supplement market, which has become a major arena for numerous generic knockoffs and designer analogs intended to avoid detection by regulatory agencies. Since dietary supplements do not need FDA approval for marketing, a large and ever-changing list of suppliers increase the possibility for adulteration with drugs or similar analogs. To give a pass/fail result of herbal supplements screening for the presence of synthetic ED drugs and their analogs, an IMS operated in the positive mode equipped with a thermo desorber was evaluated [11]. Analyte molecules were vaporized and carried from the heated inlet to the ionization region by carrier gas. ^{63}Ni source was used for ionization. With an internal nicotinamide calibrant, protonated ion peaks for ED drug reference materials evaluated were observed at drift times ranging from 13.5 to 21.5 ms, corresponding to reduced mobilities ranging from 1.29 to 0.83 $cm^2/(V\ s)$.

The method was successfully applied to the analysis of 26 commercially available herbal dietary supplements. Among the 26 samples tested, 15 were found to contain adulterants. Herbal supplements containing adulterants were classified with 100% accuracy by a comparison of the reduced mobilities of the ED drug adulterant in the herbal product to the ED drug reference materials.

In a further investigation, a portable IMS was evaluated as a rapid screening tool for the detection of new structures of PDE-5 inhibitors such as acetildenafils, sildenafils, and avanafil within adulterated herbal supplement matrices [63]. Again, IMS appeared to be a rapid and useful screening tool that could expand the detection scope of PDE-5 inhibitor family with adequate accuracy and sensitivity. All acetildenafils afford a common fragment peak that can be served as a marker for its analogs.

ESI provided IMS the capability of analyzing nonvolatile compounds in liquid solutions. However, traditional ESI requires the cleaning of the residue from the last analysis before to start the next work cycle, decreasing the ability of high-throughput ability of IMS consequently. By using a replaceable sample syringe, a Directspray ESI (shown in Figure 10) source can overcome the cross-over effect of injection and improve the sample throughput. With Directspray, the ESI source voltage was directly applied to the needle and the electrospray process develops from the Taylor cone formed at the end of the syringe needle. The blunt-tipped needle was held a few millimeter away from the aperture to the interface region, thus samples can be analyzed by a "load and spray" way with a work cycle less than 1 min [12,13].

Phthalate ester is a group of compounds that has been widely used as plaster in industry, and were found used as illicit clouding agents in some beverages in

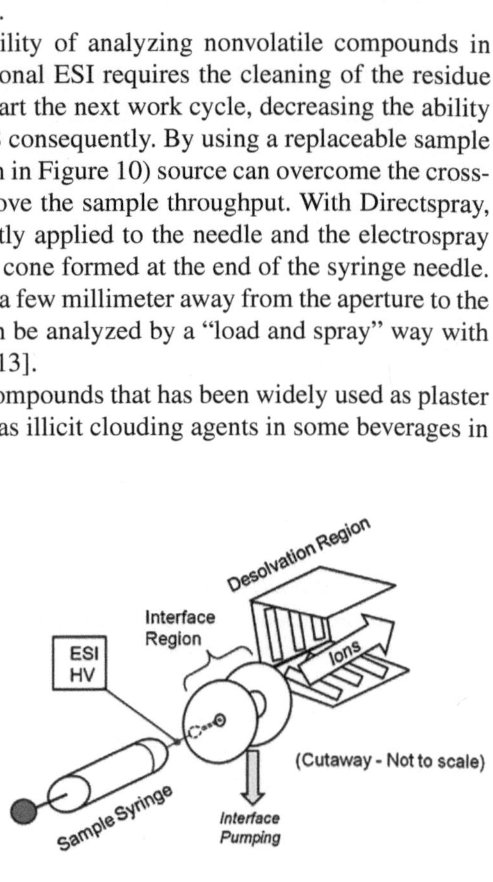

FIGURE 10 Schematic of directspray electrospray ionization source by excellims.

Taiwan. The phthalate plasticizers are used in many product pumping system and package material and thus aroused great concern of those compounds released to drinks, food, and other products. More importantly, they are considered endocrine disruptors because exposure to them has been reported to increase the incidence of developmental abnormalities and to impair reproduction and development. With Directspray ESI source, phthalates could effectively analyzed for ion mobility spectrometry (IMS). The IMS spectra of dicyclohexyl phthalate (DCHP) in Cola using ESI source appeared as a single peak separated from the other peaks of Cola alone. It worth noting that no sample pretreatment was used before analysis, the Cola sample was just mixed with methanol and injected directly to the IMS. Similar protocols have been expanded for the detection of phthalates contained in soy-based bubble tea, a popular drink in Asia. As little as 10 ng/μL of dibutyl phthalate can be detected in the bubble tea filtrate with 120 s acquisition time, a level 12.5 times lower than the detection limits of the Chinese standard method using LC. Even in a complex matrix of the bubble tea product, ESI-HPIMS has shown rapid, sensitive response to phthalate detection. These results indicate that ESI-IMS could be used as a routine on-site analytic method for phthalates in drinks distributed by street sellers [12].

Other phthalate compounds, including those that are banned by the European Union (EU) and regulated by China also can be analyzed directly by ESI-IMS. A linear response range of 1.5 orders of magnitude has been found using ESI-IMS with tested phthalates. The correlation coefficient is very good for dimethyl phthalate (DMP) plotted as the total mobility spectrum peak area versus sample concentration ($R^2 = 0.993$), which means this method is also capable for rapid and accurate quantitative analysis.

5. ANALYSIS OF FOOD CONTAMINANTS AND RESIDUE WITH ION MOBILITY SPECTROMETRY

Food contaminant refers to harmful chemicals and microorganisms in food. Because the impact of food contaminants can exhibit long-term effects and that some contaminants are often present at very low levels in food, the qualitative and quantitative analysis of food contaminants are growing challenges for food safety regulation institutions.

IMS has also have been applied to the field of food safety analysis related to pesticides and herbicide residue, veterinary drug residues, microbial toxins, and other food contaminants. The application of IMS in food safety analysis is mainly because of two reasons: the simplicity of instrument and speed of analysis.

5.1 The Detection of Pesticides

Pesticides and herbicides represent a major source of food contaminant for fruits, vegetables, and drinking water. Since thermal desorption is the dominant sample introduction method used in IMS that has been the initial method used with IMS for the analysis of pesticides from imported fruits [64]. Pesticides were detected by wiping spiked and unspiked apples and oranges, using both fiber glass and Teflon filters, and then heat desorbed into an IMS instrument. Also, solid phase desorption was used to sample volatiles from fruit using an add-on module to a GC-IONSCAN system. Beta-BHC, gamma-BHC, parathion, *trans*-chlordane, *cis*-chlordane, and endrin ketone were investigated from the surface of oranges and apples. The transfer efficiency varied from 1 to 5.9% for different pesticides. The detection limits for different pesticides analyzed in the negative ion mode ranged from 10 to 300 pg, with linear ranges of 10–6000 pg. Processing time per sample is typically under 10 min.

Pesticide intoxications often occur in agriculture, especially in developing countries. Acute pesticide poisonings are a major public health problem. There are approximately 3,000,000 and 100,000 cases of acute pesticide and OP intoxications, respectively, and more than 300,000 deaths per year in the whole world caused by acute occupational health problems and self-poisoning with pesticides [65]. For this reason, rapid, simple, and sensitive field methods are required to check the presence of these pesticides. By using an aspiration-type IMS, sulfotep, propoxur, and nicotine from a liquid matrix have been successfully detected [66]. Samples were introduced with a ceramic evaporator and ionized by an ^{241}Am radioactive source. The presence of pesticide molecules caused changes in the ion mobility distribution and thus could be detected in the presence of 2-propanol background. Visualized by an artificial neural network method, all pesticides could be separated from each other. The instrumentation required was simple and inexpensive, and the analysis time was rapid compared to other traditional detection techniques such as gas or liquid chromatography. Thus, IMS seems to be potentially valuable for the analysis of pesticides even at residue levels.

By combining ambient pressure laser desorption (APLD) and drift tube ion mobility spectrometry, a noninvasive rapid detection method for pesticides on fruit surfaces has been explored [17]. With HP-IMS, various groups of agricultural chemicals including organophosphorous pesticides, pyrethroids, carbamates and different areas of applications, such as insecticides, fungicides, and herbicides were investigated over different fruit surfaces, including plum, peach, grape, tomato, and apple (Figure 11). Drift tube ion mobility spectrometry provided better separation and qualitative results than DMS [18]. The laser irradiance power was adjusted to avoid matrix signals and only pesticides were vaporized from the fruit surfaces without damage to the fruit. The detection limits for 20 tested pesticides were in the range of some tens of ng/kg. The combinations of UV laser desorption and ion mobility spectrometry has the potential to rapidly detect different types of pesticides on a variety of fruit surfaces. The detection limits that could be achieved with such a hybrid instrument were sufficient for field screening. Another ambient pressure laser-induced acoustic desorption with a 532 nm Yd:YAG Q switched laser seemed to be a more gentle surface desorption method compared to laser desorption for there are no heat or laser light impacting the analytes directly [16].

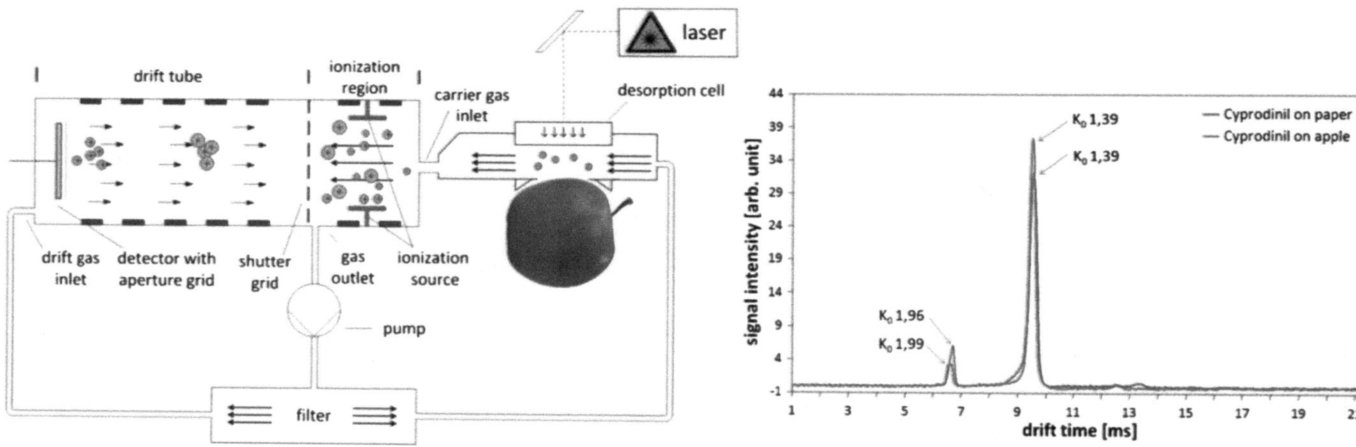

FIGURE 11 Left: Experimental setup for pesticide detection on fruits comprising an ion mobility spectrometer with tritium ion source and a laser desorption device. Right: Ion mobility spectra of individual pesticides laser desorbed from paper and an apple skin respectively, Cyprodinil appeared a single characteristic peak. *Reproduced from Ref. [17], Figures 1 and 2, p. 58.*

One major challenge of direct analysis of complex samples such as fruits, meat, and juice is the extraction and purification of trace level of pesticides and veterinary drugs from the complex matrices. Ionization efficiency can be seriously suppressed due to competitive ionization, charge exchange reaction, and other side effects during the ionization process. It is beyond the capability of IMS to quantitatively determining one specific trace level compound in the presence of hundreds of high concentration compounds unless selective sample extraction and selective ionization method is used. Traditionally, SPE and SPME have been widely used in GC- and LC-based analytical methods to purify and preconcentrate analytes. Most recently, molecularly imprinted polymer SPE (MIP-SPE), hollow fiber liquid phase microextraction (HF-LPME) provide more convenient and selective sample preparation techniques.

HF-LPME is relatively low in cost and requires only a small amount of organic solvent to extract analytes from aqueous matrices. The use of inexpensive disposable membranes can minimize carryover effects. Due to sample-to-acceptor volume ratio, HF-LPME also provides a high enrichment factor of analysts within a relatively short time during the extraction procedure. These features make HF-LPME a very sensitive technique suitable for determination of the pesticides at trace levels. The low solvent consumption also makes HF-LPME an environment-friendly extraction method. This high-throughput sample preparation method combined with ion mobility spectrometry makes it an attractive method for the determining of pesticides from cucumbers [67]. Homogenized cucumber samples were extracted with water under sonication. After the extract was centrifuged, the supernatant was taken for HF-LPME and ready for IMS analysis. The IMS instrument is equipped with a ^{63}Ni ionization source and a thermo desorption sampler. Seven pesticides could be identified by their different drift times. The results showed that the seven pesticides could be detected with detection limits in the range of 5–50 ng. The repeatability based on RSD values for peak height corresponding to the pesticide samples with appropriate concentrations was less than 12%, indicating that the instrument could provide repeatable analysis.

More specific ion mobility measurements of herbicides were performed with ESI-IMS hyphenated to a quadrupole mass spectrometer [68]. Eight sulfonylurea herbicides including sulfometuron-methyl, prosulfuron, metsulfuron-methyl, triasulfuron, rimsulfuron, tribenuron-methyl, primisulfuron-methyl, and thifensulfuron-methyl are analyzed under positive ionization mode, afforded molecular ion adducts with sodium and potassium in addition to protonated molecules. Sodium adduct ions appeared in different drift time from protonated ions, thus add extra complexity to the identification of an individual species within a complex mixture. With the IMS resolving power up to 102, it was possible to distinguish each sulfonylurea compound by mass-selected IMS spectra. To explore the potential of IMS as a rapid screening technique, an unfiltered river water sample spiked with 25 ppm sulfometuron-methyl was analyzed using developed method. The IMS-MS-SIM spectrum of the spiked sample revealed a single peak demonstrating the capability of the system to identify individual sulfonylurea herbicides species within an environmental matrix. The sensitivity of various sulfonylurea ranged from 850 ppb to 5 ppm. An added benefit of ESI-IMS aside from the ability to analyze aqueous samples with minimal preparation is the comparatively short data acquisition times. Traditional methods of analysis (LC-MS) acquire data on the minute timescale, whereas this method using ESI-IMS-QMS required only milliseconds to obtain results.

5.2 Analysis of Veterinary Drugs

With a C18 SPE cartridge, furazolidone, chloramphenicol, and enrofloxacin were effectively extracted from poultry meat extracts and then determined by an IMS [69]. The purified solution was injected into an IMS instrument equipped with thermal desorption with corona discharge ionization source. All samples were analyzed both by IMS and HPLC and the results show that there was good agreement between the results of two methods for chloramphenicol and enrofloxacin. The advantage of corona discharge as an ionization source over the conventional ^{63}Ni is an order of magnitude higher ion current. The higher current also allows using narrower ion pulse widths, which might improve the resolving power of ion mobility peaks. The signal-to-noise ratio is also clearly much higher than that of the ^{63}Ni ionization source for a single-run spectrum. Though furazolidone showed a single peak in the IMS spectrum, it was not a stable compound and was converted to its metabolites. Thus, the analyses of those compounds should be focused on the residue of metabolites.

Tetracyclin is a broad spectrum antibiotic produced by the *Streptomyces* genus of Actinobacteria, widely used in veterinary medicine to treat and prevent microbial infections, such as respiratory tract infections, arthritis, and other animal disease. Aptamers have been applied in preconcentration, extraction, purification of targets on solid phase extraction (SPE) column, and magnetic beads. The combination of antitetracycline aptamer-based sorbent with ESI-IMS was successfully applied for the analysis of tetracycline from human plasma and urine samples with good recovery and sensitivity [70]. Aptamers have excellent selectivity and recovery factor and they are much cheaper than antibodies. The combination of using aptamer as a selective preseparation and enrichment technique with ESI-IMS demonstrated a selective, simple, fast, and derivatization-free method for analysis of tetracycline in human plasma and urine samples with a compatible detection limit and a wide dynamic range relative to the reported methods. Though this method only investigated human urine and plasma, it can be expected to be successfully applied for the detection of tetracycline residue in animal food directly.

5.3 Analysis of Nitrite and Nitrate in Food

Nitrite (NO_2^-) and nitrate (NO_3^-) are both naturally occurring compounds in plant food and animal food. However, nitrite is also a kind of artificial preservative added to food intentionally to keep freshness of food. Nitrite has a harmful impact on human health by reaction with secondary amines to form toxic products in food and human digestive systems. Most developed method for determining nitrite and nitrate are ion chromatography (IC) separation with electrochemistry detection or spectrophotometer methods with chemical derivatization, which is low efficient and labor intensive. Ion mobility spectrometry with negative electrospray ionization was first reported for the separation and detection of various anions in aqueous solution with good separation and sensitivity. Nitrite and nitrate could be separated from chloride, formate, acetate, sulfate, phosphate effectively [71]. The detection limits for nitrate and nitrite was as low as 10 ppb and 40 ppb, respectively. Further, Jafari and coauthors reported the simultaneous determination of nitrite and nitrate in potato with the same protocol [72]. A linear dynamic range of about two orders of magnitude, and relative standard deviations below 5% were obtained by the proposed method for the analysis of both ions. For the analysis of nitrite and nitrate in potato samples, a simple extraction procedure was developed from potato tissue with only centrifuge and dilute involved. In this method, the analysis time was much shorter than that of methods such as IC, CE; also, the reliability was not lower that of the other methods. All samples including bottled water, tap water, and potato evaluated for this method have afforded good validation results.

5.4 Analysis of Microbial Toxins in Food

Aflatoxins belong to a group of naturally occurring mycotoxins that are produced by food spoilage fungi found on a variety of foods such as nuts, grains, dried fruits, and some vegetable oils such as peanut oil. Aflatoxins are important to monitor because they are very toxic and among the most potent carcinogenic compounds known. IMS methods applied for the determination of aflatoxin B1 (AFB1) and aflatoxin B2 (AFB2) on pistachio samples using corona discharge ionization is evaluated to be a sensitive and simple method for screening purposes [73]. This method involves an extraction step and a determination step using methanol as extraction solvent. Methanol extracts of samples were introduced by an injection port and the evaporated sample were transferred to ionization region by a carrier gas. The developed work demonstrates the capability of positive corona discharge ion mobility spectrometry for determination of total AFB1 and AFB2 in real samples at an acceptable detection limit. Although other methods may be more sensitive, they can be time-consuming due to requirements of converting aflatoxins to more intense-fluorescent derivatives. Fast response, low cost, and portability of the IMS make it competitive with other low-detection limit techniques.

Ochratoxin A (OTA) is another carcinogenic and mutagenic mycotoxin produced by *Aspergillus ochraceus*, *Aspergillus carbonarius*, or *Penicillium verrucosum*. Licorice root and its extract are intensively used in food industry for natural sweetener and antioxidant. IMS with corona discharge ionization demonstrate good performance to detect OTA in licorice root using immunoaffinity columns for the cleaning up [74]. The recoveries resulted in 111%, 109%, and 111%, respectively, with about 1% relative standard deviation (RSD%) in spiking blank licorice samples with OTA. The amount of OTA in licorice root detected by IMS method is agreeable with other reported method, which shows that IMS could be applied to fast screening purposes and accurate quantitative analysis.

6. CONCLUSIONS

Though IMS has shown good performance for speed and sensitivity, reported applications in this field is still limited and most of applications have been developed by academic institutions. Compared with conventional applications in explosive detection, chemical-warfare agent determining, and illicit drug screening, the attempt of IMS in food science is just at the beginning and full of challenges and opportunities.

The advantages of IMS include reliability and quantitative performance, speed of response, durability in accepting sample with little preparation, and low detection limits. The discussion and applications in this chapter illustrate a brief profile of applications of IMS in the food industry in recent years. Various IMS methods are applied in process control, monitoring of aging and storage stage, quality control, determination of food compositions, screening for illicit food adulterant, and food contaminant analysis; not just as a kind of routine laboratory analysis methods. Now that the versatility and capability of IMS have been demonstrated for field detection of food contaminants, the determination of trace level residues, process monitoring and the determination of food, IMS appears to have a bright future as an analytical method of choice in the food industry.

For more complex samples, IMS combined with MS offers a number of advantages. The first is that it IMS provides a rapid separation prior to mass spectrometry. IMS separates ions based on their size while MS separates ions based on their mass. With both a size and mass measurement of an ion, unique ion density information is provided for an ion, which can lead to its structural identification. For the most part, MS is blind to isomers. The addition of IMS to MS can permit the analysis of isomers. Finally, the separation of ions in mobility space prior to mass spectrometry enables the reduction of chemical noise and the ability to discern isotope ratios for ions in complex mixtures.

REFERENCES

[1] R.G. Ewing, D.A. Atkinson, G.A. Eiceman, G.J. Ewing, Talanta 54 (2001) 515–529.
[2] D.S. Moore, Rev. Sci. Instrum. 75 (2004) 2499–2512.
[3] P. Dugourd, R.R. Hudgins, D.E. Clemmer, M.F. Jarrold, Rev. Sci. Instrum. 68 (1997) 1122–1129.
[4] G.R. Asbury, C. Wu, W.F. Siems, H.H. Hill, Anal. Chim. Acta 404 (2000) 273–283.

[5] H.H. Hill, S.J. Martin, Pure Appl. Chem. 74 (2002) 2281–2291.
[6] G.R. Asbury, J. Klasmeier, H.H. Hill, Talanta 50 (2000) 1291–1298.
[7] J. Yinon, TrAC, Trends Anal. Chem. 21 (2002) 292–301.
[8] M.A. McCooeye, Z. Mester, B. Ells, D.A. Barnett, R.W. Purves, R. Guevremont, Anal. Chem. 74 (2002) 3071–3075.
[9] C. Wu, W.F. Siems, H.H. Hill, Anal. Chem. 72 (2000) 396–403.
[10] J.D. Dunn, C.M. Gryniewicz-Ruzicka, J.F. Kauffman, B.J. Westenberger, L.F. Buhse, J. Pharm. Biomed. Anal. 54 (2011) 469–474.
[11] C.M. Gryniewicz, J.C. Reepmeyer, J.F. Kauffman, L.F. Buhse, J. Pharm. Biomed. Anal. 49 (2009) 601–606.
[12] A.J. Midey, A. Camacho, J. Sampathkumaran, C.A. Krueger, M.A. Osgood, C. Wu, Anal. Chim. Acta 804 (2013) 197–206.
[13] A.J. Midey, C.A. Krueger, M.A. Osgood, J. Wu, C. Wu, Am. Lab. 45 (2013) 10–15.
[14] R. Guevremont, J. Chromatogr. A 1058 (2004) 3–19.
[15] V. LopezAvila, H.H. Hill, Anal. Chem. 69 (1997) R289–R305.
[16] S. Ehlert, A. Walte, R. Zimmermann, Anal. Chem. 85 (2013) 11047–11053.
[17] C. Weickhardt, N. Kaiser, H. Borsdorf, Int. J. Ion Mobility Spectrom. 15 (2012) 55–62.
[18] H. Borsdorf, S. Roetering, E. Nazarov, C. Weickhardt, Int. J. Ion Mobility Spectrom. 12 (2009) 15–22.
[19] Y.H. Chen, W.F. Siems, H.H. Hill, Anal. Chim. Acta 334 (1996) 75–84.
[20] C. Wu, H.H. Hill, A.P. Gamerdinger, Field Anal. Chem. Technol. 2 (1998) 155–161.
[21] H.M. Dion, L.K. Ackerman, H.H. Hill, Talanta 57 (2002) 1161–1171.
[22] C.L. Crawford, H.H. Hill, Talanta 107 (2013) 225–232.
[23] R.B. Cody, J.A. Laramee, H.D. Durst, Anal. Chem. 77 (2005) 2297–2302.
[24] Z. Takats, J.M. Wiseman, B. Gologan, R.G. Cooks, Science 306 (2004) 471–473.
[25] J.A. Castro, C. Koster, C. Wilkins, Rapid Commun. Mass Spectrom. 6 (1992) 239–241.
[26] G.A. Harris, M. Kwasnik, F.M. Fernandez, Anal. Chem. 83 (2011) 1908–1915.
[27] C. Wu, W.E. Steiner, P.S. Tornatore, L.M. Matz, W.F. Siems, D.A. Atkinson, H.H. Hill, Talanta 57 (2002) 123–134.
[28] C. Wu, W.F. Siems, G.R. Asbury, H.H. Hill, Anal. Chem. 70 (1998) 4929–4938.
[29] A.B. Kanu, P. Dwivedi, M. Tam, L. Matz, H.H. Hill, J. Mass Spectrom. 43 (2008) 1–22.
[30] D.A. Barnett, R. Guevremont, R.W. Purves, Appl. Spectrosc. 53 (1999) 1367–1374.
[31] G.A. Eiceman, Z. Karpas, H.H. Hill, Ion Mobility Spectrometry, third ed., CRC Press, Boca Raton, FL, 2013.
[32] P.J. Rauch, P.D. Harrington, D.M. Davis, Food Technol. 50 (1996) 83–85.
[33] F. Roeck, N. Barsan, U. Weimar, Chem. Rev. 108 (2008) 705–725.
[34] J.S. Vestergaard, M. Martens, P. Turkki, LWT Food Sci. Technol. 40 (2007) 1095–1101.
[35] O. Gursoy, P. Somervuo, T. Alatossava, Int. J. Food Eng. 92 (2009) 202–207.
[36] R. Alonso, V. Rodriguez-Estevez, A. Dominguez-Vidal, M.J. Ayora-Canada, L. Arce, M. Valcarel, Talanta 76 (2008) 591–596.
[37] I. Márquez-Sillero, S. Cárdenas, S. Sielemann, M. Valcárcel, J. Chromatogr. A 1333 (2014) 99–105.
[38] F.J.N.N.J.C. Strachan, I.D. Ogden, Anal. Chim. Acta 313 (1995) 5.
[39] J.I.B.V. Ruzsanyi, G.A. Eiceman, Int. J. Ion Mobility Spectrom. 6 (2003) 6.
[40] T.H. Carlo Tiebe, B. Koch, U. Ritter, I. Stephan, Int. J. Ion Mobility Spectrom. 13 (2010) 8.
[41] T. Vandendriessche, J. Keulemans, A. Geeraerd, B.M. Nicolai, M.L.A.T.M. Hertog, Food Microbiol. 32 (2012) 406–414.
[42] A.R. Shalaby, Food Res. Int. 29 (1996) 675–690.

[43] M.H.S. Santos, Int. J. Food Microbiol. 29 (1996) 213–231.

[44] Z. Karpas, B. Tilman, R. Gdalevsky, A. Lorber, Anal. Chim. Acta 463 (2002) 155–163.

[45] G.M. Bota, P.B. Harrington, Talanta 68 (2006) 629–635.

[46] Z. Hashemian, A. Mardihallaj, T. Khayamian, Talanta 81 (2010) 1081–1087.

[47] Z. Karpas, A.V. Guaman, A. Pardo, S. Marco, Anal. Chim. Acta 758 (2013) 122–129.

[48] G. Barnard, E. Atweh, G. Cohen, M. Golan, Z. Karpas, Int. J. Ion Mobility Spectrom. 14 (2011) 207–211.

[49] M. Camara, N. Gharbi, A. Lenouvel, M. Behr, C. Guignard, P. Orlewski, D. Evers, J. Agric, Food Chem. 61 (2013) 1036–1043.

[50] I. Marquez-Sillero, E. Aguilera-Herrador, S. Cardenas, M. Valcarcel, Anal. Chim. Acta 702 (2011) 199–204.

[51] Z. Karpas, A.V. Guaman, D. Calvo, A. Pardo, S. Marco, Talanta 93 (2012) 200–205.

[52] R. Garrido-Delgado, F. Mercader-Trejo, S. Sielemann, W. de Bruyn, L. Arce, M. Valcarcel, Anal. Chim. Acta 696 (2011) 108–115.

[53] R. Garrido-Delgado, L. Arce, M. Valcarcel, Anal. Bioanal.Chem. 402 (2012) 489–498.

[54] R. Garrido-Delgado, F. Mercader-Trejo, L. Arce, M. Valcarcel, J. Chromatogr. A 1218 (2011) 7543–7549.

[55] G.R. Asbury, H.H. Hill Jr, J. Chromatogr. A 902 (2000) 433–437.

[56] P.V. Johnson, H.I. Kim, L.W. Beegle, I. Kanik, J. Phys. Chem. A 108 (2004) 5785–5792.

[57] L.W. Beegle, I. Kanik, L. Matz, H.H. Hill, Anal. Chem. 73 (2001) 3028–3034.

[58] B.H. Clowers, H.H. Hill, J. Mass Spectrom. 41 (2006) 339–351.

[59] M.T. Jafari, B. Rezaei, M. Javaheri, Food Chem. 126 (2011) 1964–1970.

[60] R. Fernández-Maestre, H. Hill Jr, Int. J. Ion Mobility Spectrom. 12 (2009) 91–102.

[61] C. Krueger, A. Midey, T. Kim, M. Osgood, J. Wu, C. Wu, Int. J. Ion Mobility Spectrom. 14 (2011) 71–79.

[62] J.D. Dunn, C.M. Gryniewicz-Ruzicka, D.J. Mans, L.C. Mecker-Pogue, J.F. Kauffman, B.J. Westenberger, L.F. Buhse, J. Pharm. Biomed. Anal. 71 (2012) 18–26.

[63] D.J. Mans, R.J. Callahan, J.D. Dunn, C.M. Gryniewicz-Ruzicka, J. Pharm. Biomed. Anal. 75 (2013) 153–157.

[64] A.G.R. DeBono, R. Jackson, R. James, F. Kuja, A. Loveless, T. Le, S. Nacson, A. Rudolphl, S. Yin, Int. J. Ion Mobility Spectrom. 4 (2001) 4.

[65] K. Flemming, Dan. Med. Bull. 54 (2007) 58–59.

[66] K. Tuovinen, M. Kolehmainen, H. Paakkanen, Anal. Chim. Acta 429 (2001) 257–268.

[67] J. Wang, Z. Zhang, Z. Du, W. Sun, Anal. Methods 5 (2013) 6592–6597.

[68] B.H. Clowers, W.E. Steiner, H.M. Dion, L.M. Matz, M. Tam, E.E. Tarver, H.H. Hill, Field Anal. Chem. Technol. 5 (2001) 302–312.

[69] M.T. Jafari, T. Khayamian, V. Shaer, N. Zarei, Anal. Chim. Acta 581 (2007) 147–153.

[70] S.N. Aslipashaki, T. Khayamian, Z. Hashemian, J. Chromatogr. B 925 (2013) 26–32.

[71] P. Dwivedi, L.M. Matz, D.A. Atkinson, H.H. Hill, Analyst 129 (2004) 139–144.

[72] M.T. Jafari, F. Torki, M. Saraji, Anal. Sci. 28 (2012) 391–395.

[73] A. Sheibani, M. Tabrizchi, H.S. Ghaziaskar, Talanta 75 (2008) 233–238.

[74] M. Khalesi, M. Sheikh-Zeinoddin, M. Tabrizchi, Talanta 83 (2011) 988–993.

Part II

Mass Spectrometry Applications within Food Safety and Quality

Chapter 6

Food Proteins and Peptides

Roberto Samperi,* Anna Laura Capriotti, Chiara Cavaliere, Valentina Colapicchioni, Riccardo Zenezini Chiozzi and Aldo Laganà
Dipartimento di Chimica, Università di Roma "La Sapienza", Rome, Italy
**Corresponding author: E-mail: roberto.samperi@uniroma1.it*

Chapter Outline

1. Introduction **309**
2. Proteomic Techniques **313**
2.1 Top-Down Proteomics 315
2.2 Bottom-Up Proteomics 316
2.2.1 In-Gel Bottom-Up Proteomics 316
2.2.2 Shotgun Proteomics 318
3. Food Security **326**
3.1 Plant Proteomics 326
3.2 Stress Effects on Gene Expression 327
3.2.1 Abiotic Stress 327
3.2.2 Biotic Stress 329
3.2.3 Postharvest Proteomics 330
3.3 What about Livestock? 330
4. Food Safety, Quality, and Authentication **332**
4.1 Genetically Modified Organisms 332
4.2 Food Processing and Quality Control 332
4.2.1 Meat 333
4.2.2 Farm Animal Milk, Dairy Products, Eggs 337
4.2.3 Fish and Aquatic Invertebrates 338
5. Food Peptidomics and Digestomics **341**
5.1 Bioactive Peptides 341
5.1.1 Discovery of Bioactive Peptides Encrypted in a Protein Amino Acid Sequence 342
5.1.2 In vitro and In vivo Bioactive Peptides Production 344
5.1.3 Bioactive Peptides Sources 345
6. Conclusions and Outlook **346**
References **348**

1. INTRODUCTION

What is proteomics? In an interesting article, Lederberg and McCray [1] attributed the origin of the term "genom(e)" to the German botanist Hans Winkler (1920), and the origin of the suffix "ome" from the Sanskrit "OM," which signifies fullness, completeness as in divinity. McKusick and Ruddle, as the title of the journal they founded in 1987, introduced the term "genomics" about

Comprehensive Analytical Chemistry, Vol. 68. http://dx.doi.org/10.1016/B978-0-444-63340-8.00006-6

66 years later. The PhD Marc Wilkins, in 1994, first used the word *proteome* to indicate the protein complement of the genome. Finally, James first coined the term "proteomics" [2], although the first definition of the word is generally attributed to Pandey and Mann [3] as the large-scale study of proteins from a given cell line or organism. Today, this definition appears restrictive, and proteomics could be defined as the study, by high-throughput techniques, of the whole gene expression of a biological system in a certain condition, including posttranslational modifications (PTMs) and protein–protein interaction.

The recent evolution of "omic sciences" has introduced integrated approaches in the field of food research, which broadly overcomes the limited descriptive approaches of the early omics era. Currently, omics studies are emerging as an important part of the holistic approach that pursues a global attempt of knowledge about food, which covers the assessment of its composition, the effects of the processes of production, the modifications over time, the digestion processes, and the impact on human health.

The qualitative and quantitative determination of proteins and peptides in raw or processed food is experiencing a growing interest and importance from both the scientific and economic point of view. Proteomics and peptidomics are relatively new entries in the field of food security, safety and authenticity, and themselves can contribute to the emergence of new branches of the science of food, such as foodomics and the newly born nutriomics, digestomics, and gut metagenomics/metaproteomics.

From a historical point of view, food security can be considered the first argument concerning food approached with the tools of proteomics. In the World Food Summit of 1996, food security was defined as existing "when all people at all times have access to sufficient, safe, nutritious food to maintain a healthy and active life." This definition implies both physical and economic access to food and rests its foundation on three pillars (1) food availability, (2) food access, and (3) appropriate food utilization. Therefore, as widely recognized, food security is a complex issue, linked to sustainable trade, economy and environmental impact. The issue of food security can be summarized in four questions:

- Is there enough food in the world to feed all its inhabitants?
- Can the current levels of production meet the demand for food in the future?
- In that way, is it possible to increase food production without resorting to overexploitation of the environment that may undermine food security in the long term?
- How can we ensure that the food traded in the global market is of acceptable quality and safe to eat?

The green revolution, started worldwide in the 1960s under the auspices of Food and Agriculture Organization of the United Nations (FAO), favored the planting of crops with high yield, which, however, cannot grow successfully without the help of large amounts of fertilizers, pesticides, and irrigation to produce this result. However, the sum of factors such as depletion and climate change has contributed to the collapse of productivity of lands subjected to extreme climate. With the publication of the report "*Save and Grow*" in 2011 [4], FAO itself proposed a new paradigm of intensive crop production that conjugates highly productive and environmentally sustainable. The scientific community had already become aware for at least two decades of the fragility of the results obtained with the first green revolution and the need to find alternative ways, not only to maintain the existing levels, but also to increase production for future needs without threatening natural resources. To achieve this purpose it would be necessary in the future to find more efficient systems of food production, to reduce waste, pollution, and the consumption of energy and other resources such as soil and water. In this scenario, the increased demand for protein could not be solved with an increase in livestock, but in improving and increasing sustainable agricultural production, especially cereals. Crop yield improvement by enhancing biotic and abiotic stress resistance of plants is essential in order to accomplish such increase in food production. Another problem that seriously affects the availability of food is food losses. For different reasons, in both developed and developing countries, this loss can be roughly estimated as 25–30% of the production. In developed countries, this occurs at the level of retail sale and consumption, while in developing countries, food loss occurs mainly at postharvest and processing stages [5]. Reducing postharvest losses and increasing shelf life would be the most efficacious and affordable actions to increase food security. Selection of species and varieties suited to the climate and soils of the crops, genetically modified organisms (GMOs), intelligent agricultural practices, optimized harvesting and post harvesting strategies, suitable storage methods, are all complementary actions aimed at increasing food production without heavily depleting natural resources.

Foodomics has been defined as "a new discipline that studies the food and nutrition domains through the application of advanced omics technologies to improve consumer's well-being, health, and confidence" [6]. The complex food science domain comprises food genomics, transcriptomics, proteomics, peptidomics, metabolomics, and lipidomics. Indeed, the interaction of food with the organism enters the circle of system biology generating the nutriomics sciences. Finally, the interaction between the intestinal microflora, digested food, and proteins (peptides) secreted by the host organism into the gut will represent an environment which is located at the apex of a complex biochemical pyramid that science has just begun to climb [7]. This assemblage of a variety of research fields is at the service fundamental knowledge that already has a strong impact and will have even more in the future, not only on human health, but also in the world economy.

Food quality, safety and authenticity are characteristics closely related to each other and complementary to food security (see Figure 1). Food safety is a broad issue which includes the concept of food edibility (free of dangerous

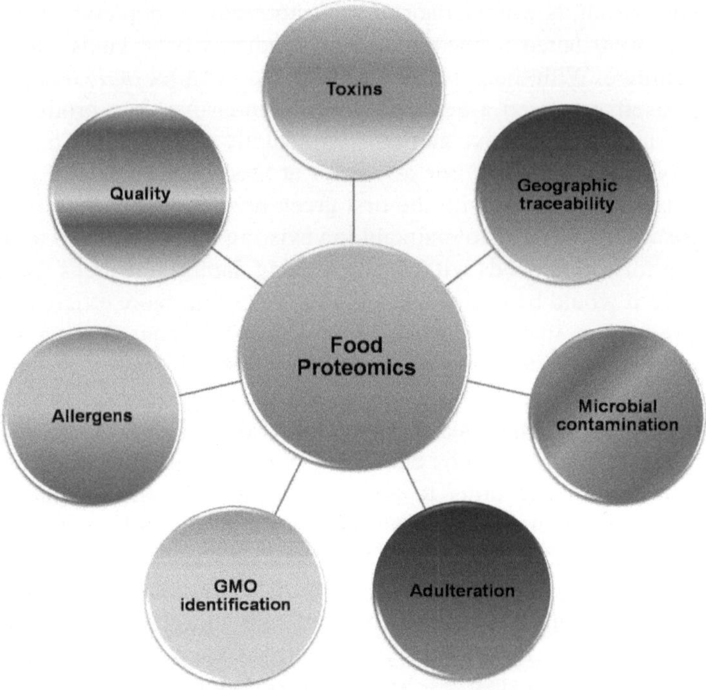

FIGURE 1 Proteomics can find several applications in food safety, security, and authenticity fields, such as geographic traceability, determination of microbial and parasite or toxin contamination, and identification of allergens, adulteration, or GMOs.

compounds, such as toxins, pathogens, allergens, and spoilage products xenobiotics), but also of salubrity (macronutrient composition, nutraceutical characteristics, subtle health effects). Finally, food traceability and authenticity (geographic origin, ingredients in processed foods, processing) are fundamental in determining both nutritional and commercial values of foodstuff. A separate discussion is needed for products that contain GMOs as, although the balance of pros and cons is neither homogeneous nor definitive, many consumers regard their presence in foods negatively; therefore, it is part of the perceived security and quality.

Modern nutritional research has a great potential in contributing to improve human health through the application of the techniques of molecular and system biology to nutritional issues. Nutriproteomics has been defined as "the large-scale analysis of the structure and function of proteins as well as of protein–protein interactions in a cell to identify the molecular targets of diet components" [8]. The proteome is a highly dynamic network, where dietary components can interfere with protein expression, modification, and interaction in different physiological or pathological conditions. Personalized nutrition could be as important as personalized medicine in maintaining or restoring homeostasis, and understanding the interaction of a nutrient with an organism at cellular and molecular level is of pivotal importance.

Proteins are extensively hydrolyzed along the stages of gastrointestinal (GI) digestion up to free amino acids (AAs) or short oligopeptides (2–20, rarely up to 40 AAs), which are released from food proteins, escape further digestion, and can be absorbed in the intestinal lymphatic system. Many of these peptides are bioactive, exerting physiological activities, for example, they exert agonist or antagonist action for the same targets of endogenous counterparts with which they often share common structural traits. The term "protein digestomics" has been coined recently [9] to indicate these peptides, but they can be also produced during food processing or in vitro enzymatic hydrolysis.

The human GI tract contains a complex society of commensal microbes (approximately 10^{11} g^{-1} feces), the so-called microbiota, which, by the expression of their genes (microbiomes), plays a key role in a wide range of host-related processes. *Inter alia*, it modulates the expression of genes involved in several intestinal functions, including nutrient transformation and absorption. Recent findings suggest that each human being has a unique and relatively stable gut microbiota, unless disrupted by external factors [10]. Microbiota expresses its own metaproteome and secretes bioactive factors, including proteins and peptides [11]. In addition, it participates in food digestion processes and may produce or destroy bioactive peptides (BPs).

2. PROTEOMIC TECHNIQUES

Mass spectrometry (MS), used in combination with a wide variety of separation methods and bioinformatic tools, is the principal methodology for proteomics. There are two fundamental strategies currently employed for protein identification and characterization in proteomics: bottom-up and top-down approaches (see Figure 2). In bottom-up approach, purified proteins, or complex protein mixtures, are subjected to proteolytic cleavage, and the resulting peptides are analyzed by MS or MS/MS. In top-down proteomics, intact protein ions or large protein fragments are subjected to gas phase fragmentation for MS/MS analysis [12]. There is some ambiguity about the employment of the terms "top-down" and "bottom-up" in proteomics. In one lexicon, they designate the entity that is subjected to the primary separation technique. In this regard, separation of proteins by gel electrophoresis followed by in-gel proteolytic digestion and liquid chromatography (LC)-MS analysis of the resulting peptide mixture can be considered a top-down approach, because the initial entities separated are proteins. In another lexicon, adopted by Reid and McLuckey [13] in their review of top-down protein characterization, top-down and bottom-up refer to the entities introduced into the mass spectrometer. As the majority of the scientists involved in proteomics research agree with the second definition, this terminology will be the one adopted in the present chapter.

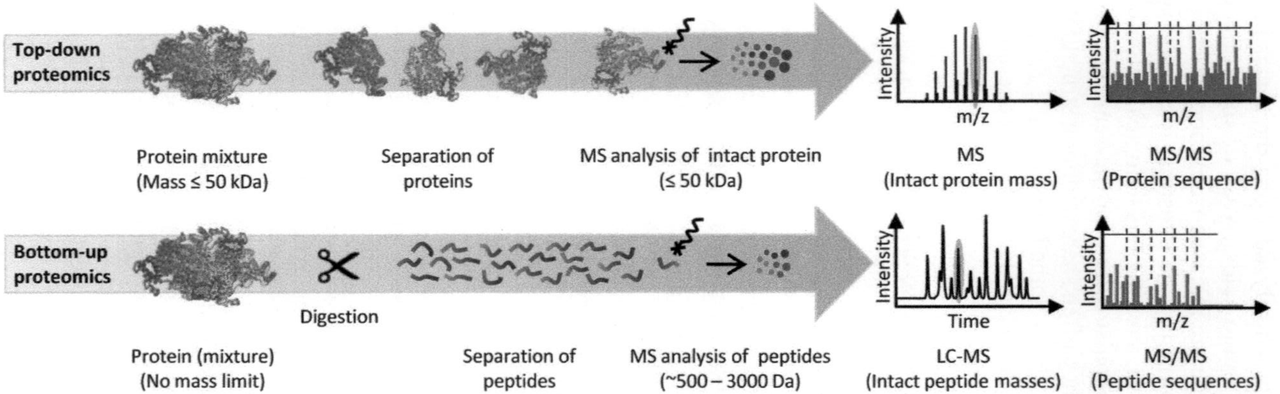

FIGURE 2 Schematic diagrams of top-down (upper panel) and bottom-up (lower panel) proteomics experiments. *Adapted with permission from Linda Switzar, Martin Giera, and Wilfried M.A. Niessen, Protein digestion: an overview of the available techniques and recent developments, Journal of Proteome Research 12 (2013) 1067–1077, Figure 1, p. 1068. Copyright 2013 American Chemical Society.*

2.1 Top-Down Proteomics

Top-down proteomics experiments consist in introducing intact proteins into the mass spectrometer. The ions generated by electrospray ionization (ESI) or, rarely, by matrix-assisted laser desorption/ionization (MALDI) are then subjected to gas phase separation, fragmentation, fragment separation, and automated interpretation of mass spectrometric and chromatographic data yielding both monoisotopic molecular weight (MW) of the intact protein and the protein fragmentation ladders. A schematic diagram of generic top-down experiment is shown in Figure 2 (upper panel). Depending on its complexity and the information required, the protein-containing sample can be submitted to one-dimension (1-D) or two-dimension (2-D) separation before MS and/or MS/MS. In turn, separations can operate online or off-line between them and with MS [14,15]. The top-down strategy can integrate and support information obtained from bottom-up analysis, in particular regarding complete protein sequence and PTM localization as well as any combination possibly existing between modifications on distinct parts of the protein sequence [15,16].

The main limitations of this approach are:

1. the need of high accuracy mass measurement analyzers able to separate the multicharged isotopic cluster of proteins as large as 229 kDa [17,18];
2. the difficulty in determining the product ion masses from multiply charged product ions (with possible ambiguity in the interpretation of top-down MS/MS spectra) [13]; and
3. the lack of suitable software for reconstructing protein identity. Moreover, differently from bottom-up experiments, full automation in data treatment is possible only depending on instrumentation capability.

Coupling LC with fourier transform ion cyclotron resonance (FTICR) instruments is not straightforward: a quadrupolar ion trap (IT), usually linear IT (LIT) is necessary to perform selected ion accumulation before the ICR cell, reducing also the space charge effect. This implies that a relatively long duty cycle is needed for a sufficiently sensitive acquisition of MS/MS spectra, as the LIT-FTICR instrument requires greater ion population than the simpler LIT instrument. Therefore, owing to the difficulties in coupling online to an FTICR instrument much of the early top-down work has been done with infusion of isolated proteins or simple protein mixtures [13]. The introduction of devices able in enhancing ion transmission and/or ion desolvation [19,20] alleviates the problem reducing the filling time of LIT.

Recently, also lower resolution instruments with a faster duty cycle such as a quadrupole-time of flight (Q-TOF) in combination with ESI in infusion mode [21], and MALDI-TOF/TOF [22] have been used. Despite the availability of high-performance mass spectrometers, methods for MS-compatible separation and high-throughput identification of intact proteins are yet underdeveloped.

2.2 Bottom-Up Proteomics

A schematic diagram of a generic bottom-up experiment is shown in Figure 2 (lower panel); see also Figure 3 for details.

2.2.1 *In-Gel Bottom-Up Proteomics*

Bottom-up strategy is the most widely used for protein identification and characterization. The first developed bottom-up strategy is achieved by the 2D polyacrylamide gel electrophoresis (2D-PAGE) in combination with MALDI-TOF MS. 2D-PAGE was introduced more than 35 years ago [23,24] as an orthogonal highly resolving separation technique, and still plays a remarkable role in the analysis of proteome samples. The technique is based on the separation of proteins in the first dimension according to their isoelectric points, pI, by isoelectric focusing (IEF), either in a pH gradient maintained dynamically with ampholytes [24] or with a pH gradient covalently immobilized in the gel matrix [25]. The proteins are, then, transferred to the second dimension sodium dodecyl sulfate (SDS)-PAGE where they are further separated according to their molecular mass [26].

A fundamental problem with IEF is that proteins tend to precipitate at their pI. Urea is a common solubilizer agent for proteins and must be added at concentration 7–9 mol/L to avoid aggregation of proteins. The addition of a nonionic detergent also increases protein solubility, while the addition of a thiol reducing agent breaks disulfide bridges. The denaturing ambient is maintained in the second dimension, where the anionic detergent is added to the buffer. SDS

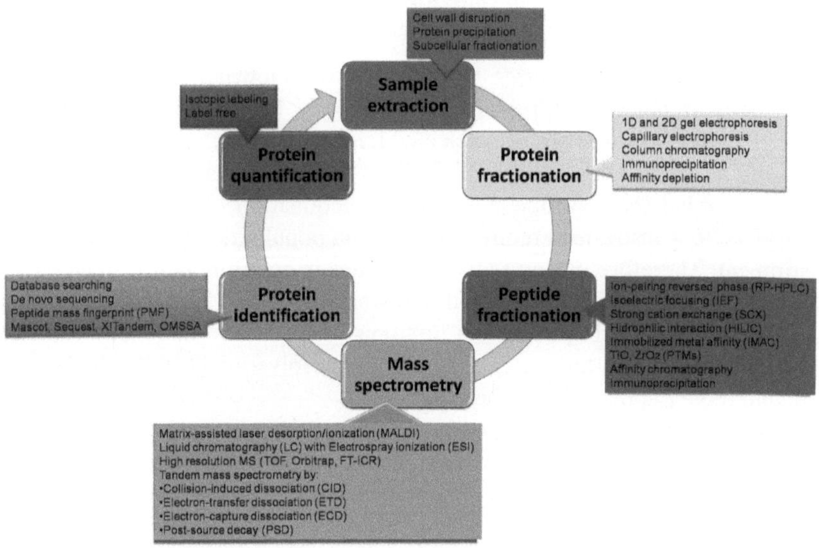

FIGURE 3 Main analytical steps and techniques used in bottom-up proteomics analysis.

binds the majority of proteins at a ratio of 1.4:1 *w/w* to form negatively charged complexes. During SDS-PAGE, separation depends on the effective molecular radius, which approximates roughly to molecular size, because of the sieving effect of the gel matrix.

Depending on the gel size and pH gradient range, 2D-PAGE can resolve more than 5000 protein spots simultaneously (1500–2000 routinely) [27]. Furthermore, it delivers a map of intact proteins, which reflects changes in protein expression level, including isoforms and PTMs. The number of spots separated can be maximized by using larger gel strips (IEF) or plates (SDS-PAGE). If the available amount of sample is not a problem, IEF can be subdivided into strips with a narrow pH range, and each strip is then coupled to an individual SDS-PAGE gel plate [28,29].

After separation, proteins in the 2D-map have to be visualized by a staining method and, possibly, quantified by computerized image analysis [30]. It has been suggested that concentration of individual proteins in a sample can span up to 12 orders of magnitude [31]: these dramatic variations in protein concentrations are a major challenge for all currently available protein detection methods [32,33]. The most important properties of protein visualization methods are low detection limit, high linear dynamic range, reproducibility, and compatibility with the subsequent protein identification procedures. Universal detection methods of proteins on gels consist mainly in staining with the anionic dyes Coomassie Blue, or silver staining, with fluorescence staining or labeling, and radioactive isotopes as a possible alternative; in any case currently no staining method meets all these requirements.

In spite of being a very powerful tool for protein analysis, 2D-PAGE is characterized by insufficient reproducibility. It is challenging to ensure direct spot-to-spot comparison between two separate gels, as the complexity of the specimen and instrumental technique adopted to obtain the final electrophoretic maps often hindered different images from being perfectly superimposable. Therefore, computer-aided image analysis is frequently not straightforward and very time-consuming. To overcome this difficulty appropriate statistical tools can be employed [34]. A particularly useful technique to measure relative abundance of proteins between comparative samples is 2-D differential in-gel electrophoresis, in which two equal amounts of samples are labeled in vitro using two different fluorophores showing different excitation and/or emission wavelengths [35]. Then, samples are mixed and separated in a single 2-D gel and the two images obtained at different wavelengths are normalized, whereby differences between the two samples are easily evidenced due to comigration. More recently a third fluorophore has been added to further control normalization [36]. However, even this approach is not without problems and several attempts have been made to overcome it, which were only in part successful [37].

Sample preparation, as far as sample prefractionation, is also a very important issue in 2D-PAGE, although some problems are not very different from that encountered in-gel-free methods and will be discussed in the next paragraph.

Other limitations widely recognized and described of 2D-PAGE arise with respect to separation of very acidic, very basic, low-abundance, integral membrane proteins, limited range in expression levels and differences in solubility, as well as the separation of low-MW (<15 kDa) and high-MW (>150 kDa) proteins [38–41]. For a very comprehensive treatment of the subject, readers are referred to the reviews by Görk et al. [27,42].

After the 2D-PAGE separation, staining, and computerized image analysis, spots of interest are excised from the gel, destained, and submitted to in-gel digestion (usually employing trypsin), then the digested gel is prepared for MS identification of peptides followed by database search. Peptide mass fingerprinting by means of MALDI-TOF MS existed before the term proteomics was coined and genome sequentiation was in an advanced stage [43–47]. This approach is now rarely employed because it gives the peptide composition, but not the aminoacidic sequence. Approximately, at the same time, but using a very different approach, Mann et al. [48] and Eng et al. [49] demonstrated that a short aminoacidic sequence, along with the fragmentation mass spectra (a sequence tag) was useful for protein identification. This observation can be considered the starting point for shotgun proteomics (see next paragraph), however, it was also used in combination with 2D-PAGE [50] when, after the introduction of Q-TOF [51] and TOF–TOF [52] instruments, MALDI MS/MS became possible.

Bioinformatics tools for processing experimental data are of paramount importance. The search engines such as Sequest [49] and Mascot [50,53], continually upgraded, match and score the peptide assignment to a protein.

2.2.2 Shotgun Proteomics

In this approach, a proteolytic enzyme, usually trypsin, digests the protein sample to produce peptides. Then, the complex peptide mixture is separated by 1-D or 2-D coupled separation method and detected by MS/MS. Reliable protein identification is accomplished by database searching for unique peptide sequence matching.

2.2.2.1 Sample Preparation

Extraction of proteins from the raw sample is an essential, critical step for obtaining good protein identification and quantification. Some samples are not a ready source of proteins and need specific protocols. The majority of samples is very complex, and/or may contain large amounts of nonprotein compounds, therefore a prefractionation method is needed to reduce the complexity or isolate the proteins of interest. Moreover, it can be difficult to identify low-abundance proteins in the presence of a large excess of relatively abundant proteins. Therefore, for effective proteome analysis it becomes critical to enrich the sample to be analyzed in subfractions of interest. Many strategies, based on physical–chemical, cellular, or immunological properties, have been developed over the years for fractionation of proteins into subproteomes.

Sequential extraction based on different solubility of protein classes in buffers is a method particularly suited [54–62]. A subcellular fractionation of tissue sample may be essential before extraction to simplify the protein identification as well as to retain their primary localization.

Such fractionation can be obtained by differential centrifugation [63] or selective immunoseparation [62,64]. Protein precipitation methods are employed to isolate proteins from other tissue components, the most popular being performed with trichloroacetic acid-acidified acetone [65,66] and methanol/chloroform/water (4:1:3, v/v/v) [67,68]. The two protein precipitation methods are not equivalent: in both cases, the precipitation is partial, and after resolubilization gives rise to different compositions [69]. Extraction of membrane proteins and even more their separation from other cell compartment proteins remain difficult and need a careful maintenance of the solvent environment [70].

Low-abundance proteins are difficult to detect in the presence of higher abundance ones due to a competition for ionization during ESI process [71]. To overcome this difficulty, a pretreatment of the sample is necessary and several strategies have been proposed for the selective depletion of high-abundance proteins and enrichment of low-abundance ones. These strategies include: precipitation with organic solvents [72] or salt solutions [73], ultrafiltration [74–76], solid phase extraction [76–79], chromatography [80], immunoaffinity depletion [81–83], peptide library beads [77,84], dual size exclusion–affinity hydrogel nanoparticles [76,85–87]. All of these strategies allow to determine low-abundance proteins and peptides and immunodepletion appears to be preferred [88]. However, some issues have yet to be overcome, in particular the partial removal of high-abundant proteins that remain in the sample and the codepletion of low-abundant proteins. Finally, PTMs occurring on proteins in response to intra- and extracellular stimula play a crucial role in protein biological functions. However, because PTMs are usually present in very low amount, specific enrichment techniques are needed, which are applied at both protein and peptide level [89,90].

In shotgun proteomics, precipitated proteins are redissolved in a denaturing medium, reduced, alkylated, and digested after dilution with an appropriate buffer [91,92]. Usually trypsin is the enzyme of choice, but sometime other proteases could be preferred [91].

2.2.2.2 Peptide Separation

Enzymatic digestion of complex protein samples produces a huge number of peptides that should be separated before MS to obtain in deep proteome coverage and reliable quantitative data. Recently, the introduction of very long monolithic columns (up to 4 m) [93], or very efficient, relatively long (40–70 cm) columns, packed with 3 or 2 μm particle size [94–96], together with very long gradient times and coupled with the last generation of mass spectrometers have improved much better the performance. Although these recent progresses in LC column technology enable the identification of about 5000 proteins in a single run, 2-D separations are still widely used. There are important factors to be considered for developing multidimensional protein identification technology, such as the time required for analysis, compatibility with MS of the buffer used for the chromatographic separation, and the effective orthogonality of the two dimensions. For the last stage of separation, which is the step directly interfaced to a mass spectrometer, reversed phase (RP)-LC is usually preferred, as it can provide high resolution, desalting of samples, and the compatibility of the phases with the ESI source and MS detection.

An important consideration for the development of multidimensional separations is the orthogonality of coupled techniques. Peak capacity can be maximized by combining separation methods based on different principles of action. A variety of first dimensions, some of which nonchromatographic, such as SDS-PAGE [97] and off-gel IEF [98,99], or chromatographic ones, including size-exclusion chromatography [100], strong cation exchange (SCX) [101,102], strong anion exchange [103], RP/RP high/low pH [104] are employed. Although SCX is the predominant first dimension used upstream of RP, comparative study demonstrates that RP/RP high/low pH showed the best peak capacity and protein identification number [105,106]. A comprehensive discussion of this topic can be found in Motoyama et al. and Tang et al. [107,108].

2.2.2.3 Mass Spectrometry

Generally, nanoESI-MS/MS is employed to analyze the proteolytic mixtures separated by chromatographic [109] or electrophoretic techniques [110]. Developments and improvements in MS instrumentation have played a key role and facilitated the MS-based proteomics. The first analyzer used for ESI-MS/MS acquisition was the Paul ion trap [109,111], which showed limited mass accuracy, sensitivity, and linear range. The more recently introduced LIT, thanks to the lower space charge effect, shows better sensitivity and larger dynamic range, therefore has completely substituted the traditional IT in the most recent application [112]. However, the problems about mass accuracy remain. To improve sensitivity, dynamic range, comprehensiveness, and throughput, LC-ESI FTICR with the strategy denominated "accurate mass tag," based on high resolution LC separation and high mass measurement accuracy (<1 ppm), has been used [113], but this approach necessitates a preventive identification by traditional IT instruments. The Q-TOF instrument, which is capable of accurate mass measurements, has also been employed [114,115], but the restricted dynamic range limited its usefulness in proteomics.

The MALDI ion source was also used in combination with TOF–TOF [51,115–117], Q-TOF [51,118], and IT [119] mass spectrometers, with the limitation of the necessity of an off-line setting, that could be realized with a robotic device to fractionate an LC separation into discrete spots on the MALDI sample plate [115]. The MALDI source gives better ionization of hydrophobic peptides than ESI, therefore it could be used to obtain complementary information in a proteomic study [115].

Nevertheless, in recent years the MS for proteomics has experienced thc dominance of the Orbitrap-based mass spectrometers. Orbitrap technology, developed by Alexander Makarov, was introduced on the market by Thermo around the mid-2000s and is still evolving toward models with improved performance [120–126]. The Orbitrap is a relatively simple IT device where only an electrostatic high potential field is applied. The ions oscillate inside the trap with a frequency depending only on the mass-to-charge ratio, which is measured and converted by an FT to the individual frequencies and intensities, yielding the mass spectrum. The key capabilities of this analyzer are high resolution and accurate mass, similar to those achievable with the FTICR instruments. Although it is possible to fragment ions in the Orbitrap, better performances can be obtained manipulating the ions fragmented in a collision cell (or other fragmentor devices) in another mass analyzer to which the Orbitrap can be linked as a high-resolution detector of precursor ions. Several ideas have been tried out to create powerful hybrid systems with the quadrupolar technology (transmission quadrupole, LIT). These hybrid instruments exploit both the characteristics of high resolution (Orbitrap) and rapid scan time (quadrupolar analyzer). In proteomics analyses, hybrid Orbitrap MS/MS instruments are commonly used in a data dependent acquisition mode, after parameter optimization [127].

2.2.2.4 Proteome Informatics Tools and Quantitative Proteomics

Bioinformatic tools are very important to handle and analyze the huge amount of data generated even by a single MS/MS-based proteomics experiment. During the 1980s, when genomics data were not available, various strategies were developed for the de novo sequencing of peptides from MS/MS spectra. The availability of protein databases derived from genome sequencing made possible in the late 90s the so-called peptide fragmentation fingerprint (PFF) identification method which, continuously implemented, is now routinely used. In the PFF data analysis, the experimental MS/MS spectra are compared with the theoretical ones, computed from peptide sequences, and stored in databases, in order to find the most similar (highest score) candidate peptide. Theoretical spectra were derived for peptide sequences obtained by in silico digestion of proteins contained in the database and indicating the enzyme employed in the experiment. A confident identification does not require that all the peptides of a given protein have to be confirmed: some peptide sequences are unique and characterize a group of proteins, differing from each other for PTMs, isoforms, etc. Before spectral comparison, raw data are treated to be more suitable for database comparison, with operations like signal filtering and background subtraction, deisotoping, charge deconvolution, and peak matching dataset alignment and information selected by the operator, like proteolytic enzyme, number of missed cleavages, fixed modifications and variable modifications derived from the sample preparation procedure. Nowadays, Mascot [53] and SEQUEST [49] search engines, originally developed during the 1990s, are now commercially available and commonly used in proteomics. Moreover, also some free access online search engines, for example, the recent Andromeda [128], can be used. A very important question is to establish the confidence of the obtained identification [129]. This function can be exploited with validation tools such as peptide prophet and protein prophet [130,131], Qscore [132], iProphet [133], and Percolator [134]. Therefore, the matching with known sequences uses genomics/proteomics annotated sequence database such as Swiss-Prot [135,136] (see Figure 4). The current bioinformatics facilities automate in a pipeline all processes involved in the MS/MS-based identification, maximizing the number of identified spectra with high level of confidence (see, for example [137,138]). A comprehensive treatment of these subjects can be found in some reviews [138–144].

Proteomic studies do not focus only on the identification of proteins in a given sample, but aim at an accurate quantification of them, as the investigation of variable protein expression profiles can yield useful information. Over the time, many absolute and relative quantitative proteomic methodologies have been developed. For relative quantification, stable isotopic labeling to create a specific mass tag that can be recognized by a mass spectrometer was introduced in 1999 both in the form of metabolic coding in isotopically enriched media [145,146], and introducing isotopic coding with an affinity tag into peptides in a single step [147] (see Figure 5). In this second case, two samples that need to be compared react with both light and heavy forms of a reagent able to react with the cysteine residues, then are combined and analyzed by LC-MS/MS. The main disadvantage of this class of reagents was that some proteins did not have any cysteine. Thereafter, various labeling strategies relying on the labeling of samples from different conditions with stable isotopes (^{2}H, ^{13}C, ^{15}N, ^{18}O) have been introduced [148,149]. The most popular of the nonisotopic methods is the isobaric tags for relative and absolute quantification [150] having the specific advantage of being able to quantify several samples in a single run. These amino-reactive reagents consist of two parts: a balancing group and a reporter group. The reporter group exists in up to eight different forms with different masses. The balancer group compensates the different weights so that the sum of the masses of these two groups is always the same (isobaric tags), and every peptide precursor ion contains contributions from all the mixed samples. When such a peak is selected for fragmentation, the two groups fall apart and reveal the relative contributions in the fragmentation spectrum.

The observation of a correlation between protein abundance and peak areas [151,152] or number of MS/MS spectra achieved [153] opened the door to label-free quantification of proteins. Label-free quantification can be divided into two distinct approaches: signal intensity measurement based on precursor ion spectra and spectral counting (see Figure 6). In the first approach, the intensity of the selected peak can be visualized in an extracted ion chromatogram as a function of the retention time and its area can be determined. Although this approach may appear straightforward, several points must be taken into account to ensure reproducible and accurate quantification between individual sample

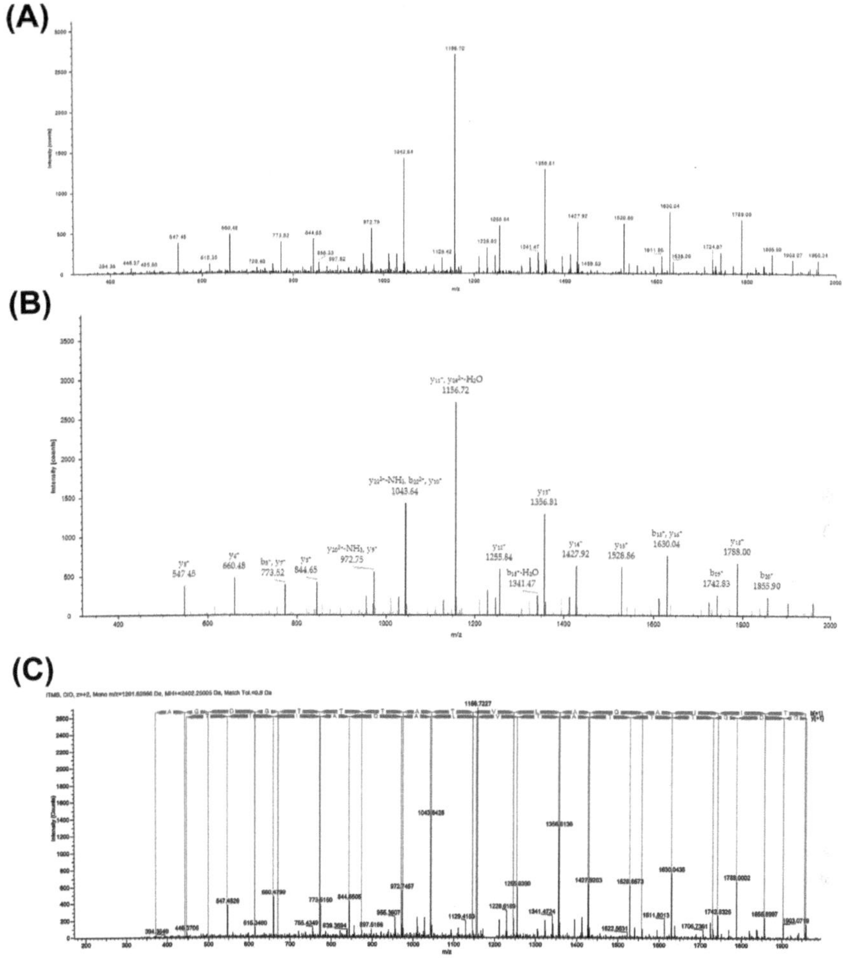

FIGURE 4 Peptide identification in a sample of tryptic digest of *Escherichia coli* analyzed by nanoliquid chromatography–tandem mass spectrometry with a hybrid linear ion trap—FT Orbitrap instrument; Mascot was used as search engine and Swiss-Prot as sequence database. (A) Raw tandem mass spectrum obtained by collision-induced dissociation of a peptide of m/z 1201.6287 (charge +2). (B) Matching between the experimental spectrum and the theoretical spectrum obtained by in silico digestion of proteins contained in the database. (C) Automatic assignment of aminoacidic sequence of the peptide.

runs when analyses of multiple samples are performed, thus a suitable computational approach is mandatory [154]. In addition, the right balance between acquired survey (precursor) and fragment ion spectra needs to be found: an increased number of acquired fragment ion spectra leads to a higher proteome coverage, but increases the cycle time. This in turn results in less acquired MS spectra, which are needed to describe the chromatographic peptide ion peak.

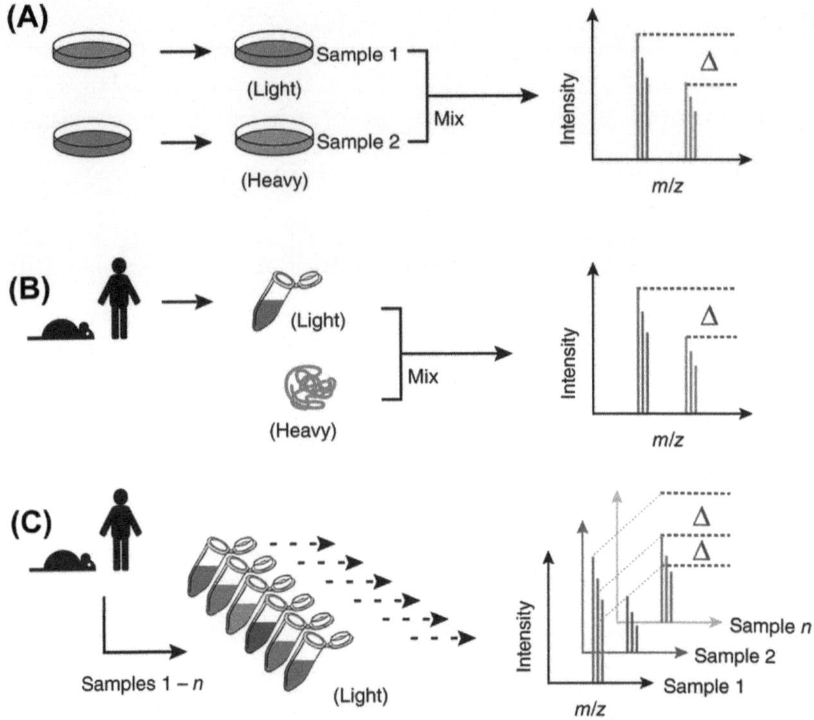

FIGURE 5 MS-based quantitative proteomics. (A) In a typical SILAC experiment, cells are grown in medium containing amino acids labeled with different isotopes to incorporate the isotopes into cellular proteins. Samples from the conditions to be compared are combined before analysis. (B) For absolute quantification light isotope-labeled sample, primary tissue is spiked with known amounts of protein with heavy isotope label. For A and B, samples are prepared after mixing the samples or spiking in the standard, respectively, and peptides are analyzed together by LC-MS. SILAC quantification is based on the differential intensities of coeluting peptides with identical amino acid sequence but with distinct isotope labels. (C) In a label-free approach, peptides derived from any biological sample are prepared separately and analyzed by LC-MS sequentially. Label-free quantification is based on the differential intensities of eluting peptides with identical amino acid sequence and mass between separate LC-MS runs. δ indicates quantitative peptide differences. *Reproduced with permission from Felix Meissner, Matthias Mann, Nat. Immunol. 15 (2014) 112–117, Figure 3, p. 115. Copyright 2014 Macmillan Publishers Ltd: Nature Immunology.*

The spectral counting approach is based on the empirical observation that the more of a particular protein is present in a sample, the more MS/MS spectra are collected for peptides of that protein. Hence, relative quantification can be achieved by comparing the number of such spectra between a set of experiments. Although intuitive and attractive in practical terms, the spectrum counting approach is still controversial. In fact, the spectrum count response is different for every peptide because, e.g., the chromatographic behavior (retention time, peak width) varies for every peptide. Therefore, quantification requires the observation of many spectra for a given protein. Both label-free

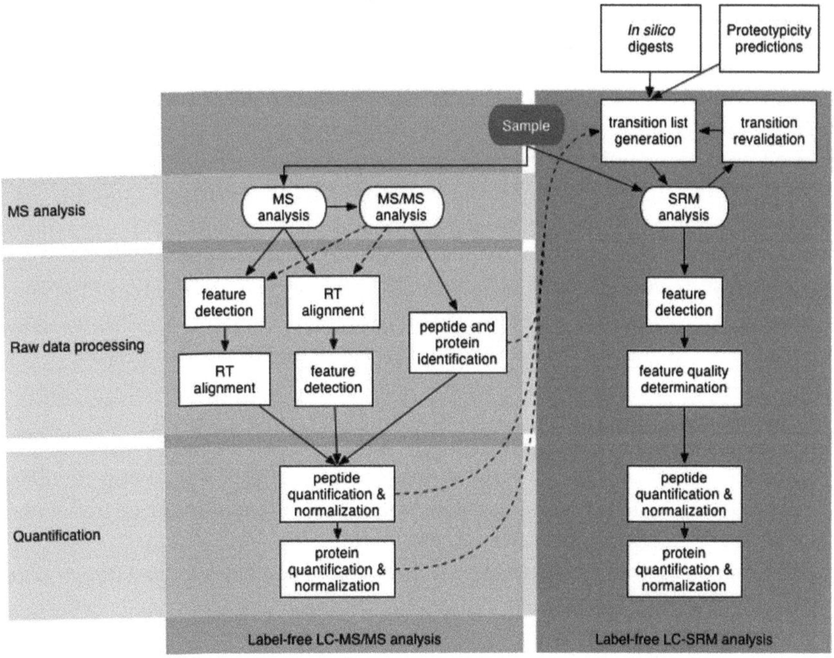

FIGURE 6 Information flow during data processing in label-free LC-MS/MS and LC-SRM. *Reproduced with permission from Ref. [158], Figure 1, p. 30. Copyright 2013 Elsevier.*

quantification methods would benefit from the faster acquisition cycle available in the last generation mass spectrometers. For further details on these subjects, the reader may consult some specific publications [140,155–158].

In many applications, quantitative relative proteomics is the first step of biomarker development (see Figure 6). After selection of candidate biomarkers, there is the need to develop analytical methods to provide protein absolute quantitative data of controlled accuracy and precision. Usually, LC-MS/MS quantitative methods for small molecules use triple quadrupole (QqQ) instruments operated in selected reaction monitoring (SRM) acquisition mode and internal standards (ISs) to obtain optimum selectivity, sensitivity, and precision in the detection and quantification of analytes. In shotgun proteomics, the most convenient solution is to select one (or more, if available) proteotypic peptide and a relative IS. With the last generation QqQ instruments it is possible to select tens of SRMs per run without loss in sensitivity, thus it is possible to analyze many selected proteotypic peptides contemporaneously in the same sample. Synthesized isotopic ISs, although expensive, are the best way to obtain very reliable data. These peptides are typically added to the extracts after enzymatic digestion, therefore, do not correct for systematic or accidental errors in sample processing during crucial steps of the analysis, such as protein extraction, sample purification, and trypsin digestion [159].

3. FOOD SECURITY

As a consequence of the pressure on the environment that results from the impact of climate change and expanding population, there is a decrease in the amount of land suitable for crop production. Currently, frontier research in agriculture has focused on understanding the influence of genetics on production capacity and phenotypes through the elucidation of the respective genomic sequences and annotation. However, understanding the gene functions is only the beginning: genes are transcribed and translated to proteins that have a greater impact on the phenotype of the biological system, then the phenotype of the plant or animal is influenced by the genotype and by the environment, they live in. The equation: genotype + environment = phenotype [160] applies to every form of life.

3.1 Plant Proteomics

Proteomics contributes to the food security issue mainly by studying the effect of biotic and abiotic stresses on gene expression in plants and to understand postharvest events and food spoilage processes.

Proteomics of plants is a complex matter, due to their genomic extension and polyploidy as well as to the fact that only a few plant genomes have been extensively sequenced. The model plants are species for which the whole genome sequence has been characterized; therefore, databases contain almost all their protein sequences, allowing a large-scale proteomic study [161]. The most studied plant models in proteomics are *Arabidopsis thaliana*, whose small genome was fully sequenced in 2000 [162], followed by rice (*Oryza sativa*), whose genome was fully sequenced in 2002 [163,164]. For many other economically important species, such as maize, wheat, soybean, tomato, potato, banana, etc., several proteomic studies have been performed even if their genomes have not yet been fully sequenced, and their proteomes, largely due to the transfer of proteomics data obtained from model plants, are now available [165]. Since proteins are mostly conserved between related species, studying proteome of plants lacking genome or sufficient expressed sequence tag, sequence information for protein identification is to some extent still possible. The National Center for Biotechnology Information (NCBI, http://www.ncbi.nlm.nih.gov/) and the Universal Protein Resource (UniProt, http://www.uniprot.org/) have developed the major, most comprehensive, and freely available sequence databases. However, these databases differ greatly both in the total number of available protein sequences and in the number of entries available for the same species with a known genome sequence. For some plants, species-specific sequence databases are also available [165]. For example, The Arabidopsis Information Resource (TAIR, http://www.arabidopsis.org/) maintains a database of genetic and molecular biology data for the model higher plant *A. thaliana.* What happens for most of the plants is that only a limited number of peptides identified may be assigned to a protein based on homology and that most of the proteins can be identified with a single peptide sequence [67].

Relative quantitative proteomics in plants do not substantially differ from that of other organisms. Nevertheless, challenges in protein analysis of plant tissues are the typically low protein concentrations, the requirement of cell wall disruption to extract proteins, high content of polysaccharides, nucleic acids, and polyphenols that need to be removed since they can affect protein extraction and fractionation. Removal of these compound groups can be achieved by selective precipitation of proteins, which can also help to increase the protein concentration in the samples [166]. The 2D-PAGE, as well as isotopic labeling or label-free shotgun proteomics method could be employed [167], indeed, the use of label-free shotgun proteomics is usually limited to model and well-annotated species.

3.2 Stress Effects on Gene Expression

Plant growth is greatly affected by both biotic (pathogens) and environmental (drought, salt, oxidative, pollution, etc.) stresses and this is a particular concern in agriculture, where stress-related alterations in plant development, growth, and productivity can determine economic losses.

In order to maintain adequate proliferation, plants have developed certain adaptive responses to external stresses, including a network of enzymes involved in the detoxification of excess reactive oxygen species (ROS) which accumulate during an imbalance of cellular homeostasis. Among the species, model plants, such as *A. thaliana* and *O. sativa*, are the most extensively studied, followed by *Triticum estivum, Triticum durum, Zea mays, Glycine max, Hordeum vulgaris,* and *Sorghum bicolor*, which are economically important plants, phylogenetically near to *O. sativa*, and with published genomes [168].

3.2.1 Abiotic Stress

An extended overview of the effect of many abiotic stresses on plant proteomes can be found in some recent reviews [169–174]. There are some common aspects of plant response to abiotic stresses at proteome level. Under optimum conditions, plant metabolism is aimed at biosynthesis of cellular components and cell division, which are associated with active growth and development. Stress conditions redirect cellular metabolism to the establishment of a new cellular homeostasis underlying plant stress acclimation. The process of acclimation is associated with a biosynthesis of several stress-protective compounds, therefore the acclimation causes increased strain on energy metabolism. As a consequence, an increased tolerance to a given stress factor is usually observed in those plants able to maintain enhanced photosynthesis and carbon assimilation rates. Photosynthesis- and metabolism-related proteins could be down- or upregulated in different ways and extent in stress-sensitive or stress-resistant species and varieties. Another characteristic feature of various stresses is an increased risk of oxidative damage. Enhanced ROS formation induces enhanced expression of ROS scavenging enzymes, mainly enzymes participating in ascorbate–glutathione cycle or glutathione-S-transferase. Another protein class conferring abiotic stress tolerance is involved in signal transduction, like calmodulins and related family members.

Drought and salinity are the major environmental constraints on plant growth and productivity. Desertification and salinization are rapidly increasing on a global scale and currently affect more than 10% of arable land, which results in a decline of the average yields of major crops greater than 50%. Although irrigation is often used as a remedy to supplement inadequate rainfall in drought prone areas, overirrigation of arable lands may increase soil salinity in the long term, thus worsening the situation. Increasing salinity leads to a reduction and/or delay in germination of plants and death of seeds before germination. Moreover, salt stress affects a wide variety of physiological and metabolic processes in plants in their vegetative stages leading to growth reduction. Stress because of salinity generally involves osmotic, ionic, and oxidative stress. The initial growth reduction is due to osmotic effect of salt outside roots and subsequent growth reduction is due to ionic stress, which is caused by the inability to prevent salt from reaching toxic levels in transpiring leaves. Plants growing naturally on saline soils have evolved various mechanisms to cope with negative effects of ionic stress. These mechanisms rely on controlled uptake, exclusion, compartmentalization, and increased extrusion of salts.

In addition to the proteins that are overexpressed in general condition of stress, such as those involved in photosynthesis and detoxification, some cellular processes are altered in specific manner. Drought and soil salinity have the common trait of reduced water uptake by roots; besides, both are common in several hot and dry semiarid regions where agriculture is dependent on irrigation. In addition, increased concentrations of salts in soil water also cause an osmotic stress and an unbalancing in intracellular ion homeostasis.

The largest drought/salinity stress-responsive group of proteins includes several wall hydrolases whose abundance decreases in cell wall upon stress, indicating a reduced cell wall elongation activity, whereas osmotic adjustment also leads to enhanced sequestration of inorganic ions in central vacuole [169–174]. Varieties tolerant to drought [175] and salinity [66] have also been examined. Tolerance appears to be associated with a more efficient defense and an improvement of protection from mechanical stress by increased cell wall lignifications. Moreover, drought tolerance seems to be related to the capacity of maintaining primary and secondary metabolism efficiency as well as cell duplication and cellular transport, while salinity tolerance seems to be governed by a higher capacity for osmotic homeostasis.

Seeds are particularly vulnerable to stress encountered between sowing and seedling establishment while salt tolerance in plants usually increases with plant ontogeny. Successful rooting of plants largely depends on successful seed germination. Rapid and uniform field emergence is a fundamental requisite for a good crop establishment, especially under adverse environmental conditions. Seed priming was defined as presowing treatments in water or in an osmotic solution that allows seed to imbibe water to proceed to the first stage of germination, but prevents radicle protrusion through the seed coat. Seed priming with antioxidant compounds such as salicylic acid and ascorbic acid seems to be an efficient method to overcome seed germination problems and to improve seedling growth in the field under salinity. The major processes that presumably play an important role during seed priming can be described as: cell cycle-related event, endosperm weakening by hydrolase activities, mobilization of storage proteins, lipid and starch mobilization, protein synthesis, and methyl cycle. Stress tolerance induced by priming is probably achieved through two strategies: seed priming sets in motion germination-related processes that facilitate the transition of quiescent dry seeds into germinating state and lead to improved potential of germination [58].

3.2.2 Biotic Stress

Various biotic stresses, such as bacteria, viruses, fungi, and insects, contribute to serious economic losses. Plant–pathogen interactions reveal a very complex pathophysiological context, in which resistance, susceptibility, and direct-induced defense reactions interplay to trigger expression responses of hundreds of genes. Most of the work on biotic stress has been done, both in vitro and in vivo, in the model plant *A. thaliana* by using fungi and bacteria as pathogen elicitors [176]. Although not completely applicable to crop species, these studies throw light on the biochemical cross-talking between host and guest, whose secretomes, during attack/defense, will determine the extent of crop damage. Clearly, proteomics has a role in elucidating parts of the defense response, especially in identifying PTMs in signaling pathways. Studies on crop plants such as rice [171] and maize [173] substantially confirmed these findings: briefly, three main categories of proteins appear to be involved in maize kernel resistance: (1) storage proteins, (2) stress-related proteins, and (3) antifungal proteins. At a first glance, the situation is more complex in rice, but this probably depends on more detailed investigations on rice than maize. Other nonmodel, economically relevant plant response to fungal and/or bacterial infections has been studied by proteomics, including tomato [177], banana [178], Chinese cabbage [179], fruit trees [180], etc., but this issue appears to be still in its infancy.

Viruses are relatively diffuse among plants and cause significant losses in economic terms. Symptoms of the infection vary with the host species and/or virus strain and include mosaic, stunt, chlorosis, dwarfing, leaf malformation, and systemic necrosis. This subject, recently reviewed, has not been widely addressed by proteomics tools [181]. The most important and widespread plant viruses are caused by the genus *Tobamovirus, Sobemovirus, Cucumovirus*, and *Potyvirus*. In all, proteomics reveals a widespread repression of proteins associated with the photosynthetic apparatus, while energy metabolism, protein synthesis, and turnover are typically upregulated, indicating a major redirection of cell metabolism. Other common features include the modulation of metabolisms concerning sugars, cell wall, and ROS as well as virus-related capside proteins.

Insect infestation is another source of stress for crops. Surprisingly, although some studies using genomics and trascriptomics are present in the literature, only two [182,183] integrate polymerase chain reaction (PCR)-based protein expression change with proteomics-based ones. Therefore, this field appears to be still open to exploration.

3.2.3 Postharvest Proteomics

About one-third of the food produced is wasted in both developed and developing countries. While in high-income countries, most of losses occur both at the retail and consumer levels, in developing countries, postharvest losses prevail, mainly due to inadequate postharvest and processing practices, and they are a significant portion of the total production. Reduction of postharvest losses would be the easiest and most environment friendly method to increase food availability in poor countries. Although staple crops (cereals, soybean, cassava, etc.) represent the greatest contribution to the food supply of billions of people, also tropical fruits are of great importance in the economy of many developing countries as goods for export. Much knowledge on protein expression change during vegetable development, maturation, and ripening has been acquired using tomato as model [184–186]. More recently, some proteomic studies, conducted on ripening of some classic tropical fruits, have revealed the increase in some defense and stress-related proteins, as well as antioxidant enzymes in mature fruits [187–189]. This information might be useful for selecting both, varieties that are more resistant and appropriate storage/transport conditions. Indeed, one of the most common problems is that after harvesting, fruits and vegetables are subjected to various stresses, both biotic and abiotic, which may compromise the quality and shelf life [17,184]. Understanding the physiopathological change during postharvest processes, would enable the selection of optimal strategies for minimizing product losses. As many commodities are stored at low temperature to retard ripening the cold stress effect has been studied in susceptible and resistant species and varieties. It has been demonstrated, for example, that peach warmed before chilling overexpressed some proteins, about half of them stress related, and that increased cold resistance [190]. Seeds are the staple in the diet of many people. Seeds are usually handled after harvesting in big lots, thus treatment to improve shelf life is difficult to apply, and the perishable nature of the product depends mainly on the conditions of storing.

Senescence and mold development are the diseases that affect the seeds during storage and transport. Appropriate grain and seed storage conditions should guarantee good seed quality and low incidence of mycotoxins. Seed aging proteomics has been studied in *A. thaliana* [191], but translation to other plant species is not obvious and, substantially, seed trade and industry are in need of seed quality markers for characterization of lots and monitoring their state of conservation.

3.3 What about Livestock?

Currently, there is a tendency to consider the breeding of animals a danger rather than a resource for food security. There is no doubt that intensive farming has a significant impact on the environment and consumption of natural resources, as well as planet warming caused by ruminant emissions [192, and references there-in]. In developed countries, animal farming is a large industrial sector, while animals bred in the wild are an exception. The farm level requires the intensive use of feed, of which cereals are a relevant part, and many people believe that these food resources could be better used to feed people directly. While in about 10,000 years of domestication, mother nature-selected animals that are more productive consistent with the environmental conditions, in the last century breeds were optimized either for milk or meat production. This has generated new issues like accumulation of recessive genetic variants in animal populations and health problems, such as udder problems in dairy cattle [193]. Recently, *Journal of Proteomics* published a special issue on farm animal proteomics [194], where in the introducing editorial the guest Editors made an important consideration. They say: "Although there have been enormous advances in the technology of proteomics… there has been limited application of proteomics in farm animal science" [195]. This statement is still valid today, and advanced proteomic studies would help to overcome this gap (see Figure 7).

Another area relating to livestock is aquaculture. Aquaculture experienced rapid growth in the last 30 years, and proteomics has been increasingly used over the last decade as a powerful comparative tool to address different questions in aquaculture [196]. Fish is considered much more healthy food than

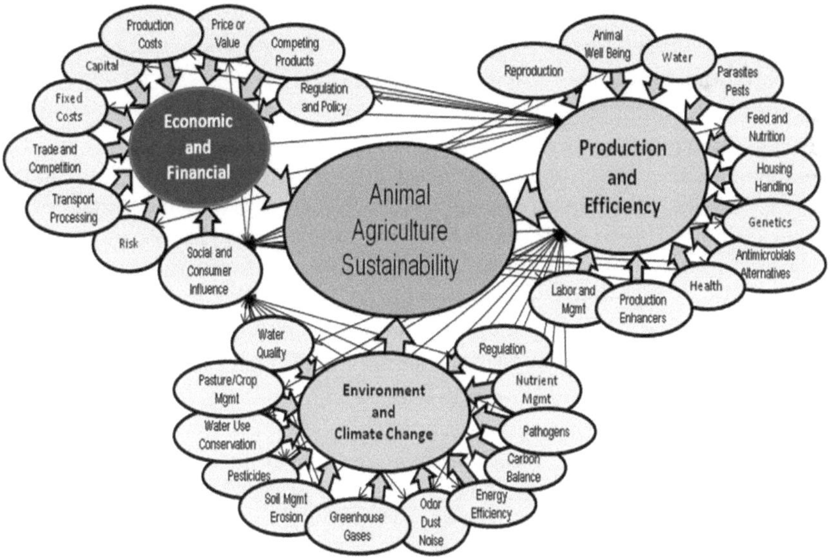

FIGURE 7 Factors related to animal agricultural sustainability. Sustainable animal production is affected by complex factors relating to numerous economic, production, and environmental factors. These factors must be understood to appreciate the complexity of modern agricultural systems and consequently the relationship between genomics–proteomics and production traits. Many of these factors also interact directly or indirectly further adding to the complexity of sustainable agriculture systems. *Reproduced with permission from [160], Figure 1, p. 22. Copyright 2013 Elsevier.*

terrestrial animal meat and fish farming encounters less hostility than intensive livestock farming. However, aquaculture intensification has significant drawbacks, including increased environmental impact due to the large amounts of waste discharged as effluent [197].

4. FOOD SAFETY, QUALITY, AND AUTHENTICATION

Safety, quality, and authentication of food are strictly connected to each other. As many subjects can relate to at least two of these three categories, they will be treated together in this section. Moreover, the very important issues of food allergy and intolerance, and food pathogens, which represent the most important concern of food safety, are treated in Chapters 8 and 14, respectively, of this book. It is noteworthy that the large majority of proteomic studies in these fields has been performed with gel-based techniques.

4.1 Genetically Modified Organisms

The safety of transgenic food derived from GMOs is still controversial, not only within consumers, but also within the scientific community. The potential presence of unplanned changes, which could produce secondary effects of gene expression caused by the insertion of alien genes, is a major concern. The Organisation for Economic Co-operation and Development in 1993 introduced the concept of "substantial equivalence." According to this concept, GMOs should be compared to the traditional varieties to evaluate whether they have the substantial equivalent components. This strategy is based on the assumption that traditional varieties having a history of safe use can serve as comparison for safety assessments of new GMOs derived from that variety. After several years of intense debate, strict regulations on different aspects of GMOs, including risk assessment, marketing, labeling, and traceability have been currently established in the European Union and other countries. The principle of maximum prudence has been adopted by these nations and consumers have now the right to know if the product they are purchasing contains GMOs. In this context, proteomics, able to quantify hundreds of proteins in parallel, has emerged as a very useful approach in the study of differentially expressed proteins in transgenic food and has raised interest in safety assessment [168,198]. All the techniques and strategies for protein analysis reported in Section 1 have been employed, however, 2D-PAGE to separate complex protein mixtures, image analysis to compare gels, and MS to determine the identity of the differentially expressed proteins are the most commonly used [199].

4.2 Food Processing and Quality Control

At first, the application of proteomics to food science was characterized by exploratory analyses of food and beverages of various origins, instead, more recently, it has been successfully applied to the study of quality control in food production processes and also of food safety. Thermal processes (refrigeration, freezing, heating, spray drying, pasteurization, and sterilization) are the most widely used operations among different processing treatments that are applied in food manufacturing. Other treatments include mechanical (homogenization, filtration, clarifying), chemical (blanching), biological (fermentation), and passive (maturation, ripening). Such treatments aim to improve the safety and to extend the shelf life of food, as well as to improve the organoleptic and nutritional characteristics of foodstuffs. Sometime, due to the variability of raw materials and processes used, unwanted chemical reactions can take place among the main components (proteins, lipids, and carbohydrates) and, depending on the severity and time of the treatment, detrimental effects on the quality can occur [200] and it is at this stage that proteomics studies are fundamental.

4.2.1 Meat

The crucial step in meat production is the transformation of live animal tissues into an edible meat product. The conversion from muscle to meat is a complex biochemical process and, because meat maturation and processing are accomplished through many biochemical steps and through changes in protein abundance and structure, the use of proteomics is of paramount relevance for understanding the conversion of muscle to meat [193]. These changes, referred to as *postmortem* changes, occur when an active cellular metabolism in absence of oxygen supply and metabolite removing by blood circulation take place. Nevertheless, other factors, related to animal genotype, sex, age, husbandry, nutrition, weight, stress status, etc., concur to significantly influence the conversion of muscle to meat, and then its tenderness and meat quality [201].

4.2.1.1 Beef Cattle Meat

D'Alessandro and Zolla in their review [202] divided the *postmortem* biochemical processes, which involve both sarcoplasmic and myofibrillar proteins, into seven steps, although biochemical investigations suggest that, the boundaries between these steps appear more labile than previously believed. In the first step, blood supply is lost, thus oxygen and nutrients no longer reach the cells, the lactate, the end product of anaerobic metabolism, accumulates, and pH decrease. In the second step, phosphocreatine, which acts as substrate for ATP reconstitution, wears off quickly and energy is mainly produced through degradation of glycogen. Some proteomic studies reported a positive correlation between some glycolytic enzyme levels and meat tenderness, though, it has been also proposed that this might reflect a technical bias due to altered protein solubility resulting from lower pH in *postmortem* muscles. Moreover, only certain enzyme isoforms have been shown to correlate with meat tenderness. As a consequence of lactate accumulation, pH should further decrease. However, this is probably an oversimplification, and other phenomena, such as phospholipids transformation, altered membrane potential through phosphatidylserine externalization, redistribution of ions, which, in turn would reflect the ongoing of apoptosis-like phenomena, and enzyme phosphorylation takes place. The third step is mainly characterized by some heat shock proteins (chaperons) which protect proteins from denaturation/digestion, enhancement, and activation by means of a high level of phosphorylation. This and other proteins' level/activity enhancement, low pH and calcium dysregulation lead, ultimately, to diffuse cell apoptosis. In the step four, there is the onset of rigor mortis, as a consequence of the end of energetic metabolisms, and the apoptotic cascade described above and caused by the formation of cross-bridges between myosin and actin filaments. If the process stopped at this point, the meat quality would be scarce. From a proteomics point of view, the tenderization (fifth step) is less defined. It is a multienzymatic process, which implies the activation of some proteolytic systems, many of these are Ca^{2+} dependent, such as calpains, calpases, etc. The effect is to some extent counterintuitive and is based on the effect (myofibril fragmentation) observation more than on proteomics evidence. This fact does not necessarily mean that proteomics is not suited to directly investigate this phenomenon; it is likely that more sophisticated technologies than those employed till now will be applied to point out the complexity of the mechanism. Step six takes into consideration the effect of oxidative stress in promoting the myofibril degradation. The anaerobic *postmortem* ambient favors the formation of ROS species that in turn can oxidize proteins resulting in browning of the meat. The presence of free radical scavenger positively influences both color and tenderness, probably exerting a protective effect also on proteolytic enzyme activity. In step seven, the degradation process of myofibrillar proteins continues; especially troponine T isoforms are cloven to small peptides that could serve as a biomarker of meat maturation. Further storage results in liquid loss and darkening of meat color.

4.2.1.2 Swine Meat

Pig meat is the most widely consumed throughout the world [201]. The use of proteomics in pork production was highly exploited and among the functions related to farm, animal production is the most developed sector. Proteomic applications range from biomarkers of muscle development to the establishment of indicators of stress at transport and slaughter or the detection of illegal growth promoters or the administration of antibiotics before their preslaughter mandatory withdrawal [203]. As for beef, proteomics have also been largely exploited to interpret the molecular mechanisms associated with the *postmortem* biochemical modifications, and findings did not differ substantially, though some differences are reported for different breeds [201].

Unlike beef, in many countries pork is mainly used for the production of cooked and of cured, uncooked products to be consumed after quite a long aging. While pork meat for fresh consumption should have the same characteristics as of beef, the manufactured products have completely different requirements.

Recently, some proteomic studies have been devoted to the determination of change in protein pattern during the transformation process of pork meat in cooked or dry-cured ham to support the industrial processing. Cooked ham industrial manufacturing involves three steps: (1) brine solution imbibitions by a multineedle syringe; (2) homogenization in a refrigerated stainless steel container; (3) assembly of meat and cooking in steam chambers to approximately 70 °C, followed by slow cooling, packaging, and pasteurization. Variables may be brine salt concentration and additive, as well as process times and temperatures. Ham texture depended on the process variables and was found to depend on the different types and quantity of myofibrillar proteins (myosin, tropomyosin, and actin related) which were extracted. Dry-cured ham is prepared by treatment with dry salt that can vary from region to region. External, semimembranous, and internal, biceps femoris, muscles were influenced differently by salt treatment and proteolysis as well as protein extraction depending on the ion strength influenced the protein composition. The ripening process has also been studied by proteomics [201]. This phase is characterized by an intense proteolysis activity of both myofibrillar and sarcoplasmic proteins, which affects several quality parameters of the final product. Changes in protein composition depend on the processing of the product, but also on the genotype of the pig. In particular, the PRKAG3 gene mutation that causes glycogen hyperaccumulation in muscle dramatically affects the *postmortem* metabolism and thereby the ultimate muscle pH decreases.

4.2.1.3 Other Farm Animals

There is a clear disparity between the importance of chicken from the point of view of the food consumption and the amount of proteomic studies devoted to it. The soluble fraction (metabolism related) of proteins was investigated during a growth period of about a month and, not surprisingly, a steady increase of glycolytic enzymes was reported [204]. Another study focused on the characterization of the chicken muscle response to restraint and transport, by means of transcriptomics, proteomics, and metabolomics: the response to stress resulted in reduced levels of glycolytic and gluconeogenic metabolites and reinforcement of muscle myofibrils [205]. The problem of transformation of carcass into meat is less important than in other animals; notwithstanding, as for beef and pork, proteomics reveal a role of glycolytic enzymes in meat quality, and, by 2D-PAGE fingerprint and principal component analysis, it is possible differentiate three chicken breeds [201].

As for rabbit (*Oryctolagus cuniculus*), proteomics has been extensively used as a model for larger species of mammals, including humans, for biomedicine and physiology studies. Nevertheless, little information can be found in the literature on the use of proteomics to study rabbit production for food. The first proteome mapping of a rabbit muscle using 2D-PAGE and peptide mass fingerprinting was effected in 2009, followed by a more detailed study on the proteome variation on response to a restricted diet and weight loss [206]. Results show that some enzymes of the energy metabolism are differentially expressed in groups of experimental animals subject to a restricted diet. As regards small ruminants, proteomics has had very limited applications and only two examples of classical muscle proteomics are to be found for sheep, whereas for goat, no proteomics studies have been described. In order to investigate the molecular basis of muscle fiber type-related composition, 2D-PAGE and MALDI-TOF have been used to characterize the expression of sarcoplasmic proteins in four different skeletal muscles in sheep. It was determined that muscles may be differentiated according to the expression profile of proteins involved in oxidative metabolism, oxidative stress, and protein turnover. Finally, a hypertrophic ovine breed showed upregulation of enzymes involved in glycolytic and oxidative metabolism, while also some stress-related proteins were overexpressed [201].

4.2.1.4 Processed Meat and Fraud

Food authentication is one of the major areas concerning food quality and safety. Several regulations have been implemented to assure correct information and to avoid species substitutions.

A form of fraud widespread is to add ingredients of lower commercial value to products based on meat of high quality. Adding chicken or turkey to beef or pork is a common practice for some products; in this case, the fraud could be to declare on the label values that do not correspond to reality. Although it is both an economic and ethical issue, the literature in this field is surprisingly scarce. Detection of frauds in meat products with DNA-based methods is not suitable for all products or types of frauds; for example, when meat has been subjected to thermal processes, DNA-based methods fail. Similarly, methods based on antibody recognition may lose specificity and sensitivity due to thermal modification of the epitopes. Thermal processes involve protein alteration to a greater or lesser extent: oxidative modifications, such as carbonylation, thiol oxidation, aromatic hydroxylation, and reaction of sugars with AA side chains are the protein modifications most frequently reported in foodstuffs that have been subjected to thermal processing. Condensations and eliminations of side chains or peptide backbone breakdown have also been described [207].

Proteomic studies on raw and processed meat of different origin make possible the discovery of protein biomarker useful for meat authentication. In a study conducted on both raw and cooked meat, it was found that three myosin light chains are species-specific for both fresh and cooked meat, and two proteotypic peptides can be selected for the accurate quantitative determination of chicken meat present at concentration as low as 0.5%, by using a stable isotopic IS [56]. Typical proteins that can be used to discover adulteration of meat with soybean proteins have also been reported [208]. In a more recent study, proteins differentiating the six species (cattle, pig, chicken, turkey, duck, and goose) and relatively stable during the meat aging and only slightly degraded in readymade products were investigated. Some observed species-specific differences in protein expression in raw meat were retained in processed products after finishing the entire technological process. Regulatory proteins, metabolic enzymes, some myofibrillar and blood plasma proteins were identified, but applications were not reported [209]. However, in principle, the quantitative measurement of an undeclared product present in a certain food may be performed using the system of the proteotypical peptides selection and isotopic dilution.

4.2.2 Farm Animal Milk, Dairy Products, Eggs

Milk contributes to postnatal development of the newborn, however, due to lactase persistency as an adaptive evolution in humans, apart from the cases of intolerance, animal milk and dairy products are nutrients for the entire human life cycle. Since the protein fraction constitutes the most nutritionally relevant component in milk, proteomics application to milk studies has acquired major importance. Among milks, bovine milk has the broadest commercial interest, but other animal milks are largely employed in the dairy product industry, especially to make cheese. Bovine and ovine milk contain about 32 g/L of proteins, of which about 80% are caseins. Sheep milk contains more fats and more proteins than cow milk. Human milk has a comparatively lower protein concentration (about 16 g/L) of which caseins are only 35%. From the protein content point of view, donkey milk is the most similar to human. Caseins are present in micellar form, while the other proteins are water soluble (whey proteins) or fat soluble milk fat globules (MFGs). By centrifugation at about 3000 rpm, it is possible to separate fats and MFGs, while by ultracentrifugation, caseins and soluble proteins can be separated from each other [210].

Most of the proteomic investigations on milk have been done by 2D-PAGE followed by MALDI-TOF identification, probably because 2D-PAGE is particularly suitable for separation of protein isoforms and PTMs. However, more recently shotgun proteomics has also been employed. The two techniques can be considered complementary, as hydrophobic peptides with larger masses were preferentially detected by RP-LC-ESI/MS, whereas smaller and basic peptide ionization was favored by MALDI [210]. Shotgun proteomics has been used to characterize the MFG membrane proteome of two enriched fractions (from whey and from butter) [211]. In this case, the purpose was to obtain the in deep proteome characterization of samples that very likely generate a consistent fraction of hydrophobic peptides without paying attention to PTMs.

Caseins are the most abundant proteins in milk and their composition and PTMs are very important for the dairy industry. For this reason casein fraction composition of milks has been extensively studied by proteomics [210, and references there in]. MFGs, produced by the mammary gland, have a membrane composed of phospholipids and 25–75% of proteins. Composition of MFG membrane proteins are of high interest because their protein content and composition differ between farm animal species and may contain information about the health status of the animal. Whey milk proteins have been extensively studied in humans, but some works have also been done for milk producing farm animals. The protein composition of whey is different for different species, reflecting the requirement of the newborn. These differences might be important in determining the composition of BPs (see next section).

Mastitis, inflammation of the mammary gland caused by various etiological agents is a very diffuse disease in lactating farm animals, which cause a qualitative and quantitative variation in milk production and a consequent economic loss. Many efforts have been made to discover early biomarkers, and proteomics has been applied to the differentiation of healthy and mastitic bovine [211] and ovine [212] milk, however, putative biomarkers still need to be validated.

Adulteration of milk, i.e., the total or, more common, partial substitution of a more valuable product with a less expensive one is a diffuse practice in making cheese. This form of fraud can in most cases be detected by DNA-based methods. However, also methods based on proteomics could be used [211]. In the case of substitution of fresh milk with UHT or powdered milk, proteomics is the most suitable technique, as thermal treatments cause glycation, lactosylation, oxidation, deamidation, dehydration of proteins, modification that can be easily revealed by simple proteomic analysis [213].

Although egg, especially egg white, is a rich and inexpensive font of high-nutritional value proteins, the literature dealing with this subject is scarce. It is well known that samples, such as egg white, in which few proteins (ovalbumin, ovotransferrin, and ovomucoid) account for approximately 75% of the total protein content, are traditionally difficult to analyze in depth by both gel-based and shotgun proteomics. In 2D-PAGE, their spots overlay those of many less abundant proteins, while MS fails, because the peptides of the few abundant proteins tend to dominate the full mass spectra and are selected for fragmentation by MS/MS over and over again, thereby preventing the detection of less abundant peptides that coelute. This difficulty has been overcome by depleting the most abundant proteins by the so-called "combinatorial peptide ligand library" technology [214]. However, this technology is only amenable to soluble proteins and, in addition, the composition of the proteome results modified in an unknown and unpredictable way, which makes it impossible to determine the absolute quantity of the proteins and introduce a bias in relative quantification. The most recent advancements in MS, with the introduction of instruments showing increased sensitivity and scan speed, enable the determination of about the same number of proteins without the need of depletion [215]. However, neither of the two papers addresses the issue of egg quality. In a recent paper, the effects of temperature and time of storage on albumen protein composition was studied [216]. Some proteins resulted degraded and among them, a putative biomarker of egg quality, i.e., lipocalin protein family decrease, was proposed. In principle, authenticity of chicken eggs is an issue. However, differences in the proteome of chicken breed might be an indication of the egg-based product origin and also the origin of lyophilized eggs used as raw material in the confectionery industry.

4.2.3 Fish and Aquatic Invertebrates

Fish are a substantial portion of the diet in many countries, providing about 40% of the proteins consumed by nearly two-thirds of the world's population [217]. Despite this relevance, the physiology of only a few fish species has been examined extensively. These are small laboratory species established as experimental models for studies in developmental biology, genetics, and environmental toxicology, such as zebra fish (*Danio rerio*), and some species of interest in aquaculture. Zebra fish has been selected as a model organism for biological research to a similar extent as other model species (*A. thaliana, Drosophila melanogaster*, etc.), and has been mainly employed as an experimental model for developmental biology of vertebrates, as well as for human diseases, like cancer or neurodegenerative disorders [218]. Currently, in addition to zebra fish, genomes of the following species are available: cavefish (*Astyanax mexicanus*), cod (*Gadus morhua*), coelacanth (*Latimeria chalumnae*), fugu (*Takifugu rubripes*), medaka (*Oryzias latipes*), platyfish (*Xiphophorus maculatus*), spotted gar (*Lepisosteus oculatus*), stickleback (*Gasterosteus aculeatus*), Tetraodon (*Tetraodon nigroviridis*), Nile tilapia (*Oreochromis niloticus*) (http://www.ensembl.org). Among them, only cod and Nile tilapia are valuable as food.

Due to the depletion of fish stocks, the fishing is not enough to meet the market demand. As a result, aquaculture, which already accounts for about 50% of aquatic animal food for human consumption, is likely to expand even more in the future in order to meet world's health requirements of fish protein. The widespread awareness about the use of scientific knowledge, as well as the emerging technologies to obtain both better farmed organisms and sustainable production, has enhanced the importance of research on seafood biology. Proteomics, as a powerful technique for the study of biological systems and their dynamics in different conditions, has been increasingly used during the last decade to address many different questions related to fish biology. However, with respect to the relevance of the topic, relatively few proteomic studies are reported in the literature. Topics which have been addressed in farmed fish using proteomics include aspects directly related to fertility (quality of gametes, growth of alevins, etc.), fish health (diminishing infections, the avoidance of stress), muscle growth [196,217, and references there in]. The aim of these research studies was to increase the production yield. Proteomic approaches have been also applied in aquatic toxicology by investigating model species of fish and species of economical relevance [219,220].

Farming of some invertebrates, mainly molluscs, is growing in recent years. Aquatic invertebrates are also popular sentinel species used in estuarine and coastal monitoring programs. Nevertheless, to date there are only few proteomic studies regarding these species: aspects such as taxonomy, shell evolution, stress, antibiotic administration and immune response [219], and climatically contrasting environment [221] have been investigated.

Almost all of the proteomic studies reported above have been performed with the 2D-PAGE approach, only three of the cited works used the shotgun strategy [220,222,223]. It can be expected that, with the grooving genomic data available, the second strategy will be increasingly employed, being intrinsically more suitable for less abundant protein identification, as demonstrated in a recent work on the sarcoplasmic fish proteome [224].

Authentication of fish species is one of the major areas concerning seafood quality and safety.

Conventional identification of fish by the examination of their morphological features, a difficult task in the case of closely related fish species, becomes almost impossible between wild and farmed fish of the same species and in processed products. DNA-based techniques suffer from the same limitation, whereas in principle these differentiations could be possible with protein-related techniques. Electrophoretic and immunological methods have been used, but these classical protein-based methods are laborious, time-consuming, and prone to error when some proteins are degraded during food processing. Proteomic-based methodologies have been recently proposed to assess fish authenticity. Differences in aminoacidic sequence of the same protein in different species would be expected, and differences of some protein expression in wild/farmed fish might be possible. To devise a fast, accurate method, a two-phase process has been used: the discovery phase and the target-driven phase [225]. In the discovery phase, 2D-PAGE permits the detection of potential species-specific profiles or protein spots for the discrimination between species, then selected spots are subjected to tryptic digestion and the recovered peptides are analyzed by MS.

MS fingerprints of the proteins define a set of molecular fish authentication markers, relying in the presence or absence of species-specific peptide masses. It is possible the selection of a target protein that presents high thermostability that can be used also for processed food. In the second phase, when peptide biomarkers have been characterized they are directly searched in the digested samples by SRM in a QqQ instrument, or its variant when working with an IT analyzer.

A central objective in fish quality research is to understand the mechanisms involved in quality changes. This can be due to the preharvest biological aspect, to *postmortem* conditions, or to a combination of both. The conditions which were found to influence the fish quality include feed composition, temperature, stress, muscle activity, the slaughter method, storage temperature, and time [196 and references there in]. However, the use of proteomics to get light at a molecular level of the involved mechanisms that are responsible for a certain quality in aquaculture species has been exploited in few published studies [226–229]. Muscle texture is a very important quality aspect of fish that is connected to proteins. *Postmortem* proteolytic tenderization of muscle and bacterial spoilage is one of the most adverse changes related to fish freshness. Tenderization affects mainly myofibrillar and cytoskeletal proteins. The biochemical processes involved in this postmortem texture changes during chilled as well as frozen storage in muscle of different fish species have been extensively studied by proteomics, mainly by using the 2D-PAGE approach [196,225]. There is a consensus on the fact that different proteolytic and glycolytic systems are involved in this process, and several structural proteins, such as myosin and α-actinin, are affected as substrates for these enzymes. However, many important details remain to be discovered, for example, the role of the oxidation by ROS that react with proteins forms carbonyls.

5. FOOD PEPTIDOMICS AND DIGESTOMICS

Similarly, to living organisms, food peptidome may be defined as the whole peptide pool present in a raw material for food preparation or peptides obtained during technological processes for food preparation and/or storage. Food peptidomics covers research concerning the origin of peptidome, its changes during processing and/or storage, the composition and changes in the pool of peptides on the properties of food as well as the methods applied in research dealing with this group of compounds. The area of interests of food peptidomics includes also functional properties and biological activity of peptides.

5.1 Bioactive Peptides

The discovery of naturally occurring functional compounds in raw or processed food is emerging as a subject of topical interest in food sciences and technologies. The interest for these compounds resides in their several biological and nutritional properties, which also include potential health benefits. Typical functional compounds include ω3-fatty acids, vitamins, glucosinolates, phenolic acids, flavonoids, but also proteins and peptides, which, despite their strong effect on the functional and biological activities of food, remained inaccessible before the advent of proteomic technologies.

Food-derived BPs are made up of short AA chain with a known sequence (mainly ranging from 2 to 20 AAs, with some up to 40 residues) that are inactive inside the parent protein. These peptides can be released by GI digestion or during food processing (ripening, fermentation, cooking) and also by storage or in vitro hydrolysis by proteolytic enzymes [230,231]. BPs possess a wide range of biological functions, such as antimicrobial activities [232], blood pressure-lowering effects [233–235], cholesterol-lowering ability [236], antiinflammatory [237], antithrombotic, and antioxidant activities [238,239], etc., and may be used as components in functional foods, nutraceuticals, food grade, biopreservatives, cosmetics, pharmaceuticals, etc. [240–243]. BPs can exert more than one of the 43 biological activities reported in BIOPEP database (http://www.uwm.edu.pl/biochemia/index.php/pl/biopep).

The effects of BPs are typically exerted at the protein level, perhaps due to peptide–protein interactions that can perturb the structural conformation and enzymatic activities. Moreover, peptides can regulate the expression of genes responsible for abnormal signaling pathways but it is unclear if these activities are based on direct peptide–nucleic acid interactions or binding and if inactivation of protein transcription factors that regulate such genes.

Understanding the nature and bioactivity of nutritional peptides means comprehending an important level of environmental regulation of the human genome, because diet is the environmental factor with the most profound lifelong influence on health. In addition, based on their original structures, several of them have been engineered to produce peptide derivatives characterized by peculiar aggregation resistance and/or increased biological activity [244].

5.1.1 Discovery of Bioactive Peptides Encrypted in a Protein Amino Acid Sequence

The analysis of food-derived BPs represents a relatively new subfield of proteomic studies applied to food sciences [245–248]. Natural BPs are ubiquitous in all life kingdoms, nonetheless there are some difficulties in discovering their presence: a food BP may be rare or unique in terms of sequence and modification, and many food genomes are not well annotated. A proteomic and peptidomic approach applied to the study of BPs allows finding peptides of interest, optimizing their production and contributes to understand the interaction mechanisms between receptor and BPs. BPs can be discovered either empirically or by prediction: both the classical hydrolysis strategy and the bioinformatics-driven reversed genome engineering can be used [249].

BPs identification relies on the detection of one particular sequence in its full length. This imposes a challenge because it is not possible to select proteotypic peptides for identification, as usually performed in the case of protein biomarkers. In shotgun proteomics, specific enzymes, usually trypsin, are applied in protein biomarker identification at peptide level, thereby a restrict range of peptides is generated by a protein. Otherwise, BPs are released in food processing, or proteins are left intact up to the final product and the peptides are liberated in situ by the host digestive system or gut microbial enzymes or a combination of both. Therefore, the release becomes a very complex and unspecific mechanism in which several enzymes coming from food processing (e.g., Alcalase, Protamex, Flavorzyme) or from digestive system (e.g., pepsin, trypsin, chymotrypsin, elastase) and having various activities are involved [230]. As a consequence, the typical MS-based proteomic approach used for protein identification must be adapted to the increased complexity of food-derived peptides which can be smaller or larger than the tryptic peptides.

Most of today's known food-derived BPs have been identified by the rather empirical "fractionate-and test" peptide discovery strategy [230,244]. Nevertheless, this approach is very labor-intensive and even extensive fractionation and purification not always enables unambiguous identification of single BP. Moreover, most known BPs are short peptides (2–6 AAs). Such molecules sit at the interface of both the world of proteomics and small molecules and hence represent a challenge in terms of analysis when using current proteomics techniques [250].

In order to circumvent the limitation of the empirical approach, computer-based ("in silico") simulation has been recently applied to discovery bioactive encrypted peptides [245,251,252]. The in silico approach involves the use of information accrued in databases, such as BIOPEP, to determine the occurrence of cryptic BPs in the primary structure of food protein sequences that can be obtained from the protein databases. Bioinformatics software can also be used to generate profiles of peptides by simulating in silico proteolytic enzyme specificities. The peptides resulting from simulated proteolysis can then be matched with BPs in databases for predetermined bioactivities. Given the complex nature of protease–protein interactions it is not guaranteed that the theoretical peptides can be reproduced experimentally; nevertheless, this targeted process can reduce the time required to screen for BPs present in diverse protein sources using several proteases, and can lead to the discovery of new precursors of known BPs. However, as the currently existing databases are often not regularly updated, there could be missed opportunities. Besides, the applicability of this method lies in the availability of genomic data and, for example, for many fishes the genome is still not sequenced.

Therefore, an integrated bioinformatics process has been proposed for the discovery of nonrecognized BPs [244], as shown in Figure 8. After the identification of BPs from protein sets in databases completed in accordance with the empirical approach, the remainder supposed inactive peptides are analyzed in silico for structural patterns, which have been previously associated with known bioactivities.

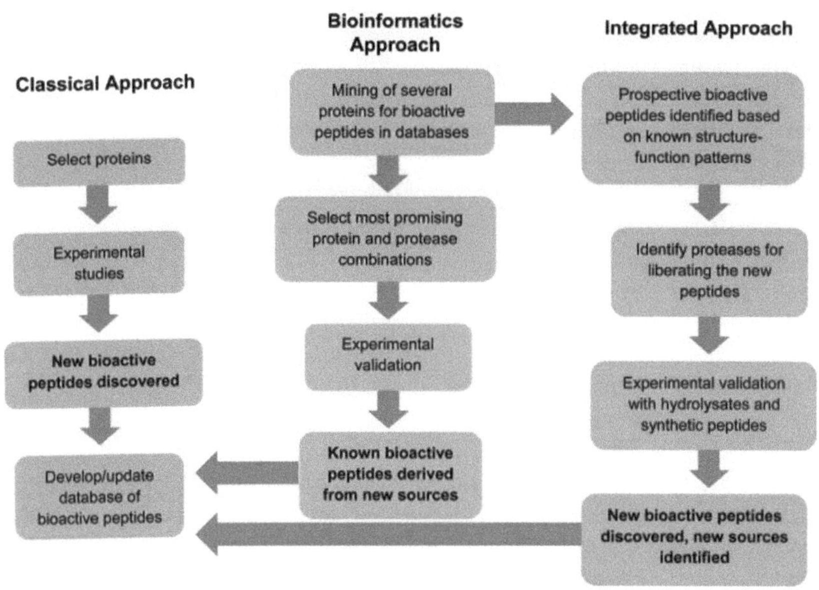

FIGURE 8 Classical, bioinformatics, and integrated approaches toward the discovery of bioactive peptides from food proteins. *Reproduced with permission from [244], Figure 2, p. 139. Copyright 2014 Elsevier.*

Open access PeptideRanker server (http://bioware.ucd.ie/wcompass/biowareweb) is a predictive tool that has the capability to rank large peptide sets, assigning scores (0–1) according to structural patterns that indicate the probability of showing bioactivity [253]. The success of this approach lies on the establishment of structure–activity correlation and should be considered the first step to select the most promising sequence, which would be tested experimentally.

5.1.2 In vitro and In vivo Bioactive Peptides Production

The human GI tract is capable to digest a wide range of proteins of different sources. Normally, the cascade of GI proteolytic and peptidolytic enzymes very efficiently degrades proteins into single AAs or very short peptides. However, there is also evidence that some dietary peptides, which survive luminal digestion and peptidases of the brush border membrane of enterocyte, can be detected in measurable amounts in the peripheral blood and urine. Further hydrolyzed vascular endothelial tissue peptidases and soluble plasma peptidases degrade absorbed peptides and, consequently, the plasma half-life for most peptides is limited to minutes [254]. Bioaccessibility of food-derived peptides involves the assessment of the identity, the relative amount of components released in the GI tract, and reaching the intestinal lumen before absorption or excretion; the bioavailability, instead, represents the proportion of specific components that reach the systemic circulation [9]. The determination of peptides in blood is outside the purpose of this chapter, therefore the subject of this paragraph will deal with bioaccessibility.

The majority of the current knowledge on the digestibility of proteins derives from in vivo or in vitro experiments. Compared to in vivo experiments on laboratory animals and humans, which are less feasible because of intrinsic technical difficulties and ethical constraints, in vitro tests are more versatile and easily practicable.

Human digestion is a complex process wherein ingested food is broken into nutrients and both mechanical and enzymatic processes take place. In the static in vitro models, proteins are sequentially exposed to conditions that simulate mouth, stomach, and intestine environments (different pH values and enzymes) [255]. Static models are an oversimplification of the reality, in which many of the physical processes that occur in vivo are not taken into account. To overcome these limitations, several dynamic gastric models have been developed. Dynamic monocompartmental models simulate gastric digestion, thus just provide a partial insight into GI digestion. The bicompartmental models aim to simulate the luminal conditions of the stomach and proximal small intestine, better simulating the in vivo conditions [255]. In vivo digestion experiments are very few in number and nonhuman model, usually piglets, have been preferably used. Stomach, duodenum, and jejunum (middle small intestine tract) could be sampled by aspiration [256], whereas the inclusion of ileal sample requires the sacrifice of the animals [257].

In this approach, peptides are released from food proteins under conditions, which are very different from each other. In addition, bioactivity of peptides is evaluated by in vitro experiments and in very isolated cases, with suitable human trials. As a result, data about the bioaccessibility of food peptides are often conflicting.

The identification of peptides, from which the bioactive ones can be identified by known sequences, is performed by the widely used method for peptidomics, that include sample preparation, usually by solid phase extraction, and LC followed by LIT MS/MS, Q-TOF MS/or Orbitrap MS/MS.

5.1.3 Bioactive Peptides Sources

Most of BPs have a positive impact on health, whereas some of the longest BPs that escape digestion in GI tract are allergens [258–260], or related to bowl diseases [261]. Until now, almost all of the BPs identified are produced by hydrolysis of the most abundant proteins [9,262,263].

In this context, it is important to consider the BP sources. For example, milk is a very important protein source for humans and, in particular, it is the unique food for newborns, so that they have received particular attention in this field. Milk and dairy products are a rich source of BPs, which may have a wide range of beneficial effects on health.

Fishes are rich sources of structurally diverse bioactive compounds, including BPs. Nonetheless, only few works have been devoted to their research. Some antioxidant and antimicrobial peptides have been isolated from fish sources, but some sequences have not yet been identified [264]. In a recent work, from sarcoplasmic-annotated protein sequences of 15 fish species, 183 encrypted peptides with potential biological activity were determined in silico, but their effective activity should be demonstrated [224]. In another recent work, both sarcoplasmic and myofibrillar proteins were extracted from a sea bass (*Dicentrarchus labrax*) sample; four different sample preparations were tested and a shotgun proteomic approach for peptide identification was used. In the best conditions, from 473 (sarcoplasmic) and 398 (myofibrillar) tryptic peptides, 44 and 18 peptides, respectively, having antimicrobial activity were identified in silico. Moreover, this work demonstrated the importance of sample preparation in BP discovery, especially for peptides, which are encrypted in peptide sequences of less abundant protein [265].

Other sources investigated are meat and meat-based food [266,267], soya milk [268], and hen egg [269], but it is very probable that BPs might be produced from any food.

Foltz and coworkers have done a strong criticism regarding the in vitro testing of BP [254] that underlines that this approach "neglects the poor absorption, distribution, metabolism, and excretion (ADME) properties of peptides resulting in low peptide bioavailability." To this criticism must be added the observation that the gut contains a microbiota codevelop with the host from birth and which interact with the guest gut metabolism [270]. Therefore, bioavailability of nutrients and nutraceuticals is a characteristic that may vary from individual to individual and, as for medicine, there is the possibility to integrating omics platforms toward personalized dietary recommendations [271].

6. CONCLUSIONS AND OUTLOOK

Boggess et al. [160] reported that three obstacles restrain a largest diffusion of proteomics in food science: (1) lack of access to advanced technology, (2) lack of training, (3) lack of adequate funding. Indeed, the first two obstacles may be considered the result of the third: more funding means access to advanced research method and technical facilities, as well as greater appeal in attracting young scientists. With increasing general awareness that "food is the first medicine," we can venture the hypothesis that this obstacle will be gradually removed in the near future. Besides, a further improvement of proteomics and bioinformatics tools can also be expected in a near future, which could overcome the problem of low throughput and the bottleneck of data mining in proteomics.

Protein identification by MS relies heavily on sequence databases. Deep proteomic analysis is therefore more efficient when a large repertoire of gene or protein sequences is available for the species studied. Although sequenced and functionally annotated genomes have recently increased considerably, the number of organisms for which high-quality genome sequence and annotation are available is still lacking and is expected to keep increasing due to progress in high-throughput sequencing techniques [161].

Undoubtedly, the first priority of food research is to ensure sustainable food security for the next generation. Climate change and the decline of fertile arable land contribute in making the problem difficult to deal with, but also a fascinating scientific challenge. An interdisciplinary approach, as well as enterprise funding for public research is needed to ward off the specter of hunger from the future of humanity. Proteomic studies on stress effects and adaption has the potential to contribute actively to promote "the best use of the worst environment" for food production. This objective is far from being reached, i.e., translate proteomics information into agricultural application (see Figure 9). Many works in the field lack validation of the results and there is still no consensus about the standard of quality in proteomic research [272], therefore, there are some contradictory data in the literature and new, high quality data are essential to realize this translation.

Foodomics can be considered as the system biology of food and, as in the system biology, proteomics can play a pivotal role. Food safety, quality, authenticity, and origin issues will benefit of peptide or protein biomarker discovery that enables the selection of targets for developing simpler, high-throughput assays.

Food processing technology and food stability is another field that will benefit of increasing use of proteomics. Protein change during processing,

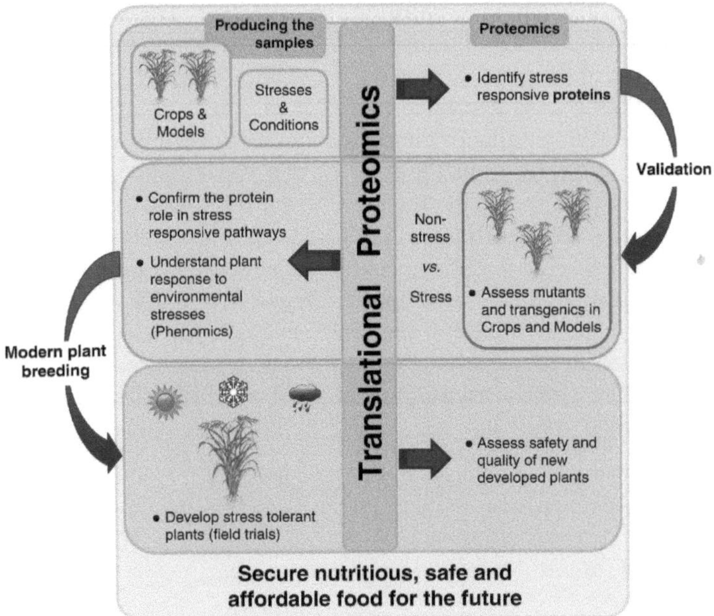

FIGURE 9 Workflow for abiotic stress tolerance improvement in crops, identifying the three main stages: first the experiments leading to protein discovery, then the validation of protein function, and finally the application in modern plant breeding. *Reproduced with permission from [170], Figure 1, p. 148. Copyright 2013 Elsevier.*

fermentation and ripening, enzyme deactivation, and food spoilage are matters of tremendous economic and health impact which are actively looking for new and more effective tools. Processing residues with high protein content (meat and fish cut out, whey, etc.) could be the starting material for production of BPs for dietary supplements. Moreover, antioxidant and antimicrobial natural peptides could revolutionize the industry of food preservation and this fact, avoiding synthetic preservatives, might have a positive effect on health.

Personalized nutrition (adapting food to individual needs) which was almost unthinkable a few years ago appears affordable nowadays. Advanced nutritional research has among its priorities the well-being promotion of the individual by focusing on how diet and dietary components could assess the risk or prevent the onset of diseases, and improve performance [70]. Diet responses vary between individuals due to their genotype, lifestyle, and environment, and the achievement of such information represents one of the most important tasks of nutrigenomics and nutriproteomics.

Intestinal functions, such as digestion, nutrient absorption, barrier integrity, motility, and mucosal immunity, are regulated in a complex fashion. Microbiota develop with the host from birth and is subject to a complex interplay that depends on the host genome, nutrition, and lifestyle. The microbiome has a major impact on human GI digestive tract (as well as immune) function: bacterial colonization of the gut has been shown to alter intestinal physiology of the host by modulating gene expression implicated, inter alia, in nutrient absorption. A deeper understanding of these interactions is a prerequisite for optimizing therapeutic strategies to manipulate the gut microbiome to combat disease and improve health. Gut metaproteomics, which studies the complex proteome of the gut environment, is still in its infancy. In vitro simulation by using intestinal cell culture, in vivo sampling, sample preparation for proteomics and peptidomics, chromatographic/electrophoretic separation, mass spectrometry, and transproteomic pipeline represent a stimulating challenge for the community of scientist involved in food and nutrition. Results are expected for disease prevention, including serious disease typical of GI tract, like Crohn's disease and ulcerative colitis.

REFERENCES

[1] J. Lederberg, A.T. McCray, Scientist 15 (2001) 8–9.
[2] P. James, Q. Rev, Biophys. 30 (1997) 279–331.
[3] A. Pandey, M. Mann, Nature 405 (2000) 837–846.
[4] FAO, Save and Grow. A Policymaker's Guide to the Sustainable Intensification of Smallholder Crop Production, Rome, 2011. http://www.fao.org/ag/save-and-grow/.
[5] FAO, Global Food Losses and Food Waste. Extent, Causes and Prevention. Food, Agriculture Organization of the United Nations, Rome, 2011. http://www.fao.org/docrep/014/mb060e/mb060e00.pdf.
[6] A. Cifuentes, J. Chromatogr. A 1216 (2009) 7109–7110.
[7] M. Herrero, C. Simó, V. García-Cañas, E. Ibáñez, A. Cifuentes, Mass Spectrom. Rev. 31 (2012) 49–69.
[8] G. Panagiotou, J. Nielsen, Annu. Rev. Nutr. 29 (2009) 329–339.
[9] G. Picariello, G. Mamone, C. Nitride, F. Addeo, P. Ferranti, Trends Anal. Chem. 52 (2013) 120–134.
[10] J.I. Gordon, Science 336 (2012) 1251–1253.
[11] J.B. Ewaschuk, H. Diaz, L. Meddings, B. Diederichs, A. Dmytrash, J. Backer, M. Looijer-van Langen, K.L. Madsen, Am. J. Physiol. Gastrointest. Liver Physiol. 295 (2008) G1025–G1034.
[12] N.L. Kelleher, Anal. Chem. 76 (2004) 196A–203A.
[13] G.E. Reid, S.A. McLuckey, J. Mass Spectrom. 37 (2002) 663–675.
[14] A.L. Capriotti, C. Cavaliere, P. Foglia, R. Samperi, A. Laganà, J. Chromatogr. A 1218 (2011) 8760–8776.
[15] M. Černý, J. Skalák, H. Cerna, B. Brzobohatý, J. Proteomics 92 (2013) 2–27.
[16] J.C. Tran, L. Zamdborg, D.R. Ahlf, J.E. Lee, A.D. Catherman, K.R. Durbin, J.D. Tipton, A. Vellaichamy, J.F. Kellie, M. Li, C. Wu, S.M.M. Sweet, B.P. Early, N. Siuti, R.D. Leduc, P.D. Compton, P.M. Thomas, N.L. Kelleher, Nature 480 (2011) 254–258.
[17] X. Han, M. Jin, K. Breuker, F.W. McLafferty, Science 314 (2006) 109–112.
[18] B. Macek, L.F. Waanders, J.V. Olsen, M. Mann, Mol. Cell. Proteomics 5 (2006) 949–958.
[19] J.S. Page, K. Tang, R.D. Smith, Int. J. Mass Spectrom. 265 (2007) 244–250.

[20] R.B. Dixon, D.C. Muddiman, A.M. Hawkridge, A.G. Fedorov, J. Am. Soc. Mass Spectrom. 18 (2007) 1909–1913.
[21] J. Liu, T.Y. Huang, S.A. McLuckey, Anal. Chem. 81 (2009) 1433–1441.
[22] C.K. Fagerquist, B.R. Garbus, W.G. Miller, K.E. Williams, E. Yee, A.H. Bates, S. Boyle, L.A. Harden, M.B. Cooley, R.E. Mandrell, Anal. Chem. 82 (2010) 2717–2725.
[23] J. Klose, J. Hum. Genet. 26 (1975) 231–243.
[24] P.H. O'Farrell, J. Biol. Chem. 250 (1975) 4007–4021.
[25] B. Bjellqvist, K. Ek, P.G. Righetti, E. Gianazza, A. Görg, R. Westermeier, W. Postel, J. Biochem. Biophys. Meth. 6 (1982) 317–339.
[26] J.A. Reynolds, C. Tanford, J. Biol. Chem. 245 (1970) 5161–5165.
[27] A. Görg, W. Weiss, M.J. Dunn, Proteomics 4 (2004) 3665–3685.
[28] S. Hoving, H. Voshol, J. van Ostrum, Electrophoresis 21 (2000) 2617–2621.
[29] R. Wildgruber, A. Harder, C. Obermaier, G. Boguth, W. Weiss, S.J. Fey, P.M. Larsen, A. Görg, Electrophoresis 21 (2000) 2610–2616.
[30] A.W. Dowsey, M.J. Dunn, G.Z. Yang, Proteomics 3 (2003) 1567–1596.
[31] N.L. Anderson, N.G. Anderson, Mol. Cell. Proteomics 1 (2002) 845–867.
[32] M.R. Wilkins, E. Gasteiger, J.C. Sanchez, A. Bairoch, D.F. Hochstrasser, Electrophoresis 19 (1998) 1501–1505.
[33] G.L. Corthals, V.C. Wasinger, D.F. Hochstrasser, J.C. Sanchez, Electrophoresis 21 (2000) 1104–1115.
[34] E. Marengo, E. Robotti, M. Bobba, Curr. Proteomics 4 (2007) 53–66.
[35] M. Ünlü, M.E. Morgan, J.S. Minden, Electrophoresis 18 (1997) 2071–2077.
[36] A. Alban, S.O. David, L. Bjorkesten, C. Andersson, E. Sloge, S. Lewis, I. Currie, Proteomics 3 (2003) 36–44.
[37] J. Shaw, R. Rowlinson, J. Nickson, T. Stone, A. Sweet, K. Williams, R. Tonge, Proteomics 3 (2003) 1181–1195.
[38] A. Görg, W. Weiss, Methods Mol. Biol. 112 (1999) 189–195.
[39] A. Görg, C. Obermaier, G. Boguth, A. Harder, B. Scheibe, R. Wildgruber, W. Weiss, Electrophoresis 21 (2000) 1037–1053.
[40] S.P. Gygi, G.L. Corthals, Y. Zhang, Y. Rochon, R. Aebersold, Proc. Natl. Acad. Sci. 17 (2000) 9390–9395.
[41] A. Panchaud, M. Affolter, P. Moreillon, M. Kussmann, J. Proteomics 71 (2008) 19–33.
[42] A. Görg, O. Drews, C. Lück, F. Weiland, W. Weiss, Electrophoresis 30 (2009) S122–S132.
[43] T.M. Billeci, J.T. Stults, Anal. Chem. 65 (1993) 1709–1716.
[44] M. Mann, P. Hojrup, P. Roepstorff, Biol. Mass Spectrom. 22 (1993) 338–345.
[45] D.J. Pappin, P. Hojrup, A.J. Bleasby, Curr. Biol. 3 (1993) 327–332.
[46] P. James, M. Quadroni, E. Carafoli, G. Gonnet, Biochem. Biophys. Res. Commun. 195 (1993) 58–64.
[47] J.R. Yates III, S. Speicher, P.R. Griffin, T. Hunkapiller, Anal. Biochem. 214 (1993) 397–408.
[48] M. Mann, M. Wilm, Anal. Chem. 66 (1994) 4390–4399.
[49] J.K. Eng, A.L. McCormack, J.R. Yates III, J. Am. Soc. Mass Spectrom. 5 (1994) 976–989.
[50] K. Kubota, T. Kosaka, K. Ichikawa, J. Chromatogr. B 815 (2005) 3–9.
[51] A. Shevchenko, A. Loboda, A. Shevchenko, W. Ens, K.G. Standing, Anal. Chem. 72 (2000) 2132–2141.
[52] K.F. Medzihradszky, J.M. Campbell, M.A. Baldwin, A.M. Falick, P. Juhasz, M.L. Vestal, A.L. Burlingame, Anal. Chem. 72 (2000) 552–558.
[53] D.N. Perkins, D.J. Pappin, D.M. Creasy, J.S. Cottrel, Electrophoresis 20 (1999) 3551–3567.

[54] W.J. Hurkman, C.K. Tanaka, J. Cereal Sci. 40 (2004) 295–299.
[55] J.K.C. Rose, S. Bashir, J.J. Giovannoni, M.M. Jahn, R.S. Saravanan, Plant J. 39 (2004) 715–733.
[56] M.A. Sentandreu, P.D. Fraser, J. Halket, R. Patel, P.M. Bramley, J. Proteome Res. 9 (2010) 3374–3383.
[57] B.J. Barkla, R. Vera-Estrella, M. Hernández-Coronado, O. Pantoja, Plant Cell 21 (2009) 4044–4058.
[58] A. Fercha, A.L. Capriotti, G. Caruso, C. Cavaliere, H. Gherroucha, R. Samperi, S. Stampachiacchiere, A. Laganà, J. Proteomics 91 (2013) 486–499.
[59] G. Ephritikhine, M. Ferro, N. Rolland, Plant Physiol. Biochem. 42 (2004) 943–962.
[60] A. Bodzon-Kulakowska, A. Bierczynska-Krzysik, T. Dylag, A. Drabik, P. Suder, M. Noga, J. Jarzebinska, J. Silberring, J. Chromatogr. B 849 (2007) 1–31.
[61] C. Cavaliere, F. Cucci, C. Guarino, R. Gubbiotti, R. Samperi, A. Laganà, Rapid Commun. Mass Spectrom. 22 (2008) 1159–1167.
[62] N. Ahmed, G.E. Rice, J. Chromatogr. B 815 (2005) 39–50.
[63] B. Cox, A. Emili, Nat. Protoc. 1 (2006) 1872–1878.
[64] R.R. Burgess, Methods Enzymol. 463 (2009) 331–342.
[65] M. Mohayeji, A.L. Capriotti, C. Cavaliere, S. Piovesana, R. Samperi, S. Stampachiacchiere, M. Toorchi, A. Laganà, J. Proteomics 99 (2014) 101–110.
[66] A.L. Capriotti, G.M. Borrelli, V. Colapicchioni, R. Papa, S. Piovesana, R. Samperi, S. Stampachiacchiere, A. Laganà, Anal. Bioanal. Chem. 406 (2014) 1423–1435.
[67] A.L. Capriotti, G. Caruso, C. Cavaliere, P. Foglia, S. Piovesana, R. Samperi, A. Laganà, Anal. Bioanal. Chem. 405 (2013) 9301–9309.
[68] A.L. Capriotti, G. Caruso, C. Cavaliere, S. Piovesana, R. Samperi, A. Laganà, Anal. Chim. Acta 776 (2013) 57–63.
[69] A.L. Capriotti, C. Cavaliere, P. Foglia, S. Piovesana, R. Samperi, S. Stampachiacchiere, A. Laganà, Anal. Chim. Acta 800 (2013) 36–42.
[70] V. Ganesh, N.S. Hettiarachchy, Biochim. Biophys. Acta 1824 (2012) 1107–1117.
[71] P. Zhu, P. Boweden, D. Zhang, J.G. Marshall, Mass Spectrom. Rev. 30 (2011) 685–732.
[72] O. Chertov, A. Biragyn, L.W. Kwak, J.T. Simpson, T. Boronina, V.M. Hoang, D.A. Prieto, T.P. Conrads, T.D. Veenstra, R.J. Fisher, Proteomics 4 (2004) 1195–1203.
[73] H.B. Krishnan, S.S. Natarajan, Phytochemistry 70 (2009) 1958–1964.
[74] X. Zheng, H. Baker, W.S. Hancock, J. Chromatogr. A 1120 (2006) 173–184.
[75] R.S. Tirumalai, K.C. Chan, D.A. Prieto, H.J. Issaq, T.P. Conrads, T.D. Veenstra, Mol. Cell. Proteomics 2 (2003) 1096–1103.
[76] A.L. Capriotti, G. Caruso, C. Cavaliere, S. Piovesana, R. Samperi, A. Laganà, Anal. Chim. Acta 740 (2012) 58–65.
[77] J.M. Koomen, D. Li, L. Xiao, T.C. Liu, K.R. Coombes, J. Abbruzzese, R. Kobayashi, J. Proteome Res. 4 (2005) 972–981.
[78] L. Hu, K.S. Boos, M. Ye, R. Wu, H. Zou, J. Chromatogr. A 1216 (2009) 5377–5384.
[79] R. Tian, H. Zhang, M. Ye, X. Jiang, L. Hu, X. Li, X. Bao, H. Zou, Angew. Chem. Int. Ed. 46 (2007) 962–965.
[80] J. Marshall, A. Jankowski, S. Furesz, I. Kireeva, L. Barker, M. Dombrovsky, W. Zhu, K. Jacks, L. Ingratta, J. Bruin, E. Kristensen, R. Zhang, E. Stanton, M. Takahashi, G. Jackowski, J. Proteome Res. 3 (2004) 364–382.
[81] Y. Gong, X. Li, B. Yang, W. Ying, D. Li, Y. Zhang, S. Dai, Y. Cai, J. Wang, F. He, X. Qian, J. Proteome Res. 5 (2006) 1379–1387.
[82] R. Pieper, Q. Su, C.L. Gatlin, S.T. Huang, N.L. Anderson, S. Steiner, Proteomics 3 (2003) 422–432.

[83] J. Brand, T. Haslberger, W. Zolg, G. Pestlin, S. Palme, Proteomics 6 (2006) 3236–3242.

[84] L. Sennels, M. Salek, L. Lomas, E. Boschetti, P.G. Righetti, J. Rappsilber, J. Proteome Res. 6 (2007) 4055–4062.

[85] A. Luchini, D.H. Geho, B. Bishop, D. Tran, C. Xia, R.L. Dufour, C.D. Jones, V. Espina, A. Patanarut, W. Zhou, M.M. Ross, A. Tessitore, E.F. Petricoin, L.A. Liotta, Nano Lett. 8 (2008) 350–361.

[86] A. Rainczuk, K. Meehan, D.L. Steer, P.G. Stanton, D.M. Robertson, A.N. Stephens, Proteomics 10 (2010) 332–336.

[87] D. Tamburro, C. Fredolini, V. Espina, T.A. Douglas, A. Ranganathan, G. Muto, A. Luchini, J. Am. Chem. Soc. 133 (2011) 19178–19188.

[88] J.R. Whiteaker, H. Zhang, J.K. Eng, R. Fang, B.D. Piening, L.C. Feng, T.D. Lorentze, R.M. Schoenherr, J.F. Keane, T. Holzman, M. Fitzgibbon, C. Lin, H. Zhang, K. Cooke, T. Liu, D.G. Camp II, L. Anderson, J. Watts, R.D. Smith, M.W. McIntosh, A.G. Paulovich, J. Proteome Res. 6 (2007) 828–836.

[89] J.V. Olsen, M. Mann, Mol. Cell. Proteomics 12 (2013) 3444–3452.

[90] A.M.N. Silva, R. Vitorino, M.R.M. Domingues, C.M. Spickett, P. Domingues, Free Radic. Biol. Med. 65 (2013) 925–941.

[91] B.R. Fonslow, J.R. Yates III, Proteolytic digestion methods for shotgun proteomics, in: J. Pawliszyn (Ed.), Comprehensive Sampling and Sample Preparation, vol. 3, Elsevier, 2012, pp. 261–276.

[92] A.L. Capriotti, G. Caracciolo, C. Cavaliere, C. Crescenzi, D. Pozzi, A. Laganà, Anal. Bioanal. Chem. 401 (2011) 1195–1202.

[93] M. Iwasaki, N. Sugiyama, N. Tanaka, Y. Ishihama, J. Chromatogr. A 1228 (2012) 292–297.

[94] M. Pirmoradian, H. Budamgunta, K. Chingin, B. Zhang, J. Astorga-Wells, R.A. Zubarev, Mol. Cell. Proteomics 12 (2013) 3330–3338.

[95] C. Tu, M.J. Mammen, J. Li, X. Shen, X. Jiang, Q. Hu, J. Wang, S. Sethi, J. Qu, J. Proteome Res. 13 (2014) 627–639.

[96] E.J. Hsieh, M.S. Bereman, S. Durand, G.A. Valaskovic, M.J. MacCoss, J. Am. Soc. Mass Spectrom. 24 (2013) 148–153.

[97] L.M.F. de Godoy, J.V. Olsen, G.A. de Souza, G. Li, P. Mortensen, M. Mann, Genome Biol. 7 (2006) R50–R50.15.

[98] B.J. Cargile, J.R. Sevinsky, A.S. Essader, J.L. Stephenson Jr., J.L. Bundy, Biomol. Tech. 16 (2005) 181–189.

[99] J. Chen, C.S. Lee, Y. Shen, R.D. Smith, E.H. Baehrecke, Electrophoresis 23 (2002) 3143–3148.

[100] G.J. Opiteck, J.W. Jorgenson, R.J. Anderegg, Anal. Chem. 69 (1997) 2283–2291.

[101] M.P. Washburn, D. Wolters, J.R. Yates III, Nat. Biotechnol. 19 (2001) 242–247.

[102] D.A. Wolters, M.P. Washburn, J.R. Yates III, Anal. Chem. 73 (2001) 5683–5690.

[103] K.G. Mawuenyega, H. Kaji, Y. Yamuchi, T. Shinkawa, H. Saito, M. Taoka, N. Takahashi, T. Isobe, J. Proteome Res. 2 (2003) 23–35.

[104] H. Toll, H. Oberachera, R. Swart, C.G. Huber, J. Chromatogr. A 1079 (2005) 274–286.

[105] M. Gilar, P. Olivova, A.E. Daly, J.C. Gebler, Anal. Chem. 77 (2005) 6426–6434.

[106] L. Callipo, A.L. Capriotti, C. Cavaliere, R. Gubbiotti, R. Samperi, A. Laganà, Biomed. Chromatogr. 25 (2011) 594–599.

[107] A. Motoyama, J.R. Yates III, Anal. Chem. 80 (2008) 7187–7193.

[108] J. Tang, M. Gao, C. Deng, X. Zhang, J. Chromatogr. B 866 (2008) 123–132.

[109] A.J. Link, J. Eng, D.M. Schieltz, E. Carmack, G.J. Mize, D.R. Morris, B.M. Garvik, J.R. Yates III, Nat. Biotech. 17 (1999) 676–682.

[110] D. Figeys, A. Ducret, J.R. Yates III, R. Aebersold, Nat. Biotech. 14 (1996) 1579–1583.
[111] G. Choudhary, S.-L. Wu, P. Shieh, W.S. Hancock, J. Proteome Res. 2 (2003) 59–67.
[112] R.E. Higgs, M.D. Knierman, V. Gelfanova, J.P. Butler, J.E. Hale, J. Proteome Res. 4 (2005) 1442–1450.
[113] R.D. Smith, G.A. Anderson, M.S. Lipton, L. Pasa-Tolic, Y. Shen, T.P. Conrads, T.D. Veenstra, H.R. Udseth, Proteomics 2 (2002) 513–523.
[114] H. Chassaigne, J.V. Nørgaard, A.J. Van Hengel, J. Agric. Food Chem. 55 (2007) 4461–4473.
[115] M.D. Stapels, D.F. Barofsky, Anal. Chem. 76 (2004) 5423–5430.
[116] R.J. Cotter, W. Griffith, C. Jelinek, J. Chromatogr. B 855 (2007) 2–13.
[117] S. Pan, R. Aebersold, R. Chen, J. Rush, D.R. Goodlett, M.W. McIntosh, J. Zhang, T.A. Brentnall, J. Proteome Res. 8 (2009) 787–797.
[118] W. Chen, P.J. Lee, H. Shion, N. Ellor, J.C. Gebler, Anal. Chem. 79 (2007) 1583–1590.
[119] A.N. Krutchinsky, M. Kalkum, B.T. Chait, Anal. Chem. 73 (2001) 5066–5077.
[120] Q. Hu, R.J. Noll, H. Li, A. Makarov, M. Hardmanc, R.G. Cooksa, J. Mass Spectrom. 40 (2005) 430–443.
[121] J.V. Olsen, L.M.F. de Godoy, G. Li, B. Macek, P. Mortensen, R. Pesch, A. Makarov, O. Lange, S. Horning, M. Mann, Mol. Cell. Proteomics 4 (2005) 2010–2021.
[122] J.V. Olsen, J.C. Schwartz, J. Griep-Raming, M.L. Nielsen, E. Damoc, E. Denisov, O. Lange, P. Remes, D. Taylor, M. Splendore, E.R. Wouters, M. Senko, A. Makarov, M. Mann, S. Horning, Mol. Cell. Proteomics 8 (2009) 2759–2769.
[123] N. Nagaraj, N.A. Kulak, J. Cox, N. Neuhauser, K. Mayr, O. Hoerning, O. Vorm, M. Mann, Mol. Cell. Proteomics 11 (2012) M111.013722-1–M111.013722-11.
[124] A.S. Hebert, A.L. Richards, D.J. Bailey, A. Ulbrich, E.E. Coughlin, M.S. Westphall, J.J. Coon, Mol. Cell. Proteomics 13 (2014) 339–347.
[125] M.W. Senko, P.M. Remes, J.D. Canterbury, R. Mathur, Q. Song, S.M. Eliuk, C. Mullen, L. Earley, M. Hardman, J.D. Blethrow, H. Bui, A. Specht, O. Lange, E. Denisov, A. Makarov, S. Horning, V. Zabrouskov, Anal. Chem. 85 (2013) 11710–11714.
[126] R.A. Zubarev, A. Makarov, Anal. Chem. 85 (2013) 5288–5296.
[127] A. Kalli, S. Hess, Proteomics 12 (2012) 21–31.
[128] J. Cox, N. Neuhauser, A. Michalski, R.A. Scheltema, J.V. Olsen, M. Mann, J. Proteome Res. 10 (2011) 1794–1805.
[129] R. Higdon, J.M. Hogan, N. Kolker, G. Van Belle, E. Kolker, Omics 11 (2007) 351–365.
[130] A. Keller, A.I. Nesvizhskii, E. Kolker, R. Aebersold, Anal.Chem. 74 (2002) 5383–5392.
[131] A.K. Yocum, K. Yu, T. Oe, I.A. Blair, J. Proteome Res. 4 (2005) 1722–1731.
[132] R.E. Moore, M.K. Young, T.D. Lee, J. Am. Soc. Mass Spectrom. 13 (2002) 378–386.
[133] L. Käll, J.D. Canterbury, J. Weston, W.S. Noble, M.J. MacCoss, Nat. Methods 4 (2007) 923–925.
[134] C. O'Donovan, M.J. Martin, A. Gattiker, E. Gasteiger, A. Bairoch, R. Apweiler, Brief. Bioinform. 3 (2002) 275–284.
[135] B. Boeckmann, A. Bairoch, R. Apweiler, M.C. Blatter, A. Estreicher, E. Gasteiger, M.J. Martin, K. Michoud, C. O'Donovan, I. Phan, S. Pilbout, M. Schneider, Nucleic Acids Res. 31 (2003) 365–370.
[136] A. Garcia Castro, S. Thoraval, L.J. Garcia, M.A. Ragan, BMC Bioinform. 6 (2005) 87.
[137] A. Keller, J. Eng, N. Zhang, X. Li, R. Aebersold, Mol. Syst. Biol. 1 (2005) 0017.
[138] P.M. Palagi, P. Hernandez, D.Walther, R.D. Appel, Proteomics 6 (2006) 5435–5444.
[139] J.E. Elias, W. Haas, B.K. Faherty, S.P. Gygi, Nat. Methods 2 (2005) 667–675.
[140] Y. Perez-Riverol, R. Wang, H. Hermjakob, M. Müller, V. Vesada, J.A. Vizcaíno, Biochim. Biophys. Acta 1844 (2014) 63–76.

[141] D. Shteynberg, A.I. Nesvizhskii, R.L. Moritz, E.W. Deutsch, Mol. Cell. Proteomics 12 (2013) 2383–2393.
[142] M.R. Hoopmann, R.L. Moritz, Curr. Opin. Biotechnol. 24 (2013) 31–38.
[143] J. Listgarten, A. Emili, Mol. Cell. Proteomics 4 (2005) 419–434.
[144] H. Lam, R. Aebersold, Methods 54 (2011) 424–431.
[145] L. Pasa-Tolic, P.K. Jensen, G.A. Anderson, M.S. Lipton, K.K. Peden, S. Martinovic, N. Tolic, J.E. Bruce, R.D. Smith, J. Am. Chem. Soc. 121 (1999) 7949–7950.
[146] Y. Oda, K. Huang, F.R. Cross, D. Cowburn, B.T. Chait, Proc. Natl. Acad. Sci. 96 (1999) 6591–6596.
[147] S.P. Gygi, B. Rist, S.A. Gerber, F. Turecek, M.B. Gelb, R. Aebersold, Nat. Biotechnol. 17 (1999) 994–999.
[148] M. Wilm, Proteomics 9 (2009) 4590–4605.
[149] K. Gevaert, F. Impens, B. Ghesquière, P. Van Damme, A. Lambrechts, J. Vandekerckhove, Proteomics 8 (2008) 4873–4885.
[150] P.L. Ross, Y.N. Huang, J.N. Marchese, B. Williamson, K. Parker, S. Hattan, N. Khainovski, S. Pillai, S. Dey, S. Daniels, S. Purkayastha, P. Juhasz, S. Martin, M. Bartlet-Jones, F. He, A. Jacobson, D.J. Pappin, Mol. Cell. Proteomics 3 (2004) 1154–1169.
[151] P.V. Bondarenko, D. Chelius, T.A. Shaler, Anal. Chem. 74 (2002) 4741–4749.
[152] D. Chelius, P.V. Bondarenko, J. Proteome Res. 1 (2002) 317–323.
[153] H. Liu, R.G. Sadygov, J.R. Yates III, Anal. Chem. 76 (2004) 4193–4201.
[154] M.M. Matzke, J.N. Brown, M.A. Gritsenko, T.O. Metz, J.G. Pounds, K.D. Rodland, A.K. Shukla, R.D. Smith, K.M. Waters, J.E. McDermott, B.J. Webb-Robertson, Proteomics 13 (2013) 493–503.
[155] M. Bantscheff, M. Schirle, G. Sweetman, J. Rick, B. Kuster, Anal. Bioanal. Chem. 389 (2007) 1017–1031.
[156] K.A. Neilson, N.A. Ali, S. Muralidharan, M. Mirzaei, M. Mariani, G. Assadourian, A. Lee, S.C. van Sluyter, P.A. Haynes, Proteomics 11 (2011) 535–553.
[157] D.A. Megger, T. Bracht, H.E. Meyer, B. Sitek, Biochim. Biophys. Acta 1834 (2013) 1581–1590.
[158] M. Sandin, J. Teleman, J. Malmström, F. Levander, Biochim. Biophys. Acta 1844 (2014) 29–41.
[159] J. Villanueva, M. Carrascal, J. Abian, J. Proteomics 96 (2014) 184–199.
[160] M.V. Boggess, J.D. Lippolis, W.J. Hurkman, C.K. Fagerquist, S.P. Briggs, A.V. Gomes, P.G. Righetti, K. Bala, J. Proteomics 93 (2013) 20–39.
[161] A. Champagne, M. Boutry, Proteomics 13 (2013) 663–673.
[162] S.C. Carpentler, B. Panis, A. Vertommen, R. Swennen, K. Sergeant, J. Renaut, K. Laukens, E. Witters, B. Samyn, B. Devreese, Mass Spectrom. Rev. 27 (2008) 354–377.
[163] J. Yu, S. Hu, J. Wang, G.K. Wong, S. Li, B. Liu, Y. Deng, L. Dai, Y. Zhou, X. Zhang, M. Cao, J. Liu, J. Sun, J. Tang, Y. Chen, X. Huang, W. Lin, C. Ye, W. Tong, L. Cong, J. Geng, Y. Han, L. Li, W. Li, G. Hu, X. Huang, W. Li, J. Li, Z. Liu, L. Li, J. Liu, Q. Qi, J. Liu, L. Li, T. Li, X. Wang, H. Lu, T. Wu, M. Zhu, P. Ni, H. Han, W. Dong, X. Ren, X. Feng, P. Cui, X. Li, H. Wang, X. Xu, W. Zhai, Z. Xu, J. Zhang, S. He, J. Zhang, J. Xu, K. Zhang, X. Zheng, J. Dong, W. Zeng, L. Tao, J. Ye, J. Tan, X. Ren, X. Chen, J. He, D. Liu, W. Tian, C. Tian, H. Xia, Q. Bao, G. Li, H. Gao, T. Cao, J. Wang, W. Zhao, P. Li, W. Chen, X. Wang, Y. Zhang, J. Hu, J. Wang, S. Liu, J. Yang, G. Zhang, Y. Xiong, Z. Li, L. Mao, C. Zhou, Z. Zhu, R. Chen, B. Hao, W. Zheng, S. Chen, Z. Guo WLi, L. Mao, C. Zhou, Z. Zhu, R. Chen, B. Hao, W. Zheng, S. Chen, W. Guo, G. Li, S. Liu, M. Tao, J. Wang, L. Zhu, L. Yuan, H. Yang, Science 296 (2002) 79–92.

[164] S.A. Goff, D. Ricke, T.H. Lan, G. Presting, R. Wang, M. Dunn, J. Glazebrook, A. Sessions, P. Oeller, H. Varma, D. Hadley, D. Hutchison, C. Martin, F. Katagiri, B.M. Lange, T. Moughamer, Y. Xia, P. Budworth, J. Zhong, T. Miguel, U. Paszkowski, S. Zhang, M. Colbert, W.L. Sun, L. Chen, B. Cooper, S. Park, T.C. Wood, L. Mao, P. Quail, R. Wing, R. Dean, Y. Yu, A. Zharkikh, R. Shen, S. Sahasrabudhe, A. Thomas, R. Cannings, A. Gutin, D. Pruss, J. Reid, S. Tavtigian, J. Mitchell, G. Eldredge, T. Scholl, R.M. Miller, S. Bhatnagar, N. Adey, T. Rubano, N. Tusneem, R. Robinson, J. Feldhaus, T. Macalma, A. Oliphant, S. Briggs, Science 296 (2002) 92–100.

[165] G.K. Agrawal, R. Pedreschi, B.J. Barkla, L.V. Bindschedler, R. Cramer, A. Sarkar, J. Renaut, D. Job, R. Rakwal, J. Proteomics 75 (2012) 4588–4601.

[166] H. Thiellement, M. Zivy, C. Damerval, V. Mechin (Eds.), Plant Proteomics: Methods and Protocols, Humana Press, Totowa, New Jersey, 2007.

[167] L.V. Bindschedler, R. Cramer, Proteomics 11 (2011) 756–775.

[168] G.K. Agrawal, A. Sarkar, P.G. Righetti, R. Pedreschi, S. Carpentier, T. Wang, B.J. Barkla, A. Kohli, B.K. Ndimba, N.V. Bykova, C. Rampitsch, L. Zolla, M.S. Rafudeen, R. Cramer, L.V. Bindschedler, N. Tsakirpaloglou, R.J. Ndimba, J.M. Farrant, J. Renaut, D. Job, S. Kikuchi, R. Rakwal, Mass Spectrom. Rev. 32 (2013) 335–365.

[169] K. Kosová, P. Vítámvás, I.T. Prášil, J. Renaut, J. Proteomics 74 (2011) 1301–1322.

[170] I.A. Abreu, A.P. Farinha, S. Negrão, N. Gonçalves, C. Fonseca, M. Rodrigues, R. Batista, N.J.M. Saibo, M.M. Oliveira, J. Proteomics 93 (2013) 145–168.

[171] S.T. Kim, S.G. Kim, G.K. Agrawal, S. Kikuchi, R. Rakwal, Proteomics 14 (2014) 593–610.

[172] R. Singh, N.S. Jwa, J. Proteome Res. 12 (2013) 4652–4669.

[173] O. Pechanova, T. Takáč, J.Samaj, T. Pechan, Proteomics 13 (2013) 637–662.

[174] Z. Hossain, A. Khatoon, S. Komatsu, J. Proteome Res. 12 (2013) 4670–4684.

[175] S. Alvarez, S.R. Choudhury, S. Pandey, J. Proteome Res. 13 (2014) 1688–1701.

[176] C. Rampitsch, N.V. Bykova, Proteomics 12 (2012) 673–690.

[177] M.F. Mazzeo, G. Cacace, F. Ferriello, G. Puopolo, A. Zoina, M.R. Ercolano, R.A. Siciliano, Plant Physiol. Biochem. 74 (2014) 42–49.

[178] X. Li, T. Bai, Y. Li, X. Ruan, H. Li, Proteome Sci. 11 (2013) 41.

[179] C. Sun, L. Wang, D. Hu, A.R. Maquina Riquicho, T. Liu, X. Hou, Y. Li, J. Proteomics 98 (2014) 15–30.

[180] A. Molassiotis, G. Tanou, P. Filippou, V. Fotopoulos, Proteomics 13 (2013) 1871–1884.

[181] M. Di Carli, E. Benvenuto, M. Donini, J. Proteome Res. 11 (2012) 4765–4780.

[182] G. Corrado, F. Alagna, M. Rocco, G. Renzone, P. Varricchio, V. Coppola, M. Coppola, A. Garonna, L. Baldoni, A. Scaloni, R. Rao, BMC Plant Biol. 12 (2012) 86.

[183] N. Ferry, S. Stavroulakis, W. Guan, G.M. Davison, H.A. Bell, R.J. Weaver, R.E. Down, J.A. Gatehouse, A.M. Gatehouse, Proteomics 11 (2011) 1985–2002.

[184] R. Pedreschi, S. Lurie, M. Hertog, B. Nicolaï, J. Mes, E. Woltering, Proteomics 13 (2013) 1772–1783.

[185] M. Lemaire-Chamley, J. Petit, V. Garcia, D. Just, P. Baldet, V. Germain, M. Fagard, M. Mouassite, C. Cheniclet, C. Rothan, Plant Physiol. 139 (2005) 750–769.

[186] M. Rocco, C. D'Ambrosio, S. Arena, M. Faurobert, A. Scaloni, M. Marra, Proteomics 6 (2006) 3781–3791.

[187] S.B. Nogueira, C.A. Labate, F.C. Gozzo, E.J. Pilau, F.M. Lajolo, J.R. Oliveira do Nascimento, J. Proteomics 75 (2012) 1428–1439.

[188] J. De Magalhaes Andrade, T. Torres Toledo, S.B. Nogueira, B.R. Cordenunsi, F.M. Lajolo, J.R. Olivera do Nascimento, J. Proteomics 75 (2012) 3331–3341.

[189] T. Torres-Toledo, S. Beserra Nogueira, B.R. Cordenunsi, F.C. Gozzo, E.J. Pilau, F.M. Lajolo, J.R. Olivera do Nascimento, Postharvest Biol. Technol. 70 (2012) 51–58.
[190] L. Zhang, Z. Yu, L. Jiang, J. Jiang, H. Luo, L. Fu, J. Proteomics 74 (2011) 1135–1349.
[191] L. Rajjou, Y. Lovigny, S.P. Groot, M. Belghazi, C. Job, D. Job, Plant Physiol. 148 (2008) 620–641.
[192] M. Herrero, P.K. Thornton, Proc. Natl. Acad. Sci. 110 (2013) 20878–20881.
[193] E. Bendixen, M. Danielsen, K. Hollung, E. Gianazza, I. Miller, J. Proteomics 74 (2011) 282–293.
[194] J. Proteomics 75 (14) (2012).
[195] P.D. Eckersall, A.M. de Almeida, I. Miller, J. Proteomics 75 (2012) 4187–4189.
[196] P.M. Rodrigues, T.S. Silva, J. Dias, F. Jessen, J. Proteomics 75 (2012) 4325–4345.
[197] D. Hlaváč, Z. Adámek, P. Hartman, J. Másílko, Aquac. Int. 22 (2014) 299–320.
[198] A. Valdés, C. Simó, C. Ibáñez, V. García-Cañas, Trends Anal. Chem. 52 (2013) 2–15.
[199] V. García-Cañas, C. Simó, C. León, E. Ibáñez, A. Cifuentes, Mass Spectrom. Rev. 30 (2011) 396–416.
[200] J.M. Gallardo, I. Ortea, M. Carrera, Trends Anal. Chem. 52 (2013) 135–141.
[201] G. Paredi, S. Raboni, E. Bendixen, A.M. de Almeida, A. Mozzarelli, J. Proteomics 75 (2012) 4275–4289.
[202] A. D'Alessandro, L. Zolla, J. Proteomics 78 (2013) 558–577.
[203] A.M. de Almeida, E. Bendixen, J. Proteomics 75 (2012) 4296–4314.
[204] M.K. Doherty, L. McLean, J.R. Hayter, J.M. Pratt, D.H.L. Robertson, A. El-Shafei, S.J. Gaskell, R.J. Beynon, Proteomics 4 (2004) 2082–2093.
[205] D. Hazard, X. Fernandez, J. Pinguet, C. Chambon, F. Letisse, J.C. Portais, Z. Wadih-Moussa, H. Rémignon, C. Molette, J. Anim. Sci. 89 (2011) 2717–2730.
[206] A.M. Almeida, A. Campos, R. Francisco, S. van Harten, L.A. Cardoso, A.V. Coelho, Anim. Genet. 41 (2010) 260–272.
[207] M. Pischetsrieder, R. Baeuerlein, Chem. Soc. Rev. 38 (2009) 2600–2608.
[208] A. Leitner, F. Castro-Rubio, M.L. Marina, W. Lindner, J. Proteome Res. 5 (2006) 2424–2430.
[209] M. Montowska, E. Pospiech, Food Chem. 136 (2013) 1461–1469.
[210] P. Roncada, C. Piras, A. Soggiu, R. Turk, A. Urbani, L. Bonizzi, J. Proteomics 75 (2012) 4259–4274.
[211] M. Affolter, L. Grass, F. Vanrobaeys, B. Casado, M. Kussmann, J. Proteomics 73 (2010) 1079–1088.
[212] E. Chiaradia, A. Valiani, M. Tartaglia, F. Scoppetta, G. Renzone, S. Arena, L. Avellini, S. Benda, A. Gaiti, A. Scaloni, J. Proteomics 83 (2013) 144–159.
[213] C.D. Calvano, A. Monopoli, P. Loizzo, M. Faccia, C. Zambonin, J. Agric. Food Chem. 61 (2013) 1609–1617.
[214] C. D'Ambrosio, S. Arena, A. Scaloni, L. Guerrier, E. Boschetti, M.E. Mendieta, A. Citterio, P.G. Righetti, J. Proteome Res. 7 (2008) 3461–3474.
[215] K. Mann, M. Mann, Proteome Sci. 9 (2011) 7.
[216] N. Qiu, M. Ma, L. Zhao, W. Liu, Y. Li, Y. Mine, J. Agric. Food Chem. 60 (2012) 7746–7753.
[217] I. Forné, J. Abián, J. Cerdá, Proteomics 10 (2010) 858–872.
[218] G.J. Lieschke, P.D. Currie, Nat. Rev. Genet. 8 (2007) 353–367.
[219] B.C. Sanchez, K. Ralston-Hooper, M.S. Sepúlveda, Environ. Toxicol. Chem. 30 (2011) 274–282.
[220] K.J. Ralston-Hooper, M.E. Turner, E.J. Soderblom, D. Villeneuve, G.T. Ankley, M.A. Moseley, R.A. Hoke, P.L. Ferguson, Environ. Sci. Technol. 47 (2013) 1091–1100.

[221] S. Artigaud, R. Lavaud, J. Thébault, F. Jean, Ø. Strand, T. Strohmeier, M. Milan, V. Pichereau, J. Proteomics, in press.
[222] S. Lemeer, C. Jopling, J. Gouw, S. Mohammed, A.J.R. Heck, M. Slijper, J. den Hertog, Mol. Cell. Proteomics 7 (2008) 2176–2187.
[223] X. Zhong, Y. Xu, Y. Liang, T. Liao, J. Wang, Aquat. Toxicol. 71 (2005) 85–95.
[224] M. Carrera, B. Cañas, J.M. Gallardo, J. Proteomics 78 (2013) 211–220.
[225] M. Carrera, B. Cañas, J.M. Gallardo, Food Res. Int. 54 (2013) 972–979.
[226] C.-I. Kolditz, G. Paboeuf, M. Borthaire, D. Esquerré, M. SanCristobal, F. Lefèvre, F. Médale, BMC Genomics 9 (2008) 506.
[227] R.N. Alves, O. Cordeiro, T.S. Silva, N. Richard, M. de Vareilles, G. Marino, P. Di Marco, P.M. Rodrigues, L.E.C. Conceição, Aquaculture 299 (2010) 57–66.
[228] M. Morzel, C. Chambon, F. Lefevre, G. Paboeuf, E. Laville, J. Agric. Food Chem. 54 (2006) 2997–3001.
[229] E. Veiseth-Kent, H. Grove, E.M. Faergestad, S.O. Fjaera, Aquaculture 309 (2010) 272–279.
[230] A. Panchaud, M. Affolter, M. Kussmann, J. Proteomics 75 (2012) 3546–3559.
[231] C.C. Udenigwe, R.E. Aluko, J. Food Sci. 77 (2012) R11–R24.
[232] C. Signoretto, P. Canepari, M. Stauder, L. Vezzulli, C. Pruzzo, Curr. Opin. Biotechnol. 23 (2012) 160–167.
[233] A.M. Turpeinen, S. Järvenpää, H. Kautiainen, R. Korpela, H. Vapaalo, Ann. Med. 45 (2013) 51–56.
[234] W.-Y. Huang, S.T. Davidge, J. Wu, Crit. Rev. Food Sci. Nutr. 53 (2013) 615–630.
[235] C. Marques, M.M. Amorim, J. Odila Pereira, M. Estevez Pintado, D. Moura, C. Calhau, H. Pinheiro, Curr. Pharm. Des. 18 (2012) 4706–4713.
[236] S. Lordan, R.P. Ross, C. Stanton, Mar. Drugs 9 (2011) 1056–1100.
[237] N. D'Orazio, M.A. Gammone, E. Gemello, M. De Girolamo, S. Cusenza, G. Riccioni, Mar. Drugs 10 (2012) 812–833.
[238] O. Power, P. Jakeman, R.J. Fitzgerald, Amino Acids 44 (2013) 797–820.
[239] A.G.P. Samaranayaka, E.C.Y. Li-Chan, J. Funct. Food 3 (2011) 229–254.
[240] C. Zhang, X. Li, K. Se-K, Food Sci. Biotechnol. 21 (2012) 625–631.
[241] F. Toldrá, M.-C. Aristoy, L. Mora, M. Reig, Meat Sci. 92 (2012) 290–296.
[242] A. Zambrowicz, M. Timmer, A. Polanowski, G. Lubec, T. Trziszka, Amino Acids 44 (2013) 315–320.
[243] R. Hartmann, H. Meisel, Curr. Opin. Biotechnol. 18 (2007) 163–169.
[244] C.C. Udenigwe, Trends Food Sci. Technol. 36 (2014) 137–143.
[245] J. Carrasco-Castilla, A.J. Hernandez-Alvarez, C. Jimenez-Martınez, G.F. Gutierrez-Lopez, G. Davila-Ortiz, Food Eng. Rev. 4 (2012) 224–243.
[246] H. Korhonen, A. Pihlanto, Curr. Pharm. Des. 9 (2003) 1297–1308.
[247] N.P. Möller, K.E. Scholz-Ahrens, N. Roos, J. Schrezenmeir, Eur. J. Nutr. 47 (2008) 171–182.
[248] B. Walther, R. Sieber, Int. J. Vitam. Nutr. Res. 81 (2011) 181–192.
[249] M.G. Grigorov, P.J. Van Bladeren, Curr. Opin. Drug Discov. Dev. 10 (2007) 341–346.
[250] S.L. Lahrichi, M. Affolter, I.S. Zolezzi, A. Panchaud, J. Proteomics 88 (2013) 83–91.
[251] C.C. Udenigwe, M. Gong, S. Wu, Process. Biochem. 48 (2013) 1794–1799.
[252] I.M. Lacroix, E.C. Li-Chan, J. Funct. Food 4 (2012) 403–422.
[253] C. Mooney, N.J. Haslam, G. Pollastri, D.C. Shields, PLoS One 7 (2012) e45012.
[254] M. Foltz, P.C. van der Pijl, G.S. Duchateau, J. Nutr. 140 (2010) 117–118.
[255] A. Guerra, L. Etienne-Mesmin, V. Livrelli, S. Denis, S. Blanquet-Diot, M. Alric, Trends Biotechnol. 30 (2012) 591–600.

[256] C. Bauchart, M. Morzel, C. Chambon, P.P. Mirand, C. Reynès, C. Buffière, D. Rémond, Br. J. Nutr. 98 (2007) 1187–1195.
[257] K. Bouzerzour, F. Morgan, I. Cuinet, C. Bonhomme, J. Jardin, I.L. Huërou-Luron, D. Dupont, Br. J. Nutr. 108 (2012) 2105–2114.
[258] E. Sewekow, D. Bimczok, T. Kähne, H. Faber-Zuschratter, L.C. Kessler, A. Seidel-Morgenstern, H.J. Rothkötter, Br. J. Nutr. 108 (2012) 1603–1611.
[259] V. Cavatorta, S. Sforza, G. Aquino, G. Galaverna, A. Dossena, E.A. Pastorello, R. Marchelli, Mol. Nutr. Food Res. 54 (2010) 1452–1457.
[260] M. Rosmilah, M. Shahnaz, J. Meinir, A. Masita, A. Noormalin, M. Jamaluddin, Int. Arch. Allergy Immunol. 162 (2013) 299–309.
[261] G. Mamone, P. Ferranti, M. Rossi, P. Roepstorff, O. Fierro, A. Malorni, F. Addeo, J. Chromatogr. B Analyt. Technol. Biomed. Life Sci. 855 (2007) 236–241.
[262] J. Choi, L. Sabikhi, A. Hassan, S. Anand, Int. J. Dairy Technol. 65 (2012) 1–12.
[263] V. Raikos, T. Dassios, Dairy Sci. Technol. 94 (2014) 91–101.
[264] L. Najafian, A.S. Babji, Peptides 33 (2012) 178–185.
[265] A.L. Capriotti, C. Cavaliere, P. Foglia, S. Piovesana, R. Samperi, R. Zenezini Chiozzi, A. Laganà, Anal. Bional. Chem. 407 (2015) 845–854.
[266] P. Castellano, M.C. Aristoy, M.Á. Sentandreu, G. Vignolo, F. Toldrá, J. Proteomics 89 (2013) 183–190.
[267] E. Escudero, L. Mora, P.D. Fraser, M.C. Aristoy, F. Toldrá, Food Chem. 138 (2013) 1282–1288.
[268] M. Tomatsu, A. Shimakage, M. Shinbo, S. Yamada, S. Takahashi, Food Chem. 136 (2013) 612–616.
[269] S. Rao, J. Sun, Y. Liu, H. Zeng, Y. Su, Y. Yang, Food Chem. 135 (2012) 1245–1252.
[270] J.K. Nicholson, E. Holmes, J. Kinross, R. Burcelin, G. Gibson, W. Jia, S. Pettersson, Science 336 (2012) 1262–1267.
[271] J.W. Lampe, S.L. Navarro, M.A.J. Hullar, A. Shojaie, Proc. Nutr. Soc. 72 (2013) 207–218.
[272] Q. Gu, L.-R. Yu, J. Proteomics 96 (2014) 353–359.

Chapter 7

Mass Spectrometry in Food Allergen Research

Linda Monaci,[1,*] Rosa Pilolli,[1] Elisabetta De Angelis[1] and Gianfranco Mamone[2]

[1]*Institute of Sciences of Food Production (ISPA-CNR), Bari, Italy;* [2]*Institute of Food Sciences (ISA-CNR), Avellino, Italy*

**Corresponding author: E-mail: linda.monaci@ispa.cnr.it*

Chapter Outline

1. Food Allergy 359
- 1.1 Overview 359
- 1.2 IgE versus Non-IgE-Mediated Food Allergy 360
- 1.3 Food Proteins Triggering Allergic Reactions 362
- 1.4 A Non-IgE-Mediated Food Allergy: The Case of Celiac Disease 363
- 1.5 Food Allergy Prevalence and Legislative Frame Worldwide 365

2. The Role of Mass Spectrometry Applied to Food Allergen Research 367
- 2.1 Mass Spectrometry in Allergen Characterization 367
- 2.2 Mass Spectrometry in Celiac Disease 370
- 2.3 Features of Different Mass Spectrometers Applied to Food Allergen Detection 372

3. Advances in MS Methods for Multiple Detection of Food Allergens 375
- 3.1 Detection of Intact Food Allergens 375
- 3.2 Detection of Peptides Tracing for Allergenic Proteins 376

4. Concluding Remarks 386

References 387

1. FOOD ALLERGY

1.1 Overview

Food allergy (FA) is a relatively new food safety issue only being effectively recognized as such in the last 25–30 years, with an increasing prevalence worldwide and an increasing number of food allergens identified. Despite the advances in biochemical and clinical research, the mechanisms underlying the onset, development, and outcomes of this pathology FA are still far from being fully elucidated.

Comprehensive Analytical Chemistry, Vol. 68. http://dx.doi.org/10.1016/B978-0-444-63340-8.00007-8

FA is due to an anomalous interaction between food components and immune system causing a broad spectrum of disorders extremely variable among individuals. This makes this pathology one of the most difficult to enclose in a comprehensive scheme. The primary public health concern is related to sensitized individuals who have produced IgE antibodies specific to food proteins with the risk of developing immediate-type reactions on subsequent exposure. Such reactions can vary from very slight to severe and occasionally fatal, depending on the dose, the individual sensitivity, and other factors [1,2] and may manifest as itching and/or swelling of the oral cavity or pharynx, skin rashes (urticaria, angioedema, eczema flare), gastrointestinal upset (diarrhea, vomiting, and abdominal pain), asthmatic symptoms (wheezing, breathlessness, coughing), or in extreme circumstances life-threatening anaphylaxis (a rapidly evolving multisystem allergic reaction that involves respiratory or circulatory compromise) [3].

Although FA-related diseases are restricted to a small category of individuals, the number of patients concerned is steadily growing in the last 10 years, making this field of primary importance for clinicians as well as for food industry. The exact prevalence of FA is difficult to ascertain due to the imprecision of laboratory tests, but recent reviews of the literature estimate that FA affects around 2–4% of the population [4–6]. The prevalence of FA peaks at 6–8% during the first few years of life and is most often due to milk, egg, peanut, fish, and shellfish, although all foods may induce allergic reactions. Most children "outgrow" their allergy to milk, egg, wheat, or soy during their first decade, whereas allergies to peanut, tree nuts, fish, and shellfish are often retained for life. Milk and egg allergies are in general the most prevalent worldwide in children below 4 years of age, whereas peanut allergy is highly prevalent in the USA, Canada, Australia, and United Kingdom. According to a study carried out within the EuroPrevall project, the most prevalent allergenic groups across the European Union (EU) population were fruits, tree nuts, vegetables, and peanut, while milk and egg were the most prevalent foods in small children [7].

Despite recent progress in immunotherapy, no actual cure currently exists for FA and therefore food allergic consumers can only manage their condition by carefully avoiding products which contain the allergenic food(s). In order to address this effectively, food allergic consumers absolutely rely on the availability, accuracy, and reliability of information provided on foods they intend to buy, resulting in the urgent need for issuing specific labeling legislation.

1.2 IgE versus Non-IgE-Mediated Food Allergy

"Non toxic" adverse reactions to food ingestion can be subdivided into "food intolerance" and "Food Allergy" (Figure 1). The former (also termed food sensitivity) is a reproducible adverse reaction with no established immunologic mechanism which may be caused by the lack of a digestive enzyme, as is the case with lactose intolerance, but also by the addition of certain organic chemicals to food products such as colorants, preservatives, flavors, sulfites, or dye [8].

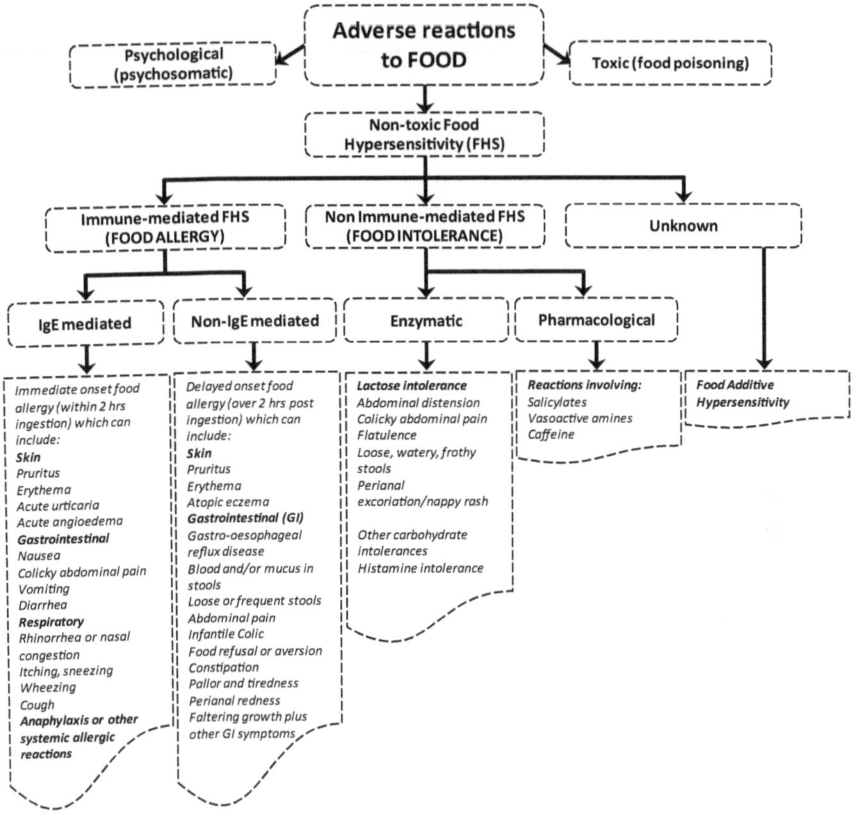

FIGURE 1 Classification of food hypersensitivities.

Biogenic amines (i.e., tyramine and histamine) contained in some foods may indeed be responsible for symptoms associated with "food intolerance" [9]. A number of articles have covered this subject [10,11] therefore it will not be treated in the present chapter.

FA arises from specific immune responses triggered by the ingestion of some food components, usually proteins. Despite the progress in understanding the molecular mechanisms of FA it is not still clear how incoming food proteins may activate immunological pathways. In normal conditions, potential allergens present in the ingested food are suppressed because digested by gastric pancreatic proteases and by gut brush border membrane (BBM) enzymes along the gastrointestinal tract. Gut epithelium tight junctions hamper the passage of proteins and peptides across the intestinal wall to reach the gut associated lymphoid tissue (GALT). However, a small amount of ingested food antigens (less than 2%) is normally absorbed and transported across the wall in an immunologically intact form [12,13]. In infants, the immaturity of both gut barrier and immune system may reduce the efficiency of the mucosal barrier [14]. However, in physiological conditions, when intact proteins do cross the gut barrier the immune system induces oral tolerance [15]. The antigens adsorbed through the digestive tract may activate the immunological pathway which can be attributed to IgE-mediated and non-IgE-mediated cellular mechanism (Figure 1). In IgE-mediated food allergies (also known as "type I" hypersensitivity reactions), specific IgE antibodies are produced in response to exposure to a food allergen. The mechanisms leading to sensitization and to production of IgE antibodies are complex and some aspects are still debated. Incoming allergens taken up by the antigen-presenting cells (APCs) and small peptide fragments (T-cell epitopes) are presented to T cells in conjunction with MHC class II molecules. T cells activate chemotactic signaling that induces conversion of B cells into IgE antibody-producing plasma cells. Following further exposure, the antigen is recognized by IgEs which are linked to specific FC3RI receptors present on the mast cells, basophils, macrophages, and other APCs. These cells release mediators (i.e., histamine, tryptase, cysteinyl-leukotrienes, and prostaglandin D2) that trigger allergic inflammatory responses [3].

Despite IgE-mediated immune responses, non-IgE-mediated allergic reactions are mediated by non-IgE antibodies (i.e., IgG, IgM, and IgA) and/or cellular immune responses [16]. Cell-mediated FA is only rarely life-threatening and symptoms are generally limited to gastrointestinal discomfort. Nevertheless, they are cause of morbidity in infants and young children. Several evidences attribute to the polarization of T cell toward the Th1 subtype, which is most likely ruled by regulatory T cells, a key role in the development of non-IgE-mediated allergies.

1.3 Food Proteins Triggering Allergic Reactions

More than 150 foods have been implicated in allergic reactions, even though the majority of these reactions are induced by a small number of foods. In 1995, the Food and Agriculture Organization (FAO) technical consultation identified eight food groups as the most common causes of allergy worldwide. These foods are known as the "big eight" and include, milk, egg, wheat, soy, peanuts, tree nuts (e.g., hazelnut, walnuts, pecans, almonds, and cashews), fish, and shellfish and are recognized as allergenic foods of public health importance [17,18]. However, the same food may contain several allergenic proteins (also termed food allergens) capable of activating an immune response in predisposed individuals. Currently more than 600 food allergens are known at molecular level, 206 of which are officially registered by the allergen nomenclature subcommittee established by the International Union of Immunological Societies (IUS) [19]. In addition, different food sources are known to contain more than one allergenic protein. As reported by the Allergome database [19], the number of food allergens doubled the number of species the allergen derived from [20].

Several suspected food allergens present in foods have not been identified yet and the structural traits that make a protein an allergen and the relationships between allergenic determinants and disease patterns are not fully understood.

Several food allergens share structural analogies such as glycosylation, molecular weight (between 10 and 70kDa), stability to food processing, and resistance to digestion. An example is shown by plant allergens: N-glycans containing α-1,3-fucose and β-1,2-xylose are known to be most frequently involved in the structures of IgE epitopes [21,22].

The widespread occurrence of fucose and xylose on N-linked glycans in plants, but also in invertebrates such as crustaceans (to less extent), explains several cross-reactive sensitization profiles toward proteins not structurally related [23]. Carbohydrate moieties can indirectly enhance the allergenic potential by conferring stability against proteolytic degradation [24,25]. A nonenzymatic glycosylation (i.e., Maillard reaction) caused by food technological process enhances the potential allergenicity of food allergens [26–28]. Very recently, Heilman et al. demonstrated that glycation of ovalbumin induced higher IgE production in mice [29].

The stability to heat treatment and the resistance to gastrointestinal digestion represent other structural analogies and properties shared among many animal- and plant-derived allergens [30–32]. These properties allow allergens to reach the epithelium gut in an immunogenic intact form. Typical examples are the 2S albumins [33], nonspecific lipid transfer proteins (nsLTPs) [34] and peanut Ara h 2 [35]. The stability to either gastrointestinal enzymes or thermal denaturation of these allergens, has been loosely attributed to a conserved skeleton of eight cysteine residues engaged in disulphide bonds [36].

1.4 A Non-IgE-Mediated Food Allergy: The Case of Celiac Disease

Among non-IgE-mediated food allergies, celiac disease (CD) is the most common food-sensitive enteropathy in humans [37]. CD is a complex chronic immune-mediated disorder of the small intestine, triggered in genetically susceptible individuals after ingestion of wheat gliadin and/or related prolamins from oat (avenin), rye (secalin), and barley (hordein) [38]. Epidemiologic studies consistently report a prevalence of CD ranging from 0.5% to 1.0% in Europe and United States [39–41]; however the true prevalence is suspected to be much higher as many affected people are asymptomatic or show mild symptoms and signs [42,43].

Genetic factors play a key role in disease as proven by the well-established familiar aggregation of diagnosed cases (5–15%) and by the concordance of approximately 85% in homozygous twins [44]. Over 97% of CD patients express at least one of the two human leukocyte antigen (HLA) class II genes DQ2 and/or DQ8 which are the most important and best characterized genetic risk factors. The majority of individuals (>90%) carry the DQ2 gene, while the rest express the DQ8 gene. These HLA haplotypes are necessary but not sufficient for the development of CD, as only about 4% of DQ2/8 positive individuals exposed to gluten developed CD, indicating that other genetic factors are also involved [45–47]. More recent epidemiological studies suggested that the timing of gluten introduction, and the pattern of breast feeding, may strongly influence the subsequent development of CD [48].

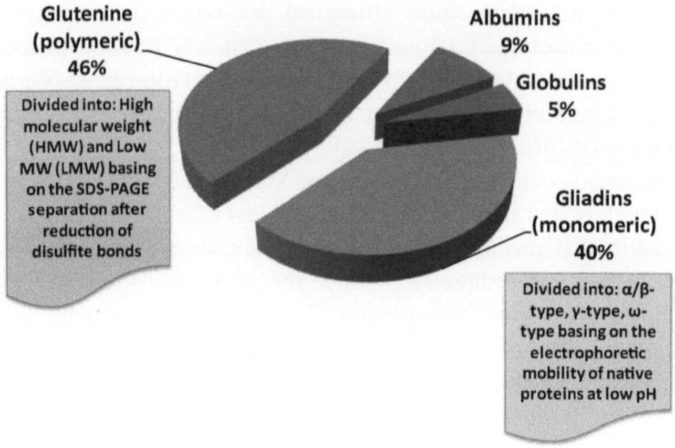

FIGURE 2 Wheat proteins distribution.

However, CD is triggered only upon the introduction of gluten-containing foods in the diet of susceptible individuals [38]. Gluten proteins are proline- and glutamine-rich and responsible for the technological characteristics of wheat flour-derived products. They are conventionally grouped into monomeric gliadins and polymeric glutenins (Figure 2). These latter are able to form intra- or inter-molecular disulfide bonds [49]. Gliadins are classified in α/β, γ, and ω subunits, according to their electrophoretic mobility in acid polyacrylamide gel electrophoresis. Glutenins consist of low molecular weight glutenin subunits (LMW-GS) and high molecular weight GS (HMW-GS), linked through inter-chain disulfide bonds. Gliadin and glutenin proteins are both rich in the amino acids glutamine and proline [50]. For this reason, during gastrointestinal digestion, a family of Pro- and Gln-rich polypeptides that are responsible for the inappropriate immune response are released. Long gliadin fragments can reach high concentration levels in the gut epithelium and reach the lamina propria either via the paracellular (as a consequence of an increased permeability of the intestinal epithelium due to an upregulation of zonulin) or via the transcellular pathways (by using enterocytic vesicles) [51]. Several fragments are recognized as toxic and immunostimulating triggering an innate or adaptive immune response respectively.

The binding of gliadin fragments to HLA-DQ2/DQ8 molecules triggers an innate immune response [38]. The formed complexes are expressed on the cellular surface of APC to be recognized by a specific population of $CD4^+$ T cells. The binding became more relevant when a single glutamine residue is modified by enzyme tissue transglutaminase (TG2) [52,53]. This enzyme catalyzed at low pH, a deamidation reaction, transforming neutral glutamine residues at specific positions into negatively charged glutamic acids [54] which increase the binding with basic amino acids located in the anchor positions of HLA-DQ2/DQ8 molecules. The interaction of deamidated peptides with HLA-DQ2/DQ8 strongly activates

CD4+ T cells, triggering enterocyte apoptosis and secretion of proinflammatory cytokines, in particular γ-interferon, leading to profound tissue remodeling. Stimulated CD4+ T cells are also able to induce B-lymphocyte differentiation into plasma cells producing high levels of antibodies to gliadin and IgA autoantibodies directed against endomysium, reticulin, and jejunum [55–57].

In concert with the adaptive immunity, gliadin peptides elicit an innate immune response by the stimulation of both dendritic cells and lamina propria mononuclear cells to produce IL-15 [58,59]. This cytokine causes an expression of NKG2D receptors on intraepithelial T cells and its ligand MICA on epithelial cells. The subsequent NKG2D–MICA interaction may result in destruction of the intestinal epithelium [60].

1.5 Food Allergy Prevalence and Legislative Frame Worldwide

Although an increasing number of foods are implicated in allergic reactions, only a few of them are involved in the most popular types of food allergies. A total of eight big food groups have been identified including milk, egg, wheat, soy, peanuts, tree nuts (e.g., hazelnut, walnuts, pecans, almonds, and cashews), fish, and shellfish and are recognized as allergenic foods of public health importance [17,18]. Therefore they are included in regulatory allergen lists worldwide. In general, milk and egg allergies are the most widespread types of allergy in children below 4 years of age. Because tolerance is developed spontaneously in the majority of patients, the prevalence is reduced in schoolchildren and adolescents. Peanut allergy is instead highly prevalent in the USA, Canada and Australia and United Kingdom (UK) affecting approximately 2.9% of one-year-old infants in Australia. The prevalence of tree nut allergy as confirmed by oral challenge ranged from 0.1% (almond in UK) to 4.3% (hazelnut in adolescents in Germany), whereas the prevalence of allergies to wheat and soy was 0–0.5% and 0–0.7%, respectively. Despite the Western countries, wheat allergy is instead very prevalent in Japan, inducing 10% of all immediate reactions to foods and it was ranked third for inducing allergic reactions after egg (29%) and milk (23%). Differently, the prevalence of fish allergy in UK, US, and Canada was found to be ranging between 0.2% and 0.6%, while in Japan 5% of allergic reactions was due to fish ingestion. As far as allergy to fruit is concerning, this type of FA has been primarily investigated in Europe and calculated to be between 0.1% and 4.3% in oral food challenge studies. Results obtained in the recent EuroPrevall project and reporting epidemiological surveys carried out in the European population, confirmed the importance of plant food allergies in schoolchildren and adults. The most prevalent four groups identified were fruits, tree nuts, vegetables, and peanut [7]. Given the several studies accomplished and the wide consumption of allergenic food products around the globe, other food products might be considered as good candidates for entering the list of allergen-containing foods to be regulated by the respective governments.

Legislation requires food labeling to protect the health of allergic consumers. On this regard, the International Codex Alimentarius Commission has provided the basis for a correct labeling in several countries. Different international authorities had recognized the importance of properly conveying the information of the presence of allergenic ingredients in foods to consumers through food labeling. Since hundreds of foods have been regarded allergenic [61], the focus in the beginning was placed on a limited set of "priority allergens." Therefore in the past years, national governments and international regulatory bodies have actively worked in this direction issuing specific regulations and guidelines about the labeling of allergenic food ingredients [62]. Initially, a list of priority allergens was drawn down by the Codex Alimentarius [63] and this was further utilized as starting document by the European Commission and other regulatory bodies, to introduce legislation about the correct declaration on the food label of a number of food categories considered to be allergenic [64]. A part from Japan requiring the only labeling of crustaceans, milk, egg, and wheat, and Korea that extended the list also to fish and soy, the legislative frame of other countries including US, China, Hong Kong, and Mexico require in general the mandatory labeling of the "big eight" food categories. Worthy to be mentioned is Australia/New Zealand also including sesame and Canada that extended also to shellfish/mollusks and mustard the foods considered allergenic accounting for a total of 11 food categories regulated by the enforced legislation.

It turns out that EU is more restrictive in the field of food allergen labeling, with the last Directive 2007/68/EC requiring the mandatory labeling of a total of 14 allergenic ingredients, whenever used and irrespective of the amounts, on the respective food label [65]. The list of allergenic ingredients includes peanut, milk, eggs, fish, crustaceans, mollusks, tree nuts (almonds, hazelnuts, walnuts, cashew nuts, pecan nuts, Brazil nuts, pistachio nuts, macadamia nuts, and Queensland nuts), soybeans, cereal containing gluten (wheat, rye, barley, and related grains), lupin, celery, sesame seed, mustard, and sulfur dioxide/sulfites (Figure 3). Nevertheless, some ingredients exempted from allergen declaration are also listed in the same Directive.

This regulatory framework is intended to provide consumers with information about allergens when they occur in foods as ingredients, meaning when they are intended components of the final food product. However, allergens may also become unintended components of a food when manufacturing processes and controls are not adequate to prevent cross-contact between allergen-containing and allergen-free foods. Over 600 alerts due to the presence of undeclared allergens in foods have been recorded between 2004 and 2012 by the Rapid Alert System for Food and Feed in the EU [66]. Contamination of food by these "hidden allergens," at the moment, represents the major health problem for allergic consumers that might lead to food recalls, which are expensive for the industry [67]. Moreover, in the European "General Food Law" it is also specified that food manufacturers are responsible for the safety of food products brought onto the market [68]. Consequently, food manufacturers need to take extra measures to prevent or control cross-contamination to protect the allergic consumers and their own reputation. Still, no specific

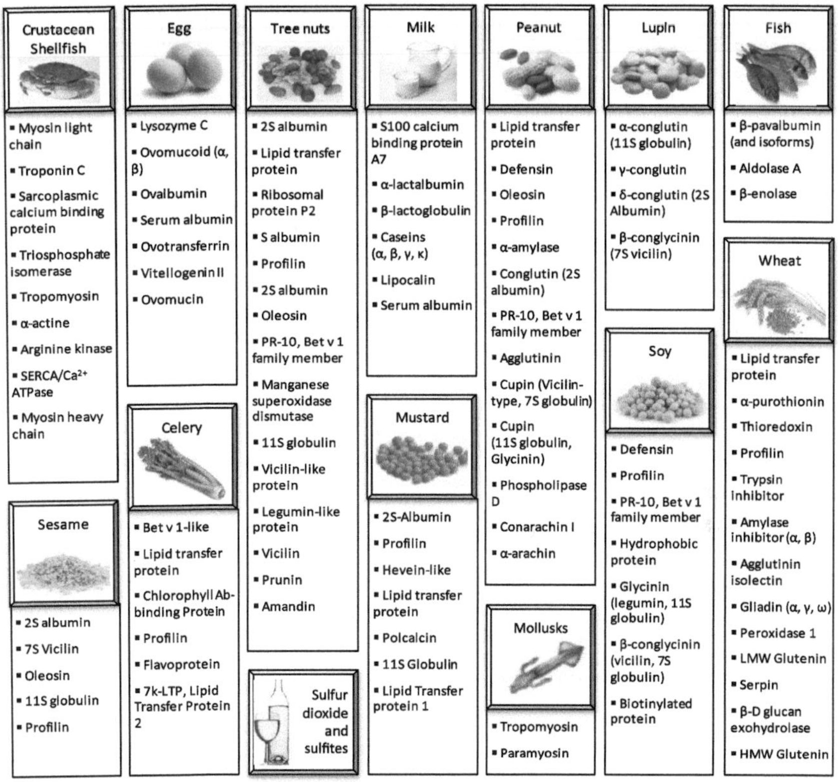

FIGURE 3 Allergenic proteins belonging to the 14 allergenic ingredients issued in the Directive 2007/68/EC of the European Commission.

legislative framework or quantitative guidelines are available for the labeling of potentially contaminated food products; often it is not clear for food producers how to manage this issue and they prefer to use some form of precautionary labeling to inform allergic consumers about the likelihood of inadvertent presence of allergenic food constituents. While the intention of such warnings is to help allergic consumers to make safe food choices, their widespread use made them ineffective [69–71]. In fact, allergic consumers often end to believe that precautionary labels are overused and thus can be ignored or mistrusted, leading to risk-taking behaviors [72–74].

2. THE ROLE OF MASS SPECTROMETRY APPLIED TO FOOD ALLERGEN RESEARCH

2.1 Mass Spectrometry in Allergen Characterization

One of the most efficient MS-based methods for allergen identification is based on electrophoresis separation (one- or two-dimensional version) followed by

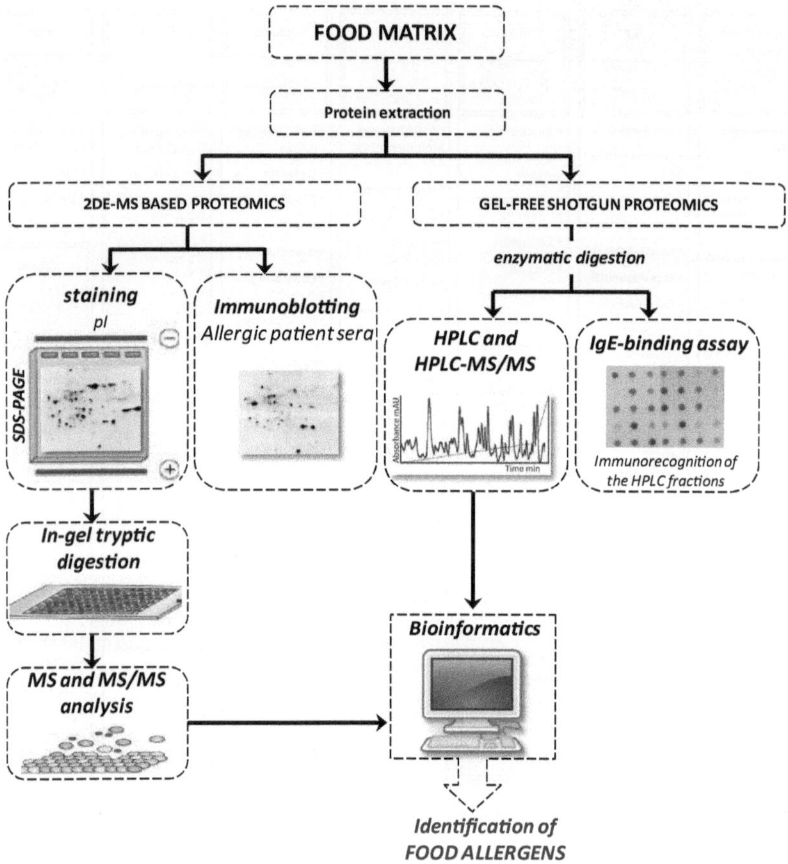

FIGURE 4 Typical proteomic-based approaches for identifying allergens or epitome in complex food matrices.

western blotting (WB) detection using sera from allergic patients, combined with an additional gel used for immune-reactive proteins characterization (Figure 4) [75]. Validation of the IgE-binding properties may be provided by immunoblot-inhibition assays typically employing sera preincubated with purified candidate allergens. These strategies have been successful used in the identification of a large number of food allergens. A total of 20 potential allergenic wheat proteins, including α-amylase inhibitor, β-amylase, profilin, serpin, and β-D-glucanexohydrolase were identified by this way [76–78]. Some authors also demonstrated that nsLTP and endochitinase allergens isolated from maize showed cross-reactivity with homologous forms in grape [79]. WB and immunoblotting experiments by using allergic patients sera followed by MS analysis also demonstrated the high immunogenicity of the following bovine milk proteins: 90% αs2-casein, 50% κ-casein, 15% β-casein, 5% αs1-casein, 45% β-lactoglobulin, 45% bovine serum albumin (BSA), 95% IgG-heavy chain, 50%

lactoferrin [80]. A muscle arginine kinase (Pen m 2) was identified as a novel crustacean allergen [81]. More recently, liquid chromatography online-coupled to tandem mass spectrometry (LC-MS/MS) analysis was applied for determining the sequences of tryptic peptide from a new IgE-binding protein of hazelnut, not reported in databases [82]. However, a large number of other allergens identified in this way, are represented by less common allergenic foods such as peach [83], barley [34], lettuce [84], chestnut [85], banana [86], and tomato [87].

The role of MS has been important for precise identification of immunogenic epitopes. This is a key aspect for designing opportune strategies of either therapeutic intervention (i.e., peptide-based vaccines) or technological reduction of the allergenic potential. Epitope mapping of α-lactalbumin, one of the major milk whey proteins, has been attempted by use of synthetic peptides combined with immunological and matrix-assisted laser ionization time-of-flight (MALDI-TOF) [88]. A tryptic dodecapeptide of ovalbumin was identified as the relevant allergenic epitope [89,90] by using magnetic immunoaffinity carrier and MALDI-TOF-MS detection.

Furthermore, stability to gastric and pancreatic digestion is a basic criterion for predicting allergenic potential of food proteins. Some allergens are only partially, or are not affected at all by digestion and are capable of crossing the gastrointestinal barrier in an intact form, thus preserving intact IgE-binding domains. As a sake of example, the nsLTP, almost ubiquitous in plants, and 2S albumin from many seeds, are highly resistant to gastroduodenal digestion, remaining intact and retaining unmodified the major epitope regions [36,91,92].

MS has been used to identify the regions of milk whey proteins, α-lactalbumin and β-lactoglobulin surviving the in vitro gastrointestinal digestion [93]. Similarly, peptides arising from an allergenic 32-kDa avocado endochitinase (Prs a 1) upon simulated gastric digestion were identified by MALDI-MS and assayed for their binding properties to IgE from individuals with a clinical history of latex-fruit allergy syndrome [94].

Another consideration fundamental in determining the allergenic potential is the uptake of food-derived allergens. Monolayers of Caco-2 cells is considered a useful model to simulate the transport of peptides across the epithelial barrier [95]. Native and pepsin-hydrolyzed forms of ω5-gliadin and nsLTP (among the most important wheat allergen) showed to cross the Caco-2 cell monolayer by the transcellular route [96]. Similarly, transport experiments had already been performed with unhydrolyzed bovine whey proteins, α-lactalbumin and β-lactoglobulin [97,98]. The gastrointestinal digestion of prefractionated milk proteins released potential β-lactoglobulin epitopes, able to translocate Caco-2 monolayers [99].

In conclusion, the combination of 2DE and/or coupled to MALDI-MS experiments proved to be successful for the identification of a large number of food allergens. In addition, the coupling between HPLC and MS/MS detection was applied for characterization and sequences elucidation of tryptic peptides from new IgE-binding proteins.

2.2 Mass Spectrometry in Celiac Disease

MS-based techniques have been applied to address many biochemical, enzymatic, and immunological issues related to the pathogenesis of CD. In particular, such approaches have been applied to characterize toxic and immunogenic gliadin peptides, to identify antigens recognized by serum antibodies of CD patients or changes in protein expression specific to gut tissue from patients with CD [100–102].

Intestinal antigens recognized by IgA serum of patients with active CD were identified combining proteomics and immunochemistry techniques. In particular, four main autoantigenic determinants including actin, ATP synthase β-chain, and two charge variants of enolase were identified [103]. However TG2 was found to be the major antigen target of IgA antiendomysial antibodies [104]. To date, TG2-specific antibodies represent a hallmark of the disease. As mentioned above, TG2 plays a key role in the adaptive immune system. In addition, consequently to catalyzing deamidation reactions of gliadin peptides, tissue Transglutaminase (tTG) may also generate additional antigenic neoepitopes by cross-linking gliadin peptides with extracellular matrix proteins (transamidation reaction) [105]. Gliadin peptides have also been demonstrated to be cross-linked to TG2 [106–108]. In this case, the tTG–gliadin complex would act as a hapten carrier, which could explain the typical antibody response against tTG in untreated CD patients [43].

MS-based approach has also been producing a massive impact on understanding the complex nature of toxic and immunogenic gliadin peptides released upon gastrointestinal digestion. By using a proteomic strategy, it has been shown that peptide toxic 31–55 [109] survived gastrointestinal digestion of both recombinant α-gliadin and whole gliadin extracted from a common wheat cultivar [110]. Iacomino et al. demonstrated by LC-MS analysis, that 22% of the 31–55 peptide translocate across monolayers of Caco-2 cells, without being degraded by hydrolysis. Interestingly, treatment of Caco-2 cells with whole gliadin digest extracted from a common wheat cultivar increased the epithelial 31–55 translocation by 35% [111]. Peptide 31–55 encompasses sequences 31–43 and 31–49 which represent the unique model peptide for the study of innate immune system in CD. The toxicity of gliadin peptide 31–43 was demonstrated in organ culture of treated biopsies [109], and in vivo feeding studies [112]. Similar results have been obtained in vivo on small intestinal [113] and oral [114] mucosae with the peptide 31–49. The number and sequence type of toxic peptides able to induce an immune response are mainly restricted to 31–43, 31–49, and 31–55 peptides. By contrast, a very high number of "immunogenic" gliadin peptides have been reported [51]. The combination of MS analytical approaches with cell biology, immunology, synthetic chemistry has been applied to identify the HLA-binding motifs and T-cell recognition patterns in gliadin-derived peptide sequences [115,116].

Among gliadin peptides, 33-meris is considered the most immunogenic as it includes six overlapping epitopes. Basing on LC-MS analysis, Shan et al. demonstrated that a 33-mer peptide is released by proteolytic hydrolysis of recombinant α2-gliadin, and was resistant to BBM enzymes degradation [117]. Hence, it has been suggested that the 33-mer can reach the underlying lamina propria and, following deamidation by TG2, trigger the pathogenic cascade of CD. A similar approach was applied to the identification of another multivalent 26-mer peptide generated from hydrolysis of recombinant γ-gliadin. 26-mer that was found to be a good substrate of TG and displayed markedly enhanced T-cell antigenicity [117]. At least 60 putative peptides that share common structural and functional features of the 33-mer and the 26-mer peptides were also identified by in silico analysis of the gluten proteome [118].

An interesting peptidomic approach has been applied for the rapid identification of gliadin peptides susceptible to deamidation by TG2 in complex mixtures of digested gliadins. By tagging the TG2-susceptible glutamine residues with a probe such as mono-dansyl-cadaverine [106] or 5-biotinamido-pentylamine [119], a restricted number of gliadin peptides sharing specific amino acid motifs for TG2 were selectively fished among a myriad of unmodified peptides. Glutamine residues target by TG2 were then identified by LC-MS/MS. The majority of gluten T-cell epitopes identified are originated from α- and γ-gliadins, and some of these sequences actually stimulated intestinal T-cell responses after tTG deamidation [120].

Currently, a strictly gluten-free diet is the only medical treatment for CD patients. An interesting research in food science has focused on preventing or reducing gluten toxicity in wheat flour. To date, two approaches are being tested: masking or proteolytic degradation of gliadin epitopes. For both, the contribution of MS analysis has been proven essential. Approaches based on enzymatic pretreatment of flour were studied for abolishing the immunogenicity and toxicity of gluten. Selected probiotic bacteria added during fermentation are able to proteolyze the proline/glutamine-rich gluten peptides and thus decrease immunotoxicity [121–124]. In vitro and in vivo studies driven by MS [125] confirmed in part the effectiveness of prolyl-endoprotease, an oral supplement to reduce gluten intake in patients. Alternative approaches for gluten degradation are based on gluten fermentation with different microbial media, including probiotic preparations [126] or sourdough lactobacilli [127]. 2DE, MALDI-TOF-MS, and LC-MS techniques were used to monitor the effective disappearance of the toxic epitopes.

Another interesting enzymatic approach consisted of inactivation of immunogenic gluten epitopes by masking the gluten motif specificity of TG2. In particular, MALDI-TOF and MS/MS analysis proved the ability of microbial TG to catalyze cross-linking of lysine methyl ester to gliadin, with subsequent blockage of the T-cell-mediated gliadin activity [128].

2.3 Features of Different Mass Spectrometers Applied to Food Allergen Detection

In the last two decades, mass spectrometry has played a pivotal role in proteomic research being the election method for protein identification in complex mixtures. Such MS approach has been subsequently exported to the food allergen field proving to be a viable alternative confirmatory tool capable of multiplex analysis. In addition, compared to the methods available based on DNA detection or antibody recognition, MS-based technology can provide a comparable sensitivity and a higher reliability in the final identification, since it enables a direct detection of the allergen itself (as intact protein) or through reliable markers. On the other hand, MS techniques overcome the limitations shown by enzyme-linked immunosorbent assay (ELISA) due to a possible modification of the epitope after processing of the containing food. The several advantages offered over the immunoassay methods rely on the possibility to run multiallergen analysis in one shot, quantitative analysis, structural protein elucidation, characterization of protein modifications, and epitope mapping. MS has also the potential to be semi or fully automated, potentially allowing the high-throughput analysis of food samples though the high cost required for the equipment.

MS methods typically employed in food allergen analysis use either matrix-assisted laser ionization (MALDI) or electrospray ionization (ESI). MALDI typically coupled to TOF-MS has been the first technique exploited for protein mass determination [129,130]. In general, the performances of TOF analyzers have greatly improved over the years in terms of resolution and mass accuracy offered. In the last decade, ESI has become the standard ionization method used for allergens detection by high performance liquid chromatography (HPLC) or micro/nano LC and MS. One of the most promising MS techniques for food allergen characterization studies and quantitative analysis was the combination between HPLC and a hybrid quadrupole-time-of-flight (QqTOF) mass spectrometer through ESI interface. Thanks to the generation of a multicharged pattern displayed by the majority of proteins upon ESI-MS analysis, it was also possible to detect either multiple-protonated peptides or middle size intact proteins provided that the envelope of multicharged ions generated entered the instrument mass range. Such multiprotonated pattern might be further deconvoluted by suitable softwares leading to unambiguous identification of the original protein. Deep investigations have been undertaken in this direction aimed at investigating, food allergen modifications induced by application of thermal processing to foods by ESI-MS [131–135].

In general, MS method scan uses two different format for protein identification namely the bottom-up and top-down approach (Figure 5). The

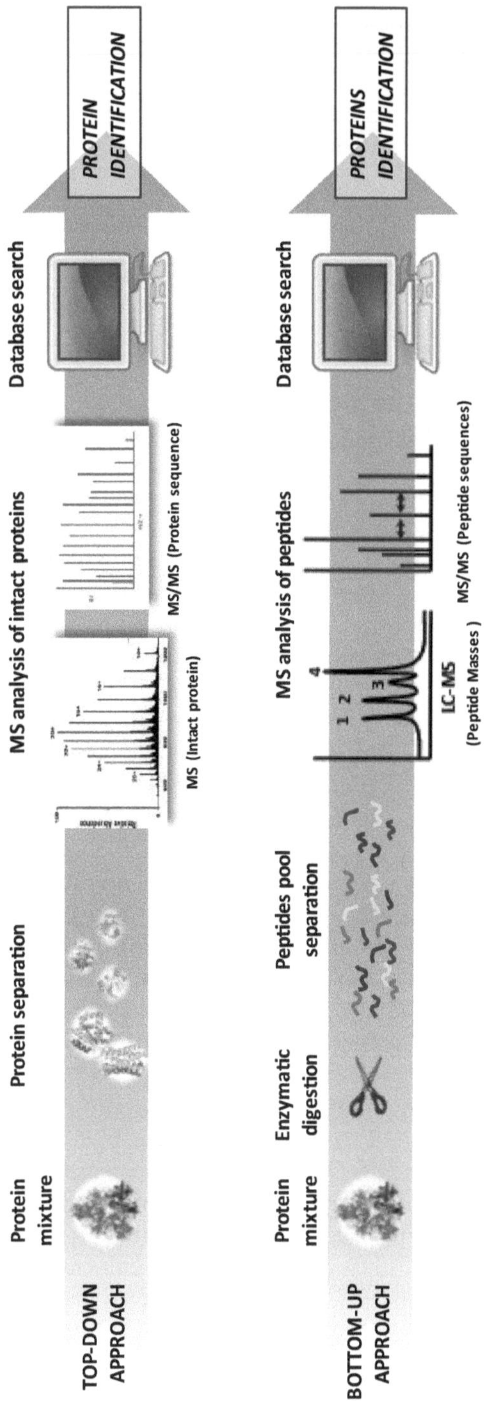

FIGURE 5 Overview of the proteomic approaches.

bottom-up is the most widespread and preferred route used in food allergen analysis and can be performed in different modes. Protein identification can be achieved by peptide mass fingerprinting through detection of a high number of peptides resulting from the specific cleavage operated by specific proteases using MALDI-TOF-MS detection. Differences in chemical modification of allergenic proteins were also highlighted using this instrumental set-up. With the enlargement of the available databases, criteria for protein identification became more stringent also requiring more accurate mass measurements. In the last decade, MALDI-TOF-based analysis has been superseded by the more advanced tandem mass spectrometers available on the market merging the double-stage format with an increased resolution offered. These instruments like the latest triple quadrupole MS, in general, boast excellent scan speeds and allow to run complex MS acquisition schemes. More insights might in general be provided by using the cutting-edge technology. Despite the old low-resolution tandem mass spectrometers where different charge states cannot be easily distinguished due to the low-resolving power offered, with the high-resolution instruments placed on the market, different acquisition schemes can be designed thus offering the additional advantage of readily determining polypeptides charge states. The most common high-resolution instruments used in food allergen research area are represented by QqTOF-MS, Orbitrap™-based MS systems and last generation of linear ion trap (LIT) MS capable of working in ultrazoom scan mode. As a result, to increase the confidence in protein assignment, henceforth fragment mass fingerprinting is the typical mode used for allergen identification in complex matrices. In this perspective, a first survey experiment is typically carried out for identifying suitable candidate peptide markers. Final data collected, consisting of a multitude of collision-induced dissociation (CID)-MS/MS spectra generated upon fragmentation of precursor peptides, could subsequently be searched against large databases using a number of different algorithms (e.g., Sequest, Mascot, etc.). A list of recognized peptides is finally compiled into a protein "hit list." By means of appropriate bioinformatic tools, a statistically based identification can be achieved basing protein attribution on a probability-based matching score at a certain confidence level. Consequently, the higher the ion score and the greater the number of peptides matching to any protein, the greater the sequence coverage and the probability of a correct protein assignment [136]. In theory, any protein target to a foodstuff might be used to infer the presence of that component itself. Once the most representative protein has been chosen, specific transitions (coupling peptide/fragments) should be designed, that identify the protein of interest. To uniquely identify the protein, the first step is to perform a BLASTing (basic local alignment search tool) search against the proteome of that species or other species that might be contained in that food. Besides, in order to ensure that the most appropriate peptides are chosen and to avoid variability of results, specific criteria should be applied for peptides selection.

3. ADVANCES IN MS METHODS FOR MULTIPLE DETECTION OF FOOD ALLERGENS

3.1 Detection of Intact Food Allergens

MS is not intrinsically a quantitative analytical technique because the ionization efficiency of a protein/peptide can vary considerably according to the molecular structure of the protein. For this reason, mass spectrometry has been mainly used in the past for protein characterization. Nonetheless, thanks to progress done by this technology over the years and to the features offered by the latest generation of mass analyzers available, efforts have been directed in the course of time toward the development of quantitative MS methods capable of delivering both qualitative and quantitative information of allergenic proteins contained in a food. As a result, the focus of the scientific community has moved to this powerful technology capable of performing a multiplex detection of allergens at the highest confidence level within a single chromatographic run, turning a potential high-throughput screening tool able to quantify allergen traces in foods.

In general, for accurate quantification the protein/peptide amount must be referred to that of a suitable standard either the whole protein or a derived peptide. In the direct quantification of intact proteins, according to a "top-down" approach, the intensity of analyte multicharged ions is compared with that of internal or external intact standard proteins. The major advantage in using a protein standard is that brought through the whole extraction procedure any issues of recovery, incomplete digestion and extraction derived modification can be nullified, assuming the process is identical for the sample and standard alike. Nevertheless, the sensitivity of "top-down" mass spectrometry methods is limited by the high complexity shown by the native protein and the wide distribution of protein charge states. Moreover the use of a specific intact protein standard is hampered by the commercial availability of "purified" proteins. However, although using isotopically labeled proteins for accurate and traceable quantification of a protein, the structure of the labeled standard should be also analyzed to better highlight eventual differences with the target protein. However due to the prohibitive costs of using isotopically labeled equivalent proteins for quantitative purposes, only a few papers have been reported in the literature showing HPLC-MS methodology applied to the detection and quantification of allergenic proteins by using isotopically labeled equivalents [137]. Alternatively, label-free methods based on HPLC-MS detection operating in selected ion-monitoring mode might be implemented for the detection of intact proteins although such approach had objective limitations when applied to highly processed foods, especially for the high limits of detection (LODs). With this aim, a few papers reported the use of MS systems for the identification and characterization of intact proteins in food commodities especially for studying modifications induced by thermal treatments [132,133,138,139]. Among them, lactoglobulins were the highly investigated proteins due to the good multiprotonated features shown by this class of proteins generating a reproducible multicharged ions enveloped by electrospray ionization [131,134,135,140].

3.2 Detection of Peptides Tracing for Allergenic Proteins

The most widespread approach used in proteomics is the "bottom-up" strategy also known as "shotgun" approach (Figure 5). This is the preferred way for allergen detection at the highest sensitivities by MS. The use of "bottom-up" methodologies for quantitative analysis in proteomics is steadily increasing and has proved to adapt also for food allergen quantitative analysis. Quantitative methodologies such as SILAC (stable isotope labeling with amino acids in cell culture), ICAT (isotope-coded affinity tag), iTRAQ (isobaric tag for relative and absolute quantification) are typically used for large-scale quantification in proteomics, since capable of exploring the dynamics of whole proteomes through the relative quantification among multiplexed samples, particularly suitable for differentially expressed proteins [141,142]. On the other hand, label-free approaches are the preferred routes for food allergen quantification as they do not involve stable isotopes.

The workflow underpinning bottom-up strategy is based on protein extraction from food, proteolytic digestion operated by specific cleavage enzymes and analysis of the final peptide mixture for the identification of stable and unique peptide markers tracing for the target proteins. Care should be taken inappropriately designing suitable and reliable peptide markers for each target protein [143]. With the aim to harmonize MS methods for food allergen detection, a task force working on the harmonization of analytical methods for food allergens issued some guidelines to help scientists in properly identifying protein and peptide markers [143]. In Table 1 are reported some criteria used for selecting

TABLE 1 Basic Criteria Used for Proper Protein and Peptide Identification in MS-Based Food Allergen Analysis

Criteria for Protein Target Selection	Criteria for Peptide Selection
Knowledge of full protein sequence	Peptide and respective fragments should be highly reproducible
Uniqueness of the protein to the desired foodstuff to be detected	Peptide and respective fragments should be uniquely originated from the target protein
High abundance of the selected protein marker	Peptide lacking of cys, met, and glutamic acid residues
Absence of protein modifications in the containing food	Reproducible retention times along the chromatographic run
Protein stability	Preferably +2 and +3 peptide ions
High extraction efficiency	Typically peptide lengths comprised between 6 and 12 residues
High efficiency and reproducibility in digestion	

suitable protein and peptide markers. On this regard, the advent of last-generation hybrid mass spectrometers has pushed in the direction of developing sensitive hyphenated methods, coupling separation techniques, and MS detection, for absolute quantification of allergens in complex food matrices. This opened a new era in the food allergen area with efforts aimed at developing sharp and high-throughput analytical MS schemes for the multiple detection of allergens in foodstuffs within a single run. Such peptide-based allergen quantification strategy can be based on multiple selected reaction monitoring (SRM) MS mode that monitors characteristic precursor ion → product ion transitions of selected peptide markers generated upon proteolytic digestion on triple quadrupole instruments. The targeted monitoring of mass and transitions of selected "proteotypic peptides" ameliorates in general the analysis since it overcomes the bias due to the presence of a large number of dominant components. The most common applications of SRM in proteomics rely on the principles of stable isotope dilution (SID) methods [144] that provide the highest possible analyte specificity for quantitative determinations. The term SID typically refers to the use of a stable isotope-labeled internal standard spiked into a sample at a known concentration. The final response ratio between the analyte and labeled compound can be interpolated on a standard curve to calculate the absolute amount of analyte in unknown samples without being affected by matrix effects. SRM and SID combination in proteomics is finding exponentially increasing application so that they can be now considered the "golden standard" for absolute quantification. This approach has been considered a tool for the precise quantitative determination of targeted proteins in complex samples [145]. In general, a reliable quantification requires the use of internal reference peptides. In the very early beginning, selected ion monitoring (SIM) scheme also involving MS/MS spectra corresponding to each fragmented precursor ion, has been used for food allergen monitoring in different food commodities by using ESI-QqTOF-MS systems coupled with either ultra high performance liquid chromatography (UHPLC) or microHPLC separation. The first applications of such QqTOF approach in food allergen research date back to the early 2000 where for the first time a capillary HPLC system coupled to QqTOF-MS was used for tracing peanuts or milk allergens in food products, assessing the potentials offered by such technology for qualitative and quantitative analysis [146–150]. The SIM method developed, was based on the detection of precursor ion peptides and the respective fragment patterns displayed in the MS/MS spectra obtained. This approach proved to be applicable to food allergen detection even though the sensitivity reached was not very challenging. With the advent of the latest highly sensitive triple quadrupole mass spectrometers, the routine quantitative analysis usually accomplished on small molecules was transferred to the protein field. This was attained by monitoring multiple transitions of the best peptide markers identified for each allergenic category. A multiallergen SRM method capable of tracing seven allergenic ingredients in a single run was for the first time described by Heick et al. [151]. Once suitable peptides and transitions were properly selected for the seven allergenic foods investigated in that work, the method was applied to bread incurred with these seven allergenic ingredients to assess the quantitative capabilities of the method. According to their findings, LODs ranging from 3 to 70 μg/g were obtained in the incurred bread matrix, depending on the specific allergen. Another interesting work following this scheme described the development of an LC-MS method and its validation for the accurate quantification of milk traces in different food products [152]. Upon proper selection of suitable markers, quantification was attained by using internal standard peptides containing isotopically labeled amino acids. Such method enabled to reach LODs down to 0.2–0.5 mg/kg comparable to the limits obtained with ELISA kits, therefore proving that LC-SRM-MS approach could be intended as a sensitive and quantitative tool for milk allergens detection in selected food matrices. In the same period, other papers appeared in literature mostly enhancing potentials of the linear ion trap MS for the multitarget analysis of nuts and fish allergens in diverse food matrices [153–155]. Some authors also described advantages and limitations of multitarget allergen analysis by using MS^3 acquisition mode [156]. A method duly optimized for the simultaneous detection of soy, egg, and milk allergens in a cookie food matrix by microHPLC-ESI-MS/MS, was recently, proposed [157]. Thanks to the innovative configuration and the versatility shown by the dual cell linear ion trap MS used, the most intense and reliable peptide markers were first identified by untargeted survey experiment, and subsequently employed to design an ad hoc multitarget SRM method (Figure 6). A total of 11 peptides were monitored simultaneously for confirmation purposes of each allergenic contaminant and the two most sensitive peptide markers/protein were selected in order to retrieve quantitative information. Relevant LODs were found to range from 0.1 μg/g for milk to 0.3 μg/g for egg and 2 μg/g for soy.

An innovation in this regard was represented by a study demonstrating for the first time the feasibility of a single stage-Orbitrap™-mass spectrometer for the fast and high-throughput screening of egg and milk allergens in wine samples [158]. In general, high resolution mass spectrometry (HRMS) offers many benefits over the classical unit-mass-resolution tandem mass spectrometry [159–162]. Hence, the combined difficulties associated with managing and monitoring a large number of SRM traces motivated the shift toward full-scan-based HRMS screening. The collection of full-scan spectra provides greater insights into the identity and chemical structure of a food component. In addition, thanks to the excellent scan speed and resolving power of the HRMS systems, a full-scan untargeted analysis in both, precursor MS and product ion MS mode at the highest resolving power and mass accuracy might be accomplished for multiplex analysis of food allergens. By following this approach, challenging LODs can be obtained. Thanks to the post acquisition accurate mass filtration of the selected peptide ions operated on the total ion current traces, thus representing a valid alternative to the SRM-based methods for quantification purposes [163]. The method was also tested on other food matrices e.g., cookies incurred with milk powder, proving to be a

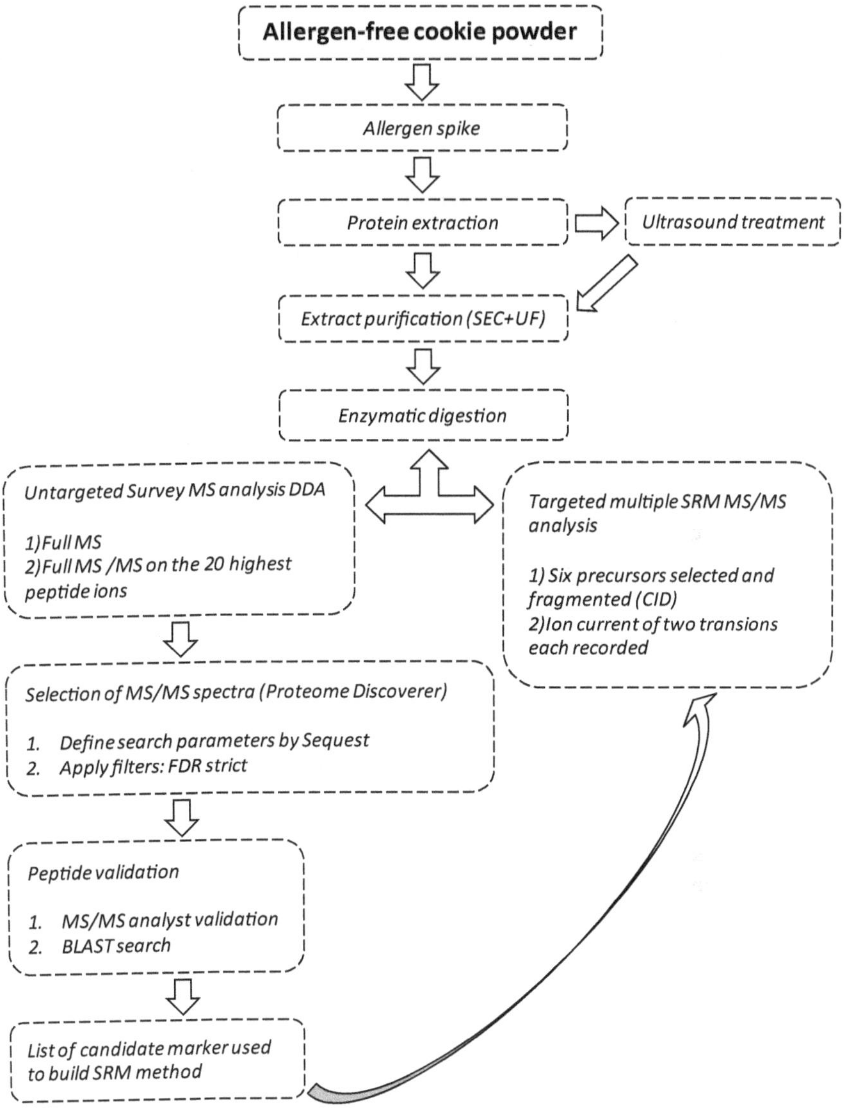

FIGURE 6 Experimental workflow and double MS acquisition scheme designed for peptide markers discovery.

good screening tool for monitoring food allergen contamination also in complex food products [158]. Potentials and features of SRM- and HRMS-based methods for the multiplex screening of egg- and milk-related proteins were recently compared choosing a type of wine as a reference matrix. Quantitative and confirmative capabilities of linear ion trap and an Orbitrap™ were assessed on the specific case study [164]. Tables 2 and 3 resume the HPLC-MS

TABLE 2 Overview of the LC-MS/MS Quantitative Methods Devised for Animal-Origin Allergen Detection in Food Products Based on Detection of Peptide Markers

Animal-Origin Allergenic Compound	Foodstuff	Target Protein	LOD	MS Method	References
Egg (ovalbumin)	Red wine	Ovalbumin	0.8 μg/mL	HPLC/LTQ linear ion trap-SRM	[166]
Egg (egg white powder)	Red wine	Ovalbumin Ovotransferrin Lysozyme Ovomucin Serum albumin	50 μg/mL	HPLC/quadrupole-time-of-flight-SIM	[167]
Egg (egg white powder)	Bread	Ovalbumin	42 μg/g	HPLC/triple quadrupole-linear ion trap-MRM	[151]
Egg (egg white powder)	White wine	Ovalbumin Lysozyme	0.4–1.1 μg/mL	HPLC/Orbitrap™ MS	[159]
Egg (whole egg powder)	Biscuits	Ovalbumin	0.3 μg/g	MicroHPLC/linear ion trap-SRM	[157]
Egg (ovalbumin + lysozyme)	White wine	Ovalbumin Lysozyme	0.3 μg/mL 0.18 μg/mL	HPLC/Orbitrap™ MS	[164]
	White wine	Ovalbumin Lysozyme	0.19 μg/mL 0.19 μg/mL	MicroHPLC/linear ion trap-SRM	
Milk (skim milk powder)	Cookie	αS1-casein	1.25 μg/g	Capillary HPLC/quadrupole-time-of-flight-SIM	[146]
Milk (caseinate)	White wines	αS1-casein αS2-casein β-casein	0.09–0.29 μg/mL	HPLC/3D ion trap-MS^2	[168]

Milk (milk powder)	Cookie	αS1-casein αS2-casein BSA	100 μg/g	Capillary HPLC/quadrupole-time-of-flight-SIM	[149]
Milk (caseinate)	White wine	αS1-casein β-casein	50 μg/mL	Capillary HPLC/quadrupole-time-of-flight-SIM	[169]
Milk (caseinate)	White wine	αS1-casein β-casein	39–51 μg/mL	HPLC/Orbitrap™ MS	[163]
Milk (caseinate)	White wine	αS1-casein αS2-casein β-casein	0.4–0.9 μg/mL	HPLC/Orbitrap™-MS	[159]
Milk (α-casein, β-casein)	Red wine	α-casein β-casein	0.5 μg/mL 0.01 μg/mL	HPLC/LTQ linear ion trap-SRM	[166]
Milk (skim milk powder)	Soy-based infant formula	β-casein β-lactoglobulin αS2-casein k-casein	20 μg/g 5–20 μg/g 5–50 μg/g 20 μg/g	HPLC/triple quadrupole-MRM	[152]
	Breakfast cereals	β-casein β-lactoglobulin αS2-casein k-casein	1 μg/g 1–2 μg/g 1–2 μg/g 2 μg/g	HPLC/triple quadrupole-MRM	
	Infant cereals	β-casein β-lactoglobulin αS2-casein k-casein	1 μg/g 1–2 μg/g 1–2 μg/g 2 μg/g	HPLC/triple quadrupole-MRM	
	Baby food	β-casein β-lactoglobulin αS2-casein k-casein	<1 μg/g 1 μg/g 1 μg/g 1 μg/g	HPLC/triple quadrupole-MRM	

Continued

TABLE 2 Overview of the LC-MS/MS Quantitative Methods Devised for Animal-Origin Allergen Detection in Food Products Based on Detection of Peptide Markers—cont'd

Animal-Origin Allergenic Compound	Foodstuff	Target Protein	LOD	MS Method	References
Milk (skim milk powder)	Bread	αS1-casein	5 μg/g	HPLC/triple quadrupole-linear ion trap-MRM	[151]
Milk (nonfat dry milk)	Biscuits	αS1-casein	3 μg/g	HPLC/triple quadrupole-linear ion trap-MRM	[137]
Milk (skimmed milk powder)	Biscuits	αS1-casein	0.1 μg/g	MicroHPLC/linear ion trap-SRM	[157]
Milk (caseinate)	White wine	αS1-casein	0.3 μg/mL	HPLC/Orbitrap™ MS	[164]
		αS1-casein	0.2 μg/mL	MicroHPLC/linear ion trap-SRM	

TABLE 3 Overview of the LC-MS/MS Quantitative Methods Devised for Plant-Origin Allergen Detection in Food Products Based on Detection of Peptide Markers

Plant-Origin Allergenic Compound	Foodstuff	Target Protein	LOD (μg/g)	MS Method	References
Hazelnut (flour)	Bread	11S globulin	5 μg/g	HPLC/triple quadrupole-linear ion trap-MRM	[151]
Hazelnut (flour)	Biscuits Biscuits	Cor a 9 Cor a 9	30 μg/g 35 μg/g	HPLC/linear ion trap-SRM HPLC/linear ion trap-SRM-MS^3	[156]
Hazelnut (flour)	Biscuits Dark chocolate	Cor a 9 Cor a 9	1.3 μg/g 14 μg/g	HPLC/linear ion trap-SRM-MS^2	[170]
Cashewnut (flour)	Biscuits Biscuits	Ana o 2 Ana o 2	14 μg/g 30 μg/g	HPLC/linear ion trap-SRM HPLC/linear ion trap-SRM-MS^3	[156]
Cashewnut (flour)	Biscuits Dark chocolate	Ana o 2 Ana o 2	0.5 μg/g 15 μg/g	HPLC/linear ion trap-SRM	[170]
Almond (flour)	Biscuits Biscuits	Pru 1 Pru 1	17 μg/g 25 μg/g	HPLC/linear ion trap-SRM HPLC/linear ion trap-SRM-MS^3	[156]
Almond (flour)	Biscuits Dark chocolate	Pru 1 Pru 1	0.9 μg/g 9 μg/g	HPLC/linear ion trap-SRM	[170]
Almond (flour)	Bread	Prunin	3 μg/g	HPLC/triple quadrupole-linear ion trap-MRM	[151]
Peanut (flour)	Biscuits Biscuits	Ara h 3/4 Ara h 3/4	10 μg/g 27 μg/g	HPLC/linear ion trap-SRM HPLC/linear ion trap-SRM-MS^3	[156]

Continued

TABLE 3 Overview of the LC-MS/MS Quantitative Methods Devised for Plant-Origin Allergen Detection in Food Products Based on Detection of Peptide Markers—cont'd

Plant-Origin Allergenic Compound	Foodstuff	Target Protein	LOD (μg/g)	MS Method	References
Peanut (flour)	Biscuits Dark chocolate	Ara h 3/4 Ara h 3/4	0.1 μg/g 7 μg/g	HPLC/linear ion trap-SRM	[170]
Peanut (flour)	Bread	Ara h 1	11 μg/g	HPLC/triple quadrupole-linear ion trap-MRM	[151]
Peanut (Ara h 1)	Vanilla ice cream	Ara h 1	10 μg/g	HPLC/quadrupole time-of-flight-SIM	[147]
Peanut (incurred material)	Dark chocolate	Ara h 1	2 μg/g	HPLC/quadrupole time-of-flight-SIM	[171]
Peanut (flour)	Chocolate/rice crispy based snacks	Ara h 2 Ara h 3/4	5 μg/g 1 μg/g	HPLC/triple quadrupole-MRM	[154]
Peanut (flour)	Breakfast cereals	Ara h 3/4	3 μg/g	HPLC/linear ion trap-SRM	[172]
Peanut (flour, IRMM-481f peanut test material)	Cookie	Ara h 1 Ara h 2 Ara h 3/4	100 μg/g Not detected 10 μg/g	HPLC/triple quadrupole-MRM	[173]
Walnut (flour)	Biscuits Biscuits	Jug r 4 Jug r 4	55 μg/g 50 μg/g	HPLC/linear ion trap-SRM HPLC/linear ion trap-SRM-MS^3	[156]

Walnut (flour)	Biscuits Dark chocolate	Jug r 4 Jug r 4	0.8 μg/g 5 μg/g	HPLC/linear ion trap-SRM	[170]
Walnut (flour)	Bread	Jug r 1	70 μg/g	HPLC/triple quadrupole-linear ion trap-MRM	[151]
Wheat gluten (commercial gluten)	*Red wine*	*Gliadin* *Glutenin*	10 mg/L	Chip-based chromatographic interface/quadrupole time-of-flight-MS^2	[174]
Wheat gluten (standard peptide)	*Corn flour*	*Gluten*	1–30 ng/g	HPLC/triple quadrupole-MRM	[175]
Lupin (flour)	Pasta	β-conglutinin α-conglutinin δ-conglutinin γ-conglutinin	1 μg/g 4 μg/g 5 μg/g 13 μg/g	HPLC/linear ion trap-SRM	[153]
	Biscuits	β-conglutinin α-conglutinin δ-conglutinin γ-conglutinin	1 μg/g 2 μg/g 18 μg/g 24 μg/g	HPLC/linear ion trap-SRM	
Soy (flour)	Bread	Glycinin	24 μg/g	HPLC/triple quadrupole-linear ion trap-MRM	[151]
Soy (flour)	Biscuits	β-Conglycinin-α-chain	2 μg/g	MicroHPLC/linear ion trap-SRM	[157]

methods so far developed for food allergen detection in different foods by employing the shotgun strategy.

More recently, a feasibility study on the application of metrologically traceable MS-based reference procedures has been described by Cryar et al. [165]. The approach employed a proteolytic digestion step of wine spiked with lysozyme standard, followed by isotope dilution MS analysis. Results showed that the MS method was feasible for quantitative analysis proving to be suitable for the production of allergens in food certified reference materials. Confirmation of the feasibility of MS methods for absolute quantification of an allergenic protein in a food matrix is a first step toward comparison of results for allergen proteins collected among different laboratory for validation purposes.

4. CONCLUDING REMARKS

The present chapter describes the topic of food allergy, distribution worldwide, and the legislative frame along with potentials of MS-based strategy applied to the food allergen field. MS in general proved to be a valid tool for the analysis of food allergens both for structural characterization of peptides, proteins, epitopes as well as for detecting traces of allergens in different food commodities. It offers several advantages over the traditional antibody-based methods and polymerase chain reaction (PCR)-based kits, with the additional possibility to identify an allergenic protein at the highest confidence level. As a consequence, in the last two decades the interest of the analysts has been placed on the development of MS-based strategies tailored to the high-throughput multitarget screening of allergens in food. Investigations have also been directed to assess quantitative potentials of the MS methodologies developed. The proteomics workflow typically utilized for allergens analysis employed a protein extraction step followed by proteolytic digestion of the partially purified extract and HPLC-MS/MS analysis of the resulting peptide mixture. Detection of the offending allergen is achieved through detection of unique previously selected peptide markers tracing for each target allergen. In perspective, still few aspects need to be tackled in order to increase confidence in the analytical measurement performed. In particular, the use of isotopically labeled equivalents (proteins or peptides) is strongly encouraged for quantitative analysis. Besides, in order to assure reliability and accuracy in the analytical measurement, the method under evaluation should undergo validation among different laboratories. On the other hand, sample preparation still represents a crucial issue and deserves to be in-depth investigated. Efforts should be directed toward a better understanding of the effects of sample treatment and processing, and the influence of the matrix on the target proteins to minimize the bias and increase the confidence in the measurements obtained by different laboratories. On this regard, the use of reference materials commercially available would help scientists in checking performances of the MS method under development.

REFERENCES

[1] A. Cianferoni, J.M. Spergel, Allergol. Int. 58 (2009) 457–466.

[2] S.A. Cochrane, L.J. Salt, E. Wantling, A. Rogers, J. Coutts, B.K. Ballmer-Weber, P. Fritsche, M. Fernandez-Rivas, I. Reig, A. Knulst, T.M. Le, R. Asero, K. Beyer, M. Golding, R. Crevel, E.N. Clare Mills, A.R. Mackie, Allergy 67 (2012) 107–113.

[3] J. Wang, H.A. Sampson, J. Clin. Invest. 121 (2011) 827–835.

[4] R. Rona, T. Keil, C. Summers, D. Gislason, L. Zuidmeer, E. Sodergren, S. Sigurdardottir, T. Lindner, K. Goldhahn, J. Dahlstrom, D. McBride, C. Madsen, J. Allergy Clin. Immun. 120 (2007) 638–646.

[5] L. Zuidmeer, K. Goldhahn, R.J. Rona, D. Gislason, C. Madsen, C. Summers, E. Sodergren, J. Dahlstrom, T. Lindner, S.T. Sigurdardottir, D. McBride, T. Keil, J. Allergy Clin. Immunol. 121 (2008) 1210–1218.

[6] C.B. Madsen, S. Hattersley, J. Buck, S.M. Gendel, G.F. Houben, J.O. Hourihane, A. Mackie, E.N. Mills, P. Norhede, S.L. Taylor, R.W. Crevel, Food Chem. Toxicol. 47 (2009) 480–489.

[7] S. McClain, C. Bowman, M. Fernández-Rivas, G.S. Ladics, R. van Ree, Clin. Transl. Allergy 4 (2014) 11.

[8] M. Kurek, Alerg. Astma Immun. 3 (1998) 66–75.

[9] B. Rymarczyk, B. Rogala, Gastroenterol. Pol. 8 (2001) 277–282.

[10] O. Ozdemir, E. Mete, F. Catal, D. Ozol, Dig. Dis. Sci. 54 (2009) 8–14.

[11] S. Guandalini, C. Newland, Curr. Gastroenterol. Rep. 13 (2011) 426–434.

[12] S. Husby, J. Jensenius, S. Svehag, Scand. J. Immunol. 22 (1985) 83–92.

[13] A. Host, Pediatr. Allergy Immunol. 5 (1994) 5–36.

[14] L.T. Weaver, M.F. Laker, R. Nelson, A. Lucas, J. Pediatr. Gastr. Nutr. 6 (1987) 351–358.

[15] M. Chehade, L. Mayer, J. Allergy Clin. Immun. 115 (2005) 3–12.

[16] H. Jyonouchi, Inflamm. Allergy Drug Targets 7 (2008) 173–180.

[17] H.A. Sampson, J. Allergy. Clin. Immun. 113 (2004) 805–820.

[18] S.H. Sicherer, H.A. Sampson, Annu. Rev. Med. 60 (2009) 261–277.

[19] Available at: http://www.allergen.org. Last accession on June 19, 2014.

[20] M. Koeberl, D. Clarke, A.L. Lopata, J. Proteome Res. 13 (2014) 3499–3509.

[21] K. Fotisch, S. Vieths, Glycoconj. J. 18 (2001) 373–390.

[22] R. van Ree, M. Cabanes-Macheteau, J. Akkerdaas, J.-P. Milazzo, C. Loutelier-Bourhis, C. Rayon, M. Villalba, S. Koppelman, R. Aalberse, R. Rodriguez, L. Faye, P. Lerouge, J. Biol. Chem. 275 (2000) 11451–11458.

[23] H. Malandain, F. Giroux, Y. Cano, Eur. Ann. Allergy Clin. Immunol. 39 (2007) 216–220.

[24] S.J. Maleki, R.A. Kopper, D.S. Shin, C.W. Park, C.M. Compadre, H. Sampson, A.W. Burks, G.A. Bannon, J. Immunol. 164 (2000) 5844–5849.

[25] S.J. Koppelman, C.A. Bruijnzeel-Koomen, M. Hessing, H.H. de Jongh, J. Biol. Chem. 274 (1999) 4770–4777.

[26] S.J. Maleki, O. Viquez, T. Jacks, H. Dodo, E.T. Champagne, S.Y. Chung, S.J. Landry, J. Allergy Clin. Immunol. 112 (2003) 190–195.

[27] K. Lehmann, K. Schweimer, G. Reese, S. Randow, M. Suhr, W.M. Becker, S. Vieths, P. Rosch, Biochem. J. 395 (2006) 463–472.

[28] L. Perrocheau, B. Bakan, P. Boivin, D. Marion, J. Agric. Food Chem. 54 (2006) 3108–3113.

[29] M. Heilmann, A. Wellner, G. Gadermaier, A. Ilchmann, P. Briza, M. Krause, R. Nagai, S. Burgdorf, S. Scheurer, S. Vieths, T. Henle, M. Toda, J. Biol. Chem. 289 (2014) 7919–7928.

[30] R. Bredehorst, K. David, J. Chromatogr. B: Biomed. Sci. Appl. 756 (2001) 33–40.

[31] H. Breiteneder, E.N. Mills, J. Allergy Clin. Immunol. 115 (2005) 14–23.

[32] F.J. Moreno, L.A. Rubio, A. Olano, A. Clemente, J. Agric. Food Chem. 54 (2006) 8631–8639.

[33] J. Lin, P.R. Shewry, D.B. Archer, K. Beyer, B. Niggemann, H. Haas, P. Wilson, M.J. Alcocer, Int. Arch. Allergy Immunol. 141 (2006) 91–102.

[34] G. Garcia-Casado, J.F. Crespo, J. Rodriguez, G. Salcedo, J. Allergy Clin. Immunol. 108 (2001) 647–649.

[35] S.J. Koppelman, S.L. Hefle, S.L. Taylor, G.A. de Jong, Mol. Nutr. Food Res. 54 (2010) 1711–1721.

[36] R. Wijesinha-Bettoni, Y. Alexeev, P. Johnson, J. Marsh, A.I. Sancho, S.U. Abdullah, A.R. Mackie, P.R. Shewry, L.J. Smith, E.N. Mills, Biochemistry 49 (2010) 2130–2139.

[37] P.H. Green, C. Cellier, N. Engl. J. Med. 357 (2007) 1731–1743.

[38] L.M. Sollid, Ann. Rev. Immunol. 18 (2000) 53–81.

[39] J.A. Murray, C. Van Dyke, M.F. Plevak, R.A. Dierkhising, A.R. Zinsmeister, L.J. Melton 3rd, Clin. Gastroenterol. Hepatol. 1 (2003) 19–27.

[40] C. Catassi, E. Fabiani, I.M. Ratsch, G.V. Coppa, P.L. Giorgi, R. Pierdomenico, S. Alessandrini, G. Iwanejko, R. Domenici, E. Mei, A. Miano, M. Marani, G. Bottaro, M. Spina, M. Dotti, A. Montanelli, M. Barbato, F. Viola, R. Lazzari, M. Vallini, G. Guariso, M. Plebani, F. Cataldo, G. Traverso, A. Ventura, Acta Paediatr. 412 (1996) 2–35.

[41] J. West, R.F. Logan, C.J. Smith, R.B. Hubbard, T.R. Card, BMJ 329 (2004) 716–719.

[42] B. Jabri, D.D. Kasarda, P.H. Green, Immunol. Rev. 206 (2005) 219–231.

[43] M. Setty, L. Hormaza, S. Guandalini, Mol. Diagn. Ther. 12 (2008) 289–298.

[44] P. Margaritte-Jeannin, M.C. Babron, M. Bourgey, A.S. Louka, F. Clot, S. Percopo, I. Coto, J.P. Hugot, H. Ascher, L.M. Sollid, L. Greco, F. Clerget-Darpoux, Tissue Antigens 63 (2004) 562–567.

[45] L.M. Sollid, Nat. Rev. Immunol. 2 (2002) 647–655.

[46] D.A. van Heel, L. Franke, K.A. Hunt, R. Gwilliam, A. Zhernakova, M. Inouye, M.C. Wapenaar, M.C. Barnardo, G. Bethel, G.K. Holmes, C. Feighery, D. Jewell, D. Kelleher, P. Kumar, S. Travis, J.R. Walters, D.S. Sanders, P. Howdle, J. Swift, R.J. Playford, W.M. McLaren, M.L. Mearin, C.J. Mulder, R. McManus, R. McGinnis, L.R. Cardon, P. Deloukas, C. Wijmenga, Nat. Genet. 39 (2007) 827–829.

[47] K.A. Hunt, A. Zhernakova, G. Turner, G.A. Heap, L. Franke, M. Bruinenberg, J. Romanos, L.C. Dinesen, A.W. Ryan, D. Panesar, R. Gwilliam, F. Takeuchi, W.M. McLaren, G.K. Holmes, P.D. Howdle, J.R. Walters, D.S. Sanders, R.J. Playford, G. Trynka, C.J. Mulder, M.L. Mearin, W.H. Verbeek, V. Trimble, F.M. Stevens, C. O'Morain, N.P. Kennedy, D. Kelleher, D.J. Pennington, D.P. Strachan, W.L. McArdle, C.A. Mein, M.C. Wapenaar, P. Deloukas, R. McGinnis, R. McManus, C. Wijmenga, D.A. van Heel, Nat. Genet. 40 (2008) 395–402.

[48] M. Silano, C. Agostoni, S. Guandalini, World J. Gastroenterol. 16 (2010) 1939–1942.

[49] F. Bonomi, S. Iametti, G. Mamone, P. Ferranti, Cereal Chem. 90 (2013) 358–366.

[50] H. Wieser, Food Microbiol. 24 (2007) 115–119.

[51] R. Ciccocioppo, A. Di Sabatino, G.R. Corazza, Clin. Exp. Immunol. 140 (2005) 408–416.

[52] Ø Molberg, S.N. Mcadam, R. Körner, H. Quarsten, C. Kristiansen, L. Madsen, L. Fugger, H. Scott, O. Norén, P. Roepstorff, K.E. Lundin, H. Sjöström, L.M. Sollid, Nat. Med. 4 (1998) 713–717.

[53] E.H. Arentz-Hansen, S.N. McAdam, O. Molberg, C. Kristiansen, L.M. Sollid, Gut 46 (2000) 46–51.

[54] B. Fleckenstein, Ø Molberg, S.W. Qiao, D.G. Schmid, F. von der Mulbe, K. Elgstoen, G. Jung, L.M. Sollid, J. Biol. Chem. 277 (2002) 34109–34116.

[55] S. Krupicková, L. Tucková, Z. Flegelová, M. Michalak, J.R. Walters, A. Whelan, J. Harries, J. Vencovský, H. Tlaskalová-Hogenová, Gut 44 (1999) 168–173.

[56] S.M. Stenman, K. Lindfors, I.R. Korponay-Szabo, O. Lohi, P. Saavalainen, J. Partanen, K. Haimila, H. Wieser, M. Mäki, K. Kaukinen, BMC Immunol. 9 (2008) 6.

[57] R. Shaoul, A. Lerner, Autoimmun. Rev. 6 (2007) 559–565.

[58] M. Londei, C. Ciacci, I. Ricciarelli, L. Vacca, S. Quarantino, L. Maiuri, Mol. Immunol. 42 (2005) 913–918.

[59] L. Maiuri, C. Ciacci, I. Ricciardelli, L. Vacca, V. Raia, S. Auricchio, J. Picard, M. Osman, S. Quaratin, M. Londei, Lancet 362 (2003) 30–37.

[60] S. Hüe, J.J. Mention, R.C. Monteiro, S. Zhang, C. Cellier, J. Schmitz, V. Verkarre, N. Fodil, S. Bahram, N. Cerf-Bensussan, S. Caillat-Zucman, Immunity 21 (2004) 367–377.

[61] S. Hefle, J. Nordlee, D. Taylor, Food Sci. Nutr. 36S (1996) S69–S89.

[62] S. Hattersley, R. Ward, A. Baka, R.W.R. Crevel, Food Chem. Toxicol. 67 (2014) 255–261.

[63] Codex Alimentarius, Codex General Standard for the Labelling of Pre-packaged Foods, FAO/WHO, 1999.

[64] S.M. Gendel, Regul. Toxicol. Pharm. 63 (2012) 279–285.

[65] European Parliament and Council, Directive 2007/68/EC of 27 November 2007 amending Annex III to Directive 2000/13/EC as regards certain food ingredients, Off. J. Eur. Union L 310 (2007) 11–14.

[66] RASFF Rapid Alert System for Food and Feed, European Commission, Brussels, Belgium, 2012.

[67] K. Vierk, K. Falci, C. Wolyniak, K.C. Klontz, J. Allergy Clin. Immun. 109 (2002) 1022–1026.

[68] EU Regulation 178/2002, Regulation (EC) No 178/2002, Off. J. Eur. Communities 31 (2002) 21–24.

[69] Regulation EC No 178/2002 of the European Parliament and of the Council.

[70] A. Sakellariou, A. Sinaniotis, L. Damianidou, N.G. Papadopoulos, E. Vassilopoulou, Allergy 65 (2010) 534–535.

[71] G.A. Zurzolo, M.L. Mathai, J.J. Koplin, K.J. Allen, Curr. Allergy Asthma Rep. 12 (2012) 292–296.

[72] EU Food Information for Consumers Regulation (No. 1169/2011) http://eur-lex.europa.eu/LexUriServ/LexUriServ.do?uri=OJ:L:2011:304:0018:0063:EN:PDF.

[73] S.L. Hefle, T.J. Furlong, L. Niemann, H. Lemon-Mule, S. Sicherer, S.L. Taylor, J. Allergy Clin. Immunol. 120 (2007) 171–176.

[74] J. Barnett, J. Leftwich, K. Muncer, K. Grimshaw, R. Shepherd, M.M. Raats, M.H. Gowland, J.S. Lucas, Allergy 66 (2011) 969–978.

[75] G. Picariello, G. Mamone, C. Nitride, F. Addeo, P. Ferranti, TrAC-Trend Anal. Chem. 52 (2013) 120–134.

[76] M. Akagawa, T. Handoyo, T. Ishii, S. Kumazawa, N. Morita, K. Suyama, J. Agric. Food Chem. 55 (2007) 6863–6870.

[77] I. Sander, M. Raulf-Heimsoth, M. Duser, A. Flagge, A.B. Czuppon, X. Baur, Int. Arch. Allergy Immunol. 112 (1997) 378–385.

[78] P. Sotkovský, M. Hubalek, L. Hernychova, P. Novak, M. Havranova, I. Setinova, A. Kitanovicova, M. Fuchs, J. Stulik, L. Tuckova, Proteomics 8 (2008) 1677–1691.

[79] E.A. Pastorello, L. Farioli, V. Pravettoni, J. Scibilia, A. Conti, D. Fortunato, L. Borgonovo, S. Bonomi, L. Primavesi, B. Ballmer-Weber, Anal. Bioanal. Chem. 395 (2009) 93–102.

[80] M. Natale, C. Bisson, G. Monti, A. Peltran, L.P. Garoffo, S. Valentini, C. Fabris, E. Bertino, A. Coscia, A. Conti, Mol. Nutr. Food Res. 48 (2004) 363–369.

[81] C.J. Yu, Y.F. Lin, B.L. Chiang, L.P. Chow, J. Immunol. 170 (2003) 445–453.
[82] C. Nitride, G. Mamone, G. Picariello, C. Mills, R. Nocerino, R. Berni Canani, P. Ferranti, J. Proteomics 86 (2013) 16–26.
[83] A. Palacin, L. Tordesillas, P. Gamboa, R. Sanchez-Monge, J. Cuesta-Herranz, M.L. Sanz, D. Barber, G. Salcedo, A. Diaz-Perales, Clin. Exp. Allergy 40 (2010) 1422–1430.
[84] G.C. Hartz, M.M. San Miguel-Moncin, A. Cistero-Bahima, K. Fotisch, K.J. Metzner, D. Fortunato, J. Lidholm, S. Vieths, S. Scheurer, Mol. Immunol. 44 (2007) 2820–2830.
[85] S.K. Lee, S.H. Yoon, S.H. Kim, J.H. Choi, H.S. Park, J. Korean, Med. Sci. 20 (2005) 573–578.
[86] M. Grob, J. Reindl, S. Vieths, B. Wuthrich, B.K. Ballmer-Weber, Ann. Allergy Asthma Immunol. 89 (2002) 513–516.
[87] O.Y. Bassler, J. Weiss, S. Wienkoop, K. Lehmann, C. Scheler, S. Dolle, D. Schwarz, P. Franken, E. George, M. Worm, W. Weckwerth, J. Proteome Res. 8 (2009) 1111–1122.
[88] H. Hochwallner, U. Schulmeister, I. Swoboda, M. Focke-Tejkl, V. Civaj, N. Balica, M. Nystrand, A. Harlin, J. Thalhamer, S. Scheiblhofer, W. Keller, T. Pavkov, D. Zafred, B. Niggemann, S. Quirce, A. Mari, G. Pauli, C. Ebner, N.G. Papadopoulos, U. Herz, E.A.F. van Tol, R. Valenta, S. Spitzauer, J. Allergy Clin. Immunol. 125 (2010) 1279–1285.
[89] Z. Bilkova, R. Stefanescu, R. Cecal, L. Korecka, S. Ouzka, J. Jezova, J.L. Viovy, M. Przybylski, Eur. J. Mass Spectrom. 11 (2005) 489–495.
[90] B. Jankovicova, S. Rosnerova, M. Slovakova, Z. Zverinova, M. Hubalek, L. Hernychova, P. Rehulka, J.L. Viovy, Z. Bilkova, J. Chromatogr. A 1206 (2008) 64–71.
[91] F.J. Moreno, B.M. Maldonado, N. Wellner, E.N. Mills, Biochim. Biophys. Acta 1752 (2005) 142–153.
[92] C.P. Mattison, C.C. Grimm, R.L. Wasserman, Mol. Nutr. Food Res. 58 (2014) 884–893.
[93] G. Picariello, P. Ferranti, O. Fierro, G. Mamone, S. Caira, A. Di Luccia, S. Monica, F. Addeo, J. Chromatogr. B Analyt. Technol. Biomed. Life Sci. 878 (2010) 295–308.
[94] A. Diaz-Perales, C. Blanco, R. Sanchez-Monge, J. Varela, T. Carrillo, G. Salcedo, J. Allergy Clin. Immunol. 112 (2003) 1002–1007.
[95] F. Delie, W. Rubas, Crit. Rev. Ther. Drug Carrier Syst. 14 (1997) 221–286.
[96] M. Bodinier, M.A. Legoux, F. Pineau, S. Triballeau, J.P. Segain, C. Brossard, S. Denery-Papini, J. Agric. Food Chem. 55 (2007) 4576–4583.
[97] I. Caillard, D. Tomé, Reprod. Nutr. Dev. 35 (1995) 179–188.
[98] J. Rytkönen, K.H. Valkonen, V. Virtanen, R.A. Foxwell, J.M. Kyd, A.W. Cripps, T.J. Karttunen, J. Agric. Food Chem. 54 (2006) 1500–1507.
[99] G. Picariello, G. Iacomino, G. Mamone, P. Ferranti, O. Fierro, C. Gianfrani, A. Di Luccia, F. Addeo, Food Chem. 139 (2013) 203–212.
[100] P. Ferranti, G. Mamone, G. Picariello, F. Addeo, J. Mass Spectrom. 42 (2007) 1531–1548.
[101] G. Mamone, G. Picariello, F. Addeo, P. Ferranti, Expert Rev. Proteomics 8 (2011) 95–115.
[102] V. De Re, M.P. Simula, V. Canzonieri, R. Cannizzaro, Dig. Dis. Sci. 55 (2010) 3041–3046.
[103] J. Stulík, L. Hernychová, S. Porkertová, O. Pozler, L. Tucková, D. Sánchez, J. Bures, Proteomics 3 (2003) 951–956.
[104] W. Dieterich, T. Ehnis, M. Bauer, P. Donner, U. Volta, E.O. Riecken, D. Schuppan, Nat. Med. 3 (1997) 797–801.
[105] S. Orrù, I. Caputo, A. D'Amato, M. Ruoppolo, C. Esposito, J. Biol. Chem. 278 (2003) 31766–31773.
[106] G. Mamone, P. Ferranti, D. Melck, F. Tafuro, L. Longobardo, L. Chianese, F. Addeo, FEBS Lett. 562 (2004) 177–1782.
[107] J.L. Piper, G.M. Gray, C. Khosla, Biochemistry 41 (2002) 386–393.

[108] B. Fleckenstein, S.W. Qiao, M.R. Larsen, G. Jung, P. Roepstorff, L.M. Sollid, J. Biol. Chem. 279 (2004) 17607–17616.

[109] L. Maiuri, R. Troncone, M. Mayer, S. Coletta, A. Picarelli, M. De Vincenzi, V. Pavone, S. Auricchio, Scand. J. Gastroenterol. 31 (1996) 247–253.

[110] G. Mamone, P. Ferranti, M. Rossi, P. Roepstorff, O. Fierro, A. Malorni, F. Addeo, J. Chromatogr. B 855 (2007) 236–241.

[111] G. Iacomino, O. Fierro, S. D'Auria, G. Picariello, P. Ferranti, C. Liguori, F. Addeo, G. Mamone, J. Agric. Food Chem. 61 (2013) 1088–1096.

[112] M.N. Marsh, S. Morgan, A. Ensari, T. Wardle, R. Lobley, C. Mills, S. Auricchio, Gastroenterology 108 (1995) A871.

[113] P.J. Ciclitira, H.J. Ellis, Dig. Dis. 16 (1998) 337–340.

[114] H. Lahteenoja, M. Maki, M. Viander, I. Raiha, P. Vilja, I. Rantala, A. Toivanen, S. Syrjänen, Am. J. Gastroenterol. 95 (2000) 2880–2887.

[115] M.F. Mazzeo, B. De Giulio, S. Senger, M. Rossi, A. Malorni, R.A. Siciliano, Protein Sci. 12 (2003) 2434–2442.

[116] S. Senger, F. Maurano, M.F. Mazzeo, M. Gaita, O. Fierro, C.S. David, R. Troncone, S. Auricchio, R.A. Siciliano, M. Rossi, J. Immunol. 175 (2005) 8087–8095.

[117] L. Shan, Ø. Molberg, I. Parrot, F. Hausch, F. Filiz, G.M. Gray, L.M. Sollid, C. Khosla, Science 297 (2002) 2275–2279.

[118] L. Shan, S.W. Qiao, H. Arentz-Hansen, Ø. Molberg, G.M. Gray, L.M. Sollid, C. Khosla, J. Proteome Res. 4 (2005) 1732–1741.

[119] S. Dørum, M.Ø. Arntzen, S.W. Qiao, A. Holm, C.J. Koehler, B. Thiede, L.M. Sollid, B. Fleckenstein, PLoS One 5 (2005) e14056.

[120] G. Mamone, A. Camarca, O. Fierro, J. Sidney, G. Mazzarella, F. Addeo, S. Auricchio, R. Troncone, A. Sette, C. Gianfrani, J. Agric. Food Chem. 61 (2013) 747–754.

[121] R. Di Cagno, M. De Angelis, P. Lavermicocca, M. De Vincenzi, C. Giovannini, M. Faccia, M. Gobbetti, Appl. Environ. Microbiol. 68 (2002) 623–633.

[122] R. Di Cagno, M. De Angelis, G. Alfonsi, M. De Vincenzi, M. Silano, O. Vincentini, M. Gobbetti, J. Agric. Food Chem. 53 (2005) 4379–4402.

[123] C.G. Rizzello, M. De Angelis, R. Di Cagno, A. Camarca, M. Silano, I. Losito, M. De Vincenzi, M.D. De Bari, F. Palmisano, F. Maurano, C. Gianfrani, M. Gobbetti, Appl. Environ. Microbiol. 73 (2007) 4499–4507.

[124] M. De Angelis, R. Coda, M. Silano, F. Minervini, C.G. Rizzello, R. Di Cagno, O. Vicentini, M. De Vincenzi, M. Gobbetti, J. Cereal Sci. 43 (2006) 301–314.

[125] T. Marti, O. Molberg, Q. Li, G.M. Gray, C. Khosla, L.M. Sollid, J. Pharmacol. Exp. Ther. 312 (2005) 19–26.

[126] M. De Angelis, C.G. Rizzello, E. Scala, C. De Simone, G.A. Farris, F. Turrini, M. Gobbetti, J. Food Prot. 70 (2007) 135–144.

[127] M. Gobbetti, C.G. Rizzello, R. Di Cagno, M. De Angelis, Food Microbiol. 24 (2007) 187–196.

[128] C. Gianfrani, R.A. Siciliano, A.M. Facchiano, A. Camarca, M.F. Mazzeo, S. Costantini, V.M. Salvati, F. Maurano, G. Mazzarella, G. Iaquinto, P. Bergamo, M. Rossi, Gastroenterology 133 (2007) 780–789.

[129] M. Guilhaus, D. Selby, V. Mlynski, Mass Spectrom. Rev. 19 (2000) 65–107.

[130] S.C. Moyer, R.J. Cotter, Anal. Chem. 74 (2002) 468A–476A.

[131] F. Fenaille, F. Morgan, V. Parisod, J.C. Tabet, P.A. Guy, Rapid Commun. Mass. Spectrom. 17 (2003) 1483–1492.

[132] R. Grandori, J. Mass Spectrom. 38 (2003) 11–15.

[133] J. Leonil, D. Molle, J. Fauquant, J.L. Maubois, R.J. Pearce, S. Bouhallab, J. Dairy Sci. 80 (1997) 2270–2281.
[134] I. Losito, T. Carbonara, L. Monaci, F. Palmisano, Anal. Bioanal. Chem. 389 (2007) 2065–2074.
[135] L. Monaci, A.J. Van Hengel, J. Agric. Food Chem. 55 (2007) 2985–2992.
[136] L. Monaci, A. Visconti, TrAC-Trend Anal. Chem. 28 (2009) 581–591.
[137] G.A. Newsome, P.F. Scholl, J. Agric. Food Chem. 6 (2012) 5659–5668.
[138] F.K. Yeboah, V.A. Yaylayan, Nahrung 45 (2001) 164–171.
[139] R. Siciliano, B. Rega, A. Amoresano, P. Pucci, Anal. Chem. 72 (2000) 408–415.
[140] L. Monaci, A.J. van Hengel, J. Chromatogr. A 1192 (2008) 113–120.
[141] W.W. Wu, G. Wang, S.J. Baek, R.F. Shen, J. Proteome Res. 5 (2006) 651–658.
[142] A.C. Gingras, M. Gstaiger, B. Raught, R. Aebersold, Nat. Rev. Mol. Cell Biol. 8 (2007) 645–654.
[143] P. Johnson, S. Baumgartner, T. Aldick, C. Bessant, V. Giosafatto, J. Heick, G. Mamone, G. O'Connor, R. Poms, B. Popping, A. Reuter, F. Ulberth, A. Watson, L. Monaci, E.N.C. Mills, J. AOAC Int. 94 (2011) 1026–1033.
[144] V. Mayya, D.K. Han, Expert Rev. Proteomics 3 (2006) 597–610.
[145] P. Picotti, O. Rinner, R. Stallmach, F. Dautel, T. Farrah, B. Domon, H. Wenschuh, R. Aebersold, Nat. Methods 7 (2010) 43–46.
[146] D. Weber, P. Raymond, S. Ben-Rejeb, B. Lau, J. Agric. Food Chem. 54 (2006) 1604–1610.
[147] K.J. Shefcheck, S.M. Musser, J. Agric. Food Chem. 52 (2004) 2785–2790.
[148] H. Chassaigne, J.V. Nørgaard, A.J. van Hengel, J. Agric. Food Chem. 55 (2007) 4461–4473.
[149] L. Monaci, J.V. Nørgaard, A.J. van Hengel, Anal. Methods 2 (2010) 967–972.
[150] S. Azarnia, J.I. Boye, V. Mongeon, H. Sabik, Food Res. Int. 52 (2013) 526–534.
[151] J. Heick, M. Fischer, B. Popping, J. Chromatogr. A 1218 (2011) 938–943.
[152] P. Lutter, V. Parisod, H. Weymuth, J. AOAC Int. 94 (2011) 1043–1059.
[153] M. Mattarozzi, C. Bignardi, L. Elviri, M. Careri, J. Agric. Food Chem. 60 (2012) 5841–5846.
[154] M. Careri, A. Costa, L. Elviri, J.B. Lagos, A. Mangia, M. Terenghi, A. Cereti, L. Perono Garoffo, Anal. Bioanal. Chem. 389 (2007) 1901–1907.
[155] M. Carrera, B. Cañas, J.M. Gallardo, J. Proteomics 75 (2012) 3211–3220.
[156] C. Bignardi, L. Elviri, A. Penna, M. Careri, A. Mangia, J. Chromatogr. A 1217 (2010) 7579–7585.
[157] L. Monaci, R. Pilolli, E. De Angelis, M. Godula, A. Visconti, J. Chromatogr. A 1358 (2014) 136–144.
[158] L. Monaci, I. Losito, F. Palmisano, A. Visconti, J. AOAC Int. 94 (2010) 1034–1042.
[159] L. Monaci, I. Losito, E. De Angelis, R. Pilolli, A. Visconti, Rapid Commun. Mass Spectrom. 27 (2013) 2009–2018.
[160] A. Kaufmann, Anal. Bioanal. Chem. 403 (2012) 1233–1249.
[161] A. Kaufmann, M. Widmer, K. Maden, Rapid Commun. Mass Spectrom. 24 (2010) 2162–2170.
[162] A. Makarov, M. Scigelova, J. Chromatogr. A 1217 (2010) 3938–3945.
[163] L. Monaci, I. Losito, F. Palmisano, M. Godula, A. Visconti, Food Addit. Contam. A 28 (2011) 1304–1314.
[164] R. Pilolli, E. De Angelis, M. Godula, A. Visconti, L. Monaci, J. Mass Spectrom. 49 (2014) 1254–1263.
[165] A. Cryar, C. Pritchard, W. Burkitt, M. Walker, G. O'Connor, D.T. Burns, M. Quaglia, J. AOAC Int. 96 (2013) 1350–1361.

[166] M. Mattarozzi, M. Milioli, C. Bignardi, L. Elviri, C. Corradini, M. Careri, Food Control 38 (2014) 82–87.
[167] S. Tolin, G. Pasini, A. Curioni, G. Arrigoni, A. Masi, F. Mainente, B. Simonato, Food Control 23 (2012) 87–94.
[168] I. Losito, B. Introna, L. Monaci, S. Minella, F. Palmisano, J. Agric. Food Chem. 61 (2013) 12436–12444.
[169] L. Monaci, I. Losito, F. Palmisano, A. Visconti, J. Chromatogr. A 1217 (2010) 4300–4305.
[170] C. Bignardi, M. Mattarozzi, A. Penna, S. Sidoli, L. Elviri, M. Careri, A. Mangia, Food Anal. Method 6 (2013) 1144–1152.
[171] K.J. Shefcheck, J.H. Callahan, S.M. Musser, J. Agric. Food Chem. 54 (2006) 7953–7959.
[172] M. Careri, L. Elviri, J.B. Lagos, A. Mangia, F. Speroni, M. Terenghi, J. Chromatogr. A 1206 (2008) 89–94.
[173] R. Pedreschi, J. Nørgaard, A. Maquet, Nutrients 4 (2012) 132–150.
[174] B. Simonato, F. Mainente, S. Tolin, G. Pasini, J. Agric. Food Chem. 59 (2011) 3101–3110.
[175] J.A. Sealey-Voyksner, C. Khosla, R.D. Voyksner, J.W. Jorgenson, J. Chromatogr. A 1217 (2010) 4167–4183.

Chapter 8

Lipidomics

Paola Donato,[1,*] Francesco Cacciola,[1] Marco Beccaria,[2] Paola Dugo[2,3] and Luigi Mondello[2,3,4]

[1]"Scienze dell'Ambiente, della Sicurezza, del Territorio, degli Alimenti e della Salute" Department, University of Messina, Messina, Italy; [2]"Scienze del Farmaco e Prodotti per la Salute" Department, University of Messina, Messina, Italy; [3]Integrated Research Center, University Campus Bio-Medico of Rome, Roma, Italy; [4]Chromaleont s.r.l., c/o "Scienze del Farmaco e Prodotti per la Salute" Department, University of Messina, Messina, Italy

**Corresponding author: E-mail: padonato@unime.it*

Chapter Outline

1. Introduction and Overview 396
2. Lipids: Definitions and Classification 397
3. Structures, Functions, and Occurrence of Lipids 398
3.1 Fatty Acids 399
3.2 Glycerolipids 401
3.3 Glycerophospholipids 402
3.4 Sphingolipids 404
3.5 Sterol Lipids 406
3.6 Prenol Lipids 406
3.7 Saccharolipids 407
3.8 Polyketides 407
4. Extraction Methods of Lipids 408
4.1 Sampling and Sample Preparation 408
4.2 Soxhlet Extraction 410
4.3 Method of Folch, Lees, and Stanley 410
4.4 Method of Bligh and Dyer 410
4.5 Supercritical Fluid Extraction (SFE) 411
4.6 Accelerated Solvent Extraction (ASE) 412
4.7 Microwave-Assisted Extraction (MAE) 412
4.8 Solid Phase Extraction (SPE) 412
5. Mass Spectrometry in Lipidomics Research 413
5.1 Ionization Modes 413
5.1.1 Electrospray Ionization (ESI) 414
5.1.2 Atmospheric Pressure Chemical Ionization (APCI) and Atmospheric Pressure Photoionization (APPI) 416
5.1.3 Matrix-Assisted Laser Desorption Ionization (MALDI) 416
5.1.4 Electron Ionization (EI) and Chemical Ionization (CI) 417
5.1.5 Fast Atom Bombardment (FAB) 418
5.1.6 Other Ionization Techniques 418
5.2 Advances in Mass Spectrometry for Lipidomics 419

Comprehensive Analytical Chemistry, Vol. 68. http://dx.doi.org/10.1016/B978-0-444-63340-8.00008-X

6. Chromatographic Separations in Lipidomics Research **420**
6.1 Thin-Layer Chromatography (TLC) 421
6.2 High-Performance and Two-Dimensional TLC 422
6.3 High-Performance Liquid Chromatography 422
6.3.1 Normal-Phase Liquid Chromatography (NP-LC) 422
6.3.2 Silver Ion Liquid Chromatography (Ag-LC) 423
6.3.3 Nonaqueous Reversed-Phase Liquid Chromatography (NARP-LC) 424
6.3.4 Other HPLC Techniques 425
6.4 Recent Advances in LC/MS Techniques for Lipidomics 425
6.5 Multidimensional Liquid Chromatography (MDLC, LC × LC) 428
6.6 Gas Chromatography 429
7. Conclusions and Future Perspectives **433**
References **434**

1. INTRODUCTION AND OVERVIEW

Lipidomics, as a novel branch of metabolomics, was first put forward more than 10 years ago [1], and was defined in a very comprehensive way as "the full characterization of lipid molecular species and of their biological roles with respect to expression of proteins involved in lipid metabolism and function, including gene regulation" [2]. Such a definition encompasses two main implications, the first being the requirement for a complete determination of all the molecular species within each lipid class in a biological sample. The analytical information obtained is then to be related to the biological roles and functions of lipids in a living organism, with respect to both physiological processes and diseases.

Unlike other biological molecules, lipids are not characterized by a certain individual structure, and their great structural diversity is reflected in a variety of biochemical functions. Apart from being used as energy stores to fuel metabolism, lipids and their interactions play a crucial role in living organisms as structural components in cell membranes; furthermore every single lipid class has a highly specific function, regardless of its role as an energetic or plastic molecule. Lipids are involved in many cellular functions and metabolic processes such as signal transduction, morphogenesis, secretion, and protein trafficking [3–6]; they participate in cell growth, reproduction, and death, and their production is very extensive by either bacterial or mammalian organisms. Abnormalities in lipid structure, functions, or metabolism are related to the pathogenesis of a number of diseases, including insulin-resistant diabetes [7], obesity [8], atherosclerosis [9], schizophrenia [10], Alzheimer's disease [11], cancer [12], as well as viral and bacterial infections [13].

Profiling of the individual lipids in biological fluids is therefore of great concern with respect to human health and disease; moreover these molecules are often used for biomarker discovery, to assess the effects of candidate pharmaceuticals of metabolic pathways. As a consequence, the task of lipid analysis and profiling has become ever increasingly important, and lipidomics has become the target of postgenomic research [14], attracting a great deal of attention from biological and analytical scientists, and the application fields of lipidomics are becoming ever wider. The structural diversity of lipids challenges analytical chemists, which have been on the forefront of separation science, and has greatly spurred the development of new technologies aimed at a full characterization of these biomolecules.

Two main strategies for the analysis of lipids have been used in most of the works reported so far: on the one hand is the direct infusion "shotgun" method, which relies on a direct-inlet (separation-free) mass spectrometry (MS) analysis of a crude lipid extract. On the other hand, the so-termed "top-down" lipidomic approach, based on hyphenated techniques, which relies on a chromatography-based separation, prior to the MS measurement for the lipid analysis. A comprehensive description of the analytical methods in lipidomics and their applications can be found in some recent reviews [15,16]. Both approaches present advantages and disadvantages, which will be illustrated in this chapter, mainly focusing on the use of different ionization techniques and MS analysis modes, as well as on the coupling of liquid chromatography (LC) to MS detection. Other separation technologies, such as thin-layer chromatography (TLC), supercritical fluid chromatography (SFC), and capillary electrophoresis have also proven to be useful tools in lipidomics, as well gas chromatography (GC); these however are less widespread, and will be described more generally, in the discussion that follows.

The most common used extraction methods for lipids will be also introduced, as they represent preliminary steps crucial for the success of any analytical technique that follows.

The subject of lipid classification and nomenclature has been extensively treated in a number of comprehensive volumes, dealing with the structure, occurrence, chemistry, and biochemistry of lipids [17–21]; herewith, the main lipid categories will be only introduced, along with their representative chemical structures, with the aim to facilitate the reader's understanding of the contents of the following sections.

2. LIPIDS: DEFINITIONS AND CLASSIFICATION

Lipids, earlier termed as "lipoids," consist in a huge number of distinct molecular species, spanning from apolar (e.g., sterol esters) to polar (e.g., phospholipids, PLs), including neutral (e.g., triacylglycerols), molecules. Almost one century ago, Bloor [22] classified lipids into three groups: simple lipoids (fats and waxes), compound lipoids (phospholipoids and glycolipoids), and derived lipoids (fatty acids, alcohols, and sterols). However, this classification did not include compounds such as terpenes, carotenoids, and steroidal hormones, also termed as lipids, which do not satisfy the former definition. In a more general way, lipids are described as biological molecules characterized by poor solubility in water and high solubility in nonpolar solvents, such as hydrocarbons, ethers, and alcohols; however such a definition can be misleading, since a number of compounds that are commonly regarded as lipids actually do not show such behavior.

The development of a comprehensive classification system for lipids has always challenged scientists in the field, due to the great heterogeneity of these molecules, differing in both their structures and properties; lipids actually comprise a wide range of unique chemical entities, which, in turn, encode for unique biological functions into living organisms. Despite the numerous efforts put so far, the creation of an integrated platform into which analytical data generated by different techniques and research laboratories can be merged is still an issue.

According to Christie [23], the term lipids should only be restricted to "fatty acids and their derivatives, and substances related biosynthetically or functionally to these compounds." Furthermore, lipids are usually divided into two groups, on the basis of their structural complexity: simple lipids and complex lipids. Triacylglycerols (TAGs) are an example of the first group, since on saponification they would yield at most two primary hydrolysis products, i.e., in the specific case, glycerol and fatty acids. In contrast glycerophospholipids would yield on saponification three or more primary hydrolysis products per mole, i.e., glycerol, fatty acids, and a polar phosphorus moiety and therefore are considered as complex lipids.

Aiming to provide an unambiguous classification scheme, which would consider lipids also from a structural and biosynthetic perspective, Fahy et al. classified lipids into well-defined categories, covering both prokaryotic and eukaryotic sources, as well as archaea and synthetic compounds [24,25]. Such a classification would suit the needs of chemists, biologists, and biomedical researchers, and also facilitate dealing with the massive amounts of data generated by the international community of scientists active in the field of lipids. The LIPID MAPS Lipid Classification System is comprised of eight lipid categories, each with its own subclassification hierarchy, based and driven by the distinct hydrophobic and hydrophilic elements that constitute a lipid. Each unique lipid molecule into the LIPID MAPS Structure Database (LMSD) is characterized by lipid class, common name, systematic name or synonym, mass, InChIKey, or LIPID MAPS ID. The latter is a 12-digit identifier which reflects their position in the classification hierarchy, providing a common platform that allows handling the existing and emerging arrays of lipid structures, as well as the development of systematic methods of data management; moreover it is compatible with informatics requirements and other macromolecular databases.

The recommendations for a system-based lipid approach, as well as a number of additional documents, have been released by the International Union of Pure and Applied Chemists and the International Union of Biochemistry and Molecular Biology (IUPAC-IUBMB) Commission on Biochemical Nomenclature, and placed on the IUPAC Web site (http://www.chem.qmul.ac.uk/iupac/).

3. STRUCTURES, FUNCTIONS, AND OCCURRENCE OF LIPIDS

In this section, the readers will be acquainted with the main lipid categories, with the aim to facilitate understanding the contents of the sections that will follow. A detailed and comprehensive description of the structure, occurrence, chemistry, and biochemistry of lipids can be in several specialized texts, to which the reader is referred [26–30].

3.1 Fatty Acids

The fatty acids (FA) are characterized by a repeating series of methylene groups, and represent the major lipid building block of complex lipids. This group of molecules are synthesized by chain elongation of an acetyl-CoA primer with malonyl-CoA (or methylmalonyl-CoA) groups, and may also contain a cyclic functionality and/or heteroatoms. The FAs of plant, animal, and microbial origin usually contain even numbers of carbon atoms in straight chains, with a carboxyl group at one end and with double bonds (DBs) of the *cis* or *Z* configuration in a specific position.

The most abundant subclass in animal tissues is that of the straight-chain *saturated FAs*, usually varying in chain length from 14 to 22 carbon atoms. In the official nomenclature, they are named systematically from the saturated hydrocarbon with the same number of carbon atoms, the final "e" being changed to "oic." The saturated FA with 18 carbon atoms (having chemical formula $CH_3(CH_2)_{16}COOH$) is named "octadecaenoic acid," accordingly. Also trivial names are commonly found in the literature ("stearic acid"), as well as shorthand designations ("18:0," in which the number after the column specifies the number of double bonds). In Table 1 the most common FAs are listed, with their systematic and trivial names, as well as shorthand designations.

Branched chain FAs are more complex molecules, bearing one or more methyl substituents at the chain length, are not commonly found in natural organisms, except than in bacteria (e.g., the mycolic acids in mycobacteria).

FAs containing one double bond are also common in animal tissues, and they are designated as *monoenoic FAs or monounsaturated FAs (MUFA)*. These compounds have even-numbered straight chains of 10–30 (and more) carbon atoms; the position of the double bond is often noted in the form (n–x), being n the FA chain length, and x the number of carbon atoms between the double bond and the carboxyl group. The double bond is usually in the *cis* configuration, like in the structure of "*cis*-9-octadecenoic" acid in Figure 1 (having chemical formula $CH_3(CH_2)_7CH{=}CH(CH_2)_7COOH$), which is the most abundant one. The trivial name for this molecule is "oleic acid," while in the shorthand nomenclature it is termed as "18:1," or "18:1($n-9$)."

Monoenoic FAs with double bonds of the *trans* configuration are also found in nature, mainly in plant lipids. However, they can also be found in human tissues, due to ingestion by ruminant animals, and the presence in meat and diary products. Trans isomers may also be formed during industrial processing of fats, oils, and production of margarine, as a result of the hydrogenation process.

Polyunsaturated FAs (PUFAs) have two to six methylene-interrupted double bonds in the *cis* configuration; the most common compound of this subclass is named systematically "*cis*-9,*cis*-12-octadecadienoic acid," or trivially "linoleic acid," and shorthand designated as 18:2($n-6$). This compound is essential for their

TABLE 1 Most Common Fatty Acids (Systematic Names, Trivial Names, and Shorthand Designations)

Systematic Name	Trivial Name	Shorthand Designation
Butanoic acid	Butyric acid	4:0
Hexanoic acid	Caproic acid	6:0
Octanoic acid	Caprylic acid	8:0
Decanoic acid	Capric acid	10:0
Dodecanoic acid	Lauric acid	12:0
Tetradecanoic acid	Myristic acid	14:0
Hexadecanoic acid	Palmitic acid	16:0
Octadecanoic acid	Stearic acid	18:0
Eicosanoic acid	Arachidic acid	20:0
Docosanoic acid	Behenic acid	22:0
Tetracosanoic acid	Lignoceric acid	24:0
9-Hexadecenoic acid	Palmitoleic acid	16:1($n-7$)
9-Octadedecenoic acid	Oleic acid	18:1($n-9$)
trans-9-Octadedecenoic acid	Elaidic acid	18:1($n-9$)
11-Octadecenoic acid	Vaccenic acid	18:1($n-7$)
9,12-Octadecadienoic acid	Linoleic acid	18:2($n-6$)
9,12,15-Octadecatrienoic acid	α-Linoleic acid	18:3($n-3$)
6,9,12-Octadecatrienoic acid	γ-Linoleic acid	18:3($n-6$)
5,8,11,14-Eicosatetraenoic acid	Arachidonic acid	20:4($n-6$)
5,8,11,14,17-Eicosapentaenoic acid	EPA, timnodonic acid	20:5($n-3$)
4,7,10,13,16,19-Docosahexaenoic acid	DHA, cervonic acid	22:6($n-3$)

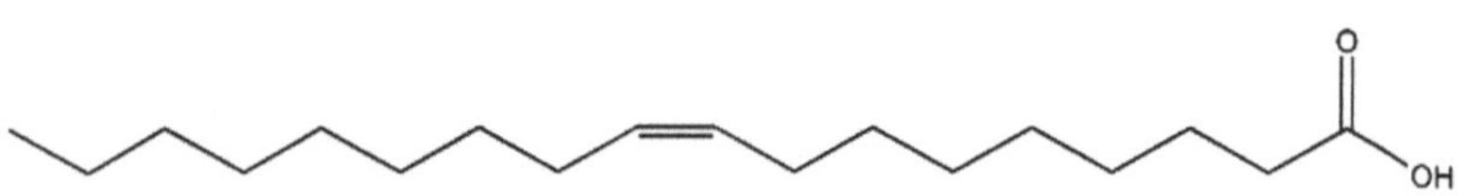

FIGURE 1 The chemical structure of oleic acid (*cis*-9-octadecenoic acid).

optimal growth and development of animal tissues; yet animal enzymes are not able to synthesize it, hence the term "essential fatty acid." This compound plays an important role in humans since, after dietary assumption, it serves as a precursor of a constituent of the membrane phospholipids, namely "arachidonic acid" (having chemical formula 20:4(n − 6), and of a series of n − 6 FAs with desaturated and elongated chains, which serves in turn as a precursor of eicosanoids involved in important biological and pharmacological effects, like prostaglandins.

Another important subclass of essential FAs is represented by the n − 3 PUFAs, commonly found in fish and marine organisms, which also serve as plastic components of animal tissues and precursors of prostanoids. The precursor of this further family is "*cis*-9,*cis*-12,*cis*-15-octadecatrienoic acid," trivially named as "linolenic acid," shorthand designated as 18:3(n − 3).

Conjugated dienoic FAs, having acetylenic bonds or conjugated double bonds in their structure, are characterized the presence of one of more double bonds of the *trans* configuration; they have shown favorable effects for human health.

Further specific subclasses are *cyclic FAs* (containing three to six carbon atoms as well as heterocyclic rings containing oxygen or nitrogen); *oxygenated FAs* (such as the 4- and 5-hydroxy FAs of cow's milk, which take part in biosynthetic pathway of eicosanoids like prostaglandins); 2-hydroxy acids found in skin and wool wax as components of the sphingolipids; specific FA subclasses containing heteroatoms of oxygen, halogen, nitrogen, and sulfur linked to the carbon chains; the so-called *thia FAs*, containing one or more sulfur atoms in their structure, like lipoic acid and biotin.

Other major lipid classes in the fatty acyl category include *FA esters*, such as wax monoesters, diesters, lactones, as well as compounds such as fatty acyl thioester-CoA derivatives, fatty acyl thioester-acyl carrier protein (ACP) derivatives, fatty acyl carnitines, which are important biochemical intermediates. *Fatty alcohols*, *fatty aldehydes*, and *fatty amides* represent further subclasses, the latter sometimes showing interesting biological activities in various organisms.

3.2 Glycerolipids

Glycerolipids (GL) are mono-, di-, and trisubstituted glycerols, the most well known being the *TAGs*, fatty acid esters of glycerol, formerly termed as triglycerides, whose structure is depicted in Figure 2. TAGs represent the most abundant lipid class in oils and fats of animal origin, and comprise the bulk of storage fat in mammalian tissue. These molecules exist as enantiomers, since a center of asymmetry is created upon enzymatic biosynthesis at carbon 2 of the glycerol backbone. During the biosynthesis and digestion of TAGs, *diacylglycerols* (diglycerides, DAGs) and *monoacylglycerols* (monoglycerides, MAGs) are formed as intermediates, with two and one FA substitution at the glycerol backbone, respectively.

According to the recommended numbering system, the prefix *sn* is placed before the name of the compound, provided that its stereochemistry is known and/or that the compound is not racemic.

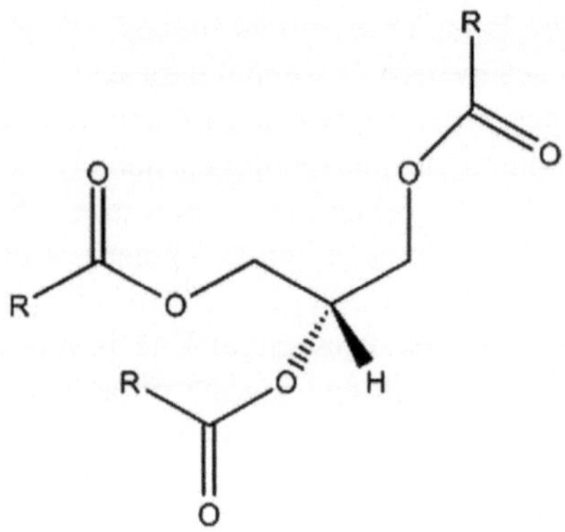

FIGURE 2 The chemical structure of a triacyl-*sn*-glycerol (triacylglycerol).

One or more sugar moieties can be linked through a glycosidic bond to glycerol, in a further subclass named as *glyceroglycolipids*.

3.3 Glycerophospholipids

Glycerophospholipids (GP), also termed as *phospholipids*, are quite ubiquitous in nature; they play a crucial role as components cellular membranes and cell lipid bilayer, and furthermore they act as binding sites for intracellular and intercellular proteins. In eukaryotic cells, they are involved in metabolism and signaling, serving as second messengers themselves or as precursors of membrane-derived second messengers.

Glycerophospholipids are classified into distinct subclasses, based on the nature of the polar head group linked at the *sn*-3 position of the glycerol backbone in eukaryotes and eubacteria, or the *sn*-1 position in the case of archaebacteria.

According to the systematic nomenclature, *monoglycerophospholipids* and *diacylglycerophospholipids* with different acyl or alkyl substituents are grouped according to the common polar moieties (or head groups), the latter being in turn further differentiated on the basis of the *sn*-1 and *sn*-2 substituents on the glycerol backbone.

Most important members of this class, for their biological role, are *phosphatidic acid* (1,2-diacyl-*sn*-glycerol-3-phosphate, see Figure 3), which serves as a precursor of most GPs; *diphosphatidylglycerol* (also known as "cardiolipin," see Figure 4), which is a constituent of the heart muscle; *lysophosphatidic acid* (1-acyl-*sn*-glycerol-3-phosphate), which is a marker of cancer.

The prefix "lyso" is added to the name of the corresponding GL, to designate those compounds which contain only one FA moiety (usually in the *sn*-1 position) in their structures; well-known examples are *lysophosphatidylcholine* (Lys-PC) and *lysophosphatidylserine* (Lys-PS).

The most abundant glycerolipid or phospholipid classes in animal tissues is *phosphatidylcholine* (PC, also known as "lecithin," see Figure 5), accounting for most of the lipid content of the external plasma membrane monolayer.

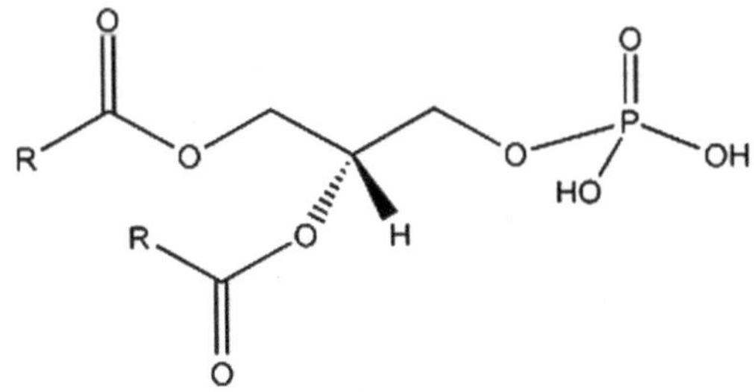

FIGURE 3 The chemical structure of phosphatidic acid (1,2-diacyl-*sn*-glycerol-3-phosphate).

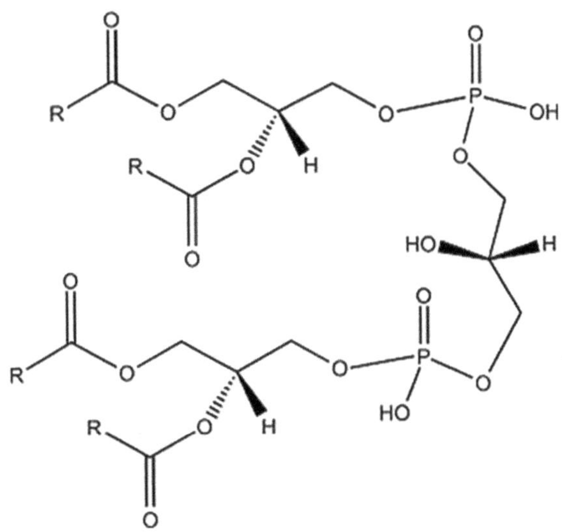

FIGURE 4 The chemical structure of cardiolipin (1′,3′-bis-(1,2-diacyl-*sn*-glycero-3-phospho)-*sn*-glycerol).

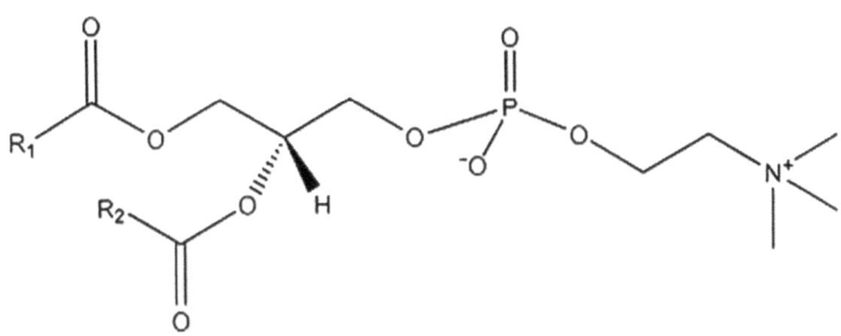

FIGURE 5 The chemical structure of phosphatidylcholine (1,2-diacyl-*sn*-glycero-3-phosphocholine).

Phosphatidylethanolamine (PE, also termed as "cephalin," see Figure 6) represents the second most abundant GP class, followed by *phosphatidylserine* (PS, see Figure 7). A further class is represented by *phosphatidylinositol* (PI, see Figure 8), which plays an important role in the regulation of some vital functions.

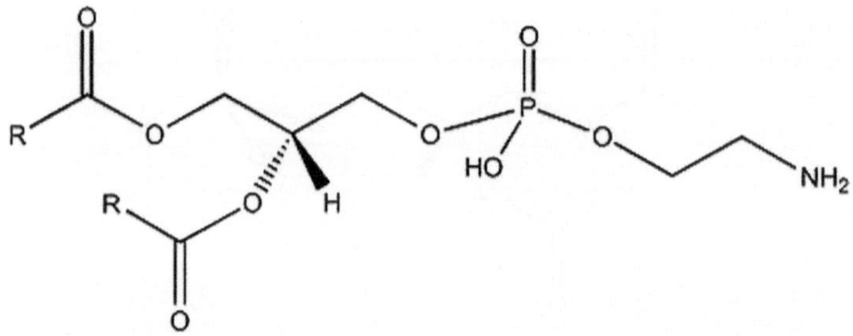

FIGURE 6 The chemical structure of phosphatidylethanolamine (1,2-diacyl-*sn*-glycero-3-phosphoethanolamine).

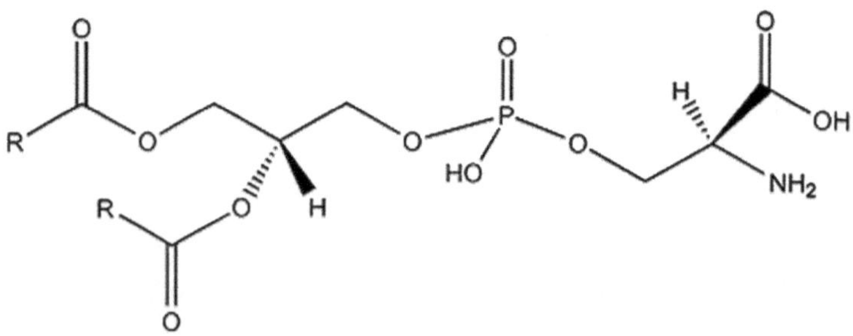

FIGURE 7 The chemical structure of phosphatidylserine (1,2-diacyl-*sn*-glycero-3-phosphoserine).

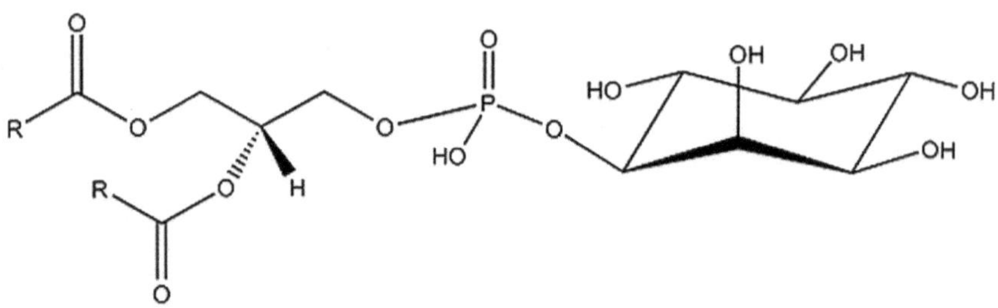

FIGURE 8 The chemical structure of phosphatidylinositol (1,2-diacyl-sn-glycero-3-phospho-(1′-myo-inositol).

Glycerophosphoglycerols and *glycerophosphoglycerophosphates* are characterized by the presence of a second glycerol unit as part of the head group, whereas for the *glycerophosphoglycerophosphoglycerols*, a third glycerol unit is acylated at the *sn*-1 and *sn*-2 positions.

3.4 Sphingolipids

Sphingolipids (SP) are characterized by a sphingoid base backbone that is synthesized de novo from serine, and a long-chain fatty acyl-CoA, converted into

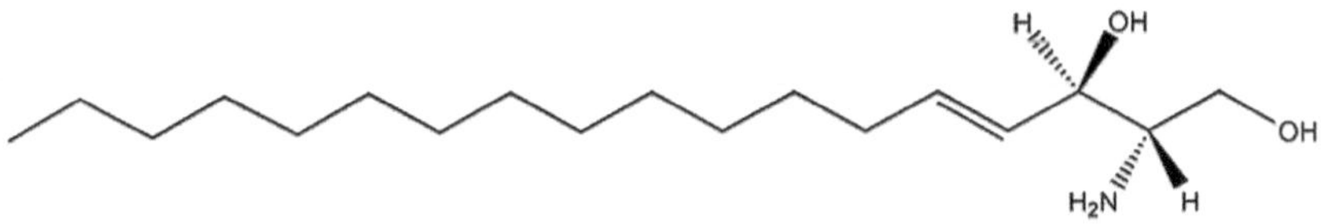

FIGURE 9 The chemical structure of sphingosine (sphing-4-enine).

species of different complexity. Main SP classes include sphingoid bases and their derivatives, ceramides, glycosphingolipids, and phosphosphingolipids.

Sphingoid bases play a functional role in living organisms, by influencing the permeability barrier of skin; structural variations such as the alkyl chain length and branching, the number and positions of double bonds, and the presence of additional hydroxyl groups critically affect this role. Among sphingoid bases, the most common in mammalians is "sphingosine," systematically named as (2S,3R,4E)-2-amino-4-octadecen-1,3-diol, also termed as "D-erythro-sphingosine" and "sphing-4-enine" (see Figure 9).

Among *sphingoid bases derivatives* are the *lysosphingolipids* (e.g., sphingosine 1-phosphocholine and sphingosine 1-glycosides), 1-phosphates, and *N*-methyl derivatives (*N*-methyl, *N,N*-dimethyl, and *N,N,N*-trimethyl).

Ceramides are *N*-acyl-sphingoid bases in which FAs are linked through an amide bond to the amino group of a long-chain base (see Figure 10). The FAs can be saturated or monounsaturated, they typically span in the 14–26 carbon atom range, and commonly also bear a hydroxyl group on carbon 2. Ceramides are major components of skin and serve as precursors of more complex sphingolipids, characterized by the presence of different polar head groups; these include the *phosphosphingolipid*, in which the head group is attached through a phosphodiester linkage, and the *glycosphingolipids*, in which the head group is attached through a glycosidic bond.

Sphingomyelins are ceramidephosphocholines, consisting of a ceramide unit linked at position 1 to phosphorylcholine (see Figure 11); they are the major phosphosphingolipids of mammals (not present in plants and microorganisms).

Glycosphingolipids are further divided into two subclasses, on the basis of the carbohydrate composition: "neutral glycosphingolipids" contain one or more uncharged sugar residues such as glucose, galactose, *N*-acetylglucosamine, *N*-acetylgalactosamine, and fucose; different families are characterized by different glyco-substituents. "Acidic glycosphingolipids" contain an ionized functional group, i.e., phosphate or sulfate, attached to neutral sugars or charged sugar residues. Among them are the *ganglioside*, highly complex acidic glycosphingolipids containing one or more sialic acid groups (*N*-acetyl or *N*-glycoloylneuraminic acid). Gangliosides are quite ubiquitous in most animal tissues; their name derived from the ganglion cells of the central nervous system, where they exert an important role in the immune response system.

Further classes are represented by the "basic glycosphingolipids" and the "amphoteric glycosphingolipids."

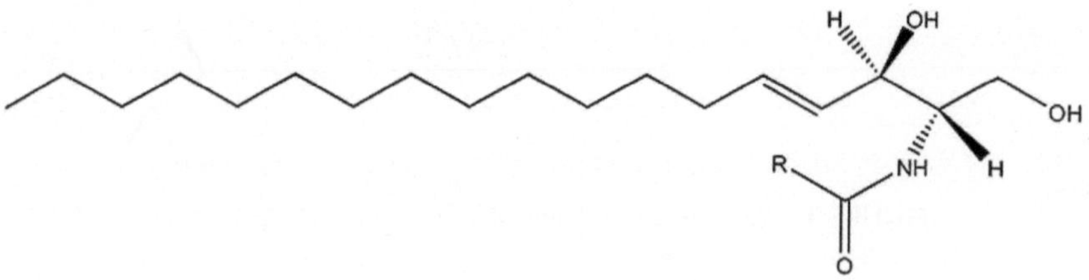

FIGURE 10 The chemical structure of ceramide (*N*-acyl-sphing-4-enine).

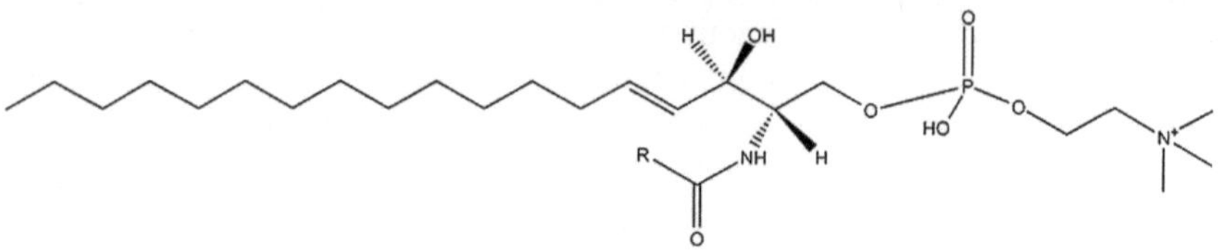

FIGURE 11 The chemical structure of sphingomyelin (*N*-acyl-sphing-4-enine-1-phosphocholine).

Hydroxyceramides and *glucosylceramides* are protein adducts formed by some sphingolipids, for example, when they are covalently attached to surface proteins of skin.

3.5 Sterol Lipids

Sterol lipids (ST) are characterized by a tetracyclic ring system; these lipid molecules play very important biological roles. The most abundant sterol in animal tissues is "cholesterol," bearing a double bond and a free hydroxyl group attached to the fused four-ring core structure (see Figure 12). Together with its derivatives, cholesterol plays a vital role as component of membrane lipids and in maintaining membrane fluidity.

The *steroids* also exert a crucial role as hormones and signaling molecules, and are divided into families based on the length of the carbon chain: the *estrogen* family (C18 steroids), the *androgens* (C19 steroids, such as testosterone and androsterone), the *progestogens*, the *glucocorticoids*, and *mineralocorticoids* (C21 steroids).

Secosteroids, characterized by the cleavage of the B ring of the core structure, comprise various species of vitamin D.

3.6 Prenol Lipids

Prenol lipids (PR) are synthesized from the 5-carbon precursors isopentenyldiphosphate and dimethylallyldiphosphate.

The *simple isoprenoids* are linear alcohols formed by the successive addition of C5 units, and are classified according to the number of these terpene units. This class includes the *carotenoids*, which exert antioxidant effects and also serve as precursors of vitamin A. The *polyterpenes* are a PR class containing more than 40 carbon atoms.

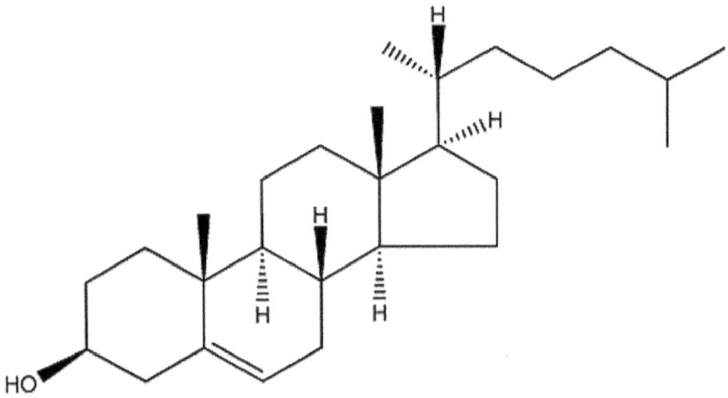

FIGURE 12 The chemical structure of cholesterol (cholest-5-en-3β-ol).

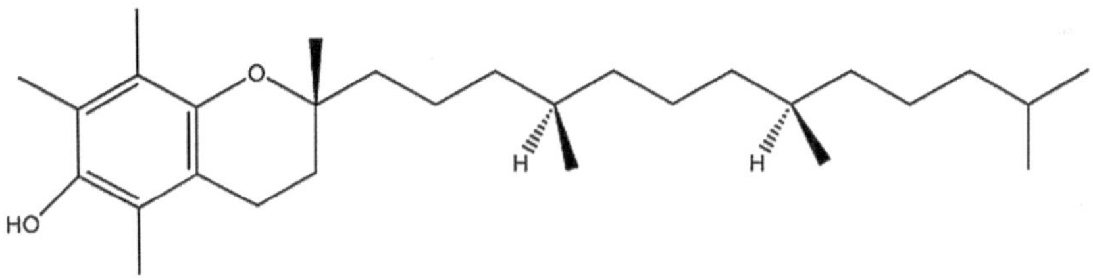

FIGURE 13 The chemical structure of α-tocopherol or vitamin E ((2R)-2,5,7,8-tetramethyl-2-[(4R,8R)-4,8,12-trimethyltridecyl]-3,4-dihydro-2H-chromen-6-ol).

The *quinones*, the *ubiquinones*, and the *hydroquinones* are another biologically important class of molecules, comprising vitamin K and vitamin E (see Figure 13).

3.7 Saccharolipids

Saccharolipids (SL) are a structural variant of glycerolipids and glycerophospholipids, in which a sugar backbone which substitutes for the glycerol backbone, to which FAs are directly linked. An example of this class of lipids is the *acylated glucosamine* serving as a precursor of the Lipid A component of the lipopolysaccharides in Gram-negative bacteria (Figure 14).

3.8 Polyketides

Polyketides (PK) comprise a large number of secondary metabolites and natural products, greatly differing in their cyclic structures, which are synthesized by enzymatic polymerization of acetyl and propionyl subunits from animal, plant, bacterial, and fungal sources. Among them are important pharmacological agents like erythromycins and tetracyclines that have found widespread use as antiparasitic, antimicrobial, and anticancer agents (see Figure 15).

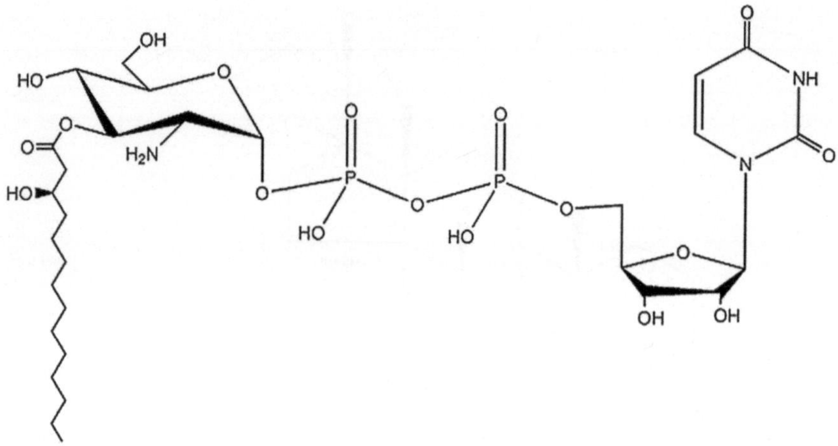

FIGURE 14 The chemical structure of a saccharolipid (UDP-3-(3R-hydroxy-tetradecanoyl)-αD-glucosamine).

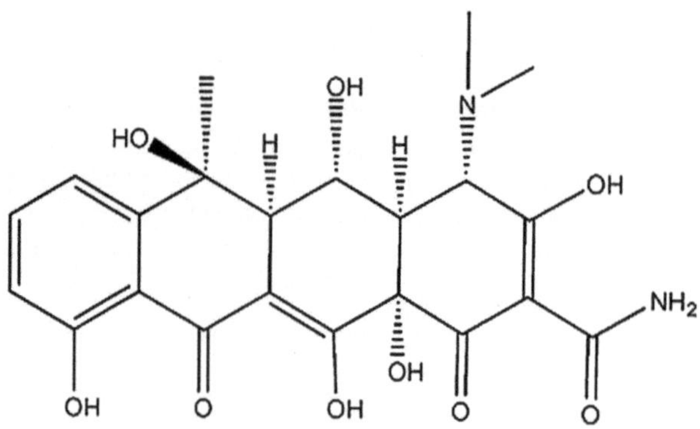

FIGURE 15 The chemical structure of a polyketide (oxytetracycline).

4. EXTRACTION METHODS OF LIPIDS

Sample extraction represents the first and often the most crucial step in analytical methods, since a proper sampling and sample handling preparation prior to analysis is mandatory to obtain satisfactory results. Since lipids are part of complex biological matrixes, an efficient extraction is needed before lipidomics analysis, in order to remove any interfering agents such as proteins, saccharides, etc. A recent book by Pawlyszin has recently treated this task, in a very comprehensive manner [31].

4.1 Sampling and Sample Preparation

Sampling and sample preparation are crucial for the success of any analytical technologies, and are critically dependent on the type of sample being analyzed, and a proper knowledge of the nature of the lipid fraction, as well as the analytical procedure applied, is needed. To know the physical structure and location of the major lipids in the sample is crucial to get a successful sampling step because the sample analyzed should represents as closely as possible the lipid composition of the whole matrix from which it was taken. Different official methods have been developed, which recommend the most adequate sample preparation and extraction procedures, for a specific type of sample. Samples like vegetable oils, which are almost 100% lipids, do not require any sample preparation prior to analysis. In contrast more complex biological samples or dietary products require extraction and purification of the lipid fraction, prior to analysis. Obviously, sampling and sample preparation should be carried out in a way to minimize any changes in lipid properties, prior to analysis.

Generally, two extraction methods are used for sample preparation in lipidomics, i.e., liquid–liquid extraction and solid phase extraction (SPE). Solvent selection is, obviously, a decisive step prior to solvent extraction of lipids from a matrix, which is aimed to separate lipids from water-soluble food components, prior to chromatographic analysis. The solvent used should ideally extract the sole lipid fraction, while leaving proteins, carbohydrates, and any other components of the matrix behind. Also the solvent polarity should be carefully selected, with respect to the polarity of the specific lipids to be extracted; obviously there is no single solvent that apply to this scope, since the different lipids in a sample may widely differ in their polarities. Polar lipids like phospholipids are more soluble in polar solvents, like alcohols. In contrast nonpolar lipids such as TAGs are more soluble in hexane and other nonpolar solvents.

Additional characteristics of the extraction solvent include inertness toward the lipid fraction, low boiling point, to be easily removed by evaporation at low temperatures. The "ideal" solvent should also be nonflammable and nontoxic.

Often the sample to be extracted is finely grounded at low temperatures, prior to solvent extraction, in order to obtain a more homogeneous sample. Particle size reduction will also afford an increased lipid surface area exposed to a solvent, and this will in turn increase the yield of an extraction procedure. The latter can also be maximized by drying the samples, prior to extraction with an organic solvent, in order to increase the degree of the solvent penetration into the matrix.

In the case of complex lipids, like saccharolipids or glycolipids, a hydrolytic step is needed, prior to extraction, to release bound lipids; this is usually performed by digestion of the sample with hydrochloridric acid upon heating.

Further treatments like filtering, centrifuging, or drying may be needed after solvent extraction, to remove any residual moisture; sometimes the use of antioxidant agents, a nitrogen atmosphere, as well as control of the light and temperature exposure are mandatory, to keep auto-oxidation to a minimum, in the case of unsaturated lipids, and thus prevent degradation and formation of off-flavors and harmful products.

4.2 Soxhlet Extraction

Soxhlet extraction was originally designed for the lipid extraction from a solid material [32]. This procedure is based on sample drying and grounding into small particles, before being placed in a porous cellulose membrane (thimble). The latter is then placed in an extraction chamber (with a distillation arm and a siphon), which is in turn suspended above a heated flask containing the solvent and below a condenser. As the solvent evaporates, it moves up into the condenser, and back into a liquid that drips down to the extraction chamber containing the sample, penetrates the sample, extracts the lipids and carries them into the bottom flask. After some time, the solvent in the flask containing the extracted fraction is evaporated; the mass of the remaining lipid can be measured, and the percentage of lipid in the initial sample calculated, to obtain a recovery yield.

Such method still represents one of the most widespread extraction techniques for isolation of fats from solid lipidic matrices, even though it presents a number of disadvantages, e.g., the large volumes of organic solvents and long time required, hazards of boiling solvents, poor extraction of polar lipids. Modern instrumentation placed is now commercially available, allowing automated or semiautomated Soxhlet extraction that affords better extraction kinetics and shorter extraction times.

4.3 Method of Folch, Lees, and Stanley

The original extraction procedure was developed by Folchand coworkers in 1957 [33]. Since then, a number of variants have been introduced, to the basic extraction. Some of them are best suited for the quantitative extraction of major lipid classes from matrices, whereas others allow the complete recovery of minor components of biological importance.

According to the most commonly used modified procedure [34]: the tissue (1 g) is homogenized for 1 min with methanol (10 mL), then chloroform is added (20 mL) and the mixture is homogenized for 2 min more. After filtration, a chloroform–methanol mixture (30 mL, 2:1 v/v) is added to the solid, which is homogenized for 3 min. The mixture is filtered again and washed with fresh solvent; the combined extracts are shaken thoroughly with one-fourth its volume of a potassium chloride solution (0.88% in water). Once the mixture is settled out, the upper aqueous layer is drawn off by aspiration, and then one-fourth the lower layer of methanol-saline solution is added (1:1, v/v), to repeat the washing procedure. The bottom layer is finally filtered, and the solvent evaporated to obtain the purified lipid fraction, which is stored in a small volume of nonpolar solvent at −20 °C until analysis.

4.4 Method of Bligh and Dyer

Originally developed as a solvent-saving, more economic alternative to the Folch procedure, the Bligh and Dyer method [35] is generally best suited for the extraction of large lipidic samples, with a high content of endogenous water. The procedure, first applied to extract phospholipids from a fish muscle tissue, is carried out as follows. The tissue (100 g) is homogenized with a solvent mixture consisting of chloroform and methanol (300 mL, 1:2 v/v), for 4 min. In case that a two-phase system is formed, more solvent is added until a single phase is obtained. After extraction, the mixture is filtered while the tissue residue is rehomogenized with more chloroform (100 mL) and filtered once more. The combined filtrates are shaken thoroughly with one-fourth its volume of a potassium chloride solution (0.88% in water). Once the mixture is settled out, the upper aqueous layer is drawn off by aspiration, and then one-fourth the lower layer of methanol–saline solution is added (1:1, v/v), to repeat the washing procedure. The bottom layer is finally filtered, and the solvent evaporated to obtain the purified lipid fraction, which is stored in a small volume of nonpolar solvent at −20 °C until analysis.

One major drawback of this method, compared to the original extraction procedure, is that it may lead to significant underestimation of the lipid content, proportionally to the lipid content of the sample.

4.5 Supercritical Fluid Extraction (SFE)

Supercritical fluid extraction (SFE) relies on the use of supercritical fluids as the solvent, rather than the organic solvents normally used, for the extraction of lipid compounds from a matrix [36]. A supercritical fluid is any substance at a temperature and pressure above its critical point, where distinct liquid and gas phases do not exist.

The most common supercritical fluid is carbon dioxide (CO_2), which affords very good recoveries for nonpolar lipids, unlike polar lipids are scarcely soluble in this fluid, and may remain partially unextracted. To this purpose, the addition of organic modifiers as cosolvents to the primary fluid is very common, to enhance the extraction efficiency. The addition of 1–10% of methanol or ethanol to CO_2 typically expands the extraction range to include more polar lipids [37].

SFE brings a number of clear advantages over organic solvent-based extraction methods, the major are solvating powers similar to liquid organic solvents, but with higher diffusivities, lower viscosity, and lower surface tension. The solvating power can be further adjusted by pressure or temperature tuning, in order to speed up the separation of the analytes from the solvent. The addition of organic modifiers to CO_2 modifies its polarity and, therefore, changes the selectivity of the extraction. With respect to organic solvents, SCFs are generally cheap, simple, and comparatively safe. This is of special attractiveness for industrial processes since toxic hazards deriving from solvent manipulation and disposal cost are greatly reduced.

A special apparatus is needed to carry out SFE, basically consisting of a solvent pump, a pressure cell to contain the sample, a means of maintaining pressure in the system, and a collecting vessel. Carbon dioxide is pumped to a heating zone, where it reaches supercritical conditions, then it passes to the extraction vessel, and, behaving as a gas, can rapidly diffuse into the solid matrix and, as a liquid, dissolve a large quantity of lipids from the extracted material. The dissolved lipids are removed from the extraction cell into a separator at lower pressure, and the extracted material settles out.

4.6 Accelerated Solvent Extraction (ASE)

Accelerated solvent extraction (ASE), or pressurized liquid extraction (PLE), uses solvents at elevated temperatures and pressures, to enhance extraction of the components of interest from sample matrices [38]. The use of the high pressure, in keeping the solvent below its boiling point, enables a high penetration of the solvent in the sample, while the high temperature allows for higher solubility and higher diffusion rate of lipid solutes in the solvent.

The main advantages of ASE are the high extraction yield, compared to those obtained with Soxhlet and other solvent-extraction techniques in use, as well as shorter extraction time, and reduced solvent consumption, as demonstrated for the extraction of γ-linolenic acid [39]. On the other hand, the employment of high temperatures precludes the use of this technique for extraction of thermolabile lipid materials; whenever if the sample contains unsaturated lipids that are prone to auto-oxidation during extraction, the solvents should be degassed to avoid or at least minimize any possible degradation.

4.7 Microwave-Assisted Extraction (MAE)

Microwave-assisted extraction (MAE) can significantly reduce the extraction time, as compared to Soxhlet, when applied to lipids, as well as the need for solvent usage. Extraction times usually range from few seconds to few minutes (15–20 min), and only a few milliliters of solvent is required. Other major advantages of this technique are improved extraction yield, better accuracy and precision upon automation, suitability to extraction of thermolabile constituents, capability for extraction of minor/trace components from few milligram of sample. Such a procedure was successfully applied for the fast extraction of yeast lipids in *Saccharomyces cerevisiae* [40].

Two main types of MAE systems are commercially available, in both of them agitation is provided during extraction, to improve the mass transfer phenomenon. In the *closed extraction vessels*, extraction is performed under controlled pressure and temperature [41]. In the *focused microwave-assisted solvent extraction* (FMASE), only a part of the extraction vessel containing the sample is irradiated with microwaves.

4.8 Solid Phase Extraction (SPE)

SPE is a sample preparation process by which compounds that are dissolved or suspended in a liquid mixture are separated from other compounds in the mixture according to their physical and chemical properties. SPE uses the affinity of solutes dissolved or suspended in a mobile phase for a stationary phase through which the sample to attain the separation of a mixture. The portion that passes through the stationary phase is collected or discarded, depending on whether it contains the desired analytes or undesired impurities. If the portion retained on the stationary phase includes the desired analytes, they can then be removed from the stationary phase, by rinsing it with an appropriate eluant.

SPE for lipidomics analysis uses bared or boned silica gel with cyano, amino, or diol groups, as normal-phase stationary materials, and methanol, hexane, and chloroform as elution solvents [42,43]. Compared to liquid–liquid extraction, SPE brings the advantage of reduced solvents and time consumption; however, it suffers from lower peak capacity, thus the extraction recovery will be significantly reduced when dealing with a large volume of samples. In general, liquid–liquid extraction methods may be preferred for nontargeted lipidomics, to extract the analytes more exhaustively, while SPE methods may afford better specificity and extraction efficiency for targeted lipidomics; a fair comparison was made by Ferreiro-Vera and coworkers for the extraction of phospholipids from human serum [44].

5. MASS SPECTROMETRY IN LIPIDOMICS RESEARCH

The huge combinatorial and structural diversity of lipids poses a great challenge for lipidomics analysis. The theoretical number of distinct lipid isoforms was estimated to be as high as 200,000 [45]. To this regard, the development of robust and affordable MS equipments has greatly expanded the analytical capabilities and the number and types of lipids that can be analyzed. Technological innovations and advances in MS-based approaches offer unsurpassed resolution and sensitivity for the determination of the global lipid composition of a given sample, thus substantially spurring the evolution of lipidomics research.

Mass spectrometry can be used alone or as a detection method, after a front-end (usually, chromatographic) separation step. Therefore there exist two complementary approaches in MS lipidomics: "shotgun lipidomics" and "top-down lipidomics"; the first approach relies on direct infusion (inlet) of a sample into an MS instrument, after ionization in suitable proper source, the second includes a preliminary online or off-line separation of the lipid sample, prior to detection.

A schematic diagram of an overall lipidomics approach is presented in Figure 16.

5.1 Ionization Modes

Current MS-based methods for lipid analysis include ionization by electrospray (ESI), atmospheric pressure chemical ionization (APCI), atmospheric pressure photoionization (APPI), matrix-assisted laser desorption ionization (MALDI), electron ionization (EI), chemical ionization (CI), and fast atom bombardment (FAB). While ESI and MALDI are the most widespread ion sources used in direct infusion MS analysis, the other are usually employed

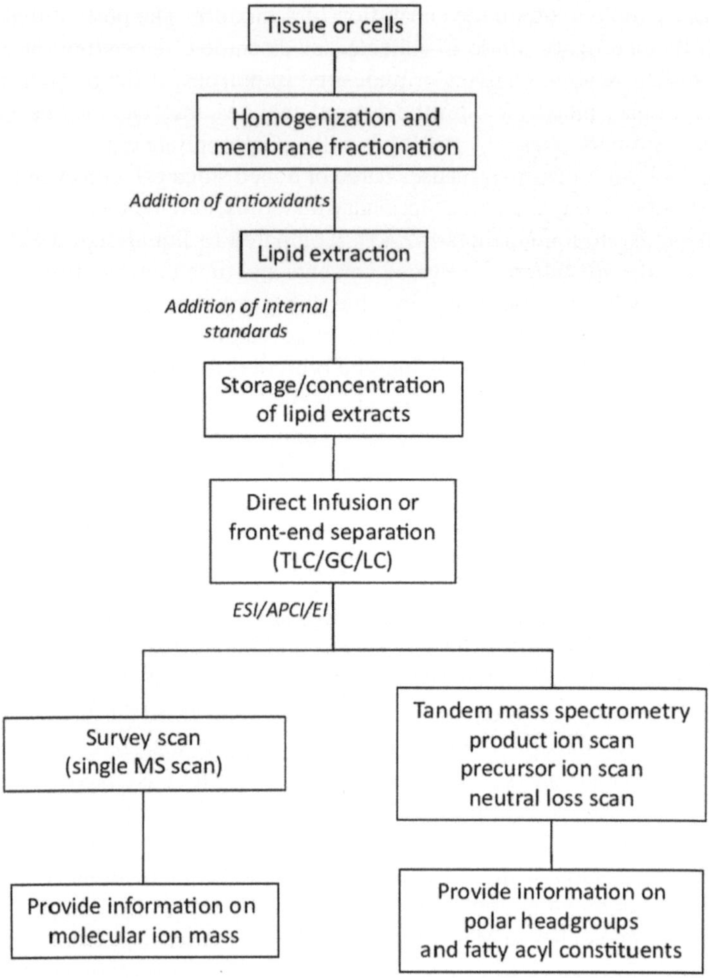

FIGURE 16 A schematic diagram of an overall lipidomics approach. *Reprinted from Ref. [15] with permission from Wiley.*

to generate gas-phase ions out of LC or gas GC effluents. In addition to these, new "soft" ionization techniques have been implemented, aiming to promote the development of lipidomics investigations; these will be described under the title "Other Ionization Techniques."

5.1.1 Electrospray Ionization (ESI)

Developed by Fenn in 1989 [46], electrospray ionization (ESI) is currently the most common ionization technique for lipid analysis. In this technique, gas-phase ions are produced directly from molecules in a solution, which is sprayed at the end of a fine capillary needle held at a high electric potential (several kV), forming small charged droplets. During the process of solvent vaporization, the sample ions are produced via either the charge-residue model (CRM), or the ion desorption model (IDM). ESI is a "soft" ionization technique, producing little or no molecular fragmentation, especially developed for macromolecules (lipids, proteins) since it is able to preserve the lipid molecular structure after ionization by minimizing the fragmentation associated with the excess of potential energy added by impact. In contrast, "hard" ionization techniques, such as, fast atom bombardment and direct chemical ionization, cause the loss of such information due to the high degree of fragmentation, molecular ions are almost replaced by fragment ions in the spectra. Ionization in ESI can occur in positive as well as negative mode, commonly yielding protonated and deprotonated molecules, respectively; it is used both in shotgun approaches, and in top-down lipidomics.

Major innovations in ESI ionization for lipid analysis consisted in adduction of metal ions, ozone electrospray, and intrasource separation. The first consists in the formation of sodiated, lithiated, and copper-containing adducts, through the adduction of doped-in metal cations, to create distinct molecular ion species, from which, complementary sequence, branching, and linkage information can be deduced [47]. In ozone electrospray (OzESI), ozone is introduced together with the desolvation gas into the ion source, being reactive toward unsaturated lipids and thus allowing to determine double-bond positions in glycerolipids, glycerophospholipids, sphingomyelins, etc. [48]. ESI with intrasource separation was first introduced by Han and Gross [1], and has become the most frequently used MS method in lipids profiling from complex biological samples, such as blood [49], cells [50] biofluids [51] and biological tissues [52].

In shotgun lipidomics, this ionization technique is used for direct quantification of lipids from "crude" extracts [53], enabling a comprehensive profiling of the cellular lipidome by direct infusion of crude lipid extracts, without front-end chromatographic separation. By adequately tuning the ESI experimental conditions, it is in fact possible to preferentially ionize anionic lipids, weak anionic lipids, and neutral lipids, by employing both positive and negative ion modes, as well as adjusting the pH of the lipid extract.

Further achievements in the MS ionization technology consisted in the implementation of nano-ESI and microfluidics. Both are much suitable for high-throughput analysis of trace volume samples [54], especially the latter, due to its ability to integrate different functions on one single chip, e.g., lysis of cells, capture of lipids, and elution of captured lipids from a solid phase for microscale purification) [55].

Of the various ionization methods used in lipid analysis, ESI offers the advantages of high sensitivity and good quantification, with less need of substantial structure-dependent response factors. Typical limits of detection of ESI-MS methods for lipidomics are in the femtomole range.

5.1.2 Atmospheric Pressure Chemical Ionization (APCI) and Atmospheric Pressure Photoionization (APPI)

Besides ESI, other soft ionization techniques are suitable for lipid analysis, i.e., atmospheric pressure chemical ionization (APCI) [56] and APPI [57]. Both, APCI and APPI are easily hyphenated to LC for lipidomics analysis, as well. While ESI is best suited for the analysis of polar lipids (like phospholipids), APCI and APPI work better for the analysis of nonpolar lipids.

APCI is a type of chemical ionization performed at atmospheric pressure, in which the LC effluent containing the analyte is heated to high temperatures and sprayed with a nebulizing gas (nitrogen), creating an aerosol of fine droplets that readily evaporate. Afterward, the electrons of a corona discharge produced primary ions are by electron impact. These primary ions in turn form secondary reactant gas ions through collision with vaporized solvent molecules, and ultimately the analyte ions are formed through gas-phase ion–molecule reactions between the analyte molecules and these solvated ions [58]. APCI-MS works best for qualitative and/or quantitative analyses of complex mixtures of relatively nonpolar compounds like glycerolipids, phospholipids, ceramides, carotenoids, and steroids [59,60].

APPI consists in a modification of APCI, in which the liquid effluent is vaporized in a heated nebulizer to generate gas-phase analytes. Ionization is triggered by a beam of photons emitted from a UV lamp, as a replacement for the corona discharge in APCI. Since most lipid molecules have high ionization energies (>10 eV), the ionization is achieved in dopant-assisted mode. The dopant radical cations are produced by absorption of photons by compounds with a low ionization energy (e.g., acetone, benzene), the analytes are then ionized by direct charge transfer from dopant radical cations solvent-mediated ionization. Used in hyphenation with LC separation, APPI provides higher signal intensities, as well as higher signal-to-noise ratios compared to APCI and ESI, and usually affords lower detection limits for the analysis of lipid molecular species such as fatty acyls and glycerolipids [61], sphingolipids [62–64], and glycerophospholipids [65].

5.1.3 Matrix-Assisted Laser Desorption Ionization (MALDI)

Matrix-assisted laser desorption ionization (MALDI) is another "soft" ionization technique [66] early introduced for lipid analysis, and still successfully used, especially for lipidomics-related disease investigations [67,68]. In MALDI, sample-matrix crystals are irradiated by a pulsed laser beam (usually, a nitrogen laser), triggering vaporization and ionization of both the target sample, and matrix molecules. The process of sample embedding in a dry crystalline matrix is crucial for the whole technique, and new compositions and applications of matrix for MALDI are continuously reported [69].

Unlike ESI, MALDI is capable to ionize the analyte directly from the solid phase, and it is also largely used in imaging mass spectrometry, for studying the lipid profile on intact tissue sections [70]. A distinct feature of MALDI-MS methods consists in the capability for simultaneous detection of different chemicals, allowing to study the interrelation between different lipids or lipids with related proteins [71,72]. The potential of rapidly revealing the overall pattern of changes in both lipidome and peptidome signatures could be of valuable interest for handling large numbers of samples for the purpose of finding new biomarkers.

Two major drawbacks of this ionization technique are the quantitation capability and difficulty to hyphenate with other techniques. The first limitation may be overcome by adding an internal standard, as demonstrated by Niklas et al. [73] for phospholipids quantitation (the limit of detection for PC was stated as below 2 μg). Strong efforts have been put in attempting to hyphenate separation techniques with MALDI-MS, which may significantly expand the applications of MALDI-MS in related areas, as in the off-line (LC)-MALDI-MS method developed by Li et al. [74], in which the undesired effects of signal suppressing and overlapping due to PC were avoided by front-end LC separation of phospholipids.

In contrast, MALDI has been easily coupled with TLC for the analysis of different lipid classes (e.g., phosphatidylcholines, sphingomyelins, neutral glycosphingolipids, and gangliosides) [75].

5.1.4 Electron Ionization (EI) and Chemical Ionization (CI)

Electron ionization (EI), formerly known as electron impact, is an ionization method in which energetic electrons interact with gas-phase atoms or molecules to produce ions. This technique was early introduced by Dempster in 1918 [76], and is widely used for gases and volatile molecules.

In an EI ion source, the sample molecules vaporized in the gas phase are bombarded with a beam of energetic electrons (70 eV) at low pressure, as the sample under investigation that contains the neutral molecules is introduced into the ion source in a perpendicular direction to the electron beam. This causes large fluctuations in the electric field around the neutral molecules and induces ionization and fragmentation, hence the term "hard" ionization technique. Radical cations products formed by the ejection of an electron from the target molecules are then directed toward the mass analyzer. The ionization efficiency and production of fragment ions depend strongly on the chemistry of the analyte and the energy of the electrons.

EI is much suitable for the ionization of volatile lipid molecules. In contrast, an additional derivatization step will be needed to convert nonvolatile lipid molecules into the corresponding methyl, trimethylsilyl, or *t*-butyldimethylsilyl derivatives. Usually, extensive fragmentation and double bond migration is observed at high energy [77].

In chemical ionization (CI), ions are produced in the ion source through the collision of the analyte with ions of a reagent gas (usually, methane, ammonia, and isobutane). Since the gas is present in large excess compared to the analyte, the collisions with other reagent gas molecules will create an ionization plasma. Positive and negative ions of the analyte are formed by reactions with this plasma [78].

CI is a lower energy process than EI, and thus will yield less fragmentation, and usually a simpler spectrum in which the lipid analytes are easily detected as intact molecular species, e.g., protonated molecules.

5.1.5 Fast Atom Bombardment (FAB)

Fast atom bombardment (FAB) uses a liquid phase as a matrix, and the sample to be analyzed is mixed with a low volatility and inert liquid matrix (e.g., glycerol) prior to ionization [79]. The sample deposited is then bombarded with an electrically neutral atom beam (Ar or Xe) of high energy (usually several keV), under vacuum.

FAB-MS, now largely replaced by APCI and ESI, is much suitable for lipid analysis after TLC separation. This technique is capable to afford useful information on intact lipid molecules, such as molecular species, FA composition, and FA position in the glycerol backbone; especially in the negative ion mode it has been successfully employed to distinguish between *cis* and *trans*-isomers of monounsaturated FAs [80].

FAB can be considered as an improved version of secondary ion mass spectrometry (SIMS), a surface characterization technique earlier developed [81] for the analysis of solid surfaces or films. In SIMS, a high-energy cation beam is used to bombard the sample surface directly, resulting in sputtering of secondary ions, which are then measured. However, the difficulty to desorb ions with *m/z* over 500 precluded its use for the analysis of large lipid molecules.

5.1.6 Other Ionization Techniques

Desorption electrospray ionization (DESI) is a technique capable to directly analyze biological samples, such as tissues or cells, without complex sample pretreatment [82]. It allows to perform untargeted analysis and structural characterization of lipids under ambient conditions. First introduced by Cooks et al., in 2005 [83], it has proven to be a powerful tool for two-dimensional and three-dimensional imaging of lipids from unmodified complex biological samples [84], as it delivers high throughput and sensitivity.

Nanospray DESI has been introduced more recently [85], as a valuable tool comprehensive lipidomics analysis, capable to afford simultaneous detection of several lipid classes and other metabolites in a single biological tissue [86,87].

Another novel soft ionization technique is continuous flow extractive desorption electrospray ionization (CF-EDESI), especially useful when dealing with nonpolar solvents such as hexane and chloroform, which are not amenable to ESI ionization. This ambient ionization technique was very recently developed and successfully applied in lipidomics analysis especially for the analysis of FAs [88]. In contrast to DESI, from which it is derived, CF-DESI is easily hyphenated to front-end LC separation techniques.

Surface acoustic wave nebulization (SAWN), first reported in 2012 [89], is a novel method to transfer nonvolatile analytes directly from the aqueous phase to the gas phase for subsequent MS analysis. This technique is easy to use and delivers high throughput, since ionization occurs from a planar device and does not require any chemical matrix; the fragmentation occurs moderately, as ions of low internal energy are produced. Phospholipids and lipid A were successfully analyzed by this technique.

5.2 Advances in Mass Spectrometry for Lipidomics

Undoubtedly, the progress made in MS technology has greatly spurred the research in the lipidomics field, with the development of the orbitrap MS [90–92] and Fourier transform ion cyclotron resonance mass spectrometry (FTICRMS) [93–95] which facilitated direct infusion ESI-MS for the simultaneous analysis of multiple lipid classes without the need for prior separation and even for extensive MS/MS analysis [96].

Lipid ions with the same nominal *m/z* values [97,98] can be separated by ultrahigh resolution (≥100,000) MS. On the other hand, identification of lipid molecular species is more confident through high-resolution MS in combination with data obtained by higher energy collision dissociation (HCD), and also quantification of low abundant lipids is more accurate.

Identification and quantification of individual lipid molecular species is feasible by using a robust analytical platform called multidimensional mass spectrometry-based shotgun lipidomics (MDMS-SL), or 2D MS as analogue to two-dimensional nuclear magnetic resonance (NMR) spectroscopy [99–101]. This technique combines four different MS/MS modes, i.e., product ion scan, precursor ion scan, neutral loss scan, and selected reaction monitoring [102]. In 2D lipid maps obtained by this technique, the first dimension is the molecular ions (*x*-axis), whereas the second dimension is the neutrally lost fragments or the monitored fragments ions (*y*-axis). The cross peaks of a primary molecular ion in the first with the second dimension represents the fragments of a given molecular ion, a pseudo product ion spectrum of the molecular ion. The structure of a given molecular ion, as well as its isobaric substituents, can be inferred through the analysis of these cross peaks.

Imaging mass spectrometry (IMS) is a very popular and powerful tool for lipidomics analysis, allowing the direct acquisition of spatial distribution (maps) of lipids from complex tissue sections, and creating images from individual mass spectra of a biological sample [103]. This technique is able to provide the visualization and distribution information of an individual molecule, useful to investigate biological processes involving the interaction and dynamic spatial distribution.

The most common ionization techniques in IMS are MALDI and SIMS. MALDI IMS is able to provide very useful clinical and biological outputs, as it is able to detect almost all classes of lipids, and to offer rich lipid signature in both positive and negative ionization mode. The first MALDI images of lipid distribution in tissues were obtained in 2004 [104]; since then, the technique has expanded its applications to focus on different health- or disease-related tissues, such as mice brain [105], cottonseed [106], human brain [107], normal and ischemic rat brain, [108] and human skin [109].

In this technique, N_2 UV laser beams are usually rastered over the tissue section, with a typical matrix of 2,5-dihydroxybenzoic acid. However, new matrixes are constantly developed, with improved features for the analysis of a specific class of lipids, since this step is crucial for the whole technique. Major drawbacks deriving by the matrix consist in the limited lateral resolution (typically, above 20 μm) due to physical limitations and chemical noise observed at the low masses [110].

Compared to MALDI, SIMS IMS affords much cleaner spectra at the low masses, and much higher (submicrometer) resolution which makes single cell detection feasible. This is a crucial step for lipidomics analysis, as it allows to attain more details on biological process while, on the other hand, giving more information about intracellular correlation among different molecules [111].

SIMS is able to determine the molecular composition and individual compound localization on a tissue section with very high spatial resolution, and without the need for any prior sample treatment, which makes the analyses easier and more straightforward and provides the closest possible to physiological conditions [112]. In 2013, Passarelli and coworkers mapped the localization of various intact lipid species across the surface of a single neuron exploiting the high spatial resolution of C60-SIMS [113]. A region of high colocalization between the vitamin E signal and compiled PC (16:0e/18:1) lipid signal could be seen on the top portion of the neurons soma, which might be caused by some unknown correlation between these two chemicals.

Also DESI and more recently, nanospray DESI imaging technology has been implemented [114], which can significantly increase the spatial resolution and sensitivity, and used in combination with multivariate statistical analysis [115], made significant contribution to lipidomics investigation. This technique has been successfully applied to diagnose several kinds of cancers (such as renal cell cancer [116], prostate cancer [117], hepatocellular carcinoma [118], and brain cancer [119]), demonstrating the potential to identify the histology type of tumors, and also to realize a near real-time tumor surgery guide by providing rapid diagnosis and tumor margin assessment.

6. CHROMATOGRAPHIC SEPARATIONS IN LIPIDOMICS RESEARCH

The first application of chromatography to lipid analysis dates back to 1952, when James and Martin attained the separation of FAs by GC [120]. Since then, the analysis of lipids has been performed by a variety of distinct chromatographic approaches, reflecting the distinct chemical classes and subclasses. These include TLC, HPLC, capillary zone electrophoresis (CZE), micellar electrokinetic chromatography (MEKC), capillary electrochromatography (CEC), SFC, and GC.

6.1 Thin-Layer Chromatography (TLC)

Thin-layer chromatography (TLC) was successfully applied to sterol and glycolipid analysis already half-century ago [121]. TLC is a development of paper chromatography, in which the stationary phase is coated onto glass plates in a thin layer. Most applications employ the adsorption mode, and the so-termed "silica gel G" is the most widespread used material. Coated plates can be prepared in laboratory, or purchased as precoated plastic, glass, or aluminum TLC plates. For analytical purposes, typically 0.25 mm thick layers of adsorbent are employed, with as little as 0.5 mg of lipid spotted, to achieve the maximum resolution. On a preparative scale, much higher sample amounts can be applied to thicker TLC plate, even at the price of a considerable loss in resolution; typically, as much as 50 mg of sample onto 20 × 20 mm, 0.5 mm plate.

For the separation, small amounts of the lipid sample are spotted onto the TLC plate by means of a syringe, around 2 cm from the bottom of the plate, which is then placed into a tank of an appropriate solvent that only touches the bottom part of the plate without reaching the sample spots, which serves as the mobile phase. Due to capillary forces, with time the solvent moves up the plate taking the sample components with it at different rates, depending on their affinity to the adsorbent. Various combinations of aqueous stationary phases and organic mobile phases can be used to obtain differential migration of distinct lipid classes. The quality of the separation will be further affected by the degree of hydration of the adsorbent. This in turn depends on a number of other factors, like the temperature used for activation of the TLC plate, the time of exposure, the storage conditions, and the humidity it takes from the atmosphere. The process is complete when the solvent nears the top of the plate; the latter is removed and dried in a stream of air or nitrogen, the latter minimizing possible auto-oxidation of susceptible lipids, if further use is needed. The plate is then usually sprayed with a dye, to make the spots visible under UV light. The lipids can be identified by comparing the distance that the spots move, with that of standards of known composition. The spray is a chemical reagent, which may render all lipids visible, or it may be specific for a certain class of compounds; detection and quantification in TLC will be discussed later on.

TLC can be successfully applied to the separation of some simple lipid classes, such as mono-, di-, and triacylglycerols, and their derivatives. Typical mobile phases for the separation of simple lipids contain hexane, diethyl ether, and formic or acetic acid in various proportions (e.g., 80:20:2, v/v); the latter will ensure successful migration of the free FAs.

Complex lipids (e.g., phospholipids and glycosphingolipids) will not migrate under these conditions, and they can then be only detected or quantified as if they were a single lipid class.

6.2 High-Performance and Two-Dimensional TLC

More recently, high-performance (HP) TLC plates have become commercially available, based on a very uniform layer of silica gel, with small particle size, and therefore capable to afford excellent separation power with reasonably elution times. This technique also achieves more accurate quantification. These plates are of special usefulness for the separation of different types of lipids, e.g., containing phospholipids or glycosphingolipids, as much more resolution can be attained, while greatly reducing the time needed for the separation.

If a complex lipid sample cannot be separated within a single TLC run, it may be resolved by rechromatography in a second direction. In the so-called "two-dimensional TLC" techniques, the sample is first applied in the bottom corner at the left hand of the plate, and developed in a first selected solvent. After the first development is completed, the plate is removed and dried, then turned counterclockwise at right angles, and chromatographed again with a second mobile phase, of different selectivity.

In the work by Yu and Ariga [122], HP-TLC plates were first given a double development for part of the plate in a polar solvent to separate the phospholipids, and then for the full length of the plate with a less polar mobile phase, in order to resolve each of the simple lipids (in spite of the triple development, the total elution time was only 30 min).

6.3 High-Performance Liquid Chromatography

Distinct HPLC methods have been used for identification and quantification of different classes of lipids, including normal-phase LC, silver-ion LC, and non-aqueous reversed-phase LC. However, none of them achieves the identification and quantification of all different classes of lipids from a cell or tissue within a single experiment. A combination of two or more separation forms must be used, either in the off-line or in the online mode.

6.3.1 Normal-Phase Liquid Chromatography (NP-LC)

Various types of forces may be involved in the mechanism of any chromatographic analysis; and in the case of lipids, hydrophobic interactions play a major role, while ionic, dispersive, and polar interactions occur in a less extent [123].

Under normal-phase conditions, lipid separation occurs on the basis of their polar head group, regardless their FA composition. Little, if any, retention time differences will therefore occur between individual molecules within a given lipid class, thus each chemical class will tend to coelute as a single chromatographic peak. Commonly, this technique uses a silica support as adsorbent, on which different organic moieties are linked by means of a short spacer. However, some free silanol groups on the silica surface remain as residue from the manufacturing process, and affect the separation, to some extent. Silica stationary phases have been earlier used as adsorbents in TLC and low-pressure column chromatography, but nowadays the majority of applications regard the LC separation of complex lipids, according to type and number of their polar functional substituents into classes, contain amino, phosphate, hydroxyl groups, etc. [124].

Isocratic elution is sometimes used but in general, and especially for complex mixtures, the better results are obtained by using a gradient elution in which the polarity of the mobile phase is gradually increased, but it much depends on detector used. NP-LC has been widely used in conjunction with MS detection in many applications, in spite of the long analysis times (typically, in the 30–90 min range), and the poor peak resolution obtained (average peak widths were in the 20–30 s order) [125]. Mobile phases used for NP-LC/MS of polar lipids typically consist in ammonium formate (e.g., 5 mM) in A: MTBE/methanol/isopropanol (IPA) (80:10:7, v/v) and B: MeOH/IPA/water (90:7:3, v/v), or pure MeOH. For the determination of nonpolar lipids, typical solvents are A: heptane, B: 10% IPA in MTBE or ammonium acetate in MeOH/water (9:1, v/v). Addition of salts, buffers, or little amounts of acids to solvents reservoir may affect the ionization and, thus, elution of the more polar compounds.

6.3.2 Silver Ion Liquid Chromatography (Ag-LC)

In silver ion liquid chromatography (Ag-LC), separation is based on the weak interaction between the silver ions and the π-electrons of double bonds; cation exchange silver ion LC, sometimes referred to as "argentation chromatography" is considered the most efficient [126].

Silver ions impregnated on the silica or bounded to the ion-exchange stationary phase, acting as electron acceptors, interact reversibly with the *pi* electrons of DBs of unsaturated lipids, in turn electron donors, to form polar complexes. The strength of such charge-transfer type bonds will depend on the number, position, and geometry of DBs (the *cis* isomers interacting more strongly than the *trans*), and will ultimately affect the retention time. This technique has been also used in TLC applications, but now finds widespread diffusion for the separation of molecular species of simple lipids, such as TAGs.

A TAG molecule can be described according to the following features:

- Total carbon number (CN), given by the sum of the carbon atoms in the three alkyl chains;
- Total number of double bonds (DBN);
- FA alkyl chain length;
- FA position on the glycerol backbone (*sn*-1/3, *sn*-2);
- Position and configuration of DBs in each fatty acid;
- "Partition number" (PN), given by the number of carbon atoms in the aliphatic residues, minus twice the number of double bonds (=CN – 2 DBN); it is also termed as "equivalent carbon number" (ECN).

Ag-LC separates TAG species according to their number of double bonds, and it is also affected by the position and geometry of DBs, as well as alkyl chain length. Briefly, retention of TAGs increases with increasing number of double bonds and decreases with increasing alkyl chain length. Due to the nonpolar nature of TAGs, atmospheric pressure chemical ionization is the technique of choice for their ionization; Ag-LC/APCI holds a potential to separate geometrical isomers of TAGs and regioisomers of unsaturated TAGs as well. Such a distinction relies on the relative abundances of fragment ions arising from the neutral losses of FAs from *sn*-1/3 and *sn*-2 positions. Such an approach has been often applied for assignment of the most abundant FA in the *sn*-2 position, for the quantitative determination of *sn*-2 occupation calibration curves have prior obtained, for mixtures of both regioisomers [127].

6.3.3 Nonaqueous Reversed-Phase Liquid Chromatography (NARP-LC)

Reversed-phase liquid chromatography (RP-LC) techniques find widespread application today, especially in conjunction with MS (RP-LC/MS), or tandem MS (RP-LC/MS/MS), techniques. Mobile phases typically employed in RP-LC are straightforward compatible with biological and lipophilic molecules, such as lipids.

In RP-LC, the degree of polarity, alkyl chain saturation, and chain length, will concur to the separation mechanism, thus enabling separation of lipids, which differ in their FA composition [128]; NARP-LC separation of molecular species of neutral lipids is based mainly on partition between the nonpolar bonded stationary and the polar mobile phases. Separation of molecular species of TAGs, DAGs, and MAGs in reversed-phase columns has been successfully achieved.

In NARP-LC, the elution order is based on increasing PN, thus components will elute in ascending order of chain length, while a double bond in any of the FA chain lengths reduces the retention time, roughly by the equivalent of two carbon atoms. The number and position of double bonds also affect retention, as well as their chain length. As a consequence, TAGs with identical equivalent carbon number tend to coelute, and are thus called critical pairs. However, under optimized conditions, by employing efficient stationary phases, nowadays commercially available, it is possible to separate a great number of TAGs with same PN. The separation of compounds with the same PN group is feasible, such as OOO, OOP, OPP, and PPP (O: oleic acid, P: palmitic acid), as well as the critical pair LLL/OLL$_n$ (L$_n$: linolenic acid), having the same PN, CN, and DBN. Also TAGs differing in the position(s) of double bonds have been resolved, such as LnLnγLn–LnLnLn, which have the same PN, CN, and DBN but differently located along the same hydrocarbon chain, i.e., Δ6 or Δ9 [129].

Although many different nonpolar stationary phases, the so-called "C$_{18}$" or "ODS" is by far the most used one; it consists of octadecylsilyl bounded covalently to the surface of spherical silica. Stationary phases containing octanyl (C_8), butyl (C_4), or ethyl (C_2) hydrocarbon chains are also occasionally used, and are less retentive than the long-chain hydrocarbon one. New columns today available can be operated at extreme pH values (i.e., <2 or >8), at high temperatures (>100 °C), or under high-pressure conditions (900 bar). For most lipid separations, mobile phases typically consist in mixtures of 2-propanol/acetonitrile, 2-propanol/methanol, acetonitrile/chloroform, etc. Elution is generally accomplished under gradient conditions; when DAGs and MAGs are part of the lipid mixture to separate, an initial step of aqueous-organic solvent is sometimes accomplished at the beginning of the run, to improve the resolution of the more polar sample components.

It is straightforward that NARP-LC fairly complements Ag-LC, for the separation of complex lipid mixtures, and this feature has been used in a number of multidimensional approaches (discussed later).

6.3.4 Other HPLC Techniques

In addition to the ones described above, other liquid chromatography techniques have also been applied to lipid analysis, even if in a more limited number of applications. These include size-exclusion chromatography, also known as "gel-permeation" or "gel-filtration" chromatography; chiral-phase chromatography; ion-pair chromatography. Very recently, hydrophilic-interaction liquid chromatography (HILIC) has been exploited for lipid separation. In this type of chromatography, individual lipid classes of lipids are separated on silica or diol columns according to their polarity [130,131].

6.4 Recent Advances in LC/MS Techniques for Lipidomics

Various chromatography-based separation methods in combination with various detection systems, have been developed for lipid analysis and considerable progress has been made in both LC hardware and column technology. However, in many cases HPLC alone is not capable to afford the separation/information required to fully characterize complex samples, such as biological matrix containing several lipids, which often exhibit similar retention behavior. To this purpose, MS detection can act as a second separation dimension, by discriminating components based on their molecular masses and fragments [132].

In terms of mass spectrometric analyzers, TAG analysis has been developed and implemented successfully with tandem quadrupoles and linear ion traps; tandem mass spectrometry (MS/MS) techniques allowed for unambiguous structural characterization for mixtures of isobaric species, yielding product ions from both positive and negative fragmentation processes. Later on, the triple quadrupole mass spectrometer was found to be well suited for lipid analysis through MS/MS operation, including product ion scanning, and selected reaction monitoring. High-resolution mass analysis of molecular ion species and product ions after collision-induced dissociation (CID) became routinely possible with the second-generation Time-of-flight (ToF) analyzers, Fourier transform ion cyclotron resonance (FT-ICR)-MS, and orbitrap mass spectrometer. Hybrid mass spectrometers such as the Q-ToF afford superior spectral resolution and mass accuracy, also overcoming duty cycle issues typically associated with other scanning instruments; the so-called "MS^E" acquisition mode in fact allows for many precursor and neutral loss acquisitions within a single experimental run.

Ultimate generation single quadrupoles allow for high speed scanning (up to 15,000 amu/s) and ultrafast polarity switching; the small size and the possibility to perform tandem MS make them ideal for benchtop LC-MS. On the other hand, ToF instruments present a number of advantages: high speed (up to 20,000 Hz), high resolution (using a reflectron), virtually no limit on mass range, femtogram-level sensitivity, sub-ppm mass accuracy, improved in-spectrum dynamic range without loss in sensitivity, high mass resolution, feasibility to use as a second stage in tandem MS experiments, in combination with either an ion trap (IT-ToF) or a quadrupole (Q-ToF) [133].

From the HPLC counterpart, recent advances in column technology (sub-2 μm and partially porous particles) and hardware (allowing operating pressures up to 15,000 psi) have arrived to meet the expected performance in terms of resolution, speed, and sensitivity with respect to conventional LC analysis. UHPLC/MS platform combining ultra high-performance liquid chromatography with an orthogonal accelerated time of flight (oa-ToF) spectrometer offers high-throughput sample analysis, providing narrow chromatographic peaks (<3 s) with good ion statistics from accurate mass measurement (<2 ppm) [134].

Electrospray ionization (ESI) and atmospheric pressure chemical ionization (APCI) have been very popular as ionization techniques, for LC/MS determination of lipids [135–137]. Based on their ionization mechanisms, ESI is more suitable for ionization of polar and ionic compounds and is capable of ionizing both small and large biomolecules. A very powerful Jetstream ESI-QqToF mass spectrometry approach in combination with high-resolution reversed-phase liquid chromatography, applied on human blood plasma, have been recently reported by Sandra et al. Figure 17 shows an 2D LC-MS feature maps, colored by abundance, of the blood plasma lipid extract in positive (A) and negative (B) ionization mode. In this work, the application of UHPLC (up to 1200 bar) allowed to greatly extend the peak capacity and productivity [138].

APCI can ionize less polar and neutral compounds more efficiently than ESI. Consequently, APCI-MS coupled to HPLC has represented the most used tool for TAG identification, because of the full compatibility with common NARP-LC conditions, easy ionization of nonpolar TAGs, and the attainment of both protonated molecules $[M + H]^+$ and fragment ions $[M + H–R_iCOOH]^+$. On the other hand, ESI is usually employed for the more polar phospholipids.

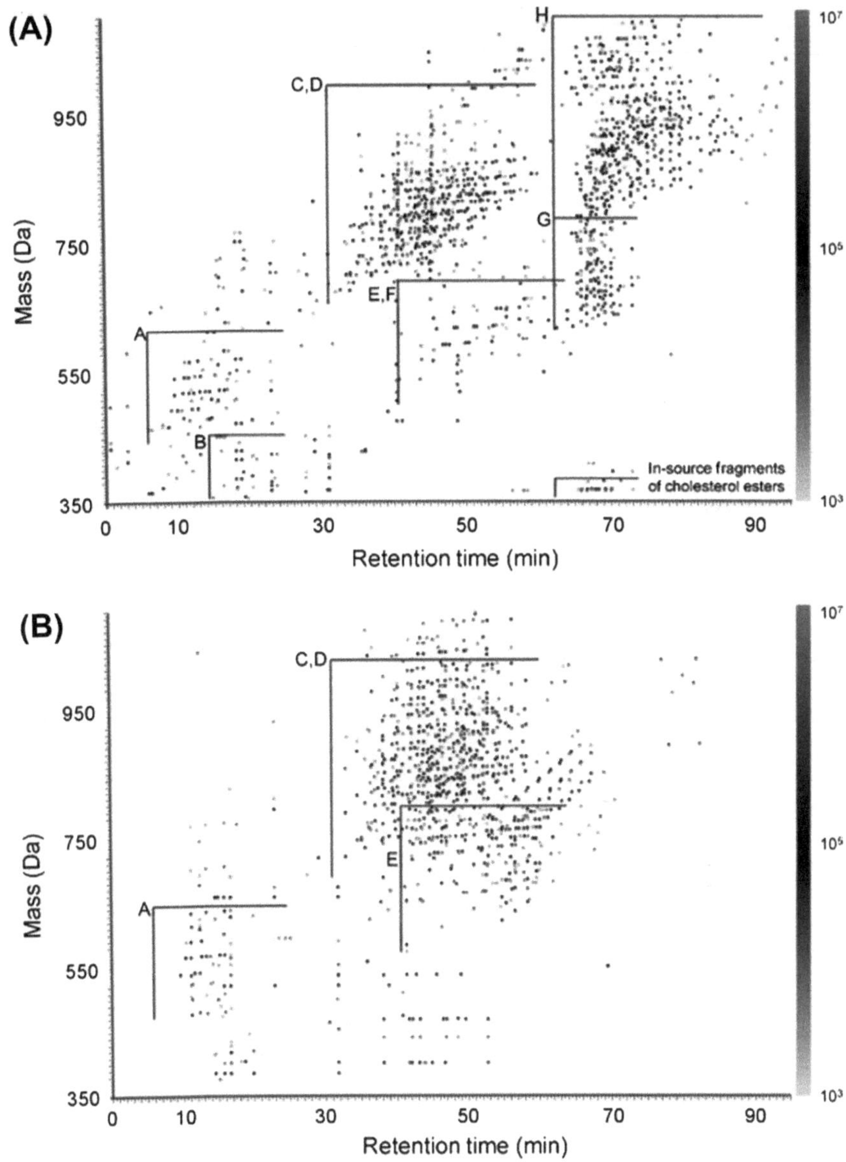

FIGURE 17 2D LC-MS feature maps, colored by abundance, of the blood plasma lipid extract in positive (A) and negative (B) ionization mode. *Reprinted from Ref. [138] with permission from Elsevier.*

However, ESI or matrix-assisted laser desorption/ionization (MALDI) has been used for TAGs, as well [139,140].

In spite of the considerable advantages to be gained from the coupling of HPLC and MS, hyphenation cannot be considered straightforward, yet [141].

6.5 Multidimensional Liquid Chromatography (MDLC, LC × LC)

In spite of the substantial progress in the field of HPLC column and instrumentation, and the considerable efforts made by analysts to develop chromatographic techniques taking full advantage of these developments, the separation efficiency required in many cases overwhelms the capability of any single-column method. As a consequence, much effort has been directed to the development of multidimensional HPLC (MDLC) systems, especially in the comprehensive mode (LC × LC), aiming to attain increased peak capacity, n_c (namely the total number of solutes that can be potentially separated, with regularly spaced adjacent peaks filling the available separation space). Several strategies have been implemented by researchers in designing comprehensive HPLC systems for lipid analysis, which exhibit an immediate and evident advantage over other multidimensional HPLC techniques: the whole sample is, in fact, subjected to an additional separation dimension. On the other hand, in the so-termed "heart-cutting" LC-LC approaches, only selected fractions are transferred to the second dimension.

The transfer of fractions, from the first to the second dimension, can be performed either in the off- or online mode; both techniques are widely employed and both present distinct *pros* and *cons*. In off-line methods (LC/LC), fractions from the first chromatographic step are collected via a fraction collector, separated from the solvent by evaporation, redissolved and then reinjected onto the second column. The most striking advantage of such an approach derives from the great flexibility in coupling any separation modes, in terms of mobile phases and buffer compatibility, time and duration. The separation power can be maximized by independent optimization of each dimension, while sensitivity can be tuned by regulating the sample concentration injected in both dimensions.

Main drawbacks consist in the potential loss of sample or degradation during solvent evaporation, possible sample contamination or artifacts formation, and difficulties in automation. Moreover, such methods are obviously more time-consuming and offer low analytical repeatability. Online approaches are more technically challenging and difficult to operate, with respect to off-line methods; on the other hand, they bring in the great advantage of fully automated solute transfer between the two separation dimensions, with no flow interruption. As a consequence, they allow for greater analytical repeatability. Additionally, the reduced sample handling makes operation with unstable compounds more feasible and minimizes losses, an issue of particular relevance when dealing with limited sample amounts. A number of factors have to be taken into consideration, when developing an online 2D method, since the more stringent requirements render the coupling of different separation modes far more complicated with specific interfaces and software needed. Instrumental set-ups and method development, capabilities, and requirements of mass spectrometry detection in multidimensional HPLC have been recently reviewed, for various target analytes, including lipids [142].

In the last few years, comprehensive two-dimensional systems in combination with mass spectrometry have been investigated for TAG separation in various food samples, such as rice oil [143], soybean and linseed oils [144], donkey milk [145], and corn oil [146], using Ag^+-HPLC in the first dimension (^{1}D) and NARP-LC in the second dimension (^{2}D), respectively. The two separation modes are based on different retention mechanisms, as discussed earlier, and hence, the column combination can be considered as entirely orthogonal. For ^{1}D separation, either a microbore (1 mm I.D.) or a narrow-bore column (2.1 mm I.D.) are usually employed; *n*-hexane modified with a small amount of acetonitrile was used in ^{1}D, whereas isopropanol and acetonitrile, in a gradient program, were used in the ^{2}D. The issues related to solvent incompatibility and analyte focusing were solved by using a microbore column with a very low flow rate in the ^{1}D and by decreasing the percentage of the weaker solvent (acetonitrile), in the ^{2}D. The 2D chromatograms were characterized by the formation of group-type patterns, with TAGs located in characteristic positions in relation to their PN and DB values. With regards to MS conditions, a data acquisition rate of 5 Hz and a mass scan range of 400–900 amu, allowed the acquisition of approximately 10 spectra for peaks of approximately 2-s duration; the spectral production frequency was sufficient for reliable peak identification.

Off-line coupling of HILIC and reversed-phase was realized by Holčapek and coworkers, for lipidomic profiling of biological tissues [147] using both UV and MS detection (namely, ESI ionization for polar phospholipid species, and APCI interface or nonpolar TAG species).

A novel online HILIC × RP-LC-ESI-MS system for phospholipid analysis have been recently reported by Dugo et al. [148]. In this work, a silica HILIC in the ^{1}D and a C18 column in the ^{2}D, based on stop-flow mode, in combination with ESI-MS detection was established to separate PLs in a Folch-extracted cow's milk and plasma samples. The enlargements of the HILIC × RP-LC-ESI-MS contour plot, and the 2D raw data, for the separations of PC molecular species, in the plasma sample are shown in Figure 18. The PC class turned out to be the richest in terms of molecular species, allowing to separate a total of 11 PCs, over two 15-min modulation cycles. The outcome of this work was to highlight the great separation power of the LC × LC setup tested, due to the orthogonal column combination, which can be used also in presence of inexpensive detectors, e.g., quadMS or evaporative light scattering, for the analysis of other lipid classes, of different origin. The only drawback of such a system was the long analysis time (roughly 6 h), even though well compensated by the enhanced resolving power, and by the greater amount of analyte information obtained.

6.6 Gas Chromatography

Since most lipids are not volatile and some of them can be easily prone to degradation at high temperatures, compared to TLC and HPLC, GC presents much more issues. In addition, GC methods are very sensitive to compound

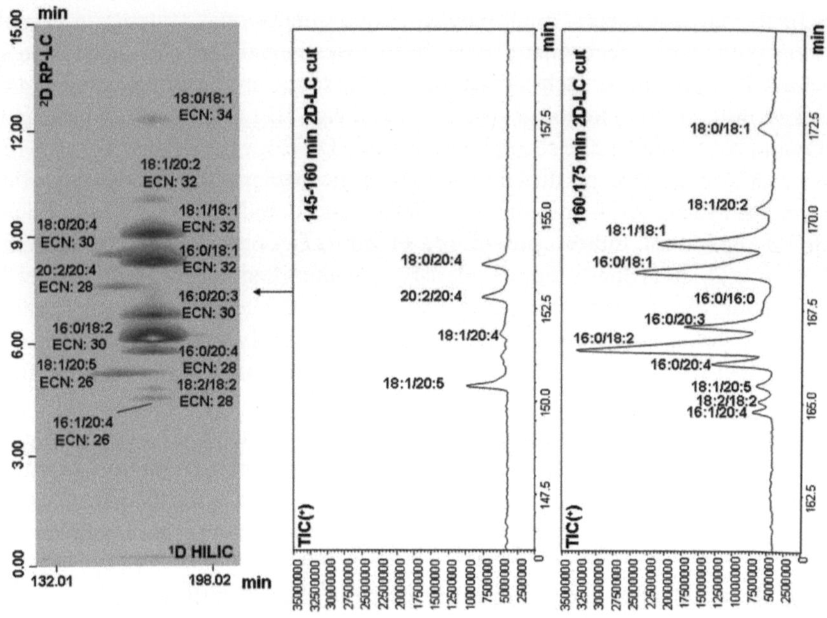

FIGURE 18 Enlargement of the HILIC × RP-LC-ESI-MS contour plot for separation of the PC molecular species, along with the corresponding 2D raw data of contained in the plasma sample. *Reprinted from Ref. [148] with permission from Elsevier.*

polarity; in most cases, in order to improve volatility, they require an essential derivatization step prior to GC separation which may significantly affect both sample recovery and most structural information about lipid molecular species. Although derivatization can solve the problems of the analyte in terms of volatility and lability when dealing with chemically simple molecules, for complex biomolecules, the necessary derivatization may be difficult and may involve multiple reaction steps [149].

Usually, GC is used to analyze several classes of lipids and may present high resolution and sensitivity, especially when focusing on one class. Therefore, a very specific extraction, a preseparation by TLC or HPLC, and a derivatization procedure are important parts of GC-based lipidomic methods. Sample preparation can include preseparation of lipid classes, hydrolysis, derivatization, or pyrolysis [150].

Before analysis of any lipid samples, it is first necessary to extract them from their tissue, plant, or food matrices in a relatively pure state. The authors have dealt with these aspects in some detail earlier in this chapter (Paragraph 4. "Extraction Methods of Lipids"), so only essential details will be discussed here, as preliminary steps to GC analysis.

While the extraction method will be chosen by the analyst according to the specific needs of the GC analysis, apart from the type of sample, some common rules need to be followed. First, lipid extraction is carried out as soon as possible after the removal of the tissue from a living organism, to avoid degradation, while plant tissues are commonly preextracted with isopropanol to minimize artifactual by tissue enzymes. Moreover, wherever possible, lipids are handled in an atmosphere of nitrogen and, in the case of PUFAs, antioxidant are often added to prevent auto-oxidation which would occur very rapidly in air.

Simple lipids are extracted with relative ease, since they generally occur in storage tissues as part of large aggregates; on the other hand, the complex lipids which usually occur as constituents of membranes, in a close association (and interaction) with proteins and polysaccharides, are not extracted so readily. More than one extraction may be needed, in the case of difficult samples, and lyophilized tissues may need to be rehydrated prior to carrying out the extraction. Furthermore, lipids extracted from tissues may contain appreciable amounts of nonlipid contaminants, such as sugars, amino acids, urea, and salts, which must be removed before the lipids are analyzed.

Once the lipid samples are extracted, usually some prefractionation is desirable, prior to GC analysis. It is common to separate lipids into simpler groups or distinct classes, by means of TLC, HPLC, or preparative HPLC (these techniques have been described earlier, in this chapter). The GC separation method adopted will differ, depending on whether simple molecules, such as the fatty acids, are to be analyzed, or the molecular species of more complicated lipids; often, more chromatographic procedures are combined, to achieve effective separations.

Lipids are hydrolyzed by heating them under reflux with an excess of dilute aqueous ethanolic alkali. The free FAs and any water-soluble hydrolysis products (saponifiable fraction), and the nonpolar, nonsaponifiable fraction (containing hydrocarbons, fatty alcohols, and sterols) are recovered separately, for further GC analysis.

Before being analyzed, the polar FAs need to be converted to low-molecular weight volatile derivatives, such as methyl esters. Alkali- or acid-catalyzed transesterification in a methanolic medium can be used to obtain FA methyl esters (FAMEs) from the saponifiable fraction, unless hydrolysis has been carried out previously: in this case, only acid-catalyzed transesterification may be applied. Peak shape and resolution will be also benefit from the derivatization of any other polar functional groups in the molecules prior to analysis, in a similar manner [151].

The separation of FAMEs is usually achieved on highly polar liquid phases, and the analytes are separated according to their chain length and degree of saturation; however, for the separation of *cis* and *trans* FAs usually GC only will be not sufficient, and the combination with other types of chromatography is applied [152]. In principle, it would be possible to identify FAMEs by GC retention times alone, however since not all fatty acid standards are commercially available, GC is commonly used in combination with derivatization and chemical degradative or spectroscopic techniques, such as Fourier transform infrared (FTIR) and, especially, MS, as further means of characterization.

Despite the lack of sensitivity, GC-FTIR may be conveniently applied as a complementary tool to GC-MS, for isomer discrimination and structure elucidation of compounds with very closely related structures. Unlike MS, the coupling of IR to GC poses the need for a specific interface; three types are commercially available: matrix isolation, direct deposition, and light pipe [153].

It must be noticed, that the reliability of the data obtained may be affected by artifact formation, structural modifications, and/or nonquantitative formation of the ester derivatives. Furthermore, major drawbacks of this commonly applied procedure of sample derivatization and separation are the loss of any structural information related to the lipid composition, and the cost in time (typically, more than an hour). On the other hand, fast and high-resolution GC methods are also available, capable to afford satisfactory separations in few minutes [154].

High oven temperature of GC is essential for analysis of high molecular weight compounds, and high temperature GC (HTGC) is a routine technique for the analysis of high boiling compounds which are eluted from the column with oven cycling up to >400 °C. Recently, an HTGC-ToF-MS system has been assembled by Sutton et al. [155]. This system using readily available equipments has been demonstrated to be viable for the online MS analysis (at up to 430 °C) of a wide range of high molecular weight compounds (*m/z* 1850), which is suitable for the analysis of very long-chain lipids [156].

Recently, the ionic liquid has been used as a novel stationary of GC column to separate lipids, especially the isomers of lipids. Delmonte et al. discussed the separation of *cis* and *trans* isomers of FAs, showing successful separation of a series of geometric and positional isomers of monounsaturated FAs [157]. Ionic liquid GC capillary column could be recommended as the most suitable column for the analysis of total FAs or FA methyl esters. It may offer more molecular species information for lipidomics analysis [158].

A great deal of effort has been applied to TAGs analysis, since these molecules represent the main component of virtually all the fats and oils of commercial importance. GC can also be used for the direct separation of TAGs into family groups based on the carbon chain length (carbon number, CN). Because of the low volatility of intact lipids, the analysis is normally performed at high temperatures on nonpolar or medium-polar low-bleeding columns. If a more polar stationary phase is used, further resolution of the CN family group peaks can be achieved [159].

In recent years, the multidimensional GC system becomes more and more widely used to analyze complex samples. As two columns with different polarities are coupled together, the main advantage of comprehensive two-dimensional GC is improvement of the chromatographic resolution due to the increase of peak capacity. The two-dimensional GC has been applied to analyze FAs profiles in many different real samples, such as wastewater [160], human plasma [161], mouse heart [162], and various cells [163,164], and becomes more and more popular in lipidomics analysis. The contour plot relative to the GC × GC separation of the most complex plasma sample is illustrated in Figure 19.

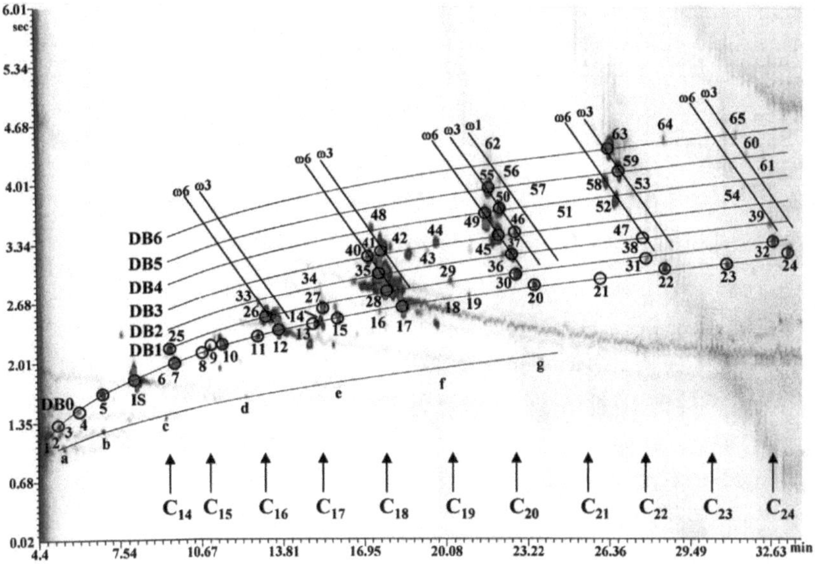

FIGURE 19 Result of a plasma FAMEs GC × GC-FID analysis. *Reprinted from Ref. [161] with permission from Wiley.*

7. CONCLUSIONS AND FUTURE PERSPECTIVES

Successfully investigated in the last decade, lipidomics can be considered an essential tool for investigation of many physiological and pathological processes, such as cancers and Alzheimer's disease. Due to the diversity of structures, several distinct techniques have been exploited for lipidomics analysis aiming to separate and determine the various lipid classes. Liquid–liquid extraction is the most used one for lipid extraction, despite the use of SPE turned out to be a valuable tool for such a task.

Electrospray mass spectrometry and matrix-assisted laser desorption ionization mass spectrometry are the most widely used techniques for both comprehensive and targeted analysis; MS imaging technology is becoming even more popular for the study of biological kinetics and processes, as they can provide quick information on interactions and dynamic spatial distribution of biological samples.

Various chromatographic technologies are used for the separation of lipids in complex sample matrixes, especially for comprehensive analysis. Thin layer chromatography is a simple, rapid, and inexpensive technology that can achieve the separation of different lipid classes; LC is definitely an essential tool for lipid separation, and shows great applicability when it is coupled to mass spectrometry or tandem mass spectrometry. Supercritical fluid chromatography is suitable for nontargeted lipid profiling due to high resolution, whereas GC has the advantage of isomer separation and highly sensitive quantitation, although derivatization of lipids is usually needed.

Undoubtedly, multidimensional techniques like LC × LC and GC × GC have significantly spurred the research in the lipidomics field. In the near future, new advances in analytical technologies for lipidomics studies, such as 2D NMR, new MS ionization modes, the improvement of MS imaging technology, and the realization of more powerful comprehensive platforms are likely to significantly promoted the investigation of lipid profiling and other aspects in lipidomics research.

In particular, innovative methods with higher sensitivity are urgently needed for probing of low abundance lipids, which pose a great challenge for bioanalytical chemists working in life science areas.

REFERENCES

[1] X.L. Han, R.W. Gross, J. Lipid Res. 44 (2003) 1071–1079.
[2] F. Spener, M. Lagarde, A. Géloën, M. Record, Eur. J. Lipid Sci. Technol. 105 (2003) 481–482.
[3] M.R. Wenk, Nat. Rev. Drug Discov. 4 (2005) 594–610.
[4] G. van Meer, EMBO J. 24 (2005) 3159–3165.
[5] M. Orešič, V.A. Hänninen, A. Vidal-Puig, Trends Biotechnol. 26 (2008) 647–652.
[6] J.E. Vance, D.E. Vance (Eds.), Biochemistry of Lipids, Lipoproteins and Membranes, Elsevier, Amsterdam, 2002.
[7] C. Wang, H. Kong, Y. Guan, J. Yang, J. Gu, S. Yang, G. Xu, Anal. Chem. 77 (2005) 4108–4116.
[8] Y. Shi, J. Biomed. Res. 24 (2010) 6–15.
[9] A.J. Lusis, Nature 407 (2000) 233–241.
[10] R. Kaddurah-Daouk, J. McEvoy, R.A. Baillie, D.J. Lee, K. Yao, P.M. Doraiswamy, K.R.R. Krishnan, Mol. Psychiatry 12 (2007) 934–945.
[11] T. Hartmann, J. Kuchenbecker, M.O.W. Grimm, J. Neurochem. 103 (2007) 159–170.
[12] J.A. Menendez, R. Lupu, Nat. Rev. Cancer 7 (2007) 763–777.
[13] N.J. Haughey, J. Steine, A. Nath, J.C. McArthur, N. Sacktor, C. Pardo, V.V. Bandaru, et al., Front. Biosci. 13 (2008) 5120–5130.
[14] J.M. Castro-Perez, J. Kamphorst, J. De Groot, F. Lafeber, J. Goshawk, K. Yu, R.J. Shockcor, J.P. Vreeken, T. Hankemeier, J. Proteome Res. 9 (2010) 2377–2389.
[15] M.B. Khalil, W. Hou, H. Zhou, F. Elisma, L.A. Swayne, A.P. Blanchard, Z. Yao, S.A.L. Bennett, D. Figeys, Mass Spectrom. Rev. 29 (2010) 877–929.
[16] M. Li, L. Yang, Y. Bai, H. Liu, Anal. Chem. 86 (2014) 161–175.
[17] M.I. Gurr, J.L. Harwood, K. Frayn (Eds.), Lipid Biochemistry, Blackwells, London, U.K, 2002.
[18] I. Hanin, G. Pepeu (Eds.), Phospholipids, Plenum Press, New York, 1990.
[19] K. Larsson (Ed.), Lipids. Molecular Organization, Functional Properties and Technical Applications, The Oily Press, Dundee, 1993.
[20] D.E. Vance, J.E. Vance (Eds.), Biochemistry of Lipids, Lipoproteins and Membranes, fourth ed., Elsevier Science, New York, 2002.
[21] L. Feng, G.D. Prestwich (Eds.), Functional Lipidomics, Dekker-CRC, New York, 2005.
[22] W.R. Bloor, Proc. Soc. Exp. Biol. Med. 17 (1920) 138–140.

[23] W.W. Christie, X. Han (Eds.), Lipid Analysis – Isolation, Separation, Identification and Lipidomic Analysis, fourth ed., The Oily Press, Bridgwater, U.K, 2010.
[24] E. Fahy, S. Subramaniam, H.A. Brown, C.K. Glass, A.H. Merrill Jr., R.C. Murphy, C.R. Raetz, D.W. Russell, Y. Seyama, W. Shaw, T. Shimizu, F. Spener, G. van Meer, M.S. VanNieuwenhze, H.S. White, J.L. Witztum, E.A. Dennis, J. Lipid Res. 46 (2005) 839–862.
[25] E. Fahy, S. Subramaniam, R.C. Murphy, M. Nishijima, C.R. Raetz, T. Shimizu, F. Spener, G. Van Meer, M.J.O. Wakelam, E.A. Dennis, J. Lipid Res. 50 (2009) S9–S14.
[26] M.I. Gurr, J.L. Harwood, K. Frayn (Eds.), Lipid Biochemistry, Blackwells, London, U.K, 2002.
[27] I. Hanin, G. Pepeu (Eds.), Phospholipids, Plenum Press, New York, 1990.
[28] K. Larsson (Ed.), Lipids. Molecular Organization, Functional Properties and Technical Applications, The Oily Press, Dundee, 1993.
[29] D.E. Vance, J.E. Vance (Eds.), Biochemistry of Lipids, Lipoproteins and Membranes, fourth ed., Elsevier Science, New York, 2002.
[30] L. Feng, G.D. Prestwich (Eds.), Functional Lipidomics, Dekker-CRC, New York, 2005.
[31] F.J. Señorans, P. Luna, Sample preparation techniques for the determination of fats in food, in: J. Pawlyszin (Ed.), Comprehensive Sampling and Sample Preparation, Elsevier, Amsterdam, 2012, pp. 203–211.
[32] F. Soxhlet, Dingler's Polytechnisches J. 232 (1879) 461.
[33] J. Folch, M. Lees, G.H.S. Stanley, J. Biol. Chem. 226 (1957) 497–509.
[34] P. Ways, D.J. Hanahan, J. Lipid Res. 5 (1964) 318–328.
[35] E.G. Bligh, W.J. Dyer, Can. J. Biochem. Physiol. 37 (1959) 911–917.
[36] P. Laakso, in: W.W. Christie (Ed.), Advances in Lipid Methodology – One, The Oily Press, Ayr, 1992, pp. 81–119.
[37] Y. Tanaka, et al., J. Oleo Sci. 53 (2004) 417–424.
[38] M. Cooney, G. Young, N. Nagle, Sep. Purif. Rev. 38 (2009) 291–325.
[39] M.T. Golmakani, J.A. Mendiola, K. Rezaei, E. Ibáñez, J. Supercrit. Fluids 62 (2012) 109–115.
[40] S. Khoomrung, P. Chumnanpuen, S. Jansa-Ard, M. Stahlman, I. Nookaew, J. Boren, J. Nielsen, Anal. Chem. 85 (2013) 4912–4919.
[41] V. Mandel, Y. Mohan, S. Hemalatha, Pharmacogn. Rev. 1 (2007) 7–18.
[42] S. Hauff, W. Vetter, Anal. Chim. Acta 636 (2009) 229–235.
[43] D. Tsikas, A. Zoerner, A. Mitschke, Y. Homsi, F.M. Gutzki, J. Jordan, J. Chromatogr. B 877 (2009) 2895–2908.
[44] C. Ferreiro-Vera, F. Priego-Capote, M.D. Luque de Castro, J. Chromatogr. A 1240 (2012) 21–28.
[45] L. Yetukuri, K. Ekroos, A. Vidal-Puig, M. Oresic, Mol. Biosyst. 4 (2008) 121–127.
[46] J.B. Fenn, M. Mann, C.K. Meng, S.F. Wong, C.M. Whitehouse, Science 246 (1989) 64–71.
[47] J.C. Smith, W. Hou, S.N. Whitehead, M. Ethier, S.A. Bennett, D. Figeys, Rapid Commun. Mass Spectrom. 22 (2008) 3579–3587.
[48] M.C. Thomas, T.W. Mitchell, D.G. Harman, J.M. Deeley, R.C. Murphy, S.J. Blanksby, Anal. Chem. 79 (2007) 5013–5022.
[49] A. Demirkan, A. Isaacs, P. Ugocsai, G. Liebisch, M. Struchalin, I. Rudan, J.F. Wilson, P.P. Pramstaller, U. Gyllensten, H. Campbell, G. Schimitz, B.A. Oostra, C.M.J. van Duijn, Psychiatr. Res. 47 (2013) 357–362.
[50] M.K. Dymond, C.V. Hague, A.D. Postel, G.S. Attard, J. R. Soc. Interface 10 (2013) 1–13.
[51] Y. Chen, M. Ogundare, C.M. Williams, Y. Wang, Y. Wang, G.W. Sewell, P.J. Smith, F.Z. Rahman, N. O'Shea, A.W. Segal, W.J. Griffiths, Biochimie 95 (2013) 461–463.

[52] S.K. Abbott, A.M. Jenner, T.W. Mitchell, S.H.J. Brown, G.M. Halliday, B. Garner, Lipids 48 (2013) 307–318.
[53] X. Han, K. Yang, H. Cheng, K.N. Fikes, R.W. Gross, J. Lipid Res. 46 (2005) 1548–1560.
[54] T. Iven, C. Herrfurth, E. Hornung, M. Heilmann, P. Hofvander, S. Stymne, L.-H. Zhu, I. Feussner, Plant Methods 9 (2013) 24–37.
[55] T. Sun, S. Pawlowski, M.E. Johnson, Anal. Chem. 83 (2011) 6628–6634.
[56] H. Farwanah, T. Kolter, K. Sandhoff, Biochim. Biophys. Acta 1811 (2011) 854–860.
[57] L. Imbert, M. Gaudin, D. Libong, D. Touboul, S. Abreu, P.M. Loiseau, O. Laprévote, P. Chaminade, J. Chromatogr. A 1242 (2012) 75–83.
[58] C. Dass, Modes of ionization, in: D.M. Desiderio, N.M. Nibbering (Eds.), Fundamentals of Contemporary Mass Spectrometry, Hoboken, John Wiley & Sons, Inc, New Jersey, 2007, pp. 15–60.
[59] W.C. Byrdwell, Lipids 36 (2001) 327–346.
[60] W.C. Byrdwell, Lipids 40 (2005) 383–417.
[61] S.S. Cai, J.A. Syage, J. Chromatogr. A 1110 (2006) 15–26.
[62] A. Delobel, S. Roy, D. Touboul, K. Gaudin, D.P. Germain, A. Baillet, F. Brion, P. Prognon, P. Chaminade, O. Laprevote, J. Mass Spectrom. 41 (2006) 50–58.
[63] S. Roy, A. Delobel, K. Gaudin, D. Touboul, D.P. Germain, A. Baillet, P. Prognon, O. Laprevote, P. Chaminade, J. Chromatogr. A 1117 (2006) 154–162.
[64] A. Munoz-Garcia, J. Ro, J.C. Brown, J.B. Williams, J. Exp. Biol. 211 (2008) 447–458.
[65] A. Delobel, D. Touboul, O. Laprevote, Eur. J. Mass Spectrom. 11 (2005) 409–417.
[66] K. Tanaka, H. Waki, Y. Ido, S. Akita, Y. Yoshida, T. Yoshida, Rapid Commun. Mass Spectrom. 2 (1988) 151–153.
[67] G.Y. Liu, F. Han, Y. Yang, Y. Xie, H. Jiang, Y.Y. Mao, H.P. Wang, M.M. Wang, R.J. Chen, J. Yang, J.H. Chen, Nephrol. Dial. Transplant. 26 (2011) 1493–1502.
[68] L. De Oliverira, N.O.S. Câmara, T. Bonetti, E.G. Lo Turco, R.P. Bertolla, A.F. Moron, N. Sass, I.D.C.G. Da Silva, Clin. Biochem. 45 (2012) 852–855.
[69] W. Dong, Q. Shen, J.T. Baibado, Y. Liang, P. Wang, Y. Huang, Z. Zhang, Y. Wang, H.-Y. Cheung, Int. J. Mass Spectrom. 343–344 (2013) 15–22.
[70] J. Schiller, R. Suss, B. Fuchs, M. Muller, O. Zschornig, K. Arnold, Front. Biosci. 12 (2007) 2568–2579.
[71] I.D. van der Werf, C.D. Calvano, F. Palmisano, L. Sabbatini, Anal. Chim. Acta 718 (2012) 1–10.
[72] R. Angelini, R. Vitale, V.A. Patil, T. Cocco, B. Ludwig, M.L. Greenberg, A.J. Corcelli, Lipid Res. 53 (2012) 1417–1425.
[73] J. Niklas, K. Hollemeyer, E. Heinzle, Anal. Biochem. 419 (2011) 351–353.
[74] G.-C. Li, Y.-H. Wang, R.-A. Wu, S.-C. Wang, Chem. Res. Chin. Univ. 27 (2011) 371–374.
[75] U. Distler, M. Hulsewig, J. Souady, K. Dreisewerd, J. Haier, N. Senninger, A.W. Friedrich, H. Karch, F. Hillenkamp, S. Berkenkamp, J. Peter-Katalinic, J. Muthing, Anal. Chem. 80 (2008) 1835–1846.
[76] A.J. Dempster, Phys. Rev. 11 (1918) 316.
[77] E.K. Long, I. Smoliakova, A. Honzatko, M.J. Picklo, Lipids 43 (2008) 765–774.
[78] M.S. Munson, F.H. Field, J. Am. Soc. Mass Spectrom. 88 (1966) 2621–2630.
[79] M. Barber, R.S. Bordoli, G.J. Elliott, R.D. Sedgwick, A.N. Tyler, Anal. Chem. 54 (1982) 645A–657A.
[80] H. Ji, V.G. Voinov, M.L. Deinzer, D.F. Barofsky, Anal. Chem. 79 (2007) 1519–1522.
[81] S. Handa, Y. Kushi, H. Kambara, K. Shizukuishi, J. Biochem. 93 (1983) 315–318.
[82] N.E. Manicke, J.M. Wiseman, D.R. Ifa, R.G. Cooks, J. Am. Soc. Mass Spectrom. 19 (2008) 531–543.

[83] Z. Takats, J.M. Wiseman, R.G.J. Cooks, Mass Spectrom. 40 (2005) 1261–1275.

[84] C.R. Ferreira, V. Pirro, L.S. Eberlin, J.E. Hallett, R.G. Cooks, Anal. Bioanal. Chem. 404 (2012) 2915–2926.

[85] P.J. Roach, J. Laskin, A. Laskin, Analyst 135 (2010) 2233–2236.

[86] J. Laskin, B.S. Heath, P.J. Roach, L. Cazares, O. Semmes, Anal. Chem. 84 (2012) 141–148.

[87] I. Lanekoff, O. Geydebrekht, G.E. Pinchuk, A.E. Konopka, J. Laskin, Analyst 138 (2013) 1971–1978.

[88] L. Li, S.H. Yang, K. Lemr, V. Havlicek, K.A. Schug, Anal. Chim. Acta 769 (2013) 84–90.

[89] S.H. Yoon, Y. Huang, J.S. Edgar, Y.S. Ting, S.R. Heron, Y. Kao, Y. Li, C.D. Masselon, R.K. Ernst, D.R. Goodlett, Anal. Chem. 84 (2012) 6530–6537.

[90] K. Schuhmann, R. Almeida, M. Baumert, R. Herzog, S.R. Bornstein, A. Shevchenko, J. Mass Spectrom. 47 (2012) 96–104.

[91] M. Carvalho, J.L. Sampaio, W. Palm, M. Brankatschk, S. Eaton, A. Shevchenko, Mol. Syst. Biol. 8 (2012) 600–616.

[92] C.J. Fhaner, S. Liu, H. Ji, R.J. Simpson, G.E. Reid, Anal. Chem. 84 (2012) 8917–8926.

[93] F. Li, X. Qin, H. Chen, L. Qui, Y. Guo, H. Liu, G. Chen, G. Song, X. Wang, F. Li, S. Guo, B. Wang, Z. Li, Rapid Commun. Mass Spectrom. 27 (2013) 24–34.

[94] L. Becker, A. Poutaraud, G. Hamm, J.-F. Muller, D. Merdinoglu, V. Carré, P. Chaimbault, Anal. Chim. Acta 795 (2013) 44–51.

[95] Y. Guo, X. Wang, L. Qui, X. Qin, H. Liu, Y. Wang, F. Li, X. Wang, G. Chen, G. Song, F. Li, S. Guo, Z. Li, Clin. Chim. Acta 414 (2012) 135–141.

[96] D. Schwudke, K. Schuhmann, R. Herzog, S.R. Bornstein, A. Shevchenko, Cold Spring Harb. Perspect. Biol. 3 (2011) a004614.

[97] K. Schuhmann, R. Herzog, D. Schwudke, W. Metelmann-Strupat, S.R. Bornstein, A. Shevchenko, Anal. Chem. 83 (2011) 5480–5487.

[98] S.S. Bird, V.R. Marur, M.J. Sniatynski, H.K. Greenberg, B.S. Kristal, Anal. Chem. 83 (2011) 6648–6657.

[99] X. Han, R.W. Gross, Expert Rev. Proteomics 2 (2005) 253–264.

[100] X. Han, R.W. Gross, Anal. Biochem. 295 (2001) 88–100.

[101] X. Han, J. Yang, H. Cheng, H. Ye, R.W. Gross, Anal. Biochem. 330 (2004) 317–331.

[102] X. Han, K. Yang, R.W. Gross, Mass Spectrom. Rev. 31 (2012) 134–178.

[103] A.H. Merrill, R.C. Murphy, Biochim. Biophys. Acta 1811 (2011) 635–636.

[104] D. Touboul, H. Piednoel, V. Voisin, S. De La Porte, A. Brunelle, F. Halgand, O. Laprevote, Eur. J. Mass Spectrom. 10 (2004) 657–664.

[105] S.N. Jackson, B. Colsch, T. Egan, E.K. Lewis, J.A. Schultz, A.S. Woods, Analyst 136 (2011) 463–466.

[106] P.J. Horn, A.R. Korte, P.B. Neogi, E. Love, J. Fuchs, K. Strupat, L. Borisjuk, V. Shulaev, Y.-J. Lee, K.D. Chapman, Plant Cell 24 (2012) 622–636.

[107] A. Veloso, E. Astigarraga, G. Barreda-Gómez, I. Manuel, I. Ferrer, M.T. Giralt, B. Ochoa, O. Fresnedo, R. Rodríguez-Puertas, J.A. Fernández, J. Am. Soc. Mass Spectrom. 22 (2011) 329–338.

[108] H.-Y. Wang, H.-W. Wu, P.-J. Tsai, C.B. Liu, Anal. Bioanal. Chem. 404 (2012) 113–124.

[109] P.J. Hart, S. Francese, E. Claude, M.N. Woodroofe, M.R. Clench, Anal. Bioanal. Chem. 401 (2011) 115–125.

[110] A. Carado, M.K. Passarelli, J. Kozole, J.E. Wingate, N. Winograd, A.V. Loboda, Anal. Chem. 80 (2008) 7921–7929.

[111] P.A. Sutton, S.J. Rowland, J. Chromatogr. A 1243 (2012) 69–80.

[112] S. Mas, R. Perez, R. Martinez-Pinna, J. Egido, F. Vivanco, Proteomics 8 (2008) 3735–3745.

[113] M.K. Passarelli, A.G. Ewing, N. Winograd, Anal. Chem. 85 (2013) 2231–2238.
[114] I. Lanekoff, B.S. Heath, A. Liyu, M. Thomas, J.P. Carson, J. Laskin, Anal. Chem. 84 (2012) 8351–8356.
[115] V. Pirro, L.S. Eberlin, P. Oliveri, R.G. Cooks, Analyst 137 (2012) 2374–2380.
[116] A.L. Dill, L.S. Eberlin, C. Zheng, A.B. Costa, D.R. Ifa, L. Cheng, T.A. Masterson, M.O. Koch, O. Vitek, R.G. Cooks, Anal. Bioanal. Chem. 398 (2010) 2969–2978.
[117] L.S. Eberlin, A.L. Dill, A.B. Costa, D.R. Ifa, L. Cheng, T. Masterson, M. Koch, T.L. Ratliff, R.G. Cooks, Anal. Chem. 82 (2010) 3430–3434.
[118] R.H. Perry, D.I. Bellovin, E.H. Shroff, A.I. Ismail, T. Zabuawala, D.W. Felsher, R.N. Zare, Anal. Chem. 85 (2013) 4259–4262.
[119] L.S. Eberlin, I. Norton, D. Orringer, I.F. Dunn, X. Liu, J.L. Ide, A.K. Jarmusch, K.L. Ligon, F.A. Jolesz, A.J. Golby, S. Santagata, N.Y.R. Agar, R.G. Cooks, Proc. Natl. Acad. Sci. U.S.A. 110 (2013) 1611–1616.
[120] A.T. James, A.J.P. Martin, Biochem. J. 50 (1952) 679–690.
[121] R.D. Bennet, E. Heftmann, J. Chromatogr. 9 (1962) 359–362.
[122] R.K. Yu, T. Ariga, Methods Enzymol. 312 (2000) 115–134.
[123] P. Donato, P. Dugo, L. Mondello, Separation of lipids, in: S. Fanali, P.R. Haddad, P.R. Poole, P. Schoenmakers (Eds.), Liquid Chromatography: Applications, Elsevier, Amsterdam, The Netherlands, 2013.
[124] P. Tranchida, P. Donato, P. Dugo, G. Dugo, L. Mondello, TRAC 26 (2007) 191–205.
[125] C. Wang, H. Kong, Y. Guan, J. Yang, J. Gu, S. Yang, G. Xu, Anal. Chem. 77 (2005) 4108–4116.
[126] W.W. Christie, J. High Resolut. Chromatogr. Chromatogr. Commun. 10 (1987) 148–150.
[127] H. Leskinen, J.P. Suomela, H. Kallio, Rapid Commun. Mass Spectrom. 21 (2007) 2361–2373.
[128] J. McHowat, J.H. Jones, M.H. Creer, J. Chromatogr. B 702 (1997) 21–32.
[129] M. Lísa, M. Holčapek, J. Chromatogr. A 1198–1199 (2008) 115–130.
[130] M. Pulfer, R.C. Murphy, Mass Spectrom. Rev. 22 (2003) 332.
[131] S. Harrabi, W. Herchi, H. Kallel, P.M. Mayer, S. Boukhchina, Food Chem. 114 (2009) 712.
[132] R. Harkewicz, E.A. Dennis, Annu. Rev. Biochem. 80 (2011) 301–325.
[133] J.M. Castro-Perez, J. Kamphorst, J. De Groot, F. Lafeber, J. Goshawk, K. Yu, R.J. Shockcor, J.P. Vreeken, T. Hankemeier, J. Proteome Res. 9 (2010) 2377–2389.
[134] P.D. Rainville, C.L. Stumpf, J.P. Shockcor, R.S. Plumb, J.K. Nicholson, J. Proteome Res. 6 (2007) 552–558.
[135] P. Dugo, M. Herrero, T. Kumm, D. Giuffrida, G. Dugo, L. Mondello, J. Chromatogr. A 1189 (2008) 196–206.
[136] P. Dugo, D. Giuffrida, M. Herrero, P. Donato, L. Mondello, J. Sep. Sci. 32 (2009) 973–980.
[137] T. Hajek, V. Skerikova, P. Cesla, K. Vynuchalova, P. Jandera, J. Sep. Sci. 31 (2008) 3309–3328.
[138] K. Sandra, A. dos Santos Pereira, G. Vanhoenacker, F. David, P. Sandra, J. Chromatogr. A 1217 (2010) 4087–4099.
[139] N.L. Leveque, S. Heron, A. Tchapla, J. Mass Spectrom. 45 (2010) 284–296.
[140] M. Malone, J.J. Evans, Lipids 39 (2004) 273–284.
[141] P. Donato, F. Cacciola, P.Q. Tranchida, P. Dugo, L. Mondello, Mass Spectrom. Rev. 31 (2012) 523–559.
[142] P. Dugo, F. Cacciola, L. Mondello, P. Donato, Comprehensive two-dimensional liquid chromatography combined with mass spectrometry, in: L. Mondello (Ed.), Comprehensive Chromatography in Combination with Mass Spectrometry, John Wiley & sons Inc, Hoboken, NJ, 2011.

[143] L. Mondello, P.Q. Tranchida, V. Stanek, P. Jandera, G. Dugo, P. Dugo, J. Chromatogr. A 1086 (2005) 91–98.
[144] P. Dugo, T. Kumm, M.L. Crupi, A. Cotroneo, L. Mondello, J. Chromatogr. A 1112 (2006) 269–275.
[145] P. Dugo, T. Kumm, B. Chiofalo, A. Cotroneo, L. Mondello, J. Sep. Sci. 29 (2006) 1146–1154.
[146] E.J.C. van der Klift, G. Vivo-Troyols, F.W. Claassen, F.L. van Holthoon, T.A. van Beek, J. Chromatogr. A 1178 (2008) 43–55.
[147] M. Lísa, E. Cifkova, M. Holčapek, J. Chromatogr. A 1218 (2011) 5146–5156.
[148] P. Dugo, N. Fawzy, F. Cichello, F. Cacciola, P. Donato, L. Mondello, J. Chromatogr. A 1278 (2013) 46–53.
[149] A. Carrasco-Pancorbo, N. Navas-Iglesias, L. Cuadros-Rodríguez, TRAC 28 (2009) 263–278.
[150] W.W. Christie (Ed.), Gas Chromatography and Lipids. A Practical Guide, The Oily Press, Bridgewater, England, 1989.
[151] W.W. Christie (Ed.), Lipid Analysis. Isolation, Separation, Identification and Structural Analysis of Lipids, The Oily Press, Bridgewater, England, 2003.
[152] R.E. McDonald, M.M. Mossoba, Methods for *trans* fatty acid analysis, in: C.C. Ahoh, D.B. Min (Eds.), Food Lipids, Marcel Dekker, New York, NY, USA, 2002.
[153] M.M. Mossoba, R.E. McDonald, M.P. Yurawecz, J.K.G. Kramer, Eur. J. Lipid Sci. Technol. 103 (2001) 826–830.
[154] L. Mondello, A. Casilli, P.Q. Tranchida, R. Costa, B. Chiofalo, P. Dugo, G. Dugo, J. Chromatogr. A 1035 (2004) 237–247.
[155] P.A. Sutton, S.J.J. Rowland, Chromatogr. A 1243 (2012) 69–80.
[156] P.A. Sutton, M.J. Wilde, S.J. Martin, J. Cvačka, V. Vrkoslav, S.J. Rowland, J. Chromatogr. A 1297 (2013) 236–240.
[157] P. Delmonte, A.-R.F. Kia, J.K.G. Kramer, M.M. Mossoba, L. Sidisky, J.I. Rader, J. Chromatogr. A 1218 (2011) 545–554.
[158] F. Destaillats, M. Guitard, C. Cruz-Hernandez, J. Chromatogr. A 1218 (2011) 9384–9389.
[159] I. Haddad, M. Mozzon, R. Strabbioli, N.G. Frega, Dairy Sci. Technol. 92 (2012) 37–56.
[160] M.S. Beldean-Galea, J. Vial, D. Thiébaut, V. Coman, Anal. Methods 5 (2013) 2315–2323.
[161] P.Q. Tranchida, R. Costa, P. Donato, D. Sciarrone, C. Ragonese, P. Dugo, G. Dugo, L. Mondello, J. Sep. Sci. 31 (2008) 3347–3351.
[162] L.C. Marney, S.C. Kolwicz Jr., R. Tian, R.E. Synovec, Talanta 108 (2013) 123–130.
[163] A.L. Payeur, M.A. Lorenz, R.T. Kennedy, J. Chromatogr. B 893–894 (2012) 187–192.
[164] P. Dittwald, J. Claesen, T. Burzykowski, D. Valkenborg, A. Gambin, Anal. Chem. 85 (2013) 1991–1994.

Chapter 9

Food Forensics

Maurizio Aceto
Dipartimento di Scienze e Innovazione Tecnologica, Università degli Studi del Piemonte Orientale "Amedeo Avogadro", Alessandria, Italy
E-mail: maurizio.aceto@mfn.unipmn.it

Chapter Outline

1. Introduction **442**
- 1.1 Definitions 445
- 1.2 The Role of Chemometrics in Food Forensics 448
- 1.3 The Principal Foods Subjected to Fraud 449
  - 1.3.1 Wine, Fermented Drinks, and Other Beverages 449
  - 1.3.2 Milk and Dairy Products 451
  - 1.3.3 Vegetable Oils 452
  - 1.3.4 Meats 453
  - 1.3.5 Fish 454
  - 1.3.6 Fruits and Vegetables 455
  - 1.3.7 Animal Products 457
  - 1.3.8 Products from Cereals 458
  - 1.3.9 Organic Food 459

2. Mass Spectrometric Analysis for Food Forensics **460**
- 2.1 Analytical Methods in Food Forensics 460
- 2.2 Isotope Ratios as Chemical Markers 461
  - 2.2.1 Wine, Fermented Drinks, and Other Beverages 465
  - 2.2.2 Milk and Dairy Products 468
  - 2.2.3 Vegetable Oils 468
  - 2.2.4 Meats 469
  - 2.2.5 Fish 470
  - 2.2.6 Fruits and Vegetables 471
  - 2.2.7 Animal Products 473
  - 2.2.8 Products from Cereals 474
  - 2.2.9 Organic Food 474
- 2.3 Trace Elements as Chemical Markers 476
  - 2.3.1 Wine, Fermented Drinks, and other Beverages 479
  - 2.3.2 Milk and Dairy Products 480
  - 2.3.3 Vegetable Oils 480
  - 2.3.4 Meats 481
  - 2.3.5 Fish 482
  - 2.3.6 Fruits and Vegetables 482
  - 2.3.7 Animal Products 485
  - 2.3.8 Products from Cereals 486
  - 2.3.9 Organic Food 486
- 2.4 Molecular Ions as Chemical Markers 486
  - 2.4.1 Strategies of Molecular Analysis 486

Comprehensive Analytical Chemistry, Vol. 68. http://dx.doi.org/10.1016/B978-0-444-63340-8.00009-1

2.4.2 Hyphenated Systems 489
2.4.3 Stand-Alone Systems 489
2.4.4 Wine, Fermented Drinks, and Other Beverages 490
2.4.5 Milk and Dairy Products 493
2.4.6 Vegetable Oils 494
2.4.7 Meats 494
2.4.8 Fish 495
2.4.9 Fruits and Vegetables 496
2.4.10 Animal Products 497
2.4.11 Products from Cereals 497
2.4.12 Organic Food 497
3. Conclusions and Future Trends 498
Acknowledgments 499
References 499

1. INTRODUCTION

The term "food forensics" involves the possibility of using powerful scientific methods for the authentication and traceability of foodstuffs, in a way similar to popular TV series such as *CSI* in which scientific methods are used to solve forensic problems. Food forensics must be considered a discipline of primary importance. The global food import bills have been estimated at USD 1,29 trillion by Food and Agriculture Organization of the United Nations [1]; it is the fifth year in succession that the world bill has surpassed USD 1 trillion. Considering the high numbers expressed, it must not be surprising how high the fraud rate is. In 2007, John Spink, Director of the Packaging for Food and Product Protection Initiative at Michigan State University (USA) estimated the total value of food fraud \$49 billion, while according to the UK Food Standards Agency, the level of fraud accounted for 10% of food market, i.e., around \$7 billion. Examples are innumerable; to name a few among the most recent cases, the 2013 meat adulteration scandal, also known as *Horsegate*, with concern to foods advertised as beef that actually contained undeclared horse meat [2]; the latest case of wine fraud in Italy [3] in which 30,000 bottles were sold labeled as *Brunello di Montalcino*, *Chianti*, and other valuable wines, but indeed they contained false or low quality wines; finally, the so-called *Honey Laundering Conspiracy* [4] that has been dubbed as the "largest food fraud in the USA history," where honey produced in China has been introduced in the USA market using a network of Asian countries to "wash" Chinese origin product with new packaging and false documents.

Italy is with no doubt among the countries most damaged by food frauds. The list of frauds perpetrated to the detriment of Italian food brands is endless:

- cow buffalo *mozzarella* made with freeze-dried cow milk, with milk from Southern America or with curd from Eastern Europe;
- wine made synthetically with water, sugar, and tartaric acid;
- rapeseed oil or olive pomace oil from Mediterranean countries colored with chlorophylls and artificially flavored, sold as olive oil;
- Chinese eels and moribund mussels from Turkey revived with local seawater;
- *pummarola*, i.e., tomato sauce, produced in China then diluted and reworked in Italy;

- capers from Northern Africa sold as *Pantelleria capers*;
- *Aceto balsamico tradizionale di Modena* produced in Germany.

According to a 2007 report by Coldiretti [5], the Italian organization of agricultural entrepreneurs, in foreign food export one product in four is fake; the global amount of food fraud to the detriment of Italian export can be estimated in 60 billions of Euros [6]. In some countries, i.e., USA, only 2% of cheese labeled as Italian is based on original products. It must be noted that if a *made in Italy* product can be sometimes recognized (e.g., wines, cheeses, salami, etc.), much more complex is recognizing the culinary product, once that foodstuffs have been worked and transformed into recipes (*cannoli siciliani, pasta e fagioli*, etc.).

These are with no doubt impressive figures and high costs to both producers and consumers. In addition, a lot of technical resources must be spent in the inspection; at present, in fact, only a small percentage of food supply can be controlled. US Food and Drug Administration reported in its 2013 annual report on food facilities and food imports [7] that a mere 14% of domestic food facilities were inspected and an even lower percentage of foreign food facilities.

Indeed, fraud accompanied commercial transactions since ancient times possibly due to a reflection of human nature. Crooked traders always were all the rage, maximizing incomes by diluting their products with cheaper raw materials. In biblical times, rules and suitability standards were issued for inspection of meat, which can be found in the *Book of Leviticus*, possibly the most ancient health codex ever written. Galen, considered among the fathers of modern medicine, warned against adulteration of herbs and spices. In ancient Rome and Athens, there were laws concerning the adulteration of wines with flavors and colors [8]. In the first century AD, Pliny the Elder, in his *Naturalis Historia* [9], Book XII, Chapter XIV, tells that "*…piper lungum facillime adulteratur Alexandrino sinapi…*" (long pepper is very easily adulterated with Alexandrian mustard) and also that "*…adulteratur iunipiri bacis…*" (pepper is adulterated with juniper berries). In the nineteenth-century text, *A treatise on adulterations of food and culinary poisons* by German chemist Frederick Accum [10], possibly the first written account of food fraud, the author exposed culinary sharp practice in London, detailing how bakers cut their flour with alum and chalk to make loaves whiter, tipped in plaster and sawdust to make them heavier; how brewers added substances like strychnine to beer to make it taste bitter and save money on hops; and, perhaps worst of all, how lead, copper, or mercury salts were used to make brightly colored sweets and jellies that would be attractive to children.

In those times only simple, empirical methods were available to inspectors in order to recognize adulterations, so that the most clamorous frauds could be committed. Nowadays, both sides have increased their talents: impostors have developed more subtle ways to bypass regulations and inspectors have at their disposal more powerful analytical methods to investigate.

A great difference among ancient and modern times is also the huge amount of information available at present. Moore et al. [11] reviewed the information concerning food frauds issued on scholarly journals and general media and developed a database published in the US Pharmacopeial Convention's Food Chemicals Codex [12], labeled USP Food Fraud Database; this resource is also freely accessible at the Web site http://www.foodfraud.org.

Food fraud can be resumed in a simple concept: the will of selling a cheaper product at the price of a more valuable one. Three main strategies can be singled out: *adulteration*, i.e., the illicit variation in composition of a food product by means of addition or, most commonly, subtraction of some of its components; *sophistication*, i.e., the illicit variation in the natural or legal composition of a food product adding an external, unauthorized compound; *counterfeiting*, i.e., substitution of a food product with a similar but cheaper one. Inside these strategies, different types of frauds can be described:

- *partial or complete substitution of a product with similar but cheaper alternatives*: this kind of fraud is relatively easy to be identified because adulterants or sophisticants may contain substances acting as natural markers;
  - sea trout sold as salmon;
  - orange juice sophisticated with apple juice;
  - olive oil blended with other vegetable oils;
  - pork salami blended with donkey or horse meat;
- *false geographic provenance of the product*: it is well known that foodstuffs coming from certain regions may result more attractive to consumers with respect to equivalent products coming from less renowned regions: this kind of counterfeit can be discovered only if markers of the respective regions are identified;
  - truffles coming from Piedmont (Italy) versus truffles coming from other countries;
  - saffron coming from Tuscany or Abruzzi (Italy) versus saffron coming from other countries;
- *false declaration of process*: several foodstuffs have an added quality value when produced with specific processes that are usually time and resources consuming; fraud is carried out when a quality brand is used for a food obtained with cheaper production methods;
  - *extra virgin olive oil* is a valuable brand that can be applied only to a product obtained by means of physical or mechanical methods, while chemical extraction with solvents is not allowed;
  - food labeled as *organic* but produced with conventional methods;
- *processes not allowed*: several kinds of fraud concern with illicit practices carried out in the view of improving the quality features of foodstuffs;
  - addition of undeclared sugars to fruit juices in order to increase their taste;
  - addition of glycerine to wine to improve its body;

   - addition of natural but not allowed dyes (e.g., flavonoids from berries other than grapes) to wine to improve its hue;
- *sale of spoiled products*: this fraud can easily be identified.

Authentication of foodstuffs is strictly bound to labeling. For example, a product labeled "vegetable oil" will be any edible oil of vegetal origin, whereas a product labeled "olive oil" must be only the one obtained from olives. Labels imply rules imposing specific features from the technical point of view.

Problems arise frequently whereas rules are different from country to country. Some countries pursue more strict rules in the commercialization of foodstuffs, other are less strict; this may request observing more or less restrictive analytical parameters. One example is the addition of sugar to must in winemaking: in most jurisdictions (e.g., Italy) this practice is forbidden, while in others (i.e., Germany) its use in chaptalization is regulated but permissible in lower quality wines only and in France it is permitted, though under strict regulation.

1.1 Definitions

Food forensics implies the possibility to verify whether a foodstuff be authentic or not. It is useful at this stage to define some terms which will be used in the following. In particular, the difference among *authentication* and *traceability* must be pointed out [13]. *Authentication* is a procedure useful to verify the features declared by a product label and to reveal if a product has been adulterated or counterfeit (Figure 1). *Traceability* is a procedure useful to verify the link among a foodstuff and the raw materials with which it has been produced (Figure 2). Authentication studies look for chemical parameters useful to discriminate authentic products from nonauthentic products, using one or more groups of variables whose distribution must be evaluated with reference to geographic provenance, botanic or animal variety, and production technology of samples of a particular food. Instead, traceability studies look for chemical parameters useful to find tracks of the different stages inside a production chain. If a food product can be traced—linked to its raw materials—it can be certainly authenticated, but if a food product can be authenticated (i.e., has all the features declared in its label) not necessarily can be traced. Authentication and traceability are not, therefore, synonyms, but are concepts pointing in the same direction, i.e., working at the consumers' advantage by guaranteeing the quality of foodstuffs. For the sake of simplicity, in the following text we will use the term *classification* referring to the whole of authentication and traceability, while we will use either *authentication* or *traceability* when the specific meaning will be implied.

Considering the scientific literature in the field of food forensics, most classification studies are focused on food authentication. This is due to the fact that by selecting the proper variables, almost every food can be authenticated,

FIGURE 1 Food authentication.

FIGURE 2 The concept of food traceability.

i.e., discriminated by its substitutes. Traceability has more strict rules and only a selection of foods can be actually traced. The following cases can be singled out:

- *short and very short production chains*: foodstuffs which withstand very few or no chemical treatments in the passage from the raw materials to the table (Figure 3, top):
  - very good possibilities of authenticating;
  - good possibilities of tracing;
    - fruit and vegetables;

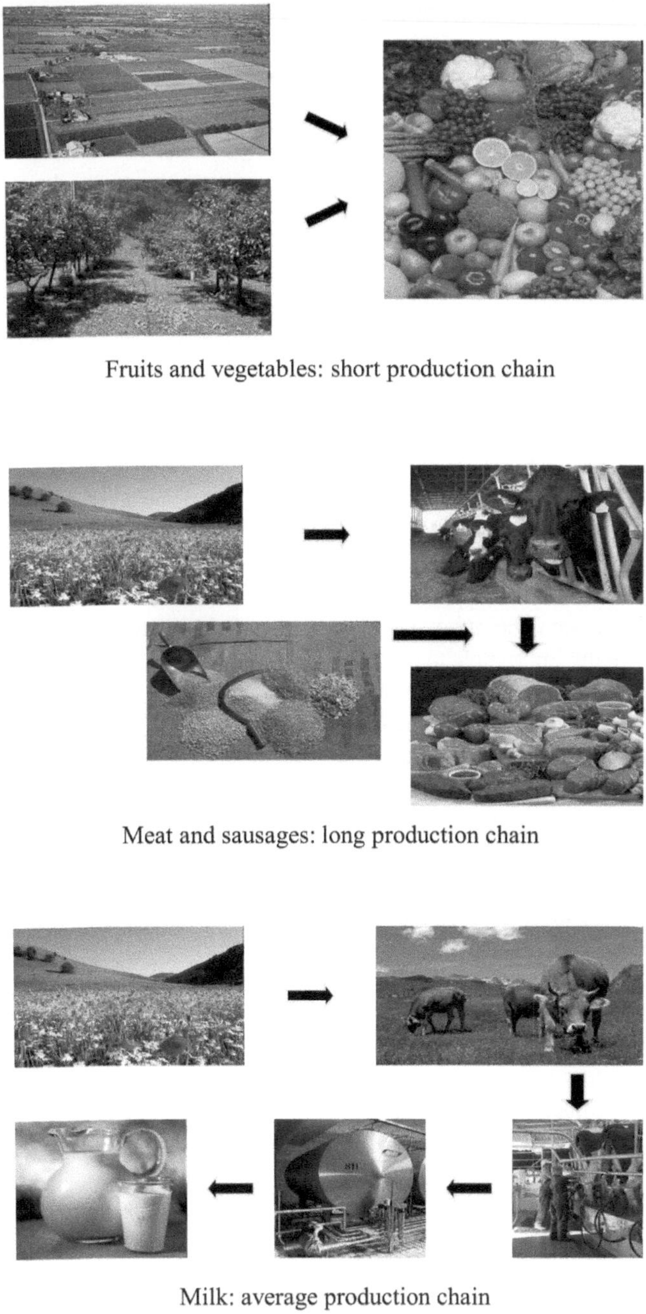

FIGURE 3 Examples of food production chains.

- *average production chains*: foodstuffs which withstand some recognizable chemical treatments in the passage from the raw materials to the table (Figure 3, bottom):
  - good possibilities of authenticating;
  - traceability to be verified;
    - vegetable oils;
    - wines;
    - milk;
    - honey;
    - products from cereals;
- *long and complex production chains*: foodstuffs which withstand many or unchecked chemical treatments in the passage from the raw materials to the table (Figure 3, middle):
  - fairly good possibilities of authenticating;
  - scarce possibilities of tracing;
    - dairy products;
    - meat and sausages.

1.2 The Role of Chemometrics in Food Forensics

Also, it is important to point out the relevant role of *chemometrics* in food classification. It is virtually impossible to define a food in terms of a single scientific description: very few products are made of pure substances (e.g., salt, sucrose) while the most are mixtures containing some main well-defined compounds and minor ingredients less known but maybe decisive for organoleptic features. The same product will show slight but significant differences depending on the provenance of raw materials, the way they were processed, and the animal/botanical species of origin. In order to be defined as authentic, a foodstuff must be characterized taking into account the possible variations along different directions. In many cases, therefore, classifying foodstuffs implies determination of several variables, e.g., trace elements or molecular ionic fragments. Since in very few cases a single parameter is sufficient, determining a large number of variables plays a key role in the possibility of authenticating or tracing a product. The variables determined can be used to individuate groups of samples, or *classes*, with homogeneous chemical features, with particular concern to *authentic* and *nonauthentic* samples, and to assign membership to unknown samples. In this instance, equal importance must be given in the chemical characterization of both authentic and nonauthentic samples, because only when the difference is clear the authenticity of unknown samples can be recognized. This implies, though, the need of managing large sets of data. This procedure can be carried out with *pattern recognition* (or *classification*) multivariate mathematical methods, using well-known techniques such as principal components analysis (PCA), cluster analysis (CA), discriminant analysis (DA), or soft independent modeling of class analogy (SIMCA). Pattern recognition methods have been applied to food forensics since at least 1975 and in the following decades, the number of applications has grown considerably [14,15]. The applications of chemometrics in food forensics have been reviewed in several instances [16–18], putting particular focus on unsupervised [19] or supervised [20] methods of pattern recognition.

1.3 The Principal Foods Subjected to Fraud

In the development of methods for food classification, there is a strong difference among authentication and traceability. In order to trace a food product, a perfect knowledge of the whole production chain, from raw materials to the final product, is mandatory because every single passage can influence the possibility to follow the chain by means of analytical measurements. This feature is less essential in authentication, where it can be sufficient distinguishing the authentic finished product from the nonauthentic one, irrespective of the production chain. In this case also, however, the knowledge of the various passages inside the production chain allows individuating the suitable analytical tests in a more efficient way. For this reason, it is well timed to illustrate shortly the production chains of the most important foodstuffs and to understand, for each category, what the more common frauds are.

1.3.1 Wine, Fermented Drinks, and Other Beverages

Wine is with no doubt one among the foodstuffs mostly subjected to frauds. High-quality wines or particular vintages may reach high prices, becoming ideal objectives for illicit commerce. Although regulations on wines are among the most complete, the extent of frauds is so high that could hardly be quantified.

The role of analytical chemistry in wine classification is definitely strategic. In the past, the only way to verify the authenticity of a wine was the tasting by experts, but today powerful instrumental methods are available. Not all frauds, though, can be revealed.

The production chain of wine is complex and many passages occur among grapes and bottled wine. Trace elements and stable isotope ratios have been suggested as traceability markers, as it will be detailed further on. In the case of authentication, several classes of organic compounds, typical of specific passages in the production chain, can be used as markers.

Saurina [21] and Versari et al. [22] recently reviewed classification methods of wine with different analytical techniques.

The most common frauds in the wine and spirits market are the following:

- adulteration:
  - addition of water;
- sophistication:
  - use of sugars or alcohols from other plants different from grapes;
  - use of additives, flavors, and dyes not allowed;
  - use of vinous products made with table grapes;
- counterfeiting:
  - wines obtained totally with musts and/or with table grapes marketed as products derived from wine grapes;

- wines marketed under registered names, i.e. Italian *Denominazione di Origine Controllata* (DOC) and/or *Denominazione di Origine Controllata e Garantita* (DOCG), without having the prescribed chemical, physical, organoleptic, and documental requisites;
- Single malt whisky replaced by blended one.

The previous considerations have valid for all other fermented drinks, among which:

- spirits from grapes (e.g., cognac);
- spirits from other vegetable species (e.g., rhum, grappas, whisky, tequila);
- beer (from barley);
- cider (from apples);
- sakè (from rice).

In all these cases, raw materials undergo at least one chemical conversion due to alcoholic fermentation. Moreover, production and conservation can include other specific treatments. In addition, in some cases more than a single raw material is used: beer, as an example, is made from both barley and hop, which makes it more difficult to go back to their origin.

A particular fermented drink is *Aceto balsamico di Modena*, which stands out for quality and commercial value. It is a product obtained from cooked grapes must, fined in barrels for at least 12 years; along this period it is subjected to progressive concentration in a series of casks of different woods and sizes, called *acetaia*, until it becomes a dense juice, rich in flavors. The chemistry underlying the aging of *Aceto balsamico* is still partially unknown, but it is certain that all passages involved in the production (alcoholic and acetic fermentations, conservation in six different types of barrels) heavily modify the raw matters. A traceability study can hardly be figured, while authentication could be based on the development of typical flavoring compounds.

Besides alcoholic drinks, another drink of high consumption is of course mineral water. The production chain of mineral water is for sure the simplest. To be marketed as mineral, water must not withdraw any treatment from source to bottle, which is a striking difference among mineral and tap water (i.e., delivered by domestic water systems). Water would then seem to be the ideal food for classification studies, being the chemical features untouched all along the production chain. Indeed, the commercial value of mineral water does not justify application of powerful but expensive analytical techniques such as mass spectrometry or nuclear magnetic resonance (NMR).

The most common frauds in the mineral water market are the following:

- adulteration:
  - dilution with tap water;
- sophistication:
  - use of purification agents;
- counterfeiting:
  - bottles labeled as mineral water but containing tap water.

1.3.2 Milk and Dairy Products

Inside dairy products, two groups are present:

- products obtained from mechanical and physical transformations (milk);
- products with more elaborate production chains, including chemical transformation passages (yoghurt, cheeses, butter).

It is obvious that the first group is more suitable to be studied for traceability. Milk traceability is relatively simple as its composition reflects instantaneously the conditions at which, the producing animal is exposed to. Transfer of nutrients and/or contaminants from grass to milk, passing through the animal stomach, is fast and it is then possible to use different vegetal biomarkers such as carotenoids, terpenes, and polyphenolic compounds to yield information on the animal's diet, while determination of stable isotopes ratio can yield information of the provenance of milk. As for water, though, the market value of the product can hardly justify traceability studies.

For what concerns authentication, it is possible to identify organic markers allowing discrimination of different productions. The chemical composition of dairy products reflects that of the milk they come from, but it also depends on other factors such as processing, aging, and quality of microbial flora. Moreover, most of them are alive products, hosting microbial species which breed on a substrate rich in nutrients and whose metabolism generates chemical compounds continuously evolving. It is apparent that traceability studies can be difficult while authentication schemes can be easily developed based on typical organic compounds.

Zachar et al. [23] recently reviewed classification methods of milk and dairy products with different analytical techniques.

The most common frauds in the dairy products market are the following:

- adulteration:
  - addition of water to milk;
- sophistication:
  - preparation of cheeses with fraudulent use of milk powder, casein or caseinates instead of natural liquid milk;
  - use of milk different from the one declared in the label (e.g., the cheaper cow's milk in place of buffalo, sheep, or goat's milk, all more expensive);
  - use of expired cheeses in the preparation of *pasta filata* fresh cheeses;
  - use of casein and butter in the production of *pasta filata* fresh cheeses;
- counterfeiting:
  - use of milk of animal origin and/or geographic source different from those prescribed in product specifications;
  - production of butter from buttermilk of buffalo, sheep, and goat's milk and its marketing as butter obtained from cream or from buttermilk of cow's milk;
  - use of not allowed additives and colorants;

- use of animal and/or vegetal fats and/or butter produced in EU in the production of fresh butter:
- false declaration of cheese made from heat-treated milk instead of raw milk.

1.3.3 Vegetable Oils

Extra virgin olive oils are among the highest quality products in Europe, which attract counterfeit. Adulterated olive oil is possibly the biggest agricultural fraud in the EU. In the production of vegetable oils, and in particular of olive oil, there is difference among authentication and traceability. In fact, it is relatively easy, on the basis of several organic compounds, distinguishing oils from different botanical sources (e.g., olive, hazelnut, sunflower, etc.) and therefore identifying fraudulent additions to higher quality oils; on the other hand, it is more difficult in tracing the production chain as the passages from olives to the final product can be many, with both physical and chemical transformations, even if the higher quality products (i.e., extra virgin olive oils) must be produced with physical methods only. It must be considered, moreover, that olive oil in particular is a complex system bound to several variables such as olive cultivars, climatic features, modality of handling the raw matter and the final product, etc.

Ben-Ayed et al. [24] recently reviewed classification methods of oils with different analytical techniques.

The most common frauds in the vegetable oils market are the following:

- adulteration:
  - olive oil illicitly subjected to deodorization to obtain products deprived of organoleptic defects, passed off as extra virgin oil;
  - olive oil illicitly subjected to disacidification to obtain products with low acidity, passed off as extra virgin oil;
- sophistication:
  - blending of olive oil with seed oil unaltered or previously subjected to specific treatments (e.g., desterolization) in order of hiding specific compounds that could reveal its addition under an analytical check;
  - oil declared as extra virgin olive oil illicitly obtained by mixing refined oil with virgin oil;
- counterfeiting:
  - seed oil colored with chlorophyll, marketed as virgin olive oil;
  - virgin and extra virgin oil made in foreign countries sold as local product;
  - virgin oil introduced in one country with merceological denomination of seed oil or foreign olive oil, temporarily imported to be refined or packaged and then given back to the original country, which instead, by means of triangulation transactions, is input on the national market at low price thanks to fiscal advantages.

1.3.4 Meats

Meat has obviously a central role in the diet of most regions, with particular concern to Western countries. At the same time, in recent years, there have been several issues of security concern in the meat market, e.g., bovine spongiform encephalopathy (BSE), human variant Creutzfeldt–Jakob disease (CJD), and more recently the use of equine meat in products labeled as beef meat. Therefore, classification of meat products has potentially a strategic importance, even considering that most consumers address their choice of purchase according to the geographic provenance of animals. Yet, the whole traceability system, which should guarantee origin and so quality of meat, is actually based on brands, tattoos, and animal passports, that is ultimately on paper documentation, not on chemical analysis.

On the other hand, in order to develop a classification scheme, the complexity of the livestock system is from the traceability point of view must be considered. Several variables, such as environment, race, diet, drinking water features, conjugal conditions, etc., can influence. Moreover, analytical strategies are necessarily different for global or microregional scales. The feed used to feed livestock can come from different sources, the chemical features (e.g., isotopic ratios, trace metals, etc.) of which will be mixed while fixing into beasts' muscle fibers, a process that in addition occurs for long periods, so that the original fingerprints of diet foods are lost. It is also possible that animals had been bred in different farms along their lives, maybe with different feeding methods. Finally, the biological and physical factors influencing the isotopic composition of animal tissues are not entirely clarified, unlike the case of wine in which stable isotope analysis is by now a powerful, straightforward method of authentication.

Information that can be easily yielded from chemical analysis is the type of livestock feeding, which is reflected on the isotope fingerprint of slaughtered meat. Extensive systems of production, such as organic farming, are strongly related to the local environment. This livestock is let to spend free outdoors, sometimes even during winter. Feeding needs few supplements from sources external to farms as prescribed by EEC Regulation 2092/91 (and further amending acts) for organic farming [25]. As a consequence, animals could incorporate mineral substances and isotope profiles typical of a local, restricted area. For livestock subjected to intensive systems, the situation is more complicate: feeding systems can change from free grazing and assumption of local feeds to whole indoors housing with assumption of feeds produced externally, possibly from different sources. In particular, soy-based protein supplements are marketed all over the world. In meats slaughtered from animals fed with these systems, the assumed mineral substances reflect a mixture of provenances, which is not an ideal situation for classification and specifically for traceability. Particularly complex is the case of poultry, where animals are hardly bred with extensive systems so that their meats have weak links with the territory.

In the case of sausages, the possibility of traceability is even worst, due to the fact that their preparation, by definition, calls for addition of different substances to favor conservation: salt, spices, additives, and sometimes selected microorganisms, all instances heavily influencing the original chemical features in terms of trace elements or isotope ratios. Authentication could nevertheless be possible thanks to identification of specific organic compounds; this is particularly important when high-quality handcrafted productions are involved, i.e., *Pata Negra raw ham*, *culatello di Zibello*, etc., which have relevant commercial values.

The analytical methods for meat classification have been recently reviewed by Vlachos et al. [26] and by Sentandreu and Sentandreu [27].

The most common frauds in the meat market are the following:

- adulteration:
  - addition of water in order to increase meat weight;
  - addition of disproportionately high amounts of extenders and fillers of non-meat origin;
- sophistication:
  - undeclared addition of offal;
  - undeclared addition of mechanically separated meat (MSM);
- counterfeiting:
  - meat produced in foreign countries sold as local meat;
  - meat of lower quality animals passed off as meat of higher quality animals (i.e., horse meat passed off as beef meat);
  - meat from animals stolen and/or illegally slaughtered;
  - meat from poached wild animals;
  - previously frozen meat labeled as fresh.

1.3.5 Fish

The term *fish* is collective term encompassing all water-dwelling animals of interest for human consumption; in the market sense, it comprises any fish, mollusc, or crustacean species, which is harvested, either from sea or from internal water sources. Fish is the most traded food commodity in the world; this means it is an ideal subject for falsification. Frauds in the fish market are very common: a recent survey [28] carried out in 21 states in the USA estimated that one-third of the analyzed fish samples were mislabeled, according to USA Food and Drug Administration (FDA) guidelines, with rates hitting as high as 52% in Southern California. This is not surprising if we consider that, once a fish is filleted and skinned, can be difficult to determine what species it actually is. From the analytical point of view, fish is not so different from meat: classification studies must start from diet of individuals destined to consumption, and this is valid only for what concerns bred individuals, while it is meaningless for what concerns caught individuals, whose diets cannot be controlled. Indeed, analytical controls on fishes are limited to sanitary check or to the possibility of distinguishing fresh items from preserved items. Processed fish products (i.e., canned products, pastes, etc.) pose additional difficulties as some chemical markers can be easily degraded.

The analytical methods for fish classification have been recently reviewed by Lavilla et al. [29]; a comprehensive list of methods is also reported in food and agriculture organization (FAO) Fisheries Technical Paper n.455 [30].

The most common frauds in the fish market are the following:

- adulteration:
  - addition of water in order to increase weight;
  - overglazing, i.e., use of excess ice in preservation;
  - soaking, i.e., use of excess additives in preservation;
- counterfeiting:
  - fish caught or farmed in foreign countries sold as local product;
  - fish of lower quality passed off as of higher quality (i.e., wild salmon passed off as farmed salmon);
  - meat from fish stolen and/or illegally caught;
  - transshipping, i.e., fish exported through different countries to avoid duties and tariffs;
  - previously frozen fish labeled as fresh.

1.3.6 Fruits and Vegetables

Fruits and vegetables are the ideal subjects for classification studies. The link between soil and the marketed product is almost free of intermediate passages that could alter chemical and physical features, in particular for what concerns inorganic parameters. Treatments with agrochemicals can introduce some organic and/or inorganic compounds; nevertheless it is not difficult to find chemical markers behaving as tracers. In most raw products, it could be possible to find the elemental of isotopic profile of the soil on which the plants grew. Basically, however, the geographical origin of most fruit and vegetable products does not appear as a real commercial argument and few labels mention their origin. As a result, classification studies on these matrices are scarce and limited to items of notable value. Particular productions, located in small areas and strictly regulated, could be easily subjected to counterfeit. In Italy, this can be the case of typical productions such as the *cardo gobbo of Nizza Monferrato* (near Asti, Piedmont), the *cherry tomato of Pachino* (near Siracusa, Sicily), or the *pistachio nut of Bronte* (near Catania, Sicily). A relevant product is the hazelnut, mostly employed in the confectionary industry; the most prized variety, the renowned *Tonda Gentile delle Langhe*, is easily counterfeit by substituting it with cheaper varieties.

The considerations made above are valid for the most common products. Some specific categories of fruit and vegetable products can draw particular attention for their commercial value. Truffle is the natural foodstuff *par excellence*, so that it is hard to obtain high-quality products by cultivation. It is possibly the foodstuff most suitable for classification studies: (1) its price is quite high (the variety *Tuber Magnatum pico*, the highly renowned white truffle from Alba costs 1500–2500 €/Kg), with prices strongly varying according to geographic provenance and variety; (2) in the passage from soil to table there is no transformation, so that its composition is only bound to soil and plant metabolism; (3) it has a complex composition, which is an advantage in order to develop classification schemes; (4) it has a fairly high content in metals, which is an advantage for traceability studies based on the distribution of trace elements. Despite all these features, for market reasons very few classification studies have been carried out since now with concern to truffles. Similar considerations can be drawn on mushrooms: even these natural products are suitable subjects for classification studies, with a considerably lower commercial value.

Another class of relevant vegetable items include spices and herbs. Traditionally used worldwide, these items cover a large part of the food market. Saffron is unquestionably the world's most expensive spice [31] and definitely the foodstuff most suitable for classification studies. The real product, prepared with traditional, manual methods from stigmas of *Crocus sativus* can be sold at 15.000–20.000 €/Kg (gold costs 30.000 €/Kg!). This outstanding quotation is justified by the fact that for 1 g of saffron it takes 150 fresh blossoms, each giving three stigmas. Second in price comes vanilla, which is commonly counterfeit by introducing synthetic vanillin into less quality products, while authentic vanilla contains mostly 4-hydroxybelzaldehyde.

From the perspective of classification studies, tea and coffee also can be considered fruit and vegetable foodstuffs, ranking very high for what concerns the impact on the market. Their production is however more complex, as the final products intended for use withstand some manipulations (e.g., toasting of coffee) that can alter the original composition of raw matter.

Arvanitoyannis and Vaitsi [32] have reviewed the application of analytical and chemometrics methods for the classification of vegetables, with particular concern to tomato.

The most common frauds in the market of fresh fruits and vegetables are the following:

- adulteration:
  - addition of water or low value materials to increase weight of products;
- sophistication:
  - addition of substitutes such as maltodextrins in coffee;
- counterfeiting:
  - products marketed under registered names (i.e. (PDO) or protected designation of origin) without having the prescribed chemical, physical, organoleptic, and documental requisites;
  - incorrect botanical declaration.

Far more complex, from the classification point of view, is the matter of fruit and vegetable juices and conserves. Inside the food market these products have a highly relevant importance, due to the fact that transformation of fresh products into conserved products is particularly interesting to food industry: in this way less perishable products (juices or conserves) are generated instead of others (fresh fruits or vegetables) whose organoleptic features are more difficult to preserve. Of course, the physicochemical operations (pressing, clarification, addition of preservatives, etc.) involved render hard traceability, leaving some possibilities for authentication based on identification of chemical markers of the original fruit or vegetable raw matters.

The most common frauds in the juices and conserves market are the following:

- sophistication:
  - addition of glucose and complex sugars for lowering acidity index;
  - addition of thickening agents (e.g., gelatin, agar–agar, fecula, etc.) for improving stability and consistency of products and for hiding the effect of the use of defective raw matters or of undesired or noncompliant production techniques;
  - addition of natural and synthetic dyes to not-fully ripe raw matter;
  - addition of antiseptic substances in order to guarantee conservation of products of unsuitable matter;
  - extreme squeezing of fruits;
  - preparation of synthetic juices;
  - addition of fruit and vegetable products different from those declared;
- counterfeiting:
  - declaration in labeling unequal to what is inspected inside confections, in terms of net weight, drained weight, residues, nutritional values, ingredients, etc.;
  - reuse of expired products, changing expiry date.

1.3.7 Animal Products

Several foodstuffs are produced by animals but consumed by man. Among these, particularly relevant are *eggs* and *honey*.

The composition of eggs should strongly reflect chicken's diet, since there is no chemical transformation in the passage from collection to marketing (at least for correctly preserved eggs). It is therefore a case similar to meat, for what concerns classification studies. A different matter is, on the contrary, for egg-based foodstuffs, whose production can involve addition of several compounds. The commercial value of eggs and egg-based products cannot justify the need of traceability.

The most common frauds in the market of eggs and egg-based products are the following:

- counterfeiting:
  - sale of broken eggs delivered to food industry;
  - marketing of broken eggs;
  - marketing of loose eggs lacking the prescribed labeling systems;
  - apposition of wrappers on packages in the days following the day of classification and selection of egg or the day of recommended sale, which corresponds to 21 days after picking up;
  - marketing of eggs belonging to weight categories different to the one declared;
  - marketing of eggs reporting a deposition date successive to the real one;
  - labeling indicating a chicken breeding system different from the one adopted in the farm from which eggs come (e.g., battery farming in place of free range).

Honey is a highly concentrated solution of carbohydrates in water. It is with no doubt a product strongly linked to the territory, being its composition and organoleptic features deriving mostly from the type of flora gathered by bees; EU regulations, moreover, specify that no additive can be added to the product. In addition to botanic variations, other relevant parameters linked to territory are the type of soil and the human activities that can have consequences on the quality.

Considering honey, the geographical parameter does not allow establishing absolute rankings, but it is well known the fact that some origins are more recognizable in terms of composition and quality. For example, honeys from Ethiopia and Italy are renowned for being among the best in the world.

At present, the most suitable method for obtaining the botanical and geographical origin of honey is melissopalynology, which is the identification of honey pollens. Also promising are techniques that allow determining the distribution of sugars and other organic compounds. For what concerns traceability, the production chain is simple, as from the collection of bees' secretion to the packaging, no chemical process are included. Therefore, the composition of packed honey reflects the inorganic parameters of the area of origin, but it must be considered that it is frequently hard to establish the territory on which bees stand on. The matter is even more complicate for *thousand flower-type* honeys, i.e., honey produced from a variety of flowers.

Arvanitoyannis et al. [33] and Camina et al. [34] have recently reviewed the application of analytical and chemometrics methods for the classification of honey.

The most common frauds in the honey market are the following:

- adulteration:
  - honey produced with illicit addition of water;
- sophistication:
  - honey produced with illicit addition of sugars (mostly glucose) or molasses;
- counterfeiting:
  - honey labeled as unifloral (e.g., *Acacia honey*) which is actually *thousand flower honey*, i.e., with botanical origin different from the one declared;
  - honey made in foreign countries sold as local product.

1.3.8 Products from Cereals

Cereals have been used since antiquity for both human and animal nutrition. The most common cereals are corn, wheat, rice, barley, rye, and oats. From milling of some cereals, flours are obtained which are the bases for production of several common derivatives, such as bread, pasta, and bakery products. The industrial processing of cereals comes through milling and other mechanical procedures, which make flours and other intermediate products easily traceable. The same holds for cereals, which are consumed after collection and refining, as in the case of rice. In the case of bread and bakery products, instead, additional substances are involved such as yeasts, fats, salt, eggs, etc., which complicate the possibility of tracing.

It must be considered that the commercial value of cereals and bakery products can hardly justify classification studies; nevertheless, valuable niche productions, such as the *pane nero di Castelvetrano* (Sicily) or some rice varieties (Indian–Pakistani basmati rice), could take advantage from authentication studies.

Vlachos and Arvanitoyannis recently reviewed the methods for classification of rice [35] and maize [36].

The most common frauds in the cereals market are the following:

- sophistication:
  - production of pasta labeled *durum wheat semolina*, made from durum wheat *semolato*;
  - production of pasta labeled *durum wheat semolina*, made from soft wheat flour;
- counterfeiting:
  - partial or total substitution of higher quality products, i.e., *basmati* rice with lesser value rice cultivars;
  - products of similar cultivars sold as original productions, i.e., *basmati* rice brand assigned to rice produced from countries other than Himalayan foothill regions of India and Pakistan.

1.3.9 Organic Food

The term *organic food* is limited to foodstuffs and agricultural productions obtained in the respect of biological equilibria, of human and animal health, and in the safeguard of environmental resources. This statement includes use of natural technical means, exclusion of chemical products and excessive exploitation of natural resources. Organic farming systems involve using organic fertilizers, such as animal manure and extensive cultivation of nitrogen-fixing plants. Conventional farming systems, on the contrary, allow to use agrochemicals and synthetic fertilizers.

From the classification point of view, the sector of organic foodstuffs is delicated. In this case, more than individuating the link between soil and product, it is important to verify analytically whether an organic product can be distinguished from a conventional or industrially made one. Differences in farming systems should be reflected in the chemical composition of plants growing on different crops, but it is difficult to evidence chemical markers able to trace or even authenticate really organic products. Several analytical tests exist based on the presence/absence of prohibited compounds (e.g., agrochemicals, pesticides, heavy metals, etc.). The content of pesticide residues is systematically lower in organic plant products than in conventional products; the same holds with concern to drugs in organic meat versus conventional meat. However, the number of undesired compounds is too large to be wholly checked. At present, the discrimination among organic and conventional products is troublesome on an analytical base. Better chemical markers are needed.

The analytical methods for classification of organic foodstuffs have been recently reviewed by Capuano et al. [37].

The most common frauds in the organic food market are the following:

- sophistication:
  - foodstuffs produced with illicit addition of various compounds;

- counterfeiting:
  - production of foodstuffs using technical means not allowed (fertilizers, pesticides, etc.);
  - marketing of agricultural products declared as obtained from organic farming, with a label quoting certification, while they come from conventional farming;
  - importing, packaging, or marketing of products labeled as organic, lacking the mandatory certificate;
  - production and manipulation of products labeled as organic without the producer had submitted to the requested control system;
  - improper use of captions such as *natural*, *bio-*, *eco-*, incorrectly recalling an organic method for conventional products;
  - undeclared use of genetically modified food.

2. MASS SPECTROMETRIC ANALYSIS FOR FOOD FORENSICS

2.1 Analytical Methods in Food Forensics

In the last 15–20 years, a large number of highly sophisticated techniques, not specifically developed for food classification, have been found to be suitable in food forensics. Spectroscopic techniques such as nuclear magnetic resonance (NMR) and infrared spectrophotometry (FT-IR) have been used in several applications on all kinds of foodstuffs [38,39]. Other techniques, usually developed for determination of compounds of biological interest, resulted to be useful; this is the case of chromatographic, electrophoretic, and enzymatic analysis. More recently, DNA analysis has proven to be one of the most powerful techniques in food forensics [40] as it is in all sectors of forensic analysis. It is no doubt, however, that the lion's share in food forensic is played by mass spectrometric (MS) techniques. The key role of MS in individuating elemental and molecular parameters that can be used as markers in the classification of foodstuffs cannot definitely be underestimated. Due to the great diagnostic potential in classification of foodstuffs, the application of MS techniques in food forensics has received much attention in the scientific literature; the subject has been recently reviewed by Aiello et al. [41] and by Drivelos and Georgiou [42] with concern to the geographical origin of foods in the European Union. Further references to applications of MS can be found in more general discussions on the subject of food forensics [40,43–48].

MS techniques can be used in food forensics with concern to three main classes of chemical markers (Figure 4):

- *isotope ratios*: the most powerful markers for food classification, a fact certified by several hundred of scientific publications;
- *trace elements*: mass spectrometry coupled with plasma techniques can determine elements at trace and ultratrace level, which are known to act as geochemical markers;
- *molecular ions of ionic fragments*: whole molecular ions or low molecular weight ions can be used to identify typical patterns.

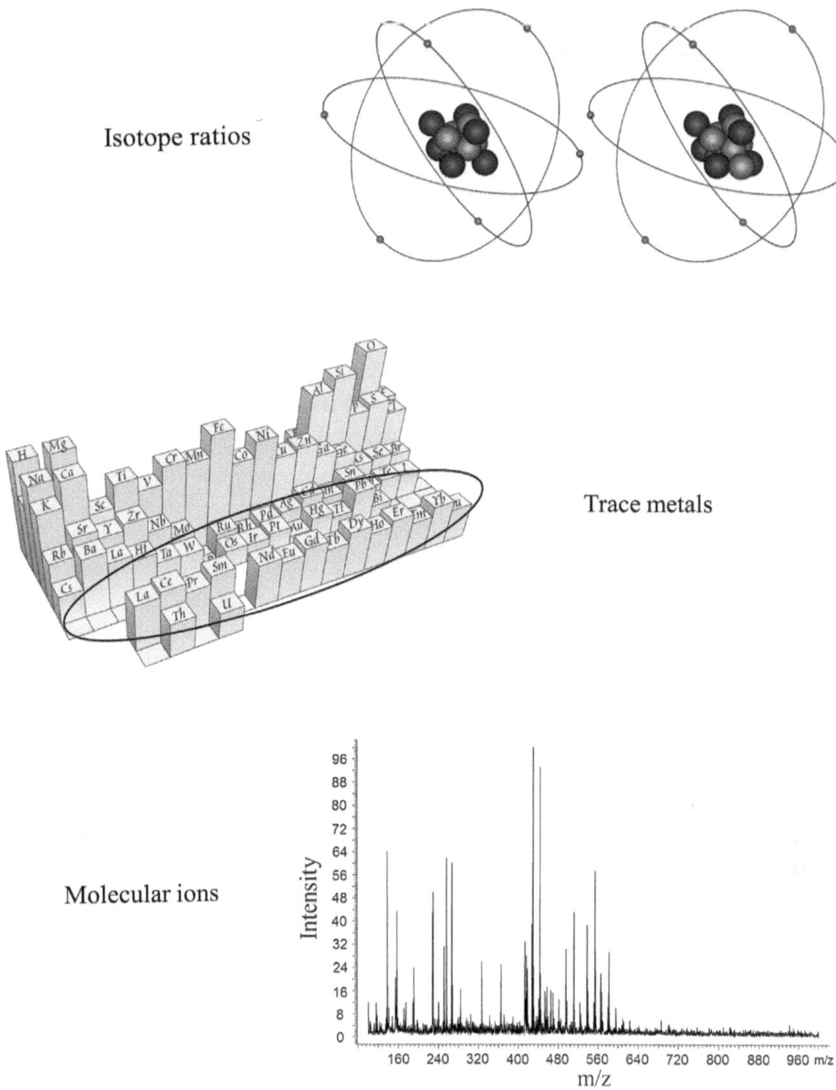

FIGURE 4 Chemical markers useful in food classification.

2.2 Isotope Ratios as Chemical Markers

The role of isotope ratio analysis is well known and consolidated in scientific field, in particular with concern to stable isotopes analysis (SIA). Its potential has been soon recognized in the field of food forensics, thanks to the comprehension of the mechanisms ruling isotopic fractionation of light elements, i.e., the elements constituting living matter. On the basis of ultraslight variations in isotope ratios, SIA allows identifying molecules in foodstuffs with similar chemical structure but coming from different raw materials or produced with different processes, e.g., for biological versus industrial synthesis. With concern to determination of hydrogen isotope ratio, the most powerful SIA method is site-specific natural isotope fractionation–nuclear magnetic resonance (SNIF-NMR), developed by Martin and Martin [49] at the beginning of 1980s; further on, Caer et al. showed the potential of SNIF-NMR for determination of $^{13}C/^{12}C$ ratio also [50]. The use of mass spectrometry to perform isotope ratio measurements is much older, as it has been initiated in the field of geosciences in the 1940s and 1950s. This soon has led to the development of the powerful technique known as isotope ratio mass spectrometry (IRMS). Technical details on SIA and IRMS can be found in several publications [51–53].

The high diagnostic power of IRMS has been recognized at legislative level by indication of its use, together with SNIF-NMR, inside national and international laws ruling food analysis. In the European Union legislation, EC Regulation No 2676/90 [54] and successive amendment in EC Regulation No 822/97 [55], ruling Community methods for the analysis of wines, indicated IRMS for determination of $^{18}O/^{16}O$ isotopic ratio of water content in wines, useful for identifying fraudulent addition of water. A further amendment in EC Regulation No 440/2003 [56] indicated IRMS also for determination of $^{13}C/^{12}C$ isotopic ratio in wine alcohol in order to identify fraudulent addition of sugar from cane or beet. Similar solutions have been adopted by organisation international de la vigne et du vin (OIV) in Resolution OENO 17/2001 [57] concerning IRMS determination of $^{13}C/^{12}C$ ratio on wine ethanol, Resolution OENO 7/2005 [58] concerning IRMS determination of $^{13}C/^{12}C$ ratio on CO_2 in sparkling wines, and Resolution OENO 353/2009 [59] concerning IRMS determination of $^{18}O/^{16}O$ ratio on water in wine and must. In the UK, the application of SIA techniques for food forensics has been strongly supported by the British Food Standards Agency (FSA) since the mid-1990s. In a similar way, the European Office for Wine, Alcohol and Spirit Drinks (BEVABS), established by EU in 1993, recommended the use of SIA to combat major fraud in the beverage sector since 1997. In the USA, the Association of Official Analytical Chemists indicated IRMS as method of choice for determination of $^{13}C/^{12}C$ isotopic ratio in honey [60] and fruit juices and maple syrups [61].

IRMS applied to whole samples is usually known as *bulk SIA*. Since the middle of 1980s, a new type of instrument, resulting from the interface of IRMS to a gas chromatographic system, became available. This hyphenated system allowed performing what is commonly known as *compound-specific isotope analysis* (CSIA), i.e., determination of isotope signatures at the molecular level. Despite the main applications of CSIA are in environmental studies, this technique has also proved to be a powerful method for food forensics. CSIA can be also carried out by extraction of single compounds from the sample.

IRMS has been used for classification of several different types of foodstuffs. Once again, it is necessary to distinguish between authentication and traceability. Isotope ratios of light elements, such as hydrogen, boron, carbon, nitrogen, oxygen, and sulfur, are strongly influenced by chemical, physical, and biological phenomena. Instead of using the simple ratio, it is more common to compare δ value that is the ratio of isotopes in the sample to the ratio in a standard. δD values in plant products originate from the water taken up by the roots and are higher in regions with relatively low humidity. $\delta^{13}C$ values are more bound to the metabolism of plants, with particular concern to the pathway followed in fixation of CO_2: the main discrimination is between C3 (the first product of photosynthesis is 3-phosphoglycerate) and C4 (the first product of photosynthesis is a C4 unit) plants [62]. C3 plants account for 95% of edible species, among which sugar beet, potato, rice, barley, wheat, soy, sunflower, olive, rye, apricot, orange, grapes, and peanut. C4 plants are 1%, examples are sugarcane, sorghum, maize, and millet. A minor category, to which pineapple, vanilla, agave, and cactus belong, is the one associated with Crassulacean acid metabolism (CAM). $\delta^{18}O$ values originate from CO_2, H_2O, and O_2 and correlate with water, reflecting the isotopic composition of groundwater and the average precipitations in the region (related to latitude, distance from the sea and altitude) and the extent of evapotranspiration, mainly influenced by humidity and temperature [63]. The geographical signature from water is transferred into plant and animal products [64]. $\delta^{15}N$ values have relatively shorter ranges but vary according to different factors: the nitrogen cycle (fixation, uptake, mineralization, nitrification, and denitrification), trophic ecology (an increase in trophic level correlates with an increase in $\delta^{15}N$ value), and anthropogenic activity [65]. Finally, $\delta^{34}S$ values are bound to sulfur sources such as bedrock weathering, atmospheric deposition, and microbiological activity. When plants are converted into food for man or animals, geographic provenance, animal/botanic species, or way of processing are all factors that can cause strong fractionation in the ratios of light elements in products, providing therefore the basis of authentication schemes. At the same time, high rates of fractionation are a negative feature if traceability, where constant parameters are sought for, is the objective; therefore, light elements isotope ratios usually are not efficient tracers. On the contrary, isotope ratios of heavy elements, such as strontium or lead, are not influenced by biological phenomena, since the corresponding elements are usually kept off from biological cycles; therefore, once established into rocks, they are potentially maintained unaltered in the passage from soil to food. Isotope ratios of Sr and Pb change strongly from rock to rock. In the case of strontium, ^{87}Sr is a radiogenic nuclide generated by the radioactive decay of ^{87}Rb, so its content in rocks (and consequently in soils derived from them) increases with age; $^{87}Sr/^{86}Sr$ values are therefore lower in younger rocks such as carbonaceous, volcanic, and basaltic rocks and higher in older rocks such as magmatic and metamorphic rocks. The case of lead is more variegated: this element has three radiogenic nuclides, ^{206}Pb, ^{207}Pb, and ^{208}Pb, plus one nonradiogenic, ^{204}Pb. Different ratios are possible with a strong link to the age of rocks. The role of the isotopic signatures of strontium and lead as suitable markers for geologic and therefore geographic provenance of foodstuffs has been generally discussed in many studies [42,66–68]. Of particular interest is a recent, fundamental study by Voerkelius et al. [69] in which authors showed that an $^{87}Sr/^{86}Sr$ database of surface waters could be used to predict the geographic origin of some types of food, such as honey and wheat.

To resume, isotope ratios of light elements are more useful to study authentication of foodstuffs, while isotope ratios of heavy elements are more useful to study traceability (and therefore authentication also). The main information yielding by means of IRMS analysis is listed in Table 1.

A fascinating application of IRMS analysis in food forensics is the characterization of *palaeodiets* in order to obtain information concerning diets of ancient people and cultures. Carbon and nitrogen isotopic signatures can help in identifying the types and amounts of proteins and plant matter consumed by individuals [70] following the fractionation mechanisms in edible matter described before. $\delta^{13}C$ values measured in bone collagen mainly indicate the protein component of the diet [71] while in bone apatite they provide a picture of dietary energy, including carbohydrates and lipids [70]. Accordingly to what it was cited before, $\delta^{13}C$ values can help in distinguishing a diet based on C3 plants (i.e., temperate grasses such as wheat and barley) from C4 plants (i.e., millet and sorghum). $\delta^{13}C$ values from bone collagen and apatite provides additional information concerning the individual's dietary energy source (C3, C4, or mixed) and protein source (C3, C4, or marine) [72]. Stable isotopes of nitrogen discriminate more efficiently between aquatic and terrestrial protein [70]; moreover, $\delta^{15}N$ values are correlated with the trophic position of the edible organism in the food chain: body tissues

TABLE 1 List of Main Isotope Ratios of Interest in Food Forensics

Isotope Ratio	Fractionation	Information
$^{2}D/^{1}H$[a]	Origin of water, biochemical	Geographical, botanical, natural versus synthetic
$^{13}C/^{12}C$[a]	Fixation of CO_2 (C3 vs C4 plants)	Botanic origin, diet
$^{15}N/^{14}N$	Trophic level, marine and terrestrial plants	Diet, agricultural practices
$^{18}O/^{16}O$	Origin of water	Geographical
$^{34}S/^{32}S$	Atmospheric deposition, biochemical	Diet, geographical
$^{87}Sr/^{86}Sr$	Underlying geology	Geographical
^{208}Pb, ^{207}Pb, ^{206}Pb, ^{204}Pb	Underlying geology	Geographical

[a]*It can also be determined by means of SNIF-NMR.*

are generally 3–4‰ higher than the $\delta^{15}N$ of the diet [73]. Examples of recent palaeodietary studies were those by Killgrove and Tycot [74] who investigated human skeletons found in two cemeteries in Rome dating to the Imperial period (first–third centuries AD) and by Gregoricka and Sheridan [75] who analyze skeletons from an early medieval Byzantine monastery in Jerusalem.

In the following paragraphs, examples of applications of IRMS analysis to food forensics are cited with concern to different food categories. This matter has been reviewed by Rossmann [66].

2.2.1 Wine, Fermented Drinks, and Other Beverages

A major part of IRMS applications in food forensics is devoted to wine. Three main objectives are sought for: (1) identification of addition of water; (2) identification of addition of sugars coming from plants other than grapes; (3) recognition of the geographic provenance.

The first objective can be obtained from the determination of $^{18}O/^{16}O$ isotopic ratio of water [76]. A close correlation between the oxygen isotope signature of must and the related wine water must be found in genuine wines. Actually, there is a very low chance that water illicitly added to wine had the same isotopic signature of the original water, which confirms the potentiality of the method in recognizing this kind of fraud.

Addition of sugars obtained from plants other than grapes needs a two-step approach. The discriminating factor is the metabolic pathway followed by plants to fix CO_2: C3 plants have $^{13}C/^{12}C$ isotopic ratio values more negative in terms of $\delta^{13}C$ [77]. Sugars from cane or maize can be easily detected, being them C4 plants while grape is a C3 plant. Addition of sugar from beet needs a further step, being beet a C3 plant as well; in this case determination of D/H ratio by means of SNIF-NMR helps in further discrimination between grapes and beet, being δD values higher in sugars from grapes.

Also of interest in enological field is the identification of addition of synthetic ingredients in winemaking. Moreno Rojas et al. [78] investigated the possibility of frauds committed using natural external L-tartaric acid in wines by means of IRMS. Particular focus was given to L-tartaric acid extracted from tamarind in place of the natural compound present in grapes, which is the only source allowed for acidification treatments according to both EU and OIV regulations. Samples of L-tartaric extracted from musts in Romania were compared with samples extracted from commercial tamarind pulp of four different origins (India, Mexico, Colombia, and Thailand). The different botanical sources (grapes vs tamarind) were discriminated on the basis of δD, $\delta^{13}C$, and $\delta^{18}O$ values. Combining hydrogen, carbon, and oxygen isotopic signatures, it was also possible to distinguish between natural L-tartaric acid and synthetic L-tartaric acid (petroleum by-products).

For what concerns the determination of the geographic provenance, the situation is more complicate. EU regulation 2676/90 [54], apart from establishing rules for identification of fraudulent water addition to wine, set the rules for creation of a database of isotopic ratio values in wines in order to allow the possibility of recognizing the geographic provenance of a wine. Wines from the main European wine-producing countries and also from some foreign countries have been analyzed since then, collecting every year hundreds of samples and determining D/H values of ethanol by ^{2}H-NMR, $\delta^{13}C$ values of ethanol, and $\delta^{18}O$ values of wine water. Conventional wine analysis data are included. Due to the different possibilities of adulteration and to the relevant seasonal and regional variability, though, it is hard to obtain a reliable assessment of geographic provenance [66]. This is a consequence of the intrinsic variability of light elements isotopic signature.

On the contrary, the isotopic signatures of heavy elements are more constant, because they are not subject to relevant seasonal variability and to biological cycles. Several studies showed that strontium and lead isotope ratios are powerful markers of geographic provenance. Horn et al. [79] were among the first suggesting that the isotopic signature of strontium could be a tracer for the geographic provenance of wine. Almeida and Vasconcelos [80] demonstrated that, while total Sr concentration changes dramatically along the passage from soil to bottled wine, $^{87}Sr/^{86}Sr$ ratio keeps identical, providing a tool for the identification of the geographic provenance of a wine. This result is justified by the fact that the isotopic signature of strontium, once set into rocks, has not influenced biological or chemical phenomena so that it can be considered a powerful geographic marker. Several studies thereafter have confirmed these results [81–83]. Similar but not identical considerations can be drawn on the use of lead isotope ratios. Contrarily to strontium, which is almost totally an element of natural origin, lead in wine has two major sources [84]: the primary source is linked to soil, as for strontium, while the secondary source is anthropogenic and it involves fertilizers, pesticides, road traffic (the introduction of leaded gasoline in 1939 by Ford has caused an increasing lead pollution in the environment, reversed in the 1970s due to the introduction of unleaded fuel), and the equipment used during vinification. Therefore, the isotopic signature of anthropogenic lead can overlap to the geochemical one. Despite this, several studies have investigated the possibility of determining the geographical origin of wine samples on the basis of lead isotopic signature [85,86]. An additional feature is that in the case of lead, more than one isotope ratio can be exploited. The key factor in the determination of lead isotope ratio is the precision of the instruments used. Conventional quadrupole inductively coupled plasma–mass spectrometry (ICP-MS) has a limited potential for distinguishing very small differences between isotope ratios, while higher precision of measurement is yielded by ICP-Time of Flight-MS, thermal ionization mass spectrometry, and multicollector sector-field-ICP-MS.

Recently, the use of boron as $^{11}B/^{10}B$ isotope ratio has been suggested as additional means for wine traceability; the results obtained by Vorster et al. [87] in discrimination of wines from South Africa seems promising, but at present this element has not found many applications in food forensics, though its natural isotopic variation could go as wide as 90‰ [88].

In food forensics, a relatively lower amount of research has been devoted to spirits. In this field also, however, do exist authenticity issues such as deliberate and illegal substitution of a prised brand with a cheaper one. Simpkins and Rigby [89] were among the first to apply IRMS for identification of counterfeit spirits. Among hard liquors, whisky is a traditional drink with a strong commercial value, which production is defined by both EU and local regulations. Prices for whisky bottles can vary dramatically according to purity (*single malt* vs *blended*), vintage, geographic provenance (Scotland vs Ireland vs Northern American countries), and raw matters (barley vs corn/wheat). For this reason, several studies have tried to develop methods for whisky classification according to the variables cited. Using compound-specific isotope analysis of hydrogen in ethanol, Hilkert et al. [90] discriminated *single malt* whiskies made from barley (a C3 plant) from *blended* whiskies made from barley and corn (a C4 plant) and from whiskies made from corn alone. This classification was made possible due to the high rate of variation of δD among C3 and C4 plants. Parker et al. [91] used compound-specific isotope analysis of carbon in volatile congeners, classifying whiskies of different brands. More recently, Meier-Augenstein et al. [92] found a good correlation among δD and $\delta^{18}O$ values for waters used in the production of Scottish whiskies and the corresponding bottled whiskies, allowing to develop a method for geographic classification and for identification of counterfeit products. Again, with application of compound-specific isotope analysis of carbon, Rhodes et al. [93] developed a method for identification of illicit addition of neutral alcohol to whisky, based on internal isotopic correlations for ethanol and the congeners from the same sample.

Bottled water accounts for a relevant segment of the food market, with particular concern to the Italian market. Frauds are always possible in terms of product misrepresentation. As an example, carbonated water can contain natural CO_2 coming from sources or industrial CO_2 artificially added. IRMS can be applied to the classification of mineral waters, investigating the correlation among the isotopic signature of hydrogen and oxygen in bottled water with estimated mean annual precipitation isotope ratios for source or bottling locations. Carbon isotopic signature in the dissolved inorganic carbon of carbonated bottled water can yield information on the origin of CO_2; extreme negative values of $\delta^{13}C$ in bottled waters can be due to exogenous (industrial) CO_2 sources. Raco et al. [94] investigated samples of bottled water from Italy, among which some sparkling waters (i.e., naturally carbonated). δ^2D and $\delta^{18}O$ values of samples were found to vary within the reported natural variations for Italian waters as samples lied along the global and Italian meteoric water line. According to the authors' opinion, these values might reflect relatively unaltered source water signatures. Voerkelius et al. [69] carried out a large-scale investigation of strontium isotope ratios on 650 different European natural mineral waters as part of the food traceability project "TRACE" funded by the EU. Isotopic data were combined with a GIS-based geological map of Europe in order to elaborate a novel spatial prediction for strontium isotopic composition of groundwater and thus the composition of bioavailable strontium, which is available for uptake by plants and subsequently transferred into the food chain.

2.2.2 Milk and Dairy Products

The classification of milk and some dairy products can be obtained by IRMS analysis. Scampicchio et al. [95] analyzed raw, pasteurized (HTST), and ultrapasteurized (UHT) milk from different Italian origins by determining $\delta^{13}C$ and $\delta^{15}N$ for milk and milk fractions (fat, casein, and whey). Significant changes in $\delta^{13}C$ and $\delta^{15}N$ were observed as a function of heat processing, which allowed identifying markers useful for pattern recognition methods. It was possible to discriminate simultaneously the geographical origin and type of processing of milk samples.

Of particular concern is the possibility of verifying the provenance of cheeses, which in several cases are high-quality products. Intermediate processes in the production chain play, of course, a relevant role in fixing the isotopic signatures of light elements. $^{13}C/^{12}C$ ratio in milk fat and in cheese protein reflects the type of forage fed to the cows mainly constituted of C3 plants. Also $^{15}N/^{14}N$ ratio results from forage, but the use of organic fertilizers and intensive farming methods can increase this value; other influent factors are climate and soil conditions. Finally, nitrogen-fixing plants, such as leguminosae, have lower $\delta^{15}N$ than nonnitrogen-fixing plants and can be administered in the cows' fodder. $^{18}O/^{16}O$ ratio of milk depends on the water ingested and the proportion of fresh versus dry fodder; ultimately, climatic features play a primary role.

Mozzarella cheese is a highly renowned Italian product made from buffalo milk, characterized by a protected designation of origin (PDO). Brescia et al. [96] analyzed buffalo milk and mozzarella cheese from two sites in Southern Italy; authors found that milk could be distinguished on the basis of $^{13}C/^{12}C$ versus $^{15}N/^{14}N$ ratios, probably due to a difference in the diets of buffalos in the two locations, while in order to distinguish mozzarella samples, D/H data determined by means of SNIF-NMR must be included. Pillonel et al. [97] distinguished Emmental cheeses produced in six European regions: Allgau (Germany), Bretagne and Savoie (France), Finland, Switzerland and Vorarlberg (Austria). Different variables were used in the study: stable isotope ratios of light elements ($^{13}C/^{12}C$, $^{15}N/^{14}N$, $^{18}O/^{16}O$, and D/H) and of $^{87}Sr/^{86}Sr$ and concentrations of major (Ca, Mg, Na, K), trace (Cu, Mn, Mo, I), and radioactive elements (^{90}Sr, ^{234}U, ^{238}U) elements. A complete discrimination was obtained only by using all the variables determined, since isotope ratios yielded a partial classification.

2.2.3 Vegetable Oils

Olive oil is a noble product with a very long tradition, with particular concern to the Mediterranean area (olive tree was already cultivated in Syria at least 6000 years ago). It is commonly recognized that the highest quality oils are produced in Italy and Spain. Frauds in the olive oil market are frequent since it is not easy to identify products with features different from those declared in labels, at least for what concern the geographic provenance of the raw matter. For this reason, several studies have investigated the possibility to classify samples. Most of the studies concern *extra virgin* olive oils, the products with highest quality. Camin et al. [98] carried out a thorough analytical study within the framework of the food traceability project "TRACE" funded by EU, in which authors analyzed 267 olive oils coming from eight European sites. Both isotopic (H, C, and O stable isotope ratios) and elemental concentrations were determined. In addition, the same parameters were measured in 314 fresh surface waters from the same sites. The main scope of the project was verifying whether olive oils could be discriminated according to the climatic and geological characteristics of the sites on which they were produced, and finding a possible relation with the corresponding surface waters. δD and $\delta^{18}O$ values in oils resulted to be correlated to each other and to corresponding waters. $\delta^{13}C$ and $\delta^{18}O$ showed significant correlation with geographical parameters, such as latitude and distance from the coast, and climatic parameters, such as temperature, relative humidity, and amount of precipitation; δD correlated with latitude, distance from the coast, altitude, and temperature. Pattern recognition carried out with canonical discriminant analysis (CDA) allowed a good classification on a geographic scale; this result was improved adding elemental data as variables in the classification.

Also, it is possible to verify the addition of lesser quality oils to olive oil with compound-specific IRMS (even if this issue can be addressed much more easily by GC-MS). Blending of olive oil with edible oils with slightly different fatty acid composition (olive pomace, sunflower, hazelnut) might be detected by using $\delta^{13}C_{16:0}$ versus $\delta^{13}C_{18:1}$ covariations combined with molecular information and carbon isotopic composition of the bulk oil [99].

2.2.4 Meats

It is not surprising, due to the importance of meat in the diet of Western countries, that stable isotopes, together with trace element distribution, had been frequently used for meat classification. As discussed before, the production chain of meat is complex due to several confounding factors such as the variegated diet of livestock, the movement of animals among different farms (or even farming systems) before slaughtering, and the metabolic turnover times of animal tissues [100]. Studies have been carried out on beef [101–105], lamb [106,107], and pork meat [108]. More recently, Heaton et al. [109] carried out an extensive study in which they analyzed beef samples from the major cattle producing regions of the world (Europe, USA, South America, Australia, and New Zealand) with both IRMS and ICP-MS. The results were promising for what concerns the geographic classification. Measurements carried out on defatted dry mass (DDM) showed that $\delta^{13}C$ values were clearly lower in beef samples from England, Ireland, and Scotland than in American samples (Brazil and USA); similar results have been reported by Schmidt et al. [105]. It is well known that differences in $\delta^{13}C$ values are bound to the proportion of C3 and C4 plant materials in the diet of livestock. Extensive C3 pasture feeding is the prevailing system in Northern Europe, while feeding with maize or grain (C4) is more common in American countries, which accounts for the less negative values. Beef samples from Central and Southern Europe show an intermediate behavior, indicating possibly prolonged periods of maize feeding. $\delta^{15}N$ values have smaller variations, but higher values may indicate the use of marine products (fishmeal, seaweed) in the cattle feed [110] or feeding with other animal products: it is known that increases in the trophic level can raise $\delta^{15}N$ values of around 3%; this practice, however, is forbidden in Europe because of the risks of BSE infection [111]. Low values of $\delta^{15}N$ may be due to the use of synthetic nitrogen-based fertilizers, whose isotopic signature can pass through crops, cereals, and eventually cattle diet, or to foraging of leguminous clover or concentrates. Of particular interest is the isotopic signature described by $^{18}O/^{16}O$. This parameter was measured on beef lipid in place of beef DDM due to instrumental problems. It was found that a significant correlation existed among $\delta^{18}O$ values and latitude of the production region. Data from $^{2}D/^{1}H$ analysis gave similar responses and a plot of $\delta^{2}D$ versus $\delta^{18}O$ mean values in beef lipid showed even better this feature, with samples from higher latitudes (Scotland, New Zealand, Shetland, England, etc.) ranging at lower values and samples from warmer climates and lower latitudes (Southern Africa, Australia, Brazil, and Uruguay) ranging at the opposite side. A supervised pattern recognition analysis with CDA, applied to isotopic and elemental data, allowed classifying samples into three broad groups, i.e., European, South American, and Australasian. Finally, Guo et al. [112] showed that information on the dietary intake of livestock and geographical origin could be obtained with isotopic analysis of C and N from defatted beef as well as from crude fat and tail hair.

The classification of meat and meat products with isotopic analysis has been reviewed by Montowska and Pospiech [113] and by Zhao et al. [114].

2.2.5 Fish

In the classification of fish using isotopic (as well as elemental) parameters, it is usual analyzing otoliths and/or scales, while soft tissues and bones are less useful. Otoliths have metabolic inertness and their composition can reflect the environmental history of organisms. Determination of isotopic signature can help establishing fish origin. $^{18}O/^{16}O$ and $^{2}D/^{1}H$ ratios are strongly latitude dependent, while $^{13}C/^{12}C$ ratio is bound to metabolism and diet [115]. $^{87}Sr/^{86}Sr$ has great potential in determining the geographic origin [116].

Salmon is a fish typically subjected to frauds, with particular concern to the actual geographical provenance and whether the fish is caught or farmed. Wild or line-caught salmon can cost three times farmed salmon. A comprehensive study on salmon classification was carried out by Thomas et al. [117], who analyzed a large number of samples of wild and farmed Atlantic salmon from 32 origins within Europe, North America, and Tasmania. Hydrogen, carbon, oxygen, and nitrogen isotope ratios were determined in different parts of animals (muscle, oil, and body water) and in different molecular fractions extracted (glycerol, fatty acids, lipids). It can be assumed that the isotopic signatures in salmon reflected both the environment in which it is grown and the composition of the diet consumed. Results of IRMS analysis evidenced that $\delta^{15}N$ values measured on choline and $\delta^{18}O$ values measured on total oil were the best variable for discrimination between wild and farmed fish, while it was more difficult to discriminate between different geographical origins. These results were confirmed by Anderson et al. [118] who analyzed for its elemental and isotopical composition of a large number of salmon muscle tissues from the Pacific Ocean.

Isotopic methods of classification of fish have been recently reviewed by Lavilla et al. [29].

2.2.6 Fruits and Vegetables

This category comprised fruits, vegetables, spices, flavors, herbs, tea, and coffee. A large part of food forensics studies involving IRMS are devoted to classification of these products, due to low complexity of their production chains. Zhao et al. [114] recently reviewed the applications of IRMS in classification of agroproducts from plant sources. Bontempo et al. [63] carried out a vast research on characterization of tomato and its derivatives, using isotopic ($\delta^{13}C$, $\delta^{15}N$, $\delta^{18}O$, $\delta^{34}S$, δD), elemental (trace elements), and anionic parameters (major anions), verifying how these variables change along the production chain and evaluating their usefulness as geographical origin markers. Tomato samples were from three Italian regions: Piedmont, Emilia-Romagna, and Apulia. Juice, *passata*, and paste were analyzed in all cases. Analysis of variance evidenced the statistically significant variables in the discrimination: for what concerns the differently processed tomato products δD, $\delta^{18}O_{bulk}$, $\delta^{18}O_{water}$, B, Cr, Sr, Ba, phosphate, sulfate, and chloride were highlighted, while for the different geographical origin $\delta^{13}C$, $\delta^{15}N$, $\delta^{18}O_{bulk}$, $\delta^{34}S$, $\delta^{18}O_{water}$, Li, Mg, P, K, Cr, Fe, Co, Ni, Ga, As, Se, Mo, Cd, Sn, Sb, Ba, La, Eu, Gd, Tm, Tl, Pb, and U were found. Classification with linear discriminant analysis (LDA) allowed a good discrimination on the basis of geographic provenance, with selection of 10 variables: Gd, La, Tl, Eu, Cs, Ni, Cr, Co, $\delta^{34}S$, and $\delta^{15}N$.

Spices can be considered quality foods in cases in which they express traditional values and are obtained with highly selected raw matters and difficult processes. Among spices, the most valuable is saffron, obtained from stigmas of *C. sativus*. Saffron of the highest quality can be sold at 15.000–20.000 €/Kg, which renders a good idea of the importance of classification. Martin et al. [119] used the carbon isotopic signature to distinguish between safranal extracted from authentic saffron versus that produced by chemical synthesis, but $\delta^{13}C$ values were not so different; application of SNIF-NMR for determination of $^{2}D/^{1}H$ ratio was more successful. More recently, Maggi et al. [120] analyzed 28 saffron samples from leading producers (Greece, Iran, Italy, and Spain) and from the same harvesting year (2006). Analyses were carried out on defatted dry matter. Using data from hydrogen, carbon, and nitrogen isotopic analysis treated with CDA, authors succeeded in discriminating samples with concern to geographical origin, with better results than using conventional chemical parameters. A further valuable example of this category is paprika (dried and powdered red pepper), a traditional spice in the Hungarian diet but a well-known and widely used product all over the world. In Hungary, the so-called *Szegedi Füszerpaprika* was acknowledged of PDO. Brunner et al. [121] investigated the production chain of *Szegedi Füszerpaprika* in order to verify whether processing could alter the isotopic signatures provided by soil. Results of both isotopic and elemental analysis (ICP-MS) showed that the production process has no significant influence on the Sr isotopic signature and on most of the investigated elements, with the exception of Al, Ti, Cr, and Fe, which were mainly introduced through the cutting process of the dried pods. A unique fingerprint for PDO *Szegedi Füszerpaprika* could be created in order to discriminate it from foreign paprika products.

With concern to flavors, vanilla is one of the most important and widely used products in the food industry. Natural vanilla flavor, extracted from the pods of the tropic orchid *vanilla*, is much more expensive than synthetic vanillin: the difference is \$1200–\$4000/kg for the natural versus \$15/kg for its synthetic equivalent. A particular feature of the vanilla orchid is that it is a CAM-type plant, which is advantageous from the diagnostic point of view since its $\delta^{13}C$ values are in a characteristic range (around −21.0‰), so they are distinguishable from C3 and C4 plants. However, measurement of $\delta^{13}C$ values of bulk vanillin is no longer sufficient for classification purposes because it is possible to enrich synthetic vanillin in ^{13}C on both the methoxyl and aldehydic groups and thereby produce synthetic material with the same $\delta^{13}C$ value as that of natural vanillin. Greule et al. [122] determined carbon and hydrogen ratios of vanillin molecule and vanillin methoxyl groups of samples of different origins, including authentic, synthetic, and semisynthetic samples, the last ones obtained biosynthetically by wood lignin and eugenol. Results showed that the combined isotopic approach with ^{13}C and $\delta^{2}H$ values of both vanillin bulk and vanillin methoxyl groups allowed to fully differentiate between natural vanillin extracted from orchid, synthetic ex guaiacol, and semisynthetic ex lignin, providing a rapid and reliable authenticity assessment of vanillin.

Fruits juices were among the first foodstuffs to be subjected to SIA for identification of frauds. As for wine, measurement of $^{13}C/^{12}C$ ratio by means of IRMS soon proved to be useful to identify sugars produced from C4 plants, such as corn or cane, in C3 plant products, such as citrus, apple, or grape juices [123]. This measurement is also useful in order to identify fraudulent addition of L-ascorbic acid to juices [124]; most commercially available L-ascorbic acid is produced starting from C4 plant carbohydrates (e.g., maize starch) whereas authentic L-ascorbic acid comes, of course, from C3 fruits so that after isolation from sample juice exhibits $\delta^{13}C$ values of 10–14%, range from 19% to 23%.

Measurement of $^{18}O/^{16}O$ ratio is used in order to distinguish among directly pressed and rediluted single strength juices. Values of $\delta^{18}O$ are higher in authentic juices since tap water, used as diluent, is relatively depleted in heavy isotopes [125]. Guyon et al. [126] used an hyphenated system composed by HPLC linked to IRMS via an interface allowing the chemical oxidation of organic matter (HPLC-co-IRMS); in this way they were able to determine $\delta^{13}C$ values of organic acids, glucose, and fructose in lime and lemon juices from various geographical origins. These samples constituted a basis to define a confidence domain to which comparing commercial samples of uncertain origin. Among these samples, several of them presented $\delta^{13}C$ values outside the defined range revealing the unlabeled addition of C4-type organic acids or sugars (lemons are C3-type plants), in disagreement with EU regulation.

Tea and coffee are two products of vegetable origin consumed all over the world. For these items, the price is strongly determined by quality, flavor, and the reputation of the producing area; classification is therefore strategic for the market value. Pilgrim et al. [127] classified tea samples from different Asian and African countries by isotopic (δD, $\delta^{13}C$, and $\delta^{15}N$) and elemental (trace elements) data. Geographical classification with pattern recognition methods yielded good results, unaffected by tea type (black, green, or oolong), quality, or harvest year. For what concerns coffee, the isotopic composition of oxygen and strontium [128] and of boron and strontium [129] in samples from different countries were found to be useful in the geographic classification.

Rummel et al. [130] discriminated orange juice samples from different regions in North and SouthAmerica, Africa, and Europe. Isotopic signature of δD, $\delta^{13}C$, $\delta^{15}N$, and $\delta^{87}Sr$ allowed a good classification of samples. In addition, authors showed the possibility of recognizing fraudulent addition of orange juice concentrate by comparing $^{87}Sr/^{86}Sr$ of soluble and insoluble components of the juices.

2.2.7 *Animal Products*

Studies on animal products are mainly focused on eggs and honey. Rogers [131] determined carbon and nitrogen isotope values in whole and delipidized yolk, albumen and egg membrane in eggs samples from 18 different brands of chicken eggs laid under caged, barn, free range, and organic farming regimes. It was possible to distinguish free range and organic egg from caged and barn laid eggs on the basis of higher $\delta^{15}N$ values, possibly due to a higher animal protein (trophic) contribution to the chicken's diet in contrast to pure plant-based foods, as suggested by De Niro and Epstein [132]. $\delta^{13}C$ values, instead, did not elucidate laying regimen. Rock [133] recently reviewed the applications of IRMS techniques for eggs classification.

An important study by Schellenberg et al. [134], within the framework of the food traceability project "TRACE" funded by EU, was focused on the possibility to classify honey samples produced in different European regions by the isotopic signatures of light elements. 516 authentic honeys from 20 European regions were analyzed. Floral types of samples were acacia, brambles, buckthorn, chestnut, citrus, clover, dandelion, eucalyptus, heather lavender, rape, rosemary, sunflower, thyme, forest, and honeydew, plus mixed floral honeys and samples of unknown floral origin. On all samples the stable isotope ratios of hydrogen, carbon, nitrogen, and sulfur were determined. Analyses were made upon precipitation of honey proteins in order to increase the concentrations of nitrogen and sulfur. Results showed the high potential of SIA in honey geographic classification. A significant correlation was found between δD values in honey protein and climatic conditions in the production regions. $\delta^{13}C$ values also were found to be influenced by climate, with higher values in samples from warmer and drier regions; regional differences were more significant than botanical origin. Nitrogen isotopic composition was not clearly correlated with geographical or botanical features of the samples. Finally, $\delta^{34}S$ values were correlated with soil sulfur isotopic ratio and eventually with geographical locations of the plant's growing areas, with strong influence of sea spray for regions close to the sea. Chemometric analysis with LDA identified $^{13}C/^{12}C$ and $^{34}S/^{32}S$ as the most discriminant variables.

A further field of forensic investigation on honey is the identification of illicit use of sugar syrups either by addition to honey or by feeding the bee colonies. $\delta^{13}C$ values from honey protein, as indicated by the official White method [135], can reveal honey manipulation with C4 plant sugars such as high-fructose corn syrup (HFCS), which derives mainly from maize, while the main nectar-providing sources of the honeybees are C3 plants; the difference between $\delta^{13}C$ value of protein and whole honey should be lower than 1‰ [136]. The addition of sugars from C3 plants (beet, wheat), however, cannot be proved by this method. To overcome this drawback, Daniele et al. [137] proposed honey organic acids as internal standards in order to detect honey adulteration.

2.2.8 Products from Cereals

Cereals are a major part in the diet of most cultures all over the world. Among cereals, rice has attracted most of the attention in classification studies, due to the fact that in many countries the geographic provenance of this product is an important issue to consumers. Several studies indicated that SIA can help in distinguishing the provenance of rice samples [138,139]. Ariyama et al. [140] used isotope signatures of heavy elements, i.e., strontium and lead, to distinguish rice samples from Japan, China, Thailand, and USA. $\delta^{87}Sr$ was found to be more effective in the discrimination, also coupled to total Sr and Rb concentration, while lead isotope ratios were found to be more influenced by anthropogenic sources of pollution and therefore less useful in the classification.

2.2.9 Organic Food

The possibility of discriminating organic foodstuffs from the equivalent items obtained by means of conventional systems has been frequently discussed in the last two decades. Particular interest has been given to the application of $\delta^{15}N$ values, due to the fact that the isotopic signature of nitrogen is strongly influenced by the type of fertilizers. Synthetic nitrogen fertilizers, commonly used in conventional agriculture, generate low $\delta^{15}N$ values (from –6‰ to 6‰) in crops while manures and fertilizers permitted in organic agriculture have much higher values (from 1‰ to 37‰). Camin et al. [141] in their recent study involving a large set of fruits grown in Italy, verified that $\delta^{15}N$ is the only isotopic parameter suitable for discrimination of organic fruits from conventional fruits, being less influenced by variables such as cultivar, year, and site of production. However, this isotopic parameter becomes unreliable in cases in which organic production is managed with crop rotation or derived from green manure based on leguminous (N-fixing) plants. Laursen et al. [142] investigated the possibility of distinguishing vegetables grown with organic agriculture from conventionally grown products, using the isotopic signatures of H, C, N, O, Mg, and S to analyzed winter wheat, spring barley, faba bean, and potato. Vegetables were obtained with organic (two different systems, using respectively animal and green manure) and conventional farming systems under similar controlled conditions in two different harvests, 2007 and 2008. Results from SIA analysis showed that none of the isotope ratios measured could allow discrimination of organic versus conventional samples for all the products studied. δD values discriminate organic versus conventional cereals, while $\delta^{15}N$ was useful to identify the use of animal manure in organic wheat, barley, and potato. On the contrary, compound-specific isotope analysis of nitrogen and oxygen isotopes in isolated nitrate allowed a good discrimination of organic and conventional plants [143]. These studies suggest that discrimination with $\delta^{15}N$ is not valid in all instances and nitrogen isotopic analysis should be combined with other analytical approaches (e.g., other stable isotope ratios or secondary metabolic profiling).

Schmidt et al. [105] investigated the isotopic signatures of carbon, nitrogen, and sulfur in samples of Irish beef coming from organic and conventional farming. Conventional beef samples showed slightly less negative and more variable $\delta^{13}C$ values, higher $\delta^{15}N$ and lower $\delta^{34}S$ values, as a result of different diet regimes of cattle. In the case of poultry meat, Rhodes et al. [144] developed a method based on determination of $\delta^{13}C$ to verify whether animals had been corn-fed, a label assigned by EU legislation to chicken fed with a diet containing at least 50% corn for the greater part of the fattening period. $\delta^{13}C$ value in protein resulted to be a reliable marker of the dietary status of the chicken, depending on the difference in the ratio of C3 versus C4 plants assumed.

Chung et al. [145] investigated the isotopic signatures of carbon and nitrogen in organic and conventional milk. Organic milk samples presented on average $\delta^{13}C$ values higher and $\delta^{15}N$ values lower than those of conventional milk samples, as expected considering dairy cow's diet and the use of natural versus synthetic fertilizers, but significant variations were found as a function of the collection date and the milk brand. Much better results were obtained by a combination of $\delta^{13}C$ and $\delta^{15}N$ values which was effective in the discrimination among organic and conventional milk.

2.3 Trace Elements as Chemical Markers

Geochemists used trace and ultratrace elements (called for simplicity *trace elements* further on) since long time ago for their role as markers of rock origin [146]. This is due to several reasons:

- variations of concentration in trace elements are wider than in major and minor elements;
- every system has many more trace elements than major and minor elements, which means a larger set of variables and therefore a potentially higher discrimination power;
- trace elements are collectively sensitive to processes to which major and minor elements are not.

These reasons are still valid when trace elements are used for food classification. In fact, these analytes are considered as suitable variables for geographic classification because they act as territory markers, providing information linked to the geochemical origin of the raw matters with which a food is produced. They can be called *primary geographic indicators*, differently from organic parameters (e.g., volatile compounds, polyphenols, etc.), which can be considered as *secondary geographic indicators* when they reflect technologies typical of regional or national traditions.

Of course, not all elements can play this role: chemical transformation of raw matters and addition of external substances all along the production chain can change the original distribution of some elements. From food to food, adequate elemental markers must be individuated.

Determination of trace elements can be performed with different elemental techniques, such as graphite furnace atomic absorption spectrometry (GFAAS), inductively coupled plasma–atomic emission spectrometry (ICP-AES), or anodic stripping voltammetry (ASV), but the performance features of the plasma–MS system cannot be matched. ICP-MS is at present the most suitable elemental technique in food forensics. This is due to its features of sensitivity, accuracy and precision, speed of analysis, and ability of determining a large amount of variables in short time. The last feature implies that large set of data can be produced and further on treated with pattern recognition methods for classification studies. Table 2 lists the main analytical features of elemental techniques.

The application of ICP-MS analysis to food forensics is relatively straightforward since at least 25 years. At the beginning of 1990s, the work by McCurdy et al. [147] on wine analysis was one of the first suggesting that trace element distribution could be utilized for traceability studies of foodstuffs. Since then, several publications have deepened this point.

As for isotope ratios determination, in the case of trace elements determination, the difference between authentication and traceability must be taken into account. In order to set up an authentication scheme for a particular food, a number of elements are determined, from major to trace and ultratrace. The

TABLE 2 Analytical Features of Elemental Techniques

	ICP-MS	ICP-AES	Flame AAS	GFAAS
Detection limits	Excellent for most elements	Very good for most elements	Very good for some elements	Excellent for some elements
Sample throughput	All elements/2–6 min/ sample	5–30 elements/ min/sample	15 s/ element/ sample	4 min/ element/ sample
Linear dynamic range	10^5 (10^8 with range extension)	10^5	10^3	10^2
Precision				
Short term	1–3%	0.3–2%	0.1–1%	1–5%
Long term (4 h)	<5%[a]	<5%[a]		
Interferences				
Spectral	Few	Common	Almost none	Few
Chemical (matrix)	Moderate	Almost none	Many	Many
Ionization	Minimal	Minimal	some	Minimal
Mass effects	High on low	NA	NA	NA
Isotopes	Yes	No	No	No
Dissolved solids				
(Maximum tolerable concentration)	0.1–0.4%	2–25%	0.5–3%	>20%
No. of elements	>75	>73	>68	>50
Sample usage	Low	High	Very high	Very low
Semiquantitative analysis	Yes	Yes	No	No
Isotope analysis	Yes	No	No	No
Routine operation	Easy	Easy	Easy	Easy
Method development	Skill required	Skill required	Easy	Skill required
Unattended operation	Yes	Yes	No	Yes

Continued

TABLE 2 Analytical Features of Elemental Techniques—cont'd

	ICP-MS	ICP-AES	Flame AAS	GFAAS
Combustible gases	No	No	Yes	No
Operating cost	High	High	Low	Medium
Capital cost	Very high	High	Low	Medium/ high

ICP-MS, inductively coupled plasma–mass spectrometry; ICP-AES, inductively coupled plasma–atomic emission spectrometry; AAS, atomic absorption spectrometry; GFAAS, graphite furnace atomic absorption spectrometry.
[a]*Precision improves with use of internal standards.*

distribution of elements is determined in groups or *classes* of food samples having particular features, according to differences in geographic provenance, animal/botanical species, processing method, or any other discriminating scheme. Using pattern recognition methods, a comparison is made between the different classes and the possibility of discrimination is evaluated. In this instance, the highest number of variables is obtained, the highest possibilities are of succeeding in discrimination among different classes, i.e., authentic versus nonauthentic. For this reason, all elements are potentially useful, regardless of their concentration, so that some authentication schemes could rely mostly on major and minor elements, some other could rely on trace and ultratrace elements. As a role of thumb, major and minor elements can be more useful when discrimination depends on animal/botanical species or process methods, which can cause larger variations in elemental compositions, while trace and ultratrace elements are more useful when geographic provenance is involved.

Traceability of foodstuffs based on element determination involves a different approach. It is not a matter of comparison between different classes of samples of the same type, but a comparison of different samples inside the same production chain. If in the case of isotopic analysis the feature to look for is the similarity of isotopic fingerprinting (mostly of heavy elements), in the case of elemental analysis the feature to look for is the similarity of elemental *distribution*. This can be done with simple statistic methods. Geochemists are used to evaluate relative data by dividing the concentration of each element by its concentration in a set of normalizing values, such as those found in chondritic meteorites [148,149]. This is particularly useful in order to eliminate the saw tooth pattern caused by the Oddo–Harkins rule, according to which even-numbered nuclides are more stable than odd-numbered nuclides. If the distribution of the elements considered is kept unaltered along the production chain, i.e., if there is no *fractionation*, then the whole chain is traceable.

Of particular interest in traceability of foods are considered the rare earth elements (REE). These elements are popular among geochemists as geological markers, because they can help identifying the origin of rocks. The potential of REE as markers is due to their similar chemical behavior. In biological systems, they do not seem to participate actively to plant metabolism: it is suggested that REE 3+ ions can compete with calcium in some instances but with no discrimination among different REE. Plants tend to assume REE ions from soil with little or no fractionation of the original distribution [150–152]. It has been proposed, therefore, that REE distribution in soil could be reflected on plants [13]. It can be hypothesized that only slight or no variations of the original distribution of REE occur when passing from soil to products, in foods with favorable production chains (see Section 1.1).

In the following paragraphs, examples of applications of ICP-MS analysis to food forensics are cited with concern to different food categories.

2.3.1 Wine, Fermented Drinks, and other Beverages

Authentication studies on wine using trace elements distribution are possibly among the first classification works on foodstuffs. Several works evidenced that elemental patterns can be used to classify wines because these patterns may reveal their geographic provenance [153–156]. Factors such as soil chemistry and regional geology influence the elemental composition of crops and can be exploited positively in the classification. On the contrary, anthropogenic factors such as viticultural practices and processing methods, which can have a strong effect, are more difficult to elucidate.

In 1994, Latorre et al. [157] differentiated the prised *Rias Baixas* Spanish wine from Galicia—Certified Brand of Origin—from its imitations. Pattern recognition analysis, performed on ICP-MS data, revealed that Li and Rb were the most discriminating variables. Similar studies were carried out by Baxter et al. [158] on wines from different regions of Spain and England. Marengo and Aceto [159] analyzed with ICP-MS samples of five different DOC and DOCG wines obtained from Nebbiolo variety: *Barolo*, *Barbaresco*, *Nebbiolo d'Alba*, *Roero*, and *Langhe Nebbiolo*. These wines are different with concern to aging (respectively 3 years, 2 years, 1 year, 6 months, and no aging) and partially to geographic provenance (different areas inside the province of Cuneo, Piedmont). It was possible to discriminate the five brands by means of pattern recognition methods. The variables with the most discriminating power resulted to be Si, Mg, Ti, Mn, and Mo, which could be related to the aging method more than to the geological features of soil, considering that samples come from a narrow area.

Relatively, fewer studies have been carried on the traceability of wines through determination of elements. Taylor et al. [160] studied soils and wines from the Okanagan Valley and the Niagara Peninsula, the Canada's two major wine-producing regions. They found that, among trace elements, strontium was able to differentiate both soils and wines from the two regions. Oddone et al.

[161] analyzed the samples of soil, grapes, must and wine in four enological production chains from Piedmont (Italy): Gavi, Barbera, Brachetto d'Acqui, and Freisa. Particular concern was given to the role of REE. Data obtained by ICP-MS showed that REE distribution was clearly maintained unaltered in the passage from soil to must; from must to wine some fractionation occurred on heavier analytes (Gd-Lu), possibly as a consequence of winemaking processes, while lighter analytes (La-Eu) seem to remain unaltered. Similar results were obtained by Aceto et al. [162] in a study on Moscato d'Asti white wine in which samples were analyzed at every step of the production chain. The fingerprint of REE was kept unaltered in the passage soil–grapes–must, while fractionation occurred in wine after the clarification with bentonites. In addition, analysis of Moscato musts from 102 samples showed that is possible to classify their geographic origin, building a basis for identification of possible addition of foreign musts.

Classification of wines by means of elements distribution has been recently reviewed by Gonzálvez and de la Guardia [163].

2.3.2 Milk and Dairy Products

The classification of milk samples by means of trace elements distribution must take into account two different sources: the metabolism of the animal species producing milk and the geographic location of farms, a factor bound to local geology and hydrology features. Benincasa et al. [164] studied the effect of animal species on milk composition. Milk samples were obtained from cows and water buffaloes fed in an Italian farm with identical forage and water, herded in the same field and managed with similar regimes of veterinary medicinal care. Sixteen elements were determined by ICP-MS. Treatment of data with PCA evidenced that cow and water buffalo milk samples could be clearly discriminated; in particular, Ca, P, Ga, Zn, Mn, Ba, and S were higher in the water buffalo milk samples while K and Rb were higher in cow milk samples. The final purpose was to identify "biomarker" elements that could be used to check fraudulent labeling of milk and associated by-products, e.g., *mozzarella* cheese.

In authentication of dairy products based on trace metals, it can be hypothesized that most of them, with few exceptions (e.g., Cu), are not influenced by milk transformation processes and therefore, they can reflect the geochemical features of soil-cows' milk chain. Pillonel et al. [97] in the already cited classification study on Emmental cheeses, determined trace elements with ICP-MS together with and radioactive elements activity. Interesting elements for discrimination were molybdenum and sodium.

2.3.3 Vegetable Oils

Classification studies on edible oils based on elemental analysis require the superior sensitivity of ICP-MS, since the lipophilic environment of oils keeps the metal content obviously at a minimum. In addition, the high carbon content is a drawback for most spectroscopic techniques and requires to be addressed with particular devices, such as addition of oxygen to the plasma and low temperature inside the nebulization chamber. Nevertheless, using ICP-MS, several studies have been issued that allowed to classify oil samples coming from different countries. Trace metals in oil can originate from soil, environment, genotype of the plant, fertilizers and/or metal-containing pesticides, manipulation of olives during the manufacturing, or contamination from the metal-processing equipment. If suitable variables (i.e., elements of natural origin only) are selected, the trace elements distribution in oil samples should reflect the geographic provenance. Benincasa et al. [165] developed an ICP-MS method to determine 18 trace elements in organic extra virgin olive oils coming from the Italian regions Apulia, Calabria, Umbria, and Abruzzo. Oil samples were obtained from two different cultivars, *Carolea* and *Coratina*. A model with LDA was built that allowed optimal discrimination of geographic provenances. The most discriminating variables resulted to be Fe, Mg, Sr, Ca, and As. Llorent-Martínez et al. [166] focused their study on the possibility of distinguishing different types of oils produced in Spain. They determined 18 elements by ICP-MS in samples of virgin olive, olive, pomace olive, corn, sunflower, and soybean oils. Application of PCA allowed a good discrimination among the different categories, in particular olive oils from other lesser quality oils. Cr, Cu, Fe, and Mn resulted to be the most discriminating variables, with Cr and Fe mostly abundant in vegetable oils while Cu and Mn were higher in olive oils. In the already cited study by Camin et al. [98] in which authors analyzed several samples of European olive oils obtaining both isotopic and elemental data, the contribution of trace elements variables in the geographic classification was a relevant one. First at all, authors classified samples on a geological basis, i.e., dividing the whole samples set into three groups according to soil type: shale/clay, limestone, or acid magmatic. Analysis of variance (ANOVA) procedure highlighted statistically significant differences in the content of 16 elements (Mg, Al, K, Ca, V, Mn, Ni, Zn, Rb, Sr, Ce, Sm, Cs, La, Eu, U) among the olive oils produced in the three different geological zones; this classification was confirmed by means of CDA. Similar results were obtained by combining elemental and isotopic data in order to perform geographic classification.

2.3.4 Meats

Trace elements in meat and meat products may derive from different sources. The occurrence of minerals in soils is obviously a primary source, with particular concern to livestock herded on pasture lands; other relevant sources are feed supplements and environmental pollution. Franke et al. [104] reviewed the information concerning the sources of trace elements in meat. They concluded that selenium and rubidium could be interesting geochemical markers. According to some studies these elements, more than to mineral supplementation, i.e., feeding practices, could be related to geographic origin, reflecting the differences of these elements in soils [167]. Other studies [168], instead, suggest a more relevant role to feeding supplements. Pollution from industry, mining or occasional events such as disasters (i.e., Chernobyl) can contribute to soil composition and ultimately to meat. In some cases, it is possible to use polluting elements as markers of geographic origin of meat. A study on tissues of livestock grazing nearby Kidston Gold Mine (North Queensland, Australia) showed that liver, muscle, and blood of animals were enriched in As and Zn [169]. In contrast, Chessa et al. [170] did not find statistically significant differences in Pb, Zn, and Cd between muscle tissues of sheep from a polluted area in Southwest Sardinia and sheep herded on unpolluted areas.

The already cited study by Heaton et al. [109] used trace element distribution in addition to isotopic data to classify beef samples from different countries. The most important elements in the classification were strontium, iron, rubidium, and selenium. The classification into three broad groups, i.e., European, South American, and Australasian was obtained with CDA.

2.3.5 Fish

In the classification of fish, several factors can influence the elemental distribution: metabolism of organisms, geomorphology, lithology, food availability, contamination from external sources, etc. An example of high-quality fish is the Galician mussel (*Mytilus galloprovincialis*), the first fish acknowledged by European PDO. Costas-Rodriguez et al. [171] established the geographical origin of mussels from different areas of Galician Rías (Vigo, Pontevedra, Arousa, Muros-Noia, Ares-Betanzos). For this, the distribution of 40 elements was determined by ICP-MS in 158 samples of Galician origin and in samples from two Mediterranean regions representative as non-Galician samples. Different supervised pattern recognition techniques (LDA, SIMCA, and ANN) were used that allowed to classify correctly PDO Galician mussels versus non-Galician samples and Galician mussels according to the ría of origin. Cubadda et al. [172] developed an ICP-MS method for the determination of trace elements distribution in marine organisms. The aim of the study was, apart from identifying pollution hot spots, verifying whether it was possible to have an elemental fingerprinting of seafood destined to the fishing market. In the case of mussels, results evidenced that a good separation was achieved between samples collected in three different farming sites. A recent study by Guo et al. [173] evidenced that the trace element distribution in fish muscles can be used as a fingerprint to discriminate among fishes captured in different sea regions. The geographic variability, in fact, was larger than the species variability. Element data were treated with partial least squares discriminant analysis (PLS-DA) and probabilistic neural network (PNN) supervised pattern recognition methods.

Classification of fish by means of elements distribution has been recently reviewed by Lavilla et al. [29].

2.3.6 Fruits and Vegetables

The soil elemental composition can serve as a fingerprint for different crops, providing a useful marker for geographical classification. Fruits and vegetables, which are the foodstuffs more directly linked to soil, reflect well this utility.

Several studies have already shown that the trace element distribution is a powerful marker for classification of fruit and vegetable foodstuffs. The elemental fingerprint given by soil can be affected by various factors such as botanic varieties, fertilization, climatic conditions, and agricultural practices, but several studies [174,175] suggest that the variations of element composition induced by these factors are smaller than the differences between production places when proper elemental markers are selected. As a consequence, in every classification study the proper variables must be selected.

In the category of fruits and vegetables, spices rank among the first places in terms of cost. It has already been cited the fact that saffron from *C. sativus* is perhaps the most expensive food in the world and therefore the ideal candidate for a classification study. D'Archivio et al. [176] analyzed saffron samples coming from three Italian regions, Abruzzo (one among the most important regions in the world producing saffron), Umbria, and Sardinia. Analysis with ICP-MS and LDA allowed discriminating the three geographical provenances; the most significant variables were Li, B, Na, Ga, Rb, Sr, Zr, Nb, Cs, Ba, Sm, and Hf with particular concern to B, Na, Sr, and Rb. These elements can be considered mostly of geochemical origin and therefore characteristic of soil, so that their distribution is not fractionated in the passage from soil to plant. In the already cited study by Brunner et al. [121] on *Szegedi* Hungarian paprika, authors distinguished authentic samples labeled with PDO from foreign paprika products on the basis of both isotopic signature of strontium and trace elements distribution, with main concern to REE.

Bettinelli et al. [177] investigated the link among soil and the different parts of tomato plant for what concerns the distribution of REE. Bontempo et al. [63], in the already cited study on tomato and its derivatives, highlighted the importance of trace elements and in particular of REE in tracing the geographic provenance of these products. Tomato samples from three Italian regions could be discriminated and several trace elements, namely Gd, La, Tl, Eu, Cs, Ni, Cr, and Co, were found to be highly relevant in the classification. Lo Feudo et al. [178] analyzed with ICP-MS tomato samples cultivated in four Italian regions (Apulia, Calabria, Basilicata, and Emilia-Romagna) along two growing seasons; also, a tomato paste sample made in Italy was compared with similar products coming from California, China, and Greece. In both instances, a good discrimination result was obtained on a geographical base, using 32 trace elements as variables and classification with supervised pattern recognition methods (LDA, KNN, and SIMCA).

Classification of hazelnuts from Piedmont was achieved with ICP-MS by Oddone et al. [179] with ICP-MS, using REE as chemical tracers. Authors demonstrated that no fractionation occurred to the original distribution of REE in the passage from soil to fruits (Figure 5), and it was possible to evaluate the differences among hazelnuts from Piedmont, other Italian regions, and Turkey.

Several classification studies on fruits and vegetables based on trace elements determination have been recently issued by a research group working at

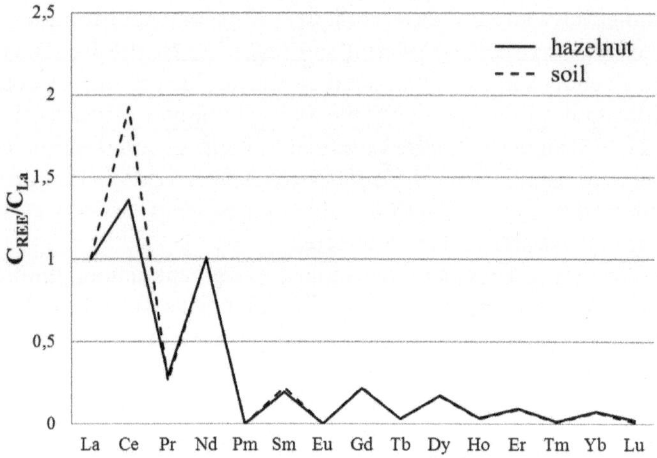

FIGURE 5 Distribution of rare earth elements (REEs) in soil and hazelnuts.

Dipartimento di Chimica, Università della Calabria (Southern Italy). They studied the distribution of trace elements in Clementine (*Citrus clementina* Hort. ex Tan.) mandarins, one of the most important cultivated varieties in the Mediterranean basin [180]. This variety is cultivated in many countries on different continents, but a world renowned production is the one located in Calabria (Southern Italy), awarded with protected geographical indications (PGI) certification by the European Union as "Clementine di Calabria." Authors were able to discriminate PGI samples of four different Calabrian zones, from non-PGI samples coming from Spain, Tunisia, and Algeria, using supervised pattern recognition methods, such as SIMCA, PLS-DA, and LDA. In another study, they classified samples of Tropea red onion (*Allium cepa* L. var. Tropea), one among the most highly appreciated Italian products, awarded with PGI certification as "Cipolla Rossa di Tropea Calabria," from non-PGI onion samples, i.e., samples coming from cultivation areas located outside the distribution area specified in the production regulations [181]. Authors evidenced the importance of the contribution of REE in the classification.

Tea and coffee are both foodstuffs of vegetable origin with a relevant commercial value. The classification of tea and coffee samples has been attempted by means of trace elements distribution using ICP-MS (in some cases also using ICP-AES). Moreda-Piñeiro et al. [182] were able to discriminate tea samples from Asian and African countries on the basis of element distribution as determined with ICP-AES and ICP-MS. Data treated with supervised pattern recognition techniques (LDA and SIMCA) allowed to distinguish also among teas from China, India, and Sri Lanka. More recently, Pilgrim et al. in the already cited study [127] combined data from ICP-MS and SIA to classify tea samples from different Asian and African countries. Twenty trace elements, among which four REE, were used for application of LDA method.

2.3.7 Animal Products

A very particular animal product is caviar, i.e., roe descending from certain fish species. Caviar is an extremely exclusive and greatly demanded delicacy and therefore one of the most expensive—and most counterfeit—foodstuffs in the world. The most prised caviar is recognized to be that produced from sturgeons of the Caspian Sea, the renowned *caviar Volga*. Perhaps due to the high incomes involved, few analytical studies have been devoted to the classification of caviar; Ebert and Islam [183] used a histological method to identify caviar imitations of Russian sturgeon roe. Rodushkin et al. [184] used the elemental distributions determined by ICP-MS to classify vendace and whitefish caviar samples according to the geographic origin (Sweden, Finland, and USA) and to the type of water in which fish grow (brackish vs freshwater, with brackish caviar being more expensive). Seventy-two elements, from μg/g to pg/g concentrations, were determined. The element distribution in caviar is influenced by its main constituents, i.e., roe and salt, but impurities originating from the caviar preparation and packaging processes can also contribute. Furthermore, contamination during the various analytical stages is possible. To check for this possibility, authors analyzed also samples of unprocessed vendace roe and salt used in caviar production. Elements potentially allowing to establish the sample provenance were As, Ba, Br, I, Li, Mo, Se and Sr while Fe, Al, Ti, and V, even though useful in the discrimination, were reputed to arise during processing and packaging. Particularly useful were element ratios, i.e., Sr/Mg, Sr/Ca, and Sr/Ba that achieved a good discrimination of geographical provenance. In fact, Sr/Ca ratio in otoliths reflects their relative proportion in the ambient water, as showed Bath et al. [185]. Differentiation between caviar from brackish and freshwater sources was accomplished using Sr isotope ratio measurements.

The classification of honey based on geographical or botanical origin has been studied using trace elements distribution in several studies. Minerals could be useful for a classification system, since they can be associated with the soil where Melliferous flora grows. In the passage from soil to honey, however, different sources can contribute to the final elemental composition. Environmental pollution can cause increase of heavy and transition metals concentration. The use of particular fertilizers, rich in REE, in crops can result in anomalous values of these elements' amounts [186]. Chudzinska and Baralkiewicz [187] analyzed with ICP-MS samples of honey of three types (honeydew, buckwheat, and rape) produced in 16 areas of Poland. Data obtained from determination of 15 elements were treated with supervised pattern recognition methods and allowed a good classification according to the regional provenance and to the honey type. Al, Mg, and Zn were the best markers of geographical origin while K and Mn were the markers of honey type. Similar results were obtained by Batista et al. [186] in a study on Brazilian honey samples. In that case Pb, Tl, Pt, Ho, and Er were the most significant variables in the geographical classification. Chen et al. [188] determined 12 elements in 163 Chinese honey samples in order to classify samples according to the botanical origin obtaining a good discrimination between samples of linden, vitex, rape, and acacia honeys.

2.3.8 Products from Cereals

Raw cereals, as fruits and vegetables, are ideal subjects for classification studies using trace elements distribution. Ariyama et al. [140] determined trace elements in rice samples from Japan, China, Thailand, and USA together with isotope ratios of heavy elements. A good classification was obtained, where the most discriminating variables were Rb, Sr, Ba, and Co. Shen et al. [189] studied the correlation between the distributions of trace elements in rice and soil, based on determination of the available fraction of metals. Samples were from four provinces of China. Analysis by means of ICP-MS revealed that most of the elements determined (Mg, K, Ca, Na, Be, Mn, Ni, Cu, and Cd) were significantly different in both soil and rice samples of the four regions. These results were confirmed by LDA chemometric analysis.

Zhao et al. [190] investigated the classification of wheat samples collected from four major wheat-producing regions in China in two subsequent harvests, on the basis of trace elements distribution. Application of LDA to elemental data allowed a good geographical classification.

2.3.9 Organic Food

Laursen et al. [191] investigated the possibility of distinguishing organic and conventional productions of selected vegetables by means of element distribution as determined with ICP-OES and ICP-MS. Wheat, barley, faba bean, and potato were the products studied, cultivated in 2 years at three different locations using both organic and conventional cropping systems. The study was carried out under controlled and comparable conditions for what concerns soil type and climate, so that differences could be due only to the cropping system. Using trace elements as variables, a good discrimination was obtained among organic and conventional products; Cd and Cl were the elements giving higher contribution to the classification, possibly due to trace impurities from inorganic fertilizers used in conventional cropping system. Kelly and Bateman [192] used $\delta^{15}N$ values and trace elements distributions to distinguish between organic and conventional grown tomato and lettuce samples; results were better for tomatoes and poor for lettuces. Systematic differences were found in the concentrations of certain elements, such as manganese, calcium, copper, and zinc, possibly due to the presence of elevated levels of arbuscular mycorrhizal fungi in soils run with organic systems.

2.4 Molecular Ions as Chemical Markers

2.4.1 Strategies of Molecular Analysis

The previous two paragraphs accounted for techniques relying on *elemental* parameters. A classical use of mass spectrometry, however, is the separation and identification of *molecules*. This can be done either using *hyphenated systems*, i.e., interfacing a mass spectrometer with a separative system, or with *stand-alone* or *MS-only systems*. Hyphenated systems will be the subject of the following paragraph, while stand-alone systems will be dealt within Section 2.4.3. Both systems are now more than consolidated techniques to identify molecular markers to be used in food classification. General discussions on the application of hyphenated and stand-alone systems in food forensics have been recently issued by Herrero et al. [193] and by Wang et al. [194].

An improvement in the diagnostic power of molecular analysis with MS systems applied to food forensics is the development of advanced *fingerprinting* and *profiling* methods. The concepts of fingerprinting and profiling are not specifically referred to MS since they encompass all the analytical methods that can provide multiple responses from samples; in this sense, techniques such as NMR, vibrational spectroscopies (FT-IR and Raman), electronic spectroscopies (UV–Visible–NIR spectrophotometry) are included together with MS techniques. While traditional methods are focused on identification and quantification of specific compounds in a particular food sample, which is the approach typical of *targeted* strategies, fingerprinting and profiling methods (i.e., *nontargeted* strategies) provide specific information about groups of samples based on their component distribution without need of accurate identification of all molecules. The difference between fingerprinting and profiling is that the latter requires previous knowledge of the samples because it focuses on a specific group of related compounds. Fingerprint/profiling methods call for highly reliable and powerful detection systems such as MS. Moreover, the amount of information given by mass spectral data sets asks for a mandatory chemometric treatment of data with pattern recognition methods [195]. Food forensics studies using fingerprinting/profiling methods based on hyphenated systems are strongly greatly increasing in the last years. The application of these methods has been extensively reviewed by Esslinger et al. [196].

A further improvement occurring in the very last years in the knowledge and potentiality of food forensics resulting from the development of *foodomics*. The new frontier of the application of mass spectrometry to food analysis has an English name but a Spanish father, Alejandro Cifuentes [197]. This discipline results from a powerful interaction among food and nutrition science, advanced analytical techniques (i.e., biomolecular and bioinformatics), with particular emphasis in chemometrics (Figure 6, based on Herrero et al. [193]). Further in-depth analyses on this subject can be found in recent publications [198,199]. Even if the concept of foodomics goes beyond the field of food forensics, one of its main features is the analytical study of foods for compound profiling with the aim of identifying biomarkers useful for food classification. It is clear that mass spectrometry is definitely the core technology for such a task: the depth of information achieved by MS techniques cannot be reached by other techniques. In addition, interfacing of MS with protein and peptide databases has allowed new possibilities of the characterization of biomolecules.

The definition of foodomics encompasses, among others, three main analytical sectors, which have been developed in the last 20 years and have found large

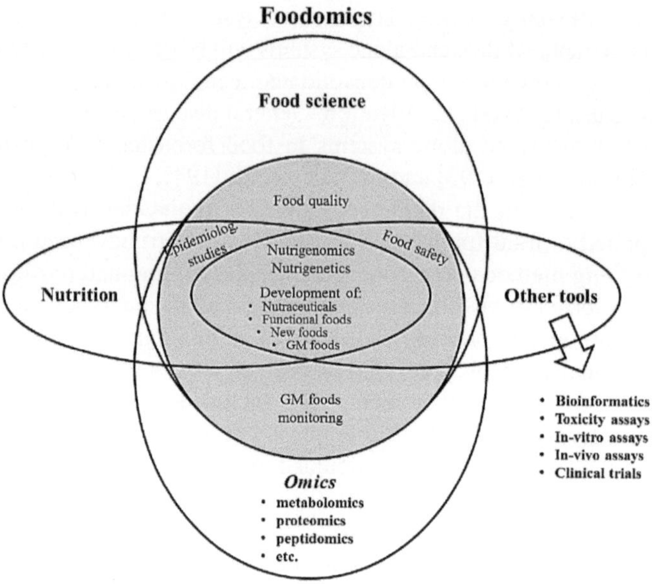

FIGURE 6 The concept of *foodomics.*

application, among several other fields, in food forensics: *proteomics*, *peptidomics*, and *metabolomics*. Proteomics is the analytical field devoted to the characterization of the proteins expressed in a particular biological system; from the analytical point of view, a major problem is due to the different physicochemical properties of proteins and their large concentration range in real samples. Peptidomics is the analysis of all peptide content within an organism, tissue, or cell, including transient products of protein degradation; in this case, also difficulties arise from the concentration range and the high number of peptidic sequences available in real samples. The applications of proteomics and peptidomics in food forensics have been recently reviewed [200,201]. Metabolomics is focused on the analysis of a metabolome, i.e., the full set of endogenous or exogenous low molecular weight compounds (<1500 Da), which means a large number of different molecules with different chemical and physical properties in a wide concentration range [202]. The term *metabolomics* has been established in the late 1990s. Oliver et al. used the expression *metabolome* for the first time in 1998 [203]. Recalling the concepts of *fingerprinting* and *profiling* expressed before, there are three different ways for analyzing metabolites in a biological system: namely, *target analysis*, *metabolic profiling*, and *metabolic fingerprinting*. Target analysis is the quantitative determination of selected, well-individuated analytes. Metabolic profiling is a nontargeted strategy requiring previous knowledge of the sample, which focuses on the study of a group of related metabolites (e.g., amino acids, alcohols, etc.) or metabolites from a specific metabolic pathway. Metabolic fingerprinting is focused on comparing patterns or *fingerprints* of metabolites among different samples, aiming not to identify all the involved metabolites but to detect those that can be considered biomarkers. The differences among these approaches have been clarified by Fiehn [204] and reviewed by Dettmer et al. [205]. Due to the great potentialities of metabolomics-based approaches, it is not surprising that several reviews on their applications in the field of food science and nutrition had been recently issued [206–209], with a large number of citations to works of interest in the field of food forensics. Oms-Oliu et al. [210] reviewed the use of metabolomics for classification of plant-derived food. A major part of the cited studies describes the application of MS techniques either in hyphenated or in stand-alone systems. Cubero-Leon et al. [211], in their review work on food classification with metabolomics, showed that the percentage of studies involving mass spectrometry was nearly 50% in the last 10 years, with NMR and vibrational spectroscopy accounting for most of the remaining part.

2.4.2 Hyphenated Systems

In the field of food forensics, a large number of applications have been developed using MS as a powerful detection system coupled with separative techniques. The composition of most foodstuffs results from a complex mixture of dozens or even hundreds of compounds. The application of separative techniques (chromatographic and/or electrophoretic) hyphenated to detection system giving full structural information makes it possible to elucidate the chemical nature of foodstuffs and to develop reliable classification schemes. The hyphenated systems GC-MS and LC-MS have, therefore, a long tradition in the field of food forensics. Interfacing of MS with gas chromatographic systems is historically older and more consolidated, even though limited to the separation of volatile or volatilizable compounds. Nowadays the development of atmospheric pressure ionization (API) sources, such as electrospray ionization (ESI) and atmospheric pressure chemical ionization (APCI), which have solved the issues of pressure difference between a liquid phase at high-pressure and low-pressure vacuum regions, has favored the spread of LC-MS-based methods. Applications have been recently reviewed [212–214]. Relatively more recent is the interfacing of MS with electrophoretic systems [215,216] with particular concern to capillary electrophoresis (CE). CE-MS systems have been applied mainly for identification of organic compounds that can play as biological markers in food classification.

2.4.3 Stand-Alone Systems

Besides hyphenated systems, only stand-alone or MS methods have been developed. A list, not necessarily extensive, of the most important stand-alone MS techniques nowadays available must include Pyrolysis (Py-MS), direct infusion of samples by means of ESI-MS or atmospheric pressure photoionization mass spectrometry (APPI-MS), matrix-assisted laser desorption/ionisation–time of flight mass spectrometry (MALDI-ToF-MS), proton transfer reaction–time of flight (PTR-ToF-MS), desorption electrospray ionization (DESI-MS), direct analysis in real time (DART-MS), and secondary electrospray ionization (SESI-MS).

Stand-alone techniques allowed the building of *molecular profiling* procedures and yielded new insight in the identification of molecular markers in food matrices that can be exploited for classification purposes, without need of prior separation. Sample profiling, therefore, can be achieved with similar performances compared to hyphenated systems but in a faster and easier way. These methods have a high potential in food forensics, in particular when coupled with chemometric data treatment. In fact, the spectra obtained can be considered as quantitative fingerprints of the organic components of food samples and, even if simple visual inspection of the spectra usually does not allow identifying prominent peaks that could be used for classification, the use of multivariate pattern recognition methods can yield optimal results. In addition, stand-alone MS methods can request minimal time consumption and low costs (a few minutes are necessary for sample preparation and mass spectrum acquisition) in contrast with traditional methods of molecular biology. The construction of molecular profile databases can integrate methods and provide high-tech tools in large-scale screening for control and authentication of foodstuffs.

A particular system, specifically devoted to the profiling of volatile compounds, is the headspace mass spectrometer (HS-MS). It is a variant of the so-called electronic nose or *e-nose* systems [21,217] in which a mass spectrometer is used as a chemical sensor in order to yield a fingerprint pattern of volatile compounds, which coupled with pattern recognition methods, allows classification of samples according to variety and to other technological features. With respect to classical e-nose systems, based on solid state sensors, the advantage of MS-based e-nose systems lies in the signal reproducibility, higher specificity, and interpretation of the results. In fact, it is possible to obtain information about complex mixtures of volatile compounds even if it is not possible to identify and/or quantify each chemical species. The synergic use of HS-MS with multivariate statistical techniques allows the relevant information for classification and discrimination of the investigated samples to be extracted. Śliwińska et al. [218] have reviewed the applications of artificial senses, including e-noses, in food analysis and classification.

In the following sections, the application of hyphenated and stand-alone MS techniques to food forensics is illustrated according to the different food categories. In each section, examples involving hyphenated systems are described firstly and then come examples involving stand-alone MS systems. As the application of classical targeted methods using hyphenated systems is relatively straightforward in food forensics, particular focus will be given to fingerprinting/profiling applications and to methods involving stand-alone MS.

2.4.4 Wine, Fermented Drinks, and Other Beverages

The application of hyphenated systems to wine analysis and classification has a long tradition. Profiling of volatile compounds, polyphenols, and other chemical classes have been used for the classification of wine samples according to geographical origin, grape variety, and aging. Marengo et al. [219] analyzed with solid phase microextraction (SPME)-GC-MS samples of five different DOC and DOCG wines obtained in Piedmont from Nebbiolo variety: *Barolo*, *Barbaresco*, *Nebbiolo d'Alba*, *Roero*, and *Langhe Nebbiolo*. These wines have different aging periods (respectively 3 years, 2 years, 1 year, 6 months, and no aging). This feature is remarkably reflected in the respective values: *Barolo* and *Barbaresco*, in fact, are among the most prised wines in the world. Using a target approach, it was possible to discriminate the five brands by means of pattern recognition methods (PCA, HCA, Kohonen self-organizing map, stepwise LDA, and SIMCA) according to the aging characteristics. Vaclavik et al. [220] used HPLC coupled to ToF-MS to differentiate red wines of three varieties (Cabernet Sauvignon, Merlot, and Pinot Noir) of various geographical origins (European and the US retail market). A nontargeted metabolomic approach was applied to data: the molecular identification of the compounds separated was not necessary, while ions with identical elution profiles and related m/z values were extracted as molecular features. Data were treated by means of multivariate statistical analysis and yielded a good classification according to varieties. A higher level of confidence for the identification process was achieved by using tandem mass spectrometry. Jaitz et al. [221] developed a rapid LC-MS/MS method for phenols and polyphenols profiling, including three isomeric pairs, in wine samples from 11 Austrian regions, including 6 grape varieties and 5 vintages. A good classification was achieved according to geographical, variety, and vintage features using CDA. Serrano et al. [222] developed an LC method with UV-Vis and fluorescence detection for the differentiation of Spanish PDO wines based on their chromatographic profiles; the most discriminating compounds were then identified by means of MS.

With concern to beer classification, Mattarucchi et al. [223] used a fingerprinting approach with LC-MS to authenticate Trappist beer samples of different brands. In this case, a total of 232 beers were fingerprinted and classified through multivariate data analysis. The selected beer samples were clearly distinguished from beers of different brands; only three samples were wrongly classified when compared with other types of beer of the same Trappist brewery.

In recent years, stand-alone MS techniques have been intensively applied to wine. Fingerprinting with protein or peptides profiles have been proposed. Among the first applications of MALDI-ToF-MS to wine protein characterization, it must be cited in the work by Szilágyi et al. [224], which dates back to 1996. The paper suggested the feasibility of MALDI as an alternative to IRMS or Py-MS for distinction of wines and musts. Chambery et al. [225] applied MALDI-ToF-MS in the analysis of peptide profiles obtained from whole wine protein tryptic digests to the classification of high-quality white wines produced in Campania (Southern Italy). MALDI spectra provided fingerprints of samples that attained their discrimination according to different grape varieties or different winemakers. The qualitative presence or absence of peptide m/z values were used as variables for both types of classification. However, Nunes-Miranda et al. [226] reviewed the methods of protein and peptide profiling, highlighting some critical points in the possibility of wine authentication. Fulcrand et al. [227] proposed a method involving direct MS with ESI-ToF and MALDI-ToF-MS to determine the profile of polyphenolic compounds and use it as a tool for wines discrimination.

An interesting application of MS in wine analysis is the use of MS-based e-nose systems for profiling/fingerprinting of volatile compounds. In the case of alcoholic drinks, e-nose can be useful in the classification of products according to the varieties or enological features (i.e., aged vs young wines) [21]. With respect to classical e-nose systems based on solid state sensors, MS-based e-nose systems have the advantage of withstanding the high concentrations of ethanol that often interfere with solid state sensors. An example of application to wine classification was issued by Cynkar et al. [228], which used an MS-based e-nose system combined with chemometrics to ascertain the geographical origin of Tempranillo wines produced in Australia and Spain. Capone et al. [229] used solid phase extraction (SPE) to extract volatile compounds from different typical wines from Apulia (Southern Italy). The analysis of the extracts by GC-MS in conjugation with e-nose identified 18 compounds over odor threshold, which were used for further pattern recognition analysis with PCA. By means of PLS-DA analysis, data from e-nose and GC-MS analysis were correlated, establishing the relationships between e-nose response and wine aroma compounds. Vera et al. [230] used MS-based e-nose to discriminate between beer samples produced in different factories. Jelen et al. [231] analyzed raw spirits made from rye, potato, and corn in order to verify the possibility of a classification based on the botanical origin. They used an SPME-MS method in which volatile compounds profiles were determined; PCA and LDA pattern recognition methods allowed a good separation between the three botanical sources. A detailed review on the application of e-noses for the analysis of alcoholic beverages has been published [232].

Other stand-alone MS systems were used on alcoholic drinks. With concern to beers, Cajka et al. [233] developed a method based on direct interfacing of SPME to DART-ToF-MS for classification of beer samples from Belgium, the Netherlands, and Czech Republic. This system, together with a more classical SPME-GC-MS system, allowed the characterization of the profile of volatile compounds, which reflect the chemical features typical of different brewing processes. Šedo et al. [234] showed how the application of MALDI-ToF-MS could be used in the classification of beers, allowing to distinguish samples of the same brand but from different breweries. The protein profiles resulted to be highly discriminating, having a strong relation to the brewing materials, technology processes, and sensory characteristics of beer samples. Protein composition of beer depends on the raw materials and the enzymatic reactions that take place in the course of the beer production process.

Among spirits, a major interest has been devoted to characterization and classification of whisky, according to its commercial and traditional value. Aylott et al. [235] applied Py-MS followed by multivariate analysis of the resulting mass spectra to distinguish authentic samples from nonauthentic, in addition to higher-alcohol congener analysis by GC-MS; Py-MS proved to be reliable as a stand-alone method for whisky classification. An improvement of stand-alone MS methods was later proposed by Møller et al. [236] who used direct infusion of samples with ESI-MS to classify whisky samples from different countries. Authors verified that analysis in negative ion mode yielded the most characteristic whisky fingerprinting mass spectra. Three groups were clearly discriminated: Scotch whiskies, American whiskies, and presumably counterfeit whiskies commercialized in Brazil. Discrimination among single malt and blended Scotch whiskies was also achieved. The proposed method resulted to be a simple and rapid way to check quality and proof of authenticity of whisky samples, with further improvement by means of data treatment with pattern recognition chemometric methods. In addition, application of ESI-MS/MS analysis allowed identifying the most diagnostic anions, which resulted to be simple sugars, disaccharides (indicating the possible use of caramel to adjust color and flavor) and phenolic compounds. Garcia et al. [237], using ESI coupled to ultra-high resolution and accuracy Fourier transform ion cyclotron resonance mass spectrometry (FT-ICR MS), extended the range of diagnostic molecules and further improved the possibility of identifying counterfeit samples. In this case, both positive and negative ion modes yielded useful information.

The Aceto Balsamico Tradizionale di Modena, a high-quality vinegar produced in Emilia-Romagna (Central Italy), is a typical case in which the composition of a foodstuff is so complex that it is hard to find chemical markers for its classification. Its characteristic flavor is the effect of different variables. In such cases, molecular fingerprinting is the best approach for classification. Cocchi et al. [238] used HS-MS, the already cited *e-nose* system, to distinguish correctly aged (and therefore marketable) Aceto Balsamico Tradizionale di Modena products from the ones, which are still undergoing maturation. Classification with supervised pattern recognition methods (SIMCA, LDA) was successful and allowed recognizing the different stages of maturation.

2.4.5 Milk and Dairy Products

Protein profiling is particularly useful in the classification of milk and dairy products. Factors such as genetic and nongenetic polymorphisms and milk processing (i.e., thermal denaturation and proteolysis) can induce variations in the structure of individual proteins, so that milk samples from different species or from different animal breeds of the same species, or resulting from different processes, can be distinguished. Nicolaou et al. [239] analyzed with MALDI-ToF-MS binary and tertiary mixtures of cow's, goat's, and sheep's milks; using supervised pattern recognition methods (linear PLS and nonlinear Kernel PLS) on the protein and peptidic profiles of the samples, they were able to identify the addition of an adulterant milk regardless of milk processing. Yang et al. [240] used two-dimensional gel electrophoresis (2-DE) coupled with MS to acquire protein profiles of milk samples from different animals (cow, goat, camel, yak, and buffalo). The different protein spots acted as specific molecular markers for identification of milk samples, with particular concern to adulteration; in fact, the animal species could be singled out even in binary mixtures. Cunsolo et al. [241] used MALDI-ToF-MS to identify the fraudulent addition of bovine and caprine milk in raw donkey's milk, a product considered as precious pharmafood thanks to its use in anallergic diets. Extraneous milks were detected by using the protein profiles of some whey proteins as molecular markers. Cunsolo et al. [242] reviewed the applications of MS-based techniques for characterization of milk protein profiles and their use in milk classification.

2.4.6 Vegetable Oils

Given the complexity of the composition of vegetable oils and the high-quality and commercial value of extra virgin olive oils, hyphenated and stand-alone MS techniques can have great potential in their classification. Di Donna et al. [243] developed a method for classification of Italian olives based on the polyphenolic profile as determined with HPLC-MS. Salter et al. [244] used pyrolysis-MS to classify extra virgin olive oils from different regions of Italy and obtained from different olive cultivars. The molecular fingerprints of samples were analyzed with artificial neural network pattern recognition method and allowed a good discrimination of olive oils on the basis of their geographical origin. Vaclavik et al. [245] used high-resolution ToF-MS coupled with DART to determine the comprehensive profiling of triacylglycerols in oils products. Authors were able to discriminate between extra virgin olive oil, olive pomace oil, and olive oil; also, adulteration of extra virgin olive oil with hazelnut oil was achieved.

With concern to e-nose MS-based systems, Oliveros et al. [246] used HS-MS for classification of olive oils of different geographical origin with a fast method of analysis. Samples were from five different Mediterranean areas. Both unsupervised and supervised pattern recognition methods, applied to the profiles of volatile compounds, yielded good results in the classification. Similar results were obtained by Cosio et al. [247] using an e-nose system in combination with artificial neural networks.

2.4.7 Meats

The classification of meat and meat products can be achieved by means of different strategies. Profiling of volatile compounds and lipids is one typical approach. Narvaez-Rivas et al. [248] determined with GC-MS the profile of volatile compounds in samples of Iberian dry-cured hams. Classification according to different pork fattening systems was achieved.

Zaima et al. [249] used a metabolomic approach with MALDI-imaging mass spectrometry in the assessment of beef authenticity. The analysis was carried out on extracted lipids and in few minutes enabled the visualization of the distribution of individual biomolecules. Peaks in the mass range 700–1500 m/z were used for a pattern recognition treatment with PCA; the results allowed obtaining a good discrimination among beef samples from three different areas of Japan.

Among cured meat products, it is highly renowned the quality of Iberian and Italian dry-cured hams. Del Pulgar et al. [250] used the recently developed proton transfer reaction time of flight mass spectrometry (PTR-ToF-MS) technique for the characterization of dry-cured hams produced in Italy (Prosciutto di San Daniele, Prosciutto di Parma and Prosciutto Toscano) and Spain (Dehesa de Extremadura) according to PDOs. Samples were analyzed by direct injection without any pre-treatment. Analysis by means of PTR-ToF-MS yielded information concerning the volatile organic profile, which reflects the different manufacturing practices (i.e., rearing system of livestock, salting, and curing process, etc.); data from molecular profiling were treated with PCA that allowed a good classification according to the production processes of the different PDOs. González-Dominguez et al. [251] studied the classification of different types of Iberian ham with other stand-alone MS techniques. Authors analyzed intramuscular fat extracts using two techniques, direct infusion electrospray mass spectrometry (ESI(+)-MS) and flow injection atmospheric pressure photoionization ionization mass spectrometry (APPI(+)-MS). Total lipid profiles (triacylglycerides, diacylglycerides, monoacylglycerides, and free fatty acids) were obtained with simple sample preparation. Data were treated with PLS-DA method that allowed classifying Iberian ham samples according to their feeding systems (traditional feeding based on natural resources vs feeding based on commercial foodstuffs) and provenance.

From a completely different point of view, molecular analysis by means of MS techniques can be used to check for *absence* of meat in particular foodstuffs. *Kosher* is the term for food that may be consumed by Jews according to *halakha* (Jewish law); in a similar way, *halal* stands for food considered permissible to Muslims according to Islamic law. Compliance of food to *kosherness* and *halalness* is very important for Jew and Muslim people and its control is therefore a relevant commercial matter considering the extension of the global Halal food market. Indrasti et al. [252] characterized with two-dimensional comprehensive GC-ToF-MS the fatty acid profile in samples containing animal fat, in order to discriminate among cattle fat, chicken fat, goat fat, and the prohibited lard. Rohman and Man [253] reviewed the analytical methods used in the detection and quantification of pig derivatives in food products to verify compliance to *kosherness* and *halalness* of food destined for consumption by Jew and Muslim people. Among the different methods listed, applications were reported using LC-MS and GC-MS.

2.4.8 Fish

Fish authentication demands for more innovative methodologies since the identification of morphological characteristics, such as the head, fins, skin, or bones, can be lost during processing so that the original fish species cannot be recognized. In addition, globalization and freer markets determined the fact that a growing number of species are used for transformed products. Sophisticated MS techniques using *molecular profiling* can address these needs. Recently, it has been proposed the use of MALDI-ToF-MS for the unequivocal identification of seafood species [254]. This method is based on the evaluation of the mass signals generated from proteins with molecular weights of about 11 kDa, which can play as specific biomarkers for species discrimination. The applicability of the method was checked by analyzing protein extracts from 25 different seafood species, selected among largely consumed products either of high commercial value or commonly involved in frauds. Authors underlined the fact that this method does not require either preliminary information on the investigated sample or preliminary identification of the proteins generating biomarkers. It was nevertheless evidenced the role of parvalbumins, calcium-binding proteins with molecular weights in the 10,000–12,000 Da range, as progenitors of several of the biomarkers individuated. With concern to processed seafood products, Barik et al. [255] developed a proteomic method for identification of the species of origin based on detection of species-specific sarcoplasmic peptides. The technique used was a combination of 2-DE gel electrophoresis and MALDI-ToF-MS; in addition, LC-MS analysis was performed on the protein spots cut from gel.

2.4.9 Fruits and Vegetables

The classification of fruit and vegetable products can be favored in cases in which foodstuffs are characterized by a wide aromatic or polyphenolic profile. Examples are the characterization of tomato, hazelnuts, coffee, and other items with a marked flavor. Lo Feudo et al. [256] classified samples of fresh tomatoes and triple concentrate tomato pastes of different geographical origin according to volatile fraction profiling as determined with HS-SPME-GC-MS. With a similar metabolomic approach, Jandric et al. [257] classified fruit juices using ultra performance liquid chromatography (UPLC)-QToF MS analysis. Of particular interest are some recently issued studies that illustrated the potentialities of an analytical platform involving sample preparation by headspace SPME, separation by two-dimensional comprehensive GC-MS (GC × GC-MS) and data processing using advanced fingerprinting–profiling approaches. This platform was applied in the classification of coffee samples [258], hazelnuts [259], flavored fruit foods [260], and hazelnuts with comparison to hazelnut-containing sweet pastes [261].

Guo et al. [262] analyzed Chinese apple juices of different varieties and geographical origin with HS-SPME-GC-MS. A fingerprinting approach was followed, using the chromatographic profiles as nonspecific signals and avoiding an a priori identification of the volatile compounds separated. The application of stepwise LDA yielded satisfactory discrimination of the samples according to variety and geographical origin.

Sabatino et al. [263] developed a hyphenated method using HPLC with both diode array and MS detectors for authentication of saffron. The high specificity of MS allowed the unequivocal identification of characteristic markers of adulterants (e.g., safflower, marigold, turmeric, etc.).

With concern to coffee, Arruda et al. [264] analyzed with HS-SPME-GC-MS coffee samples harvested in different maturation stages and treated by different processes. Profiling of volatile compounds, with more than 110 molecules, allowed to discriminate both the maturity stage and processing type used. Garrett et al. [265] analyzed green beans of *arabica* coffee of different cultivars and geographic provenance by means of direct infusion electrospray ionization Fourier transform ion cyclotron resonance mass spectrometry (ESI FT-ICR MS) and a metabolomic approach. The application of PCA and PLS-DA pattern recognition methods to data from the identified metabolites allowed discriminating samples according to both genetic and geographical features.

2.4.10 Animal Products

Radovic et al. [266] applied the potential of Py-MS for the authentication of honey samples according to the botanical origin. Ten unifloral types were tested: acacia, chestnut, eucalyptus, heather, lime, rapeseed, sunflower, citrus, lavender, and rosemary. Using discriminant function analysis (DFA) as supervised pattern recognition method, the data obtained allowed a good discrimination among the different botanical sources but not among the geographical provenances. More recently, Wang et al. [267] used MALDI-ToF-MS to determine the geographic origin of honey samples according to the protein pattern, in order to overcome the limitations of other parameters such as volatile trace compounds, saccharides, or trace elements in the classification. Protein profiling has already shown its potential in the botanical classification of honey [268]. In this work, honey samples from Hawaii and other regions in the world were analyzed and their MALDI-ToF-MS protein ion mass spectra were transformed into protein ion mass spectral barcodes, which allowed a robust and accurate comparison with the mass spectral database library generated by authors with dedicated software. Classification with PCA of protein fingerprints, expressed as barcodes, allowed separating honeys of different geographical and floral origins.

2.4.11 Products from Cereals

Studies on cereals and bakery foods are not very common. Bianchi et al. [269] characterized the profile of volatile compounds in samples of PDO Altamura bread, a renowned bakery food produced in Apulia (Southern Italy). The influence of different baking modes was also evaluated. Beleggia et al. [270] determined with HS-SPMR-GC-MS the profiles of the volatile components in semolina and pasta samples obtained from four durum wheat cultivars, evidencing differences among the different cv employed. Dinelli et al. [271] used HPLC interfaced with ESI-TOF-MS to highlight differences in the profiles of phenolic compounds in modern and old common wheat varieties. Pattemore et al. [272] used MALDI-ToF-MS for identification of cereal varieties. Authors identified single nucleotide polymorphism (SNP)-based markers.

2.4.12 Organic Food

Levandi et al. [273] analyzed different varieties of conventionally grown spring and winter wheat and of organic wheat with an untargeted metabolomic strategy using HPLC-ESI-MS/MS. The results of pattern recognition treatment with PCA allowed discriminating between conventional and organic wheat varieties.

Zörb et al. suggested an analytical approach based on 2D gel electrophoresis coupled with MALDI-ToF-MS to determine the protein profile in wheat grains and to distinguish organic and conventional wheat products [274]. A selected number of proteins were isolated that could be useful as markers.

A field of particular concern is the one of *transgenic foods*, a term which is referred to food products containing or derived from genetically modified organisms (GMO). GMOs are the well-known organisms derived from recombinant DNA technology which have found wide applications in many fields, among which agriculture. The use of GMOs is of course prohibited in the production of organic foodstuffs in legislations all over the world, therefore proper analytical strategies are needed in order to recognize their fraudulent (or in many cases involuntary) application. These strategies rely mostly on gene expression profiling, protein profiling, or metabolic profiling, recalling the use of MS techniques. Examples of applications are on maize [275,276] and soybeans [277]. This matter has been recently reviewed by Valdés et al. [278].

A slightly different concept is one of the healthy products: this includes foods, dietary supplements, and nutraceutics that can positively affect human health and wellness. Phytochemicals have been proposed as health promoters. Several commercial products containing phytochemicals are submitted for evaluation of health claim. Evaluation must be sustained by an increase in the knowledge on phytochemicals' bioactivity and their impact in health, which can be obtained by means of analytical platforms based on MS techniques [279].

3. CONCLUSIONS AND FUTURE TRENDS

The contribution of mass spectrometry in the field of food forensics has been well demonstrated by the huge number of applications developed in the last years. Several approaches have been issued, concerning hyphenated systems versus stand-alone MS systems, different strategies (targeted vs nontargeted methods) and different ways of sample introduction. The advantages of MS techniques against traditional techniques are apparent in terms of robustness, completeness of information, and fastness of application, which means reduced costs of analysis also.

The prospects of the application of MS techniques to the classification of foodstuffs are almost limitless. There is no doubt that mass spectrometry had features well-suited to become the ideal tool for food certification. In particular, a huge improvement of knowledge in this field will be given by the close interaction of MS analysis, bioinformatics, availability of large databases, and all the advanced "omics" technologies. These potentialities could now be applied to build robust and reliable procedures for insertion in national and international regulations. Unfortunately in many countries, with particular concern to those with large-scale industry, the matter of food forensics has to face the interest of great producers. It would be hard to explain why, at present, only in few cases

MS techniques were indicated as methods of choice in official national regulations. At the end, therefore, limitations in the possible developments will be rather a politic-economic matter than a technical or scientific one.

ACKNOWLEDGMENTS

Author is indebted with Dr Massimo Baldizzone and Dr. Matteo Oddone for their invaluable help in carrying out research on food classification and for being such great friends.

REFERENCES

[1] Food and Agriculture Organization of the United Nations, Food Outlook - Biannual Report on Global Food Markets, http://www.fao.org/docrep/019/i3751e/i3751e.pdf.

[2] http://www.bbc.com/news/uk-21375594.

[3] http://www.repubblica.it/cronaca/2014/05/29/news/vini_falso_brunello_di_montalcino_sequestrate_30_mila_bottiglie-87525005/.

[4] http://www.theglobeandmail.com/technology/science/honey-laundering-the-sour-side-of-natures-golden-sweetener/article562759/.

[5] http://www.coldiretti.it/docindex/cncd/informazioni/753_07.shtm.

[6] http://www.coldiretti.it/News/Pagine/94--7-Febbraio-2014.aspx.

[7] http://www.fda.gov/Food/GuidanceRegulation/FSMA/ucm376478.htm?source=govdelivery&utm_medium=email&utm_source=govdelivery.

[8] S. Sumar, H. Ismail, Adulteration of foods – past and present, Nutr. Food Sci. 95 (1995) 11–15.

[9] Pliny the Elder, Natural History, the Loeb Classical Library, Harvard University Press, Cambridge, MA, USA, 1952.

[10] F. Accum, A Treatise on Adulterations of Food and Culinary Poisons: Exhibiting the Fraudulent Sophistications of Bread, Beer, Wine, Spirituous Liquors, Tea, Coffee, Cream, Confectionery, Vinegar, Mustard, Pepper, Cheese, Olive Oil, Pickles, and Other Articles Employed in Domestic Economy, and Methods of Detecting Them, Longman, Hurst, Rees, Orme, and Brown, London, 1820.

[11] J.C. Moore, J. Spink, M. Lipp, Development and application of a database of food ingrediente fraud and economically motivated adulteration from 1980 to 2010, J. Food Sci. 77 (2012) 118–126.

[12] USP, Food Chemical Codex, 2014–2016, ninth ed., US Pharmacopeial Convention, Rockville, MD (USA), 2014.

[13] M. Aceto, M. Baldizzone, M. Oddone, Keeping the track of quality: authentication and traceability studies on wine, in: P. O'Byrne (Ed.), Red Wine and Health, Nova Science Publishers, Hauppauge, NY, 2009, pp. 429–466.

[14] M. Forina, G. Drava, Chemometrics for wine. Applications, Analusis Mag. 25 (1997) M38–M42.

[15] M. Forina, S. Lanteri, C. Armanino, Chemometrics in food chemistry, in: Chemometrics and Species Identification, Topics in current chemistry series, vol. 141, Springer-Verlag, Berlin, 1987, pp. 91–143.

[16] N.E. Tzouros, I.S. Arvanitoyannis, Agricultural produces: synopsis of employed quality control methods for the authentication of foods and application of chemometrics for the classification of foods according to their variety or geographical origin, Crit. Rev. Food Sci. Nutr. 41 (2001) 287–319.

[17] Y. Chen, Y. Ni, Application of chemical pattern recognition techniques in food quality control, Chem. Res. Appl. 21 (2009) 1–7.

[18] S. Lingxia, C. Jinping, Z. Gaiming, L. Miaoyun, Research progress in application of chemometrics in food analysis, Sci. Technol. Food Ind. 33 (2012) 444–448.

[19] L.A. Berrueta, R.M. Alonso-Salces, K. Heberger, Supervised pattern recognition in food analysis, J. Chromatogr. A 1158 (2007) 196–214.

[20] M. Kozak, C.H. Scaman, Unsupervised classification methods in food sciences: discussion and outlook, J. Sci. Food Agric. 88 (2008) 1115–1127.

[21] J. Saurina, Characterization of wines using compositional profiles and chemometrics, Trends Anal. Chem. 29 (2010) 234–245.

[22] A. Versari, V.F. Laurie, A. Ricci, L. Laghi, G.P. Parpinello, Progress in authentication, typification and traceability of grapes and wines by chemometric approaches, Food Res. Int. 60 (2014) 2–18.

[23] P. Zachar, M. Soltes, R. Kasarda, J. Novotny, M. Novikmecova, D. Marcincakova, Analytical methods for the species identification of milk and milk products, Mljekarsvo 61 (2011) 199–207.

[24] R. Ben-Ayed, N. Kamoun-Grati, A. Rebai, An overview of the authentication of olive tree and oil, Compr. Rev. Food Sci. F. 12 (2013) 218–227.

[25] Council regulation (EEC) No 2092/91 of 24 June 1991 on organic production of agricultural products and indications referring thereto on agricultural products and foodstuffs, Off. J. Eur. Communities, L198 (July 22,1991) 1–15.

[26] A. Vlachos, I.S. Arvanitoyannis, P. Tserkezou, An updated review of meat authenticity methods and applications, Crit. Rev. Food Sci. Nutr. (2013). http://dx.doi.org/10.1080/10408398.2012.691573.

[27] M.Á. Sentandreu, E. Sentandreu, Authenticity of meat products: tools against fraud, Food Res. Int. 60 (2014) 19–29.

[28] http://oceana.org/sites/default/files/National_Seafood_Fraud_Testing_Results_FINAL.pdf.

[29] I. Lavilla, M. Costas-Rodriguez, C. Bendicho, Authentication of fishery products, in: M. de la Guardia, A. Gonzalvez (Eds.), Food Protected Designation of Origin: Methodologies and Applications, Comprehensive Analytical Chemistry, vol. 60, Elsevier, Amsterdam, 2013, pp. 657–717.

[30] I. Martinez, D. James, H. Loréal, Application of Modern Analytical Techniques to Ensure Seafood Safety and Authenticity, FAO Fisheries technical paper n.455, FAO, Rome, 2005.

[31] J.P. Melnyk, S. Wang, M.F. Marcone, Chemical and biological properties of the world's most expensive spice: Saffron, Food Res. Int. 43 (2010) 1981–1989.

[32] I.S. Arvanitoyannis, O.B. Vaitsi, A review on tomato authenticity: quality control methods in conjunction with multivariate analysis (chemometrics), Crit. Rev. Food Sci. Nutr. 47 (2007) 675–699.

[33] I.S. Arvanitoyannis, C. Chalhoub, P. Gotsiou, N. Lydakis-Simantiris, P. Kefalas, Novel quality control methods in conjunction with chemometrics (multivariate analysis) for detecting honey authenticity, Crit. Rev. Food Sci. Nutr. 45 (2005) 193–203.

[34] J.M. Camina, R.G. Pellerano, E.J. Marchevsky, Geographical and botanical classification of honeys and apicultural products by chemometric methods. A review, Curr. Anal. Chem. 8 (2012) 408–425.

[35] A. Vlachos, I.S. Arvanitoyannis, A review of rice authenticity/adulteration methods and results, Crit. Rev. Food Sci. Nutr. 48 (2008) 553–598.

[36] I.S. Arvanitoyannis, A. Vlachos, Maize authentication: quality control methods and multivariate analysis (chemometrics), Crit. Rev. Food Sci. Nutr. 49 (2009) 501–537.

[37] E. Capuano, R. Boerrigter-Eenling, G. van der Veer, S.M. van Ruth, Analytical authentication of organic products: an overview of markers, J. Sci. Food Agric. 93 (2013) 12–28.
[38] I.J. Colquhoun, M. Lees, Nuclear magnetic resonance spectroscopy, in: P.R. Ashurst, M.J. Dennis (Eds.), Analytical Methods of Food Authentication, Blackie Academic & Professional, London, 1998, pp. 36–75.
[39] C.N.G. Scotter, R. Wilson, Infrared spectroscopy, in: P.R. Ashurst, M.J. Dennis (Eds.), Analytical Methods of Food Authentication, Blackie Academic & Professional, London, 1998, pp. 76–96.
[40] S. Primrose, M. Woolfe, S. Rollinson, Food forensics: methods for determining the authenticity of foodstuffs, Trends Food Sci. Tech. 21 (2010) 582–590.
[41] D. Aiello, D. De Luca, E. Gionfriddo, A. Naccarato, A. Napoli, E. Romano, A. Russo, G. Sindona, A. Tagarelli, Multistage mass spectrometry in quality, safety and origin of foods, Eur. J. Mass Spectrom. 17 (2011) 1–31.
[42] S.A. Drivelos, C.A. Georgiou, Multi-element and multi-isotope-ratio analysis to determine the geographical origin of foods in the European Union, Trend Anal. Chem. 40 (2012) 38–51.
[43] S. Kelly, K. Heaton, J. Hoogewerff, Tracing the geographical origin of food: the application of multi-element and multi-isotope analysis, Trends Food Sci. Tech. 16 (2005) 555–567.
[44] D.M.A.M. Luykx, S.M. van Ruth, An overview of analytical methods for determining the geographical origin of food products, Food Chem. 107 (2008) 897–911.
[45] D. Sun (Ed.), Modern Techniques for Food Authentication, Elsevier, Amsterdam, 2008.
[46] A. Gonzalvez, S. Armenta, M. de la Guardia, Trace-element composition and stable-isotope ratio for discrimination of foods with protected designation of origin, Trends Anal. Chem. 28 (2009) 1295–1311.
[47] S.E. Ebeler, G.R. Takeoka, P. Winterhalter (Eds.), Progress in Authentication of Food and Wine, ACS Symposium Series, American Chemical Society, Washington, DC, 2011.
[48] Food protected designation of origin, in: M. de la Guardia, A. Gonzálvez (Eds.), Methodologies and Applications, Comprehensive Analytical Chemistry, vol. 60, Elsevier, Amsterdam, 2013.
[49] G.J. Martin, M.L. Martin, Deuterium labelling at the natural abundance level as studied by high field quantitative ^{2}H NMR, Tetrahedron Lett. 22 (1981) 3525–3528.
[50] V. Caer, M. Trierweiler, G.J. Martin, M.L. Martin, Determination of site-specific carbon isotope ratios at natural abundance by carbon-13 nuclear magnetic resonance spectroscopy, Anal. Chem. 63 (1991) 2306–2313.
[51] P.A. de Groot (Ed.), Handbook of Stable Isotope Analytical Techniques, Elsevier, Amsterdam, 2004.
[52] G. Gremaud, A. Hilkert, Isotopic-spectroscopic technique: stable isotope ratio mass spectrometry (IRMS), in: D. Sun (Ed.), Modern Techniques for Food Authentication, Elsevier, Amsterdam, 2008, pp. 269–320.
[53] W. Meier-Augenstein, Stable Isotope Forensics: An Introduction to the Forensic Application of Stable Isotope Analysis, John Wiley & Sons, Chichester, UK, 2010.
[54] Commission regulation (EC) No 2676/90 determining community methods for the analysis of wines, Off. J. Eur. Communities, L272 (October 3, 1990) 1–192.
[55] Commission Regulation (EC) No 822/97 of 6 May 1997 amending Regulation (EEC) No 2676/90 determining Community methods for the analysis of wines, Off. J. Eur. Communities L117 (07/05/1997) 10–12.
[56] Commission Regulation (EC) No 440/2003 of 10 March 2003 amending Regulation (EEC) No 2676/90 determining Community methods for the analysis of wines, Off. J. Eur. Communities L66 (11/03/2003) 15–23.

[57] OIV (Organisation International de la Vigne et du Vin), Determination by Isotope Ratio Mass Spectrometry of $^{13}C/^{12}C$ of Wine Ethanol or that Obtained through the Fermentation of Musts, Concentrated Musts or Grape Sugar, Resolution OENO 17/2001, Paris, 2001.
[58] OIV (Organisation International de la Vigne et du Vin), Determination of the Carbon Isotope Ratio $^{13}C/^{12}C$ of CO_2 in Sparkling Wines Method Using Isotope Ratio Mass Spectrometry (IRMS), Resolution OENO 7/2005, Paris, 2005.
[59] OIV (Organisation International de la Vigne et du Vin), Method for $^{18}O/^{16}O$ Isotope Ratio Determination of Water in Wine and Must, Resolution OENO 353/2009, Paris, 2009.
[60] Association of Analytical Communities (AOAC), Official Methods of Analysis Method 998.12: C-4 Plant Sugars in Honey, Internal Standard Stable Carbon Isotope Ratio Method, vol. 44, AOAC International, Gaithersburg, MD, USA, 2012, pp. 25–36.
[61] Association of Analytical Communities (AOAC), Official Methods of Analysis Method 2004.01: Carbon Stable Isotope Ratio of Ethanol Derived from Fruit Juices and Maple Syrups, vol. 44, AOAC International, Gaithersburg, MD, USA, 2012, p. 1.
[62] B.N. Smith, S. Epstein, Two categories of $^{13}C/^{12}C$ ratios for higher plants, Plant Physiol. 47 (1971) 380–384.
[63] L. Bontempo, F. Camin, L. Manzocco, G. Nicolini, R. Wehrens, L. Ziller, R. Larcher, Traceability along the production chain of Italian tomato products on the basis of stable isotopes and mineral composition, Rapid Commun. Mass Spectrom. 25 (2011) 899–909.
[64] K.A. Hobson, Tracing origins and migration of wildlife using stable isotopes: a review, Oecologia 120 (1999) 314–326.
[65] W. Meier-Augenstein, Stable Isotope Forensics: An Introduction to the Forensic Application of Stable Isotope Analysis, John Wiley & Sons, New York, 2010.
[66] A. Rossmann, Determination of stable isotope rations in food analysis, Food Rev. Int. 17 (2001) 347–381.
[67] G. Fortunato, K. Mumic, S. Wunderli, L. Pillonel, J.O. Bosset, G. Gremaud, Application of strontium isotope abundance ratios measured by MC-ICP-MS for food authentication, J. Anal. At. Spectrom. 19 (2004) 227–234.
[68] H. Förstel, The natural fingerprint of stable isotopes - use of IRMS to test food authenticity, Anal. Bioanal. Chem. 388 (2007) 541–544.
[69] S. Voerkelius, G.D. Lorenz, S. Rummel, C.R. Quétel, G. Heiss, M. Baxter, C. Brach-Papa, P. Deters-Itzelsberger, S. Hoelzl, J. Hoogewerff, E. Ponzevera, M. Van Bocxstaele, H. Ueckermann, Strontium isotopic signatures of natural mineral waters, the reference to a simple geological map and its potential for authentication of food, Food Chem. 118 (2010) 933–940.
[70] M. Katzenberg, Stable isotope analysis: a tool for studying past diet, demography, and life history, in: M. Katzenberg, S. Saunders (Eds.), Biological Anthropology of the Human Skeleton, Wiley-Liss, New York, 2008, pp. 413–442.
[71] H. Krueger, C. Sullivan, Models for carbon isotope fractionation between diet and bone, in: J.R. Turnlund, P.E. Johnson (Eds.), Stable Isotopes in Nutrition, ACS Symposium Series 258, American Chemical Society, Washington, DC, 1984, pp. 205–220.
[72] C. Kellner, M. Schoeninger, A simple carbon isotope model for reconstructing prehistoric human diet, Am. J. Phys. Anthropol. 133 (2007) 1112–1127.
[73] M. Schoeninger, M. DeNiro, Nitrogen and carbon isotopic composition of bone collagen from marine and terrestrial animals, Geochim. Cosmochim. Acta 48 (1984) 625–639.
[74] K. Killgrove, R.H. Tykot, Food for Rome: a stable isotope investigation of diet in the Imperial period (1st-3rd centuries AD), J. Anthropol. Archaeol. 32 (2012) 28–38.
[75] L.A. Gregoricka, S.G. Sheridan, Ascetic or affluent? Byzantine diet at the monastic community of St. Stephen's, Jerusalem from stable carbon and nitrogen isotopes, J. Anthropol. Archaeol. 32 (2013) 63–73.

[76] A. Rossmann, F. Reniero, I. Moussa, H.L. Schmidt, G. Versini, M.H. Merle, Stable oxygen isotope content of water of EU data-bank wines from Italy, France and Germany, Z. Lebensm. Unters. Forsch. A 208 (1999) 400–407.

[77] G.D. Farquhar, J.R. Ehleringer, K.T. Hubick, Carbon isotope discrimination and photosynthesis, Annu. Rev. Plant Physiol. Mol. Biol. 40 (1989) 503–537.

[78] J.M. Moreno Rojas, S. Cosofret, F. Reniero, C. Guillou, F. Serra, Control of oenological products: discrimination between different botanical sources of L-tartaric acid by isotope ratio mass spectrometry, Rapid Commun. Mass Spectrom. 21 (2007) 2447–2450.

[79] P. Horn, P. Schaaf, B. Holbach, S. Holzl, H. Eschnauer, $^{87}Sr/^{86}Sr$ from rock and soil in vine and wine, Z. Lebensm. Unters. Forsch. 196 (1993) 407–409.

[80] C.M.R. Almeida, M.T.S.D. Vasconcelos, Does the winemaking process influence the wine $^{87}Sr/^{86}Sr$? A case study, Food Chem. 85 (2004) 7–12.

[81] C. Durante, C. Baschieri, L. Bertacchini, M. Cocchi, S. Sighinolfi, M. Silvestri, A. Marchetti, Geographical traceability based on Sr-87/Sr-86 indicator: a first approach for PDO Lambrusco wines from Modena, Food Chem. 141 (2013) 2779–2787.

[82] S. Marchionni, E. Braschi, S. Tommasini, A. Bollati, F. Cifelli, N. Mulinacci, M. Mattei, S. Conticelli, High-precision Sr-87/Sr-86 analyses in wines and their use as a geological fingerprint for tracing geographic provenance, J. Agric. Food Chem. 61 (2013) 6822–6831.

[83] P. Martins, M. Madeira, F. Monteiro, R.B. De Sousa, A.S. Curvelo-Garcia, S. Catarino, Sr-87/Sr-86 ratio in vineyard soils from Portuguese denominations of origin and its potential for origin authentication, J. Int. Sci. Vigne Vin 48 (2014) 21–29.

[84] M. Barbaste, L. Halicz, A. Galy, B. Medina, H. Emteborg, F.C. Adams, R. Lobinski, Evaluation of the accuracy of the determination of lead isotope ratios in wine by ICP MS using quadrupole, multicollector magnetic sector and time-of-flight analyzers, Talanta 54 (2001) 307–317.

[85] C.M.R. Almeida, M.T.S.D. Vasconcelos, Determination of lead isotope ratios in port wine by inductively coupled plasma mass spectrometry after pre-treatment by UV-irradiation, Anal. Chim. Acta 396 (1999) 45–53.

[86] R. Larcher, G. Nicolini, P. Pangrazzi, Isotope ratios of lead in italian wines by inductively coupled plasma mass spectrometry, J. Agric. Food Chem. 51 (2003) 5956–5961.

[87] C. Vorster, L. Greeff, P.P. Coetzee, The determination of B-11/B-10 and Sr-87/Sr-86 isotope ratios by Quadrupole-based ICP-MS for the fingerprinting of South African wine, S. Afr. J. Chem. 63 (2010) 207–214.

[88] J. Hoefs, Stable Isotope Geochemistry, Springer-Verlag, Berlin, 2009.

[89] W.A. Simpkins, D. Rigby, Detection of illicit extension of potable spiritous liquors using ^{13}C:^{12}C ratios, J. Sci. Food Agr. 33 (1982) 898–903.

[90] A.W. Hilkert, C.B. Douthitt, H.J. Schlüter, W.A. Brand, Isotope ratio monitoring gas chromatography/mass spectrometry of D/H by high temperature conversion isotope ratio mass spectrometry, Rapid Commun. Mass Spectrom. 13 (1999) 1226–1230.

[91] I.G. Parker, S.D. Kelly, M. Sharman, M.J. Dennis, D. Howie, Investigation into the use of carbon isotope ratios ($^{13}C/^{12}C$) of Scotch whisky congeners to establish brand authenticity using gas chromatography-combustion-isotope ratio mass spectrometry, Food Chem. 63 (1998) 423–428.

[92] W. Meier-Augenstein, H.F. Kemp, S.M.L. Hardie, Detection of counterfeit Scotch whisky by ^{2}H and ^{18}O stable isotope analysis, Food Chem. 133 (2012) 1070–1074.

[93] C.N. Rhodes, K. Heaton, I. Goodall, P.A. Brereton, Gas chromatography carbon isotope ratio mass spectrometry applied to the detection of neutral alcohol in Scotch whisky: an internal reference approach, Food Chem. 114 (2009) 697–701.

[94] B. Raco, E. Dotsika, A. Cerrina Feroni, R. Battaglini, D. Poutoukis, Stable isotope composition of Italian bottled waters, J. Geochem. Explor. 124 (2013) 203–211.

[95] M. Scampicchio, T. Mimmo, C. Capici, C. Huck, N. Innocente, S. Drusch, S. Cesco, Identification of milk origin and process-induced changes in milk by stable isotope ratio mass spectrometry, J. Agric. Food Chem. 60 (2012) 11268–11273.
[96] M.A. Brescia, M. Monfreda, A. Buccolieri, C. Carrino, Characterisation of the geographical origin of buffalo milk and mozzarella cheese by means of analytical and spectroscopic determinations, Food Chem. 89 (2005) 139–147.
[97] L. Pillonel, R. Badertscher, P. Froidevaux, G. Haberhauer, S. Hölzl, P. Horn, A. Jakob, E. Pfammatter, U. Piantini, A. Rossmann, R. Tabacchi, J.O. Bosset, Stable isotope ratios, major, trace and radioactive elements in emmental cheeses of different origins, Lebensm.-Wiss. Technol. 36 (2003) 615–623.
[98] F. Camin, R. Larcher, G. Nicolini, L. Bontempo, D. Bertoldi, M. Perini, C. Schlicht, A. Schellenberg, F. Thomas, K. Heinrich, S. Voerkelius, M. Horacek, H. Ueckermann, H. Froeschl, B. Wimmer, G. Heiss, M. Baxter, A. Rossmann, J. Hoogewerff, Isotopic and elemental data for tracing the origin of European olive oils, J. Agric. Food Chem. 58 (2010) 570–577.
[99] J.E. Spangenberg, N. Ogrinc, Authentication of vegetable oils by bulk and molecular carbon isotope analyses with emphasis on olive oil and pumpkin seed oil, J. Agric. Food Chem. 49 (2001) 1534–1540.
[100] F. Schwagele, Traceability from a European perspective, Meat Sci. 71 (2005) 164–173.
[101] L. Hegerding, D. Seidler, H.J. Danneel, A. Gessler, B. Nowak, Oxygen isotope-ratio-analysis for the determination of the origin of beef, Fleischwirtschaft 82 (2002) 95–100.
[102] J.P. Renou, G. Bielicki, C. Deponge, P. Gachon, D. Micol, P. Ritz, Characterization of animal products according to geographic origin and feeding diet using nuclear magnetic resonance and isotope ratio mass spectrometry. Part II: beef meat, Food Chem. 86 (2004) 251–256.
[103] M. Boner, H. Förstel, Stable isotope variation as a tool to trace the authenticity of beef, Anal. Bioanal. Chem. 378 (2004) 301–310.
[104] B.M. Franke, G. Gremaud, R. Hadorn, M. Kreuzer, Geographic origin of meat-elements of an analytical approach to its authentication, Eur. Food Res. Technol. 221 (2005) 493–503.
[105] O. Schmidt, J.M. Quilter, B. Bahar, A.P. Maloney, C.M. Scrimgeour, I.S. Begley, F.J. Monahan, Inferring the origin and dietary history of beef from C, N and S stable isotope ratio analysis, Food Chem. 91 (2005) 545–549.
[106] E. Piasentier, R. Valusso, F. Camin, G. Versini, Stable isotope ratio analysis for authentication of lamb meat, Meat Sci. 64 (2003) 239–247.
[107] D. Sacco, M.A. Brescia, A. Buccolieri, A.C. Jambrenghi, Geographical origin and breed discrimination of Apulian lamb meat samples by means of analytical and spectroscopic determinations, Meat Sci. 71 (2005) 542–548.
[108] I. González-Martin, C. González-Pérez, J. Hernández Méndez, E. Marqués-Macias, F.S. Sanz-Poveda, Use of isotope analysis to characterize meat from Iberian-breed swine, Meat Sci. 52 (1999) 437–441.
[109] K. Heaton, S.D. Kelly, J. Hoogewerff, M. Woolfe, Verifying the geographical origin of beef: the application of multi-element isotope and trace element analysis, Food Chem. 107 (2008) 506–515.
[110] F.J. Winkler, Application of natural abundance stable isotope mass spectrometry in food control, in: A. Frigerio, H. Milon (Eds.), Chromatography and Mass Spectrometry in Nutrition Science and Food Safety, Elsevier, Amsterdam, Netherlands, 1984, pp. 173–190.
[111] A. Delgado, N. Garcia, δ15N and δ13C analysis to identify cattle fed on feed containing animal proteins: a safety/quality index in meat, milk and cheese, in: Proceedings of the 6th International Symposium on Food Authenticity and Safety, Nantes, France, 2001, pp. 1–11.

[112] B.L. Guo, Y.M. Wei, J.R. Pan, Y. Li, Stable C and N isotope ratio analysis for regional geographical traceability of cattle in China, Food Chem. 118 (2010) 915–920.

[113] M. Montowska, E. Pospiech, Is authentication of regional and traditional food made of meat possible? Crit. Rev. Food Sci. Nutr. 52 (2012) 475–487.

[114] Y. Zhao, B. Zhang, G. Chen, A. Chen, S. Yang, Z. Ye, Recent developments in application of stable isotope analysis on agro-product authenticity and traceability, Food Chem. 145 (2014) 300–305.

[115] A.T. Correia, F. Barrosa, A.N. Sial, Stock discrimination of European conger eel (*Conger conger* L.) using otolith stable isotope ratios, Fish Res. 108 (2011) 88–94.

[116] R. Barnett-Johnson, T.E. Pearson, F.C. Ramos, C.B. Grimes, R.B. MacFarlane, Tracking natal origins of salmon using isotopes, otoliths, and landscape geology, Limnol. Oceanogr. 53 (2008) 1633–1642.

[117] F. Thomas, E. Jamin, K. Wietzerbin, R. Guérin, M. Lees, E. Morvan, I. Billault, S. Derrien, J.M. Moreno Rojas, F. Serra, C. Guillou, M. Aursand, L. McEvoy, A. Prael, R.J. Robins, Determination of origin of Atlantic salmon (*Salmo salar*): the use of multiprobe and multielement isotopic analyses in combination with fatty acid composition to assess wild or farmed origin, J. Agric. Food Chem. 56 (2008) 989–997.

[118] K.A. Anderson, K.A. Hobbie, B.W. Smith, Chemical profiling with modeling differentiates wild and farm-raised salmon, J. Agric. Food Chem. 58 (2010) 11768–11774.

[119] G. Martin, G. Remaud, G.J. Martin, Food flavours - Generation, analysis, and process influence, in: C. Charalambous (Ed.), Proceedings of the 8th International Flavour Conference, Elsevier, Amsterdam, 1995, pp. 355–378.

[120] L. Maggi, M. Carmona, S.D. Kelly, N. Marigheto, G.L. Alonso, Geographical origin differentiation of saffron spice (*Crocus sativus* L. stigmas) - preliminary investigation using chemical and multi-element (H, C, N) stable isotope analysis, Food Chem. 128 (2011) 543–548.

[121] M. Brunner, R. Katona, Z. Stefanka, T. Prohanska, Determination of the geographical origin of processed spice using multielement and isotopic pattern on the example of Szegedi paprika, Eur. Food Res. Technol. 231 (2010) 623–634.

[122] M. Greule, L.D. Tumino, T. Kronewald, U. Hener, J. Schleucher, A. Mosandl, F. Keppler, Improved rapid authentication of vanillin using $\delta^{13}C$ and $\delta^{2}H$ values, Eur. Food Res. Technol. 231 (2010) 933–941.

[123] J. Bricout, J.C. Fontes, Analytical distinction between cane and beet sugar, Ann. Fals. Exp. Chim. 67 (1974) 211–215.

[124] M. Gensler, A. Rossmann, H.L. Schmidt, Detection of added L-ascorbic acid in fruit juices by isotope ratio mass spectrometry, J. Agric. Food Chem. 308 (1995) 299–307.

[125] J. Koziet, A. Rossmann, G.J. Martin, P. Johnson, Determination of the oxygen-18 and deuterium content of fruit and vegetable juice water. A European inter-laboratory comparison study, Anal. Chim. Acta 302 (1995) 29–37.

[126] F. Guyon, P. Auberger, L. Gaillard, C. Loublanches, M. Viateau, N. Sabathié, M.H. Salagoïty, B. Médina, $^{13}C/^{12}C$ isotope ratios of organic acids, glucose and fructose determined by HPLC-co-IRMS for lemon juices authenticity, Food Chem. 146 (2014) 36–40.

[127] T.S. Pilgrim, R.J. Watlinga, K. Grice, Application of trace element and stable isotope signatures to determine the provenance of tea (*Camellia sinensis*) samples, Food Chem. 118 (2010) 921–926.

[128] C. Rodrigues, C. Máguas, T. Prohaska, Strontium and oxygen isotope fingerprinting of green coffee beans and its potential to proof authenticity of coffee, Eur. Food Res. Technol. 232 (2011) 361–373.

[129] H. Liu, C. You, C. Chen, Y. Liu, M. Chung, Geographic determination of coffee beans using multi-element analysis and isotope ratios of boron and strontium, Food Chem. 142 (2014) 439–445.

[130] S. Rummel, S. Hoelzl, P. Horn, A. Rossmann, C. Schlicht, The combination of stable isotope abundance ratios of H, C, N and S with (87)Sr/(86)Sr for geographical origin assignment of orange juices, Food Chem. 118 (2010) 890–900.

[131] K.M. Rogers, Stable isotopes as a tool to differentiate eggs laid by caged, barn, free range, and organic hens, J. Agric. Food Chem. 57 (2009) 4236–4242.

[132] M.J. De Niro, S. Epstein, Influence of diet on the distribution of nitrogen isotopes in animals, Geochim. Cosmochim. Acta 45 (1981) 341–351.

[133] L. Rock, The use of stable isotope techniques in egg authentication schemes: a review, Trends Food Sci. Tech. 28 (2012) 62–68.

[134] A. Schellenberg, S. Chmielus, C. Schlicht, F. Camin, M. Perini, L. Bontempo, K. Heinrich, S.D. Kelly, A. Rossmann, F. Thomas, E. Jamin, M. Horacek, Multielement stable isotope ratios (H, C, N, S) of honey from different European regions, Food Chem. 121 (2010) 770–777.

[135] J.W. White, K. Winters, P. Martin, A. Rossman, Stable carbon isotope ratio analysis of honey: validation of internal standard procedure for worldwide application, J. Assoc. Off. Anal. Chem. Int. 81 (1998) 610–619.

[136] A. Guler, H. Kocaokutgen, A.V. Garipoglu, H. Onder, D. Ekinci, S. Biyik, Detection of adulterated honey produced by honeybee (*Apis mellifera* L.) colonies fed with different levels of commercial industrial sugar (C3 and C4 plants) syrups by the carbon isotope ratio analysis, Food Chem. 155 (2014) 155–160.

[137] G. Daniele, D. Maitre, H. Casabianca, Identification, quantification and carbon stable isotopes determinations of organic acids in monofloral honeys. A powerful tool for botanical and authenticity control, Rapid Commun. Mass Spectrom. 26 (2012) 1993–1998.

[138] H. Oda, A. Kawasaki, T. Hirata, Determination of the geographic origin of brown-rice with isotope ratios of 11B/10B and 87Sr/86Sr, Anal. Sci. 17 (2001) i1627–i1630.

[139] T. Korenaga, M. Musashi, R. Nakashita, Y. Suzuki, Statistical analysis of rice samples for compositions of multiple light elements (H, C, N, and O) and their stable isotopes, Anal. Sci. 26 (2010) 873–878.

[140] K. Ariyama, M. Shinozaki, A. Kawasaki, Determination of the geographic origin of rice by chemometrics with strontium and lead isotope ratios and multielement concentrations, J. Agric. Food Chem. 60 (2012) 1628–1634.

[141] F. Camin, M. Perini, L. Bontempo, S. Fabroni, W. Faedi, S. Magnani, G. Baruzzi, M. Bonoli, M.R. Tabilio, S. Musmeci, A. Rossmann, S.D. Kelly, P. Rapisarda, Potential isotopic and chemical markers for characterising organic fruits, Food Chem. 125 (2011) 1072–1082.

[142] K.H. Laursen, A. Mihailova, S.D. Kelly, V.N. Epov, S. Bérail, J.K. Schjoerring, O.F.X. Donard, E.H. Larsen, N. Pedentchouk, A.D. Marca-Bell, U. Halekoh, J.E. Olesen, S. Husted, Is it really organic? - multi-isotopic analysis as a tool to discriminate between organic and conventional plants, Food Chem. 141 (2013) 2812–2820.

[143] A. Mihailova, N. Pedentchouk, S.D. Kelly, Stable isotope analysis of plant-derived nitrate - novel method for discrimination between organically and conventionally grown vegetables, Food Chem. 154 (2014) 238–245

[144] C.N. Rhodes, J.H. Lofthouse, S. Hird, P. Rose, P. Reece, J. Christy, R. Macarthur, P.A. Brereton, The use of stable carbon isotopes to authenticate claims that poultry have been corn-fed, Food Chem. 118 (2010) 927–932.

[145] I. Chung, I. Park, J. Yoon, Y. Yang, S. Kim, Determination of organic milk authenticity using carbon and nitrogen natural isotopes, Food Chem. 160 (2014) 214–218.

[146] W.M. White, Geochemistry, Wiley-Blackwell, New York, 2013.

[147] E. McCurdy, D. Potter, M. Medina, Trace elements in wine, Lab. News 9 (1992) 10–11.

[148] A. Masuda, Regularities in variation of relative abundances of lanthanide elements and an attempt to analyze separation index patterns of some minerals, J. Earth Sci. 10 (1962) 173–187.

[149] C.D. Coryell, J.W. Chase, J.W. Winchester, A procedure for geochemical interpretation of terrestrial rare-earth abundance patterns, J. Geophys. Res. 68 (1963) 559–566.

[150] T. Liang, S. Ding, W. Song, Z. Chong, C. Zhang, H. Li, A review of fractionations of rare earth elements in plants, J. Rare Earths 26 (2008) 7–15.

[151] P.H. Brown, A.H. Rathjen, R.D. Graham, D.E. Tribe, Rare earth elements in biological systems, in: in: K.A. Gschneider, L. Eyring (Eds.), Handbook on the Physics and Chemistry of Rare Earths, vol. 13, Elsevier, Amsterdam, 1990, pp. 423–452.

[152] G. Tyler, Rare earth elements in soil and plant systems: a review, Plant Soil 267 (2004) 191–206.

[153] J.D. Greenough, H.P. Longerich, S.E. Jackson, Trace element concentrations in wines by ICP-MS: evidence for the role of solubility in determining uptake by plants, Can. J. Appl. Spectrosc. 41 (1996) 76–80.

[154] V.F. Taylor, H.P. Longerich, J.D. Greenough, Geology and wine 5. – provenance of Okanagan Valley wines, British Columbia, using trace elements: Promise and limitations, Geosci. Can. 29 (2002) 110–120.

[155] G. Thiel, G. Geisler, I. Blechschmidt, Determination of trace elements in wines and classification according to their provenance, Anal. Bioanal. Chem. 378 (2004) 1630–1636.

[156] P. Pohl, What do metals tell us about wine? Trends Anal. Chem. 26 (2007) 941–949.

[157] M.J. Latorre, C. Garcia-Jares, B. Medina, C. Herrero, Pattern recognition analysis applied to classification of wines from Galicia (Northwestern Spain) with Certified Brand of Origin, J. Agric. Food Chem. 43 (1994) 1451–1455.

[158] M.J. Baxter, H.M. Crews, M.J. Dennis, I. Goodall, D. Anderson, The determination of the authenticity of wine from its trace element composition, Food Chem. 60 (1997) 443–450.

[159] E. Marengo, M. Aceto, Statistical investigation of the differences in the distribution of metals in Nebbiolo-based wines, Food Chem. 81 (2003) 621–630.

[160] V.F. Taylor, H.P. Longerich, J.D. Greenough, Multielement analysis of Canadian wines by Inductively Coupled Plasma Mass Spectrometry (ICP-MS) and multivariate statistics, J. Agric. Food Chem. 51 (2003) 856–860.

[161] M. Oddone, E. Robotti, M. Marengo, M. Baldizzone, M. Aceto, Studio della tracciabilità sulla filiera del vino mediante determinazione dei lantanidi con ICP-MS, in: J.D. Coisson, M. Arlorio, A. Martelli (Eds.), Proceedings of VI Italian Congress "Chimica degli Alimenti", Taro, Alessandria, 2007, pp. 573–577.

[162] M. Aceto, E. Robotti, M. Oddone, M. Baldizzone, G. Bonifacino, G. Bezzo, R. Di Stefano, F. Gosetti, E. Mazzucco, M. Manfredi, E. Marengo, A traceability study on the Moscato wine chain, Food Chem. 138 (2013) 1914–1922.

[163] A. Gonzálvez, M. de la Guardia, Mineral profile, in: M. de la Guardia, A. Gonzalvez (Eds.), Food Protected Designation of Origin: Methodologies and Applications, Comprehensive Analytical Chemistry, vol. 60, Elsevier, Amsterdam, 2013, pp. 51–76.

[164] C. Benincasa, J. Lewis, G. Sindona, A. Tagarelli, The use of multi element profiling to differentiate between cow and buffalo milk, Food Chem. 110 (2008) 257–262.

[165] C. Benincasa, J. Lewis, E. Perri, G. Sindona, A. Tagarelli, Determination of trace element in Italian virgin olive oils and their characterization according to geographical origin by statistical analysis, Anal. Chim. Acta 585 (2007) 366–370.

[166] E.J. Llorent-Martínez, P. Ortega-Barrales, M.L. Fernández-de Córdova, A. Domínguez-Vidal, A. Ruiz-Medina, Investigation by ICP-MS of trace element levels in vegetable edible oils produced in Spain, Food Chem. 127 (2011) 1257–1262.

[167] K.J. Hintze, G.P. Lardy, M.J. Marchello, J.W. Finley, J. Agric. Food Chem. 49 (2001) 1062–1067.

[168] K.J. Hintze, G.P. Lardy, M.J. Marchello, J.W. Finley, J. Agric. Food Chem. 50 (2002) 3938–3942.

[169] S.L. Bruce, B.N. Noller, A.H. Grigg, B.F. Mullen, D.R. Mulligan, P.J. Ritchie, N. Currey, J.C. Ng, A field study conducted at Kidston Gold Mine to evaluate the impact of arsenic and zinc from mine tailing to grazing cattle, Toxicol. Lett. 137 (2003) 23–34.

[170] G. Chessa, G. Calaresu, G. Ledda, M.C. Testa, A. Orrù, Lead, zinc and cadmium in biological tissues of sheep bred in a polluted area, in: B. Markert, K. Friese (Eds.), Trace Elements-Their Distribution and Effect in the Environment, Elsevier, Amsterdam, 2000, pp. 479–483.

[171] M. Costas-Rodríguez, I. Lavilla, C. Bendicho, Classification of cultivated mussels from Galicia (Northwest Spain) with European Protected Designation of Origin using trace element fingerprint and chemometric analysis, Anal. Chim. Acta 664 (2010) 121–128.

[172] F. Cubadda, A. Raggi, E. Coni, Element fingerprinting of marine organisms by dynamic reaction cell inductively coupled plasma mass spectrometry, Anal. Bioanal. Chem. 384 (2006) 887–896.

[173] L. Guo, L. Gong, Y. Yu, H. Zhang, Multi-element fingerprinting as a tool in origin authentication of four East China Marine species, J. Food Sci. 78 (2013) C1852–C1857.

[174] K. Ariyama, T. Nishida, T. Noda, M. Kadokura, A. Yasui, Effects of fertilization, crop year, variety, and provenance factors on mineral concentrations in onions, J. Agric. Food Chem. 54 (2006) 3341–3350.

[175] K. Ariyama, Y. Aoyama, A. Mochizuki, Y. Homura, M. Kadokura, A. Yasui, Determination of the geographic origin of onions between three main production areas in japan and other countries by mineral composition, J. Agric. Food Chem. 55 (2007) 347–354.

[176] A.A. D'Archivio, A. Giannitto, A. Incani, S. Nisi, Analysis of the mineral composition of Italian saffron by ICP-MS and classification of geographical origin, Food Chem. 157 (2014) 485–489.

[177] M. Bettinelli, S. Spezia, C. Baffi, G.M. Beone, R. Rocchetta, A. Nassisi, ICP-MS determination of REEs in tomato plants and related products: a new analytical tool to verify traceability, At. Spectrosc. 26 (2005) 41–50.

[178] G. Lo Feudo, A. Naccarato, G. Sindona, A. Tagarelli, Investigating the origin of tomatoes and triple concentrated tomato pastes through multielement determination by inductively coupled plasma mass spectrometry and statistical analysis, J. Agric. Food Chem. 58 (2010) 3801–3807.

[179] M. Oddone, M. Aceto, M. Baldizzone, D. Musso, D. Osella, Authentication and traceability study of hazelnuts from Piedmont, Italy, J. Agric. Food Chem. 57 (2009) 3404–3408.

[180] H. Benabdelkamel, L. Di Donna, F. Mazzotti, A. Naccarato, G. Sindona, A. Tagarelli, D. Taverna, Authenticity of PGI "Clementine of Calabria" by multielement fingerprint, J. Agric. Food Chem. 60 (2012) 3717–3726.

[181] E. Furia, A. Naccarato, G. Sindona, G. Stabile, A. Tagarelli, Multielement fingerprinting as a tool in origin authentication of PGI food products: Tropea Red Onion, J. Agric. Food Chem. 59 (2011) 8450–8457.

[182] A. Moreda-Piñeiro, A. Fisher, S.J. Hill, The classification of tea according to region of origin using pattern recognition techniques and trace metal data, J. Food Compos. Anal. 16 (2003) 195–211.

[183] M. Ebert, R. Islam, Caviar and caviar imitations - adulterations of sturgeon roe provable by histological method, Fleischwirtschaft 87 (2007) 124–126.

[184] I. Rodushkin, T. Bergman, G. Douglas, E. Engström, D. Sörlin, D.C. Baxter, Authentication of Kalix (N.E. Sweden) vendace caviar using inductively coupled plasma-based analytical techniques: evaluation of different approaches, Anal. Chim. Acta 583 (2007) 310–318.

[185] G.E. Bath, S.R. Thorrold, C.M. Jones, S.E. Campana, J.W. McLaren, J.W. Lam, Strontium and barium uptake in aragonitic otoliths of marine fish, Geochim. Cosmochim. Acta 64 (2000) 1705–1714.

[186] B.L. Batista, L.R.S. da Silva, B.A. Rocha, J.L. Rodrigues, A.A. Berretta-Silva, T.O. Bonates, V.S.D. Gomes, R.M. Barbosa, F. Barbosa, Multi-element determination in Brazilian honey samples by inductively coupled plasma mass spectrometry and estimation of geographic origin with data mining techniques, Food Res. Int. 49 (2012) 209–215.

[187] M. Chudzinska, D. Baralkiewicz, Application of ICP-MS method of determination of 15 elements in honey with chemometric approach for the verification of their authenticity, Food Chem. Toxicol. 49 (2011) 2741–2749.

[188] H. Chen, C. Fan, Q. Chang, G. Pang, X. Hu, M. Lu, W. Wang, Chemometric determination of the botanical origin for Chinese honeys on the basis of mineral elements determined by ICP-MS, J. Agric. Food Chem. 62 (2014) 2443–2448.

[189] S. Shen, L. Xia, N. Xiong, Z. Liu, H. Sun, Determination of the geographic origin of rice by element fingerprints and correlation analyses with the soil of origin, Anal. Methods 5 (2013) 6177–6185.

[190] H. Zhao, B. Guo, Y. Wei, B. Zhang, Effects of wheat origin, genotype, and their interaction on multielement fingerprints for geographical traceability, J. Agric. Food Chem. 60 (2012) 10957–10962.

[191] K.H. Laursen, J.K. Schjoerring, J.E. Olesen, M. Askegaard, U. Halekoh, S. Husted, Multi-elemental fingerprinting as a tool for authentication of organic wheat, barley, faba bean, and potato, J. Agric. Food Chem. 59 (2011) 4385–4396.

[192] S.D. Kelly, A.S. Bateman, Comparison of mineral concentrations in commercially grown organic and conventional crops - tomatoes (*Lycopersicon esculentum*) and lettuces (*Lactuca sativa*), Food Chem. 119 (2010) 738–745.

[193] M. Herrero, C. Simó, V. García-Cañas, E. Ibáñez, A. Cifuentes, Foodomics: MS-based strategies in modern food science and nutrition, Mass Spectrom. Rev. 31 (2012) 49–69.

[194] X. Wang, S. Wang, Z. Cai, The latest developments and applications of mass spectrometry in food-safety and quality analysis, Trends Anal. Chem. 52 (2013) 170–185.

[195] M.M.W.B. Hendriks, F.A. Eeuwijk, R.H. Jellema, J.A. Westerhuis, T.H. Reijmers, H.C.J. Hoefsloot, A.K. Smilde, Data-processing strategies for metabolomics studies, Trends Anal. Chem. 30 (2011) 1685–1698.

[196] S. Esslinger, J. Riedl, C. Fauhl-Hassek, Potential and limitations of non-targeted fingerprinting for authentication of food in official control, Food Res. Int. 60 (2014) 189–204.

[197] A. Cifuentes, Foodomics. Advanced Mass Spectrometry in Modern Food Science and Nutrition, John Wiley & Sons, New York, 2013.

[198] F. Toldrá, L.M.L. Nollet (Eds.), Proteomics in Foods. Principles and Applications, Springer, Berlin, 2013.

[199] S. Sforza (Ed.), Food Authentication Using Bioorganic Molecules, DEStech Publications, Lancaster, PA (USA), 2013.

[200] G. Mammone, G. Picariello, S. Caira, F. Addeo, P. Ferranti, Analysis of food proteins and peptides by mass spectrometry-based techniques, J. Chromatogr. A 1216 (2009) 7130–7142.

[201] J.M. Gallardo, I. Ortea, M. Carrera, Proteomics and its applications for food authentication and food-technology research, Trends Anal. Chem. 52 (2013) 135–141.

[202] E. Trujillo, C. Davis, J. Millner, Nutrigenomics, proteomics, metabolomics, and the practice of dietetics, J. Am. Diet. Assoc. 106 (2006) 403–413.

[203] S.G. Oliver, M.K. Winson, D.B. Kell, F. Baganz, Systematic functional analysis of the yeast genome, Trends Biotechnol. 16 (1998) 373–378.

[204] O. Fiehn, Metabolomics–the link between genotypes and phenotypes, Plant Mol. Biol. 48 (2002) 155–171.

[205] K. Dettmer, P.A. Aronov, B.D. Hammock, Mass spectrometry-based metabolomics, Mass Spectrom. Rev. 26 (2007) 51–78.

[206] J.P. Antignac, F. Courant, G. Pinel, E. Bichon, F. Monteau, C. Elliott, B. Le Bizec, Mass spectrometry-based metabolomics applied to the chemical safety of food, Trends Anal. Chem. 30 (2011) 292–301.

[207] C. Hu, G. Xu, Mass-spectrometry-based metabolomics analysis for foodomics, Trends Anal. Chem. 52 (2013) 36–46.

[208] M. Castro-Puyana, M. Herrero, Metabolomics approaches based on mass spectrometry for food safety, quality and traceability, Trends Anal. Chem. 52 (2013) 74–87.

[209] C. Ibáñez, V. García-Cañas, A. Valdés, C. Simó, Novel MS-based approaches and applications in food metabolomics, Trends Anal. Chem. 52 (2013) 100–111.

[210] G. Oms-Oliu, I. Odriozola-Serrano, O. Martín-Belloso, Metabolomics for assessing safety and quality of plant-derived food, Food Res. Int. 54 (2013) 1172–1183.

[211] E. Cubero-Leon, R. Peñalver, A. Maquet, Review on metabolomics for food authentication, Food Res. Int. 60 (2014) 95–107.

[212] A. Soria, A. Ruiz-Matute, M. Sanz, I. Martínez-Castro, Chromatographic technique: Gas chromatography (GC), in: D. Sun (Ed.), Modern Techniques for Food Authentication, Elsevier, Amsterdam, 2008, pp. 321–359.

[213] A.K. Malik, C. Blasco, Y. Pico, Liquid chromatography-mass spectrometry in food safety, J. Chromatogr. A 1217 (2010) 4018–4040.

[214] V. Di Stefano, G. Avellone, D. Bongiorno, V. Cunsolo, V. Muccilli, S. Sforza, A. Dossena, L. Drahos, K. Vekey, Applications of liquid chromatography-mass spectrometry for food analysis, J. Chromatogr. A 1259 (2012) 74–85.

[215] C. Ibáñez, C. Simó, V. García-Cañas, A. Cifuentes, M. Castro-Puyana, Metabolomics, peptidomics and proteomics applications of capillary electrophoresis-mass spectrometry in Foodomics: a review, Anal. Chim. Acta 802 (2013) 1–13.

[216] V. García-Cañas, C. Simó, M. Castro-Puyana, A. Cifuentes, Recent advances in the application of capillary electromigration methods for food analysis and Foodomics, Electrophoresis 35 (2014) 147–169.

[217] B. Plutowska, W. Wardencki, Aromagrams - aromatic profiles in the appreciation of food quality, Food Chem. 101 (2007) 845–872.

[218] M. Śliwińska, P. Wiśniewska, T. Dymerski, J. Namieśnik, W. Wardencki, Food analysis using artificial senses, J. Agric. Food Chem. 62 (2014) 1423–1448.

[219] E. Marengo, M. Aceto, V. Maurino, Classification of Nebbiolo-based wines from Piedmont (Italy) by means of solid-phase microextraction-gas chromatography-mass spectrometry of volatile compounds, J. Chromatogr. A 943 (2002) 123–137.

[220] L. Vaclavik, O. Lacina, J. Hajslova, J. Zweigenbaum, The use of high performance liquid chromatography–quadrupole time-of-flight mass spectrometry coupled to advanced data mining and chemometric tools for discrimination and classification of red wines according to their variety, Anal. Chim. Acta 685 (2011) 45–51.

[221] L. Jaitz, K. Siegl, R. Eder, G. Rak, L. Abranko, G. Koellensperger, S. Hann, LC–MS/MS analysis of phenols for classification of red wine according to geographic origin, grape variety and vintage, Food Chem. 122 (2010) 366–372.

[222] D. Serrano-Lourido, J. Saurina, S. Hernández-Cassou, A. Checa, Classification and characterisation of Spanish red wines according to their appellation of origin based on chromatographic profiles and chemometric data analysis, Food Chem. 135 (2012) 1425.

[223] E. Mattarucchi, M. Stocchero, J.M. Moreno-Rojas, G. Giordano, F. Reniero, C. Guillou, Authentication of trappist beers by LC-MS fingerprints and multivariate data analysis, J. Agric. Food Chem. 58 (2010) 12089–12095.

[224] Z. Szilágyi, G. Vas, G. Mády, K. Vékey, Investigation of macromolecules in wines by matrix-assisted laser desorption/ionization time-of-flight mass spectrometry., Rapid Commun. Mass Spectrom. 10 (1996) 1141–1143.

[225] A. Chambery, G. del Monaco, A. Di Maro, A. Parente, Peptide fingerprint of high quality Campania white wines by MALDI-TOF mass spectrometry, Food Chem. 113 (2009) 1283–1289.

[226] J.D. Nunes-Miranda, G. Igrejas, E. Araujo, M. Reboiro-Jato, J.L. Capelo, Mass Spectrometry-based fingerprinting of proteins & peptides in wine quality control: a critical overview, Crit. Rev. Food Sci. Nutr. 53 (2013) 751–759.

[227] H. Fulcrand, C. Mané, S. Preys, G. Mazerolles, C. Bouchut, J.P. Mazauric, J.M. Souquet, E. Meudec, Y. Li, R.B. Cole, V. Cheynier, Direct mass spectrometry approaches to characterize polyphenol composition of complex samples, Phytochemistry 69 (2008) 3131–3138.

[228] W. Cynkar, R. Dambergs, P. Smith, D. Cozzolino, Classification of Tempranillo wines according to geographic origin, Anal. Chim. Acta 660 (2010) 227–231.

[229] S. Capone, M. Tufariello, L. Francioso, G. Montagna, F. Casino, A. Leone, P. Siciliano, Aroma analysis by GC/MS and electronic nose dedicated to Negroamaro and Primitivo typical Italian Apulian wines, Sensor Actuat. B-Chem. 179 (2013) 259–269.

[230] L. Vera, L. Acena, J. Guasch, R. Boque, M. Mestres, O. Busto, Characterization and classification of the aroma of beer samples by means of an MS e-nose and chemometric tools, Anal. Bioanal. Chem. 399 (2011) 2073–2081.

[231] H.H. Jelen, A. Ziolkowska, A. Kaczmarek, Identification of the botanical origin of raw spirits produced from rye, potato, and corn based on volatile compounds analysis using a SPME-MS method, J. Agric. Food Chem. 58 (2010) 12585–12591.

[232] M. Pilar Marti, R. Boque, O. Busto, J. Guasch, Electronic noses in the quality control of alcoholic beverages, Trends Anal. Chem. 24 (2005) 57–66.

[233] T. Cajka, K. Riddellova, M. Tomaniova, J. Hajslova, Recognition of beer brand based on multivariate analysis of volatile fingerprint, J. Chromatogr. A 1217 (2010) 4195–4203.

[234] O. Šedo, I. Márová, Z. Zdráhal, Beer fingerprinting by matrix-assisted laser desorption-Ionisation-time of flight mass spectrometry, Food Chem. 135 (2012) 473–478.

[235] R.I. Aylott, A.H. Clyne, A.P. Fox, D.A. Walker, Analytical strategies to confirm Scotch Whisky authenticity, Analyst 119 (1994) 1741–1746.

[236] J.K.S. Møller, R.R. Catharino, M.N. Eberlin, Electrospray ionization mass spectrometry fingerprinting of whisky: immediate proof of origin and authenticity, Analyst 130 (2005) 890–897.

[237] J.S. Garcia, B.G. Vaz, Y.E. Corilo, C.F. Ramires, S.A. Saraiva, G.B. Sanvido, E.M. Schmidt, D.R.J. Maia, R.G. Cosso, J.J. Zacca, M. Nogueira Eberlin, Whisky analysis by electrospray ionization-Fourier transform mass spectrometry, Food Res. Int. 51 (2013) 98–106.

[238] M. Cocchi, C. Durante, A. Marchetti, C. Armanino, M. Casale, Characterization and discrimination of different aged 'Aceto Balsamico Tradizionale di Modena' products by head space mass spectrometry and chemometrics, Anal. Chim. Acta 589 (2007) 96–104.

[239] N. Nicolaou, Y. Xu, R. Goodacre, MALDI-MS and multivariate analysis for the detection and quantification of different milk species, Anal. Bioanal. Chem. 399 (2011) 3491–3502.

[240] Y. Yang, N. Zheng, J. Yang, D. Bu, J. Wang, L. Ma, P. Sun, Animal species milk identification by comparison of two-dimensional gel map profile and mass spectrometry approach, Int. Dairy J. 35 (2014) 15–20.

[241] V. Cunsolo, V. Muccilli, R. Saletti, S. Foti, MALDI-TOF mass spectrometry for the monitoring of she-donkey's milk contamination or adulteration, J. Mass Spectrom. 48 (2013) 148–153.

[242] V. Cunsolo, V. Muccilli, R. Saletti, S. Foti, Review: applications of mass spectrometry techniques in the investigation of milk proteome, Eur. J. Mass. Spectrom. 17 (2011) 305–320.

[243] L. Di Donna, F. Mazzotti, A. Naccarato, R. Salerno, A. Tagarelli, D. Taverna, G. Sindona, Secondary metabolites of *Olea europaea* leaves as markers for the discrimination of cultivars and cultivation zones by multivariate analysis, Food Chem. 121 (2010) 492–496.

[244] G.J. Salter, M. Lazzari, L. Giansante, R. Goodacre, A. Jones, G. Surricchio, D.B. Kell, G. Bianchi, Determination of the geographical origin of Italian extra virgin olive oil using pyrolysis mass spectrometry, J. Anal. Appl. Pyrol. 40/41 (1997) 159–170.

[245] L. Vaclavik, T. Cajka, V. Hrbek, J. Hajslova, Ambient mass spectrometry employing direct analysis in real time (DART) ion source for olive oil quality and authenticity assessment, Anal. Chim. Acta 645 (2009) 56–63.

[246] C. Oliveros, R. Boggia, R. Casale, C. Armanino, M. Forina, Optimisation of a new headspace mass spectrometry instrument discrimination of different geographical origin olive oils, J. Chromatogr. A 1076 (2005) 7–15.

[247] M.S. Cosio, D. Ballabio, S. Benedetti, C. Gigliotti, Geographical origin and authentication of extra virgin olive oils by an electronic nose in combination with artificial neural networks, Anal. Chim. Acta 567 (2006) 202–210.

[248] M. Narvaez-Rivas, I.M. Vicario, M.J. Alcalde, M. Leon-Camacho, Volatile hydrocarbon profile of Iberian dry-cured hams. A possible tool for authentication of hams according to the fattening diet, Talanta 81 (2010) 1224–1228.

[249] N. Zaima, N. Goto-Inoue, T. Hayasaka, H. Enomoto, M. Setou, Authenticity assessment of beef origin by principal component analysis of matrix-assisted laser desorption/ionization mass spectrometric data, Anal. Bioanal. Chem. 400 (2001) 1865–1871.

[250] J.S. del Pulgar, C. Soukoulis, F. Biasioli, L. Cappellin, C. García, F. Gasperi, P. Granitto, T.D. Märk, E. Piasentier, E. Schuhfried, Rapid characterization of dry cured ham produced following different PDOs by proton transfer reaction time of flight mass spectrometry (PTR-ToF-MS), Talanta 85 (2011) 386–393.

[251] R. González-Dominguez, T. García-Barrera, J.L. Gómez-Ariza, Iberian ham typification by direct infusion electrospray and photospray ionization mass spectrometry fingerprinting, Rapid Commun. Mass Spectrom. 26 (2012) 835–844.

[252] D. Indrasti, Y.B.C. Man, S. Mustafa, D.M. Hashim, Lard detection based on fatty acids profile using comprehensive gas chromatography hyphenated with time-of-flight mass spectrometry, Food Chem. 122 (2010) 1273–1277.

[253] A. Rohman, Y.B. Che Man, Analysis of pig derivatives for Halal authentication studies, Food Rev. Int. 28 (2012) 97–112.

[254] M.F. Mazzeo, B. De Giulio, G. Guerriero, G. Ciarcia, A. Malorni, G.L. Russo, R.A. Siciliano, Fish authentication by MALDI-TOF mass spectrometry, J. Agric. Food Chem. 56 (2008) 11071–11076.

[255] S.K. Barik, S. Banerjee, S. Bhattacharjee, S.K. Das Gupta, S. Mohanty, B.P. Mohanty, Proteomic analysis of sarcoplasmic peptides of two related fish species for food authentication, Appl. Biochem. Biotechnol. 171 (2013) 1011–1021.

[256] G. Lo Feudo, B. Macchione, A. Naccarato, G. Sindona, A. Tagarelli, The volatile fraction profiling of fresh tomatoes and triple concentrate tomato pastes as parameter for the determination of geographical origin, Food Res. Int. 44 (2011) 781–788.

[257] Z. Jandric, D. Roberts, M.N. Rathor, A. Abrahim, M. Islam, A. Cannavan, Assessment of fruit juice authenticity using UPLC-QToF MS: a metabolomics approach, Food Chem. 148 (2014) 7–17.

[258] C. Cordero, E. Liberto, C. Bicchi, P. Rubiolo, S.E. Reichenbach, X. Tian, Q. Tao, Targeted and non-targeted approaches for complex natural sample profiling by GC×GC-qMS, J. Chrom. Sci. 48 (2010) 251–261.

[259] C. Cordero, E. Liberto, C. Bicchi, P. Rubiolo, P. Shieberle, S.E. Reichenbach, Q. Tao, Profiling food volatiles by comprehensive two-dimensional gas chromatography coupled with mass spectrometry: advanced fingerprinting approaches for comparative analysis of the volatile fraction of roasted hazelnuts (*Corylus avellana* L.) from different origins, J. Chromatogr. A 1217 (2010) 5848–5858.

[260] C. Cagliero, C. Bicchi, C. Cordero, P. Rubiolo, B. Sgorbini, E. Liberto, Fast headspace-enantioselective GC-mass spectrometric-multivariate statistical method for routine authentication of flavoured fruit foods, Food Chem. 132 (2012) 1071–1079.

[261] L. Nicolotti, C. Cordero, C. Cagliero, E. Liberto, B. Sgorbini, P. Rubiolo, C. Bicchi, Quantitative fingerprinting by headspace-two-dimensional comprehensive gas chromatography-mass spectrometry of solid matrices: some challenging aspects of the exhaustive assessment of food volatiles, Anal. Chim. Acta 798 (2013) 115–125.

[262] J. Guo, T. Yue, Y. Yuan, Feature selection and recognition from nonspecific volatile profiles for discrimination of apple juices according to variety and geographical origin, J. Food Sci. 77 (2012) C1090–C1096.

[263] L. Sabatino, M. Scordino, M. Gargano, A. Belligno, P. Traulo, G. Gagliano, HPLC/PDA/ESI-MS evaluation of saffron (*Crocus sativus* L.) adulteration, Nat. Prod. Commun. 6 (2011) 1873–1876.

[264] N.P. Arruda, A.M.C. Hovell, C.M. Rezende, S.P. Freitas, S. Couri, H.R. Bizzo, Arabica coffee discrimination between maturation stages and post-harvesting processing types, using solid phase microextraction coupled to gas chromatography and principal components analysis, Quim. Nova 34 (2011) 819–824.

[265] R. Garrett, E.M. Schmidt, L.F.P. Pereira, C.S.G. Kitzberger, M.B.S. Scholz, M.N. Eberlin, C.M. Rezende, Discrimination of arabica coffee cultivars by electrospray ionization Fourier transform ion cyclotron resonance mass spectrometry and chemometrics, LWT-Food Sci. Technol. 50 (2013) 496–502.

[266] B.S. Radovic, R. Goodacre, E. Anklam, Contribution of pyrolysis-mass spectrometry (Py-MS) to authenticity testing of honey, J. Anal. Appl. Pyrol. 60 (2001) 79–87.

[267] J. Wang, M.K. Kliks, W. Qu, S. Jun, G. Shi, Q. Li, Rapid determination of the geographical origin of honey based on protein fingerprinting and barcoding using MALDI TOF MS, J. Agric. Food Chem. 57 (2009) 10081–10088.

[268] S.R. Won, D.C. Lee, S.H. Ko, J.W. Kim, H.I. Rhee, Honey major protein characterization and its application to adulteration detection, Food Res. Int. 41 (2008) 952–956.

[269] F. Bianchi, M. Careri, E. Chiavaro, M. Musci, E. Vittadini, Gas chromatographic–mass spectrometric characterisation of the Italian Protected Designation of Origin 'Altamura' bread volatile profile, Food Chem. 110 (2008) 787.

[270] R. Beleggia, C. Platani, G. Spano, M. Monteleone, L. Cattivelli, Metabolic profiling and analysis of volatile composition of durum wheat semolina and pasta, J. Cereal Sci. 49 (2009) 301–309.

[271] G. Dinelli, A. Segura-Carretero, R. Di Silvestro, I. Marotti, D. Arraez-Roman, S. Benedettelli, L. Ghiselli, A. Fernadez-Gutierrez, Profiles of phenolic compounds in modern and old common wheat varieties determined by liquid chromatography coupled with time-of-flight mass spectrometry, J. Chromatogr. A 1218 (2011) 7670–7681.

[272] J.A. Pattemore, N. Rice, D.F. Marshall, R. Waugh, R.J. Henry, Cereal variety identification using MALDI-TOF mass spectrometry SNP genotyping, J. Cereal Sci. 52 (2010) 356–361.

[273] T. Levandi, T. Pussa, M. Vaher, A. Ingver, R. Koppel, M. Kaljurand, Principal component analysis of HPLC-MS/MS patterns of wheat (*Triticum aestivum*) varieties, Proc. Est. Acad. Sci. 63 (2014) 86–92.

[274] C. Zorb, T. Betsche, G. Langenkamper, Search for diagnostic proteins to prove authenticity of organix wheat grains (*Triticum aestivum* L), J. Agric. Food Chem. 57 (2009) 2932–2937.

[275] M.C. García-López, V. Garcia-Cañas, M.L. Marina, Reversed-phase high performance liquid chromatography-electrospray mass spectrometry profiling of transgenic and non-transgenic maize for cultivar characterization, J. Chromatogr. A 1216 (2009) 7222–7228.

[276] T. Levandi, C. Leon, M. Kaljurand, V. Garcia-Cañas, A. Cifuentes, Capillary electrophoresis-time of flight-mass spectrometry for comparative metabolomics of transgenic vs. conventional maize, Anal. Chem. 80 (2008) 6329–6335.

[277] C. Simó, E. Domínguez-Vega, M.L. Marina, M.C. García, G. Dinelli, A. Cifuentes, CE-TOF MS analysis of complex protein hydrolyzates from genetically modified soybeans — A tool for foodomics, Electrophoresis 31 (2010) 1175–1183.

[278] A. Valdés, C. Simó, C. Ibáñez, V. García-Cañas, Foodomics strategies for the analysis of transgenic foods, Trends Anal. Chem. 52 (2013) 2–15.

[279] M.D. del Castillo, N. Martinez-Saez, M. Amigo-Benavent, J.M. Silvan, Phytochemomics and other omics for permitting health claims made on foods, Food Res. Int. 54 (2013) 1237–1249.

Chapter 10

Emerging Contaminants

Julián Campo[1,*] and Yolanda Picó[2]
[1]Environmental Forensic and Landscape Chemistry Research Group. Desertification Research Centre - CIDE (Spanish Council for Scientific Research, University of Valencia, Generalitat Valenciana) Valencia, Spain; [2]Food and Environmental Safety Research Group, Faculty of Pharmacy, University of Valencia, Valencia, Spain
*Corresponding author: E-mail: Julian.Campo@uv.es

Chapter Outline

1. Introduction 515
2. Emerging Contaminants 518
3. Extraction Techniques 521
4. Mass Spectrometry Analyzers 524
5. Gas Chromatography-Mass Spectrometry (GC-MS) 528
5.1 GC-MS/MS (GC-MS^2) 539
5.2 Comprehensive GC (GC × GC) 543
6. Liquid Chromatography-Mass Spectrometry (LC-MS) 545
6.1 LC-MS/MS (LC-MS^2) 555
6.2 LC-High Resolution Mass Spectrometry (LC-HRMS) 558
6.3 Comprehensive LC (LC × LC) 561
7. Other Methods 563
8. Conclusions 568
Acknowledgments 569
References 569

1. INTRODUCTION

According to the Health and Consumer Protection Directorate-General of the European Commission, food contaminants are substances that can be present in certain foodstuffs due to **environmental contamination, cultivation practices, or production processes**. If present above certain levels, these substances can pose a threat to human health. European Union (EU) rules ensure that food placed on the market is safe to eat and does not contain contaminants at levels, which could threaten **human health**. In addition to classical pollutants (ammonia, heavy metals, etc.), there are a number of new toxicant residues known as **emerging contaminants** that can be defined as unregulated, recently discovered and potentially of concern to human health [1]. Among these, pesticides, for example, are not new chemicals but are a problem that have existed historically, which are revisited with a current population and then, can fit the definition of emerging [2]. Consequently, in order to avoid the negative impact this pollution has on the

Comprehensive Analytical Chemistry, Vol. 68. http://dx.doi.org/10.1016/B978-0-444-63340-8.00010-8

quality of food, and particularly on **human health**, many world governments, including the EU, are taking measures to control and minimize the presence of such contaminants in foodstuffs and, in this way, improve **food safety** [3].

To ensure the food safety, one of the major trends of analytical chemistry is to **develop robust, fast, efficient, sensitive, and cost-effective procedures** for the trace analysis of target and nontarget polluting compounds in food, which help to evaluate the risks associated with **mixtures of contaminants** [4]. Additionally, recent issues about adulteration and/or contamination of food highlighted the need for both food manufacturers and regulatory agencies to utilize fast and accurate analytical techniques to proactively ensure product safety [2]. Among these emerging food contaminants, the most prominent groups are agrochemical and pharmaceutical residues (pesticides, veterinary drugs, and their metabolites); those of industrial origin (perfluoroalkyl substances (PFASs); polybrominated flame retardants (PBFRs), which include the polybrominated diphenyl ethers (PBDEs); polycyclic aromatic hydrocarbons (PAHs); and nanomaterials); as well as emerging groups of biotoxins (bacterial toxins, mycotoxins, and phycotoxins) and compounds formed during the processing of food (acrylamides, chloropropanol, furans, bisphenols, phthalates, etc.). Engineering nanomaterials would also be an issue within this topic. However, due to their impact and particular requirements, they will be the treated in a different chapter.

Most of the advances in the detection of such compounds at ultratrace concentrations are due to the development of the coupling of **chromatography and mass spectrometry (MS) techniques** [4]. Methods coupling MS with separation techniques, including gas chromatography (GC) and liquid chromatography (LC), have shown great convenience and suitability for food analysis due to their ease of automation and high-throughput capability [5]. Figure 1 classifies the emerging contaminants according to the most appropriate determination

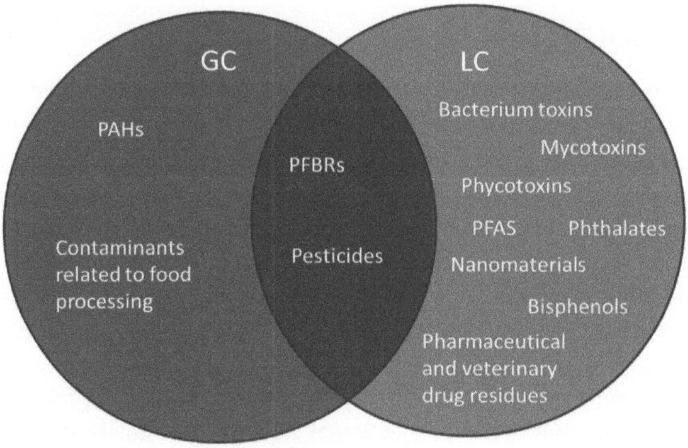

FIGURE 1 Analytical methods mostly used for separation and detection of emerging compounds in food.

technique. GC techniques coupled to MS systems [GC-MS, GC-tandem mass spectrometry (GC-MS/MS), comprehensive two-dimensional GC-MS (GC × GC-MS)] continues to be one of the most used analytical methods for the analysis of contaminant residues in food (particularly for the determination of compounds that are not so sensitive to atmospheric pressure ionization (API) (e.g., organochlorines and pyrethrins) or for certain food matrices with high fat content). Among the analytes that can be detected by GC-MS, or their possible combinations, are pesticides (e.g., herbicides, insecticides, and fungicides) [6–8], PBFRs [9–11], PAHs [12,13], and others related to food processing [14,15].

Currently, analytical methods based on LC, and more recently on ultra-high performance liquid chromatography (UHPLC), have shown a great progress for food analysis, and are applied as routine techniques in most laboratories. They can be used for a wide range of emerging compounds since the use of stable isotope internal standards have allowed to develop highly sensitive and accurate assays. These methods are applied in the assessment of contaminants such as bacterium toxins [16,17], mycotoxins [18–20], phycotoxins [21,22], veterinary drug residues [23–27], PFASs [10,28,29], nanomaterials [30], phthalates, and bisphenols [14,31,32], among others. Some families of pesticides [33,34] and some PBFRs including mostly hydroxylated metabolites [35] can also be screened by LC. In fact, LC is preferred over GC for pesticides because currently used pesticides are quite polar, thermally labile, or not easily vaporized, and consequently, worst detected by GC [36].

Finally, **ambient ionization techniques** achieve the direct analysis of sample surfaces under ambient air conditions with little or no sample pretreatment ideally suited to direct analysis without LC separation. These techniques allow ions to be created from condensed sample phases under ambient conditions and then collected and analyzed by MS [37]. They can be classified in base to the mechanism of the ionization process involved, as those related to electrospray ionization (ESI) as desorption electrospray ionization (DESI) [38,39], and those associated with atmospheric pressure chemical ionization (APCI) as direct analysis in real time (DART) [40,41]. Other methods offering minimal sample pretreatment and allowing the ionization of analyte molecules outside the MS instrument are low-temperature plasma ionization [42,43], and atmospheric solids analysis probe (ASAP) [44].

New emerging analytical techniques within this field are now being envisaged and some of them have already shown interesting prospects and promising benefits. Analytical techniques as ion mobility (IM) spectrometry provides separation of ionized samples in milliseconds, via drift time differences determined by their size and structure, reducing the total analysis time to seconds [32]. Furthermore, imaging MS (IMS) is a remarkable and novel approach used in surface-based analysis of metabolites in tissue sections or surfaces, which is also being used for contaminants assessment. 2D and 3D constructions of the chemical abundance of cells or tissues from IMS analysis provide deeper knowledge concerning the spatial organization of certain processes [45].

In relation to the MS detectors, the most commonly used are single (Q) and triple quadrupole (QqQ), ion-trap (IT), quadrupole linear ion trap (LTQ), time-of-flight (TOF), quadrupole time-of-flight (QqTOF) and Orbitrap. Their development during last years has established them as powerful analytical tools used for the detection/quantification and/or identification/confirmation of emerging contaminants and their metabolites. Recent developments in this analytical instrumentation have enabled faster and reliable analysis of a great number of compounds. The recently introduced Orbitrap mass spectrometers can provide high resolving power and mass accuracy, to tackle complex analytical problems involved in contaminant residue analyses [37,46].

In this chapter, a brief overview of the state-of-the-art in GC and LC coupled to MS techniques, multidimensional chromatography, and ambient ionization techniques for the analysis of groups of food contaminants, which are of emerging concern (emerging contaminants) is presented. Different techniques are described and illustrated with recent examples (i.e., the analysis of biotoxins (mycotoxins and phycotoxins), pesticide residues and their degradation products, some groups of pharmaceuticals and veterinary drugs, as well as some groups of products of industrial origin (PFASs and PBFRs). Most of these contaminants are of particular concern because they can be accumulated and biomagnified through the food chain and can potentially cause severe damage to human health (e.g., some of them are suspected to be carcinogenic and endocrine disruptors) [31,47].

2. EMERGING CONTAMINANTS

Emerging contaminants are a broad category of environmental and food chemicals previously unknown or unrecognized as being of concern but that are under increasing scientific scrutiny [5,48]. These contaminants can be classified in three categories of interest to food safety:

- The first category includes the less-well-studied emerging contaminants, such as degradation products, metabolites, and other breakdown products formed in the environment and/or during the treatment processes.
- The second category comprises those compounds suspected of being present in food and the environment, which are characterized by the lack of data relating to human exposure and toxicological concern.
- In the third category, it is possible to find contaminants that previously have been considered not to pose a concern but in the future could pose a risk due to changes in use patterns, food processing, etc.

It is clear that emerging contaminants in food are a broad area of research with limited information available. In some cases, there is evidence that they may affect human and environmental health (e.g., many pesticides and pharmaceutical metabolites are more toxic than the parent compound). The number of contaminants considered as emerging ones is increasing day-to-day by including compounds historically known and studied but that are reappearing as new toxicological and exposure research is carried out. This is possible due to the development of new and more efficient techniques for the trace analysis of target and nontarget analytes in food. In spite of being a huge scientific area of increasing interest, there have been few attempts to compile main advances accomplished during last years; but special mention has to be done to those of Farré and Barceló [4], Kantiani et al. [1], Petrovic et al. [49], and Picó et al. [50].

Among the emerging contaminants, those most studied and cited in literature can be classified in four different groups:

The first group comprises **residues of substances** that may remain on or in food after they are applied in a particular moment of the food processing chain. The levels of these residues in foods are often stipulated by regulatory bodies in many countries. Exposure of the general population to these residues most commonly occurs through consumption of treated food. In all cases, the levels of residues in food should not harm the consumer. The two most important groups of emerging residues are pesticides and veterinary drugs.

- Pesticides are substances or mixture of substances intended for preventing, destroying, repelling, or mitigating any pest and weeds [36,37,51]. Their toxic effects are compound-specific and include several known mechanisms of action. Examples include, but are not limited to, the inhibition of acetylcholinesterases in the central and peripheral nervous system by the organophosphorus and carbamates insecticides, and the inhibition of protein and lipid synthesis by dinitroaniline and S-alkyl dialkylcarbamothiodithioate herbicides, respectively [37]. In 2009, the EU issued new and revised maximum residue level (MRL) MRLs for the roughly 1,100 pesticides ever used in the world. The revision was intended to simplify the previous system, under which certain pesticide residues were regulated by the Commission; others were regulated by Member States, and others were not regulated at all [52].
- Veterinary drug residues have become an increasingly prominent problem and are an already well-established group of contaminants [19,26,53]. Antibiotics are used in veterinary practice, at therapeutic levels, primarily to treat diseases and to prevent infections and in animal production, at subtherapeutic levels, to increase feed efficiency and to promote growth in food-producing animals. Residues, mostly in products of animal origin, could develop allergies and resistant bacteria [50]. Tetracyclines, quinolones, macrolides, and sulphonamides are some of the veterinary drugs that have been potentially used by farmers to fight against livestock infections, while acaricides or parasiticides are often used for the treatment of endoparasitic and parasitic diseases [25].

The second group includes the contaminants, which are in the environment where the food is grown, harvested, transported, stored, packaged, processed, and consumed. Among them, the emerging contaminants included the following:

- PFASs are a large group of compounds characterized by a fully fluorinated hydrophobic linear carbon chain attached to one or more hydrophilic heads. PFASs repel both water and oil, and are therefore ideal chemicals for surface treatments and other industrial and commercial applications [1]. These compounds are biomagnified in the food chain, leading to increased levels of PFASs in animal-derived foods [29]. Animal studies have demonstrated hepatotoxicity, developmental toxicity, immunotoxicity, as well as hormonal effects [54,55].
- PBFRs are commonly used in electronics, toys, motor vehicles, plastics, and textiles to reduce flammability. This group of chemicals consists of tetrabromobisphenol A, PBDEs, polybrominated biphenyls (PBBs), and hexabromocyclododecane (HBCD). Three commercial PBDE formulations have been produced: penta-BDE, octa-BDE, and deca-BDE [56]. Different toxicological properties of PBDEs have been reported mainly related to the fact that they can act as endocrine disruptors. As a result, Penta-BDE and Octa-BDE mixtures are currently banned by the Stockholm Convention [57].
- PAHs are a group of compounds considered of priority concern by health and environment administrations all over the world on the basis of their mutagenic and carcinogenic effects. Contamination of food by PAHs arises from environmental sources (e.g., deposition from the air on the surface of plants, intake by marine organisms in polluted zones, etc.) and food preparation (e.g., grilling, roasting, or smoking) [58].

The third group is formed by natural toxins, biologically produced. Biotoxins are further classified into fungal biotoxins (mycotoxins), microbial biotoxins, plant biotoxins (phytotoxins), and animal biotoxins, including fish toxins (phycotoxins). The emerging ones included the following:

- Mycotoxins, which are secondary metabolites produced by toxigenic fungi that commonly contaminate agricultural commodities in the field, during transportation and storage under a wide range of climatic conditions [19,59–61]. Multiple mycotoxin contamination of predominantly consumed food commodities such as maize, groundnuts, and their food by-products can exert serious health problems in consumer populations [62]. About 300–400 different toxic fungal secondary metabolites have been characterized so far, being those of health and economic importance the aflatoxins, ochratoxins, fumonisins, deoxynivalenol, zearalenone, and T-2 toxin. These toxins often cause mycotoxicosis, or induce immunosuppression [18] as well as may be carcinogenic, mutagenic, hepatotoxic, nephrotoxic, and cytotoxic.
- Phycotoxins are potent organic compounds produced by dinoflagellates, other flagellated phytoplankton, and cyanobacteria that inhabit marine, brackish, or freshwater bodies or soils. The effects of different phycotoxins range from

peracute lethality from respiratory paralysis to acute liver failure and death, skin irritation, and tumor promotion [21,63,64]. The most common types of poisoning by marine biotoxins are diarrheic shellfish poisoning, paralytic shellfish poisoning, amnesic shellfish poisoning, neurologic shellfish poisoning, and azaspiracid shellfish poisoning [22,65]. Ciguatera fish poisoning and poisoning by palytoxins and spirolides are nowadays considered as the emerging group of phycotoxins [50].

The last group involves the compounds formed during the processing of food as *acrylamides* (produced when carbohydrate-rich foods are exposed to high temperatures [66]); *chloropropanols* (their formation in food is not fully elucidated but it is suggested that hydrochloric acid and residual lipids from the applied material are the precursors for these substances [67]); *furans* (historically well-known but reemerging as new toxicological studies are undertaken and new exposure data and analytical methodology become available [50]); *bisphenols* (are the core substrate to produce bisphenol A diglycidyl ether, the main monomer used in the epoxy resin industry, which can migrate from the protective lining into food and can generate different derivatives during food storage [14]); *phthalates* (mainly used as plasticizers but because of their toxicity are included in the amended list of priority substances of the EU [31], since they are endocrine disruptors with potential neurotoxic and carcinogenic effects in humans [68]).

3. EXTRACTION TECHNIQUES

Sample preparation step is necessary to isolate and concentrate contaminant residues. Extraction is traditionally considered as the bottleneck of these analyses. However, there is a clear trend within the sample preparation step to simplify extraction and to reduce or even eliminated the cleanup steps. Extraction methods as generic as possible, able to extract hundreds of compounds simultaneously are preferred in order to get cost-effective analysis. The main techniques for sample preparation remain solvent extraction (SE) and solid-phase extraction (SPE), and both can be used for the extraction and or cleanup of emerging contaminants. In the former, the most commonly used solvents are acetonitrile, ethyl acetate, in conjunction with anhydrous sodium sulfate, methanol, acetone, dichloromethane, or mixtures of these [51,69,70]. The most important advantage of SE is that it can be indistinctly used for liquid and solid samples with high efficiency. Soxhlet—a continuous reflux extractor—invented in 1879 by Franz von Soxhlet is still in use [71,72]. However, this technique is long, tedious, and requires special glass material. The most common way to perform SE is by mechanical shaken or ultrasounds.

Especial mention has to be done to the QuEChERS (Quick, Easy, Cheap, Effective, Rugged, Safe) extraction, and its modifications, since its combination of compound isolation/extraction with cleanup has given it a very high popularity for multiclass residue extraction in various food matrices [59,73,74]. This method was initially developed for pesticide residue determination and is based on acetonitrile extraction (buffered or not depending on the version) by salting out followed by dispersive SPE (dSPE) cleanup. It is a very generic extraction method, whose application fields are continuously extended to other types of contaminants, such as mycotoxins [59,75], veterinary drugs [25,53,59], acrylamide [68], BFRs [11], and PAHs [72]. QuEChERS presents several advantages in terms of speed, convenience, and low solvent consumption. Its wide applicability to many contaminant/matrix combinations makes it an ideal candidate as a universal sample preparation technique [37].

Besides the QuEChERS-based approaches [76], other alternatives within SE rely on the application of sources of energy different to the mechanical shaking or ultrasounds to favor the extraction of the analytes. The most popular among them are pressurized liquid extraction (PLE) and microwave-assisted extraction (MAE) [53,57,77]. PLE uses high pressure and temperature to favor the extraction and is usually employed as the extraction step for determination of BFR and norbonenes [57], pesticides [78], and veterinary drugs [79]. Depending on the analytes' polarity, the solvents used in PLE can be apolar as dichloromethane, hexane, and ethyl acetate or polar as water. The latter is called pressurized hot water extraction and offers the attractive to be a green extraction method fully compatible with SPE, and the disadvantage to be only useful for polar analytes. PLE with different sorbents could be used as selective pressurized extraction approach. Integration of the PLE and cleanup process is also possible by loading sorbent at the bottom of the PLE stainless cell [78]. Comparing with traditional extraction and cleanup approaches [liquid–liquid extraction (LLE)-SPE, solid phase microextraction (SPME), and PLE-gel permeation chromatography (GPC)], the selective PLE can simplify extraction step and significantly reduce extraction time and solvent consumption.

MAE is also an attractive separation method, due to the drastic reduction of extraction time and solvent consumption compared with conventional methods. Recently, MAE with closed vessels and polar solvents has been tested for melamine [80]. Total time required for the MAE procedure was 2 min, instead of the 40 min needed for the ultrasonic method. Medium-assisted nonpolar solvent dynamic MAE has been applied to extract organophosphorus pesticides from cereals. Without adding any polar solvent, graphite powders were used as microwave absorption medium to transform microwave energy into heat energy [81]. By dynamic MAE using hexane as extraction solvent, 10 organophosphorus could be extracted completely and the extract was directly analyzed by GC-MS without any cleanup process. Although PLE and MAE emerged as very promising few years ago, their use is still limited because they involve additional expenses and give similar recoveries to SE.

In addition to these, other extraction approaches used for particular cases are ion-pair extraction with quaternary ammonium cations at basic pH, widely used for PFASs [28,82], the addition of a potassium hydroxide or enzymatic digestion in order to release target analytes from covalent bonding [83], and steam distillation [30].

One important problem in both GC-MS and LC-MS is certain compounds present in the sample that can affect the initial ionization of the analyte through what is often called ion suppression/enhancement or matrix effects [82]. This is the reason why the cleanup steps have not been fully eliminated from the extraction procedures since they help to reduce or even eliminate these effects. Although the cleanup helps, it does not eliminate completely matrix effect and then, they should be taken into account for proper quantification. The calibration approaches most frequently used to reduce or compensate matrix effects are calibration using external standards prepared in sample matrix (matrix-matched standards) and calibration using internal standards. The advantage of using matrix-matched standards is the simplicity of application and economy; however, the need of obtaining a blank matrix for every sample type is mandatory, and turns it into unpractical technique for routine analysis. Otherwise, calibration with internal standards is the best option if the isotopically labeled standards of the target compounds are available. This system corrects signal suppression, as both labeled and native compounds suffer the same suppression effect. An isotopically labeled compound is the best solution for a quantitative compensation but they are sometimes not available and/or highly expensive [36]. However, although these matrix effects can be corrected for quantification, they decrease analytical sensitivity. Thus, cleanup is, many times, mandatory.

SPE can be used to extract compounds from liquid food samples or mostly performed as an additional cleanup step after SE or as a combined single isolation/cleanup step. With the packing materials available for SPE, it is possible to find reversed-phase octadecyl silica (C18), hydrophilic–lipophilic balance (HLB) phases, ionic exchangers (anionic and cationic, strong and weak), graphitized black carbon (GBC), Florisil, polystyrene–divinylbenzene materials, and carbon nanotubes [23,30]. SPE is usually performed as cleanup either off-line or online, by using of reversed-phase cartridges, usually polymeric HLB [84].

If a single cleanup step is not sufficient, tandem SPE or mixed-mode cartridges achieved cleaner extracts [85], and selective SPE, with molecularly-imprinted polymers (MIPs) or immunoaffinity columns provided unique specificity to isolate the compounds of interest. The formers have recently been used for quinolones [86], tetracyclines [87], and phthalates [88] in the SPE cartridge format. This MIP-SPE can be applied either off-line, or connected in line with an HPLC column offering possibilities for automation of the whole pretreatment/enrichment and analytical process [89]. Selective immunoaffinity columns are also used to prepare mostly mycotoxins samples [90].

In the case of samples with high fat content, additional SPE cleanup techniques able to separate fat from the analytes include GPC or counter-current chromatography (CCC) [9,71]. The latter is a chromatographic separation technique based on the partition of solutes between two immiscible liquid phases as they interact in a thin tube under a centrifugal force field. CCC has been applied to the separation of organohalogenated contaminants from fat coming from fish tissues.

In addition to those conventional techniques, miniaturized extraction techniques for emerging contaminants include the use of SPME [91], stir-bar-sorptive extraction (SBSE) [92], and liquid-phase microextraction techniques (LPME), including microextraction with supramolecular solvents (SUPRAS) [93], hollow-fiber (HF) LPME [94], ultrasound-assisted emulsification micro-extraction (USAEME) [95], matrix solid-phase dispersion (MSPD) [8,66,96], and dispersive liquid-liquid microextraction (DLLME) [75,97].

4. MASS SPECTROMETRY ANALYZERS

Nowadays, the analyzers used in the analysis of emerging contaminants in food can be classified in

- MS/MS
- High resolution mass spectrometers (HRMS)

Some of them are able to provide both features: accurate MS measurement and MS/MS. The basic principle of MS/MS is the selection of precursor ion, fragmentation of this ion in the collision cell, and measurement of the mass–charge ratio (*m/z*) of the product ions formed. There are two fundamentally different approaches to MS/MS: tandem in space and tandem in time. Tandem-in-space instruments have separate mass analyzers in physically different locations of the instrument. Tandem "in-time" is accomplished with ions trapped in the same place where they are further fragmented, with the various stages of MS conducted within the same physical trapping volume but at different times during the experiment [98]. A hybrid mass spectrometer is an instrument, which combines analyzers of different types. Examples of tandem mass spectrometers include, but are not limited to, QqQ, quadruple linear ion trap (LTQ), QqToF, and quadrupole-Orbitrap (Q Orbitrap) and linear quadrupole Orbitrap (LTQ–Orbitrap) hybrid instruments. Tandem-in-time instruments are typically ion-trapping mass spectrometers, which comprise ^{3}D quadrupole ion traps (IT), linear IT (as part of the LTQ), and Orbitrap instruments.

These different, hyphenated or not, mass analyzers can work in many different working modes depending on their characteristics. Figure 2 shows the information attainable depending on the characteristics of the mass spectrometer.

Triple quadrupole (QqQ). Methods based on QqQ geometry are the most widely used for the analysis of emerging contaminants because of their high sensitivity, selectivity and specificity, wide linear dynamic range (three orders of magnitude) and precision (the repeatability of peak areas is usually below 10%). The use of the selected reaction monitoring (SRM) mode reduces the probability of spectral interferences allowing identification by monitoring

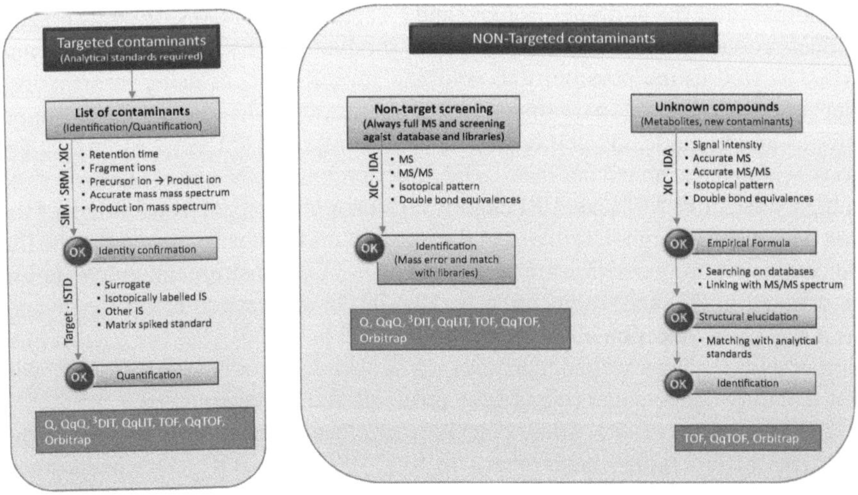

FIGURE 2 Scheme of the different working mode and the mass analyzers able to do it.

usually two transitions per target compound [37]. Disadvantages of QqQ-MS methods are that (1) they are restricted to target analysis, (2) the SRM mode limits the total number of transition-ion pairs detected for each chromatographic run, with typical monitoring of less than 50–100 transitions, and (3) interferences from matrix components may produce additional chromatographic peaks even in extracted ion chromatograms (EICs), which may result in false positive identifications. Other limitations are the lack of postacquisition reinterrogation of data other than for those analytes preprogrammed into the method, and the reliance on the availability of reference standards. SRM-MS/MS attains unambiguous identification by separating the target ions through Q1 and further detecting fragment ions of the target by Q3. SRM is able to reduce chemical noise and to detect multiple transitions (Q1/Q3 pairs). Although the complete resolution by chromatography is difficult to obtain, MS/MS detection achieves selective analysis [5].

Ion trap (IT) and linear ion trap (LIT). They are characterized by their MS^n capability that enables the sequential and multistage isolation of precursor ions, fragmentation, trapping, and mass scanning in the same space and as function of time. These features, not available in the QqQ and QqTOF instruments, allow the identification of unknown compounds and high full-scan sensitivity. There are two types of ion-traps, the ^{3}DIT (with a tridimensional form) and the linear form that uses a quadrupole. The ^{3}DIT-MS instruments present less sensitivity, smaller linear dynamic ranges, and lower quantification capability in relation to the QqQ systems in SRM mode. Despite of only ions with m/z larger than approximately one-third of the precursor ion, and above *m/z* 50, can be efficiently trapped. Significant improvements in IT technology have been achieved by the implementation of two quadrupoles preceding a LIT mass analyzer (LTQ). These hybrid platforms offer the attributes of QqQ and LIT operation modes, including SRM mode, product-ion, neutral-loss, enhanced product ion (EPI), enhanced MS and MS^3, as well as the possibility to combine several of them using information-dependent acquisitions experiments [98]. In enhanced MS scan mode, Q1 operates as an ion guide, Q2 cell is set to a low energy, and Q3 as a LIT to trap the ions, which are scanned out in an axial direction toward the ion detector to yield a highly sensitive MS scan. EPI employs the tandem-in-space capabilities of the ion path with the high-sensitivity of the ion-trap mass scan. In this mode, the ion selection and fragmentation are performed as in a QqQ instrument, precursor ion is selected in Q1, and fragments generated in Q2 are trapped in Q3 that works as a trap prior detection allowing the MS^3 mode in a single run (very useful for compounds having a second transition at low abundance or not detectable as well for reducing the potential risk of false positives and negatives) [36].

Time-of-flight (TOF). These detectors have been increasingly applied in the analyses of emerging contaminants in food [99,100]. TOF instruments measure fast and accurately *m/z* ratios based on the mass-dependent flight time differences from the entrance of the analyzer to the detector, affording fast full-spectral acquisition rates, full-spectral sensitivity, at high-mass resolution (that of the modern TOF-MS usually in the range 30,000–60,000 full width at half maximum, FWHM) with high mass accuracy (<2 mDa). The main advantage of the TOF-MS is that it enables the screening of a sample for a large number of analytes (theoretically an unlimited number), allowing a posterior process of data files for additional compounds [99]. Its principal disadvantage is that it cannot perform MS/MS. However, the application of the in-source collision-induced dissociation (CID) gives structural information through the characteristic pattern of accurate mass fragments, the isotopic pattern, and the elemental composition of the molecular ion, which are very useful when no standards are available to support the identification. Useful information to identify metabolites and/or degradation products is obtained from fragment ions because some of them are very characteristic of the molecule. If one of these characteristic fragment ions show up at a different retention time than the protonated molecule, this may indicate the presence of a possible degradation product that either has the fragment mass or contains that fragment in the molecule [36]. There are two complementary configurations for TOF-MS analyzers interfaced to GC instruments that provide exact mass measurements (sub 5 ppm) with moderate acquisition rate (20 Hz in the more recent instruments), and instruments that provide higher acquisition speed of 100–500 spectra/s but usually provide unit mass resolution. The latter is compatible with GC×GC. High-speed GC-TOF-MS was often used for quantitative analysis, whereas high-resolution GC-TOF-MS was used for qualitative analysis of organic compounds in food due to its limited dynamic range.

Recently developed LC-TOF-MS methods allow fast mass-assignment accuracy within 3 ppm. Analytical throughput was also increased by using a dual-nebulizer ESI source and an automated calibrant-delivery system [5]. Modern

LC-TOF-MS instruments are amenable to identify not only target contaminants in food, but also nontarget ones. Since quantitative aspects of TOF-MS configurations are still being investigated, most of the authors exploit their potential to screen the sample for a large number of compounds (including transformation products) and carry out the quantitative analysis with QqQ-MS or LTQ-MS platforms [101].

Quadrupole time-of-flight (QqTOF). It is a hybrid QqTOF mass spectrometer with MS/MS capability. This platform can perform a full spectral acquisition of the entire product-ion profile at a defined mass window, with high resolution (>5,000 FWHM), and high full spectral sensitivity (mass accuracy at 5 ppm for both precursor MS and product ion MS/MS modes). Its potential to identify unknown molecules, without the use of standards, can be extended to any compound in a number of matrices. Compared to QqQ-MS, LC-QqTOF-MS has the major advantage of structure identification and confirmation. However, LC-QqQ-MS offers lower limits of detection (LODs) for quantitative analysis compared to LC-QqTOF-MS, even though the two techniques provide similar linear dynamic range and repeatability. So far, a number of the latest articles and reviews have shown that LC-TOF-MS and LC-QqTOF-MS are advanced, efficient techniques for food analysis, primarily including the following aspects: (1) fast screening for target analytes, especially its capability for an increasing number of compounds in food samples; (2) identification of nontarget analytes by accurate-mass determination of the protonated or deprotonated molecules and their product-ion mass spectra by collision-induced dissociation; and, (3) quantification capability or performance, such as LOD, linearity, reproducibility, and retrospective data analysis, is often compared to LC-QqQ-MS [84,102].

Orbitrap mass spectrometer and Fourier transform ion cyclotron resonance (FT-ICR). Orbitrap-MS and FT-ICR-MS are high-resolution mass spectrometers, which leads to highly selective and accurate analysis of emerging compounds in complex food and water matrices by virtue of their high resolution (typically 100,000–1,000,000 FWHM), related to the finely tuned electrostatic field, and high mass precision (1–2 ppm, allowing discrimination between isobaric interferences and ions of interest) [5]. The Orbitrap allows the radially trapping of ions about a central spindle-like electrode with a coaxial outer cylindrical electrode. The *m/z* values are measured based on the mass-dependent frequency at which ions undergo harmonic oscillations during their trapping period. Frequencies are measured by the acquisition of the time-domain images of current transients and the subsequent Fourier transformation enables to obtain the mass spectra. Even if Orbitrap mass analyzer presents slower data acquisition rate than QqTOF instruments, it provides outstanding mass accuracy, mass resolution, and reliable high-sensitivity MS^n performance. Despite the rapid and continuing changes to instrument specifications, TOF analyzers typically still have less mass-resolving power than instruments based upon the Orbitrap, although high mass-resolving power with Orbitrap instruments comes at the expense of scan speed [98]. Other instruments as FT-ICR are considered as one of the most sensitive ion-detection methods with mass accuracy of <1 ppm, but the high cost of instrumentation limits their application [4].

To the moment, the Orbitrap mass analyzer has only been applied to LC and is nowadays available in three different configurations:

- The stand-alone Orbitrap combined with an API source, which can obtain characteristic fragment of the molecules as previously described for TOF instruments by in-source CID.
- The combination of a LIT with an Orbitrap analyzer to form a hybrid LTQ-Orbitrap mass spectrometer. This device has various advantages over the QqTOF-MS analyzers, namely higher full-scan sensitivity, greater mass resolution and accuracy, higher tandem MS sensitivity (MS^n), and a wider linear dynamic range that makes multistage MS^n possible for structural elucidation. This instrument keep all the characteristics of an Orbitrap and a LIT including the higher-energy CID.
- The Q-Orbitrap combines quadruple precursor ion selection with Orbitrap detection that is capable of precursor ion selection for both selected ion monitoring (SIM) and data-dependent MS/MS experiments and achieves higher fragmentation speeds.

The three configurations have already a range of applications within food safety demonstrating the good performance of the instruments.

5. GAS CHROMATOGRAPHY-MASS SPECTROMETRY (GC-MS)

Most of the analytical methods for separation and detection of emerging compounds in food used currently are those based on GC coupled to MS (GC-MS) including GC-MS/MS and GC×GC-MS. GC-MS is particularly suitable for volatile and semivolatile, moderate to nonpolar small molecules. However, GC are not amenable to the direct analysis of highly polar, thermally unstable, and high molecular-weight chemicals (poor chromatographic peak shape or a peak-tailing problems may affect the separation and ultimately affect the LODs) [5]. Although a derivatization procedure often helps to solve these problems and to improve the method sensitivity, this additional step can increase considerably the analysis time as well as can influence the accuracy of the results because of possible losses of analytes and/or incomplete reactions. N,O-Bis(trimethylsilyl) trifluoroacetamide (BSTFA) is still used to derivatize labile groups such as hydroxyl on melamine [80] and alkylphenols [92], with the more stable trimethylsilyl group.

Table 1 summarizes the most recent GC-MS and related methods developed and applied to assess emerging food contaminants, including acrylamide, chloropropanols, BFRs and organophosphorus flame retardants (OPFRs), PAHs, phthalates, and pesticides. Acrylamide, OPFRs, and pesticides can also be determined by LC.

TABLE 1 Emerging Contaminants and Analytical Techniques Used in GC for Their Extraction and Determination in Different Food Matrices

Food Contaminant	Matrix	Extraction	Determination	Recovery (%)	LODs ($\mu g\ kg^{-1}$)	Reference
Food Processing Contaminants or Adulterants						
Acrylamide	Fried potato, fried eggplant, taamia, minnan, ab.rae/hilumur, gorassa	QuEChERS with modified MTMOS–TEOS dSPE5	*GC-qMS* Split-less injection HPWax (30 m × 250 μm × 0.25 μm) 100% polyethylene glycol EI ionization in SIM mode	88–103	9.1–12.8	[68]
Furan	Selected Chinese food (canned ham, fruit juices, infant formula)	*DI-SPME* (CAR/PDMS, 75 μm) 1 g of solid samples + 1 g NaCl + 5 mL of H_2O 5 mL of liquid samples + 1 g NaCl	*GC-qMS* Thermodesoprtion of analytes HP5ms (30 m × 0.25 mm × 0.25 μm) 5%-phenyl-methylpolysiloxane EI in SIM mode	93–105	0.175	[91]
Melamine (derivatization with BSTFA)	Seafood	*MAE* closed-vessel 20 mL Diethylamine/Milli-Q water/Acetonitrile 10/40/50 At 70 °C, 600 W, and 1 min	*GC-qMS* Split-less injection DB-5MS (30 m × 0.25 mm × 0.25 μm) 5%-phenyl-methylpolysiloxane EI in SIM mode	>96	550	[80]
Bisphenol A (derivatization with acetic acid)	Beverage samples	*USAEME-ISD* 5 mL of sample + NH_3 buffer + 30 μL acetic acid + 500 μL $CHCl_3$. Ultrasound emulsification and separation by centrifugation	*GC-3DIT-MS* Split-less injection VF-5 ms (25 m × 0.25 mm, 0.25 μm) 5%-phenyl-methylpolysiloxane EI in SIM mode	≥82	0.038	[95]

Continued

TABLE 1 Emerging Contaminants and Analytical Techniques Used in GC for Their Extraction and Determination in Different Food Matrices—cont'd

Food Contaminant	Matrix	Extraction	Determination	Recovery (%)	LODs (μg kg^{-1})	Reference
Chloropropanols						
3-MCPD, acrylamide (ac)	Bread, fried chips, fried instant noodles, soy sauce, instant noodle flavoring	MSPD (Extrelut NT)	*GC–QqQ-MS/MS* Split-less injection Innovax (3 m × 0.32 mm × 0.25 μm) cross-bonded polyethylene glycol combined in sequence with DB-5 ms (30 m × 0.25 mm × 0.25 μm) 5%-phenyl-methylpolysiloxane EI ionization and in SRM mode	3-MCPD: 90.5–107 Ac: 81.9–95.7	5	[66]
Halogenated Contaminants						
62 PCBs, 42 PBDEs, HBCDs	Tilapia fish species	High-speed solvent extractor (hexane/acetone 1:1, v/v), GPC for fat removal and eluted with hexane/dichloromethane	*GC-qMS (PBDEs, PCBs)* Determination parameters were not indicated. LC-MS/MS (HBCD isomers)	86–115	–	[9]
8 PBDEs, 8 MeO-PBDEs, BFRs, halogenated norbornenes	55 biota samples, covering 3 trophic levels	PLE: hexane: dichloromethane (1:1). Cleanup: SPE in alumina cartridges (Al–N)	*GC-qMS* (BFRs) Split-less DB-5 (15 m × 0.25 mm × 0.1 μm) 5%-phenyl-methylpolysiloxane NCI using NH_4^+ and SIM mode *GC-QqQ-MS/MS* (norbonenes) NCI using CH_4^+ and SRM mode	50–98	0.1–2.1 l.w.	[57]
14 PCBs and PBDEs	Milk	Hexane extracts were loaded on an LC-Florisil column to isolate analytes. The elute were dried and dissolved in acetone used as the disperser solvent in DLLME	*GC-qMS* (BFRs) Split-less DB-5 (15 m × 0.25 mm × 0.1 μm) 5%-phenyl-methylpolysiloxane EI and SIM mode	74–132	0.01–0.4	[97]

PBDE, OFRs, pesticides	Mussels	*PBDEs*: PLE dried (Hydromatrix), extracted: dichloromethane inside stainless-steel cells. Purified: activated silica gel, alumina and GPC. *OFRs*: Florisil SPE. ASE (Dionex ASE 200 methylene chloride). Purified (s). Solvent exchanged to hexane, added to a solid-phase 2 g silica glass extraction column	*GC-qMS* Pesticides: EI-GC-MS PBDEs: NCI- GC-MS OFRs: EI/NCI GC-MS, LC-MS/MS Split-less DB-5HT (15 m × 0.25 × 0.1 μm) 5%-phenyl-methylpolysiloxane The MS operates in: NCI using CH_4^+ or in EI ionization and in SIM mode	*Pesticide* 69–107 *PBDEs*: 55–130 *OFRs*: 70–130	0.02–160	[30]
PCBs, PBDEs, OCPs	Fish, cephalopod, bivalve, gastropod, crustacea, cetacean	Homogenized (anhydrous Na_2SO_4), Soxhlet extractor (3:1 dichloromethane: Hexane). Lipid removed by GPC biobeads S-X3. Cleanup: multilayer silica gel column	GC-HRMS *PCBs and OCPs* DB5-MS (30 × 0.25 mm × 0.25 μm) 5%-phenyl-methylpolysiloxane *PBDEs* DB5-MS (15 m × 0.25 mm × 0.1 μm) EI mode and SIM mode	*PCBs*: 64–87 *PBDEs*: 87–107	*PCBs*: 0.04–0.08 *BDEs*: 0.1–0.5	[71]
PBDEs, PBBs, PCBs	Whole herring, sprat, eggs from common eider, eggs and livers from herring gull	Homogenized, dried in a 1-fold amount of dry Na_2SO_4. Lipids removed by GPC. Fractionation: Florisil™ column	*GC-QqQ-MS/MS* Split-less DB5-MS (15 m × 0.25 mm × 0.1 μm) 5%-Phenyl-methylpolysiloxane EI mode and SRM mode	*PBDEs*: 66–74 *PCBs*:56	*PBDEs,* *PBBs*: 0.02–0.77 *PCBs*: 0.01–0.22	[114]
22 BFRs (PBDEs, PBT, PBEB, etc.)	Trout	Modified QuEChERS (using ethyl acetate) Solvent change to hexane Cleanup in a handmade silica mini-column	*GC-QqQ-MS/MS* (EI), GC-MS (QqQ as a single Q in NCI) Split-less injection DB-XLB (15 m × 0.18 mm × 0.07 μm) (12% phenyl)-methylsiloxane EI and NCI and SRM or SIM modes	70–119	*EI-MS/MS*: 5–1000 *NCI-MS*: 5-100	[11]

Continued

TABLE 1 Emerging Contaminants and Analytical Techniques Used in GC for Their Extraction and Determination in Different Food Matrices—cont'd

Food Contaminant	Matrix	Extraction	Determination	Recovery (%)	LODs ($\mu g\ kg^{-1}$)	Reference
BFRs	Tuna fish	Soxhlet for 2 h with 100 mL of n-hexane: acetone (3:1, v/v) plus cleanup through acid silica	*GC × GC-TOF-MS* Spit-less injection First dimension: HT-8 (30 m × 0.25 mm i.d. × 0.25 μm) 5% phenylmethyl siloxane Cold tap Second dimension: BPX-50 column (1.6 m × 0.1 mm × 0.1 μm) 50% phenylmethylsiloxane EI in full scan *m/z* range 50–600	–	–	[113]
PAHs						
16 PAHs, 20 OCPs, 18 PCBs	Hen egg yolk	Automated Soxhlet extraction. Purification by GPC	*GC-QqQ-MS/MS* Split-less injection BPX5 (30 m × 0.25 mm × 0.25 μm) 5% phenyl methylpolysiloxane EI and SRM mode	74.5–104.7	–	[72]
16 PAHs, 13 FRs, 18 pesticides, 14 PCBs, 7 PBDE	Catfish muscle	QuEChERS (acetonitrile), cleanup: d-SPE(zirconium-based sorbent)	*LP-GC-QqQ-MS/MS* Split-less injection Rti-5 ms (15 m × 0.53 mm × 1 μm) 5% phenyl methylpolysiloxane, and 5 m × 0.18 mm non-coated Virtual length (5.5 m × 0.18 mm) EI and SRM mode	70–120	*PAHs, pesticides*: 0.5–5 FRs: 1–10 *PCBs*: 0.1–0.5 *PBDEs*: 0.5–10	[13]

PAHs, PCBs, PBDEs	Fish	Modified QuEChERS (using ethyl acetate) Solvent change to hexane Cleanup in a handmade silica mini-column	*GC × GC-TOF-MS* Columns used in the first and second dimension had (30 m × 0.25 mm × 0.25 μm) and (1 m × 0.1 mm × 0.1 μm), respectively. Different stationary phases LV-PTV injection	–	*PCBs*: 0.01–0.25 *PBDEs*: 0.025–5 *PAHs*: 0.025–0.5	[105]
Phthalates						
Phthalates, estrogenic hormones, polycyclic musks, bisphenol-A, alkylphenols, pesticides	Thicklip gray mullet	Enzymatic hydrolysis (Bond elute Plexa), cleanup: SPE (Florisil)	*GC-qMS* LVI-PTV injection No more conditions specified	63–122	0.04–459	[31]
Phthalates	Edible vegetable oils	Acetonitrile with precipitation under freezing and SPE cleanup using alumina	*GC-³DIT-MS* Split-less injection HP-5MS (30 m × 0.250 mm × 0.25 μm) 5%-phenylmethylpolysiloxane EI in SIM mode	62–100	20–50 1500 for DiNP and DiDP	[108]
Phthalates and alkylphenols (derivatized with BSTFA)	Vegetables	SBSE Extraction with methanol, evaporation to almost-dryness, addition of water and SBSE	*GC-qMS* Thermal desorption 3 columns (30 m × 0.25 mm × 0.25 μm) • 100% dimethylpolysiloxane • 5% diphenyl dimethylpolysi-loxane • 50% diphenyldimethylpolysi-loxaneEI and SIM mode	83–118	0.014–0.105	[92]

Continued

TABLE 1 Emerging Contaminants and Analytical Techniques Used in GC for Their Extraction and Determination in Different Food Matrices—cont'd

Food Contaminant	Matrix	Extraction	Determination	Recovery (%)	LODs (μg kg^{-1})	Reference
Phthalates	Soybean milk	MIPs pH adjustment to 4.45 (isoelectric proteins point). Elimination of proteins by centrifugation. Supernatant flow through MISPE. Phthalates eluted with EtOAc	*GC-qMS* Split-less injection DB-5 (30 m × 0.25 mm, 0.25 μm) 5% phenyl methylpolysiloxane EI in SIM mode	76–108	13–22	[88]
Pesticides						
Fenobucarb, diazinon, chlorothalonil, chlorpyrifos	Apple	5 g of apple + 5 mL of H_2O with 10% NaCl HS-SPME with polydimethylsiloxane (PDMS, 100 μm) Online thermal desorption	*GC-qMS* Split-less DM-5MS (30 m × 0.25 mm × 0.25 μm) 5% phenyl methylpolysiloxane EI with SIM mode	80–105	0.01–0.2	[6]
25 pesticides	Tea	QuEChERS: Toluene and ethanol (50:50, v/v) Purification: PSA, GCB	*GC-qMS* HP-5 (30 m × 0.25 mm × 1 μm) 5% phenyl methylpolysiloxane EI in SIM mode	76.4–115.6	5–10	[115]
OPPs	Barley, wheat, oilseed rape, meadow fescue, 4 carrot cultivars	QuEChERS (double-distilled H_2O, acetonitrile, $MgSO_4$, NaCl, dispersive extraction technique). Clean-up: PSA	*GC-qMS* PTV injection HP-5MSI (30 m × 0.25 mm × 0.25 μm) 5% phenyl methylpolysiloxane EI in SIM	60–392	10–50	[104]

86 pesticides (OPPs, OCPs, pyrethroids, carbamates)	Cold pepper, eggplant, carrot, cucumber, potato, hot pepper, squash, beans, okra, onions, cauliflower, tomato (cultivation, green house)	Acetonitrile, celite, separator funnel with petroleum ether. Saturated solution of NaCl and H_2O, washed (Na_2SO_4). Cleanup: florisil column	*GC-qMS* Split-less injection HP-5MS (30 m × 0.25 mm × 0.25 μm) 5% phenyl methylpolysiloxane EI in SIM	–	1	[116]
Bifenthrin	Milled wheat	Anhydrous sodium sulfate, methanol–acetone 1:1, ethyl acetate–acetone 4:1, sodium sulfate and Florisil	*GC-^{3}DIT-MS* Split-less injection Capillary column not specified EI in SRM	79.6–82.6	4	[69]
135 pesticides	Green, black dry tea leaves, stalks	Modified QuEChERS (acetonitrile extraction and cleanup by liquid–liquid partitioning to hexane).	*GC–QqQ-MS/MS* Split-less injection HP-5 MS (15 m × 0.25 mm × 0.25 μm) 5% phenyl methylpolysiloxane EI in SRM	*Green tea*: 53–112 *Black tea*: 42–110	1–100	[7]
52 pesticides	Medicine and food dual-purpose herbs	PLE: 5 g of sample with diatomaceous earth 2:1 Florisil + GCB, as cleanup adsorbents in the cell Flush at 50% with ethyl acetate at 120 °C and 1500 psi for 6 min in 2 cycles	*GC-QqQ-MS/MS* With column back-flush HP-5 UI column (20 m × 0.18 mm × 0.18 μm) and a restrictor column (1.3 m × 0.180 mm id., 0.18 μm) 5% phenyl methylpolysiloxane EI in SRM mode	62–127	0.2–5	[78]

Continued

TABLE 1 Emerging Contaminants and Analytical Techniques Used in GC for Their Extraction and Determination in Different Food Matrices—cont'd

Food Contaminant	Matrix	Extraction	Determination	Recovery (%)	LODs (μg kg^{-1})	Reference
416 pesticides	Fruits and vegetables	QuEChERS (acetate buffered version)	*GC-QqTOF-MS* Split-less injection DB-5 ms (30 m × 0.25 mm × 0.25 μm). 5% phenyl methylpolysiloxane APCI and MS^E and accurate mass mode to	–	–	[103]
165 pesticides	Fruits and vegetables	QuEChERS (acetonitrile + NaCl + $MgSO_4$ followed by SPE Carbon/NH_2 cartridge)	*GC-QqTOF-MS* Split-less injection DB-35MS (30 m × 0.25 mm × 0.25 μm) 5% phenyl methylpolysiloxane EI and MS and MS/MS modes	–	–	[106]
34 pesticides	Milk, cream	Acetone/cyclohexane/ethyl acetate 2:1:1, Mg_2SO_4, NaCl. Purification: GPC, cleanup: SPE (GCB, PSA)	*GC × GC-TOF-MS (high speed TOF)* Split-less injection **First dimension:** **1.** deactivated Guard column (5 m × 0.25 mm) + VF5 (30 m × 0.25 mm × 0.25 μm); or, alternatively **2.** BPX550 (2.2 m × 0.1 mm) +HP5ms (30 m × 0.25 × 0.25 μm) 5% phenyl methylpolysiloxane Cold trap **Second dimension:** **1.** BPX-50 (2.2 m × 0.1 mm) or **2.** BPX-50 (1.5 m × 0.15 mm) 50% phenyl methylsiloxane EI in full scan *m/z* 40 and 600	66–84	0.2–0.4	[112]

Pesticides	Oil seeds	MSPD Homogenized with aminopropyl silica (dispersion sorbent, DS) + florisil (cleanup sorbent) and elution with acetonitrile	*GC × GC-TOF-MS (high speed TOF)* Split-less injection: First dimension: DB-5 MS (30 m × 0.25 mm × 0.25 μm) 5% phenyl methylsiloxane Cold trap Second dimension: DB-17 (1.15 m × 0.1 mm × 0.15 μm) 50% phenyl methylsilico EI in full scan *m/z* 50 and 500	–	–	[8]
Pyrethroid pesticides	Fruit and vegetable juices	HF-LPME 480 rpm speed agitation, extraction time of 41 min and NaCl concentration of 3% (w/v). Accurel Q3/2 polypropylene hollow fiber membranes with 1-octanol	*GC-qMS* Split-less mode HP-5MS (30 m × 2.1 mm, 0.25 μm) 5% phenyl methylpolysiloxane EI and MS		0.02–0.07	[94]

MA-NPS-DMAE = medium-assisted nonpolar solvent dynamic microwave extraction.
MTMOS–TEOS = sol–gel hybrid based on methyltrimethoxysilane–tetraethoxysilane.

The combination of GC to MS is commonly due through electron impact ionization (EI). The two dominant ionization methods used for GC-MS analysis are EI and chemical ionization (CI). The EI results in more fragments of low *m/z* and few, if any, molecules approaching the molecular mass unit. Advantages of EI ionization are the small influence of molecular structure on response and the large number of characteristic fragments [50]. This facilitates comparison of generated spectra with commercial available EI library spectra such as that of the National Institute of Standards of the USA. This feature is possible by the development of automated mass spectral deconvolution software, which allows a rapid screening of the samples. However, this ionization is not very appropriate for labile molecules that are most sensitive if softer CI (positive CI or negative CI) is used. One of the main benefits of using CI is that a mass fragment closely corresponding to the molecular weight of the analyte of interest is produced. An alternative source, much less explored, is APCI that, until recently, has hardly ever been applied coupled with GC [103]. Recent studies show very good prospects for this source in combination to HRMS.

The determination of emerging contaminants in food requires the introduction of almost all the analyte present in the sample into the GC system. Splitless injection is routinely used to the monitoring of emerging contaminants in foods. Programmed temperature-vaporizing (PTV) inlets [104] are also used including large volume injection [31,105] in order to introduce the maximum amount of analyte in the sample.

Separation efficiency is in general a characteristic of a GC column and is related to the solute "band broadening," which is related to width of solute peaks and the ratio of solute retention factors. GC uses long, open tubular capillary columns and is subsequently far more efficient at separation than any other chromatographic technique [5]. The polarity of the analyte is crucial for the choice of the stationary compound, which in an optimal case would have a similar polarity as the analyte. Most common stationary phases in open tubular columns are 5% phenyl methyl siloxane. There are few examples, but low-pressure GC (LP-GC) has also been applied for the fast analysis of various pollutants in different environmental and food matrices. A typical LP-GC set-up involves the use of a short microbore column (typically 0.5–1 m × 0.10 mm internal diameter) at the injector side connected with a zero dead-volume connector to a short megabore column (typically 10 m × 0.53 mm) to be used with higher gas velocities. Although the use of LP-GC results in a loss of separation efficiency, it is 3- to 5-fold more rapid than conventional GC. This system provides fast separation of PCBs, PBDEs, PAHs, pesticides, and novel flame retardants within 9 min [13].

The MS detectors that have been used are Q and QqQ, IT, and TOF. In many methods using quadrupoles, MS detection has been optimized by developing a SIM program using the most prominent masses (i.e., *m/z* values associated with each target compound, determined by running the GC-MS analysis in scan mode). SIM detection improves sensitivity, by decreasing the background noise, for the quantitative analyses of target compounds compared to the scan mode. However, SIM is not useful to identify nontarget compounds [50]. QqQ–MS/MS is considered as one of the most powerful techniques; there are a number of references in the literature to the use of GC-QqQ-MS/MS in the analysis of emerging contaminants in food [7,9,13,66,72,78].

Given their high mass resolution and accuracy as well as their sensitivity in scan mode (they are not quadrupoles), TOF platforms are successful in the screening, identification, and structure elucidation of some emerging contaminants (e.g., pesticides and their degradation products) in food and water samples [37]. Currently, three types of TOF-MS instruments differing in their basic characteristics are available to be combined to GC:

- Unit-resolution instruments that feature high acquisition speeds (500–1,000 spectra/s) commonly used in GC×GC.
- High-resolution/accurate mass analyzers (5000–12,500 FWHM) providing only moderate acquisition speed (20–50 spectra/s),
- High-speed high-resolution/accurate mass analyzers permitting high acquisition speeds (up to 200 spectra/s) as well as high mass resolving power (50,000 FWHM) also useful in GC×GC.

In addition to these TOF-MS instruments for single MS analysis, a hybrid instrument combining quadrupole and TOF-MS has recently been introduced allowing either analysis under the conditions of HR-TOF-MS/(MS) or Q/HR-TOF-MS (MS/MS) with selection of precursor ions and monitoring of the product-ion mass spectra [103,106].

The two most applied advances within this field that are discussed in detail below are the GC-MS/MS and the comprehensive GC×GC.

5.1 GC-MS/MS (GC-MS^2)

MS^2 experiments can be performed using QqQ, ^{3}DIT, and QqTOF mass analyzers. Despite IT-MS/MS capability that enables the sequential and multistage isolation of precursor ions, fragmentation, trapping, and mass scanning in the same space and as function of time, which allow the identification of unknown compounds, this mass analyzer presents less sensitivity, smaller linear dynamic ranges, and lower quantification capability than QqQ systems using SRM mode. Consequently, the QqQ platform in SRM mode is considered the most sensitive and robust for GC interface allowing low dwell times with high scan speeds [5,37,69,107].

Few examples of ^{3}DIT application are included in Table 1, one for phthalates [108] that did not exploit the MS/MS possibilities of the analyzer, and the other for pesticides residues [69] that use SRM. However, this mass analyzer is falling into disused due to the previously reported disadvantages.

Using GC-QqQ-MS/MS, Cajka et al. [7] screened 135 pesticides in green and black dry tea leaves and stalks using a simple sample preparation procedure, avoids the use of GPC and/or dSPE cleanup steps, decreasing the overall cost of the method and the analysis time. Figure 3 shows an overlay of SRM chromatograms of the 164 pesticides in a matrix-matched standard (0.1 mg/kg) acquired using the optimized MS/MS method. Typically, two SRM transitions per analyte are used in laboratories to meet identification criteria. Du et al. [78] carried out a selective PLE and GC-QqQ-MS/MS method for simultaneous determination of 52 pesticide residues in medicine and food dual-purpose herbs by means of an ultra-inert capillary GC-MS column and column back-flush system (to decrease analysis time and extend the life of GC–MS/MS system). Employing SRM for the quantitative analysis, LODs between 0.2 and 5 μg/kg were obtained. Other compounds analyzed by GC-QqQ-MS/MS, also in SRM mode, were both 3-monochloropropane-1,2-diol (3-MCPD) and acrylamide [66]. These authors proved that a 3-m polyethylene glycol precolumn was enough to improve the peak-tailing of both analytes with less column bleed. This choice increased the separation efficiency of polar and low molecular weight compounds such as 3-MCPD and acrylamide, which were found in samples of bread, fried chips, fried instant noodles, soy sauce, and instant noodle flavoring.

The ultra-trace analysis of 22 BFRs in fish by QqQ-MS/MS has been also described providing improved selectivity and sensitivity compared to a routinely used single quadrupole MS with limits of quantification in the range of 0.005–1 mg/kg [10]. Similarly, Luzardo et al. [72] performed an analysis of 16 PAHs, 20 organochlorine pesticides, and 18 PCBs on eggs, collected from the markets of the Canary Islands (Spain), by means of GC-QqQ-MS/MS. Their results showed no influence of the method of production (conventional, free-run, and organic) in the presence of these compounds, and none of the samples surpassed the MRLs established in the EU.

There are still few applications of the QqTOF and they are mostly in the field of pesticide residues. Cervera et al. [103] developed a GC-APCI-QqTOF-MS method for around 130 pesticides in fruit and vegetable samples, including strawberries, oranges, apples, carrots, lettuces, courgettes, red peppers, and tomatoes. The screening strategy consisted in first rapid searching and detection, and then a refined identification step using the QqTOF capabilities (MS^E and accurate mass). Identification was based on the presence of one characteristic *m/z* ion (Q) obtained with the low collision-energy function and at least one fragment ion (q) obtained with the high collision-energy function, both with mass errors of less than 5 ppm, and an ion intensity ratio (q/Q) within the tolerances permitted. Following this strategy, 15 of 130 pesticides were identified in the samples. Figure 4 shows illustrative examples of pesticides found slightly over the MRL in lettuce and carrot. Zhang et al. [106] also reported a method to identify pesticides residues using the most conventional EI source instead of

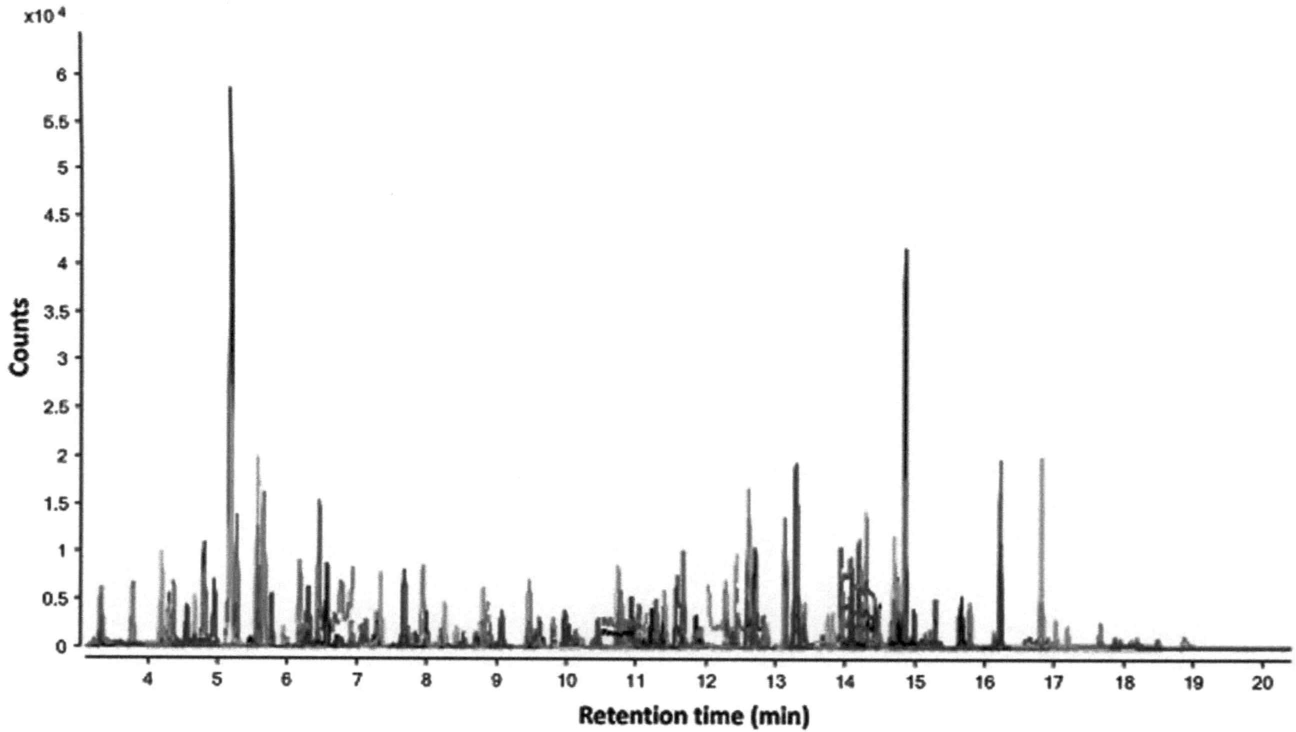

FIGURE 3 An overlay of SRM chromatograms of the 164 pesticides in the matrix (tea)-matched standard (0.1 mg/kg) acquired using the optimized MS/MS method. *Cajka et al. [7].*

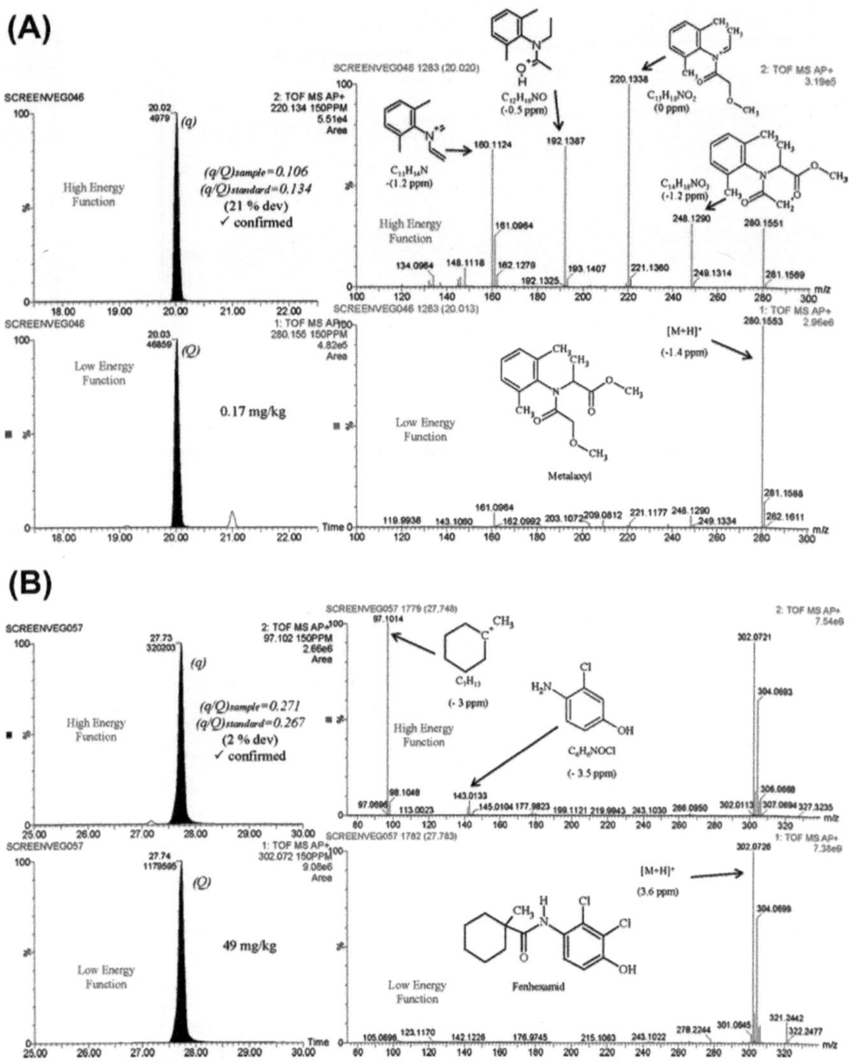

FIGURE 4 GC-APCI-TOF-MS narrow-window extracted ion chromatogram (mass window 150 ppm) showing the detection of (A) metalaxyl in one carrot sample (maximum residue level 0.1 mg/kg) and (B) fenhexamid in lettuce (maximum residue level 40 mg/kg). Experimental APCI accurate-mass spectra for the high collision-energy function and for the low collision-energy function are also shown with the chemical structures proposed for the most abundant fragment ions together with the experimental mass errors (in parts per million). *Cervera et al. [103].*

APCI and real isolation and fragmentation of the precursor ion. GC-QTOF-MS demonstrated a strong potential in pesticide residue analysis and proved to be a powerful tool for the structure elucidation of unknown compounds in complicated matrices.

5.2 Comprehensive GC (GC × GC)

GC × GC significantly enhances the separation capabilities of a one-dimensional GC. GC × GC involves the separation of target analytes by two orthogonal capillary columns, in which the second column is normally of smaller diameter and shorter in length. The resulting peaks are much sharper with smaller peak width and higher peak intensity. The coeluting peaks from the first column could be separated by the second column. In addition, it also offers enhancement in sensitivity of around one order of magnitude in terms of peak height [109]. However, some technical problems arise when GC × GC is coupled to mass analyzers since these have to be fast enough in order to differentiate all information coming from the comprehensive GC, and accordingly it is usually coupled to TOF-MS [110].

Despite its potential, the cost of the required instruments and the need for dedicated software for visualization and quantification have slowed down the spread of the development of GC × GC in food safety [111]. However, there are good examples of researches applying this approach as the analysis of 34 pesticides, isomers, and their metabolites in milk and cream [112]. This method includes intensive extraction with acetone/cyclohexane/ethyl acetate followed by GPC and SPE cleanup on a mixed bed GBC and primary/secondary amine silica gel (PSA) column before determination. Wang et al. [8] developed an oil-absorbing MSPD extraction with comprehensive GC × GC TOF-MS suitable for screening 68 pesticide residues (27 organophosphorus, 23 organic chlorines, 11 synthetic pyrethroids, and 7 carbamates) in peanut, soybean, rape seed, sesame, and sunflower seed. A 35-min orthogonal separation was performed with a nonpolar–polar column set. Identification of 68 pesticides by TOF-MS was conducted using mass spectral libraries (NIST libraries), and then confirmations were made via reanalyses and comparisons with reference standards.

GC × GC-TOF-MS has also been studied for the simultaneous analysis of several classes of organobromines (OBs), including PBDEs, polybrominated biphenyls (PBBs), methoxylated PBDEs (MeO-PBDEs), several halogenated naturally produced compounds (HNPs), eight novel brominated flame retardants, polybrominated hexahydroxanthene derivatives (PBHDs), 2,4,6-tribromoanisole, and a mixed halogenated compound in Bluefin tuna muscles. The proposed methodology maximized separation of both within and among OB families, and among these and other halogenated micropollutants detected in these samples and coextracted matrix components. Special attention has been paid to solve coelution problems observed during the analysis of OBs with one-dimensional GC-based techniques. Satisfactory separation among several relevant PBDEs and MeO-PBDEs has been obtained allowing their unambiguous determination in a single run. Several new tri- and tetra-BHD derivatives were also identified, indicating that these compounds could apparently exist as structured families in nature. In addition, a tetrabrominated dimethoxylated-biphenyl and two tetrabrominated dimethoxylated-BDEs were also tentatively identified [113].

In a different study, and in the same line, GC×GC-TOF-MS was applied to the simultaneous determination of 18 PCBs, 7 PBDEs, and 16 PAHs [105]. All target PCBs, PBDEs, and PAHs were separated in GC × GC system consisting of BPX5 (nonpolar) in the first dimension and BPX50 (polar) in the second dimension. The most critical analytes for the column selection and oven temperature program optimization were mainly isomeric groups of PAHs. The examples of chromatograms of PCB and PAHs in fish and mussel matrices are shown in Figure 5.

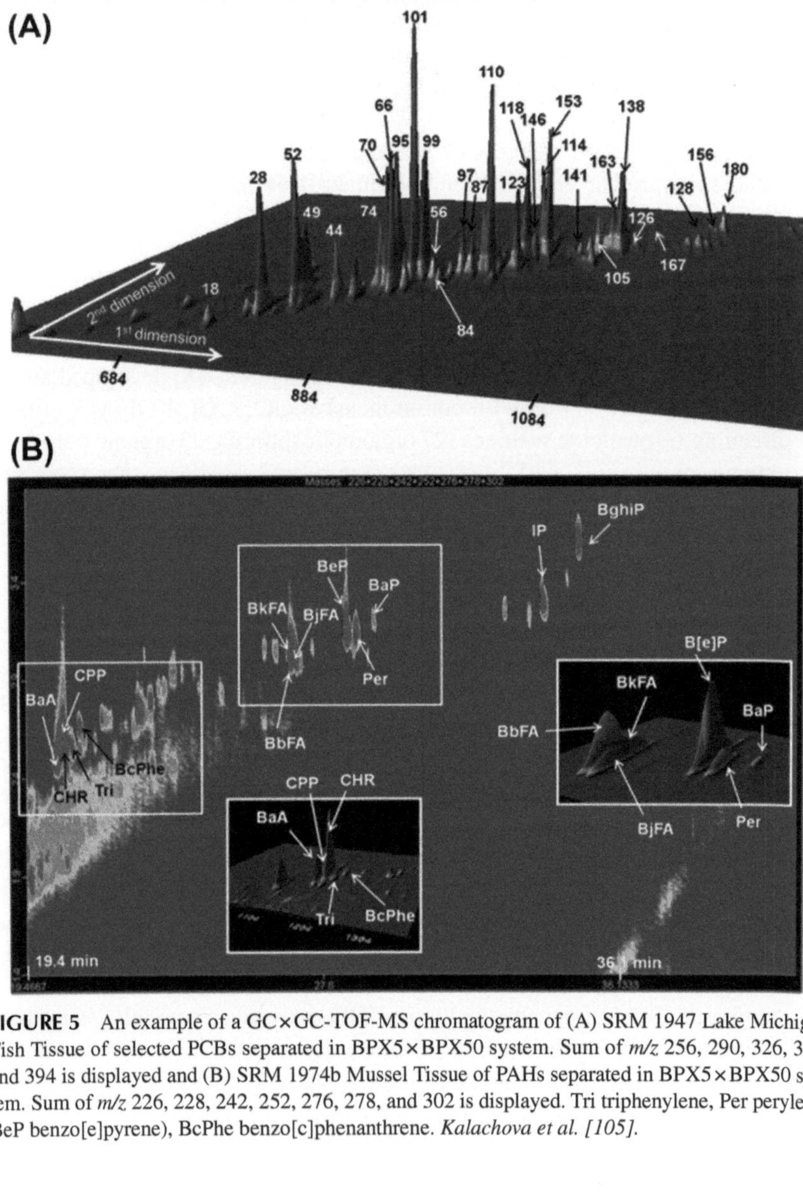

FIGURE 5 An example of a GC×GC-TOF-MS chromatogram of (A) SRM 1947 Lake Michigan Fish Tissue of selected PCBs separated in BPX5×BPX50 system. Sum of *m/z* 256, 290, 326, 360, and 394 is displayed and (B) SRM 1974b Mussel Tissue of PAHs separated in BPX5×BPX50 system. Sum of *m/z* 226, 228, 242, 252, 276, 278, and 302 is displayed. Tri triphenylene, Per perylene, BeP benzo[e]pyrene), BcPhe benzo[c]phenanthrene. *Kalachova et al. [105].*

GC × GC-TOF-MS has demonstrated to be a powerful tool for the simultaneous determination of different types of contaminants. This technique, even though complicate, saves time because allows separation of many compounds.

6. LIQUID CHROMATOGRAPHY-MASS SPECTROMETRY (LC-MS)

Recent developments in both LC and MS have resulted in very powerful instrumentation for sensitive and selective determination of more polar or ionic contaminants at trace levels in food. Accordingly, LC–MS (Q, QqQ, IT, TOF, or Orbitrap) [30,74] has increased exponentially its application within this field (in comparison to GC–MS methods, where the analyte must be volatile or derivatized to permit volatilization) [90]. Table 2 summarizes the most recent LC–MS and related methods developed and applied to assess food contamination.

The combination of API-MS/MS with separation techniques as LC or UHPLC is currently the more frequently used in the analysis of compounds of moderate to high polarity, particularly using APIs as APCI, ESI, and atmospheric pressure photo ionization (APPI). However, the use of APCI and APPI for analysis of food contaminants have been left in the wake of the ESI's overwhelming popularity. This may be related to the increasing number of analytes currently searched but may also reflect the improvements in source and probe design for ESI not yet paralleled in APCI or APPI [98].

The most widely used LC separation technique for contaminants is reversed-phase (RP) on monolithic columns (mainly silica-based monoliths). Monolithic columns currently used in LC show higher permeability and lower flow resistance than the conventional LC columns packed with small particles, providing thus significantly shorter separation times at moderate operation pressures. In fact, remarkable progress is been recently achieved in the development of monolithic stationary phases, with special attention to the RP, but also hydrophilic interaction LC (HILIC) and ion-exchange applications of monolithic LC columns. This is based in the fact that monolithic silica rods can be chemically modified by binding a variety of nonpolar, weakly or strongly polar functionalities as Chromolith C18 or Chromolith C8, which are commercially available and are widely applied [89]. However, very polar or ionic samples are weakly retained in RP columns and aqueous-organic normal-phase LC separation, named HILIC, has to be employed to allow successful separation [37]. These strategies allowed reducing the inner diameter and the particle size of the columns. The use of narrow bore columns (1–3 mm i.d.) is a common practice to determine emerging contaminants in food as can be observed in Table 2.

Recently, an important development in the field of LC has been the UHPLC, in which, columns are packed with smaller particles reaching higher flow rates and faster separations with superior resolution and improved sensitivity due to the narrow peak shape [4]. In this system, the mobile phase is chosen as a

TABLE 2 Emerging Contaminants, Analytical Techniques Used in LC for Extraction and Determination in Different Food Matrices

Food Contaminant	Matrix	Extraction	Determination	Recovery (%)	LODs (ng/g)	Reference
Mycotoxins						
Mycotoxins, veterinary drugs	Hen egg	1 g sample + 4 mL of a solution MeOH:Water:CH_3COOH (80:20:1), 0.5 g CH_3COONa and 2.0 g Na_2SO_4 anhydrous	*LC-ESI-QqQ-MS/MS* PI mode Silica C18 (150 mm × 2.1 mm × 3 μm) Water/AcN 0.1% CH_3COOH SRM	23–95	0.9–27.1	[59]
15 mycotoxins: Aflatoxins, fumonisins, trichothecenes, ochratoxin A, citrinin, sterigmatocystin, zearalenone	Milk thistle seeds, natural extract	QuEChERS + DLLME 2g sample + 8 mL 30 mM NaH_2PO_4 at pH 7.1 + 1% CH_3COOH in CAN + t (4 g $MgSO_4$, 1 g NaCl, 1 g sodium citrate, 0.5 g disodium hydrogen citrate sesquihydrate) DLLME with Cl_2CH_2	*UHPLC-QqQ-MS/MS* PI mode Zorbax Eclipse plus HRD (50 mm × 2.1 mm, 1.8 μm) Water/MeOH with 0.3% HCOOH and 5 mM $HCOONH_4$ SRM	46.1–98.9	0.45–459	[75]
Deoxynivalenol-3-glucoside	Cereal-based food	Methanol/water (80:20; v/v), clean-up: immunoaffinity columns	*LC-LIT-MS/MS* PI mode Kinetex C18 column (100 × 2.1 mm, 2.6 μm) Water/MeOH 0.5% CH_3COOH at 0.2 mL/min SRM	90	4	[20]
320 toxic And potentially toxic fungal secondary metabolites	Maize, maize-based products, nuts, nut-based products, beer, soybean	Extraction and dilution: ACN/water/glacial acetic acid 79:20:1, 20:79:1 v/v/v	*LC-ESI-QTRAP-MS/MS* PI and NI modes Gemini C18 (dimensions not provide) MeOH/water/glacial acetic acid with 5 mM CH_3COONH_4 SRM	Solids: 19–272 Liquids: 38–582%	Solids: 0.0002–5 Liquids: 0.0002–5	[18]

23 fungal, 3 bacterial metabolites	3 varieties Dried edible mushroom	Acetonitrile/water/acetic acid (79:20:1, v/v/v),	*LC-ESI-QTRAP-MS/MS* PI mode Gemini C18 (150 × 4.6 mm, 5 μm) SRM	74–114.1	0.01–12	[16]
12 mycotoxins, 8 veterinary drugs, 16 pesticides	Bakery (raw materials, finished products)	QuEChERS acetate buffered: AcN 1% CH_3COOH, stabilization CH_3COONa, anhydrous Na_2SO_4. Clean-up: dSPE	*UHPLC-ESI-Orbitrap-MS* PI mode Acquity UPLC BEH C18 (100 × 2.1 mm, 1.7 μm) Water/MeOH with 0.05% HCOOH and 1 mM $HCOONH_4$ at 0.2 mL/min *m/z* 50 to 100,000 FWHM	70–110	10	[19]
Phycotoxins						
Yessotoxin (YTX), 45 OH YTX, homo YTX, 45 OH homo YTX, carboxy YTX, carboxy homo YTX	Mussel	Acetone (Ultra-Turrax), n-hexane, methanol	*LC-ESI-QqQ-MS/MS* PI mode Pursuit XRs ultra C18 (50 × 2 mm, 2.8 μm) Water/MeOH and 2 mM HCOOH and 5.2 mM NH_4OH at 0.2 mL/min SRM	–	0.02	[22]
OA, YTX, AZA, PTX, CI, PbTx	Mussels and other bivalves	2 g of samples + 3 × 3 mL MeOH Three SPE clean-ups • Strata-X; • Oasis HLB • C18Elution: 0.5 mL Methanol, methanol with 0.3% NH_4OH, and ACN: 0.15% HCOOH in water	*UHPLC-ESI-QTRAP-MS/MS* PI mode Kinetex XB-C18 (100 × 2.1 mm, 2.6 μm) 0.15% HCOOH in water/ACN at 0.35 mL/min SRM, EPI	–	0.12– 13.6	[125]

Continued

TABLE 2 Emerging Contaminants, Analytical Techniques Used in LC for Extraction and Determination in Different Food Matrices—cont'd

Food Contaminant	Matrix	Extraction	Determination	Recovery (%)	LODs (ng/g)	Reference
Pesticides						
>250 pesticides, veterinary drugs	Animal feed (chicken, hen, rabbit, horse)	"Dilute-and-shoot" with water and acetonitrile (1% formic acid), clean-up: Florisil cartridges	*UHPLC-ESI-QqTOF-MS/MS (MS^E)* PI and NI modes Acquity UPLC BEH C18 (100 × 2.1 mm, 1.7 μm) Water/MeOH 0.1% HCOOH and 4 mM $HCOONH_4$ at 0.3 mL/min *m/z* 100–1,000, two simultaneous acquisition function LE and HE (30 eV)	10–171	1–150	[84]
24 pesticides	Fresh tomatoes	QuEChERS acetate buffered: AcN 1% CH_3COOH, stabilization CH_3COONa, anhydrous Na_2SO_4. Clean-up: dSPE	*LC-ESI-MS* PI mode Shim-pack C18 (75 × 2 mm, 2.1 μm) 5 mM CH_3COONH_4 with 0.1% HCOOH + AcN SIM mode	71.3–112.3	0.001–0.2	[73]
Azamethiphos, 3 avermectins, 2 carbamates, 2 benzoylureas	Farmed fish (turbot, panga, salmon), shellfish (scallop, clam, mussel, cockle)	MSPD. Clean-up: silica gel, elution: 0.5% Acetic acid in acetonitrile	*LC-ESI-QqQ-MS/MS* PI and Ni modes Hypersil ODS (100 × 3.2 mm, 3 μm) Water/ACN 5 mM CH_3COONH_4 at 0.4 mL/min SRM	84.9–118	1.5–3.7	[96]
98 OPPs, carbamates	Edible oil, meat, egg, cheese, chocolate, coffee, rice, tree nuts, citrics, vegetables	QuEChERS acetate buffered: AcN 1% CH_3COOH, stabilization CH_3COONa, anhydrous Na_2SO_4. Clean-up: dSPE: PSA and other material depending on the matrix	*LC-ESI-QqQ.MS/MS* PI mode RP C-12 (50 × 2 mm, 4 μm) Water/MeOH 5 mM $HCOONH_4$ at 0.25 mL/min SRM	70–120	10	[118]

54 pesticides	Apple, lemon, orange, tomato, carp, European catfish	QuEChERS (acetonitrile, $MgSO_4$, NaCl, $Na_3C_6H_5O_7 \cdot 2H_2O$, [HOC(COOH) $(CH_2COONa)_2 \cdot 1.5H_2O$dSPE: PSA, C18, graphitized black carbon	*UHPLC-ESI-LTQ-Orbitrap-MS/ddMS* PI mode XB-C18 (50 × 2.10 mm, 1.7 μm) Water/MeOH 0.1% HCOOH at 0.4 mL/min m/z 50–500; 60,000 FMWH and ddMS/MS ions >100 cps	58–120	0.1–2	[127]
53 pesticides	Raspberry, strawberry, black and red currant	QuEChERS (triphenyl phosphate, acetonitrile, NaOH. Clean-up: DSPE ($MgSO_4$, PSA, GCB, C18)	*UHPLC-ESI-QqQ-MS/MS* PI mode Zorbax plus C18 (100 × 2.1, 1.8 μm) Water/MeOH 0.01% HCOOH and 5 mM $HCOONH_4$ at 0.5 mL/min SRM	19.7–135.9	0.3–22.7	[119]
300 pesticides	Cucumber, lemon, wheat flour, rocket, and black tea	Acetonitrile	*LC × LC-ESI-QqQ-MS/MS* HILIC column: 1st Dimension YMC-Pack diol (100 × 2.1 mm; 5 μm) RP column: 2nd Dimension Poroshell 120 EC-C18 (100 × 2.1 mm; 2.7 μm) Water/CAN 5 mM 5 mM $HCOONH_4$ and 0.1% CH_3COOH at 0.2 mL/min SRM	70–120	0.001–0.01	[129]
Acetamiprid, desmethyl-, fenhexamid, -O-glucoside, iprodione, despropyl-, malathion, desmethyl-, propamocarb, desmethyl- (metabolites)	29 fruits And vegetables	QuEChERS (acetonitrile, anhydrous $MgSO_4$, sodium-acetate trihydrate. Clean-up: PSA, anhydrous $MgSO_4$)	Non-target *LC-ESI-TOF-MS* and Target *LC-ESI-QqQ-MS/MS* PI and NI modes XDB-C18 (50 × 4.6 mm; 1.8 μm) Water/AcN 0.1% HCOOH at 0.5 mL/min TOF 50–1000 *m/z* and in source CID QqQ-SRM	Only qualitative data		[33]

Continued

TABLE 2 Emerging Contaminants, Analytical Techniques Used in LC for Extraction and Determination in Different Food Matrices—cont'd

Food Contaminant	Matrix	Extraction	Determination	Recovery (%)	LODs (ng/g)	Reference
166 pesticides	Fruits, vegetables	QuEChERS (Acetonitrile, anhydrous $MgSO_4$, sodium-acetate trihydrate. Clean-up: PSA, anhydrous $MgSO_4$)	*UHPLC-ESI-Q-Orbitrap-MS* PI mode Acquity UPLC BEH C18 (100 × 2.1 mm, 1.7 μm) Water/AcN 0.1% HCOOH at 0.5 mL/min *m/z* 65 to 950 70,000 FWHM and ddMS/MS	81–110	–	[34]
Pharmaceuticals, Veterinary Drugs						
Sulphonamides, fluoroquinolones, macrolides, anthelmintics	Meat-based baby food, powdered milk-based infant formulae	LLE-based method "dilute-and shoot" (with 0.1% CH_3COOH in ACN) Acetate buffered QuEChERS	*UHPLC-ESI-QqQ-MS/MS* PI mode Acquity UPLC BEH C18 (100 × 2.1 mm, 1.7 μm) Water/MeOH 0.05% HCOOH at 0.2 mL/min SRM	46.3–134.6	0.1–5	[25]
Phenicol drugs, florfenicol-amine	Animal-derived foods (chicken, pork meat, fish, prawns, honey)	Fish, meat products: Acetic acid extraction, SPE (MCX) Honey: 1% Ammonium hydroxide solution, SPE (HLB)	*UHPLC-ESI-QqQ-MS/MS* PI mode Ascentis phenyl–hexyl (100 × 2.1 mm, 2.7 mm) Water/MeOH CH_3COOH and CH_3COONH_4 (pH 5.5) at 0.5 mL/min SRM	52–106	<0.1–1	[120]
Sulfonamides, penicillins	Calf liver, pangasius fish	Aqueous buffer, acetonitrile. Clean-up: SPE	*UPLC-ESI-Orbitrap-MS* PI mode Kinetex Core–Shell C18 (150 × 2.1 mm, 2.6 μm) Water/AcN with HCOOH at 0.4 mL/min *m/z* 140 to 1000 at 50,000 FWHM	Only qualitative		[130]

3 aminoglycosides, 3 β-lactams, 2 lincosamides, 4 macrolides, 4 quinolones, 4 sulfonamides, 3 tetracyclines, amprolium	Chicken muscle	2% Trichloroacetic acid in water–acetonitrile (1:1, v/v), removing fat with hexane, dilution, filtration	*LC-ESI-QqQ-MS/MS* PI mode ZIC–HILIC (100 × 2.1 mm, 3.5 μm) Water $HCOONH_4$ (pH2)/AcN at 0.2 mL/min SRM	53–99	0.1–20	[26]
86 pharmaceuticals and PCPs	Mussels	Acetonitrile, aqueous pH 2 buffer for the acidic ext. Or aqueous NH_4OH (pH 10) for the basic ext. Treated with Na_4EDTA. Clean-up: SPE (Oasis HLB)	*LC-ESI-QqQ-MS/MS* PI and NI modes XTerra C18 (10 cm × 2.1 mm, 3.5 μm) Water-methanol:acetonitrile 0.1% HCOOH and 10 mM $HCOONH_4$ at SRM	40–150	0.02–160	[30]
Tetracyclines, macrolides, quinolones, sulphonamides, anthelmintics (benzimidazoles, avermectins)	Eggs	**1.** *LLE*: AcN, 0.5M citric acid, 0.1M Na_2EDTA. SPE clean-up **2.** *SPE*: Na_2EDTA (pH 4). OASIS HLB washed (n-hexane), methanol, acetonitrile, 0.04% (v/v) 35% NH_3 in MeOH **3.** *MSPD*: C18 (n-hexane, dichloromethane, methanol). SPE (Florisil, methanol, acetonitrile, 0.04% (v/v) 35% NH_3 in MeOH **4.** Acetate buffered *QuEChERS*	*UHPLC-ESI-QqQ-MS/MS* PI mode Acquity UPLC BEH C18 (100 × 2.1 mm, 1.7 μm) Water/MeOH with 0.05% CH_3COOH at 0.3 mL/min SRM	1–71–97 2–3–2 4–2–96		[53]

Continued

TABLE 2 Emerging Contaminants, Analytical Techniques Used in LC for Extraction and Determination in Different Food Matrices—cont'd

Food Contaminant	Matrix	Extraction	Determination	Recovery (%)	LODs (ng/g)	Reference
12 macrolides, 7 β-lactams, 2 lincosamides morantel, orbifloxacin	Milk	AcN	*UHPLC-ESI-QqQ-MS/MS* PI mode UPLC HSS T3 (10 × 2.1 mm, 1.8 μm) 0.05% HCOOH in water/AcN at 0.3 mL/min SRM	51.5–139.0	0.1–2.5	[27]
Sulfonamides	Pork, chicken, liver, fish, milk and honey	"Stemit" QuEChERS that simultaneously extract and cleaned- up *SEP/MAC* QuEChERS performed in a single tube	*UHPLC-ESI-QTRAP-MS/MS* PI mode ZORBAX Eclipse AAA (150 × 4.6 mm, 3.5 μm) Water/AcN with 0.02% HCOOH at 0.5 mL/min SRM, PreS-IDA-EPI, NLS-IDA-EPI, SRM-IDA-EPI	67–116	≤7.5	[124]
Alkylphenols						
Alkylphenols (APs)	Mussels	Dispersion: Water. Extraction: Isooctane by steam distillation. Clean-up: SPE aminopropyl sorbent	*LC-ESI-QqQ-MS/MS* PI and NI modes XTerra C18 (10 cm × 2.1 mm, 3.5 μm) Water-methanol:acetonitrile 0.1% HCOOH and 10 mM $HCOONH_4$ at 0.4 mL/min SRM	–	0.02–160	[30]
25 non-Volatile compounds NPOs and other contaminants	Adhesives used in food packaging materials	Acetonitrile, shaken, acrylic polymer precipitated. Filtered and diluted (1/100 acetonitrile)	*UPLC-TOF-MS, UPLC-HDMS* PI and NI modes Acquity UPLC BEH C18 (100 × 2.1 mm, 1.7 μm) Water/AcN at 0.4 mL/min *m/z* 100–1,000, in source fragmentation	Qualitative identification		[126]

Flame Retardants						
HBCD stereoisomers	Fish	SUPRAS microextraction 750 mg Of fish 650 µL reverse aggregates of decanoic acid (DeA)	*LC-ESI-QTRAP-MS/MS* NI mode Chiral Nucleodex β-PM (200 × 4.0 mm, 5 µm) Water/MeOH and MeOH/AcN at 0.3 mL/min SRM	87–114	0.5–5.6	[93]
HBCD stereoisomers	Biota samples	Not described	*LC × LC-ESI-QqQ-MS/MS* NI mode RP column 1st dimension Synergi (150 × 2 mm, 4 µm) Chiral column: 2nd Dimension Nucleodex beta-PM (200 × 4 mm, 5 µm) Water/MeOH gradient SRM	–	–	[128]
PFASs						
PFBA, PFPA, PFHxA, PFHpA, PFOA, PFNA, PFDcA, PFUnA, PFDoA, PFTrA, PFBS, PFHxS, PFOS, PFDcS, PFOSA	Whole herring, sprat, eggs (common eider) eggs, livers (herring gull)	Acetonitrile, ENVI-Carb, glacial acetic acid	*LC-ESI-TOF-MS* NI mode ACE-C18 (150 × 2.1 mm, 3 µm) Water/MeOH 5 mM CH_3COONH_4 at 0.2 mL/min *m/z* 50-100 in source CID	95	0.01–1.20	[10]
PFBS, PFHS, PFOS, PFDS, PFOSA, PFHxA, PFHpA, PFOA, PFNA, PFDA, PFUnDA, PFDoDA	Mussels	Ion-pairing (0.5 M tetrabutyl ammonium hydrogen sulfate, sodium carbonate buffer, methyl tert-butyl ether)	*LC-ESI-QqQ-MS/MS* NI mode Betasil C18 (100 × 2.1 mm, 5 µm) Water/MeOH 5 mM CH_3COONH_4 at 0.3 mL/min SRM	70–130%	0.02–160	[30]

Continued

TABLE 2 Emerging Contaminants, Analytical Techniques Used in LC for Extraction and Determination in Different Food Matrices—cont'd

Food Contaminant	Matrix	Extraction	Determination	Recovery (%)	LODs (ng/g)	Reference
PFBS, PFHxS, PFOS, PFDS, PFHxA, PFHpA, PFOA, PFNA, PFDA, PFUnA, PFDoA, PFTrA, PFTeA, 6:2, 8:2, 10:2 FTUCAs, 6:2, 8:2, 10:2 FTOHs, PFOSA, NMeFOSA	Eggs (glaucous-winged, California, ringbilled, herring gulls)	10 mM KOH acetonitrile/water (80/20 v/v), clean-up, fractionation: SPE (Oasis WAX)	*HPLC-APPI-QqQ-MS/MS* NI mode ACE 3C18 (50 × 2.1 mm, 3 μm) Water/MeOH 2 mM CH_3COONH_4 at 0.3 mL/min SRM	>70%	–	[83]
PFOA, PFOS	Fish, seafood, meat, milk, cereal-based food, eggs, vegetables, honey, oils, beverages	Ion-pairing (tetrabutylammonium, hydrogen sulfate solution, sodium carbonate buffer 0.25 M, pH 10, methyl tert-butyl ether	*HPLC-QTRAP-MS/MS* NI mode Betasil C18 (50 × 2.1 mm, 5 μm) Water/MeOH 2 mM CH_3COONH_4 at 0.2 mL/min SRM	–	PFOA: 3 PFOS: 1.5	[28]
PFBA, PFPeA, PFHxA, PFHpA, PFOA, PFNA, PFDA, PFUdA, PFDoA, PFTrA, PFTeA, PFHxDA, PFODA, PFBS, PFHxS, PFOS, PFDS, FHEA, FOEA, FDEA	Cereals, fish, fruit, milk, ready-to-eat foods, oil, meat	Alkaline digestion, purification: TFC (Aria TLX-1 system), 2 extraction columns	*LC-ESI-QqQ-MS/MS* NI mode Hypersil GOLD PFP (50 × 3 mm) Water/MeOH 2 mM CH_3COONH_4 at 0.2 mL/min SRM	70–120%	0.005–0.65	[29]

SEP/MAC, fast single-tube extraction/partitioning-multifunction adsorption clean-up; OA, okadaic acid; YTX, yessotoxins; AZA, azaspiracids; PTX, pectenotoxins; CI, cyclic imines; PbTx, brevetoxins.

compromise between optimal chromatographic separation and good ionization efficiency and overall MS performance. UHPLC-MS allows not only faster analyses, which are obviously very interesting, but also multiclass, multiresidue methods with short injection cycle times and minimal sample preparation decreasing the cost per analysis and reducing the impact of these methods on the environment. Analytical strategies applied in this field have evolved to respond to statutory requirements for the scope of analytes detected, and the needed method detection limits to meet enforcement action levels, such as those stipulated by the European Commission in 2002/657/EC [117].

Summarizing, current development efforts have focused on (1) generic, versatile, and simplified extraction procedures (e.g., QuEChERS, dilute and shoot, medium polarity SE). (2) improved chromatographic performance (mainly related to the development of UHPLC technology) in terms of shorter run time and higher sensitivity, and (3) new HRMS analyzers exploiting high-performance TOF and Orbitrap technology instruments, which provide significant advantages with regard to selectivity, sensitivity, accuracy, and speed for rapid screening of organic contaminants in complicated samples and possible identification for unknown compounds [50].

6.1 LC-MS/MS (LC-MS^2)

Almost the same mass analyzers combined with GC are also combined with LC. However, the ionization techniques used in the combination of LC-MS are called "soft" ionization techniques because they produce a lack or minimal extend of analyte fragmentation. Then, the use of MS/MS is almost mandatory to carry out a proper identification of the analytes. In fact, only one study is compiled in Table 2 that used single-stage nominal mass determination [73].

Most common MS/MS instruments are QqQ and LTQ (or QTRAP), but QqTOF and different combinations of Qs and ITs with the Orbitrap have gained popularity in recent years, even though acquisition and maintenance is more costly [5]. These combinations of HRMS and MS/MS will be covered in the next section.

Commonly both, QqQ and LTQ work in SRM that is monitoring a number of precursor ion→product ion transitions. Modern MS/MS analyzers are capable of full scan acquisition rates of up to 10,000 *m/z* per second, dwell times of 1 ms, and polarity switching in 30 ms or less. Table 2 schematizes a number of studies that apply these mass analyzers to determine mycotoxins [16,20,75], phycotoxins [22], pesticides [33,96,118,119], and veterinary drugs [26,27,53,120]. However, the most efficient ways to monitor emerging contaminants in food are multiclass, multiresidue methods because they are designed to detect a large number of contaminants simultaneously [121]. MS-based multiclass analysis is the recent trend.

When coupled with multiresidue extraction techniques, LC-MS/MS on QqQ instruments in SRM mode is capable of screening a large number of target contaminants, even in difficult food matrices. Using these last generation instruments, it is possible to develop multiresidue methods able to cover between 200 and 300 analytes simultaneously. For example, a multiclass, multiresidue UHPLC–MS/MS method was successfully validated for screening 113 of the 127 veterinary drug residues tested at or below US regulatory tolerance levels in bovine muscle. A novel aspect of this method was the postcolumn infusion of mobile-phase additives, such as ammonium formate, during the elution of anthelmintic drugs to enhance their MS detection properties [121]. The final method presented was capable to identify 98 compounds and quantify 87. Similarly, Zhan et al. [122] optimized a UHPLC–MS/MS method for the analysis of 255 veterinary drug, pesticide, mycotoxins, and other potential chemical contaminants in milk, using a generic extraction technique based on a two-step precipitation attaining total analysis times below 10 min Figure 6 illustrated a standard and a real sample. Interestingly, confirmatory analysis was performed according to the revised EU criteria (2002/657/EC) [123] and the applying SRM of two transitions per compound could reduce the risk of false positives. Moreover, the ratio between different product ions and relative retention time provided additional identification and confirmation, as shown in Figure 6. A QuEChERS-like extraction method was developed by Capriotti et al. [59] for the simultaneous analysis of mycotoxins and

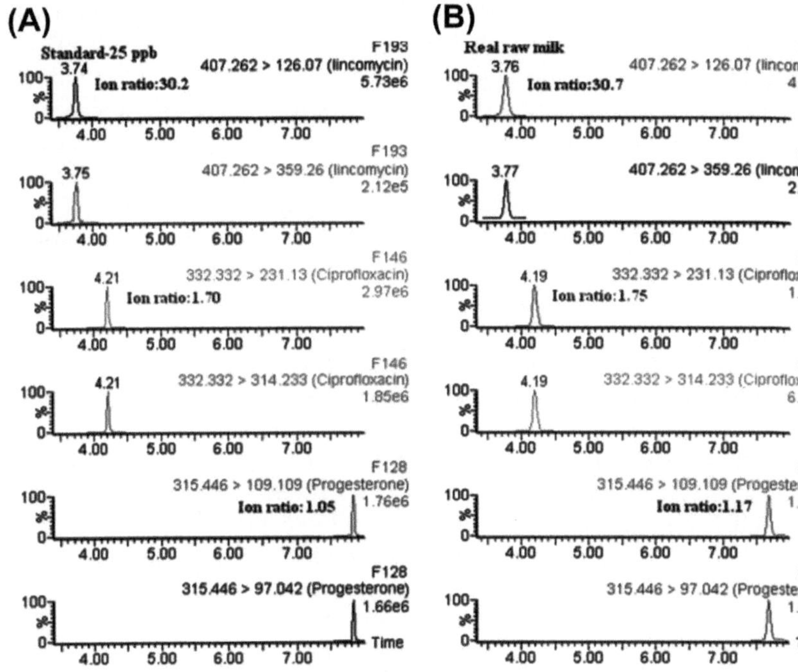

FIGURE 6 Chromatograms of standard solution (A), a real raw milk sample positive for lincomycin, ciprofloxacin, and progesterone (B). *Zhan et al. [122].*

veterinary drugs in hen eggs by LC–MS/MS with ESI source. Various classes of mycotoxins (enniatins, beauvericin, ochratoxins, aflatoxins) and antimicrobials (tetracyclines, ionophores, coccidiostats, penicillins, cephalosporins, fluoroquinolones, sulfonamides) were considered for the development of this method. Particular attention was devoted to extraction optimization: different solvents (acetone, acetonitrile, and methanol), different pH values, and different sample to extracting volume ratios were tested and evaluated in terms of recovery, RSD, and ESI signal suppression due to matrix effect. Chromatographic and MS conditions were optimized to obtain the best instrumental performances for most of the analytes. Quantitative analysis was performed by means of matrix-matched calibration, in a range that varied depending on the analyte and its established maximum limit. Abia et al. [18] reported on multimycotoxin occurrence in staple food commodities from Cameroon. Samples, including maize and maize-based products, nuts and nut-based products as well as beer, beverage, and soybean were analyzed for 320 toxic and potentially toxic fungal secondary metabolites. A total of 69 metabolites were detected in all studied commodities.

The other important trend is the determination of emerging contaminants of industrial or environmental origin that are at very low concentrations. Analysis of PFASs in food samples is normally carried out by LC-ESI-MS/MS allowing LODs in the pg–ng/g level in food samples. For example, Perez et al. [29] assessed the levels of 21 PFASs in 283 food items (38 from Brazil, 35 from Saudi Arabia, 174 from Spain, and 36 from Serbia) among the most widely consumed food stuffs in these geographical areas. The analysis of food stuffs was carried out by turbulent flow chromatography (TFC) combined to LC-QqQ-MS. For all the selected matrices, the method LOD and the method limit of quantification (LOQ) were in the range of 5–650 pg/g and 17–2,000 pg/g, respectively. The performance of LTQ has also been demonstrated for trace quantification of PFASs [82]. Recently, Guerranti et al. [28] studied the levels of perfluorooctane sulfonate (PFOS) and perfluorooctanoic acid (PFOA) in human milk and food samples (cereal-based products, meat, eggs, fish and seafood, fruit and vegetables, etc.) from the city of Siena and its province (central Italy) based on LC-LTQ-MS/MS. According to their results, PFOS was found in fish samples (7.65 ± 34.2 ng/g), and in meat, milk and dairy products (1.43 ± 7.21 ng/g and 1.35 ± 3.45 ng/g, respectively).

The improved full-spectrum sensitivity in LIT due to its high ion-accumulation capacity provides very promising modes, making the technique very powerful for identification of unknown or suspected analytes of interest, even with poor fragmentation and/or low concentrations. Owing to the fast scanning capability, another attractive possibility of LTQ for semitargeted analysis is the use of information-dependent acquisition (IDA) for automatically combining a survey scan with the dependent (enhanced trap) scan during a single-looped experiment. In the recent years, some reports in the literature have demonstrated the feasibility of using these LC-LTQ enhanced modes for tracing chemical residues, multitarget screening, and structural characterization from complex matrices. Huang et al. [124] obtained synchronous supplementary confirmation in the LC-LTQ system performing an IDA experiment using the QqQ-based scanning mode (SRM, precursor ion scan, or neutral loss scan) with dynamic background subtraction as the survey scan and an automatically triggering EPI (full-spectrum ion-trap MS/MS mode) by the built-in IDA software as dependent scan for sulphonamides. Compared with the conventional SRM-based targeted measurement and more sophisticated multidimensional platform toward exhaustive identification, the homolog-targeted screening opened up an interesting strategy for expanding analyte coverage of target category of compounds and unknown structural analogues without any compromise of the data quality. Wi et al. [125] developed an LC-LTQ-MS method with fast polarity switching and a scheduled SRM algorithm mode for multiclass screening and identification of lipophilic marine biotoxins in bivalve molluscs. An EPI library was constructed after triggered collection of data via IDA of EPI spectra from standard samples. Overall, the method exhibited excellent sensitivity and reproducibility, and it will have broad applications in the monitoring of lipophilic marine biotoxins.

6.2 LC-High Resolution Mass Spectrometry (LC-HRMS)

The use of LC-QqQ-MS may compromise confirmation and sensitivity because of the loss of qualitative information required to support the structural elucidation of analytes in the SRM mode. This qualitative information can be obtained in full-scan mode, but with a loss of sensitivity. TOF and Orbitrap can resolve this limitation by its ability to provide accurate mass measurements of full-product ions assuring accurate identification of analytes, with better sensitivity than QqQ-MS in the full-scan mode. For these purposes, LC-HRMS, using either TOF or Orbitrap analyzers, has gained wider acceptance in the last few years in the field of residue analysis due to the availability of more rugged, sensitive, and selective instrumentation.

HRMS considered as a full-scan technique would, in theory, makes possible the simultaneous analysis of an indefinite number of compounds. Both UHPLC and HRMS resulted in the development of new analytical strategies that involved compounds belonging to different groups, including antibiotics, growth promoters, pesticides, and natural toxins [76]. New HRMS instruments, working in a full-scan modality, could permit to execute successful targeted screening having at the same time the added benefits of potential nontargeted/unknown analyses and retrospective controls by postacquisition processing. In particular, this last point is attractive for future comprehensive toxicological monitoring and quantification owing to the ability to reanalyze past data without reextraction/injection. The increased resolution (between 10,000 and 100,000) and high mass accuracy (relative mass error <3 ppm) offer the ability to perform accurate mass measurements of ions of interest, facilitating the elucidation of the elemental composition for confirmation/elucidation studies. A theoretically unlimited number of compounds can be monitored in one analytical run. This feature maps well against the use of accurate-mass databases, which can foster the effective screening of known species in an automated fashion. For example, Canellas et al. [126] identified the nonvolatile compounds as potential migrants from adhesives used in food packaging. A number of the current acrylic adhesive formulations were extracted and prepared for analysis. The extracts were screened using UHPLC-TOF-MS. This approach allowed the identification of several components by a combination of exact mass and in-source CID. Polgar et al. [33] described a methodology based on LC-TOF-MS for the systematic screening of pesticides and their main metabolites in fruit and vegetable samples, using accurate-mass databases of parent compounds and their diagnostic fragment ions. The approach is based on (1) search for parent pesticides; (2) search for their metabolites in the positive samples, assuming common fragmentation pathways between the metabolites and parent pesticides; and (3) search for pesticide conjugates using the data from parent species and their diagnostic fragmentation. A typical chromatogram and mass spectrum is shown in Figure 7.

The same strategy also demonstrated that, by using an Orbitrap analyzer coupled to UHPLC, the simultaneous separation and detection of pesticides, toxins, and antibiotics can be performed as an alternative to MS/MS [19]. An example of the usefulness of this approach was reported by De Dominicis et al. [19] with a fast method for extracting and purifying bakery matrices (raw materials and finished products) for analytes including 12 mycotoxins, 8 veterinary drugs, and 16 pesticides using UHPLC coupled to an Orbitrap Exactive HRMS. Analyzing extracts of blank, naturally contaminated, and fortified minicakes, prepared through a combined use of industrial and pilot plant production lines, at different concentration levels (1–100 mg/kg) of various contaminants, authors demonstrated that this is a very useful tool with a high potential for efficient multiresidual analysis in the field of food safety obtaining LODs of 10 mg/kg for most of the compound categories analyzed.

Applications of UHPLC hyphenated QqTOF-MS for pesticide residues analyses have been also reported. Aguilera-Luiz et al. [25] developed a fast and generic method for the simultaneous monitoring of >250 pesticides and veterinary drugs in animal feed. A "dilute-and-shoot" extraction with water and acetonitrile (1% formic acid) followed by a cleanup step with Florisil cartridges was applied. The extracts were analyzed by UHPLC-QqTOF-MS/MS using both positive and negative electrospray ionization. The detection of the residues was accomplished by retention time and accurate mass using an in-house database.

The different hybrid orbitraps have also been applied to the determination, mostly of pesticide residues. Wang et al. [34] presented an application of UHPLC-ESI-Q-Orbitrap for determination of 166 pesticide residues in fruits and vegetables. UHPLC-ESI-Q-Orbitrap-MS (i.e., full MS scan) acquired full MS data for quantification, and UHPLC-ESI-Q-Orbitrap dd-MS^2 (i.e.,

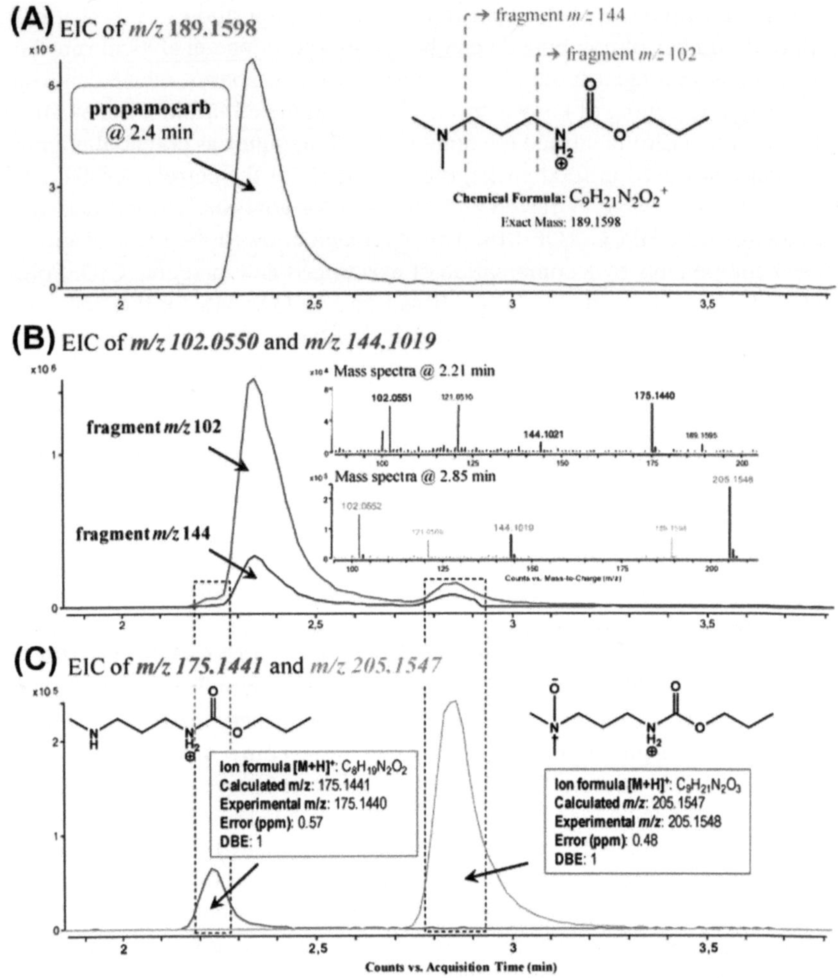

FIGURE 7 Detection of propamocarb (PRO) and its metabolites in a head lettuce sample by LC–TOF-MS: (A) EIC of the PRO parent compound (*m/z* 189); (B) EICs of fragment ions of PRO (*m/z* 102 and *m/z* 144); the presence of metabolites was indicated by the additional peaks of fragment ions (mass spectra at these retention times are shown as insets); (C) EICs and proposed structures of the two discovered metabolites (*m/z* 175 and *m/z* 205). *Polgar et al. [33].*

data-dependent scan) obtained product-ion spectra for confirmation. UHPLC-ESI-Q-Orbitrap-MS quantification was achieved using matrix-matched standard calibration curves with isotopically labeled standards or chemical analogues as internal standards. Confirmation was based on mass accuracy ≤5 ppm and LC retention time tolerance within ±2.5%. Overall, the UHPLC-ESI Q-Orbitrap has demonstrated great performance for quantification and confirmation of pesticide residues in fresh fruits and vegetables. Farré et al. [127] also analyzed pesticide residues in fruits and fish by means of UHPLC-LTQ-Orbitrap-MS, which acquired full-scan MS data for quantification, and data dependent MS^2 and MS^3 product-ion spectra for identification and/or confirmation. The regression coefficients for the calibration curves in this study were ≥0.99. The LODs for 54 validated compounds were ≤2 ng/mL (analytical standards) obtaining mass accuracies always ≤4 ppm, which corresponded to a maximum mass error of 1.6 amu. As Figure 8 illustrates, discrimination of the analytes could be only made from their $ddMS^2$ spectra at fragmentation energy of 35%. This result indicates that MS/MS is still indispensable for identification, especially regarding structural isomers that have the same calculated exact mass.

Since emerging contaminants can be unknown complex mixtures, the most advanced techniques are necessary for screening purposes to elucidate the composition. Recently, the high-definition mass spectrometer, presented as Synapt HDMS developed by Waters is a technique capable to provide much structural information. This equipment is a Q-IM-TOF-MS providing an expanded range of fragmentation protocols for structural characterization studies because it provides first- and second-generation product ions from a precursor in one experiment. The first generation of fragments is separated by ion mobility that is a gas-phase electrophoretic technique that gives rapid separations of gas-phase ions on the milliseconds timescale. This technique has been applied to analyze unknown compounds coming from the adhesive formulation in contact with food [126]. Using the MassFragment™ tool to interrogate fragmentation data, a wide series of compounds were identified, demonstrating the usefulness and importance of these tools for difficult problems.

6.3 Comprehensive LC (LC × LC)

Two-dimensional HPLC separations can be used to significantly increase the number of compounds separated in a single run in comparison to single dimension separations and to classify the sample on the basis of structural similarities. Combining different LC separation mechanisms, such as RP, ionic exchange or HILIC chromatography, allow to accomplish noncorrelated retention with maximum peak capacity. LC × LC is practiced either in online or in off-line setup. In the off-line "heart-cut" setup, only a few selected fractions from the first-dimension column are separated on a second-dimension column in alternating cycles controlled via a switching valve interface. In comprehensive LC × LC chromatographic techniques, the whole effluent from the first-dimension column is collected in fractions, which are in-line transferred to the second-dimension column. Although comprehensive LC × LC allows automatic performance avoiding tedious and time-consuming off-line fraction transfer between the separation systems, it has experienced a much slower development than GC × GC. A probable reason is that the chemistry of the stationary phase and the composition of the mobile phase in each dimension should match for real-time two-dimensional operation. For online comprehensive LC × LC practice, it is essential to

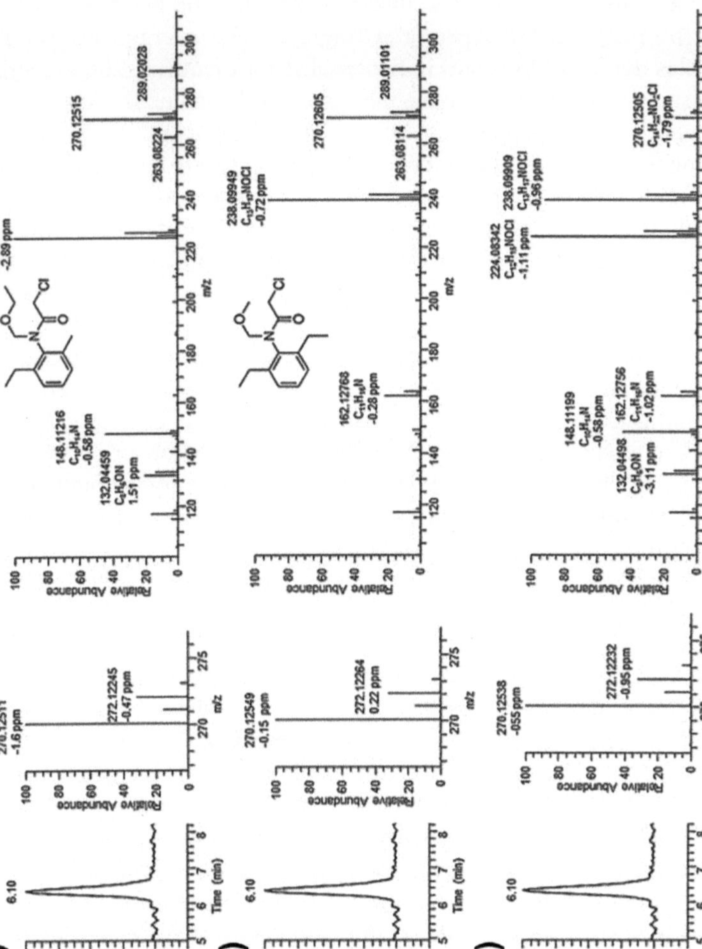

FIGURE 8 UHPLC peak as well as precursor and product-ion mass spectra obtained by LTQ-Orbitrap MS/ddMS2 for (A) alachlor, (B) acetochlor spiked separately in orange blanks at 5 ng/g, and (C) together in a nonlaboratory spiked lemon sample at 6.9 and 4.7 ng/g, respectively. *Farré et al. [127].*

accomplish the separation in the second dimension while the next fraction from the first dimension is collected, usually in less than 1 min. Such fast efficient separations require short efficient columns, fast gradients and high flow-rates of the mobile phase. This can be accomplished on columns packed with sub-2 μm particles, but at the cost of very high operation pressures [89].

LC×LC has been still little applied in the field of food safety because the most important issue in these type of analysis is sensitivity. However, there are already two recent interesting applications to determine emerging contaminants. One tackles the enantioselective analysis of isomers of HBCD, using a two-dimensional LC approach to avoid co-elution, in particular between (+) α-HBCD, (+) β-HBCD, or (+) γ-HBCD. After isomer separation on a conventional column, the single isomers are transferred to an enantioselective LC column using heart cuts. Two enantioseparations are conducted in two separate partial chromatograms: one for α-HBCD and one for β- and γ-HBCD. The result is a completely undisturbed enantioselective separation for α-HBCD at a resolution of 4.11. A peak capacity of 107 was achieved. This peak capacity is utilized by the six peaks of the three isomers with two enantiomers each by 6% [128]. The enantioseparations of the ^{12}CHBCD standard, ^{13}CHBCD standard, and HBCD isomers from an environmental sample (glaucous gull liver, collected in East Greenland in 2004) are shown in Figure 9 a, b, and c, respectively.

Using this approach, a fully automated system was developed for the determination of more than 300 different pesticides from various food commodities [129]. The samples were extracted with acetonitrile before their injection into the two-dimensional LC-system (without manual cleanup). The separation of analytes and matrix compounds was carried out by a HILIC column in the first dimension, and all analytes eluted within one small fraction at the beginning of the run. This fraction was transferred to the analytical RP separation with a packed loop interface; however, some very polar compounds with a stronger retention on the HILIC column were measured directly. The method was validated for over 300 pesticides in cucumber, lemon, wheat flour, rocket, and black tea, obtaining for the large majority of the analytes LODs of at least at 0.01 mg/kg. For over 50% of the analytes, the method presented good sensitivity even at 0.001 mg/kg showing robust results even with dirty matrices like hops and tea.

7. OTHER METHODS

In this section, some less used methods but with promising prospects are discussed. These methods include (1) miniaturized analytical techniques in their different forms, (2) ambient ionization techniques, and (3) IMS.

Food analysis is a challenging research area, in which miniaturized analytical techniques are increasingly being used in order to reduce analysis time and to have cheap methods with high-resolution and efficiency. Capillary electrophoresis (CE), capillary-LC (CLC), and recently nano-LC are among the

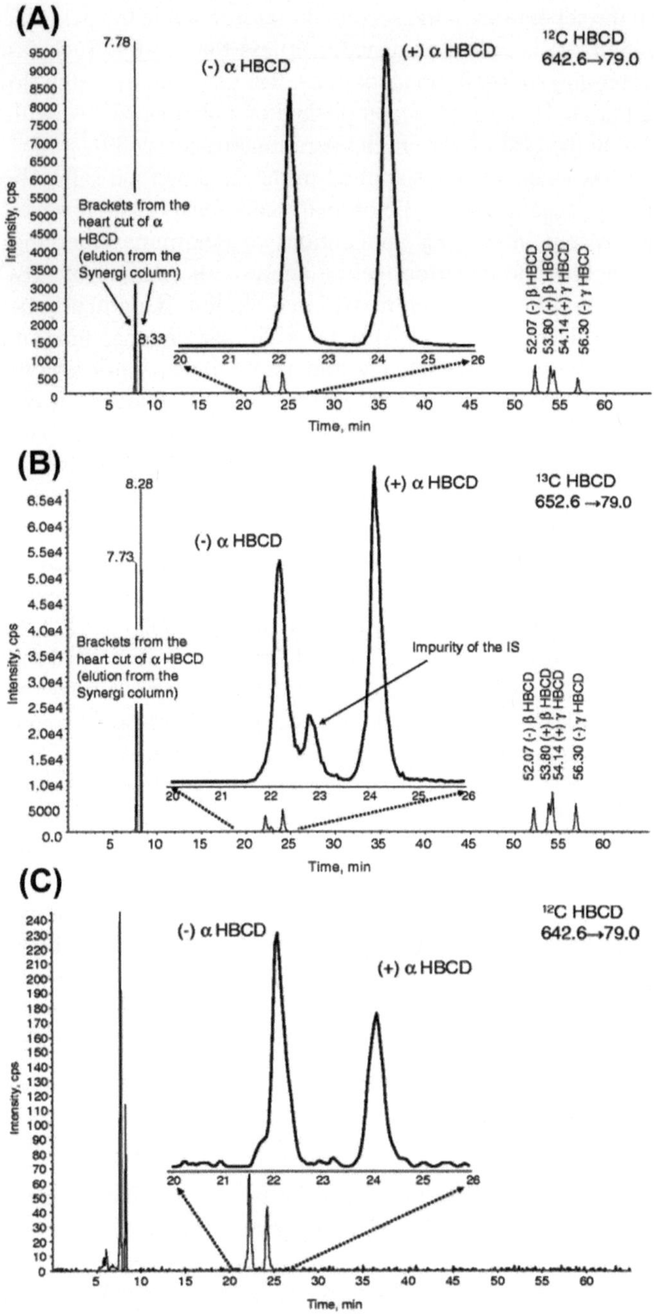

FIGURE 9 (A) Two-dimensional separation of HBCD utilizing a Synergi (L: 150 mm; i.d.: 2 mm; particle size: 4 µm; pore size: 80 Å; Phenomenex, Torrance, CA, USA); (B) and an enantioselective column (Nucleodex beta-PM; L: 200 mm; i.d.: 4 mm; particle size: 5 µm; Macherey–Nagel GmbH & Co., Düren, Germany); (C) Enantioseparation of HBCD of a sample of glaucous gull liver from East Greenland, collected in 2004. *Bester and Vorkamp [128].*

techniques most cited in literature because of their great potential in the analytical field [131]. In CLC and nano-LC, the separations are carried out in capillaries of narrow internal diameter (10–320 μm) containing selected stationary phases. Main features of these techniques are the short analysis time, high mass sensitivity, minute volumes of samples and mobile phases, and easy coupling with MS. Application of CE was performed by Alshana et al. [132] for the determination of five nonsteroidal antiinflammatory drugs in bovine milk and dairy products. To do so, DLLME was coupled with field-amplified sample stacking in capillary electrophoresis. Under optimum conditions, enrichment factors were in the range 46–229 and LODs of the analytes ranged from 3.0–13.1 μg/kg for all matrices analyzed.

Carbamate pesticides, such as oxamyl, methomyl, aldicarb, carbofuran, pirimicarb, thiocarb, and ditalimfos, were analyzed in tomato using CLC coupled to an MS detector by means of a microfabricated, heated nebulizer chip (APPI). This system obtained very high sensitivity to such compounds in comparison to ESI in which the same analytes did not produce good signals. Capillary column (50 × 0.3 mm i.d. and particle size 3.0 μm) was packed with silica C18 and a water–methanol mixture was pumped at 5 μL/min in gradient mode. Tomato samples were prepared using a buffered QuEChERS method after spiking with the seven pesticides and analyzed obtaining LODs <0.25 ng/mL (below the MRL limits). Aqai et al. [133] demonstrated the usefulness of nano-LC for the analysis of ochratoxin A and its nonchlorinated analogue, ochratoxin B, in wheat and cereal samples. Super-paramagnetic microbeads coated with monoclonal antibodies (Mabs) were used in a novel direct inhibition flow cytometric immunoassay plus immunoaffinity isolation. Analytes were detected by coupling the nanochromatographic apparatus with a QqTOF-MS to obtain a LOD as low as 0.15 ng/g for ochratoxin A, which is far below the lowest maximum level of 3 ng/g established by the EU.

In Ambient-MS, the ionization of analytes occurs under ambient conditions outside of the mass spectrometer inlet. More than 30 different Ambient-MS techniques can be mentioned depending on their dominant desorption/volatilization method [44]. At least three of these techniques known as DESI, DART, and ASAP have been commercialized for use with a wide range of MS instruments. One of the main advantages of these methods is that they permit rapid, direct measurement of compounds present on the condensed sample phase. Samples are just exposed to the ionization medium under ambient conditions and do not usually need extraction and separation processes, and that makes the food analysis extraordinarily easy and efficient [5].

DESI uses a spray of electrospray solvent composed of charged droplets as the ionizing agent (in DESI, a charged, high-velocity spray of microdroplets is directed toward the surface of interest and causes desorption of the target species by momentum transfer. Secondary off-spring droplets, which include the species of interest, migrate to the atmospheric pressure MS interface where solvent evaporation yields the ionized compounds with a mechanism similar to

ESI), whereas DART is a dry technique using primary and secondary ionization brought about by metastable atoms of helium or molecules of nitrogen (neutral metastable species are formed by electrical discharge in a gas phase with either a corona discharge (DART) or glow discharge process). These species react with ambient water and air molecules to produce the reactive ionic species, which then react with the analyte molecules to produce adduct ions that are typically protonated molecules. In the ASAP technique, samples are introduced directly into the mass spectrometer using a sealed melting-point glass capillary (probe) approximately 10 cm in length. The probe is inserted into the MS and the sample is volatilized by heated gas (usually nitrogen) [4]. The direct analysis of food samples by ASAP, while creating the opportunity for faster analysis, also gives rise to other concerns relating to interpretation and robustness of data and compliance with EU legislation on residues/contaminants in food. Accurate quantification via the direct analysis of solid samples by ASAP is not feasible due to the often heterogeneous distribution of analytes and the need to obtain a matrix-matched calibration. While ASAP is capable of producing semiquantitative data, the most likely use of ASAP in food analysis is rapid screening for the detection of analytes [44].

DART-MS has been employed by Farré et al. [40] to investigate its feasibility in combination with LTQ-Orbitrap for the direct analysis of several xenobiotics present in fruit peels. For oranges and apples, direct peel exposition demonstrated good interday precision (and proper linearity, with a dynamic range from 1 to 2,500 ng for apple. Besides the merits of fast analysis (c. 1 min) and combination of HRMS and MS/MS to elucidate nontarget or unknown compounds, the results obtained were in good agreement with those from SE DART and from UHPLC-Orbitrap analyses, showing that this method had good quantitative accuracy and capability for identification.

Multiple mycotoxins (deoxynivalenol, nivalenol, zearalenon, actyldeoxynivalenol, deepoxy-deoxynivalenol, fusarenon-X, altenuene, alternariol, alternariol-methylether, diacetoxyscirpenol, sterigmatocystin) isolated from wheat and maize by modified QuEChERS procedure was analyzed by DART ionization coupled to Orbitrap-MS [134]. Sample preparation procedure and instrument parameter settings were optimized after an initial evaluation of ionization efficiencies for the major groups of mycotoxins achievable with DART technology. The lowest calibration levels estimated for the analytes ranged from 50 to 150 μg/kg.

Since its introduction, DESI-MS has been mainly applied in pharmaceutical and forensic analysis. However, it is envisaged that DESI can be a rapid, qualitative or semiquantitative prescreening tool, ultimately being applied on site prior to sampling and sample transport to food control laboratories and/or whenever a food scandal requires results for high sample number within a very limited time span [39]. This ambient ionization technique was used for determining flunitrazepam in various alcoholic beverages [135]. Using this technique, no pretreatment of the samples was necessary and identification of the drug was accomplished in minutes (six samples were analyzed in a row within 6 min with no observable sample carryover). The acquired mass spectra provide the information of the identity of the drink based on the detected characteristic ions from the matrices. This study also demonstrated the capability of DESI-MS to perform quantitative analysis with LOQ of 3 μg/mL.

As an interesting alternative, an automated surface-sampling technique called liquid extraction surface analysis (LESA) coupled with infusion nanoelectrospray HRMS and MS/MS was applied to the qualitative determination of surface chemical residues resulting from the artificial spraying of selected fresh fruits and vegetables with representative pesticides [136]. This technique is automated and employs nanoelectrospray ionization (nanoESI) following LESA directly from the surface of fruits or vegetables (Figure 10). The selectivity afforded by LESA microliquid extraction robotically coupled with infusion nanoESI-MS (either the high-resolution capabilities of the Exactive Orbitrap or the MS/MS full-scan mass spectra available from the 5500 QTRAP) demonstrates that relatively low complexity sample extracts do not necessarily require the chromatographic separation or tedious sample preparation methodologies currently employed for the analysis of fruit and vegetable matrices.

Ambient ionization MS techniques in food analysis offer simple, high-throughput qualitative and quantitative analyses of major and trace contaminants. However, some limitations have not been solved yet: (1) the analytical results are usually expressed as pg/mm^2 obtained after direct detection of a contaminant from the surface of food, while in general MRLs are typically expressed in mg/kg of homogeneous sample; (2) contaminants are often not evenly distributed

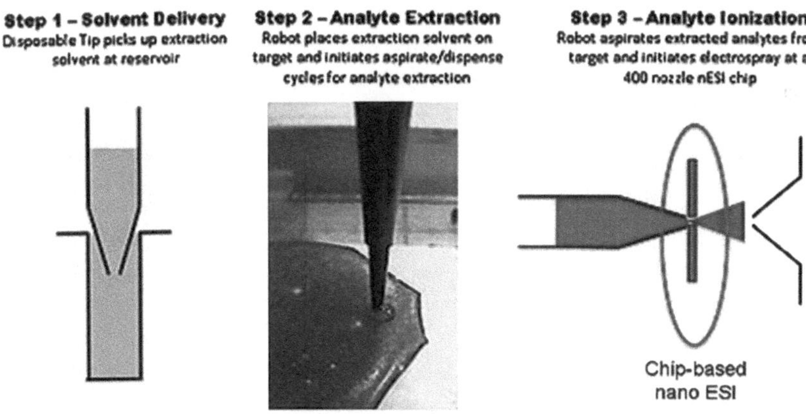

FIGURE 10 Schematic representations of LESA automation steps. Step 1: The sample robot aliquots a sample of extraction-spray solvent into a pipette tip. Step 2: The solvent in the pipette tip is dispensed/aspirated onto the surface of an apple skin to perform extraction of any chemicals on the surface of the apple. The pipette-tip diameter is 800 μm, which produces a surface area wetted with extraction solvent of c. 1–3 mm, depending on extraction volume chosen. Step 3: The pipette tip and the sample extract are robotically positioned at the inlet of the ESI chip for nano ESI-MS analysis. *Eikel and Henion [136].*

on the surface of the intact product; and, (3) the lack of separation can result in matrix interferences [4].

IMS has been a useful tool for investigating protein, peptide, drug, and metabolite distributions in human and animal tissue samples for almost 15 years. The major advantages of this method includes a broad mass range, the ability to detect multiple analytes in a single experiment without the use of labels, and the preservation of biologically relevant spatial information [137]. With a significant increase in the number and variety of applications employing this technology, there is an ongoing interest in matrix-assisted laser desorption/ionization IMS (MALDI IMS). It combines the sensitivity and selectivity of MS with spatial analysis to provide a new dimension for histological analyses to provide unbiased visualization of the arrangement of biomolecules in tissue. Advances in instrumentation have allowed for increasing spatial resolution in IMS images, although this has introduced new challenges with data acquisition times and data storage and processing. New strategies to increase the dynamic range of analysis, including Fourier transform-based methods and the incorporation of ion mobility, have been shown to be effective for broadening the application of IMS to include many new analytes as emerging contaminants [138].

8. CONCLUSIONS

The great variety of emerging contaminants summarized in this chapter, and that can be present in food, gives an indication of the importance of research and technical development carried out during last years in analytical chemistry for their analyses. Improvement in sensitivity and selectivity offered by new chromatography and MS equipment has made significant progress enabling rapid, reliable, and accurate techniques for detection, identification, and quantification of these contaminants in food commodities.

The application of LC-MS offers an improvement over GC-MS, avoiding the derivatization step for the more polar compounds. However, GC-MS continues to be a useful tool for food safety and quality control. Advanced LC-MS/MS approaches have opened up the range of contaminants that can be analyzed achieving LODs in the range μg and ng/g. The emergence and rapid development of UHPLC–MS/MS technology has reduced injection cycle times, improved the sensitivity, and increased the chromatographic resolution of compounds. Consequently, new analytical methods turned to multiclass methods, in order to face up to the large number of compounds that should be screened.

The use of HRMS as QqTOF, Orbitrap, and FT-ICR-MS allows extremely high specificity and sensitivity for the screening of large number of samples, and the possible detection of unknown compounds, being an ideal tool for screening and confirmation. But this technology is expensive and most of the current quantitative LC-MS analyses of food samples are performed on LC-QqQ-MS, which also presents good sensitivity and repeatability. Multidimensional chromatographic techniques provide extraordinary gains in separation power and resolution that make them ideal for the analysis of complex matrices. As state-of-the-art MS detectors are continuously enhanced, the identification power of these techniques in food analysis will increase extending the use of these applications.

The newly developed techniques, as ambient ionization MS, have been increasingly applied in various food analyses including chemical analysis of contaminants, with the merit of providing rapid, direct measurement of compounds present on the surface of foods with minimal, or no, sample preparation. Unfortunately, there is still a lack of specificity and, sometimes, of sensitivity needed to comply with the MRLs established by EU legislation.

Expected future developments include

- Evolution of methods to include larger number of compounds and classes of compounds.
- Reduction in sample preparation via online SPE or other techniques to increase the speed of analysis beyond current standards (with special attention to multidimensional systems).
- Improvement of LC-MS interfaces to minimize matrix effects.
- More HRMS approaches should be extensively evaluated for quantitative analysis.
- Although these direct-analysis MS tools (DESI, DART) present promising capabilities in food safety, their potential for quantitative analysis has to be further demonstrated.

ACKNOWLEDGMENTS

JC acknowledges the VALi+d postdoctoral contract (APOSTD/2014/010) of the Generalitat Valenciana for funding.

REFERENCES

[1] L. Kantiani, M. Llorca, J. Sanchis, M. Farre, D. Barcelo, Emerging food contaminants: a review, Anal. Bioanal. Chem. 398 (2010) 2413–2427.

[2] Y. Picó, M.J. Andrés, M. Farré, D. Barceló, UHPLC-QqTOF-MS/(MS) for Identifying and Quantifying Emerging Contaminants in Food 10th Annual LC/MS/MS Workshop on Environmental Application and Food Safety. Barcelona, 2014, p. 87–92.

[3] C.o.t.E. Communities, White paper on Food Safety, Commission of the European Communities, 2000, p. 52.

[4] M. Farré, D. Barceló, Analysis of emerging contaminants in food, TrAC Trends Anal. Chem. 43 (2013) 240–253.

[5] X. Wang, S. Wang, Z. Cai, The latest developments and applications of mass spectrometry in food-safety and quality analysis, TrAC Trends Anal. Chem. 52 (2013) 170–185.

[6] L.B. Abdulra'uf, G.H. Tan, Multivariate study of parameters in the determination of pesticide residues in apple by headspace solid phase microextraction coupled to gas chromatography-mass spectrometry using experimental factorial design, Food Chem. 141 (2013) 4344–4348.

[7] T. Cajka, C. Sandy, V. Bachanova, L. Drabova, K. Kalachova, J. Pulkrabova, et al., Streamlining sample preparation and gas chromatography-tandem mass spectrometry analysis of multiple pesticide residues in tea, Anal. Chim. Acta 743 (2012) 51–60.

[8] X. Wang, P. Li, W. Zhang, Q. Zhang, F. Ma, L. Yu, et al., Screening for pesticide residues in oil seeds using solid-phase dispersion extraction and comprehensive two-dimensional gas chromatography time-of-flight mass spectrometry, J. Sep. Sci. 35 (2012) 1634–1643.

[9] K.A. Asante, S. Takahashi, T. Itai, T. Isobe, G. Devanathan, M. Muto, et al., Occurrence of halogenated contaminants in inland and coastal fish from Ghana: levels, dietary exposure assessment and human health implications, Ecotoxicol. Environ. Saf. 94 (2013) 123–130.

[10] P. Carlsson, D. Herzke, M. Wedborg, G.W. Gabrielsen, Environmental pollutants in the Swedish marine ecosystem, with special emphasis on polybrominated diphenyl ethers (PBDE), Chemosphere 82 (2011) 1286–1292.

[11] K. Kalachova, T. Cajka, C. Sandy, J. Hajslova, J. Pulkrabova, High throughput sample preparation in combination with gas chromatography coupled to triple quadrupole tandem mass spectrometry (GC–MS/MS): a smart procedure for (ultra)trace analysis of brominated flame retardants in fish, Talanta 105 (2013) 109–116.

[12] K. Kalachova, J. Pulkrabova, L. Drabova, T. Cajka, V. Kocourek, J. Hajslova, Simplified and rapid determination of polychlorinated biphenyls, polybrominated diphenyl ethers, and polycyclic aromatic hydrocarbons in fish and shrimps integrated into a single method, Anal. Chim. Acta 707 (2011) 84–91.

[13] Y. Sapozhnikova, S.J. Lehotay, Multi-class, multi-residue analysis of pesticides, polychlorinated biphenyls, polycyclic aromatic hydrocarbons, polybrominated diphenyl ethers and novel flame retardants in fish using fast, low-pressure gas chromatography–tandem mass spectrometry, Anal. Chim. Acta 758 (2013) 80–92.

[14] A. Alabi, N. Caballero-Casero, S. Rubio, Quick and simple sample treatment for multiresidue analysis of bisphenols, bisphenol diglycidyl ethers and their derivatives in canned food prior to liquid chromatography and fluorescence detection, J. Chromatogr. A 1336 (2014) 23–33.

[15] M. Franek, D. Rubio, I. Diblikova, F. Rubio, Analytical evaluation of a high-throughput enzyme-linked immunosorbent assay for acrylamide determination in fried foods, Talanta 123 (2014) 146–150.

[16] C.N. Ezekiel, M. Sulyok, J.C. Frisvad, Y.M. Somorin, B. Warth, J. Houbraken, et al., Fungal and mycotoxin assessment of dried edible mushroom in Nigeria, Int. J. Food Microbiol. 162 (2013) 231–236.

[17] M. Horka, P. Karasek, J. Salplachta, F. Ruzicka, M. Vykydalova, A. Kubesova, et al., Capillary isoelectric focusing of probiotic bacteria from cow's milk in tapered fused silica capillary with off-line matrix-assisted laser desorption/ionization time-of-flight mass spectrometry identification, Anal. Chim. Acta 788 (2013) 193–199.

[18] W.A. Abia, B. Warth, M. Sulyok, R. Krska, A.N. Tchana, P.B. Njobeh, et al., Determination of multi-mycotoxin occurrence in cereals, nuts and their products in Cameroon by liquid chromatography tandem mass spectrometry (LC-MS/MS), Food Control 31 (2013) 438–453.

[19] E. De Dominicis, I. Commissati, M. Suman, Targeted screening of pesticides, veterinary drugs and mycotoxins in bakery ingredients and food commodities by liquid chromatography-high-resolution single-stage Orbitrap mass spectrometry, J. Mass Spectrom. 47 (2012) 1232–1241.

[20] M. Suman, E. Bergamini, D. Catellani, A. Manzitti, Development and validation of a liquid chromatography/linear ion trap mass spectrometry method for the quantitative determination of deoxynivalenol-3-glucoside in processed cereal-derived products, Food Chem. 136 (2013) 1568–1576.

[21] P.F. Solter, V.R. Beasley, Chapter 38-Phycotoxins, in: W.M. Haschek, C.G. Rousseaux, M.A. Wallig (Eds.), Haschek and Rousseaux's Handbook of Toxicologic Pathology, Third ed., Academic Press, Boston, 2013, pp. 1155–1186.

[22] P. Visciano, M. Schirone, R. Tofalo, M. Berti, M. Luciani, N. Ferri, et al., Detection of yessotoxin by three different methods in Mytilus galloprovincialis of Adriatic Sea, Italy, Chemosphere 90 (2013) 1077–1082.

[23] M. Camara, A. Gallego-Pico, R.M. Garcinuno, P. Fernandez-Hernando, J.S. Durand-Alegria, P.J. Sanchez, An HPLC-DAD method for the simultaneous determination of nine beta-lactam antibiotics in ewe milk, Food Chem. 141 (2013) 829–834.

[24] A. Kaufmann, M. Widmer, Quantitative analysis of polypeptide antibiotic residues in a variety of food matrices by liquid chromatography coupled to tandem mass spectrometry, Anal. Chim. Acta 797 (2013) 81–88.

[25] M.M. Aguilera-Luiz, J.L. Martinez Vidal, R. Romero-Gonzalez, A. Garrido Frenich, Multiclass method for fast determination of veterinary drug residues in baby food by ultra-high-performance liquid chromatography-tandem mass spectrometry, Food Chem. 132 (2012) 2171–2180.

[26] C. Chiaochan, U. Koesukwiwat, S. Yudthavorasit, N. Leepipatpiboon, Efficient hydrophilic interaction liquid chromatography-tandem mass spectrometry for the multiclass analysis of veterinary drugs in chicken muscle, Anal. Chim. Acta 682 (2010) 117–129.

[27] Y.-Y. Tang, H.-F. Lu, H.-Y. Lin, Y.-C. Shih, D.-F. Hwang, Multiclass analysis of 23 veterinary drugs in milk by ultraperformance liquid chromatography-electrospray tandem mass spectrometry, J. Chromatogr. B-Anal. Technol. Biomed. Life Sci. 881–882 (2012) 12–19.

[28] C. Guerranti, G. Perra, S. Corsolini, S.E. Focardi, Pilot study on levels of perfluorooctane sulfonic acid (PFOS) and perfluorooctanoic acid (PFOA) in selected foodstuffs and human milk from Italy, Food Chem. 140 (2013) 197–203.

[29] F. Pérez, M. Llorca, M. Köck-Schulmeyer, B. Škrbić, L.S. Oliveira, K. da Boit Martinello, et al., Assessment of perfluoroalkyl substances in food items at global scale, Environ. Res. 135 (2014) 181–189.

[30] N.G. Dodder, K.A. Maruya, P. Lee Ferguson, R. Grace, S. Klosterhaus, M.J. La Guardia, et al., Occurrence of contaminants of emerging concern in mussels (Mytilus spp.) along the California coast and the influence of land use, storm water discharge, and treated wastewater effluent, Mar. Pollut. Bull. 81 (2014) 340–346.

[31] C. Bizarro, O. Ros, A. Vallejo, A. Prieto, N. Etxebarria, M.P. Cajaraville, et al., Intersex condition and molecular markers of endocrine disruption in relation with burdens of emerging pollutants in thicklip grey mullets (*Chelon labrosus*) from Basque estuaries (South-East Bay of Biscay), Mar. Environ. Res. 96 (2014) 19–28.

[32] A.J. Midey, A. Camacho, J. Sampathkumaran, C.A. Krueger, M.A. Osgood, C. Wu, High-performance ion mobility spectrometry with direct electrospray ionization (ESI-HPIMS) for the detection of additives and contaminants in food, Anal. Chim. Acta 804 (2013) 197–206.

[33] L. Polgar, J.F. Garcia-Reyes, P. Fodor, A. Gyepes, M. Dernovics, L. Abranko, et al., Retrospective screening of relevant pesticide metabolites in food using liquid chromatography high resolution mass spectrometry and accurate-mass databases of parent molecules and diagnostic fragment ions, J. Chromatogr. A 1249 (2012) 83–91.

[34] J. Wang, W. Chow, D. Leung, J. Chang, Application of ultrahigh-performance liquid chromatography and electrospray ionization quadrupole orbitrap high-resolution mass spectrometry for determination of 166 pesticides in fruits and vegetables, J. Agric. Food Chem. 60 (2012) 12088–12104.

[35] D. Lankova, O. Lacina, J. Pulkrabova, J. Hajslova, The determination of perfluoroalkyl substances, brominated flame retardants and their metabolites in human breast milk and infant formula, Talanta 117 (2013) 318–325.
[36] A. Masiá, C. Blasco, Y. Picó, Last trends in pesticide residue determination by liquid chromatography–mass spectrometry, Trends Environ. Anal. Chem. 2 (2014) 11–24.
[37] H.V. Botitsi, S.D. Garbis, A. Economou, D.F. Tsipi, Current mass spectrometry strategies for the analysis of pesticides and their metabolites in food and water matrices, Mass Spectrom. Rev. 30 (2011) 907–939.
[38] L. Hartmanova, V. Ranc, B. Papouskova, P. Bednar, V. Havlicek, K. Lemr, Fast profiling of anthocyanins in wine by desorption nano-electrospray ionization mass spectrometry, J. Chromatogr. A 1217 (2010) 4223–4228.
[39] M.W.F. Nielen, H. Hooijerink, P. Zomer, J.G.J. Mol, Desorption electrospray ionization mass spectrometry in the analysis of chemical food contaminants in food, TrAc-Trends Anal. Chem. 30 (2011) 165–180.
[40] M. Farré, Y. Picó, D. Barceló, Direct peel monitoring of xenobiotics in fruit by direct analysis in real time coupled to a linear quadrupole ion Trap–Orbitrap mass spectrometer, Anal. Chem. 85 (2013) 2638–2644.
[41] A. Martinez-Villalba, L. Vaclavik, E. Moyano, M. Teresa Galceran, J. Hajslova, Direct analysis in real time high-resolution mass spectrometry for high-throughput analysis of antiparasitic veterinary drugs in feed and food, Rapid Commun. Mass Spectrom. 27 (2013) 467–475.
[42] G. Huang, W. Xu, M.A. Visbal-Onufrak, Z. Ouyang, R.G. Cooks, Direct analysis of melamine in complex matrices using a handheld mass spectrometer, Analyst 135 (2010) 705–711.
[43] J.S. Wiley, J.F. Garcia-Reyes, J.D. Harper, N.A. Charipar, Z. Ouyang, R.G. Cooks, Screening of agrochemicals in foodstuffs using low-temperature plasma (LTP) ambient ionization mass spectrometry, Analyst 135 (2010) 971–979.
[44] R.J. Fussell, D. Chan, M. Sharman, An assessment of atmospheric-pressure solids-analysis probes for the detection of chemicals in food, TrAc-Trends Anal. Chem. 29 (2010) 1326–1335.
[45] C. Ibáñez, V. García-Cañas, A. Valdés, C. Simó, Novel MS-based approaches and applications in food metabolomics, TrAC-Trends Anal. Chem. 52 (2013) 100–111.
[46] A.K. Malik, C. Blasco, Y. Pico, Liquid chromatography-mass spectrometry in food safety, J. Chromatogr. A 1217 (2010) 4018–4040.
[47] T. Schiliro, I. Gorrasi, A. Longo, S. Coluccia, G. Gilli, Endocrine disrupting activity in fruits and vegetables evaluated with the E-screen assay in relation to pesticide residues, J. Steroid Biochem. Mol. Biol. 127 (2011) 139–146.
[48] S. Panseri, A. Catalano, A. Giorgi, F. Arioli, A. Procopio, D. Britti, et al., Occurrence of pesticide residues in Italian honey from different areas in relation to its potential contamination sources, Food Control 38 (2014) 150–156.
[49] M. Petrovic, M. Farré, E. Eljarrat, M.S. Díaz-Cruz, D. Barceló, Chapter 14-Environmental analysis: emerging pollutants, in: S. Fanali, P.R. Haddad, C.F. Poole, P. Schoenmakers, D. Lloyd (Eds.), Liquid Chromatography, Elsevier, Amsterdam, 2013, pp. 389–410.
[50] Y. Picó, M. Farré, D. Barceló, Chapter 20-Emerging contaminants, in: Y. Picó (Ed.), Chemical Analysis of Food: Techniques and Applications, Academic Press, Boston, 2012, pp. 665–691.
[51] O. Akoto, H. Andoh, G. Darko, K. Eshun, P. Osei-Fosu, Health risk assessment of pesticides residue in maize and cowpea from Ejura, Ghana, Chemosphere 92 (2013) 67–73.
[52] Regulation (EC) No 1107/2009 of the European Parliament and of the Council of 21 October 2009 Concerning the Placing of Plant Protection Products on the Market and Repealing Council Directives 79/117/EEC and 91/414/EEC Official Journal of the European Community L 309 59.

[53] A. Garrido-Frenich, M.d.M. Aguilera-Luiz, J.L. Martinez Vidal, R. Romero-Gonzalez, Comparison of several extraction techniques for multiclass analysis of veterinary drugs in eggs using ultra-high pressure liquid chromatography-tandem mass spectrometry, Anal. Chim. Acta 661 (2010) 150–160.

[54] L.S. Haug, C. Thomsen, A.L. Brantsaeter, H.E. Kvalem, M. Haugen, G. Becher, et al., Diet and particularly seafood are major sources of perfluorinated compounds in humans, Environ. Int. 36 (2010) 772–778.

[55] L.S. Haug, S. Salihovic, I.E. Jogsten, C. Thomsen, B. van Bavel, G. Lindstrom, et al., Levels in food and beverages and daily intake of perfluorinated compounds in Norway, Chemosphere 80 (2010) 1137–1143.

[56] Y.R. Kim, F.A. Harden, L.-M.L. Toms, R.E. Norman, Health consequences of exposure to brominated flame retardants: a systematic review, Chemosphere 106 (2014) 1–19.

[57] E. Baron, I. Rudolph, G. Chiang, R. Barra, E. Eljarrat, D. Barcelo, Occurrence and behavior of natural and anthropogenic (emerging and historical) halogenated compounds in marine biota from the Coast of Concepcion (Chile), Sci. Total Environ. 461 (2013) 258–264.

[58] F.J. López-Jiménez, A. Ballesteros-Gómez, S. Rubio, Determination of polycyclic aromatic hydrocarbons (PAH4) in food by vesicular supramolecular solvent-based microextraction and LC–fluorescence detection, Food Chem. 143 (2014) 341–347.

[59] A.L. Capriotti, C. Cavaliere, S. Piovesana, R. Samperi, A. Lagana, Multiclass screening method based on solvent extraction and liquid chromatography-tandem mass spectrometry for the determination of antimicrobials and mycotoxins in egg, J. Chromatogr. A 1268 (2012) 84–90.

[60] E. García-Cela, A.J. Ramos, V. Sanchis, S. Marin, Risk management towards food safety objective achievement regarding to mycotoxins in pistachio: the sampling and measurement uncertainty issue, Food Control 31 (2013) 392–402.

[61] P.G. Jin, Z. Han, Z.X. Cai, Y.J. Wu, Y.P. Ren, Simultaneous determination of 10 mycotoxins in grain by ultra-high-performance liquid chromatography-tandem mass spectrometry using 13C15-deoxynivalenol as internal standard, Food Addit. Contam. Part A 27 (2010) 1701–1713.

[62] C.P. Woloshuk, W.B. Shim, Aflatoxins, fumonisins, and trichothecenes: a convergence of knowledge, FEMS Microbiol. Rev. 37 (2013) 94–109.

[63] P. Otero, P. Rodríguez, A.M. Botana, A. Alfonso, L.M. Botana, Chapter 15-Analysis of natural toxins, in: S. Fanali, P.R. Haddad, C.F. Poole, P. Schoenmakers, D. Lloyd (Eds.), Liquid Chromatography, Elsevier, Amsterdam, 2013, pp. 411–430.

[64] C. Picot, G. Limon, G. Durand, D. Parent-Massin, A.-C. Roudot, Probabilistic dietary exposure to phycotoxins in a recreational shellfish harvester subpopulation (France), J. Exposure Sci. Environ. Epidemiol. 23 (2013) 435–441.

[65] S. Suikkanen, A. Kremp, H. Hautala, B. Krock, Paralytic shellfish toxins or spirolides? the role of environmental and genetic factors in toxin production of the Alexandrium ostenfeldii complex,, Harmful Algae 26 (2013) 52–59.

[66] X.-m. Xu, H.-l. He, Y. Zhu, L. Feng, Y. Ying, B.-f. Huang, et al., Simultaneous determination of 3-monochloropropane-1,2-diol and acrylamide in food by gas chromatography-triple quadrupole mass spectrometry with coupled column separation, Anal. Chim. Acta 760 (2013) 93–99.

[67] S. Andres, K.E. Appel, A. Lampen, Toxicology, occurrence and risk characterisation of the chloropropanols in food: 2-monochloro-1,3-propanediol, 1,3-dichloro-2-propanol and 2,3-dichloro-1-propanol, Food Chem. Toxicol. 58 (2013) 467–478.

[68] M.M.A. Omar, W.A. Wan Ibrahim, A.A. Elbashir, Sol–gel hybrid methyltrimethoxysilane–tetraethoxysilane as a new dispersive solid-phase extraction material for acrylamide determination in food with direct gas chromatography–mass spectrometry analysis, Food Chem. 158 (2014) 302–309.

[69] T.M. Dordevic, S.S. Siler-Marinkovic, R.D. Durovic, S.I. Dimitrijevic-Brankovic, J.S.G. Umiljendic, Stability of the pyrethroid pesticide bifenthrin in milled wheat during thermal processing, yeast and lactic acid fermentation, and storage, J. Sci. Food Agric. 93 (2013) 3377–3383.

[70] S.E. McNamee, R. Cunningham, C.T. Elliott, Simultaneous immunochemical detection of four banned antibiotic growth promoters in raw and cooked poultry tissue, Food Addit. Contam. Part A 30 (2013) 1270–1278.

[71] G.-H. Byun, H.-B. Moon, J.-H. Choi, J. Hwang, C.-K. Kang, Biomagnification of persistent chlorinated and brominated contaminants in food web components of the Yellow Sea, Mar. Poll. Bull. 73 (2013) 210–219.

[72] O.P. Luzardo, Á. Rodríguez-Hernández, Y. Quesada-Tacoronte, N. Ruiz-Suárez, M. Almeida-González, L.A. Henríquez-Hernández, et al., Influence of the method of production of eggs on the daily intake of polycyclic aromatic hydrocarbons and organochlorine contaminants: an independent study in the Canary Islands (Spain), Food Chem. Toxicol. 60 (2013) 455–462.

[73] L.A. Arias, C.R. Bojaca, D.A. Ahumada, E. Schrevens, Monitoring of pesticide residues in tomato marketed in Bogota, Colombia, Food Control 35 (2014) 213–217.

[74] A. Yang, A.M. Abd El-Aty, J.-H. Park, A. Goudah, M.M. Rahman, J.-A. Do, et al., Analysis of 10 systemic pesticide residues in various baby foods using liquid chromatography-tandem mass spectrometry, Biomed. Chromatogr.: BMC 28 (2014) 735–741.

[75] N. Arroyo-Manzanares, A.M. Garcia-Campana, L. Gamiz-Gracia, Multiclass mycotoxin analysis in *Silybum marianum* by ultra high performance liquid chromatography-tandem mass spectrometry using a procedure based on QuEChERS and dispersive liquid-liquid microextraction, J. Chromatogr. A 1282 (2013) 11–19.

[76] B.J.A. Berendsen, L.A.M. Stolker, M.W.F. Nielen, Selectivity in the sample preparation for the analysis of drug residues in products of animal origin using LC-MS, TrAc-Trends Anal. Chem. 43 (2013) 229–239.

[77] M. Corrales, A. Fernandez, M.G. Vizoso Pinto, P. Butz, C.M.A.P. Franz, E. Schuele, et al., Characterization of phenolic content, in vitro biological activity, and pesticide loads of extracts from white grape skins from organic and conventional cultivars, Food Chem. Toxicol. 48 (2010) 3471–3476.

[78] G. Du, Y. Xiao, H.-R. Yang, L. Wang, Y-l. Song, Y.-T. Wang, Rapid determination of pesticide residues in herbs using selective pressurized liquid extraction and fast gas chromatography coupled with mass spectrometry, J. Sep. Sci. 35 (2012) 1922–1932.

[79] C. Blasco, A. Masia, F. Gabriel Morillas, Y. Pico, Comparison of the effectiveness of recent extraction procedures for antibiotic residues in bovine muscle tissues, J. Aoac Int. 94 (2011) 991–1003.

[80] C. Truzzi, S. Illuminati, C. Finale, A. Annibaldi, C. Lestingi, G. Scarponi, Microwave-assisted solvent extraction of melamine from seafood and determination by gas chromatography-mass spectrometry: optimization by factorial design, Anal. Lett. 47 (2014) 1118–1133.

[81] L.J. Wu, Y. Song, X. Xu, N. Li, M.Y. Shao, H.Q. Zhang, et al., Medium-assisted non-polar solvent dynamic microwave extraction for determination of organophosphorus pesticides in cereals using gas chromatography-mass spectrometry, Food Chem. 162 (2014) 253–260.

[82] Y. Picó, M. Farré, M. Llorca, D. Barceló, Perfluorinated compounds in food: a global perspective, Crit. Rev. Food Sci. Nutr. 51 (2011) 605–625.

[83] W.A. Gebbink, R.J. Letcher, N.M. Burgess, L. Champoux, J.E. Elliott, C.E. Hebert, et al., Perfluoroalkyl carboxylates and sulfonates and precursors in relation to dietary source tracers in the eggs of four species of gulls (Larids) from breeding sites spanning Atlantic to Pacific Canada, Environ. Int. 37 (2011) 1175–1182.

[84] M.M. Aguilera-Luiz, R. Romero-Gonzalez, P. Plaza-Bolanos, J.L. Martinez Vidal, A. Garrido Frenich, Wide-scope analysis of veterinary drug and pesticide residues in animal feed by liquid chromatography coupled to quadrupole-time-of-flight mass spectrometry, Anal. Bioanal. Chem. 405 (2013) 6543–6553.

[85] U. Błaszczyk, The application of multistep extraction and liquid chromatography with fluorescence detection for analysis of azaarenes in edible oil samples, J. Food Compos. Anal. 33 (2014) 175–180.

[86] D. Moreno-Gonzalez, F.J. Lara, L. Gamiz-Gracia, A.M. Garcia-Campana, Molecularly imprinted polymer as in-line concentrator in capillary electrophoresis coupled with mass spectrometry for the determination of quinolones in bovine milk samples, J. Chromatogr. A 1360 (2014) 1–8.

[87] Y.K. Lv, L.M. Wang, L. Yang, C.X. Zhao, H.W. Sun, Synthesis and application of molecularly imprinted poly(methacrylic acid)-silica hybrid composite material for selective solid-phase extraction and high-performance liquid chromatography determination of oxytetracycline residues in milk, J. Chromatogr. A 1227 (2012) 48–53.

[88] J. He, R. Lv, J. Zhu, K. Lu, Selective solid-phase extraction of dibutyl phthalate from soybean milk using molecular imprinted polymers, Anal. Chim. Acta 661 (2010) 215–221.

[89] P. Jandera, Advances in the development of organic polymer monolithic columns and their applications in food analysis—a review, J. Chromatogr. A 1313 (2013) 37–53.

[90] M. Bryla, M. Roszko, K. Szymczyk, R. Jedrzejczak, M.W. Obiedzinski, J. Sekul, Fumonisins in plant-origin food and fodder—a review, Food Addit. Contam. Part A 30 (2013) 1626–1640.

[91] S.J. Wu, E.T. Wang, Y. Yuan, Detection of furan levels in select Chinese foods by solid phase microextraction-gas chromatography/mass spectrometry method and dietary exposure estimation of furan in the Chinese population, Food Chem. Toxicol. 64 (2014) 34–40.

[92] J.I. Cacho, N. Campillo, P. Viñas, M. Hernández-Córdoba, Determination of alkylphenols and phthalate esters in vegetables and migration studies from their packages by means of stir bar sorptive extraction coupled to gas chromatography–mass spectrometry, J. Chromatogr. A 1241 (2012) 21–27.

[93] A.B. Lara, C. Caballo, M.D. Sicilia, S. Rubio, Enantiomer-specific determination of hexabromocyclododecane in fish by supramolecular solvent-based single-step sample treatment and liquid chromatography-tandem mass spectrometry, Anal. Chim. Acta 752 (2012) 62–68.

[94] M. Arvand, E. Bozorgzadeh, S. Shariati, Two-phase hollow fiber liquid phase microextraction for preconcentration of pyrethroid pesticides residues in some fruits and vegetable juices prior to gas chromatography/mass spectrometry, J. Food Compos. Anal. 31 (2013) 275–283.

[95] A.R. Fontana, M.M. de Toro, J.C. Altamirano, O.-S. Derivatization, P. Microextraction, Technique for determination of bisphenol a in beverage samples by Gas chromatography-mass spectrometry, J. Agric. Food Chem. 59 (2011) 3559–3565.

[96] A.M. Carro, D. Garcia-Rodriguez, P. Gonzalez-Siso, R.A. Lorenzo, Determination of chemotherapeutic agents in fish and shellfish by matrix solid-phase dispersion and liquid chromatography-tandem mass spectrometry, J. Sep. Sci. 35 (2012) 2866–2874.

[97] X. Liu, A. Zhao, A. Zhang, H. Liu, W. Xiao, C. Wang, et al., Dispersive liquid-liquid microextraction and gas chromatography-mass spectrometry determination of polychlorinated biphenyls and polybrominated diphenyl ethers in milk, J. Sep. Sci. 34 (2011) 1084–1090.

[98] S.J. Hird, B.P.Y. Lau, R. Schuhmacher, R. Krska, Liquid chromatography-mass spectrometry for the determination of chemical contaminants in food, TrAC-Trends Anal. Chem. 59 (2014) 59–72.

[99] L. Geis-Asteggiante, A. Nunez, S.J. Lehotay, A.R. Lightfield, Structural characterization of product ions by electrospray ionization and quadrupole time-of-flight mass spectrometry to support regulatory analysis of veterinary drug residues in foods, Rapid Commun. Mass Spectrom. 28 (2014) 1061–1081.

[100] C.C. Kern, J.C. Usbeck, R.F. Vogel, J. Behr, Optimization of matrix-assisted-laser-desorption–ionization-time-of-flight mass spectrometry for the identification of bacterial contaminants in beverages, J. Microbiol. Methods 93 (2013) 185–191.

[101] O. Lacina, J. Urbanova, J. Poustka, J. Hajslova, Identification/quantification of multiple pesticide residues in food plants by ultra-high-performance liquid chromatography-time-of-flight mass spectrometry, J. Chromatogr. A 1217 (2010) 648–659.

[102] A. Kaufmann, The current role of high-resolution mass spectrometry in food analysis, Anal. Bioanal. Chem. 403 (2012) 1233–1249.

[103] M.I. Cervera, T. Portoles, F.J. Lopez, J. Beltran, F. Hernandez, Screening and quantification of pesticide residues in fruits and vegetables making use of gas chromatography-quadrupole time-of-flight mass spectrometry with atmospheric pressure chemical ionization, Anal. Bioanal. Chem. 406 (2014) 6843–6855.

[104] T. Eggen, E.S. Heimstad, A.O. Stuanes, H.R. Norli, Uptake and translocation of organophosphates and other emerging contaminants in food and forage crops, Environ. Sci. Poll. Res. 20 (2013) 4520–4531.

[105] K. Kalachova, J. Pulkrabova, T. Cajka, L. Drabova, J. Hajslova, Implementation of comprehensive two-dimensional gas chromatography-time-of-flight mass spectrometry for the simultaneous determination of halogenated contaminants and polycyclic aromatic hydrocarbons in fish, Anal. Bioanal. Chem. 403 (2012) 2813–2824.

[106] F. Zhang, H. Wang, L. Zhang, J. Zhang, R. Fan, C. Yu, et al., Suspected-target pesticide screening using gas chromatography-quadrupole time-of-flight mass spectrometry with high resolution deconvolution and retention index/mass spectrum library, Talanta 128 (2014) 156–163.

[107] E. Cieslik, A. Sadowska-Rociek, J.M.M. Ruiz, M. Surma-Zadora, Evaluation of QuEChERS method for the determination of organochlorine pesticide residues in selected groups of fruits, Food Chem. 125 (2011) 773–778.

[108] I. Ierapetritis, A. Lioupis, E. Lampi, Determination of phthalates into vegetable oils by isotopic dilution gas chromatography mass spectrometry, Food Anal. Methods 7 (2014) 1451–1457.

[109] S.W.C. Chung, B.L.S. Chen, Determination of organochlorine pesticide residues in fatty foods: a critical review on the analytical methods and their testing capabilities, J. Chromatogr. A 1218 (2011) 5555–5567.

[110] J.V. Seeley, S.K. Seeley, Multidimensional Gas chromatography: Fundamental advances and new applications, Anal. Chem. 85 (2012) 557–578.

[111] M. Castro-Puyana, M. Herrero, Metabolomics approaches based on mass spectrometry for food safety, quality and traceability, TrAC-Trends Anal. Chem. 52 (2013) 74–87.

[112] D.G. Hayward, T.S. Pisano, J.W. Wong, R.J. Scudder, Multiresidue method for pesticides and persistent organic pollutants (POPs) in milk and cream using comprehensive two-dimensional capillary Gas Chromatography–Time-of-flight mass spectrometry, J. Agric. Food Chem. 58 (2010) 5248–5256.

[113] M. Pena-Abaurrea, A. Covaci, L. Ramos, Comprehensive two-dimensional gas chromatography-time-of-flight mass spectrometry for the identification of organobrominated compounds in bluefin tuna, J. Chromatogr. A 1218 (2011) 6995–7002.

[114] Y. Wu, Y. Wang, J. Li, Y. Zhao, F. Guo, J. Liu, et al., Perfluorinated compounds in seafood from coastal areas in China, Environ. Int. 42 (2012) 67–71.

[115] M. Amirahmadi, S. Shoeibi, M. Abdollahi, H. Rastegar, R. Khosrokhavar, M.P. Hamedani, Monitoring of some pesticides residue in consumed tea in Tehran market, Iran. J. Environ. Health Sci. Eng. 10 (2013).

[116] M.H. El-Saeid, M.T. Selim, Multiresidue analysis of 86 pesticides using gas chromatography mass spectrometry: II-Nonleafy Vegetables, J. Chem. (2013), Article ID 727149, 10 pages. http://dx.doi.org/10.1155/2013/727149.

[117] Commission Decision (2002/657/EC) of 12 August 2002 Implementing Council Directive 96/23/EC Concerning the Performance of Analytical Methods and the Interpretation of Results Official Journal of the European Community L 221 28.

[118] S.W.C. Chung, B.T.P. Chan, Validation and use of a fast sample preparation method and liquid chromatography-tandem mass spectrometry in analysis of ultra-trace levels of 98 organophosphorus pesticide and carbamate residues in a total diet study involving diversified food types, J. Chromatogr. A 1217 (2010) 4815–4824.

[119] E. Fornal, A. Stachniuk, Application of a truly one-point calibration for pesticide residue control by liquid chromatography-mass spectrometry, J. Chromatogr. B-Anal. Technol. Biomed. Life Sci. 901 (2012) 107–114.

[120] E. Alechaga, E. Moyano, M.T. Galceran, Ultra-high performance liquid chromatography-tandem mass spectrometry for the analysis of phenicol drugs and florfenicol-amine in foods, Analyst 137 (2012) 2486–2494.

[121] L. Geis-Asteggiante, S.J. Lehotay, A.R. Lightfield, T. Dutko, C. Ng, L. Bluhm, Ruggedness testing and validation of a practical analytical method for >100 veterinary drug residues in bovine muscle by ultrahigh performance liquid chromatography-tandem mass spectrometry, J. Chromatogr. A 1258 (2012) 43–54.

[122] J. Zhan, X.-j. Yu, Y.-y. Zhong, Z.-t. Zhang, X.-m. Cui, J.-f. Peng, et al., Generic and rapid determination of veterinary drug residues and other contaminants in raw milk by ultra performance liquid chromatography-tandem mass spectrometry, J. Chromatogr. B-Anal. Technol. Biomed. Life Sci. 906 (2012) 48–57.

[123] R.J.B. Peters, A.A.M. Stolker, J.G.J. Mol, A. Lommen, E. Lyris, Y. Angelis, et al., Screening in veterinary drug analysis and sports doping control based on full-scan, accurate-mass spectrometry, TrAc-Trends Anal. Chem. 29 (2010) 1250–1268.

[124] C. Huang, B. Guo, X. Wang, J. Li, W. Zhu, B. Chen, et al., A generic approach for expanding homolog-targeted residue screening of sulfonamides using a fast matrix separation and class-specific fragmentation-dependent acquisition with a hybrid quadrupole-linear ion trap mass spectrometer, Anal. Chim. Acta 737 (2012) 83–98.

[125] H. Wu, M. Guo, Z. Tan, H. Cheng, Z. Li, Y. Zhai, Liquid chromatography quadrupole linear ion trap mass spectrometry for multiclass screening and identification of lipophilic marine biotoxins in bivalve mollusks, J. Chromatogr. A 1358 (2014) 172–180.

[126] E. Canellas, C. Nerin, R. Moore, P. Silcock, New UPLC coupled to mass spectrometry approaches for screening of non-volatile compounds as potential migrants from adhesives used in food packaging materials, Anal. Chim. Acta 666 (2010) 62–69.

[127] M. Farré, Y. Pico, D. Barcelo, Application of ultra-high pressure liquid chromatography linear ion-trap orbitrap to qualitative and quantitative assessment of pesticide residues, J. Chromatogr. A 1328 (2014) 66–79.

[128] K. Bester, K. Vorkamp, A two-dimensional HPLC separation for the enantioselective determination of hexabromocyclododecane (HBCD) isomers in biota samples,, Anal. Bioanal. Chem. 405 (2013) 6519–6527.

[129] S. Kittlaus, J. Schimanke, G. Kempe, K. Speer, Development and validation of an efficient automated method for the analysis of 300 pesticides in foods using two-dimensional liquid chromatography-tandem mass spectrometry, J. Chromatogr. A 1283 (2013) 98–109.

[130] S. Borras, A. Kaufmann, R. Companyo, Correlation of precursor and product ions in single-stage high resolution mass spectrometry. A tool for detecting diagnostic ions and improving the precursor elemental composition elucidation, Anal. Chim. Acta 772 (2013) 47–58.

[131] C. Fanali, L. Dugo, P. Dugo, L. Mondello, Capillary-liquid chromatography (CLC) and nano-LC in food analysis, TrAc-Trends Anal. Chem. 52 (2013) 226–238.

[132] U. Alshana, N.G. Goger, N. Ertas, Dispersive liquid-liquid microextraction combined with field-amplified sample stacking in capillary electrophoresis for the determination of non-steroidal anti-inflammatory drugs in milk and dairy products, Food Chem. 138 (2013) 890–897.

[133] P. Aqai, J. Peters, A. Gerssen, W. Haasnoot, M.F. Nielen, Immunomagnetic microbeads for screening with flow cytometry and identification with nano-liquid chromatography mass spectrometry of ochratoxins in wheat and cereal, Anal. Bioanal. Chem. 400 (2011) 3085–3096.

[134] L. Vaclavik, M. Zachariasova, V. Hrbek, J. Hajslova, Analysis of multiple mycotoxins in cereals under ambient conditions using direct analysis in real time (DART) ionization coupled to high resolution mass spectrometry, Talanta 82 (2010) 1950–1957.

[135] P. D'Aloise, H. Chen, Rapid determination of flunitrazepam in alcoholic beverages by desorption electrospray ionization-mass spectrometry, Sci. Justice 52 (2012) 2–8.

[136] D. Eikel, J. Henion, Liquid extraction surface analysis (LESA) of food surfaces employing chip-based nano-electrospray mass spectrometry, Rapid Commun. Mass Spectrom. 25 (2011) 2345–2354.

[137] E.M. Weaver, A.B. Hummon, Imaging mass spectrometry: from tissue sections to cell cultures, Adv. Drug Delivery Rev. 65 (2013) 1039–1055.

[138] M.M. Gessel, J.L. Norris, R.M. Caprioli, MALDI imaging mass spectrometry: spatial molecular analysis to enable a new age of discovery, J. Proteomics 107 (2014) 71–82.

Chapter 11

Engineered Nanomaterials in the Food Sector

Ralf Greiner,* Volker Gräf, Anna Burcza, Birgit Hetzer, Johanna Milsmann and Elke Walz

Department of Food Technology and Bioprocess Engineering, Federal Research Institute of Nutrition and Food, Max Rubner-Institut, Karlsruhe, Germany

**Corresponding author: E-mail: ralf.greiner@mri.bund.de*

Chapter Outline

1. Introduction 580
2. Food Nanotechnology 581
3. Naturally Occurring Nanomaterials and Those Derived from Conventional Food Processing 582
4. Fabrication of Nanomaterials 583
5. Food Contact Materials 584
5.1 Food Packaging Systems 584
5.1.1 Barrier Properties 584
5.1.2 Mechanical Strength and Heat Resistance 586
5.1.3 Active Packaging 586
5.1.4 Intelligent/Smart Packaging 587
5.1.5 Tracking, Tracing, and Brand Protection 587
5.1.6 Biodegradable Packaging Systems 588
5.1.7 Edible Coatings 588
5.2 Coatings for Kitchen Utensils and Processing Equipment 588
5.3 Nanofiltration Membranes 589
6. Nanosized Food Ingredients and Nanosized Delivery Systems 589
6.1 Nanosized Food Ingredients 589
6.2 Nanosized Delivery Systems 590
6.2.1 Improvement of Bioavailability 591
6.2.2 Masking of Off-flavors 592
6.2.3 Protection of Bioactives Compound 592
7. Structuring of Foods 592
8. Nanosensors 593
9. Safety Aspects of Nanomaterials in Food 593
10. Legal Obligation 595
11. Analytical Approaches to Characterize Nanoparticles in Food 595
11.1 FFF-ICP-MS 600
11.2 Single-Particle ICP-MS 602
12. Conclusions 606
References 608

Comprehensive Analytical Chemistry, Vol. 68. http://dx.doi.org/10.1016/B978-0-444-63340-8.00011-X

1. INTRODUCTION

Nanoscience is a convergence of physics, chemistry, materials science, and biology, which deals with the manipulation and characterization of matter on length scales between the molecular- and the micron-size. One definition of nanoscience that is now very widely used was introduced by the 2004 Royal Society report [1]: "Nanoscience is the study of phenomena and manipulation of materials at atomic, molecular and macromolecular scales, where properties differ significantly from those at a larger scale." Nanotechnology, however, can be defined and explained in many different ways depending on which particular aspects of this complex and transformational field are being emphasized. Therefore, it is not at all surprising that there is no internationally agreed definition of nanotechnology and nanomaterial applied to the food sector. Many "official" definitions of nanotechnology exist and there is little consensus within or across governmental and nongovernmental organizations [2]. The US National Nanotechnology Initiative, for example, provides the following definition [3]: "Nanotechnology is the understanding and control of matter at the nanoscale, at dimensions between approximately 1 and 100 nm, where unique phenomena enable novel applications. Encompassing nanoscale science, engineering, and technology, nanotechnology involves imaging, measuring, modeling, and manipulating matter at this length scale." According to the 2004 Royal Society Report [1]: "Nanotechnologies are the design, characterisation, production and application of structures, devices and systems by controlling shape and size at nanometre scale" and the Organisation for Economic Co-operation and Development (OECD) defines nanotechnology as follows [4]: "Nanotechnology is the set of technologies that enables the manipulation, study or exploitation of very small (typically less than 100 nm) structures and systems. Nanotechnology contributes to novel materials, devices and products that have qualitatively different properties compared to their macroscale counterparts." The term "nanotechnology" is used for many different, not always new, technologies to design, process, and use materials at the nanometre scale.

The word "nanomaterial" is generally used when referring to materials with the size of 1–100 nm in at least one dimension. Thus, nanomaterials could be in the previously mentioned size range in three dimensions (nanoparticles, very fine powder preparations), two dimensions (nanofibres, nanowires, nanotubes), or one dimension (nanosheets, nanofilms, nanocoatings). It is inherent that these materials should also display different properties or functions from their macroscale counterparts as a result of their small size, respectively, their drastically increased surface to volume ratio. These differences include physical, mechanical, electrical, magnetic, and optical properties, chemical reactivity, and very likely of most concern, toxicity. However, it is important to distinguish between properties that change smoothly over a series of size reductions and properties that change abruptly below a certain critical size. The abrupt change of properties below a certain size is the key novelty of nanomaterials. This critical size depends on the property in question and on the material. Therefore, there is no scientific reason in support of the above indicated upper size limit. When looking at potential applications and use of nanomaterials in the food sector, it becomes obvious that the properties of most of the materials in question change smoothly when reducing the size. Therefore, the nature of properties and functions unique to the nanosize are still under discussion in the food sector.

This chapter provides an overview of potential applications of nanomaterials in the food sector and the role of mass spectrometry-based technologies for their detection and characterization. In addition to dynamic light scattering and electron microscopy, (single particle) ICP-MS and the combination of field-flow fractionation with ICP-MS are seen as very promising approaches in this context.

2. FOOD NANOTECHNOLOGY

Nanotechnologies and nanomaterials have emerged as a potential aid toward advances in the production of improved quality food with functionalized properties [5]. Research activities on applications of nanomaterials in the food sector include improved production and processing techniques; modified taste, color, flavor, texture, and consistency of food products; enhanced absorption of nutrients and bioactive compounds; reduced fat and salt content; improved shelf life and safety of food products; novel food packaging materials and nanosensors for better traceability and monitoring [5–19]. The application of nanotechnology processes and nanomaterials in the food sector fall into the following main categories:

- Development of food contact materials such as nanofilters, food packaging, kitchen utensils, processing equipment, and food containers.
- Development of nanosized food ingredients and nanosized delivery systems for biologically active compounds.
- Formation of nanostructures in food products.
- Development of nanoscale sensors and indicators.

Currently, many nanotechnology applications in the food sector are at research and developmental or near-market stages. Some years ago, only a couple of food ingredients, food additives, delivery systems for biologically active compounds, and food contact materials have found their way to the market in some countries [7,9]. However, the market for food products and food contact materials containing or consisting of nanomaterials was expected to grow rapidly worldwide [20]. Due to the lack of transparency, no clear information about the actual use of nanomaterials in the food industry is available.

The success of these products will be dependent on consumer acceptance as well as the legal framework and its applicability. Furthermore, reliable data on the benefits and risks of the application of nanomaterials in the food sector and their economic competitiveness are almost nonexisting. Many of the applications would require rigorous food safety testing before commercialization. In fact, investigations on the behavior of nanomaterials during digestion and in the human body have been started only recently and there is still a debate on appropriate testing methodologies. At present, the regulation and monitoring of nanomaterials in food regardless of global jurisdiction is a highly debated issue. In this respect, size is the only one of the parameters that should be considered. Properties such as chemical composition, unique physicochemical properties of the engineered nanomaterials, their interaction within the food matrix and with other food components, as well as their potential uptake and absorptivity in the gastrointestinal tract also need to be taken into account. All these properties will need to be examined in order to perform rigorous risk-benefit assessment. However, difficulties in the specific definition of engineered nanomaterials and the lack of standardized methods for their detection and characterization in complex matrices, such as food, as well as their potential absorption renders the effective implementation of the legal requirements presently difficult.

3. NATURALLY OCCURRING NANOMATERIALS AND THOSE DERIVED FROM CONVENTIONAL FOOD PROCESSING

Nanomaterials and nanostructures are and always have been a naturally occurring part of raw materials and foods [21]. Even unprocessed foods such as fresh fruits consist of structural components in the nanoscale. For example, proteins are generally globular structures of 1–10 nm in size. The majority of polysaccharides and lipids are linear polymers with thicknesses less than 1 nm, and are examples of one-dimensional nanostructures [22]. However, almost all food is processed in some way before it is eaten. Processing of raw materials leads to the huge variety of products available and is responsible for the various properties of a food such as texture, flavor, shelf life, and nutritional value. Many food processing operations, such as grinding, coagulation, emulsifying, or homogenizing, produce new nanostructures [23]. When foams are prepared and stabilized or emulsions are formed, two-dimensional nanostructures are created, one molecule thick, at the air–water (e.g., the foamy head on a glass of beer) or oil–water (e.g., sauces, creams, yoghurts, butter, and margarine) interface. Setting a gel, adding polymers to delay the sedimentation of dispersions or the creaming of emulsions, generally involves creating three-dimensional nanostructures by causing food biopolymers to assemble into fibrous networks. When starch is boiled to make custard, small three-dimensional crystalline lamellae, 10 nm in thickness, are melted. The texture of the paste or gel formed on cooling depends on the recrystallization of starch polysaccharides, as does the long-term staling of bakery products. When milk is homogenized, fat globules are produced that are about 100 nm in size. The dairy industry utilizes three basic micro- and nanosized structures (casein micelles, fat globules, whey proteins) to build all sorts of emulsions (butter), foams (ice cream and whipped cream), complex liquids (milk), plastic solids (cheese), and gel networks (yogurt) [24]. To distinguish naturally occurring nanomaterials and nanostructures as well as those generated by conventional food processing from intentionally manufactured nanomaterials, the term engineered nanomaterials (ENM) was introduced. By definition, an ENM needs to be intentionally produced in a defined size or size distribution for a specific purpose or function.

Even if some of the most important constituents of food such as proteins, starches, and fats undergo structural changes at the nanometre and micrometre scales during conventional food processing, the first comprehensive scientific perspective on a microstructural view of food was not published until as recently as 1987 [24]. Furthermore, imaging of the nanostructures introduced through food processing was only possible after the invention of the electron microscope and the development of tools such as probe microscopes (e.g., the scanning tunneling microscope (1982), and its more versatile offspring, the atomic force microscope (1986)). Today, foods are structured using a formulation or recipe, with structure formation (e.g., biopolymer transformation, phase creation, reactions) and stabilization (e.g., crystallization, network formation) occurring at the same time. A better understanding of the nature of nanostructures in foods will allow conventional technologies to be used rationally to improve food structure [25–27], and a better rational selection, modification, and processing of raw materials will be possible [22]. It is unlikely that the aforementioned modifications in conventional food processing and raw materials will be widely perceived as food nanotechnology. Thus, no modification to existing food regulation and no labeling will be required.

4. FABRICATION OF NANOMATERIALS

Nanomaterials can be produced by two principally different approaches called "top-down" and "bottom-up" [28]. Both approaches can be done in gas, liquid, supercritical fluids, solid states, or in vacuum. Most of the manufacturers are interested in the ability to control particle size, particle shape, size distribution, particle composition, and degree of particle agglomeration. Top-down manufacturing of nanomaterials involves breaking down larger pieces of matter using mechanical, chemical, or other form of energy. In the food sector, grinding, milling, and homogenization are examples for "top-down" approaches. Bottom-up manufacturing involves arranging atoms or molecules to larger entities via chemical reactions, electrostatic interactions, or self-assembly. This approach is capable of producing more complex molecular structures. The casein micelle is one example for a stable nanomaterial based on self-assembly of biological compounds [13]. An important limitation in the fabrication of nanomaterials for food application is due to the restriction to use food grade components only; a significant difference in terms of technological possibilities compared to, for example, pharmaceutical applications.

5. FOOD CONTACT MATERIALS

The term "food contact material" is used for various applications such as food packaging systems, food storage containers, kitchen and tableware, refrigerators, and food processing equipment. Nanostructured food contact materials and food contact materials containing nanoparticles have already been launched on the market or are currently under development.

5.1 Food Packaging Systems

Containment, protection, convenience, and communication are the four primary functions of food packaging that have been identified. These four functions are interconnected and all must be assessed and considered simultaneously in the package development process. Protection is often regarded as the primary function of the package. It protects the food from outside environmental influences such as water, water vapor, gases, light, UV radiation, odors, microorganisms, dust, shocks, vibrations, and compressive forces [29]. Therefore, food packaging provides a longer shelf life for foods and beverages by preventing, for example, spoilage, microbial contamination, and nutrient loss. Nanomaterials have the potential to improve food packaging and will be either confined in the matrix of the packaging material or form a nanolayer on its surface [30]. Food packaging has been identified as the largest area of current nanomaterial application in the food sector [7]. The application of nanomaterials in food packaging aims, for example, in prolonged shelf life of the packaged foods, improved food safety, higher quality foods, better traceability, environment-friendly packaging systems, lighter packaging systems, and ease of packaging production processes [31–41]. The main applications of nanomaterials in the area of food packaging systems were identified to be the following:

- Improvement of packaging properties such as flexibility, barrier properties for gases, moisture, volatile flavor and aroma compounds, UV radiation, as well as mechanical and heat-resistance properties.
- Packaging systems actively changing the condition of the packaged food to extend shelf-life or to improve safety or sensory properties, while maintaining the quality of the food ("active" food packaging).
- Packaging systems monitoring the conditions of packaged foods to give information about the quality of the packaged food during transport and storage ("intelligent"/"smart" packaging).
- Biodegradable packaging systems with acceptable barrier and mechanical properties (green packaging).

5.1.1 Barrier Properties

A range of nanomaterial-reinforced polymers have been developed recently. The level of nanomaterials in these nanocomposites is in general lower than 5% (w/w). Because of these relatively low levels, the nanomaterials do not significantly change properties of the packaging systems such as density, transparency, and processing characteristics [35,42]. Improved food packaging materials containing clay nanoparticles were among the first to be commercially available. The essential nanoclay raw material is montmorillonite, a widely available natural clay commonly obtained from volcanic ash/rocks. It is a 2-to-1 layered smectite clay mineral with a platelet-like structure. Individual platelet thicknesses are just 1 nm, but surface dimensions are generally 300 to more than 600 nm, resulting in an unusually high aspect ratio. Because of its hydrophilicity, the clay mineral does not disperse readily in polymers used for packaging systems. Hence, naturally occurring montmorillonite needs to be surface modified to be used in packaging systems. Nanoclay–polymer composites have been made from polyamides, nylons, starch, polyolefins, polylactide, polystyrene, ethylene–vinylacetate copolymer, epoxy resins, polyurethane, polyimides, and polyethyleneterephthalate. Incorporation of nanoclay platelets into a polymer was reported to reduce gas and moisture permeation by up to 90% and enhance gloss and stiffness [38,43–46]. The enhanced barrier properties are due to the staggered arrangement of the clay particles within the polymer matrix resulting in a more tortuous path, the gas or vapor molecules need to travel through the polymer layer. Potential applications of nanoclay-reinforced polymers in the food sector include food packaging systems for processed meats, cheese, confectionery, cereals, boil-in-the-bag foods, fruit juices, dairy products, beer, edible oils, and carbonated drinks [7,9,47]. Due to its higher aspect ratio compared to montmorillonite, laponite was studied as a further nanomaterial to be used in food packaging to improve barrier properties [48]. Laponite, a synthetic layered silicate, is manufactured from naturally occurring inorganic mineral sources.

In 2007, Imperm® (Nanocor® Inc., USA), for example, is reported to be used by Miller Brewing Co. (USA) for their multilayer PET beer bottles [7]. This technology minimizes the loss of carbon dioxide from the beer and the ingress of oxygen into the bottles, thus keeping beverages fresher and extending shelf life. In addition, the resultant bottles are both lighter and stronger than glass and are less likely to shatter. A shelf life of the brew of six months was obtained so far. Nanocor® Inc. (USA) and Southern Clay Products (USA) are now working on a plastic beer bottle that may prolong the shelf life to 18 months [7].

Besides a reduction in permeability for gases and vapors, improved barrier properties of packaging materials for UV radiation was achieved by incorporating nanomaterials into the polymer. Nanoparticles of pigments such as titanium dioxide and zinc oxide were used as UV absorbers in packaging materials and containers [7]. Titanium dioxide becomes transparent on nanoscale, but retains its UV absorption characteristics. This suggests applications in transparent wraps, films, or containers made of plastics such as polystyrene, polyethylene, and polyvinylchloride, where UV degradation needs to be avoided.

5.1.2 Mechanical Strength and Heat Resistance

Polymer–clay nanocomposites also demonstrate enhanced heat and abrasion resistance [49–51], and packaging materials incorporating titanium nitride nanoparticles, carbon nanotubes, or nanofibres were developed for increased mechanical strength without increasing the weight of the packaging material [35,38,39]. In addition, a multiwalled carbon nanotube-polyamide six composite was reported to have a better processability due to lower viscosity [52].

5.1.3 Active Packaging

Packaging systems actively changing the condition of the packaged food to extend shelf life or to improve safety or sensory properties, while maintaining the quality of the food are termed "active." For such active packaging materials, a physical contact of the active compound with the food is essential [53]. Therefore, the active compound is either designed to migrate into the packaged food or is tightly bound to the packaging material and thus the active effect is limited to the area of contact between the food and the packaging material. Currently, the most common application in active food packaging is the use of nanoscale silver embedded in the polymers as an antimicrobial agent and such packaging material is reported to be commercially available [5]. Examples include nanosilver food storage containers or nanosilver plastic bags. Several nanosilver-containing packaging products advertised 5–10 years ago were withdrawn from the Web sites or the Uniform Resource Locators (URLs) are no longer available. The mechanism of the antimicrobial effect of silver nanoparticles is still under discussion [54–56], even if the antimicrobial properties of silver have been known to cultures worldwide for many centuries. The antimicrobial activity of silver has been attributed to silver ions released from the silver surface [55,56]. Since size reduction to the nanoscale was reported to increase the antimicrobial efficiency of silver, a direct "particle-specific" toxic effect beyond the known antimicrobial activity of released silver ions was proposed [55].

Besides silver nanoparticles, nanoparticles of zinc oxide, magnesium oxide, titanium dioxide, and carbon nanotubes have been studied for their use in antimicrobial food packaging systems [57–60]. Further developments include the slow release of preservatives such as benzoic acid [35], nisin [61,62], and cinnamon oil [63] from food packaging systems, as well as the use of certain enzymes with antimicrobial activity [64].

Oxidation of food constituents due to the presence of oxygen within food packaging systems is the main cause for food deterioration. Therefore, active oxygen scavenging packaging systems have been developed [65,66]. However, some of the active packaging systems reported to contain nanomaterials consist of nanoclay composites, whereas the active component, the oxygen scavengers are molecules such as functionalized unsaturated hydrocarbons.

5.1.4 Intelligent/Smart Packaging

Packaging systems monitoring the conditions of packaged foods to give information about the quality of the packaged food during transport and storage are termed "intelligent" or "smart." Nanostructured indicators, nanosensors, antigen-detecting biosensors, or DNA-based biochips are being incorporated into food packaging materials in order to monitor microbial (e.g., food-borne pathogens or food-spoilage organisms) and/or chemical (e.g., toxins or allergens) contamination, as well as environmental conditions [67–70]. These systems respond with a color change indicating whether the food product is safe to be consumed or not. In addition to ensuring food safety, implementation of smart packaging system may also reduce food waste by replacing the shelf life expiration date (best before). Today, the expiration date is very often used as the best indication of product quality, taste, and nutrition. However, some of the food products are good for consumption at least six months after the expiration date. Therefore, smart packaging systems enable supermarkets and consumers to keep food until it is spoiled or contaminated by pathogens. Furthermore, indicators implemented in labels, printing inks, and coatings can monitor the integrity of the packaging by detecting leaks or time–temperature changes [70,71].

5.1.5 Tracking, Tracing, and Brand Protection

Nanoscale devices attached to food packaging or even the food product itself are discussed in respect to allow food to be tracked and traced from field or farm to factory to supermarkets and beyond. These approaches could help to avoid counterfeiting in the food sector. For tracking and tracing, for example, a nanobarcode consisting of stripes made of nanoparticles of gold, silver, and platinum varying in width and length was developed [72]. The nanobarcode must be read with a modified microscope for anticounterfeiting purposes. Nanodisks of gold and nickel, functionalized with chromophores could also be used as tags for tracking and tracing [73]. The chromophores emit a unique light spectrum when illuminated with a laser beam. Furthermore, a nanobarcode detection system that can be read by a computer scanner was developed [74]. Using a technique called Dip Pen Nanolithography, ink material could be deposited onto packaging materials in such a way that nanolithographic pattern are generated, which could be used for tracking and tracing [75]. A further very promising technology is the use of nanoscale Radio Frequency Identification Display (RFID) tags [76]. Unlike barcodes, which need to be scanned manually and read individually, RFID tags do not require line-of-sight for reading and it is possible to automatically read hundreds of tags in a second. The technology consists of microprocessors and an antenna that can transmit data to a wireless receiver. Retailing chains like Wal-Mart (USA), Home Depot (USA), Metro group (Germany), and Tesco (United Kingdom) have already tested this technology [77].

5.1.6 Biodegradable Packaging Systems

There is a great interest in the use of biodegradable packaging systems for fresh and processed food in order to reduce packaging waste [78,79]. Biodegradable packaging, however, often lacks the required mechanical stability and exhibits poorer barrier properties compared to plastic polymers [80]. The application of nanomaterials to biodegradable packaging systems provides advantages in respect to improving the properties of these packaging systems as well as the cost-benefit efficiency and thus promises to expand the use of biodegradable films in food packaging [33,81,82]. Currently, research activities with respect to biodegradable nanocomposites suitable for packaging applications focus on starch and derivates, polylactide, poly(butylene succinate), polyhydroxybutyrate, and aliphatic polyesters like polycaprolactone. Improvements of moisture barrier and mechanical properties of starch films [83–88] as well as improvements in tensile modulus and a reduction in oxygen permeability of polylactide films [89] has been reported after incorporation of clay nanoparticles into the polymers. Some major companies have been reported to already produce biodegradable polylactic acid and polycaprolactone nanocomposites for food packaging [90]. Furthermore, electrospinning was used to generate a biodegradable packaging system starting from chitin [91].

5.1.7 Edible Coatings

Edible coatings are discussed for meats, cheese, fruit and vegetables, confectionery, bakery goods, and fast food. They are created entirely from food-grade ingredients such as proteins, polysaccharides, or lipids and are usually 5 nm thick. Edible coatings could provide a barrier for moisture and gas or act as a vehicle to deliver colors, flavors, antioxidants, enzymes, and antibrowning agents [25,92], resulting in increased shelf life and quality of the coated product. US-based Sono-Tek Corp., for example, has commercially launched antimicrobial edible coatings for sliced meats and bakery products [93].

5.2 Coatings for Kitchen Utensils and Processing Equipment

Nanoscale coatings to create antimicrobial, scratch-resistant, antireflective, dirt-repellent, or corrosion-resistant surfaces are being developed for kitchen utensils and food processing equipment. Especially, abattoirs and meat and dairy processing plants could benefit from the introduction of surfaces aiming at reducing biofilm formation and facilitating cleaning of the surfaces [67,94]. However, currently the most widespread application is the utilization of the antimicrobial properties of silver nanoparticles, for example, by incorporation into the surface of cutting boards, baby milk bottles and drinking cups, tea pots as well as kitchen and tableware [95]. Furthermore, the US-based OilFresh Corporation has marketed a new nanoceramic product to be used in restaurant deep-frying machines. Due to its large surface area, oil use could be reduced by half and less energy is needed for cooking [77,96].

5.3 Nanofiltration Membranes

The application of nanofiltration in food processing is considered to be promising and cost effective. The nanofiltration membranes allow the passage of predominantly monovalent ions and water, but retain divalent and multivalent ions as well as more complex and larger molecules [97]. Water treatment for drinking water production is the main application of nanofiltration in food production [97–101]. Especially for the developing world, nanofiltration provides one solution to have widespread access to clean drinking water. In addition to the removal of turbidity and microorganisms from water, nanofiltration is commonly used for water desalination. Widespread application of nanofiltration technology is also seen by the dairy and sugar industry. Nanofiltration techniques are used in the dairy industry for simultaneous concentration and partial desalination [102]. Nanofiltration is also applicable for removing lactose from milk in order to make milk and milk-derived products suitable for lactose-intolerant people. Furthermore, nanofiltration could be applied by the food industry to recover lactic acid from fermentation media; to reduce color or salt of food products; to remove toxins, microorganisms, and viruses from food products, water, and beverages; to adjust flavor; and to concentrate food, dairy, and beverage products or by-products.

6. NANOSIZED FOOD INGREDIENTS AND NANOSIZED DELIVERY SYSTEMS

The efficacy of food ingredients, supplements, and additives is primarily a function of their conserved biological activity and availability. Therefore, size reduction and the use of nanosized delivery systems aim at enhanced absorption and bioavailability of the bioactive compound [103].

6.1 Nanosized Food Ingredients

The stability and solubility of valuable food ingredients can be tremendously improved by changing their size into nanoscale. A variety of nanosized food ingredients, supplements, and additives are currently reported to be commercially available. To fortify soft drinks and food products, a synthetic form of the tomato-carotenoid lycopene with a particle size of about 100 nm is produced by BASF [9]. Besides health benefits, addition of the water-dispersible lycopene to beverages also provides color. In addition, nanoselenium is being marketed as an additive to green tea [7,9]. Due to an enhanced uptake of selenium, this tea is claimed to provide a number of health benefits. Nanosized calcium, magnesium, zinc, and iron are also claimed to provide health benefits by enhancing mineral uptake and bioavailability [9]. Two further compounds that might be available as nanomaterials are silicon dioxide and titanium dioxide. Silicon dioxide is approved as an anticaking agent in certain food ingredient applications, while titanium dioxide is an approved coating and whitening agent for some foods;

both compounds, however, are not approved in their nanoform. With the aim to provide consumers a possibility to reduce their salt intake, nanoscale-sodium chloride is under development [104]. It is expected that due to its small size only small amounts are needed to cover a large area of the food surface; nevertheless inducing the same salty impression as common salt.

6.2 Nanosized Delivery Systems

Nanosized encapsulation systems have gained increased interest as potential delivery systems for nutrients, bioactive compounds, additives, supplements, and processing aids. Nanoencapsulation has emerged as an extension of micro-encapsulation technology, a well-established technology in the food industry. The potential of nanosized delivery systems is based on [25,105–113]

- improving the absorption of the encapsulated compound by modulating its solubility, bioaccessibility, membrane permeability, or stability in the gastro-intestinal tract (GIT),
- increasing the stability of the encapsulated compound by protection from external factors such as temperature, pH, moisture, oxygen, enzymes, and therefore protection from degradation during processing, storage, distribution, or in the GIT,
- overcoming incompatibilities between the encapsulated compound and the food matrix,
- improving compatibility of the encapsulated compound with other food constituents as well as its dispersibility and distribution in the food matrix,
- controlling the release of the encapsulated compound by, for example a pH-, moisture-, or enzyme trigger,
- flavor and taste masking.

Nanoencapsulation could make a significant contribution to the economic viability of food formulations because only a reduced amount of the active ingredient is needed. Furthermore, nanoscale delivery systems can be incorporated into food and beverages without affecting their taste or appearance. The choice of the encapsulation material depends on the compound to be encapsulated, the encapsulation technology applied, and the desired application. The food industry, however, is limited to the use of food-grade ingredients for encapsulation. In the meantime, a wide range of nanoscale delivery systems has been described [25,105–139]. These delivery systems differ in structures and composition. Among others, casein, α-lactalbumin, β-lactoglobulin, zein, chitosan, alginate, polylactic-co-glycolic acid have been used as encapsulate materials to generate structures such as nanosized self-assembled liquid structures, nanocochleates, nanoliposomes, archaeosomes, micelles, solid lipid nanoparticles, nanoemulsions, and biopolymeric nanoparticles.

The German-based company Aquanova®, for example, has developed a new technology called NovaSOL®, which utilizes micelles with a diameter of about

30 nm. With this technology, benzoic acid, citric acid, ascorbic acid, vitamin A, vitamin E, soybean isoflavones, β-carotene, lutein, coenzyme Q10, α-lipoic acid, and omega-3-fatty acids have been encapsulated. The loaded micelles can be easily and effectively introduced into food and beverages and provide increased intestinal absorption as well as thermal, mechanical, and pH stability of the encapsulated compound. The delivery system of the Israel-based company NutraLease Ltd. is called NutraLease™ and is based on the nanosized self-assembled liquid structures technology [7,9]. The delivery system consists of hollow spheres with a diameter of approximately 30 nm made from fats with an aqueous interior. The nanosized self-assembled structured liquids technology has been used to encapsulate compounds such as coenzyme Q10, lycopene, β-carotene, lutein, phytosterols, vitamin A, vitamin D, vitamin E, omega-3-fatty acids, and soy isoflavones [77]. The so-called Canola Active oil containing the NutraLease™ delivery system loaded with phytosterols was marketed by Shemen Industries (Israel). The oil is claimed to reduce the cholesterol intake into the body by as much as 14% by competing for bile solubilization [77]. With NanoCluster™, Royal BodyCare Life Sciences® Inc. (USA) markets a delivery system, which is, for example, used in a powdered chocolate drink, called Slim Shake chocolate [7,9,77]. The drink is claimed to be sufficiently sweet without added sugar or sweeteners by incorporating cocoa into nanoclusters. Bioral™ is a nanocochleate delivery system for micronutrients and antioxidants marketed by BioDelivery Sciences International (USA) [7,9,77]. Nanocochleates are coiled nanoparticles, crystalline in structure, derived from soy phospholipids with a size of 50 nm. In addition, more nanosized carrier systems are reported to be under development [9]. Self-assembled nanotubes with a cavity diameter of 8 nm, for example, have been developed from the hydrolyzed milk protein α-lactalbumin [119,120]. The cavity can be used to bind nutrients and bioactive compounds or to mask undesired flavors [14,119,120].

6.2.1 *Improvement of Bioavailability*

Encapsulation of bioactive compounds is aiming to improve their bioavailability. The low bioavailability of a bioactive compound might be due to its low solubility, low bioaccessibility, low membrane permeability, or low stability in the gastrointestinal tract. Thus, the reasons for the low bioavailability of the bioactive compound in question need to be identified and then the delivery system has to be carefully designed to overcome these problems. So far, the number of in vivo studies using a bioactive compound in a food-grade nanosized delivery system is very low. Studies performed with coenzyme Q10 [140–143], vitamin D [144,145], vitamin E [140,146], curcumin [147–151], iron [152–157], and calcium [158–161] resulted in general in a 2- to 10-fold higher plasma concentration of the bioactive compound under investigation. Besides bioavailability, biokinetic parameters were also changed by encapsulation. The effect on bioavailability and biokinetic parameters depends on the bioactive compound, the formulation, as well as the method used for encapsulation. Another important factor is the food matrix the delivery systems are incorporated in. However, only a very few of the studies indicated above were performed by administering the formulations incorporated in a food matrix.

6.2.2 Masking of Off-flavors

Encapsulation can also be used to mask off-flavors. Iron supplements such as iron sulfate exhibit a typical unpleasant iron taste and color. These unwanted attributes could be masked using phospholipid-coated iron phosphate nanoparticles [152]. The undesirable odor or flavor of soy products, sweeteners, and omega-3-fatty acids was reported to be significantly reduced using β-lactoglobulin nanoparticles [162] or cyclodextrins [163] for encapsulation. Nanoencapsulation has also been used by George Weston Foods, one of the leading bakeries in Western Australia, to mask the taste and odor of tuna fish oil, which is high in omega-3-fatty acids in their Tip-Top® Up™ bread [9]. The delivery system releases the tuna fish oil only when reaching the stomach and hence the unpleasant taste of fish oil can be avoided. In addition, encapsulation extends shelf life of the omega-3-fatty acid-containing food products by protecting the polyunsaturated fatty acids from oxidation.

6.2.3 Protection of Bioactives Compound

Many bioactive compounds are unstable and interact with oxygen, food components, or digestive enzymes. Therefore, protection of the bioactive compound is required during processing and storage until consumption. It was already shown that nanoencapsulation is capable of increasing the stability of bioactive compounds in a food matrix [163–169] and in the gastrointestinal tract [170,171]. The capsule material acts as a physical barrier and prevents access of, for example, oxygen, other food components, adverse pH conditions, and free radicals to the encapsulated bioactive compound. Thus, destruction of the bioactive compound in a food matrix is reduced during processing and storage resulting in an extension of the product shelf life.

7. STRUCTURING OF FOODS

The design of certain food structures may affect taste, texture, and consistency of food. As mentioned previously, functionality of raw materials and their successful processing always involved the presence, creation, and modification of forms of self-assembled nanostructures. Nanomaterials, however, such as nanoliposomes, nanoemulsions, nanoparticles, and nanofibres, have been reported as an option to create novel structures and introduce new functionalities into foods [25,172]. Examples are low-fat products with the same mouth-feel as their full fat alternatives or food products with a desirable stability, but a reduced content of certain additives. Furthermore, low-fat nanostructured mayonnaise, spreads, and ice creams are under development [7,9]. Nanostructuring of ice cream, for example, is expected to reduce its fat content from 8–16% to 1%. Introducing multiple emulsions and nanoemulsions may reduce the need for stabilizers and thickeners to achieve a desirable food texture and stability [135,173].

8. NANOSENSORS

Nanosensors to control food safety and quality have been developed recently or are still under development. Among others, nanoparticle-based sensors, array-based sensors, nanocantilevers, nano-test-strips, nanoparticles in solution, electronic noses, and electronic tongues could serve as nanosensors to detect food-borne pathogens, spoilage microorganisms, toxins, allergens, contaminants, and chemicals in foods [174–179]. The advantage of a nanosensor is the possibility to develop a portable device and to measure in real time with high sensitivity. It was argued that even one single microorganism in a food product could be recorded [180]. Portable devices have been developed in two projects funded by the European Union [9]. The device developed in the BioFinger project uses nanocantilevers to detect food-borne pathogens, toxin, and other undesired compounds in raw materials and food products, whereas an array-based sensor was developed in the GoodFood project. Further examples for nanosenors to detect food-borne pathogens use antibody-conjugated fluorescent nanoparticles [180,181] or biofunctional magnetic nanoparticles [182]. In addition, methods have been reported to control food quality using an electronic nose [183,184]. The electronic nose was shown to be capable of monitoring changes in the volatile components of fruits.

9. SAFETY ASPECTS OF NANOMATERIALS IN FOOD

Currently, the impact of using nanomaterials in the food sector on human health is a highly debated issue. Thereby insoluble, indigestible, and potentially biopersistent nanomaterials are of major concern. In respect to the impacts of nanomaterials on human health, the physicochemical properties of a nanomaterial such as particle number concentration, shape, specific surface area, chemical composition, surface chemistry, coatings, surface charge, crystal structure are perhaps more important than size or size distribution. The physicochemical properties of a nanomaterial determine its chemical reactivity, solubility, biological degradability, agglomeration, or aggregation behavior as well as its interaction with the environment. The potential for a nanomaterial to induce toxic effects depends on how the body "sees" and hence reacts with the nanomaterial at any given point in time (dynamic situation). Since properties of a nanomaterial can differ significantly from those of the corresponding bulk material or its molecular form, toxicity profiles of the bulk material or the molecular form might not apply to the nanoform. Therefore, thorough risk assessment of nanomaterials is necessary before their commercialization to protect consumers and the environment from potential hazards [185,186]. It is generally accepted that the currently used risk-assessment paradigm is considered applicable for nanomaterials [187,188]. Risk assessment is a scientifically based process consisting of the hazard identification, hazard characterization, exposure assessment, and risk characterization. In this context, risk assessment can be described as characterizing the potential hazards and the associated risks to life and health resulting from exposure of humans to nanomaterials present in foods and beverages. However, currently only very limited information about exposure of humans to nanomaterials present in foods and beverages is available. It was estimated that with a Western diet 10^{12}–10^{14} nano- and microparticles are consumed daily [189]. Furthermore, major gaps in knowledge with regard to the behavior, fate, and effects of nanosized material via the gastrointestinal route have been identified. It is not known in whether nanosized materials bind to other food components, agglomerate, or remain as free particles in the gastrointestinal tract. It is possible that they will not remain in a free form in the lumen and hence will not be available for absorption. As with other food components, interaction of nanosized materials is very likely to change during the passage through the gastrointestinal tract. Nanosized material may also affect gut function or gut microflora. Solubility and digestibility are two factors that largely determine the fate of nanomaterials in the gastrointestinal tract. When dissolved or digested, it can be assumed that nanomaterials lose their nanoscale properties. Due to the large variety of nanomaterials that could be used in the food sector, general statements on the behavior of nanomaterials in the gastrointestinal tract are not possible. For example, very small modifications in the formulation of nanoscale lipid particles strongly affected their digestibility in an in vitro model [190]. Indigestible and insoluble nanoparticles can be either excreted or absorbed intact. From studies with predominantly organic and inorganic, insoluble nanoparticles it is known that absorption of intact particles is size dependent and depending on the nature of the nanoparticles different mechanisms are involved [189]. Besides absorption of the intact nanoparticles, their interaction with the intestinal epithelium may pose a health risk. This interaction might facilitate absorption of compounds that otherwise cannot enter the body ("Trojan Horse Effect") [191]. Furthermore, a lack of knowledge regarding the effect of size reduction into the nanoscale on biokinetics has been reported [187,188]. As soon as nanomaterials have been absorbed, they can be distributed throughout the body via the blood or the lymphatic system. Interactions of nanomaterials with blood components can have a significant impact on their distribution and excretion. Nanomaterials have been reported to reach potentially sensitive target sites such as bone marrow, lymph nodes, the spleen, the brain, the liver, and the heart [192–194]. Accumulation within the human body and the possibility to overcome natural barriers such as the blood–brain barrier are seen as major issues in respect to safety [195,196].

The European Food Safety Authority (EFSA) published the first practical guidance for the risk assessment of the application of nanoscience and nanotechnologies in the food and feed chain [188]. The guidance covers risk assessments for food additives, enzymes, flavorings, food contact materials, novel foods, feed additives, and pesticides. However, two specific hurdles in performing risk assessments on nanomaterials in food and beverages have been identified: difficulties in detecting, measuring, and characterizing nanomaterials in complex matrices such as food and insufficient information on toxicology data. Therefore, risk assessment of nanomaterials in food and beverages needs to be performed on a case-by-case basis [197]. Detecting, measuring, and characterizing nanomaterials in foods and beverages as well as in organs, tissues, and cells is a fundamental requirement of risk assessment. Without appropriate routine analytical approaches to identify and characterize nanomaterials in complex matrices, it is impossible to obtain reliable data about exposure and the fate of nanomaterials in the human body following exposure and absorption.

10. LEGAL OBLIGATION

Currently, no international regulations of engineered nanomaterials in the food sector exist. However, the European Union has taken the lead in the regulation of nanomaterials in food. According to Regulation No. 1169/2011 of the European Commission [198], labeling indicating the presence of engineered nanomaterials in foods is required since December 2014. The definition for engineered nanomaterials laid down in the above mentioned regulation will need to be adjusted and adapted to the technical and scientific progress or to definitions agreed at an international level. According to a recommendation of the European Commission, ENM is defined as a nanomaterial if at least 50% of the number-based particle size distribution is in the size range between 1 and 100 nm and/or its specific surface area by volume is greater than 60 $m^2\ cm^{-3}$. Important challenges in respect to the labeling issue relate primarily to establishing validated methods and instrumentation for detection, quantification, and characterization of nanomaterials in foods and beverages.

Nanomaterials or changes caused by a modified particle size in the nanorange are also covered in some other European Regulations, such as the regulation on enzymes (Regulation No. 1332/2008) [199], on food additives (Regulation No. 1333/2008) [200], on active and intelligent food contact materials (Regulation No. 450/2009) [201], and on plastic food contact materials (Regulation No. 10/2011) [202]. In addition, it was recommended to re-examine necessary modifications of the Novel Food Directive in order to be applicable for nanomaterials or materials derived from nanotechnology.

11. ANALYTICAL APPROACHES TO CHARACTERIZE NANOPARTICLES IN FOOD

As mentioned previously, analytical methods to quantify and characterize nanomaterials in complex matrices such as food are urgently needed in respect to labeling issues of such materials in the food sector as well as to performing an adequate risk assessment. Several different analytical methods for detection and characterization of isolated nanoparticles in simple matrices such as water have been developed so far. Analyzing nanomaterials in food, however, is much more sophisticated because foods are generally very complex matrices containing many different structures and substances including nanomaterials or even a mixture of different nanomaterials. Therefore, sample preparation techniques such as centrifugation, filtration, or field-flow fractionation need to be applied prior to analysis. However, separation of the (engineered) nanomaterials from the food matrix without affecting their size/size distribution as well as their low concentrations in a food matrix represent the major challenges. The properties of a nanomaterial are very likely to change during sample preparation and analysis because these properties are a function of the environment of the nanomaterial, and even a simple procedure such as dilution of the sample with pure water will affect the ionic strength and the pH-value of the system. Thus, in respect to safety aspects of ENM incorporated into a food matrix, it is not sufficient to characterize the pristine nanomaterial because changes in composition of the surrounding medium, cooling, or heating as well as mechanical stress during food production and digestion may result in alterations of the ENM and their properties. Proteins, for example, were reported to interact with the surface of nanoparticles to form the so-called "protein corona" leading to a shift in surface charge and particle size [203,204].

When characterizing nanomaterials in food matrices, size or size distribution are not the only parameters of interest. Properties such as number or mass concentration, shape and morphology, chemical composition, purity, solubility, surface chemistry and reactivity, surface charge, crystallinity as well as aggregation and agglomeration state need to be determined. No single characterization technique is capable of providing all relevant information. Therefore, a combination of different analytical techniques is required. The European Food Safety Authority (EFSA), for example, claims that particle size should always be determined by at least two independent methods, one being electron microscopy, as the results of analytical approaches may differ because of differences in the underlying physical principles [188,205]. An overview of different characterization methods for nanomaterials is given in Table 1. All methods listed have advantages and limitations related to sensitivity and specificity in complex food matrices (concentrations/level of detection, polydispersity, chemical composition, etc.) [206–212].

For the development and validation of characterization methods and to calibrate the analytical systems, reference materials and representative test materials are needed. However, such materials are scarce, especially for the food sector. For example, titanium dioxide reference material is provided by The National Institute of Standards and Technology (NIST) [213]. The available reference materials and representative test materials were summarized recently [214].

The most commonly used methods for particle characterization are electron microscopy (EM), often combined with energy dispersive X-ray analysis

TABLE 1 Important Parameters and Analysis Methods for the Characterization of Nanomaterials

Parameter	Currently Available Methods (Examples)
Particle size (Primary/Secondary)	Microscopy methods: e.g., transmission electron microscopy (TEM), scanning electron microscopy (SEM), scanning transmission electron microscopy (STEM), atomic force microscopy (AFM), scanning transmission X-ray microscopy.
	Separation methods (flow separation and chromatography methods): e.g., field-flow fractionation (FFF), hydrodynamic chromatography (HDC), size exclusion chromatography (SEC), high performance liquid chromatography (HPLC); differential mobility analysis; (Ultra)centrifugation methods.
	Spectroscopy methods: X-ray diffraction.
	Light (laser) scattering methods: e.g., dynamic light scattering (DLS) also known as photon correlation spectroscopy (PCS); multiangle light scattering; static light scattering; photon cross-correlation spectroscopy; nanoparticle tracking analysis (NTA).
	Single-particle inductively coupled plasma mass spectrometry (sp-ICP-MS).
Chemical composition/identity	Elemental analysis: optical emission spectroscopy; atomic absorption spectroscopy; X-ray photoelectron spectroscopy; energy dispersive X-ray spectroscopy; nuclear magnetic resonance spectroscopy, mass spectrometry (MS) in particular ICP-MS.
	Molecular composition: mass spectrometry using suited ionization techniques (e.g., matrix-assisted laser desorption/ ionization coupled with separation methods (e.g., HPLC, capillary electrophoresis (CE)), nuclear magnetic resonance spectroscopy.
	Shell/core composition (for encapsulates, micelles): by a suitable method given above, after disintegration of the particles and separation of the components, e.g., by HPLC, SEC, CE, HDC.
Physical form and morphology	Microscopy methods (TEM, SEM, STEM, AFM); X-ray diffraction.
Crystalline phase	X-ray diffraction.
Particle concentration	Mainly light scattering methods (for dispersions).
	Particle concentration (in pure dry powders) may also be calculated from particle size, mass concentration, and density of the material.

Continued

TABLE 1 Important Parameters and Analysis Methods for the Characterization of Nanomaterials—cont'd

Parameter	Currently Available Methods (Examples)
Mass concentration and density	Suited methods from those listed under chemical composition, e.g., ICP-MS; analytical electron microscopy; gravimetric methods; centrifugal sedimentation (for suspensions). A possible method for measurement of density is provided by OECD TG 109.
Specific surface area	Gas adsorption (Brunauer Emmett Teller method).
Surface chemistry	Any of the suitable chemical characterization methods listed above.
Surface charge	Electrophoresis, e.g., capillary electrophoresis, laser doppler electrophoresis.
Redox potential	Potentiometric methods.
Dissolution/solubility	Standard tests for water solubility (e.g., OECD TG 105) and log P_{OW} (OECD TG 107, 117) can be used. Dissolution rate constants.
Viscosity	Methods such as OECD TG 114.
Pour density	DIN ISO 697, EN/ISO 60.
Dustiness	Methods such as EN 15051:2006, DIN 33897.
Chemical reactivity/ catalytic activity	Kinetic measurements of the chemical, biochemical, and/or catalyzed reactions.
Photocatalytic activity	Kinetic measurements of the chemical, biochemical, and/or catalyzed reactions.
pH value	pH meter/indicator.

ICP-MS, inductively coupled plasma mass spectrometry; OECD, Organisation for Economic Co-operation and Development.
Adapted from Ref. [188].

(EDX), and dynamic light scattering (DLS). The limitation of EM results from the need of elaborate sample preparation, which may induce artifacts due to particle agglomeration/aggregation during drying of the samples. Furthermore, the fact that only small amounts of samples can be analyzed has an impact on the statistical significance of the results in respect to size distribution. The determination of particle size in complex food matrices via EM might be challenging because organic coatings are difficult to consider and thus the derived size refers only to the (inorganic) core of the particles.

DLS is a widespread and rapid method for the determination of the hydrodynamic diameter of particles. It does not need any calibration, but no information on the shape and composition of the particles could be obtained. Moreover, the technique is not applicable if there is a need to discriminate between particles of different chemical composition or for polydisperse samples. The International Standards for DLS [215,216] only use two parameters to describe size distributions, an average size (harmonic intensity-weighted arithmetic average particle diameter ($\overline{x}_{DLS}$) and a polydispersity index (PI). Acceptable size distributions in particular applications could be obtained, but the mathematical models to deduce particle size distribution from the measured signals is in general not sufficiently reproducible and reliable to be incorporated in an international standard.

The European NanoLyse project [217] was set up to develop validated methods and reference materials for analyzing engineered nanoparticles in food and beverages. The original idea was to develop analytical methods for all relevant classes of engineered nanoparticles with reported or expected food and food contact material applications, i.e., metal, metal oxide/silicate, surface functionalized and organic encapsulated engineered nanoparticles. Some of the main achievements reported so far include the development of sample preparation methods such as enzymatic digestion for silver nanoparticles in chicken meat [218], acid digestion for silica nanoparticles in tomato soup [219], as well as the development and validation of a quantitative approach to detect fullerenes at ppb levels in vegetable oil based on liquid chromatography-mass spectrometry (LC/MS). A direct application of fullerenes in the food sector is not anticipated, but due to their industrial manufacture in amounts of several tons per year, fullerenes might appear as contaminants in foods. Several LC/MS-based detection approaches for fullerenes in aquatic and biological samples are already available (reviewed in Ref. [220]). Concerning the validation of analytical methods, the results of the NanoLyse project are promising, but so far only a limited number of ENM–food matrix combinations has been studied within the framework of the project. However, validation of the methods has been performed with spiked samples. Therefore, the chemical identity as well as the size of the nanomaterials to be analyzed were well known. Application of the developed analytical approaches to foods already commercialized is currently missing. In those foods, neither the chemical identity nor the size of the nanomaterial potentially present is known and, in addition, such foods may contain a mixture of nanomaterials of different chemical identities. Distinction between engineered and naturally occurring nanomaterials is a further unsolved problem of almost all analytical approaches for the detection and characterization of nanomaterials available so far.

Two promising approaches that are gaining importance in nanoparticle detection and characterization in foods at relevant concentrations are the coupling of field-flow fractionation with inductively coupled plasma mass spectrometry (FFF-ICP-MS) and single-particle ICP-MS (sp-ICP-MS). The size fractionation via FFF combined with ICP-MS as detector has the potential to achieve excellent sensitivity, elemental specificity, and enables the characterization of polydisperse systems. Thus, a major challenge in analyzing nanoparticles in complex matrices such as foods can be addressed.

11.1 FFF-ICP-MS

Coupling of ICP-MS with field-flow fractionation (FFF) is a relatively new technique for the detection, characterization, and quantification of particularly inorganic nanomaterials in biological tissues, food, cosmetics, and environmental samples.

Field-flow fractionation was developed by John Calvin Giddings in order to simplify the measurement of complex materials by separation into components [221,222]. According to Giddings, separation capability of FFF ranges from about 1 nm to more than 100 μm and the technique can be applied to both simple and complex macromaterials of biological, biomedical, industrial, and environmental relevance [223]. FFF was also shown to be useful in the fractionation of nanoparticles [218,219,224–247]. When an FFF system is used in combination with appropriate detectors, determination of particle sizes renders possible by calibrating the system with particle-size standards. Ideal standards exhibit the same chemical composition, surface properties, and shape as the particles to be analyzed. However, currently suitable standards are hardly available worldwide.

FFF is a separation technique where a fluid dispersion or solution (mobile phase) is pumped through a long and narrow channel. In order to cause separation of the particles present in the mobile phase, a field is applied perpendicular to the direction of flow. Among others, gravitational, centrifugal, thermal-gradient, electrical, magnetic forces, or an asymmetrical flow through a semipermeable membrane have been described. Separation of the particles depends on their differing "mobilities" under the force exerted by the field. Sedimentation field-flow fractionation (SdFFF), for example, was demonstrated to be a very useful tool for size characterization of silica particles, in general, and of silica particles incorporated in food matrixes [229]. Symmetrical flow field-flow fractionation was reported to successfully characterize silver nanoparticles [235].

However, the most common field-flow fractionation technique for analysis of nanoparticles is "asymmetric flow FFF" (AF4). AF4 uses a separation channel of several centimetres in length, but only 100–500 μm in height with a semipermeable membrane at the bottom of the channel (Figure 1). The mobile phase is continuously pumped through the channel and forms a parabolic flow profile (laminar flow). Furthermore, it is to some extend drawn off through the semipermeable membrane at the bottom of the channel resulting in a force perpendicular to the direction of the flow of the mobile phase (cross-flow). The sample to be analyzed is injected into the channel and focused close to the membrane. Due to the cross-flow, macromolecules or particles present in the sample to be analyzed are forced toward the membrane. The size-dependent diffusivities of particles or macromolecules lead to their arrangement in different layers

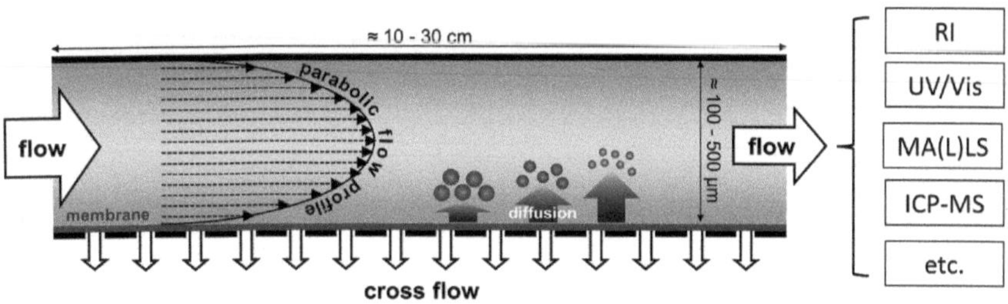

FIGURE 1 Asymmetric flow field-flow fractionation principle with different detectors. *Adapted from Postnova.*

of the parabolic flow profile inside the channel. As a result, macromolecules or particles with a small hydrodynamic diameter will elute first and larger macromolecules or particles will elute later. However, the development of a suitable AF4 approach for a certain type of nanomaterial is time consuming. Flow rate or flow profile, injection volume, the semipermeable membrane (material, molecular cutoff), and the mobile phase need to be optimized and adapted to each analyte. For example, poor recovery rates have been observed because nanoparticles tend to adsorb to the semipermeable membrane depending on their surface properties and the membrane material [224,231]. However, this kind of interaction may differ even for identical nanoparticles incorporated into different food matrices [230]. One possibility to reduce such interaction is the addition of surfactants to the mobile phase. In addition, the composition of the mobile phase needs to be chosen in a way to avoid agglomeration or other alterations of the particles. After calibrating the system with suitable standard materials, particle sizes or molecular masses of macromolecules can be determined. Detection and characterization of the macromolecules or particles eluting from an AF4 system can either be performed offline after fractionation using a fraction collector or online. The following principles have been relatively often used with an AF4 system: refractive index (RI), UV/Vis, multiangle (laser) light scattering (MA(L)LS), and dynamic light scattering (DLS). In Figure 2, the separation of a mixture of silver nanoparticles of three different sizes (10, 40, 100 nm) by AF4 with subsequent UV/vis detection at 420 nm is shown. Deionized water (pH 9.2, adjusted with NaOH) was used as a mobile phase.

Recently, coupling of AF4 with ICP-MS was considered as a promising approach to detect, quantify, and characterize inorganic nanoparticles incorporated into a food matrix [218,219]. Thereby ICP-MS could be used both in offline and online mode. Besides improvement in sensitivity, the possibility to simultaneously elucidate the size and chemical composition of the particles was reported as an advantage. When using ICP-MS in online mode, composition and flow of the mobile phase need to be adjusted properly. For example, instabilities or a decrease of signal intensity over time due to salt formation on ICP-MS cones have been reported because of a high salt concentration in the outflow of the AF4 [240]. Furthermore, polyatomic spectral interferences are

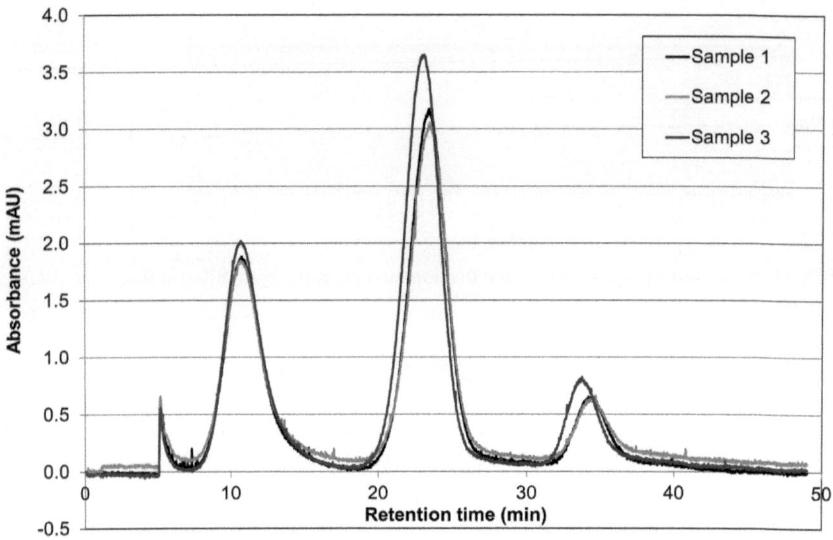

FIGURE 2 Fractogram of a mixture of silver nanoparticles of three different sizes (10, 40, 100 nm) by AF4, separation of the silver nanoparticles was performed in triplicate.

known to occur in the plasma and may bias the results [241]. Modern instruments make use of several technologies (collision/reaction cell) to reduce these interferences.

So far, AF4-ICP-MS was used to analyze and characterize silica nanoparticles in tomato soup [219], silver nanoparticles in chicken meat [218], aqueous matrices [230] or culture media [224] and gold nanoparticles in aqueous matrices [231]. The possibility to generate accurate and reliable results by applying AF4-ICP-MS was reported [230] as well as a good agreement of results obtained by AF4-ICP-MS with those obtained by transmission electron microscopy (TEM) [224]. However, only a few nanomaterials in a limited number of food matrices have been studied with FFF-ICP-MS until now. Furthermore, all studies have been performed with spiked samples or nanomaterials present in aqueous solutions. Studies on "real" food are still missing.

11.2 Single-Particle ICP-MS

Single-particle ICP-MS (sp-ICP-MS) has been introduced into the analysis and characterization of inorganic nanomaterials only recently [218,234,242–247]. The emerging technique was first described in 2003 [242] and in the meantime, it has been shown to be very promising in respect to the simultaneous determination of size and concentration of inorganic nanoparticles in environmental and food samples. Due to the state of the technical development, sp-ICP-MS allows currently only single-element detection. A specific feature of sp-ICP-MS is the possibility to distinguish between the ionic and particulate form of

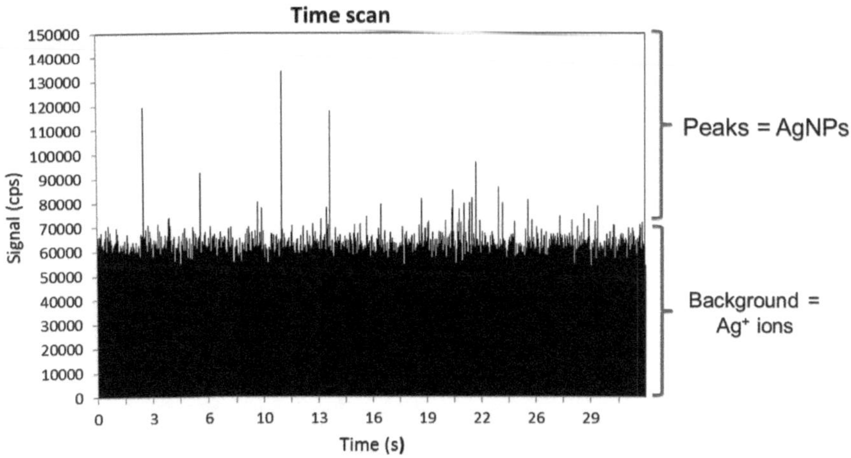

FIGURE 3 sp-ICP-MS measurement of a mixture of ionic and nanoparticulate silver. The sample contains a high background of ionic silver besides a few silver nanoparticles.

a certain analyte (Figure 3). The continuous background signal corresponds to the ionic form and the distinct pulse above the background signal to the particulate form of the analyte [242,245]. With sp-ICP-MS, nanosilver particles with a size of 20 nm or above are distinguishable from its ionic form, when present in the same sample [248].

To obtain a number-based size distribution of the analyte, the nanoparticle-containing sample needs to be introduced into the ICP-MS system in high dilution. Thus, it is guaranteed that individual particles enter the plasma. There they are ionized and each particle produces a corresponding detectable pulsed signal. Nebulization efficiency of an ICP-MS plays a significant role in the determination of particle number concentrations [247]. An accurate determination of the nebulization efficiency is not unbiased and needs at first a well-characterized reference material in respect to particle size and particle number distribution.

A further requirement for an accurate sp-ICP-MS measurement is a sufficiently short acquisition time (dwell time) in order to guarantee that an individual particle generates a single detectable pulse and therefore a distinct signal above the continuous background (Figure 4). Too-short dwell times on the other hand may cause a split of the detector pulse of an individual particle [247]. In general, dwell times in between 3 and 10 ms have been identified as optimal. Most modern ICP-MS instruments are capable of using dwell times as short as 0.1 ms. Dwell times below 1 ms, however, result in a broadening of the obtained particle size distribution and in an increase of the particle number concentration [218,244]. On the other hand, more data points per particle peak are collected with dwell times below 1 ms. Thus, it might be possible to study agglomeration or aggregation events.

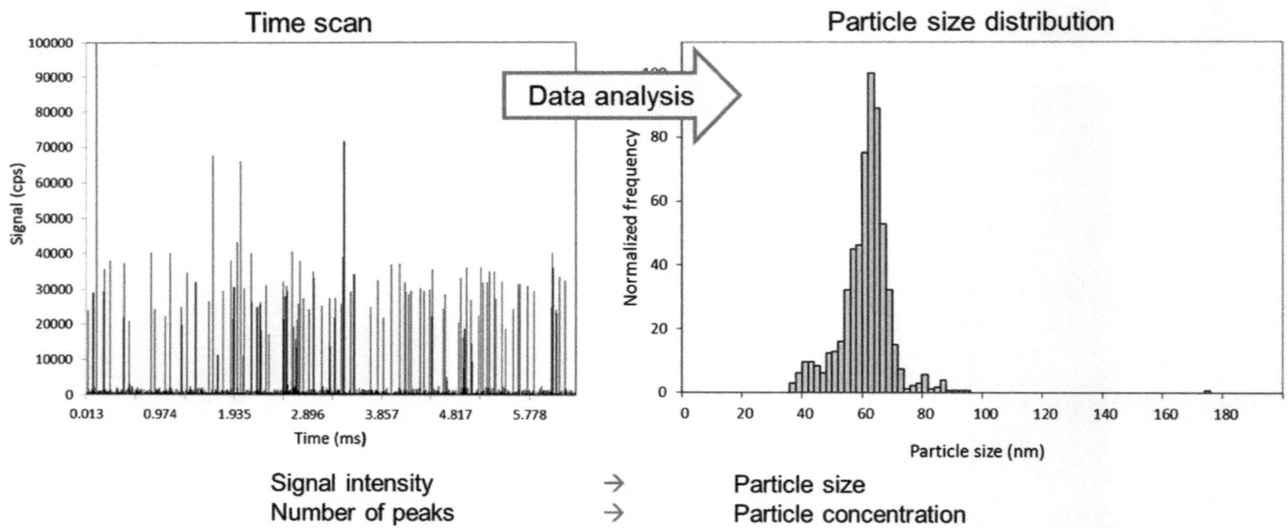

FIGURE 4 Time-resolved data acquisition of 60 nm silver nanoparticles and calculation of the corresponding particle size distribution. In this sample, the background of ionic silver is low.

According to sp-ICP-MS theory, the number of pulses generated is proportional to the particle concentration in the sample and the pulse intensities are related to particle mass. With the knowledge of the chemical composition of the particles to be analyzed and their shape, the corresponding particle diameter can be calculated [242]. After collecting the time-resolved raw data in a time-scan histogram, a cumulative distribution function is applied and signal counts versus number of events may be plotted. In order to calculate the particle sizes from the pulse intensities, an appropriate calibration an ionic ICP-MS standard of the analyzed element is needed. To discriminate individual particles from the background signal, several mathematical approaches have been proposed [243,244]. However, this process is not automated yet and thus time consuming. Nevertheless, particle size distributions of the nanoparticles present in the sample could be obtained by applying such mathematical approaches (Figure 4).

Currently only one validated sp-ICP-MS approach for sizing and quantification of nanoparticles in a food matrix is available [246]. Validation was performed with spiked samples (chicken meat spiked with nanosilver) only; no studies on "real" food samples have been published.

Since ICP-MS is capable of multielemental analysis, it is also possible to analyze particles of different elemental composition in one single sample. This could be achieved either by sp-ICP-MS in two separate runs or by coupling FFF online or offline with ICP-MS as long as the particles of different chemical composition can be separated by FFF. In terms of the detection limit for particle concentration, sp-ICP-MS was reported to be more sensitive than AF4-ICP-MS [234]. However, AF4-ICP-MS provides the possibility of a greater size resolution compared to sp-ICP-MS [234]. In Table 2, a detailed comparison of AF4-ICP-MS and sp-ICP-MS is shown. Both AF4-ICP-MS and sp-ICP-MS have their advantages and limitations in respect to sample preparation, type of nanomaterial, detection limit, etc., [238] and it has to be decided case by case which of the approaches is better suited for the specific sample to be analyzed. Very recently, it was demonstrated that by combining sp-ICP-MS and AF4 additional information about nanoparticles present in a food matrix can be obtained [218]. The determination of the silver nanoparticles present in a meat matrix was based on an external calibration of the AF4 channel. However, size determination was hampered by a nonideal behavior of the silver nanoparticles in the AF4. The number-based particle size distribution of the silver nanoparticles was determined offline with sp-ICP-MS.

Besides conventional ICP-MS, some other mass spectrometric methods have been used for the analysis of nanomaterials. For example, hydrodynamic chromatography coupled with MALDI-TOF mass spectrometry was applied to characterize and quantify liposome-type nanoparticles in a beverage matrix [249]. Furthermore, silver nanoparticles have been quantified via online species-unspecific postchannel isotope dilution in combination with AF4/sector field ICP-mass spectrometry (ICP-SF-MS) [233].

TABLE 2 Comparison of AF4-ICP-MS with sp-ICP-MS Based on Ref. [234].

Parameter	sp-ICP-MS	AF4-ICP-MS
Particle size determination • limitations	Calculated from pulse intensities (including calibration, transport/ nebulization efficiency) • affected by dwell time • only one element per particle, underestimation if different chemical composition	Estimated from retention time after calibration with particles of known sizes • affected by membrane type and mobile phase composition • only hydrodynamic diameter
Detection limit • size • concentration	~20 nm ~ng/l (depends on particle size)	~2 nm (wide size range) ~µg/l
Resolution	Better for larger particles	Depends not on size but on flow conditions (cross and detector flow)
Recovery	Losses in nebulizer	Analyte could adsorb onto the membrane
Multiform analysis	Good distinction of ionic and particulate form of the analyte	Advantageous for analysis of nanoparticle complexation and aggregation
Advantages compared to other techniques	Applicable to polydisperse samples, multielemental detection possible, high sensitivity	

12. CONCLUSIONS

Nanotechnology is expected to offer technological advantages in the production, processing, storage, transportation, traceability, safety, and security of food. However, nanotechnology-derived products need to demonstrate their economic competitiveness prior to commercialization. Up to now, information concerning the economic competitiveness of nanotechnology-derived products is almost lacking. Food packaging makes up the largest share of the current and short-term predicted market for nanotechnology applications. However, nanoscale delivery systems and nanoscale sensors and indicators are expected to catch up or even overtake in the near future.

Even if engineered nanomaterials are beginning to be used in common food products, it is extremely difficult to obtain figures on the current market situation. There is still a lack of transparency by companies on this issue. In fact, many companies that sell products containing nanomaterials commercially may not even know that nanomaterials are in their supply chain [250]. There are a few product databases for nanotechnology products from independent institutions accessible via the Internet [251–254]. Since there is no compulsory registration for nanomaterials existing, those databases may be incomplete or contain products that do not contain or consist of nanomaterials. Meanwhile, some products in the food sector that were indicated a few years ago as containing or consisting of nanomaterials were withdrawn from the market or advertising was changed in a way that no conclusions about the presence of nanomaterials in the products could be made. Validation of product entries will be required for any credible database of nanomaterials.

So far, the majority of the applications of nanomaterials in the food sector are food contact materials. Due to economic considerations, however, the market share for such packaging is rising very slowly. Dietary supplements containing nanomaterials is undoubtedly another area of high activity. Application of nanomaterials or nanostructures aims to increase the bioavailability of bioactive compounds. Marketed products are, for example, Easy Iron® (improved iron availability) or Nutri-Nano™ (improved coenzyme Q10 availability). The product micelles marketed by Aquanova® (NovaSOL®) or Frutarom (Nutralease®) and the nanoformulated lycopene (LycoVit®) developed by BASF could be found in liquid foods such as beverages. LycoVit® is authorized under the Novel Food Regulation as a novel food. Exact sales figures for all of these products or applications are not available. In dry products, such as spices, silicon dioxide is used as an anticaking agent for many years and its application is authorized as a food additive (E551). Silicon dioxide is composed of nanoscale silica particles that aggregate to bigger particles before added to foods. β-Cyclodextrin (E459), which displays a nanoscale molecule trap, is approved for the use in certain foods and finds application as a carrier for flavor compounds in beverage powders and snack items.

Data assessing risks of the use of engineered nanomaterials in the food sector are sparse. It is of most concern that nanoparticles migrate deeper into the human body, in ways we do not understand, and produce impacts we have not yet realized and are perhaps currently unable to detect. However, before nanosized materials will find widespread application in the food sector, information on potential health risks that may arise from their consumption must be available. There are major gaps in knowledge with regard to the behavior, fate, and effects of nanosized materials along the gastrointestinal route. It is not known whether nanosized materials bind to food components, agglomerate, or remain as free particles in the gastrointestinal tract. Furthermore interactions of nanosized materials with food components is very likely to change during passage through the gastrointestinal tract. Interactions with the food matrix are probably the most difficult task to consider in respect to safety assessment of engineered nanomaterials in the food sector. Nanosized material may also affect gut function or gut microflora. An important issue is whether the nanosized material is differently digested, absorbed, and metabolized compared to its macroscale equivalent. If absorption and bioavailability of the nanosized form is improved, there might be a need to establish new accepted daily intakes for these materials in the nanoform. Furthermore, nanosized materials might facilitate uptake of other substances from the intestine. Finally, almost no information on migration of nanoparticles from food packaging or surfaces used in food storage and processing into food products or beverages is currently available.

The current problems to detect and characterize nanomaterials in complex matrices such as food make their risk assessment a challenging task. Without appropriate routine analytical approaches to identify and characterize nanomaterials in complex matrices, it is impossible to obtain reliable data about exposure and the fate of nanomaterials in the human body. In addition to electron microscopy, sp-ICP-MS and field-flow fractionation coupled with ICP-MS are currently reported to be the most promising approaches for nanomaterial detection and analysis in complex matrices such as food. However, there is still a lot of work and development needed until robust, validated analytical approaches for routine use will be available. The main challenge of all detection, measurement, and characterization methods and techniques is an appropriate sample preparation.

REFERENCES

[1] Nanoscience and Nanotechnologies: Opportunities and Uncertainties, The Royal Society & The Royal Academy of Engineering, London, 2004.
[2] J.A. Shatkin (Ed.), Nanotechnology: Health and Environmental Risks, second ed., CRC Press, Boca Raton, Florida, 2012.
[3] National Nanotechnology Initiative. http://www.nano.gov/nanotech-101/what. (accessed 17.04.14).
[4] Organisation for Economic Co-operation and Development (OECD). http://www.oecd.org/sti/nano. (accessed 17.04.14).
[5] M. Cushen, J. Kerry, M. Morris, M. Cruz-Romero, E. Cummins, Trends Food Sci. Technol. 24 (2012) 30–46.
[6] P. Sanguansri, M.A. Augustin, Trends Food Sci. Technol. 17 (2006) 547–556.
[7] Q. Chaudhry, M. Scotter, J. Blackburn, B. Ross, A. Boxall, L. Castle, R. Aitken, R. Watkins, Food Addit. Contam. 25 (2008) 241–258.
[8] V.J. Morris, New Food Mag. 4 (2008) 53–55.
[9] R. Greiner, J. Brazilian Soc, Food Nutr. 34 (2009) 243–260.
[10] M.A. Augustin, P. Sanguansri, Adv. Food Nutr. Res. 58 (2009) 183–213.
[11] K.A. Abbas, A.M. Saleh, A. Mohamed, N. MohdAzhan, J. Food Agric. Environ. 7 (2009) 14–17.
[12] S. Da Pieve, S. Calligaris, M.C. Nicoli, Indust. Aliment. 48 (2009) 44–52.
[13] N. Sozer, J.L. Kokini, Trends Biotechnol. 27 (2009) 82–89.
[14] P.R. Srinivas, M. Philbert, T.Q. Vu, Q. Huang, J.L. Kokini, E. Saos, H. Chen, C.M. Peterson, K.E. Friedl, C. Dade-Ngutter, V. Hubbard, P. Starke-Reed, N. Miller, J.M. Betz, J. Dwyer, J. Milner, S.A. Ross, J. Nutr. 140 (2010) 119–124.
[15] R. Nair, S.H. Varghese, B.G. Nair, T. Maekawa, Y. Yoshida, D.S. Kumar, Plant Sci. 179 (2010) 154–163.

[16] Q. Chaudhry, L. Castle, Trends Food Sci. Technol. 10 (2011) 1–9.
[17] S. Neethirajan, D. Jayas, Food Bioproc. Technol. 4 (2011) 39–47.
[18] V. Ghormade, M.V. Deshpande, K.M. Paknikar, Biotechnol. Adv. 29 (2011) 792–803.
[19] L. Rashidi, K. Khosravi-Darani, Crit. Rev. Food Sci. Nutr. 51 (2011) 723–730.
[20] Helmut Kaiser Consultancy, Nano Food 2040, Nanotechnology in Food, Food Processing, Agriculture, Packaging and Consumption, State of Science, Technologies, Markets, Applications and Developments to 2015 and 2040, Zürich, Switzerland, 2007.
[21] B.S. Shekhon, Nanotech. Sci. Appl. 3 (2010) 1–15.
[22] International Food Safety Authorities Network (INFOSAN), INFOSAN Information Note No. 01/2008 – Nanotechnology, 2008.
[23] V.J. Morris, R. Parker, Food Nanotechnology, World of Food Science, 2008.
[24] J.M. Aguilera, D.W. Stanley, Microstructural Principles of Food Processing and Engineering, second ed., Springer-Verlag, Heidelberg, Germany, 1999.
[25] J. Weiss, P. Takhistoy, D.J. McClements, J. Food Sci. 71 (2006) R107–R116.
[26] H. Yang, Y. Wang, S. Lai, H. An, Y. Li, F. Chen, J. Food Sci. 72 (2007) R65–R75.
[27] V.J. Morris, N.C. Woodward, A.P. Gunning, J. Sci. Food Agric. 91 (2011) 2117–2125.
[28] Q. Huang (Ed.), Nanotechnology in the Food, Beverage and Nutraceutical Industries, Woodhead Publishing Limited, Cambridge, UK, 2012.
[29] G.L. Robertson, Food Packaging: Principles and Practice, third ed., CRC Press, Boca Raton, Florida, 2012.
[30] K.K. Yang, X.L. Wang, Y.Z. Wang, J. Indust. Eng. Chem. 13 (2007) 485–500.
[31] A.L. Brody, Food Technol. 57 (2003) 52–54.
[32] A.W. Hu, Z.H. Fu, Packag. Engin. 24 (2003) 22–24.
[33] J.M. Lagarón, L. Cabedo, D. Cava, J.L. Feijoo, R. Gavara, E. Giménez, Food Add. Contam. 22 (2005) 994–998.
[34] A.L. Brody, Food Technol. 60 (2006) 92–94.
[35] A. Sorrentino, G. Gorrasi, V. Vittoria, Trends Food Sci. Technol. 18 (2007) 84–95.
[36] T.H. McHugh, World Food Sci. 4 (2008) 1–3.
[37] A.L. Brody, B. Bugusu, J.H. Han, C. Koelsch Sand, T.H. McHugh, J. Food Sci. 73 (2008) R107–R117.
[38] H.M.C. de Azeredo, Food Res. Int. 42 (2009) 1240–1253.
[39] A. Arora, G.W. Padua, J. Food Sci. 75 (2010) R43–R49.
[40] B. Reddy (Ed.), Advances in Diverse Industrial Applications of Nanocomposites, InTech Europe, Rijeka, Croatia, 2011.
[41] S. Goyal, G.K. Goyal, Russ. J. Agric. Socio-Econ. Sci. 10 (2012) 14–24.
[42] S.G. Lei, S.V. Hoa, M.-T. Ton-That, Comp. Sci. Technol. 66 (2006) 1274–1279.
[43] D. Adame, G.W. Beall, Appl. Clay Sci. 42 (2009) 545–552.
[44] S. Sinha Ray, M. Okamoto, Prog. Polym. Sci. 28 (2003) 1539–1641.
[45] Z. Ke, B. Yongping, Mater. Lett. 59 (2005) 3348–3351.
[46] B. Xu, Q. Zheng, Y. Song, Y. Shangguan, Polymer 47 (2006) 2904–2910.
[47] A.L. Brody, Food Technol. 7 (2007) 102–107.
[48] Y.-L. Chung, S. Ansari, L. Estevez, S. Hayrapetyan, E.P. Giannelis, H.-M. Lai, Carbohydr. Polym. 79 (2010) 391–396.
[49] R. Kotsilkova, V. Petkova, Y. Pelovski, J. Therm. Anal. Calorim. 64 (2001) 591–598.
[50] S.S. Ray, P. Maiti, M. Okamoto, K. Yamada, K. Ueda, Macromolecules 35 (2002) 3104–3110.
[51] K.K. Wang, C.M. Koo, I.J. Chung, J. Appl. Polym. Sci. 89 (2003) 2131–2136.
[52] B. Schartel, P. Pötschke, U. Knoll, M. Abdel-Goad, Eur. Polym. J. 41 (2003) 1061–1070.

[53] L. Vermeiren, F. Devlieghere, J. Debevere, Food Addit. Contam. Part A Chem. Anal. Control Expo. Risk Assess. 19 (2002) 163–171.
[54] T.M. Tolaymat, A.M. El Badawy, A. Genaidy, K.G. Scheckel, T.P. Luxton, M. Suidan, Sci. Total Environ. 408 (2010) 999–1006.
[55] Z. Xiu, Q. Zhang, H.L. Puppala, V.L. Colvin, P.J.J. Alvarez, Nano Lett. 12 (2012) 4271–4275.
[56] M.E. Samberg, P.E. Orndorff, N.A. Monteiro-Riviere, Nanotoxicology 5 (2011) 244–253.
[57] D.S. Cha, M.S. Chinnan, Crit. Rev. Food Sci. 44 (2004) 223–237.
[58] N. Jones, B. Ray, K.T. Ranjit, A.C. Manna, FEMS Microbiol. Lett. 279 (2008) 71–76.
[59] D. Sun, T. Jin, J.Y. Su, H. Zhang, H.J. Sue, J. Food Sci. 74 (2009) M46–M52.
[60] V. Krishna, S. Pumprueg, S.H. Lee, J. Zhoa, W. Sigmund, B. Koopman, B.M. Moudgil, Proc. Saf. Environ. Prot. 83 (2005) 393–397.
[61] D.S. Cha, K. Cooksey, M.S. Chinnan, H.J. Park, LWT Food Sci. Technol. 36 (2003) 209–213.
[62] N.B. Franklin, K.D. Cooksey, K.J.K. Getty, J. Food Protect. 67 (2004) 480–485.
[63] A. Rodriguez, C. Nerin, R. Batlle, J. Agric. Food Chem. 56 (2008) 6364–6369.
[64] J. Vartiainen, M. Rättö, S. Paulussen, Packaging Technol. Sci. 18 (2005) 243–251.
[65] J. Rivett, D.V. Speer, Oxygen Scavenging Film with Good Interplay Adhesion, US Patent No. 75141512, 2009.
[66] A. Lopez-Rubio, R. Gavara, J.M. Lagar6n, Trends Food Sci. Technol. 17 (2006) 567–575.
[67] A. Garland (Ed.), Nanotechnology in Plastics Packaging. Commercial Applications in Nanotechnology, Pira International, Leatherhead, UK, 2004.
[68] S. Otles, B. Yalcin, LogForum 4 (2008) 1–9.
[69] D.A. Pereira de Abreu, J.M. Cruz, P. Paseiro Losada, Food Rev.Int. 28 (2012) 146–187.
[70] K.L. Yam, P.T. Takhistov, J. Miltz, J. Food Sci. 70 (2005) R1–R10.
[71] M. Smolander, H.-L. Alakomi, T. Ritvanen, J. Vainionpaa, R. Ahvenainen, Food Control 15 (2004) 217–229.
[72] M.J. Natan, T. Mallouk, Colloidal Rod Particles as Nano-barcodes, European Patent No. EP1227929 B1, 2008.
[73] J.M. Nam, C.S. Thaxton, C.A. Mirkin, Science 301 (2003) 1884–1886.
[74] Y. Li, C.U. Yht, D. Luo, Nat. Biotechnol. 23 (2005) 885–889.
[75] H. Zhang, R. Elghanian, L. Demers, N. Amro, S. Disawal, S. Cruchon-Dupeyrat, Direct-write Nanolithography Method of Transporting Ink with an Elastomeric Polymer Coated Nanoscopic Tip to Form a Structure Having Internal Hollows on a Substrate, US Patent No. 7491422, 2009.
[76] S. Baruah, J. Dutta, Environ. Chem. Lett. 7 (2009) 191–204.
[77] T. Joseph, M. Morrison, Nanotechnology in Agriculture & Food: A Nanoforum Report, 2006.
[78] T.P. Labuza, W.M. Breene, J. Food Proces. Preserv. 13 (1988) 1–69.
[79] L. Vermeiren, F. Devlieghere, M. van Beest, N. de Kruijf, J. Debevere, Trends Food Sci. Technol. 10 (1999) 77–86.
[80] X. Tang, S. Alavi, Carbohyd. Polym. 85 (2011) 7–16.
[81] S. Sinha Ray, M. Bousmina, Progr. Mater. Sci. 50 (2005) 962–1079.
[82] R. Zhao, P. Torley, P.J. Halley, J. Mat. Sci. 43 (2008) 3058–3071.
[83] M. Avella, J.J. de Vlieger, M.E. Errico, S. Fischer, P. Vacca, M.G. Volpe, Food Chem. 93 (2005) 467–474.
[84] B. Chen, J.R.G. Evans, Carbohydr. Polym. 61 (2005) 455–463.
[85] A.J.F. de Carvalho, A.A.S. Curvelo, J.A.M. Agnelli, Carbohydr. Polym. 45 (2001) 189–194.

[86] S.A. McGlashan, P.J. Halley, Polym. Int. 52 (2003) 1767–1773.
[87] H.M. Park, X. Li, C.Z. Jin, C.Y. Park, W.J. Cho, C.S. Ha, Macromol. Mater. Eng. 287 (2002) 553–558.
[88] H.M. Park, W.K. Lee, C.Y. Park, W.J. Cho, C.S. Ha, J. Mater. Sci. 38 (2003) 909–915.
[89] J.-W. Rhim, S.-I. Hong, C.-S. Ha, LWT Food Sci. Technol. 42 (2009) 612–617.
[90] P. Bordes, E. Pollet, L. Averou, Prog. Polym. Sci. 34 (2009) 125–155.
[91] C. Kriegel, K.M. Kit, D.J. McClements, J. Weiss, Food Biophys. 4 (2009) 213–228.
[92] H.M.C. de Azeredo, L.H. Mattoso, D.F. Wood, T.G. Williams, R.D. Avena-Bustillos, T.H. Mc Hugh, J. Food Sci. 74 (2009) N31–N35.
[93] FoodQuality news.com. http://www.foodqualitynews.com/Innovation/Sono-Tek-unveils-ultrasonic-spray-coating-tech-for-sliced-meats. (accessed 17.04.14).
[94] A.B. Kananeh, E. Scharnbeck, U.D. Kück, N. Räbiger, Food Bioprod. Process. 88 (2010) 349–356.
[95] BUND, Aus dem Labor auf den Teller: Die Nutzung der Nanotechnologie im Lebensmittelsektor, 2008.
[96] M. Pehanich, Food Process. 11 (2006) 46–48.
[97] B. Cuartas-Uribe, M.I.A. Miranda, E.S. Costa, A.B. Pia, J. Dairy Sci. 90 (2006) 1094–1101.
[98] A. Schäfer, A.G. Fane, T.D. White (Eds.), Nanofiltration. Principles and Applications, Elsevier, Oxford, 2005.
[99] M.C. Duke, A. Lim, S.C. Luz, L. Nielsen, Food Bioprod. Process. 86 (2008) 290–295.
[100] D.K. Tiwari, J. Behari, P. Sen, World Appl. Sci. J. 3 (2008) 417–433.
[101] B. Díaz-Reinoso, A. Moure, H. Domínguez, J.C. Parajó, J. Food Eng. 91 (2009) 587–593.
[102] J. Kelly, P. Kelly, Int. Dairy J. 5 (1995) 291–303.
[103] K. Oehlke, M. Adamiuk, D. Behsnilian, V. Gräf, E. Mayer-Miebach, E. Walz, R. Greiner, Food Funct. 5 (2014) 1341–1359.
[104] FAO/WHO, Expert Meeting on the Application of Nanotechnologies in the Food and Agriculture Sectors: Potential Food Safety Implications, Meeting Report, 2010.
[105] H. Chen, J. Weiss, F. Shahidi, Food Technol. 603 (2006) 30–36.
[106] L.Y. Chen, G.F. Remondetto, M. Subirade, Trends Food Sci. Technol. 17 (2006) 272–283.
[107] J. Ubbink, J. Krüger, Trends Food Sci. Technol. 17 (2006) 244–254.
[108] E. Acosta, Curr. Opin. Colloid Interface Sci. 14 (2009) 3–15.
[109] Q. Huang, P. Given, M. Qian (Eds.), Micro/Nano Encapsulation of Active Food Ingredients, Oxford University Press, Oxford, 2009.
[110] M.A. Augustin, Y. Hemar, Chem. Soc. Rev. 38 (2009) 902–912.
[111] M.X. Quintanilla-Carvajal, B.H. Camacho-Díaz, L.S. Meraz-Torres, J.J. Chanona-Pérez, L. Alamilla-Beltrán, A. Jimenéz-Aparicio, G.F. Gutiérrez-López, Food Eng. Rev. 2 (2010) 39–50.
[112] L. Sagalowicz, M.F. Curr, Opin. Colloid Interface Sci. 15 (2010) 61–72.
[113] Q. Huang, H. Yu, Q. Ru, J. Food Sci. 75 (2010) R50–R58.
[114] S.A. Agnihotri, N.N. Mallikarjuna, T.M. Aminabhavi, J. Control Release 100 (2004) 5–28.
[115] L. Chen, M. Subriade, Biomaterials 26 (2005) 6041–6053.
[116] J. Barauskas, M. Johnsson, F. Tiberg, Nano Lett. 5 (2005) 1615–1619.
[117] E. Dickinson (Ed.), Food Colloids: Interactions, Microstructure and Processing, Royal Society of Chemistry, Cambridge, 2005.
[118] R.M. Mozafari, J. Flanagan, L. Matia-Merino, A. Awati, A. Omri, E.Z. Suntres, H. Singh, J. Sci. Food Agric. 86 (2006) 2038–2045.
[119] J.F. Graveland-Bikker, C.G. de Kruif, Trends Food Sci. Technol. 175 (2006) 196–203.
[120] J.F. Graveland-Bikker, I.A.T. Schaap, C.F. Schmidt, C.G. de Kruif, Nano Lett. 6 (2006) 616–621.

[121] J. Flanagan, H. Singh, Crit. Rev. Food Sci. Nutr. 46 (2006) 221–237.
[122] E. Semo, E. Kesselman, D. Danino, Y.D. Livney, Food Hydrocoll. 21 (2007) 936–942.
[123] M.R. Mozafari, C. Johnson, S. Hatziantoniou, C. Demetzos, J. Liposome Res. 18 (2008) 309–327.
[124] N. Garti, in: Delivery and Controlled Release of Bioactives in Foods and Nutraceuticals, Woodhead Publishing, Cambridge, UK, 2008.
[125] D.J. McClements, E.A. Decker, Y. Park, J. Weiss, Crit. Rev. Food Sci. Nutr. 49 (2009) 577–606.
[126] J.F. Graveland-Bikker, R.I. Koning, H.K. Koerten, R.B.J. Geels, R.M.A. Heeren, C.G. de Kruif, Soft Matter 5 (2009) 2020–2026.
[127] A. Fernandez, S. Torres-Giner, J.M. Lagaron, Food Hydrocoll. 23 (2009) 1427–1432.
[128] D.J. McClements, J. Food Sci. 75 (2010) R30–R42.
[129] Y.D. Livney, Curr. Opin. Colloid Interface Sci. 15 (2010) 73–83.
[130] P. da Silva Malheiros, D.J. Daroit, A. Brandelli, Trends Food Sci. Technol. 21 (2010) 284–292.
[131] N. Liu, H.J. Park, Coll. Surf. B Biointerfaces 76 (2010) 16–19.
[132] U. Klinkesorn, D.J. McClements, Food Biophys. 5 (2010) 73–81.
[133] M. Hu, Y. Li, E.A. Decker, H. Xiao, D.J. McClements, J. Agric. Food Chem. 58 (2010) 1283–1289.
[134] F. Donsì, M. Annunziata, M. Sessa, G. Ferrari, LWT Food Sci. Technol. 44 (2011) 1908–1914.
[135] M. Esmaili, S.M. Ghaffari, Z. Moosavi-Movahedi, M.S. Atri, A. Sharifizadeh, M. Farhadi, R. Yousefi, J.M. Chobert, T. Haertle, A.A. Moosavi-Movahedi, LWT Food Sci. Technol. 44 (2011) 2166–2172.
[136] R. Gupta, D. Rousseau, Food Funct. 3 (2012) 302–311.
[137] D.J. McClements, Soft Matter 8 (2012) 1719–1729.
[138] P. Prombutara, Y. Kulwatthanasal, N. Supaka, I. Sramala, S. Chareonpornwattana, Food Control 24 (2012) 184–190.
[139] A.Y. Choi, C.T. Kim, H.Y. Park, H.O. Kim, N.R. Lee, K.E. Lee, H.S. Gwak, J. Agric. Food Chem. 61 (2013) 2096–2102.
[140] R. Wajda, J. Zirkel, T. Schaffer, J. Med. Food 10 (2007) 731–734.
[141] K. Terao, D. Nakata, H. Fukumi, G. Schmid, H. Arima, F. Hirayama, K. Uekama, Nutr. Res. 26 (2006) 503–508.
[142] J. Hatanaka, Y. Kimura, Z. Lai-Fu, S. Onoue, S. Yamada, Int. J. Pharm. 363 (2008) 112–117.
[143] P. Ercan, E.S. Nehir, Int. J. Food Sci. Technol. 47 (2012) 1986–1992.
[144] M. Haham, S. Ish-Shalom, M. Nodelman, I. Duek, E. Segal, M. Kustanovich, Y.D. Livney, Food Funct. 3 (2012) 737–744.
[145] F. Diarrassouba, G. Garrait, G. Remondetto, P. Alvarez, E. Beyssac, M. Subirade, Food Chem. 145 (2014) 646–652.
[146] E.I. Back, C. Frindt, E. Ocenášková, D. Nohr, M. Stern, H.K. Biesalski, Eur. J. Nutr. 45 (2006) 1–6.
[147] V. Kakkar, S. Singh, D. Singla, I.P. Kaur, Mol. Nutr. Food Res. 55 (2011) 495–503.
[148] M. Takahashi, S. Uechi, K. Takara, Y. Asikin, K. Wada, J. Agric. Food Chem. 57 (2009) 9141–9146.
[149] S. Onoue, H. Takahashi, Y. Kawabata, Y. Seto, J. Hatanaka, B. Timmermann, S. Yamada, J. Pharm. Sci. 99 (2010) 1871–1881.
[150] H. Yu, Q. Huang, J. Agric. Food Chem. 60 (2012) 5373–5379.
[151] C. Schiborr, A. Kocher, D. Behnam, J. Jandasek, S. Toelstede, J. Frank, Mol. Nutr. Food Res. 58 (2014) 516–527.

[152] F. Rohner, F.O. Ernst, M. Arnold, M. Hilbe, R. Biebinger, F. Ehrensperger, S.E. Pratsinis, W. Langhans, R.F. Hurrell, M.B. Zimmermann, J. Nutr. 137 (2007) 614–619.
[153] F.M. Hilty, M. Arnold, M. Hilbe, A. Teleki, J.T.N. Knijnenburg, F. Ehrensperger, R.F. Hurrell, S.E. Pratsinis, W. Langhans, M.B. Zimmermann, Nat. Nano 5 (2010) 374–380.
[154] N. Sakaguchi, T.P. Rao, K. Nakata, H. Nanbu, L.R. Juneja, Int. J. Vit. Nutr. Res. 74 (2004) 3–9.
[155] M. Roe, R. Collings, J. Hoogewerff, S. Fairweather-Tait, Eur. J. Nutr. 48 (2009) 115–119.
[156] M.C. Fidler, L. Davidsson, C. Zeder, T. Walczyk, I. Marti, R.F. Hurrell, Int. J. Vit. Nutr. Res. 74 (2004) 294–300.
[157] M.C. Fidler, T. Walczyk, L. Davidsson, C. Zeder, N. Sakaguchi, L.R. Juneja, R.F. Hurrell, Br. J. Nutr. 91 (2004) 107–112.
[158] H.S. Chen, J.H. Chang, J.S.B. Wu, J. Food Sci. 73 (2008) H246–H251.
[159] H.Y. Gao, H.I. Chen, W.X. Chen, F. Tao, Y.H. Zheng, Y.M. Jiang, H.J. Ruan, Food Chem. 109 (2008) 493–498.
[160] S. Huang, J.C. Chen, C.W. Hsu, W.H. Chang, Nanotechnology 20 (2009) 375102.
[161] H.S. Park, J. Ahn, H.S. Kwak, J. Med. Food 11 (2011) 454–459.
[162] P. Zimet, Y.D. Livney, Food Hydrocolloids 23 (2009) 1120–1126.
[163] G. Astray, C. Gonzalez-Barreiro, J.C. Mejuto, R. Rial-Otero, J. Simal-Gándara, Food Hydrocolloids 23 (2009) 1631–1640.
[164] A. Hentschel, S. Gramdorf, R.H. Müller, T. Kurz, J. Food Sci. 73 (2008) N1–N6.
[165] P. Zimet, D. Rosenberg, Y.D. Livney, Food Hydrocolloids 25 (2011) 1270–1276.
[166] J.P. Jee, S.J. Lim, J.S. Park, C.K. Kim, Eur. J. Pharm. Biopharm. 63 (2006) 134–139.
[167] J.S. Almeida, F. Lima, S.D. Ros, L.O.S. Bulhões, L.M. de Carvalho, R.C. Beck, Nanoscale Res. Lett. 5 (2010) 1603–1610.
[168] A. Barras, A. Mezzetti, A. Richard, S. Lazzaroni, S. Roux, P. Melnyk, D. Betbeder, N. Monfilliette-Dupont, Int. J. Pharm. 379 (2009) 270–277.
[169] D.G. Kim, Y.I. Jeong, C. Choi, S.H. Roh, S.K. Kang, M.K. Jang, J.W. Nah, Int. J. Pharm. 319 (2006) 130–138.
[170] A. Dube, K. Ng, J.A. Nicolazzo, I. Larson, Food Chem. 122 (2010) 662–667.
[171] A. Dube, J.A. Nicolazzo, I. Larson, Eur. J. Pharm. Sci. 41 (2010) 219–225.
[172] D.J. McClements, E.A. Decker, J. Weiss, J. Food Sci. 72 (2007) R109–R124.
[173] E. Dickinson, Curr. Opin. Colloid Interface Sci. 15 (2010) 40–49.
[174] M.G. Valdés, A.C. Valdés González, J.A. García Calzón, M.E. Díaz-García, Microchim. Acta 166 (2009) 1–19.
[175] K. Nachay, Food Technol. 61 (2007) 34–36.
[176] S. Bhattacharya, J. Jang, L. Yang, D. Akin, R. Bashir, J. Rapid Methods Autom. Microbiol. 15 (2007) 1–32.
[177] G. Naja, P. Bouvrette, J. Champagne, R. Brousseau, J.H.T. Luong, Appl. Biochem. Biotechnol. 162 (2010) 460–475.
[178] X. Zhang, Q. Guo, D. Cui, Sensors 9 (2009) 1033–1053.
[179] D. Tang, J.C. Sauceda, Z. Lin, S. Ott, E. Basova, I. Goryacheva, S. Biselli, J. Lin, R. Niessner, D. Knopp, Biosens. Bioelectron. 25 (2009) 514–518.
[180] X. Zhao, L.R. Hilliard, S.J. Mechrey, Y. Wang, R.P. Bagwe, S. Jin, W. Tan, Proc. Nat. Acad. Sci. 101 (2004) 15027–15032.
[181] J. Fu, B. Park, G. Siragusa, L. Jones, R. Tripp, Y. Zhao, Y.-J. Cho, Nanotechnology 19 (2008) 1–7.
[182] Y. Cheng, Y. Liu, J. Huang, K. Li, W. Zhang, Y. Xian, L. Jin, Talanta 77 (2009) 1332–1336.
[183] H. Zhang, J. Wang, S. Ye, J. Food Eng. 86 (2008) 370–378.
[184] A. Behar, D. Fugere, M. Passoff, Slipping through the Cracks: An Issue Brief on Nanomaterials in Food, As You Sow, Oakland, USA, 2013.

[185] N. O'Brien, E. Cummins, J. Environ. Sci. Health Part A 45 (2010) 992–1007.
[186] N. O'Brien, E. Cummins, Risk Anal. 31 (2011) 706–726.
[187] European Food Safety Authority, EFSA J. 958 (2009) 1–39.
[188] European Food Safety Authority, EFSA J. 9 (2011) 2140.
[189] J.J. Powell, N. Faria, E. Thomas-McKay, L.C. Pele, J. Autoim. 34 (2010) J226–J233.
[190] C. Olbrich, O. Kayser, R.H. Müller, Int. J. Pharm. 237 (2002) 119–128.
[191] L.K. Limbach, P. Wick, P. Manser, R.N. Grass, A. Bruinink, W.J. Stark, Environ. Sci. Technol. 41 (2007) 4158–4163.
[192] B. Wang, W. Feng, M. Wang, J. Nanopart. Res. 10 (2007) 263–276.
[193] J. Wang, G. Zhou, C. Chen, H. Yu, T. Wang, Y. Mad, G. Jia, Y. Gao, B. Li, J. Sun, Y. Li, F. Jiao, Y. Zhao, Z. Chai, Toxicol. Lett. 168 (2007) 176–185.
[194] B.K. Gaiser, S. Hirn, A. Kermanizadeh, N. Kanase, K. Fytianos, A. Wenk, N. Haberl, A. Brunelli, W.G. Kreyling, V. Stone, Toxicol. Sci. 131 (2013) 537–547.
[195] A. Elsaesser, C.V. Howard, Adv. Drug Deliv. Rev. 64 (2012) 129–137.
[196] B. Trouiller, R. Reliene, A. Westbrook, P. Solaimani, R.H. Schiestl, Cancer Res. 69 (2009) 8784–8789.
[197] I.C. Munro, L.A. Haighton, B.S. Lynch, S. Tafazoli, Food Add. Contam. 26 (2009) 1534–1546.
[198] Regulation (EU) No 1169/2011 of the European Parliament and of the Council of 25 October 2011 on the Provision of Food Information to Consumers. Off. J. Eur. Union.
[199] Regulation (EU) No 1332/2008 of the European Parliament and of the Council of 16 December 2008 on Food Enzymes. Off. J. Eur. Union.
[200] Regulation (EU) No 1333/2008 of the European Parliament and of the Council of 16 December 2008 on Food Additives. Off. J. Eur. Union.
[201] Commission Regulation (EC) No 450/2009 of 29 May 2009 on Active and Intelligent Materials and Articles Intended to Come into Contact with Food. Off. J. Eur. Union.
[202] Commission Regulation (EU) No 10/2011 of 14 January 2011 on Plastic Materials and Articles Intended to Come into Contact with Food. Off. J. Eur. Union.
[203] T. Cedervall, I. Lynch, S. Lindman, T. Berggård, E. Thulin, H. Nilsson, K.A. Dawson, S. Linse, PNAS 104 (2007) 2050–2055.
[204] G. Maiorano, S. Sabella, B. Sorce, V. Brunetti, M.A. Malvindi, R. Cingolani, P.P. Pompa, ACS Nano 4 (2012) 7481–7491.
[205] R.F. Domingos, M.A. Baalousha, Y. Ju-Nam, M.M. Reid, N. Tufenkji, J.R. Lead, G. Leppard, K.J. Wilkinson, Environ. Sci. Technol. 43 (2009) 7277–7284.
[206] S. Bandyopadhyay, J.R. Peralta-Videa, J.L. Gardea-Torresdey, Environ. Eng. Sci. 30 (2013) 118–125.
[207] T.P.J. Linsinger, Q. Chaudhry, V. Dehalu, P. Delahaut, A. Dudkiewicz, R. Grombe, F. von der Kammer, E.H. Larsen, S. Legros, K. Loeschner, R. Peters, R. Ramsch, G. Roebben, K. Tiede, S. Weigel, Food Chem. 138 (2013) 1959–1966.
[208] A. Lopez-Serrano, R.M. Olivas, J.S. Landaluze, C. Camara, Anal. Methods 6 (2014) 38–56.
[209] D.M.A.M. Luykx, R.J.B. Peters, S.M. van Ruth, H. Bouwmeester, J. Agric. Food Chem. 56 (2008) 8231–8247.
[210] C. Szakal, S.M. Roberts, P. Westerhoff, A. Bartholomaeus, N. Buck, I. Illuminato, R. Canady, M. Rogers, ACS Nano 8 (2014) 3128–3135.
[211] K. Tiede, A.B.A. Boxall, S.P. Tear, J. Lewis, H. David, M. Hassellov, Food Addit. Contam. Part A Chem. Anal. Control Expo. Risk Assess. 25 (2008) 795–821.
[212] A. Weir, P. Westerhoff, L. Fabricius, K. Hristovski, N. von Goetz, Environ. Sci. Technol. 46 (2012) 2242–2250.

[213] NIST. https://www-s.nist.gov/srmors/viewTable.cfm?tableid=231. (accessed 17.04.14).
[214] G. Roebben, K. Rasmussen, V. Kestens, T.P.J. Linsinger, H. Rauscher, H. Emons, H. Stamm, J. Nanopart. Res. 15 (2013) 1–13.
[215] ISO 13321, Particle Size Analysis – Photon Correlation Spectroscopy, 1996.
[216] ISO 22412, Particle Size Analysis – Dynamic Light Scattering (DLS), 2008.
[217] NanoLyse. http://www.nanolyse.eu. (accessed 17.04.14).
[218] K. Loeschner, J. Navratilova, C. Købler, K. Mølhave, S. Wagner, F. von der Kammer, E.H. Larsen, Anal. Bioanal. Chem. 405 (2013) 8185–8195.
[219] R. Grombe, J. Charoud-Got, H. Emteborg, T.P.J. Linsinger, J. Seghers, S. Wagner, F. von der Kammer, T. Hofmann, A. Dudkiewicz, M. Llinas, C. Solans, A. Lehner, G. Allmaier, Anal. Bioanal. Chem. 406 (2014) 3895–3907.
[220] B.F.G. Pycke, T.-C. Chao, P. Herckes, P. Westerhoff, R.U. Halden, Anal. Bioanal. Chem. 404 (2012) 2583–2595.
[221] J.C. Giddings, J. Sep. Sci. 1 (1966) 123–125.
[222] J.C. Giddings, F.J. Yang, M.N. Myers, Science 193 (1976) 1244–1245.
[223] J.C. Giddings, Science 260 (1993) 1456–1465.
[224] E. Bolea, J. Jimenez-Lamana, F. Laborda, I. Abad-Alvaro, C. Blade, L. Arola, J.R. Castillo, Analyst 139 (2014) 914–922.
[225] C. Augsten, K. Mäder, Int. J. Pharm. 251 (2008) 23–30.
[226] M. Baalousha, B. Stolpe, J.R. Lead, J. Chromat. A 1218 (2011) 4078–4103.
[227] E. Bolea, J. Jimenez-Lamana, F. Laborda, J.R. Castillo, Anal. Bioanal. Chem. 401 (2011) 2723–2732.
[228] L. Calzolai, D. Gilliland, C.P. Garcìa, F. Rossi, J. Chromatogr. A 1218 (2011) 4234–4239.
[229] C. Contado, L. Ravani, M. Passarella, Anal. Chim. Acta 788 (2013) 183–192.
[230] O. Geiss, C. Cascio, D. Gilliland, F. Franchini, J. Barrero-Moreno, J. Chromatogr. A 1321 (2013) 100–108.
[231] H. Hagendorfer, R. Kaegi, J. Traber, S.F.L. Mertens, R. Scherrers, C. Ludwig, A. Ulrich, Anal. Chim. Acta 706 (2011) 367–378.
[232] K. Loeschner, J. Navratilova, S. Legros, S. Wagner, R. Grombe, J. Snell, F. von der Kammer, E.H. Larsen, J. Chromatogr. A 1272 (2013) 116–125.
[233] B. Meermann, A.-L. Fabricius, L. Duester, F. Vanhaecke, T. Ternes, J. Anal. Spectrom. 29 (2014) 287–296.
[234] D.M. Mitrano, A. Barber, A. Bednar, P. Westerhoff, C.P. Higgins, J.F. Ranville, J. Anal. Spectrom. 27 (2012) 1131–1142.
[235] A.R. Poda, A.J. Bednar, A.J. Kennedy, A. Harmon, M. Hull, D.M. Mitrano, J.F. Ranville, J. Steevens, J. Chromatogr. A 1218 (2011) 4219–4225.
[236] R.N. Qureshi, W.T. Kok, Anal. Bioanal. Chem. 399 (2011) 1401–1411.
[237] B. Roda, A. Zattoni, P. Reschiglian, M.H. Moon, M. Mirasoli, E. Michelini, A. Roda, Anal. Chim. Acta 635 (2009) 132–143.
[238] A. Ulrich, S. Losert, N. Bendixen, A. Al-Kattan, H. Hagendorfer, B. Nowack, C. Adlhart, J. Ebert, M. Lattuada, K. Hungerbühler, J. Anal. Spectrom. 27 (2012) 1120–1130.
[239] A. Zattoni, B. Roda, F. Borghi, V. Marassi, P. Reschiglian, J. Pharm. Biomed. Anal. 87 (2014) 53–61.
[240] M. Costas, I. Lavilla, S. Gil, F. Pena, I. de la Calle, N. Cabaleiro, C. Bendicho, Anal. Chim. Acta 679 (2010) 49–55.
[241] R. Thomas, Characterization of Nanoparticles by Practical guide to ICP-MS, CRC Press, 2013.
[242] C. Degueldre, P.Y. Favarger, Colloids Surf. A Physicochem. Eng. Aspects 217 (2003) 137–142.

[243] F. Laborda, J. Jimenez-Lamana, E. Bolea, J.R. Castillo, J. Anal. Spectrom. 28 (2013) 1220–1232.
[244] J. Liu, K.E. Murphy, R.I. Maccuspie, M.R. Winchester, Anal. Chem. 86 (2014) 3405–3414.
[245] D.M. Mitrano, J. Ranville, A. Bednar, K. Kazor, A.S. Hering, C.C. Higgins, Environ. Sci. Nano 1 (2014) 248–259.
[246] R.J.B. Peters, Z.H. Rivera, G. Bemmel, H.J.P. Marvin, S. Weigel, H. Bouwmeester, Anal. Bioanal. Chem. 406 (2014) 3875–3885.
[247] J. Tuoriniemi, G. Cornelis, M. Hassellov, J. Anal. Spectrom. 29 (2014) 743–752.
[248] Y. Echegoyen, C. Nerin, Food Chem. Toxicol. 62 (2013) 16–22.
[249] J.P.F.G. Helsper, R.J.B. Peters, L. Brouwer, S. Weigel, Anal. Bioanal. Chem. 405 (2013) 1181–1189.
[250] S. Buratti, A. Rizzolo, S. Benedetti, D. Torreggiani, J. Food Sci. 71 (2006) E184–E189.
[251] Consumer Products Inventory, an Inventory of Nanotechnology-based Consumer Products Introduced on the Market. http://www.nanotechproject.org/inventories/consumer/. (accessed 17.04.14).
[252] ANEC/BEUC Inventory. http://nanoinformation.at/fileadmin/_migrated/content_uploads/Nanoproduktliste_BEUC_ANEC_2010.xls. Consumer Products Inventory. (accessed 17.04.14).
[253] BUND Nanoproduktdatenbank. http://www.nanowatch.de. (accessed 17.04.14).
[254] http://www.taenk.dk/. (accessed 17.04.14).

Chapter 12

Food Pathogens

Isin Akyar
Department of Medical Microbiology & Acibadem Labmed Clinical Laboratories, Acibadem University School of Medicine, Istanbul, Turkey
E-mail: isinakyar@gmail.com

Chapter Outline

1. Introduction 618
2. Common Foodborne Pathogens (Spoilage and Pathogenic Bacteria) 618
2.1 Spoilage Bacteria 618
2.2 Pathogenic Bacteria 619
3. Diagnosis of Microorganisms 620
4. Mass Spectrometry in Microbial Analysis 620
4.1 Ionization Source 622
4.2 Mass Analyzer 623
4.3 Ion Detector 624
5. Identification of Food Pathogens by Mass Spectrometry 624
5.1 Matrix-Assisted Laser Desorption Ionization Time-of-Flight-Mass Spectrometry (MALDI-TOF-MS) 624
5.1.1 Examples of MALDI-TOF-MS Studies in Foodborne Pathogens 629
5.2 Electrospray Ionization Mass Spectrometry (ESI-MS) 636
5.2.1 Examples of ESI-MS Studies in Foodborne Pathogens 637
5.3 PCR Coupled to Mass Spectrometry Using Electrospray Ionization (PCR/ESI-MS) 638
5.4 Bioaerosol Mass Spectrometry 640
5.5 Fractionation Methods Coupled with Mass Spectrometry: GC-MS and Pyrolysis–GC-MS 641
5.6 Capillary Electrophoresis-MS 642
5.7 Liquid Chromatography MS 642
5.8 Surface Enhanced Laser Desorption Ionization-Mass Spectrometry (SELDI-MS) 644
5.9 MS Data Analysis Methods 645
6. Conclusions and Future Aspects 646
References 648

Comprehensive Analytical Chemistry, Vol. 68. http://dx.doi.org/10.1016/B978-0-444-63340-8.00012-1

1. INTRODUCTION

Outbreaks of food-related illness linked to foodborne pathogens are of growing concern to public health officials and the food industry. There is an urgent need for rapid and precisely fixed methods in identifying microorganisms as agents of those outbreaks. Pathogens most commonly associated with outbreaks of foodborne illness are *Escherichia coli* O157:H7 (and other *E. coli* serotypes), *Salmonella*, *Listeria monocytogenes*, *Campylobacter*, and *Norovirus*. Rapid, accurate identification and characterization of these (and other) pathogens are required for trace-back, clinical, and forensic analysis. Because of its speed, sensitivity, and specificity, mass spectrometry (MS) is becoming more intensively used for rapid microbial identification by detection and/or identification of biomolecules that are extremely specific for the microorganism that generates them, typically proteins or DNA [1].

The objectives of this chapter are to introduce the mass spectrometry concept and principles in the detection of food pathogens. Due to the outbreaks of foodborne illnesses, there is urgency in developing precise and rapid methods to identify those pathogens. In this chapter, some information about mass spectrometry in the diagnosis of microbiological samples is reviewed, including the mass spectrometry types: Matrix-assisted laser desorption ionization time-of-flight-mass spectrometry (MALDI-TOF-MS), electrospray ionization mass spectrometry (ESI-MS), polymerase chain reaction coupled to mass spectrometry using electrospray ionization (PCR/ESI-MS), bioaerosol mass spectrometry, and chromatography coupled with mass spectrometry: gas chromatography mass spectrometry (GC-MS) and pyrolysis–GC-MS, capillary electrophoresis–MS, liquid chromatography mass spectrometry (LC-MS).

2. COMMON FOODBORNE PATHOGENS (SPOILAGE AND PATHOGENIC BACTERIA)

2.1 Spoilage Bacteria

Spoilage bacteria are microorganisms that cause the decay of food and unpleasant odors, tastes, and textures to flourish. A spoiled food has lost its previous nutritional value, texture, or flavor and can become hazardous to people and inappropriate to eat. The microbial spoilage of food products is a crucial problem, with high economic losses for the food industry, particularly under inappropriate refrigeration conditions. In this manner, spoilage bacteria are able to grow in large number in food, spoil the food, and introduce changes in the taste/smell, which influence the quality of the products. These bacteria do not cause illness under normal conditions; nevertheless, when consumed in high concentration, they can cause gastrointestinal disturbance [2]. Various bacterial species can cause decay in food products and the spoilage microbiota depends mostly on the processing and preservation method. Storage temperature also has an important key role in the growth of unacceptable microbiota in food.

Fresh foodstuffs such as fish and meat stored at refrigeration temperatures can result in the growth of *Pseudomonas* spp., including spoilage species, such as *Pseudomonas fragi* and *Pseudomonas putida*. A light preservation and atmospheric changes, e.g., by vacuum-packaging, may restrict these bacterial species and help the growth of other species, such as lactic acid bacteria, Enterobacteriaceae, *Bacillus* spp., and *Clostridium* spp. Among those bacteria, two genera, *Bacillus* spp. and *Clostridium* spp., are able to generate spores that can survive under hot conditions and germinate after a pasteurization process, being an important issue in food safety. Spoilage species may be food-specific e.g., *Erwinia* spp. has been detected in products of vegetal origin whereas seafood products are generally spoiled by species such as *Shewanella* spp. or *Photobacterium* spp. Commonly, bacteria can spoil different foods due to the physical–chemical preservation profile [3].

2.2 Pathogenic Bacteria

Foodborne diseases are caused by agents that enter the body through the ingestion of food. Food can transmit disease from person to person as well as serve as a growth medium for bacteria that cause food poisoning. It is not easy to estimate foodborne illnesses globally yet, but it was established that in 2005 alone, 1.8 million people died from diarrheal diseases. Most cases are caused by consumption of contaminated food and water. In industrialized countries, the percentage of the population suffering from foodborne diseases yearly is up to 30% [4].

Pathogenic bacteria frequently do not change the color, odor, taste, or texture of a food product, so it is hard to recognize if the product is contaminated. The bacteria in food can cause foodborne infections when food contaminated with bacteria is eaten and the bacteria continue to grow in the intestines and cause illness. Bacterial infection is frequently caused by bacterial pathogenic species such as *E. coli*, *Salmonella* spp., *L. monocytogenes*, *Staphylococcus aureus*, *Bacillus cereus*, *Clostridium perfringens*, *Campylobacter* spp., *Shigella* spp., *Streptococcus* spp., *Vibrio cholerae*, including O1 and non-O1, *Vibrio parahaemolyticus*, *Vibrio vulnificus*, *Yersinia enterocolitica*, and *Yersinia pseudotuberculosis*. Food intoxication results from ingesting toxins (or poisons) generated in food as a by-product of bacterial growth. In food intoxications, it is toxins not bacteria that cause illness. These toxins may not change the appearance, odor, or flavor of food. *S. aureus* and *Clostridium botulinum* are common bacteria that generate toxins. In some cases, such as that of *C. perfringens*, there is a mix between infection and intoxication because the disease is caused by toxins released in the gut after the intake of a large amount of vegetative cells.

An additional problem is the continuous detection of emerging foodborne pathogens that include new pathogens, pathogens that arise due to changing ecology or changing technology that connects a potential pathogen with the food chain or emerge "de novo" by transfer of mobile virulence factors [5].

Emerging foodborne pathogens include *E. coli* O157:H7, *Aeromonas hydrophila, Aeromonas caviae, Aeromonas sobria, Mycobacterium* spp., vancomycin-resistant *Enterococci*, nongastric *Helicobacter* spp., *Enterobacter sakazakii*, nonjejuni/coli species of *Campylobacter*, and non-O157 Shiga toxin-producing *E. coli* [6].

3. DIAGNOSIS OF MICROORGANISMS

Nowadays, there are different methods available to identify microorganisms in food and these are summarized in Table 1. Standard culture/colony-based methods continue to be the most reliable and precise in identifying the foodborne pathogens. However, the most important drawbacks of these methods are that they are labor-intensive and slow, often requiring 2–3 days for initial results, and up to 7–10 days for confirmation. Early detection is critical to prevent or solve the pathogen-related problems. Immunology-based methods, for example, enzyme immunoassay (EIA), enzyme-linked immunosorbent assay (ELISA), immunochromatography (ICG) strip test, and western blots, have been widely used for the detection of bacterial cells, spores, viruses, and toxins. In comparison with traditional culture techniques, immunological methods are much more rapid. However, the detection of microorganisms in real time is still challenging. PCR-based techniques can detect a single copy of a target DNA sequence and thus are less prone to producing false-positive results. The methods detect a microorganism by amplifying the target rather than the signal, and might be employed to detect a single pathogenic bacterium in food. Although rapid methods to detect pathogens based on PCR have been developed, some problems remain in the identification of atypical strains whose sequences are unknown.

For more than a hundred years, mass spectrometry has been a favorable tool to understand the chemistry of proteins and the biological processes involved in their synthesis. Nevertheless, until the invention of soft ionization technique such as MALDI and ESI (this methodology cannot be used as a routine tool in laboratories. Currently, mass spectrometry is integrated in microbiology laboratories, obtaining a fast and trustworthy identification of microorganisms based on proteomic and other biomarker analysis [7–9].

4. MASS SPECTROMETRY IN MICROBIAL ANALYSIS

Microbiological identification by mass spectrometry has tremendous advantages

1. Culture and isolation of the microorganism is not necessary, so that fastidious microorganisms can be identified.
2. Highly sensitive and accurate identification of microorganisms is achieved in a small sample amount.
3. Resistance markers and resistance profile can be determined at the same time as identification analysis [7].

TABLE 1 Characteristics of Conventional (Culture-Based), Immunological, Nucleic Acid, and Mass Spectrometry-Based Assays

Method	Advantages	Drawbacks	Detection Limit (cfu/mL or g)	Time Before Result	Specificity
Conventional culture/colony-based methods	Very sensitive, easily adaptable, cost-effective	Labor-intensive, time-consuming.	1	1–3 days	Good
Immunology-based methods	High-throughput capacity, rapid, cost-effective and simple, possibility to quantify the target pathogen	Lack of selectivity and sensitivity due to the difficulty to generate selective antibodies and the requirement of large amounts of the respective antigen to quantify bacteria	10^4	1–2 h	Moderate/good
	Antigen microarrays enable testing several pathogens at the same time				
Nucleic acid-based methods	Specific, rapid, PCR techniques have the ability to produce millions of copies of a specific portion of DNA sequence, that may be a gene or gene clusters, through the use of nucleotide sequences from 16S ribosomal RNA genes, it is now possible to identify or, to infer about the bacterial species	The use of species-specific sequences to detect and identify pathogens requires the prediction of the species presence in the clinical samples	10^3	3–6 h	Excellent
Mass spectrometry-based methods	Based on the detection of characteristic biomarkers of bacteria (e.g., proteins and peptides or nucleic acids)	The detection of candidate biomarkers from complex samples is not possible	1	10–20 min	Excellent
	Very sensitive, simple, rapid, cost-effective high-throughput capability				

There are several MS-based analyses of many classes of biomolecules and the application for each class can also be achieved through combination with chromatographic techniques, including gas chromatography (GC), capillary electrophoresis (CE), liquid chromatography (LC), and affinity methods, such as SELDI. MS approaches to the identification of microorganisms can take advantage of differences in microbial genomic sequences resulting from measurable differences in the molecular masses of PCR products. PCR assays cannot be used directly for classification of unknown microbial samples. Methods integrating PCR and MS support additional information not supplied by either technique alone [10–13].

MS has emerged as a rapid and sensitive method for accurate pathogen identification at the genus, species or subspecies level. The MS methods are based on the detection of characteristic biomarkers of bacteria (e.g., proteins and peptides, nucleic acids, or even lipids).

A mass spectrometer consists of three components: an ion source, a mass analyzer, and a detector. There are various types of each component. In the following sections, those most commonly used to identify and quantify food pathogens are described.

4.1 Ionization Source

An ionization source is where the generation of electrically charged ionized particles that gain or lose electrons in a gas phase takes place. Various ionization processes can be employed [7]. Among these processes are MALDI and ESI (electrospray ionization) that are most known.

MALDI is based on the usage of a matrix complexed with a given sample molecule that is bombarded with a laser in order for the sample molecule to form a sample ionization. The sample is normally mixed with a suitable matrix material and applied to a metal plate. Commonly the sample is mixed with as little matrix as possible as the matrix will also become excited and ablate and ionize as well. Then, a pulsed laser irradiates the sample, triggering ablation, and desorption of the sample and matrix material. The matrix itself acts as a substance that infuses the sample as well as a transformer for the laser's energy into excitation energy to allow for the vaporization of the sample ions and matrix ions from the surface of the matrix. Finally, the analyte molecules are ionized by being protonated or deprotonated in the hot plume of ablated gases, and can then be accelerated into whichever mass spectrometer is used to analyze them. There are many mass analyzers that can be combined with a MALDI but the most common one, due to its intrinsic characteristics as being able to determine the accurate mass, is time-of-flight (TOF).

Direct analyses of intact bacterial cells through MALDI-TOF-MS have been used to differentiate bacterial species and subspecies in many clinical microbiology laboratories. The protein profiles in whole-cell MALDI mass spectra have taxonomically characteristic features that can be used to differentiate bacteria at the genus, species, and strain levels, although only a small portion of the bacterial proteome is detected [9].

Surface enhanced laser desorption ionization (SELDI) is a modification of the procedure used in MALDI. Instead of mixing the UV sensitive matrix with the sample, the sample is spotted on a surface modified with a chemical functionality. Common surfaces include ionic exchangers, reverse phases, and metal binders. Biomarkers of interest in the sample bind to the surface, while the others are removed by washing. The UV matrix is then added to the spot and allowed to cocrystallize. After the ionization with the UV laser, the ions are analyzed, in the same manner as in MALDI [14].

Electrospray ionization (ESI) is especially useful in producing ions from macromolecules because it overcomes the propensity of these molecules to fragment when ionized. A sample solution is sprayed from a small tube into a strong electric field in the presence of a flow of warm nitrogen to assist desolvation. The droplets formed evaporate in a region maintained at a vacuum of several torr causing the charge to increase on the droplets. ESI produces multiple charged ions, extending the mass range of the analyzer to accommodate the kDa-MDa orders of magnitude observed in proteins and their associated polypeptide fragments [15,16].

ESI-MS has been used less frequently for direct microbial identification. Xiang and colleagues characterized microorganisms by performing global ESI–tandem mass spectrometry (MS/MS) analyses of cell lysates [17]. Vaidyanathan and colleagues reported the direct ESI-MS analyses of whole bacterial cells without prior separation [18]. The newly developed technique of desorption electrospray ionization (DESI) MS allows the direct analysis of condensed phase samples by spraying them with electrosprayed solvent droplets. This approach has been used to differentiate several bacteria species—including *E. coli*, *S. aureus*, *Enterococcus* sp., *Bordetella bronchiseptica*, *Bacillus thuringiensis*, *Bacillus subtilis*, and *Salmonella typhimurium*—based on their DESI mass spectral profiles. Distinguishable DESI mass spectra, in the mass range 50–500 Da, have been obtained from whole bacteria using both positive and negative ion modes; the analysis time can be less than 2 min [9].

4.2 Mass Analyzer

The mass analyzer is the major part of the mass spectrometer, the charged fragments (ions and radical ions), are accelerated and deflected by a strong magnetic field that influences their travel, resulting in a curvilinear path. Ions and radical ions are collected, detected, and quantified with high accuracy and sensitivity, depending on the mass/charge ratio (m/z). Although there are many analyzers such as triple quad, quadrupole linear ion trap (QTRAP), TOF-MS, QqTOF-MS, and Orbitrap the most common type is the time-of-flight (TOF) [7].

4.3 Ion Detector

Some type of electron multiplier is used, though other detectors including Faraday cups and ion-to-photon detectors are also used. Because the number of ions leaving the mass analyzer at a particular instant is typically quite small, considerable amplification is often necessary to get a signal.

Mass analyzer characterized and distinguished ions according to their mass/charge ratio (m/z). Ion detector produces mass spectrum for each compound, which is typically symbolized as a bar graph with the mass-to-charge ratio (m/z) in the X-axis, and the intensity or relative abundance of the ions in the Y-axis. The highest peak is assigned as 100% of intensity known as the base peak [7].

5. IDENTIFICATION OF FOOD PATHOGENS BY MASS SPECTROMETRY

Bacterial identification can be achieved by either identifying the specific masses of the biomarker molecule that could be compared with characteristic protein masses in databases or by correlating the whole spectral profile to a reference database [19]. In the former, bacterial strains are identified by determining the masses of biomarker and comparing them with a protein database [20]. Many studies have been accomplished with protein biomarkers by MALDI-TOF-MS for bacterial species identification [20,21] An important objection of protein database searches is the need of high mass accuracy. Identification is restricted to well-characterized microorganisms with known protein sequences available in proteome databases [22]. The second type, also named "fingerprint approach" is the most applied for bacterial identification. It depends upon spectral differences of bacterial species and identification is carried out by comparison of the spectral profile of an unknown strain to a reference database of spectral profiles [23,24]. This approach admits the differentiation of bacterial strains, due to the high specific spectral profiles, named "fingerprints," gathered. For this purpose, it is unnecessary to identify the proteins but just to determine a number of characteristic peaks that are representative of the corresponding genus and/or species [6].

5.1 Matrix-Assisted Laser Desorption Ionization Time-of-Flight-Mass Spectrometry (MALDI-TOF-MS)

One of its major challenges is the detection of candidate biomarkers in complex samples such as food. The MALDI interface is best suited to identify microbial threats, generally, in a pure culture of microorganisms [7,10].

The first applications of MALDI-TOF-MS to food pathogens isolated and separated the protein fractions from bacterial cells. However, MALDI-TOF-MS immediately turned to the analysis of whole cells directly without any sample pretreatment, called intact cell mass spectrometry (IC-MS). The desirable situation would be to establish one standardized protocol to acquire specific and reproducible spectral profiles in a rapid and cost-effective way. However, different sample preparation protocols were applied. The three most preferred are summarized below:

1. Colonies are directly picked up from the culture plates, the bacterial biomass is directly applied to the MALDI-TOF-MS sample plate. The advantage of this method is the rapidity, but the disadvantages are less homogeneous crystallization, less reproducibility, more noise, and less resolution of peaks.
2. The bacterial biomass is harvested in organic solvent or in the matrix solution to get bacterial suspensions. This method requires one to two cycles of washing/centrifugation steps. Advantages of this method are homogeneous crystallization, good reproducibility, and good resolution of peaks. Disadvantages are more noise and the need for time-consuming washing steps.
3. The bacterial biomass after harvesting in organic solvent is just centrifuged and the supernatant analyzed. This method requires just a centrifugation step. Advantages of this method are speed, homogeneous crystallization, and low noise. Best reproducibility and best resolution of peaks were demonstrated with this method [10].

Early detection and characterization of microorganisms can help to reduce health hazards and avoid the spread of microbe-related diseases.

The acquired spectra have to be representative and generative to allow the comparability of bacterial spectral fingerprints. For this purpose, a standardized protocol has to be followed, beginning from sample preparation through to instrumental parameters. Spectral profiles were less sensitive to culture conditions but can show particular variability depending on the sample preparation protocol [25,26]. The most rapid and cost-effective method is the direct application of bacterial biomass taken from culture plates to the MALDI-TOF-MS sample plate. Afterward, the bacterial cells are overlaid with the matrix solution [27]. Besides being the most rapid and cost-effective method, the direct spotting of biomass to the sample plate had several disadvantages. The difficulty in taking the correct amount of biomass impedes obtaining a homogeneous distribution of the sample and matrix. Even though this technique was favorably applied for bacterial species identification, it has been demonstrated that spectra showed more noise and less peak resolution with this fast method, making it difficult to get reproducible spectral profiles [28,29]. Most applications of MALDI-TOF-MS for bacterial identification directly analyze bacterial cell extracts, acquired by just one dilution/centrifugation step. Using whole cells or extracting the proteins, it has been demonstrated that when applying the same culture conditions, sample preparation, matrix, organic solvents, and MALDI-TOF-MS analyzer, reproducible spectral profiles are acquired [28,30].

Even though, different protocols can lead to a high variability in the resulting spectral profiles, some peaks were detected, even if different protocols were used. Such characteristic peaks could serve as biomarker proteins for the corresponding genus and/or species identification [31].

The actual data acquisition with MALDI-TOF-MS is nowadays commonly performed in an automated manner. Namely, the laser focus scans the sample in a predefined pattern and gathers a mass spectrum from a defined number of laser pulse cycles, frequently several hundreds, to yield a representative average mass spectrum. The raw spectrum is generally processed to obtain a mass fingerprint that provides the information about peak apex *m/z* values, in this manner lessening the size of individual files considerably. The important step for species identification is the comparison of the mass fingerprint of the sample to be identified to a database including reference mass fingerprints, for example, MASCOT, SWISS-PROT [12,13].

MALDI-TOF-MS fingerprinting is a rapid and reliable method for bacterial identification in food. Recently, several reports have shown the feasibility of using MALDI-TOF-MS for identifying microorganisms [23,32]. MALDI mass spectrometry-based systems for rapid characterization of microorganisms in biodefense or medical diagnostics usually detect intact proteins in the 5000–20,000 Da range. The detection and comparison of protein mass patterns has become an appropriate tool for the rapid identification of bacteria, depending on the high specific mass profiles acquired. Many bacterial identification studies by MALDI-TOF-MS fingerprinting are applied to clinical diagnostics of bacterial strains associated with human infectious diseases. On the contrary, just a few works have been done in the field of microbial food analysis by MALDI-TOF-MS for the identification of foodborne pathogens and/or spoilers. These works involved the classification and identification of various common pathogens causing foodborne diseases such as *A. hydrophila*, *Arcobacter* spp., *Campylobacter* spp., *Clostridium* spp., *Listeria* spp., *Salmonella* spp., *Staphylococcus* spp., *V. parahaemolyticus*, *Yersinia* spp., *Bacillus* spp. and species of the Enterobacteriaceae family [25–42].

Some studies simultaneously detect foodborne pathogens and food-spoilage bacteria, including genera such as *Escherichia*, *Yersinia*, *Proteus*, *Morganella*, *Salmonella*, *Staphylococcus*, *Micrococcus*, *Lactococcus*, *Pseudomonas*, *Listeria*, and *Leuconostoc* [24]. In further studies, an extensive spectral library was created, involving the essential pathogenic and spoilage bacterial species that can be present in seafood [43,44]. These works involved genera, such as *Acinetobacter*, *Aeromonas*, *Bacillus*, *Carnobacterium*, *Listeria*, *Pseudomonas*, *Shewanella*, *Staphylococcus*, *Stenotrophomonas*, *Vibrio*, and genera of the Enterobacteriaceae family [6]. IC-MALDI-TOF-MS has been favorably used to differentiate foodborne and clinically associated bacteria such as *Escherichia*, *Campylobacter*, *Salmonella*, *Yersinia*, *Helicobacter*, and *Listeria* at the genus and species levels [24,33,45,46]. Although intraspecies-level differentiation has been possible for some serovars of *Salmonella*, *L. monocytogenes* O157 versus non-O157, specific biomarkers to subspecies-level differentiation remains a challenge, in part because MS fingerprints within a species are very similar [24,33,30,47]. For instance, to obtain subspecies-level differentiation of *Salmonella*, protocols had to be optimized to generate spectra with a much higher number of reproducible protein peaks (>300) than generally obtained (usually 30). Because of this disadvantage, the utility of MALDI-TOF-MS for high-level differentiation needs further investigation [43,44]. Such singular or characteristic peak masses can serve for the rapid identification of a bacterial genus and/or species. Nevertheless, precise identification cannot be accomplished based on a single biomarker protein, but under consideration of a number of characteristic mass patterns, symbolizing the spectral fingerprint. Moreover, when working with microbial mixtures, such biomarkers become more substantial since the presence or absence of singular peak patterns could lead to a conclusion of the bacterial species present. The detection of biomarker proteins by MALDI-TOF-MS has been favorably carried out for the identification of two bacterial species isolated from contaminated water, lettuce, and cotton cloth. However, the performance of MALDI-TOF-MS fingerprinting for microbial mixtures has not been demonstrated yet. Another essential objection to MALDI-TOF-MS fingerprinting is that the classification of a bacterial genus or species, as well as the determination of singular biomarker patterns is only attainable in the frame of a spectral reference library. The number of studies on bacterial species identification by MALDI-TOF-MS in foodstuffs is steadily rising. In the whole process of bacterial identification by MALDI-TOF-MS in food products, bacteria are isolated from food samples and cultivated to have single colonies. Afterward, the bacterial cells are lysed by an organic solvent and a strong acid, the most commonly applied ones being acetonitrile and trifluoroacetic acid. Once the spectral profiles have been acquired, data analysis is carried out, involving the extraction of representative peak mass lists and the comparison of spectral data to get bacterial differentiation. Moreover, cluster analysis of the peak mass lists shows phyloproteomic relationships between bacterial species, permitting the identification of unknown strains, just like the typing of closely related strains. Several different protocols and techniques have been described for the sample preparation and data analysis [48].

The spectral profile of the strain of interest is correlated to a spectral library of reference strains for bacterial identification by MALDI-TOF-MS fingerprinting. Several private databases have been developed, involving spectral profiles of more than 500 bacterial strains, such as The Spectral Archive And Microbial Identification System (SaramisTM; AnagnosTec GmbH, Potsdam, Germany) [22,27,49]. The MALDI Biotyper 3.0 (Bruker Daltonics) searches an ample database of more than 5000 bacterial species and new spectral profiles are being added on a daily basis. The database has shown to be appropriate for the routine bacterial identification in clinical laboratories, being a rapid, cost-effective, and precise technique that identifies correctly 92% of the species [32,50]. Most of the studies, as well as these databases, are focused on human pathogens causing infectious diseases. However, the databases also involve species that play an essential role in food safety and quality providing valuable reference data for the identification of food pathogens and food-spoilage bacterial species. Nevertheless, the critical objection to these databases is the restricted availability.

From this point of view, it would be favorable to have a public database for the compliance of the spectral information for each species that would permit the comparison with results from several researches and favoring a more accurate method for identification of intact bacteria based on a large part of the available data. There have been a few efforts to start a public database, e.g., a library holding spectra of 24 foodborne bacterial species including *Escherichia* spp., *Yersinia* spp., *Proteus* spp., *Morganella* spp., *Salmonella* spp., *Staphylococcus* spp., *Micrococcus* spp., *Lactococcus* spp., *Pseudomonas* spp., *Leuconostoc* spp., and *Listeria* spp. [51]. A reference library of mass spectral fingerprints of the major pathogenic and spoilage bacterial species in seafood products was created involving more than 50 bacterial species [43,44]. In further studies, the library allowed the correct identification of unknown bacterial strains isolated from commercial seafood products [52]. The accumulated reference library of seafoodborne and spoilage bacterial species can be applied to any other foodstuff. The spectral library could easily be extended by further bacterial species and strains that are of interest in food products [6,53].

Peak-matching techniques remove the subjectivity of visual comparison. Jarman and colleagues developed an automated peak detection algorithm to extract representative mass ions from a fingerprint and to correlate spectra to fingerprints in a reference library. This algorithm accomplishes the identification of an unknown spectrum by estimating a degree of matching [31]. Some other researchers established a software (BGP-database, available on http://sourceforge.net/projects/bgp) to analyze and compare spectral profiles, permitting rapid identification. This software discovers the best match between the tested strain and the reference strains of the database [35]. In some other studies, the freely available Web-based application SPECLUST (http://bioinfo.thep.lu.se/speclust.html) was utilized to extract representative peak masses and to get final peak mass lists for each bacterial strain. Required mass lists can then be correlated and common peak masses defined. The Web interface estimates the mass difference between two peaks taken from different peak lists and determines if the two peaks are identical after allowing a certain measurement uncertainty (σ) and peak match score(s) [54]. The Web program finalized very fast, easy to use, and could be extended simply by new spectral mass lists. Although, it was not capable of investigating an unknown spectrum directly against the library, the correlation of peak mass lists could be achieved to identify the spectral profile of an unknown strain. The Web-application was successfully applied to identify pathogenic and spoilage bacterial strains, isolated from commercial seafood products. Moreover, the program permits the rapid determination of specific biomarker peaks and involves a clustering option [52]. Additional bioinformatics programs, such as Statgraphics Plus 5.1 (Statpoint Technologies, Inc., Warrenton, USA), propose several functionalities. First, spectral data have to be transformed to a binary table, displaying the presence (1) and absence (0) of a peak mass. Subsequently, various algorithms for cluster analysis can be applied, as well as principal component analysis (PCA) [44]. In a different study, the BioNumerics 6.0 software (Applied-Maths, Sint-Martens-Latem, Belgium) was utilized for data analysis and machine learning for bacterial identification by MALDI-TOF-MS [55]. Clustering of the spectral data acquired by MALDI-TOF-MS is another approach to bacterial identification and classification. Conway and colleagues introduced the term "Phyloproteomics" a novel analytical tool that solves the issue of comparability between proteomic analyses, utilizes a total spectrum-parsing algorithm, and produces biologically meaningful classification of specimens. The clustering of peak mass lists permitted a better visualization of similarities and differences of spectral comparison [56]. The creation of a dendrogram depending upon mass spectral data is a rapid technique to analyze spectral profiles and to visualize spectral relations by grouping the acquired peak mass lists of bacterial strains. In this manner, while clustering has been favorably applied for the differentiation and identification of bacterial strains at the genus and species level, at the same time clustering of mass spectral data has been applied as a typing method for the phyloproteomic study of different strains of the same species, with the aim to classify the strains [35,40,43,44,26,29]. When comparing the dendrograms symbolizing phyloproteomic relations to the phylogenetic trees, a high concordance was found by these authors. Since the peak patterns observed by MALDI-TOF-MS are generally attributed to ribosomal proteins the similarity of the MALDI-TOF-MS cluster to phylogenetic trees acquired by the analysis of ribosomal genes is not unexpected [40]. Nevertheless, in comparison to the sequence analysis of the 16S rRNA gene that is generally used for phylogenetic studies, the classification of bacterial strains by MALDI-TOF-MS fingerprinting allowed better differentiation. This is valuable for some genera, such as *Bacillus* and *Pseudomonas*, as the differentiation at the species level is complicated with 16S rRNA analysis [6]. Direct analysis of microorganisms using MALDI-MS has a number of advantages such as rapidity, low detection limits, simplified mass spectra (featuring the signals of predominantly singly charged ions), and tolerance to contaminants. The protein profiles in whole-cell MALDI mass spectra have taxonomically characteristic features that can be used to discriminate bacteria at the genus, species, and strain levels, despite only a small portion of the bacterial proteome being detected. Today, in many clinical microbiology laboratories, direct analyses of intact bacterial cells by MALDI-TOF-MS have been used [57].

5.1.1 *Examples of MALDI-TOF-MS Studies in Foodborne Pathogens*

The observed biomarkers helped researchers in not only detecting pathogenic bacteria (*Bacillus anthracis*, *Yersinia pestis*, and *Brucella melitensis*), but also in differentiating them from the corresponding nonpathogenic species. By examining a series of strains of several *Bacillus* species (*anthracis*, *thuringiensis*, *cereus*, and *subtilis*), it was possible to develop genus, species, and strain-specific biomarkers from the estimated molecular masses of the intact proteins.

The application of this technique for rapid chemotaxonomic classification of microorganisms is demonstrated [58,59].

MALDI-TOF mass spectrometry has been coupled with Internet-based proteome database search algorithms for direct microorganism identification. Demirev and colleagues applied this approach to characterize intact *Helicobacter pylori* (strain 26,695) Gram-negative bacteria by including a specific and common posttranslational modification, N-terminal Met cleavage, in the search algorithm [60]. Conway and colleagues analyzed 25 carefully selected isolates of pathogenic *E. coli* and additional Enterobacteriaceae members in order to evaluate MALDI-TOF-MS as a tool for rapid identification of common clinical bacterial isolates. Organisms were prepared according to clinical microbiological protocols and analyzed with only the slightest supplementary processing. Spectra were reproducible from preparation to preparation and consisted of 40–100 peaks primarily symbolizing intracellular proteins with masses up to 25 kDa. Spectra of 14 genetically different bacteremic isolates of *E. coli* were correlated with isolates symbolizing other genera inside the Enterobacteriaceae family. *E. coli* isolates were closely related to each other and were readily differentiable from other Enterobacteriaceae, involving *Salmonella* and *Shigella* by utilizing a new spectrum comparison algorithm. Before long, the methodology allows the analysis of 40 unknown isolates per hour per instrument. These results imply that MALDI-TOF-MS provides a rapid and reliable approach for performing phyloproteomics. It helps the identification of unknown bacterial isolates depending upon similarities within protein biomarker databases [56].

Nowadays, several clinical laboratories are establishing mass-spectrometric techniques to analyze and identify microorganisms. Nevertheless, slightest work has been done with mixtures of bacteria. For demonstrating microbial mixtures could be analyzed by MALDI-MS, mixed bacterial cultures were analyzed in a double-blind fashion. Cebula and colleagues used nine different bacterial species to generate 50 different simulated mixed bacterial cultures similar to that done for an initial blind study previously reported. The samples were analyzed by MALDI-MS with automated data extraction and several newly established analysis algorithms. The sample constituents were identified accurately to the species level in all but one of the samples. The elimination of associated organisms was challenging for the current algorithms, particularly, in differentiating *Serratia marcescens*, *E. coli*, and *Y. enterocolitica*, which have some closeness in their MALDI-MS fingerprints [61]. Leuschner and colleagues applied MALDI-TOF-MS for characterization of strains of *Salmonella enterica* subsp. *enterica*. Whole cells were analyzed by MALDI-TOF-MS. Spectra with a maximum of 500 mass peaks between (m/z) 0 and 25,000 were examined for consensus peaks manually and by a computer software algorithm. Consensus peaks were monitored by both methods for spectra of *S. enterica* serovars *Derby*, *Hadar*, *Virchow*, *Anatum*, *Typhimurium*, and *Enteritidis*. Differences in numbers of consensus peaks in spectra acquired by manual and computer correlation outlined that software including statistical analysis of peak accuracy is required. The significance of this study was to demonstrate the development of a system for peak profiles analysis in whole cell by MALDI-TOF-MS spectra to allow intra- and interlaboratory comparison [62]. Rapid identification of microorganisms utilizing MALDI is a rapidly evolving area of research due to the scarce sample preparation, speed of analysis, and broad applicability of the technique. This approach relies on expressed biochemical markers, often proteins, to identify microorganisms. As a result, variations in culture conditions that affect protein expression may limit the ability of MALDI-MS to correctly identify an organism. The efforts to investigate the effects of culture conditions on MALDI-MS signatures specifically examine the effects of pH, growth rate, and temperature. Continuous cultures prepared in bioreactors were used to get specific growth rates and pH for *E. coli* HB 101. Despite measurable morphological differences between growth conditions, the MALDI-MS data related each culture with the appropriate library entry (*E. coli* HB 101 generated using batch culture on an Luria broth (LB) media), independent of pH or growth rate. The effect of varying growth temperature on *Y. enterocolitica* was also investigated. Although, the anticipated effects on phenotype were monitored, the MALDI-MS technique supported the appropriate identification [26]. Fargerquist and colleagues identified an approximately 10-kDa protein biomarker observed in the MALDI-TOF-MS of cell lysates of five thermophilic species of *Campylobacter*: *jejuni*, *coli*, *lari*, *upsaliensis*, and *helveticus*. The biomarker was identified by genomic and proteomic sequencing as a DNA-binding protein HU. This protein was useful as a biomarker because it strongly ionizes by MALDI and its molecular weight changes between species and among strains inside a species. Intra- and interspecies variation of the HU molecular weight depends upon variations in the amino acid sequence of the HU protein and not due to co- or posttranslational modifications. The strong ionization effectiveness of HU by MALDI is probably due, in part, to four lysine residues clustered at the carboxyl end of the protein [21]. Donohue and colleagues used MALDI-MS for the characterization of 17 species of *Aeromonas* represented by 32 strains, which involved type, reference, and clinical isolates. Intact cells from each strain were used to generate a reproducible library of protein mass spectral fingerprints. Under the test conditions used, peak lists of the mass ions observed in each species showed that three mass ions were in all the 17 species tested. These common mass ions can be used as genus-specific biomarkers to identify *Aeromonas* in unknown samples. A dendrogram generated using the m/z signatures of all the strains tested showed that the mass spectral data hold sufficient information to differentiate between genera, species, and strains. There are several advantages of using MALDI-MS-based protein mass spectral fingerprinting of whole cells for the identification of microorganisms as well as for their differentiation at the subspecies level: (1) the capability to detect proteins, (2) high throughput, and (3) simple sample preparation. The precision and speed with which data can be acquired makes MALDI-MS a powerful tool for food monitoring and detection of biological hazards [34].

Methicillin-resistant coagulase-negative *Staphylococci* (MR-CNS) are of increasing importance in animal and public health, veterinary medicine, and along the meat and milk production line. Only limited data were so far available on MR-CNS characteristics. Huber and colleagues evaluated the prevalence of MR-CNS, to identify the detected *Staphylococci* to species level, and to assess the antibiotic resistance profiles of isolated MR-CNS strains. After two-step enrichment and growth on chromogenic agar, MR-CNS were detected in 48.2% of samples from livestock and chicken carcasses, 46.4% of samples from bulk tank milk and minced meat, and 49.3% of human samples. Using MALDI-TOF-MS, 414 selected MR-CNS strains belonged to seven different species (*Staphylococcus sciuri*, 32.6%; *Staphylococcus fleurettii*, 25.1%; *Staphylococcus haemolyticus*, 17.4%; *Staphylococcus epidermidis*, 14.5%, *Staphylococcus lentus*, 9.2%; *Staphylococcus warneri*, 0.7%; *Staphylococcus cohnii*, 0.5%) were identified. *S. sciuri* and *S. fleurettii* thereby predominated in livestock, bulk tank milk (BTM), and minced meat samples, whereas *S. epidermidis* and *S. haemolyticus* predominated in human samples. In addition to beta-lactam resistance, 33–49% of all 414 strains were resistant to certain non-beta-lactam antibiotics (ciprofloxacin, clindamycin, erythromycin, and tetracycline). A high prevalence of MR-CNS was found in livestock production. This was of concern in view of potential spread of mecA to *S. aureus* (MRSA). MALDI-TOF-MS proved to be a fast and reliable tool for species identification of MR-CNS isolated from different origins [63].

A rapid and reliable bacterial source tracking (BST) method is important to counter risks to human health posed by fecal contamination of surface waters. Genetic fingerprinting methods, such as repetitive sequence-based PCR (rep-PCR), have shown promise as BST tools but are time-consuming and labor-intensive. Siegrist and colleagues compared the ability of MALDI-TOF-MS to characterize and differentiate between intimately related environmental strains of *E. coli* and to classify them according to their sources to a commonly used rep-PCR-based method. Among the criteria assessed were repeatability and the ability to group *E. coli* isolates according to their respective sources. MALDI-TOF-MS-based method has a lower repeatability than rep-PCR but provides an increased ability to correctly assign *E. coli* isolates to specific source groups. Furthermore, five biomarkers that appear conserved among strains isolated from avian were identified. MALDI-TOF-MS may represent a promising, novel, and rapid approach to address the problem of fecal contamination of surface waters [64].

IC-MALDI-TOF-MS generates a reproducible spectrum depending upon growth conditions, strain, or species. This technique was applied to the identification of coagulase-negative *Staphylococci* (CoNS) using whole viable bacteria. Once a bacterium has been recognized as Micrococcaceae, the strategy was to identify peaks in the spectrum that can be used to identify the species or subspecies. MALDI-TOF-MS was carried out by utilizing bacteria acquired from one isolated colony. One reference strain for each of the 23 clinically associated species or subspecies of Micrococcaceae was chosen. For each reference strain, the MALDI-TOF-MS profile of 10 colonies obtained from 10 different passages was analyzed. For each strain, only common peaks with a relative intensity above 0.1 were retained obtained a set of 3–14 selected peaks per strain. The MALDI-TOF-MS profile of 196 tested strains was then compared with that of the set of selected peaks of each of the 23 reference strains. The best hit was always with the set of peaks of the reference strain belonging to the same species as that of the tested strain, demonstrating that the 23 sets of selected peaks can be used as a database for the rapid species identification of CoNS. Similar results were obtained using four different growth conditions. Broadening this strategy to other groups of relevant pathogenic bacteria will permit rapid bacterial identification [35].

Conventional methods used for primary detection of *L. monocytogenes* from foods and subsequent confirmation of presumptive positive samples involve prolonged incubation and biochemical testing which generally require 4–5 days to obtain a result [65]. Barbuddhe and colleagues developed a robust method to analyze the extracts of *Listeria* species and strains by MALDI-TOF-MS promising for identification of *Listeria* species, typing and even for differentiation at the level of clonal lineages among pathogenic strains of *L. monocytogenes*. For this, following a short inactivation/extraction procedure, cell material from a bacterial colony was laid on a sample plate, dried, overlaid with a matrix, and analyzed by MALDI-TOF-MS. Mass spectra derived from *Listeria* isolates showed characteristic peaks, at both the species and lineage levels. MALDI-TOF-MS fingerprinting may have potential for *Listeria* identification and subtyping and may improve infection control measures [30]. A rapid, precise differentiation between *Listeria* strains is important for proper therapeutic management and timely intervention for infection control. The method successfully compared with the pulsed-field gel electrophoresis analysis of 48 *Listeria* strains.

In the same way, Jadhav and colleagues developed a simple and rapid proteomics-based MALDI-TOF-MS in order to detect *L. monocytogenes* directly from selective enrichment broths. Milk samples spiked with single species and multiple species cultures were incubated in a selective enrichment broth for 24 h, followed by an additional 6 h secondary enrichment. As few as one colony-forming unit (cfu) of *L. monocytogenes* per milliliter of initial selective broth culture could be detected within 30 h. On applying the same approach to solid foods previously implicated in listeriosis, namely chicken pâté, cantaloupe, and Camembert cheese, detection was achieved within the same time interval at inoculation levels of 10 cfu/mL. Unlike the routine application of MALDI-TOF-MS for identification of bacteria from solid media, this study proposes a cost-effective and time-saving detection scheme for direct identification of *L. monocytogenes* from broth cultures [65].

Species differentiation is essential for the early detection and identification of pathogenic and food-spoilage microorganisms that may be present in fish and seafood products. The proteomic approach demonstrated to be a competent tool for species identification. Böhme and colleagues characterized the main 26 species of seafood-spoilage and pathogenic Gram-negative bacteria, involving *A. hydrophila*, *Acinetobacter baumannii*, *Pseudomonas* spp., and *Enterobacter* spp., by MALDI-TOF-MS of low molecular weight proteins extracted from intact bacterial cells by a rapid procedure. From the collected spectra, a library of specific mass spectral fingerprints was created. To analyze spectral fingerprints, peaks in the mass range of 2000–10,000 Da were recorded and representative mass lists of 10–35 peak masses were collected. At least one singular biomarker peak was monitored for each species, and several genus-specific peaks were detected for genera *Proteus*, *Providencia*, *Pseudomonas*, *Serratia*, *Shewanella*, and *Vibrio*. Phyloproteomic relationships based on these data were correlated to phylogenetic analysis based on the 16S rRNA gene, and a similar clustering was found. The method was also applied to the identification of three bacterial strains isolated from seafood by correlating the spectral fingerprints with the created library of reference [43]. Whole-cell fingerprinting by MALDI-TOF-MS combined with bioinformatic software tool (MALDI Biotyper 2.0) was used to identify 152 *Staphylococcal* strains corresponding to 22 *Staphylococcal* species. Dubois and colleagues analyzed spectra of the 152 isolates, formerly identified at the species level using a sodA gene-based oligonucleotide array, against the main spectra of 3030 microorganisms. A total of 151 strains out of 152 (99.3%) were correctly identified at the species level; only one strain was identified at the genus level. The MALDI-TOF-MS method demonstrated different clonal lineages of *S. epidermidis* that were of either human or environmental origin, which implies that the MALDI-TOF-MS method could be beneficial in the profiling of *Staphylococcal* strains. The topology of the dendrogram created by the MALDI Biotyper 2.0 software from the spectra of 120 *Staphylococcus* reference strains (symbolizing 36 species) was in general agreement with that inferred from the 16S rRNA gene-based analysis. Their findings signify that the MALDI-TOF-MS technology, associated with a broad-spectrum reference database, is an effective tool for the rapid and trustworthy identification of *Staphylococci* [40]. Fagerquist and colleagues studied six protein biomarkers from two strains of *E. coli* O157:H7 and one non-O157:H7. Nonpathogenic strain of *E. coli* has been identified by MALDI-TOF-TOF-MS/MS and top-down proteomics. Proteins were extracted from bacterial cell lysates, ionized by MALDI, and analyzed by MS/MS. Protein biomarker ions were identified from their sequence-specific fragment ions by comparison to a database of in silico fragment ions derived from bacterial protein sequences. In-house Web-based software was utilized to swiftly correlate the mass-to-charge (m/z) of MS/MS fragment ions to the m/z of in silico fragment ions collected from hundreds of bacterial protein sequences. A peak-matching algorithm and a *p*-value algorithm were utilized to score freely and rank identifications on the basis of the number of MS/MS-in silico matches. In this study, the six proteins identified were the acid stress chaperone-like proteins, HdeA and HdeB; the cold shock protein, CspC; the YbgS (or homeobox protein); the putative stress-response protein YjbJ (or CsbD family protein); and a protein of unknown function, YahO. HdeA, HdeB, YbgS, and YahO proteins were modified posttranslationally with removal of an N-terminal signal peptide. Gene sequencing of hdeA, hdeB, cspC, ybgS, yahO, and yjbJ for 11 strains of *E. coli* O157:H7 and 7 strains of the "near-neighbor" serotype O55:H7 demonstrated a high degree sequence homology between these two serotypes. Even though it was not likely to distinguish O157:H7 from O55:H7 from these six biomarkers, it was possible to distinguish *E. coli* O157:H7 from a nonpathogenic *E. coli* by top-down proteomics of the YahO and YbgS. In the case of the YahO protein, a single amino acid residue substitution in its sequence (resulting in a molecular weight difference of only 1 Da) was sufficient to distinguish *E. coli* O157:H7 from a non-O157: H7, nonpathogenic *E. coli* by MALDI-TOF-TOF-MS/MS, whereas this would be difficult to distinguish by MALDI-TOF-MS. As a result, a protein biomarker ion at m/z approximately 9060 monitored in the MS spectra of non-O157: H7 *E. coli* strains but absent from MS spectra of *E. coli* O157:H7 strains was identified by top-down analysis to be the HdeB acid stress chaperone-like protein consistent with former identifications by gene sequencing and bottom-up proteomics [66].

The rapid identification of food pathogenic and spoilage bacteria is critical to guarantee food quality and safety. Seafood contaminated with pathogenic bacteria is one of the leading causes of food intoxications, and the fast spoilage of seafood products cause high economic losses. Böhme and colleagues studied the main seafood pathogenic and spoilage Gram-positive bacteria by MALDI-TOF, involving *Bacillus* spp., *Listeria* spp., *Clostridium* spp., *Staphylococcus* spp., and *Carnobacterium* spp. A reference library was created, including the spectral fingerprints of 32 reference strains and the extracted peak lists with 10–30 peak masses. Genus-specific as well as species-specific peak masses were selected to serve as biomarkers for the rapid bacterial identification. Moreover, the peak mass lists were clustered with the Web-application SPECLUST to demonstrate the phyloproteomic correlations among the studied strains. Subsequently, the method identified six strains isolated from seafood by comparison with the reference library. Moreover, phylogenetic analysis based on the 16S rRNA gene was accomplished and contrasted with the proteomic approach. This study was the first that applied MALDI-TOF-MS fingerprinting to Gram-positive bacterial identification in seafood, as a fast and precise technique to guarantee seafood quality and safety [67]. Jay-Russel and colleagues reported the isolation of Campylobacter species from the same population of feral swine that was investigated in San Benito County, California, during the 2006 spinach-related *E. coli* O157:H7 outbreak. *Campylobacter* species were cultured from buccal and rectal-anal swabs, colonic feces and tonsils utilizing a combination of selective enrichment and antibiotic-free membrane filtration methods. MALDI-TOF-MS was utilized to identify species. The findings underscore the significance of protecting raw vegetable crops from fecal contamination by wild or feral animals [68]. *Cronobacter* species represent an emerging opportunistic foodborne pathogen associated with meningitis and necrotizing enterocolitis in infants. Current evidence indicates that powdered infant formula (PIF) is the main source of *Cronobacter* contamination. A total of 75 strains of *Cronobacter* spp. from different geographic regions, as well as from PIF processing environments, were identified and typed with different methods, including biochemical profiling by the API 20E system (bioMérieux, Marcy l'Etoile, France), protein profiling by MALDI-TOF-MS, and genotypic profiling by ribotype. Analysis by MALDI-TOF-MS and biochemical identification was more accurate compared with ribotype analysis. However, MALDI-TOF-MS typing and ribotype analysis showed more discriminatory ability compared with biochemical phenotyping. MALDI-TOF-MS was found to be a rapid and reliable tool to identify *Cronobacter* spp. in PIF and has the potential to trace dissemination of *Chronobacter* along the production chain [69].

There are also some MALDI-TOF-MS studies with virus pathogens. Sugarcane mosaic virus (SCMV) is an important virus pathogen in crop production, causing serious losses in grain and forage yields in susceptible cultivars. For the efficient control of this virus, a better understanding of its interactions and associated resistance mechanisms at the molecular level is required. The responses of resistant and susceptible genotypes of maize to SCMV and the molecular basis of the resistance were studied using a proteomic approach based on two-dimensional polyacrylamide gel electrophoresis (2-DE) and MALDI-TOF-MS/MS analysis. Ninety-six protein spots showed statistically significant differences in intensity after SCMV inoculation. The classification of differentially expressed proteins showed that SCMV-responsive proteins were mainly involved in energy and metabolism, stress and defense responses, and photosynthesis. Among these identified proteins, 19 have not been identified previously as virus-responsive proteins, and seven were new and did not have assigned functions. These proteins may be candidate proteins for future investigation, and they may present new biological functions and play important roles in plant–virus interactions. The behavioral patterns of the identified proteins suggest the existence of defense mechanisms operating during the early stages of infection that differed in two genotypes. In addition, there are overlapping and specific phytohormone responses to SCMV infection between resistant and susceptible maize genotypes. This study may provide important insights into the molecular events during plant responses to virus infection [70]. Attien and colleagues investigated the microbial quality of meat products in Abidjan focused on *Staphylococcus* genus and the toxin production profile of *S. aureus* isolated. Bacteria were collected from 240 samples of three meat products sold in Abidjan and 180 samples issued from clinical infections. The strains were identified by both microbiological and MALDI-TOF-MS methods. They observed that 96/240 of meat samples were contaminated by *Staphylococcus*. Eleven species were isolated from meats [71].

5.2 Electrospray Ionization Mass Spectrometry (ESI-MS)

ESI-MS, with or without earlier analyte separation by LC has also been applied to the identification and characterization of bacteria. The coupling of protein/peptide separation by LC with ESI-MS analysis has one essential benefit over the routine MALDI-TOF-MS, proteins with equal masses can present different retention times in the LC and then, can be differentiated. This improved separation provides a greater number of protein peaks, which increases the likelihood of detecting appropriate biomarkers. Disadvantages of ESI-MS involve low tolerance to contaminants in the sample and enhanced complexity of the mass spectra due to the presence of multiple charge states. Sample processing and data analysis are more time-consuming (hours as opposed to minutes) with LC-ESI-MS, and the shifting in the retention times of biomarker peaks due, for example, to column wear or subtle differences in solvent composition, requires some attention because could influence reproducibility and interlaboratory differences. A known amount of a standard of known retention time can correct the shifts and serve as a standard to establish relative peak intensity [72]. Whole-cell ESI-MS spectra could be used to distinguish between *Bacillus* species, as well as to support further differentiation between strains of *B. subtilis*. It is demonstrated in those studies that spectra are effected by changes in the cone potential of the electrospray ion source, and that the application of different voltages could increase the discriminatory ability of this method. Twine and colleagues studied the differences between *C. botulinum* flagellin proteins by ESI-MS to identify marker ions with potential for identifying and characterizing strains of *C. botulinum*. These ions had the potential for use in swift detection and differentiation of *C. botulinum* cells, demonstrating *botulinum* neurotoxin contamination. This was the first report of glycosylation of Gram-positive flagellar proteins by the "sialic acid-like" nonulosonate sugar, legionaminic acid [73]. The strength of a coupled LC-ESI-MS access was demonstrated by Everley and colleagues that apply LC-QqTOF-MS to identify biomarkers for nonpathogenic *E. coli*, O157 EHEC and non-O157 EHEC as well as for *Shigella flexneri* and Shigella *sonnei* [53,72]. Vaidyanathan and colleagues reported the direct ESI-MS analyses of whole bacterial cells without prior separation [18]. Even though ESI-MS has been used less frequently for direct microbial identification, Xiang and colleagues characterized microorganisms by carrying out global ESI–tandem mass spectrometry (MS/MS) analyses of cell lysates [17].

Desorption electrospray ionization (DESI) which is similar to direct analysis in real time (DART) mass spectrometry is an ambient ionization technique that permits the direct analysis of condensed phase samples by spraying them with electrosprayed solvent droplets. This access has been used to differentiate many bacteria species including *E. coli*, *S. aureus*, *Enterococcus* sp., *S. typhimurium* based on their DESI mass spectral profiles. Distinguishable DESI mass spectra, in the mass range 50–500 Da, have been acquired from whole bacteria utilizing both positive and negative ion modes; the analysis time can be approximately 2 min [10].

5.2.1 Examples of ESI-MS Studies in Foodborne Pathogens

Wahl and colleagues reported an analytical method to detect residual agar from a bacterial spore sample as an indication of culturing on an agar plate. The method was based on the resolubilization of agar polysaccharide from a bacterial spore sample, enzymatic digestion, followed by electrospray ionization tandem mass spectrometry (ESI-MS (n)) analysis for detection of a specific agar fragment ion. A range of *Bacillus* species and strains were selected to show the influence of this approach. The results of a proficiency test with 42 blinded samples were introduced demonstrating the benefit of this method with no false positives and only three false negatives for samples that were below the detection level of the method as recorded [74]. Vaidyanathan and colleagues investigated flow-injection electrospray ionization mass spectrometry (FI-ESI-MS) of unfractionated cell-free extracts collected from bacterial cells suspended in a solvent mixture as a fast analytical method for reproducible, high-throughput bacterial identification. Five bacterial strains (two *E. coli*, two *Bacillus* spp., and one *Brevibacillus laterosporus*) were checked. Axenically grown bacterial cells were suspended in an acidic organic solvent and the cell-free extract was gradually injected into a solvent flow stream that was sprayed into the ionization chamber of the ESI-MS. At least three classes of macromolecules, namely phospholipids, glycolipids, and proteins, contributed to the spectral information. The spectra generated involved reproducible information, which was beneficial for differentiation between the bacteria [18]. Xiang and colleagues established a method to select useful biomarkers for the identification of microorganisms based on ESI-ion trap-MS. In this method, crude cell lysates were processed using a dual microdialysis device and then, directly infused into an ion trap-MS. The low ESI flow rate and precursor ion accumulation capability of the ion trap-MS allows high-sensitivity MS/MS analyses. Precursor ions are automatically selected and analyzed using tandem MS (MS/MS) to produce "global" MS/MS surveys and processed to yield two-dimensional MS/MS spectral displays. The unique MS/MS spectral patterns can be utilized to identify mass spectrometric-detected species useful as biomarkers, which then, support a basis for reliable microorganism identification. The results demonstrated the application of this method for the identification of microorganisms, as well as for detection of bacteriophage MS^2 in the presence of a large excess of *E. coli* [17].

5.3 PCR Coupled to Mass Spectrometry Using Electrospray Ionization (PCR/ESI-MS)

PCR-electrospray ionization mass spectrometry (PCR-ESI-MS) has arisen as a technology that is capable of identifying nearly all known human pathogens either from microbial isolates or directly from clinical specimens. Assay primers are crucially designed to target one or more of the broad pathogen categories: bacterial, mycobacterial, fungal, or viral. With broad-range amplification followed by detection of mixed amplicons, the process can identify genetic proof of known and unknown pathogens. This access has advantages over traditional assays that generally target the presence or absence of one or more pathogens with known genetic composition [75]. Identification by PCR-ESI-MS using oligonucleotides is more specific for bacterial groups than for particular species, even if changeable regions are amplified between species and strains. Moreover, there are species-specific oligonucleotides used as primers that target genes for antibiotic resistance or some pathogenic characteristics. Following amplification, amplicons are subjected to mass spectrometry and the pattern acquired is correlated with those in the databases. The capability to identify an organism without earlier knowledge of the Gram, or group of microorganisms is another benefit since the stains are not needed. This technology will be advancing the identification of microorganism in infections, it has some benefits as accuracy and specificity, and one important disadvantage, the cost [7]. Pierce and colleagues together with Ibis Biosciences (a division of Abbott Molecular) developed an analysis to identify the common foodborne pathogens *Salmonella*, *E. coli*, *Shigella*, and *L. monocytogenes* using the Plex-ID biosensor system. This biosensor utilizes ESI-MS to detect the base composition of short PCR amplicons. PCR-MS in general and this assay in particular could reduce the time necessary to detect and identify foodborne pathogenic bacteria. The assay was successful at detecting all *S. enterica* subspecies and some serovars. This assay has the potential to significantly reduce the time necessary to characterize bacteria in food samples and greatly improve the response time for foodborne bacterial outbreaks [76]. *Lactobacillus reuteri* INIA P579 was used for the production and purification of reuterin. The purity of reuterin was assessed by high resolution (HR-ESI-MS) and nuclear magnetic resonance (NMR) spectroscopy. After purification, reuterin concentration obtained was 1.3 M. The inhibitory activity using *E. coli* K-12 as indicator strain was estimated to be 510 AU/ml. Survival curves in tryptic soy broth revealed that reuterin required to inhibit the growth of three *L.* monocytogenes strains was in the range of 2–4 AU/mL. Purified reuterin (10 AU/g) significantly reduced the growth of *L. monocytogenes* in cold-smoked salmon kept under moderate or strong temperature abuse conditions. After 15 days at 8 °C, cold-smoked salmon with added reuterin exhibited *L. monocytogenes* counts 2.0 log cfu/g lower than control smoked salmon with no reuterin added. At 30 °C, reuterin also controlled the growth of the pathogen, with counts 1.4 and 0.9 log cfu/g lower than those observed in control smoked salmon after 24 and 48 h, respectively. This study shows that the addition of purified reuterin might be used as a hurdle technology to improve the safety and extend the shelf-life of lightly preserved seafood products such as cold-smoked salmon [77]. Bovine mastitis, an inflammatory disease of the mammary gland, is one of the most costly diseases affecting the dairy industry. The treatment and prevention of this disease is linked heavily to the use of antibiotics in agriculture and early detection of the primary pathogen is essential to control the disease. Perreten and colleagues analyzed milk samples from cows suffering from mastitis for the presence of pathogens using PCR-ESI-MS and compared with standard culture diagnostic methods. Concurrent identification of the primary mastitis pathogens was obtained for 64% of the tested milk samples, whereas divergent results were obtained for 27% of the samples. The PCR-ESI-MS failed to identify some of the primary pathogens in 18% of the samples, but identified other pathogens as well as microorganisms in samples that were negative by culture. The PCR-ESI-MS identified bacteria to the species level as well as yeasts and molds in samples that contained a mixed bacterial culture (9%). The sensitivity of the PCR-ESI-MS for the most common pathogens ranged from 57.1% to 100% and the specificity ranged from 69.8% to 100% using culture as gold standard. The PCR-ESI-MS also revealed the presence of the methicillin-resistant gene mecA in 16.2% of the milk samples, which correlated with the simultaneous detection of *Staphylococci* including *S. aureus*. Perreten and colleagues demonstrated that PCR-ESI-MS, a more rapid diagnostic platform compared with bacterial culture, has the significant potential to serve as an important screening method in the diagnosis of bovine clinical mastitis and has the capacity to be used in infection control programs for both subclinical and clinical disease [78].

5.4 Bioaerosol Mass Spectrometry

Bioaerosol mass spectrometry (BAMS) has been set up to analyze and identify individual aerosolized microbial particles in real time. BAMS has the capability to promptly detect species-level differences between single cells without the necessity for sample work-up or preconcentration; it provides a high degree of detection specificity and analysis times in the millisecond range. Fergenson and colleagues utilized BAMS for the differentiation of the *Bacillus* spore species *B. thuringiensis* and *Bacillus atrophaeus*. The reagentless BAMS reflectron mass spectra of *Bacillus* spores and vegetative cells have been restricted to signals having masses of less than *m/z* 300. Although BAMS has supported some creation at species-level differentiation, it has been prevented by low sensitivity at high mass. Advanced high-mass sensitivity will extend the fingerprint mass range and enhance the likelihood of detecting more robust species-specific biomarkers [79]. Czerwieniec and colleagues analyzed single vegetative cells and spores of *B. atrophaeus*, utilizing bioaerosol mass spectrometry. Key biomarkers were identified from organisms grown in ^{13}C and ^{15}N isotopically enriched media. Spore spectra include peaks from dicipolinate and amino acids. The results demonstrate that compounds observed in the spectra correspond to material from the spore's core and not the exosporium. Standard compounds and mixtures were analyzed for correlation. The biomarkers for vegetative cells were apparently different from those of the spores, including mainly of phosphate clusters and amino acid fragments [80]. The high-mass analyses of airborne pathogens can be advanced by combining aerosol TOF-MS with MALDI. Commonly, in the BAMS method, the aerosol and matrix vapor are mixed and then transported into a cooling chamber, where the matrix is condensed onto the aerosol. The coated particles are then analyzed by TOF-MS. Van Wuijckhuijse and colleagues established an online system for analyzing proteins and other biological material present in single aerosol particles [81]. Protein aerosol particles having masses of up to 20,000 Da can be detected; this mass range involves those of useful biomarkers for bacteria including *Bacillus* spores [10,81].

5.5 Fractionation Methods Coupled with Mass Spectrometry: GC-MS and Pyrolysis–GC-MS

GC-MS has been applied favorably to metabolite profiling for the characterization of microorganisms. The maximum molecular weight of compounds that can be analyzed by GC-MS is approximately1000 Da. For nonvolatile analytes, conversion to volatile ones through derivatization before GC-MS analysis is frequent. Ishida and colleagues used thermally assisted hydrolysis and methylation-GC (in the presence of tetramethylammonium hydroxide) with MALDI-MS/on-probe sample pretreatment as a complementary means of directly analyzing the bacterial phospholipids in whole cells (*E. coli* K-12), without requiring any monotonous sample pretreatment steps. As a result, the bacterial lipids were absolutely characterized in terms of the acyl groups and the molecular structures by combining the results obtained by thermally-assisted hydrolysis and methylation -gas chromatography (THM-GC) and MALDI-MS [82]. Smilde and colleagues detected a very high number (93%) of commercially available metabolites of the in silico metabolomes of *B. subtilis* and *E. coli*. Such data are becoming more and more abundant, and appropriate tools for fusing these types of data sets are needed. Fusion of metabolomics data leads to an inclusive view on the metabolome of an organism or biological system [83]. GC-MS/MS can improve the detection limit and specificity of the analysis of carbohydrate biomarkers (e.g., muramic acid) in complex environmental and clinical samples. Specifically, this technique includes monitoring of the precursor ion→product ion transition in the multiple selected reaction monitoring (SRM) mode [10]. Gardner and colleagues investigated the use of comprehensive two-dimensional gas chromatography (GCxGC)-TOF-MS with multivariate analysis for the characterization of foodborne pathogen bacteria profiles. The cellular fatty acid profiles of eight strains of *Bacillus*, *Staphylococcus*, and Enterobacteriaceae (*E. coli* and *Salmonella*) were characterized by these. An original pattern method was developed to standardize the raw GCxGC retention data through the use of a chemical indexing mixture. Fatty acid profiles extracted from the templates were correlated by multidimensional scaling and PCA. Variations in the profiles of Gram-positive and Gram-negative bacteria were detected, and a series of heterogeneous mixtures consisting of different fractions (containing one Gram-positive and one Gram-negative bacteria strain) were also discerned from their homogeneous constituents [84]. Pyrolysis–MS is another method often used to gather a pyrolysate fingerprint to classify and identify bacteria. Pyrolysis is a method in which chemical or biochemical compounds are decomposed by heating. Pyrolysis products derived from many classes of compounds including lipids and proteins have all been used for bacterial differentiation [10].

5.6 Capillary Electrophoresis-MS

Capillary electrophoresis (CE)-MS is complementary to classic MS-coupled separation methods in the protein and peptide analysis. CE-MS is promising for the discovery of food pathogen biomarkers since it is able to detect both intact proteins and fragment peptides [10]. As a good example of this technique applications, Hu and colleagues identified bacterial species-specific proteins in bacterial mixtures by CE-MS/MS. Through database searching, they identified peptides that matched proteins associated to *Pseudomonas aeruginosa*, *S. aureus*, and *S. epidermidis*. The selected peptides that they identified accurately from Xcorr values ranking at the top of the search results permit them to identify the corresponding bacterial species present in the sample. The method monitor only the selected marker peptide masses in the proteolytic digests of pure bacterial cell extracts. They have successfully applied this procedure to the identification of various mixtures of the three pathogens. Even minor bacterial species present at a concentration of 1% can be identified with great certainty. This procedure for CE-MS/MS analysis of bacteria-specific marker peptides supports excellent selectivity and high accuracy when identifying bacterial species in complex systems. Moreover, Hu and colleagues have utilized this access to identify *P. aeruginosa* in a saliva sample spiked with *E. coli* and *P. aeruginosa* [85].

5.7 Liquid Chromatography MS

Ho and colleagues examined the effects of sample preparation methods on the detection of bacterial proteins by LC-MS. Several factors influence the protein patterns (i.e., number and masses of proteins). For instance, altering the polarity and pH of the extraction solvent, the number of detected proteins can be controlled. The protein patterns can change even when the total ion chromatogram appears to be similar under the same sample preparation conditions. Besides, Ho and colleagues have tested experimentally the effect of LC-MS-analyzed protein patterns (molecular masses between 2000 and 60,000) on microbial identification by protein databases. This was in contrast to the current database investigation, where only the masses of smaller proteins (<or = 20,000) from direct MALDI-MS were utilized. Despite the differences in protein patterns, results of the searching in the database show that the best matches come from the correct microorganism [86]. In addition, the influence of differences in the protein patterns on bacterial identification using a database has been investigated. Lo and colleagues defined a process utilizing a combination of LC-MS/MS of selected bacteria-specific peptides and database investigation in order to determine the species of bacteria present in a mixture. They identified the proteotypic peptides that were related with specific bacteria by searching protein databases for the LC-MS/MS data. The retention time windows for specific peptide markers were utilized as an extra compulsion in order to analyze peptide markers of many bacterial species in a single LC-selective proteotypic-peptide analysis (SPA). Lo and colleagues carried out LC-MS/MS analyses on

the proteolytic digest of cell extracts and observed only the selected marker peptide ions at given elution time windows. The corresponding bacterial species could be typified when the selected peptides that eluted at expected elution windows were identified truly from the database. At the same time, they also accomplished to identify up to eight bacterial species simultaneously during a single LC-MS/MS process, even through bacteria mixed in several amounts [87]. Moreover, a multidimensional LC-MS method has been utilized for the rapid determination and identification of bacterial proteins [10]. Although the lipid profiles may be correlated to the growth conditions, phospholipids are useful biomarkers for some microorganisms. Phospholipids can be ionized as intact entities for MS and MS/MS analyses. Individual fatty acids present in bacterial cells can be identified from product ion (MS/MS) spectra obtained in the negative ion mode. Effective enhancements in the detection limits for phospholipid analysis have been accessed utilizing ESI-MS and MS/MS [88].

Chenau and colleagues reported an immuno-LC-MS/MS method for the direct (i.e., without prior culture), sensitive, and specific detection of *Y. pestis* in complex food and environmental samples. In the first step, a bottom-up proteomics approach highlighted three relevant protein markers encoded by the *Y. pestis*-specific plasmids pFra (murine toxin) and pPla (plasminogen activator and pesticin). Suitable proteotypic peptides were thoroughly selected to monitor the three protein markers by targeted MS using SRM mode. Immunocapture conditions were optimized for the isolation and concentration of intact bacterial cells from complex samples. The immuno-LC-SRM assay has a limit of detection of 2 × 10(4) cfu/mL in milk or tap water, which compares well with those of state-of-the-art immunoassays [89]. Gram-negative and -positive bacteria elicit different response patterns by the host. The proteomic profiles of milk whey samples from cows naturally infected with *E. coli* or *S. aureus* as compared to whey from healthy cows were determined by one-dimensional, LC-MS/MS, bioinformatics processing, and pathway analyses. Since mammary epithelial cells contribute to immune responses in mammary glands, the genes of selected proteins were measured in mammary epithelial T cells (MAC-T cells) by real-time quantitative PCR (qPCR) after stimulation with heat-inactivated *E. coli* strain P4 and *S. aureus* strain Smith CP bacteria. A total of 173 proteins were identified, including 73 proteins differentially expressed among normal, *E. coli*, and *S. aureus* treatment groups. *E. coli* was more effective at significantly altering the concentration of the affected proteins. The mRNA of 23 proteins out of 24 measured by qPCR was significantly altered in MAC-T cells. Pathway analyses identified top canonical pathways significantly enriched in our samples, the most significant being the acute phase response signaling pathway. Also, top networks of genes with significant associations to identified proteins were identified. Our study has demonstrated a wider proteome profile of *E. coli* and *S. aureus* mastitic milk whey, identified more low abundant defense proteins than reported before, and has linked for the first time identified proteins to several network functions and Biocarta pathways [90].

Intact protein expression profiling has proven to be a powerful tool for bacterial subspecies differentiation. To facilitate typing, epidemiology, and trace-back of *Salmonella* contamination in the food supply, a minimum of serovar level differentiation is required. Subsequent identification and validation of marker proteins is integral to rapid screening development and to determining which proteins are subject to environmental pressure. Bacterial sequencing efforts have expanded the number of sequenced genomes available for single-nucleotide polymorphism (SNP) analyses, but annotation is often missing, start site errors are not uncommon, and the likelihood of expression is not known. McFarland and colleagues showed that the combination of intact protein expression profiles and top-down LC-MS/MS facilitates the identification of proteins that result from expressed serovar specific nonsynonymous SNPs. Combinations of these marker proteins can be used in assays for rapid differentiation of bacteria. LC-MS generated intact protein expression profiles establish which bacterial protein masses differ across samples and can be determined without prior knowledge of the sample. Subsequent top-down LC-MS/MS is used to identify expressed proteins and their posttranslational modifications (PTM), identify serovar specific markers, and validate genomic predicted orthologues as expressed biomarkers [91].

5.8 Surface Enhanced Laser Desorption Ionization-Mass Spectrometry (SELDI-MS)

Thulasiraman and colleagues, utilizing surface-enhanced SELDI ProteinChip technology, developed a fast procedure for the detection, purification, and identification of proteins in bacterial extracts. The effectiveness of this method for monitoring the expression and identification of temperature- and calcium-regulated virulence factors of *Y. pestis* is demonstrated. *Y. pestis* infection of its mammalian host is thought to be accompanied by rapid upregulation of a number of genes following a shift from 26 °C (the temperature of the flea vector) to 37 °C (the temperature of the mammalian host). To model this process, *Y. pestis* cells were grown at 26 and 37 °C in a Ca^{2+}-deficient medium. Through an initial protein profiling of the crude bacterial extract on strong anion exchange and copper affinity, ProteinChip arrays detected five proteins that were upregulated and three proteins that were down-regulated at 37 °C. Two of the proteins predominately expressed at 37 °C were semipurified in less than two days. The two proteins were identified as catalase–peroxidase and Antigen 4. Aside from its speed, a salient feature of the SELDI technique is the microgram amounts of crude sample required for analysis [92]. During *Campylobacter*-related food-borne disease outbreaks, the discrimination of closely related *Campylobacter* is important for epidemiological studies, which search for the routes of geographical spread and ways of transmission. In a pilot study, Kiehntopf and colleagues carried out differentiation of *Campylobacter* at the level of species and even strains. *Campylobacter* type-strains and isolates from various regions were cultured and afterward subjected to physicochemical lysis. Protein lysates were then applied on CM10 and IMAC30 ProteinChip Array surfaces and analyzed using a PCS 4000 SELDI Protein Chip System (Bio-Rad Laboratories). By correlating the spectra from *Campylobacter jejuni*, *Campylobacter coli*, *Campylobacter upsaliensis*, and *Campylobacter lari*, they could monitor 166 and 160 different protein peaks ($p < 0.05$) using CM10 and IMAC30 chips, subsequently. The development of classification trees, comprising two to four of these peaks, permits differentiation of *Campylobacter* species and even strains. Furthermore, species and strains could be adequately separated from each other by hierarchical cluster analysis. In this manner, according to the study that Kiehntopf and colleagues performed, SELDI-TOF-MS is a promising tool to differentiate *Campylobacter* species and even strains. Species/strain-specific ions monitored in addition to well-established markers identified by MALDI-TOF might be valuable for future *Campylobacter*-identifying algorithms [93].

5.9 MS Data Analysis Methods

Data analysis procedures for the discrimination of pathogens can depend on the simplest profiling of intact proteins or on the complicated statistical treatment of MS data. MALDI-MS analysis of whole intact cells has been utilized for protein profiling of waterborne pathogen *Aeromonas* [34]. MALDI-MS analysis is demonstrated to be a powerful tool especially suited for environmental monitoring and detection of microbial hazards in drinking water. Donohue and colleagues created a library of over 45 unique m/z signatures from 40 strains that were representative of the 17 recognized species of *Aeromonas*, as well as 3 reference strains from genus *Vibrio* and 2 reference strains from *Plesiomonas shigelloides*. The library was utilized to help speciate 52 isolates of *Aeromonas* [94]. This approach has several beneficial characteristics: the capability to detect intact biomarkers (e.g., proteins), simple sample preparation, and high throughput. The disadvantage of this technique is spectral repeatability, which is distinctly affected by differences in culture conditions and analytical parameters. Because of the complex nature of mass spectral data, many statistical algorithms have been established to couple the analyzed spectra to reference spectra or to produce biomarker spectra for microbial discrimination. A biomarker-selective algorithm has been utilized to extract key signals from spectra; the created biomarker library has been utilized to identify bacterial mixtures. The identification process included correlation of the sample spectra with the library spectra using an algorithm corresponding to a statistical test of importance. The two mathematical procedures used most frequently for the classification of spectral data are multivariate analysis (MVA) and the artificial neural network (ANN). MVA is based on multivariate statistics, which includes observation and analysis of more than one statistical variable (in this case, *m/z*) at a time. Many MVA procedures involving principal constituent analysis and cluster analysis have been applied to the discrimination of microorganisms. The ANN is an algorithm for machine learning. This type of artificial intelligence permits the identification of biomarkers, which are capable of discriminating microbial species. Multivariate statistics is frequently utilized to analyze pyrolysis mass spectra of bacterial samples. Most commonly, PCA is the utilized multivariate data analysis tool. PCA is a method for forming additional variables (i.e., principal constituents) which are linear combinations of the original variables (i.e., *m/z* values in mass spectra). The principal constituents gave reason for the discriminations among spectra. Computer algorithms must be utilized in defining useful biomarkers for bacterial identification [10]. Demirev and colleagues developed a method for rapid microorganism identification that combines MS with protein database searches. The method is based on determining the masses of a set of ions by MALDI-TOF-MS of intact or treated cells. Consecutive correlation of each ion in the set to a protein, along with the organismic source of the protein, was performed by searching an Internet-accessible protein database. The procedure has been favorably performed on *B. subtilis* and *E. coli*, two organisms with entirely sequenced genomes [95]. Probable organism sources are assigned to each mass, with the best match, theoretically, leading to identification of the correct microorganism. The sequence information of the peptides from virus and bacterial samples has been used to identify the proteins and figure out the microorganism's source in order to avoid the restricted mass accuracy of protein mass measurements [10]. Warscheid and colleagues developed a new ion trap time-of-flight hybrid mass spectrometer (qIT-TOF-MS) and applied it to peptide sequencing in proteolytic digests provided from spore mixtures of *Bacilli*. The sequence information can be acquired from fragmentation of intact proteins or proteolytic peptides of the intact proteins by investigating fragment ion (MS/MS) mass spectra against proteome databases [96]. Top-down proteomics methodology defines intact proteins without biomarker digestion, separation, and cleanup requirement. This procedure has been used favorably for microbial proteomics in the analysis of bacterial species, either pure or in a mixture [97,98]. Wynne and colleagues demonstrated that this rapid proteomics-based access to phylogenetic characterization generates similar results to the traditional techniques, and may even be applied to target organisms of undetermined taxonomy [98]. Microbial identification has been demonstrated utilizing experimentally derived databases of virus-specific tryptic peptide masses and bacterial protein masses [99]. Data analysis utilizing protein database search algorithms should be very precise and beneficial as long as the protein databases are appropriate for the pathogens of interest. Several statistical methods have been employed for discrimination among species or subspecies and proved to be beneficial [10].

6. CONCLUSIONS AND FUTURE ASPECTS

In recent years, MALDI-TOF-MS has revolutionized routine identification of bacteria. Food-spoilage bacteria, foodborne pathogens, starter cultures, and probiotics must be precisely identified. Thus, fermentation failures can be avoided and routes of contamination be revealed in order to prevent recontamination and to improve the production plants hygiene. The need to develop identification methods to get taxonomic resolution and sensitivity discriminates significantly depending on the task. Related with the legal supplies and the type of food analyzed, highly sensitive detection methods are needed for the detection of hygiene indicator bacteria. Presently, traditional culture-based methods are still utilized to identify the most of food-associated bacteria in the daily routine of food microbiology laboratories. Complete identification requires at least two days, or even more for fastidious organisms. Furthermore, discrimination of isolates with a different taxonomic background but identical physiological characteristics is restricted utilizing these phenotypic procedures. As a result, traditional phenotypic methods are—to some extent—complemented with serological and genotyping methods. Usually, serological methods do not necessitate pure cultures, but have a low level of differentiation power and are limited to species for which antisera are appropriate. On the contrary, genotyping methods allow swift identification and are highly sensitive and specific, but they still need high levels of technical expertise and remain expensive. As a result, they are not applicable for routine identification. MALDI-TOF-MS, also achieves rapid identification of bacteria. Differently, to genotyping methods, it can be efficient in routine analysis. All types of food-associated microorganisms can be identified with the same protocol. Reliable data are obtained inside a total turnaround time of 16 h. Because of short reversal times, low sample volume needed and low reagent costs, MALDI-TOF-MS has recently become a powerful tool for the identification of clinical isolates. The relevance of MALDI-TOF-MS to food-associated bacteria, and taking the specific needs of food microbiological laboratories and the food industry into account, should be investigated widely [100]. The methodology of MALDI-TOF-MS fingerprinting can be applied to various foodstuffs, since bacterial strains are isolated from the corresponding matrix before analysis and the constituted spectral library may simply be extended by adding bacterial species and strains that are of interest in the corresponding food product. The broad spectral data can be efficiently examined by searching common peak masses and by cluster analysis. Typical peak patterns can be determined and serve as biomarker proteins for the swift identification of an unknown spectrum. The generation of a dendrogram based on phyloproteomic relations permits its use for the typing of similar strains, extending conventional typing methods. While correlating phyloproteomic clustering with the classical taxonomy access depends on 16S rRNA gene analysis, high similarities were monitored in the grouping of bacterial strains. Nevertheless, genetic analysis does not always allow discrimination among species of the same genus, depending on the high similarity of DNA sequences, since spectral profiles obtained by MALDI-TOF analysis can show specificities for individual bacterial species. In this manner, proteomic tools and phyloproteomic clustering resulted in improved species differentiation ability. A disadvantage of MALDI-TOF-MS fingerprinting for bacterial identification in food is the restricted amount of reference data. In future, the leading goal should be the publication of public spectral libraries, involving strains acquired from culture collections, as well as strains isolated from food samples, to let other researchers correlate their spectra to the library and, in this manner, allow a more precise and rapid identification of bacteria in food [6].

Proteomics is a promising tool for quality control in the future. There are some studies about this subject. As an example, Klein and colleagues utilized proteomics as a quality control tool of pharmaceutical probiotic bacterial lysate products. Probiotic bacteria have a broad range of applications in veterinary and human therapeutics. Inactivated probiotics are complex samples and quality control (QC) should measure as many molecular features as imaginable. For the identification and validation of biomarkers of disease in complex biological samples such as biofluids, capillary electrophoresis coupled to mass spectrometry (CE/MS) has been utilized as a multidimensional and high-throughput method. In this study, Klein and colleagues assessed the appropriateness of CE/MS to measure the regularity of different lots of the probiotic formulation Pro-Symbioflor, which is a bacterial lysate of heat-inactivated *E. coli* and *Enterococcus faecalis*. Over 5000 peptides were detected by CE/MS in five various lots of the bacterial lysate and in a sample of culture medium. 71 to 75% of the total peptide essence was similar in all lots. Sequence analysis identified peptides of both *E. coli* and *E. faecalis* and peptides derived from the culture medium, thus affirming the presence of the strains in the formulation. Ontology analysis indicated that most of the peptides identified for *E. coli* derived from the cell membrane or the fimbrium, even though peptides identified for *E. faecalis* were improved for peptides originating from the cytoplasm. The bacterial lysate peptides as a whole are identified as highly conserved molecular patterns by the innate immune system as a microbe-associated molecular pattern (MAMP). Sequence analysis also identified the presence of soybean, yeast, and casein protein fragments that partly constitute the culture medium [101].

REFERENCES

[1] C.K. Fagerquist, In Rapid Characterization of Microorganisms by Mass Spectrometry; ACS Symposium Series, American Chemical Society, Washington DC, 2011.
[2] C.D.W. Blackburn, Food Spoilage Microorganisms, Woodhead Publishing, 2006.
[3] L. Gram, L. Ravn, M. Rasch, J.B. Bruhn, A.B. Christensen, M. Givskov, Int. J. Food Microbiol. 78 (1–2) (2002) 79–97.
[4] WHO, Food Safety and Foodborne Illness, Fact Sheet Number 237, World Health Organization, 2007.
[5] R.V. Tauxe, Int. J. Food Microbiol. 78 (1–2) (2002) 31–41.
[6] See http://www.intechopen.com/books/food-quality/species-identification-of-food-spoilage-and-pathogenic-bacteria-by-maldi-tof-mass-fingerprinting.
[7] See http://www.intechopen.com/books/common-eye-infections/diagnostics-methods-in-ocular-infections-from-microorganism-culture-to-molecular-methods.

[8] H. Gu, P.-L. Ho, K.W.T. Tsang, L. Wang, B. Xu, J. Am. Chem. Soc. 125 (2003) 15702–15703.
[9] J.C. Liu, P.J. Tsai, Y.C. Lee, Y.C. Chen, Anal. Chem. 80 (2008) 5425–5432.
[10] Y.P. Ho, M. Reddy, Clin. Chem. 56 (4) (2010) 525–536.
[11] R. Sampath, K.L. Russell, C. Massire, M.W. Eshoo, V. Harpin, L.B. Blyn, PloS One 2 (2007) e489.
[12] X. Iriart, R.A. Lavergne, J. Fillaux, A. Valentin, J.F. Magnaval, A. Berry, S. Cassaing, J. Clin. Microbiol. 50 (6) (2012) 2107–2110.
[13] M. Walker, Proteomics 11 (2011) 3143–3153.
[14] See http://en.wikibooks.org/wiki/Proteomics/Protein_Identification_-_Mass_Spectrometry/Types_Mass_Spectrometry.
[15] C.S. Ho, M.H.M. Chan, R.C.K. Cheung, L.K. Law, L.C.W. Lit, K.F. Ng, M.W.M. Suen, H.L. Tai, Clin. Biochem. Rev. 24 (1) (2003) 3–12.
[16] J.J. Pitt, Clin. Biochem. Rev. 30 (1) (2009) 19–34.
[17] F. Xiang, G.A. Anderson, T.D. Veenstra, M.S. Lipton, R.D. Smith, Anal. Chem. 72 (11) (2000) 2475–2481.
[18] S. Vaidyanathan, D.B. Kell, R. Goodacre, J. Am. Soc. Mass Spectrom. 13 (2) (2002) 118–128.
[19] B.L.M. van Baar, FEMS Microbiol. Rev. 24 (2000) 193–219.
[20] P.A. Demirev, A.B. Feldman, J.S. Lin, Johns Hopkins APL Tech. Dig. 25 (1) (2004) 27–37.
[21] C.K. Fagerquist, W.G. Miller, L.A. Harden, A.H. Bates, W.H. Vensel, G. Wang, R.E. Mandrell, Anal. Chem. 77 (15) (2005) 4897–4907.
[22] D. Dare, in: Y.-W. Tang, C.W. Stratton (Eds.), Advanced Techniques in Diagnostic Microbiology, Springer Science and Business Media, LLC, New York, 2006, pp. 117–133.
[23] R. Giebel, C. Worden, S.M. Rust, G.T. Kleinheinz, M. Robbins, T.R. Sandrin, I.L. Allen, S. Sima, M.G. Geoffrey, Academic Press, Adv. Appl. Microbiol. 71 (2010) 149–184.
[24] M.F. Mazzeo, A. Sorrentino, M. Gaita, G. Cacace, M. Di Stasio, A. Facchiano, G. Comi, A. Malorni, R.A. Siciliano, Appl. Environ. Microbiol. 72 (2) (2006) 1180–1189.
[25] K. Bernardo, N. Pakulat, M. Macht, O. Krut, H. Seifert, S. Fleer, F. Hünger, M. Krönke, Proteomics 2 (6) (2002) 747–753.
[26] D.S. Wunschel, E.A. Hill, J.S. McLean, K. Jarman, Y.A. Gorby, N. Valentine, K.J. Wahl, Microbiol. Methods 62 (3) (2005) 259–271.
[27] J.J. Bright, M.A. Claydon, M. Soufian, D.B. Gordon, J. Microbiol. Methods 48 (2–3) (2002) 127–138.
[28] C.J. Keys, D.J. Dare, H. Sutton, G. Wells, M. Lunt, T. McKenna, M. McDowall, H.N. Shah, Infect. Genet. Evol. 4 (3) (2004) 221–242.
[29] K. Böhme, I.C. Fernández-No, J. Barros-Velázquez, J.M. Gallardo, B. Cañas, P. Calo-Mata, Anal. Methods 2 (12) (2010) 1941–1947.
[30] S.B. Barbuddhe, T. Maier, G. Schwarz, M. Kostrzewa, H. Hof, E. Domann, T. Chakraborty T. Hain, Appl. Environ. Microbiol. 74 (17) (2008) 5402–5407.
[31] K.H. Jarman, D.S. Daly, K.K. Anderson, K.L. Wahl, Chemom. Intell. Lab. Syst. 69 (1–2) (2003) 61–76.
[32] P. Seng, M. Drancourt, F. Gouriet, B. La Scola, P.E. Fournier, J.M. Rolain, D. Raoult, Clin. Infect. Dis. 49 (4) (2009) 543–551.
[33] R.E. Mandrell, L.A. Harden, A. Bates, W.G. Miller, W.F. Haddon, C.K. Fagerquist, Appl. Environ. Microbiol. 71 (10) (2005) 6292–6307.
[34] M.J. Donohue, A.W. Smallwood, S. Pfaller, M. Rodgers, J.A. Shoemaker, J. Microbiol. Methods 65 (3) (2006) 380–389.
[35] E. Carbonnelle, J.-L. Beretti, S. Cottyn, G. Quesne, P. Berche, X. Nassif, A. Ferroni, J. Clin. Microbiol. 45 (7) (2007) 2156–2161.

[36] A. Grosse-Herrenthey, T. Maier, F. Gessler, R. Schaumann, H. Böhnel, M. Kostrzewa, M. Krüger, Anaerobe 14 (4) (2008) 242–249.

[37] T.H. Hazen, R.J. Martinez, Y. Chen, P.C. Lafon, N.M. Garrett, M.B. Parsons, C.A. Bopp, M.C. Sullards, P.A. Sobecky, Appl. Environ. Microbiol. 75 (21) (2009) 6745–6756.

[38] M. Alispahic, K. Hummel, D. Jandreski-Cvetkovic, K. Nöbauer, E. Razzazi-Fazeli, M. Hess, C. Hess, J. Med. Microbiol. 59 (3) (2010) 295–301.

[39] S. Ayyadurai, C. Flaudrops, D. Raoult, M. Drancourt, BMC Microbiol. 10 (1) (2010) 285.

[40] D. Dubois, D. Leyssene, J.P. Chacornac, M. Kostrzewa, P.O. Schmit, R. Talon, R. Bonnet, J. Delmas, J. Clin. Microbiol. 48 (3) (2010) 941–945.

[41] R. Stephan, D. Ziegler, V. Pflüger, G. Vogel, A. Lehner, J Clin. Microbiol. 48 (8) (2010) 2846–2851.

[42] R. Stephan, N. Cernela, D. Ziegler, V. Pflüger, M. Tonolla, D. Ravasi, M. Fredriksson-Ahomaa, H. Hächler, J. Microbiol. Methods 87 (2) (2011) 150–153.

[43] K. Böhme, I.C. Fernández-No, J. Barros-Velázquez, J.M. Gallardo, P. Calo-Mata, B. Cañas, J. Proteome Res. 9 (6) (2010) 3169–3183.

[44] I.C. Fernández-No, K. Böhme, J.M. Gallardo, J. Barros-Velázquez, C. Cañas, P. Calo-Mata, Electrophoresis 31 (6) (2010) 1116–1127.

[45] R. Kolinska, M. Drevinek, V. Jakubu, H. Zemlickova, Folia Microbiol. (Praha) 53 (5) (2008) 403–409.

[46] M.A. Winkler, J. Uher, S. Cepa, Anal. Chem. 71 (16) (1999) 3416–3419.

[47] R. Dieckmann, R. Helmuth, M. Erhard, B. Malorny, Appl. Environ. Microbiol. 74 (24) (2008) 7767–7778.

[48] R.D. Holland, F. Rafii, T.M. Heinze, J.B. Sutherland, K.J. Voorhees, J.O. Lay Jr., Rapid Commun. Mass Spectrom. 14 (10) (2000) 911–917.

[49] M. Erhard, U.C. Hipler, A. Burmester, A.A. Brakhage, J. Wöstemeyer, Exp. Dermatol. 17 (4) (2008) 356–361.

[50] A. Bizzini, C. Durussel, J. Bille, G. Greub, G. Prod'hom, J. Clin. Microbiol. 48 (5) (2010) 1549–1554.

[51] I. van den Broek, M. Blokland, M.A. Nessen, S. Sterk. See http://wileyonlinelibrary.com, http://dx.doi.org/10.1002/mas.21419.

[52] K. Böhme, I. Fernández-No, J. Gallardo, B. Cañas, P. Calo-Mata, Food Bioprocess Technol. 4 (6) (2011) 907–918.

[53] S. Brul, P.M. Fratamico, T.A. McMeekin (Eds.), Tracing Pathogens in the Food Chain, Woodhead Publishing Series in Food Science, Technology and Nutrition, Philadelphia, USA, 2011.

[54] R. Alm, P. Johansson, K. Hjerno, C. Emanuelsson, M. Ringnér, J. Häkkinen, J. Proteome Res. 5 (4) (2006) 785–792.

[55] K. De Bruyne, B. Slabbinck, W. Waegeman, P. Vauterin, B. De Baets, P. Vandamme, Syst. Appl. Microbiol. 34 (1) (2011) 20–29.

[56] G.C. Conway, S.C. Smole, D.A. Sarracino, R.D. Arbeit, P.E. Leopold, J. Mol. Microbiol. Biotechnol. 3 (1) (2001) 103–112.

[57] P.M. Murray, J. Mol. Diagn. 14 (2012) 419–423.

[58] T. Krishnamurthy, P.L. Ross, U. Rajamani, Rapid Commun. Mass Spectrom. 10 (8) (1996) 883–888.

[59] P. Lasch, W. Beyer, H. Nattermann, M. Stämmler, E. Siegbrecht, R. Grunow, D. Naumann, Appl. Environ. Microbiol. 75 (22) (2009) 7229–7242.

[60] P.A. Demirev, J.S. Links, F.J. Pineda, C. Fenselau, Anal. Chem. 73 (19) (2001) 4566–4573.

[61] K.H. Jarman, S.T. Cebula, A.J. Saenz, C.E. Petersen, N.B. Valentine, M.T. Kingsley, K.L. Wahl, Anal. Chem. 72 (2000) 1217–1223.

[62] R.G. Leuschner, N. Beresford-Jones, C. Robinson, Lett. Appl. Microbiol. 38 (1) (2004) 24–31.

[63] H. Huber, D. Ziegler, V. Pflüger, G. Vogel, C. Zweifel, R. Stephan, BMC Vet. Res. 7 (6) (2011). http://dx.doi.org/10.1186/1746-6148-7-6.

[64] T.J. Siegrist, P.D. Anderson, W.H. Huen, G.T. Kleinheinz, C.M. McDermott, T.R. Sandrin, J. Microbiol. Methods 68 (3) (2007) 554–562.

[65] S. Jadhav, D. Sevior, M. Bhave, E.A. Palombo, J. Proteomics 31 (97) (2014) 100–106.

[66] C.K. Fagerquist, B.R. Garbus, W.G. Miller, K.E. Williams, E. Yee, A.H. Bates, S. Boyle, L.A. Harden, M.B. Cooley, R.E. Mandrell, Anal. Chem. 82 (7) (2010) 2717–2725.

[67] K. Böhme, I.C. Fernández-No, J. Barros-Velázquez, J.M. Gallardo, B. Cañas, P. Calo-Mata, Electrophoresis 32 (21) (2011b) 2951–2965.

[68] M.T. Jay-Russel, A. Bates, L. Harden, W.G. Miller, R.E. Mandrell, Zoonoses Public Health 59 (5) (2012) 314–319.

[69] Y. Lu, Y. Chen, X.A. Lu, J. Lv, C.X. Man, Y.L. Chai, Y.J. Jiang, J. Dairy Sci. 97 (2) (2014) 632–641.

[70] L. Wu, S. Wang, X. Chen, X. Wang, L. Wu, X. Zu, Y. Chen, PLoS One 8 (7) (2013) e70295. http://dx.doi.org/10.1371/journal.pone.0070295.

[71] P. Attien, H. Sina, W. Moussaoui, G. Zimmermann-Meisse, T. Dadié, D. Keller, P. Riegel, V. Edoh, S.O. Kotchoni, M. Djè, G. Prévost, L. Baba-Moussa, Biomed. Res. Int. (2014). http://dx.doi.org/10.1155/2014/485620.

[72] R.A. Everley, T.M. Mott, S.A. Wyatt, D.M. Toney, T.R. Croley, J. Am. Soc. Mass Spectrom. 19 (11) (2008) 1621–1628.

[73] S.M. Twine, C.J. Paul, E. Vinogradov, D.J. McNally, J.R. Brisson, J.A. Mullen, D.R. McMullin, H.C. Jarrell, J.W. Kelly, S.M. Logan, FEBS J. 275 (17) (2008) 4428–4444.

[74] K.L. Wahl, H.A. Colburn, D.S. Wunschel, C.E. Petersen, K.H. Jarman, N.B. Valentine, Anal. Chem. 82 (4) (2010) 1200–1206.

[75] D.M. Wolk, E.J. Kaşeta, V.H. Wysocki, J. Mol. Diagn. 14 (4) (2012) 295–304.

[76] S.E. Pierce, R.L. Bell, R.S. Hellberg, C.M. Cheng, K.S. Chen, D.M. Williams-Hill, W.B. Martin, M.W. Allard, Appl. Environ. Microbiol. 78 (23) (2012) 8403–8411.

[77] R. Montiel, I. Martín-Cabrejas, S. Langa, N. El Aouad, J.L. Arqués, F. Reyes, M. Medina, Food Microbiol. 44 (2014) 1–5.

[78] V. Perreten, A. Endimiani, A. Thomann, J.R. Wipf, A. Rossano, M. Bodmer, A. Raemy, K.A. Sannes-Lowery, D.J. Ecker, R. Sampath, R.A. Bonomo, J. Dairy Sci. 96 (6) (2013) 3611–3620.

[79] D.P. Fergenson, M.E. Pitesky, H.J. Tobias, P.T. Steele, G.A. Czerwieniec, S.C. Russel, Anal. Chem. 76 (2004) 373–378.

[80] G.A. Czerwieniec, S.C. Russel, H.J. Tobias, M.E. Pitesky, D.P. Fergenson, P. Steele, A. Srivastava, J.M. Horn, M. Frank, E.E. Gard, C.B. Lebrilla, Anal. Chem. 76 (2004) 373–378.

[81] A.L. van Wujckhuijse, M.A. Stowers, W.A. Kleefsman, B.L.M. van Baar, ChE. Kientz, J.C.M. Marijnissen, J. Aerosol Sci. 36 (5–6) (2005) 677–687.

[82] Y. Ishida, K. Kitagawa, A. Nakayama, H. Ohtani, J. Anal. Appl. Pyrolysis 77 (2006) 116–120.

[83] A.K. Smilde, M.J. van der Werf, S. Bijlsma, B.J. van der Werff-van der Vat, R.H. Jellema, Anal. Chem. 77 (20) (2005) 6729–6736.

[84] J.Y. Gardner, D.E. Brillhart, M.M. Benjamin, L.G. Dixon, L.M. Mitchell, J.M. Dimandja, J. Sep. Sci. 34 (2) (2011) 176–185.

[85] A. Hu, P.S. Tsai, Y.P. Ho, Anal. Chem. 77 (5) (2005) 1488–1495.

[86] Y.P. Ho, P.H. Hsu, J. Chromatogr. A 976 (1–2) (2002) 103–111.

[87] A.A. Lo, A. Hu, Y.P. Ho, J. Mass Spectrom. 41 (8) (2006) 1049–1060.

[88] M.K. Moe, T. Anderssen, M.B. Strom, E. Jensen, J. Am. Soc. Mass Spectrom. 16 (1) (2005) 46–59.
[89] J. Chenau, F. Fenaille, S. Simon, S. Filali, H. Volland, C. Junot, E. Carniel, F. Becher, Anal. Chem. 86 (12) (2014) 6144–6152.
[90] E.M. Ibeagha-Awemu, A.E. Ibeagha, S. Messier, X. Zhao, J. Proteome Res. 9 (9) (2010) 4604–4619.
[91] M.A. McFarland, D. Andrzejewski, S.M. Musser, J.H. Callahan, Anal. Chem. 86 (14) (2014) 6879–6886.
[92] V. Thulasiraman, S.L. McCutchen-Maloney, V.L. Motin, E. Garcia, Biotechniques 30 (2) (2001) 428–432.
[93] M. Kiehntopf, F. Melcher, I. Hanel, H. Eladawy, H. Tomaso, Foodborne Pathog. Dis. 8 (2011) 875–885.
[94] M.J. Donohue, J.M. Best, A.W. Smallwood, M. Kostich, M. Rodgers, J.A. Shoemaker, Anal. Chem. 79 (5) (2007) 1939–1946.
[95] P.A. Demirev, Y.P. Ho, V. Ryzhov, C. Fenselau, Anal. Chem. 71 (14) (1999) 2732–2738.
[96] B. Warscheid, K. Jackson, C. Sutton, C. Fenselau, Anal. Chem. 75 (20) (2003) 5608–5617.
[97] P.A. Demirev, A.B. Feldman, P. Kowalski, J.S. Lin, Anal. Chem. 77 (22) (2005) 7455–7461.
[98] C. Wynne, C. Fenselau, P.A. Demirev, N. Edwards, Anal. Chem. 81 (23) (2009) 9633–9642.
[99] Z.P. Yao, P.A. Demirev, C. Fenselau, Anal. Chem. 74 (11) (2002) 2529–2534.
[100] M. Pavlovic, I. Huber, R. Konrad, U. Busch, Open Microbiol. J. 7 (2013) 135–141.
[101] G. Klein, J.P. Schanstra, J. Hoffmann, H. Mischak, J. Siwy, K. Zimmermann, PloS One 8 (6) (2013) e66682.

Chapter 13

Foodomics

Koichi Inoue and Toshimasa Toyo'oka*
Laboratory of Analytical and Bio-Analytical Chemistry, School of Pharmaceutical Sciences, University of Shizuoka, Suruga-ku, Shizuoka, Japan
**Corresponding author: E-mail: toyooka@u-shizuoka-ken.ac.jp*

Chapter Outline

1. Introduction 654
2. Definition of Foodomics 654
3. General Flowchart on Foodomics of MS 655
4. Food Matrices 656
4.1 Food Metabolites Related to Genes 657
4.2 Food Metabolites Related to the Impact of Stress 658
4.3 Food Metabolites Related to Nutritional Status 658
4.4 Food Metabolites Related to Quality Grade 659
4.5 Food Metabolites Related to Environmental Conditions 659
5. Sample Preparations 660
5.1 Liquid–Liquid Extraction, Solid-Phase Extraction, and Dispersive Solid-Phase Extraction 660
5.2 SPME and Similar Methods 661
5.3 Centrifugal Ultrafiltration 662
6. Separation 663
6.1 GC Techniques 664
6.2 LC Techniques 664
6.3 CE Techniques 666
6.4 Others Separating Techniques 667
7. MS Detection 669
7.1 MALDI-MS Techniques 669
7.2 Direct MS Techniques 670
7.3 MS Combined with Separation Techniques 670
8. Data Analysis 671
9. Data Assessment 674
10. Advanced Future Foodomics Approach with MS 676
10.1 Advanced Foodomics Approach for Human Response by Foods 676
10.2 Advanced Foodomics Approach for Food Microbiology 677
10.3 Advanced Foodomics Approach for Food Safety and Quality 677
List of Abbreviations 679
References 680

Comprehensive Analytical Chemistry, Vol. 68. http://dx.doi.org/10.1016/B978-0-444-63340-8.00013-3

1. INTRODUCTION

Today, ensuring food safety and quality is the most demanding task for food researchers, government, and the food industry. One of the most relevant topics concerning food quality is authentication of the food components. Food safety is a continuing concern not only for the prevention of poor-quality products, but also for the risk derived from pesticides, additives, veterinary medicines, and unknown substances. They can present especially hazardous health risks to humans, including allergenic or toxic effects. Regarding allergens, common foodstuffs such as shrimp, nuts, avocado, wheat, and milk can trigger allergies affecting individual health. Thus, in order to tackle the new food challenges, we expect to face in the future, we must make sure that we directly reduce any possible risk from food.

Based on the evaluation of food safety and quality, various approaches have been discussed to perform the screening of peptides, proteins, oxidized chemicals, additives, natural substances, nutrient factors, contaminants, and others compounds in food. However, in the past, the possibility to carry out these evaluations by nontargeted tactics was a concept for the future compared to the analytical strategy of conventional target screening of chemicals such as pesticides, additives, veterinary medicines, additives, etc. Target analytical methods for the determination of conventional compounds in foods are limited by the wide range of foodstuffs and the large number of substances to determine as well. As a good example of the situation, in 2008, melamine was detected for first time as an adulterant in infant foods [1,2]. Melamine contains a high percentage of nitrogen that can make the protein content of food appear higher than the actual value. It is illegal to artificially enhance protein concentrations by the addition of nitrogen-rich compounds [3]. Later on, the nitrogen-rich dicyandiamide was also detected in infant formula [4]. For food assessment, a broad vision implies not only the determination of already-known compounds, nutrient factors and/or effects, but also the possibility of resolving any unexpected contamination, degradation, expediency, multifunctionality, and undiscovered ability of foods. Little is known about the extent of changes in the nontargeted and exhaustive profiling of foods based on *omics* technologies. In this chapter, a new approach, called "foodomics" that carries out nontargeted and/or exhaustive profiling of foods based on mass spectrometry (MS) for the comprehensive and high-throughput characterization of all aspects of food is described.

2. DEFINITION OF FOODOMICS

In 2009, Cifuentes defined "foodomics" as a discipline that studies food and nutrition through the application and integration of *omics* technologies to improve consumers' well-being, health, and knowledge [5]. This discipline is expected to be not only a useful model to cover in a simple and straightforward way the abovementioned new technologies and issues within food science, but more decisively, a global discipline that comprises all of the

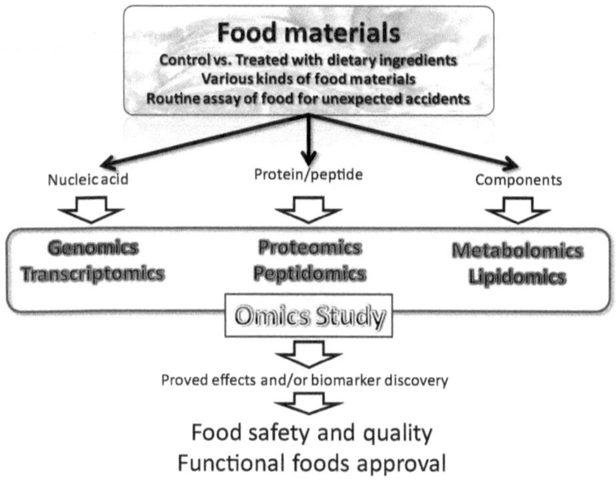

FIGURE 1 Advanced foodomics platform with mass spectrometry.

emerging working areas in which food (including nutrition), advanced analytical techniques (mainly *omics* tools), and bioinformatics are combined using, among others, MS approaches [6]. The ideal foodomics platform is shown in Figure 1 [7]. Foodomics cover a wide range of application from the development of transgenic food to the full characterization of foods using molecular tools such as genomics, transcriptomics, peptidomics, proteomics, lipidomics, and/or metabolomics for fingerprinting, profiling, authenticity, and/or biomarkers classification. Moreover, the subjects are food materials, and the final goal is that the characterization of food features using nontargeted and/or exhaustive analytical techniques. Among the advanced food analysis tools that can be used to develop the necessary *omics* study, MS is already an important tool for proteomics, peptidomics, metabolomics, and lipidomics [6]. Proteomics and peptidomics are the nontargeted large-scale analysis of proteins and/or peptides in an organism, tissue, or cell for food science including food processing and characterization of healthy ingredients described in Chapter 8, "Food proteins and peptides." Thus, in this chapter, we focus on metabolomics for the analysis of nontargeted components (molecular weight entities of approximately <1500 Da).

3. GENERAL FLOWCHART ON FOODOMICS OF MS

The starting point of a foodomic approach for food safety and quality is the complete characterization of food, including the regulation of genes, impact stress, nutritional status, quality grade, and environmental condition. The general flowchart for foodomics with MS shows the selection of the food matrices, sample preparation, separation, MS detection, data analysis, data assessment, and final goal of food safety and quality, as shown in Figure 2. The key point

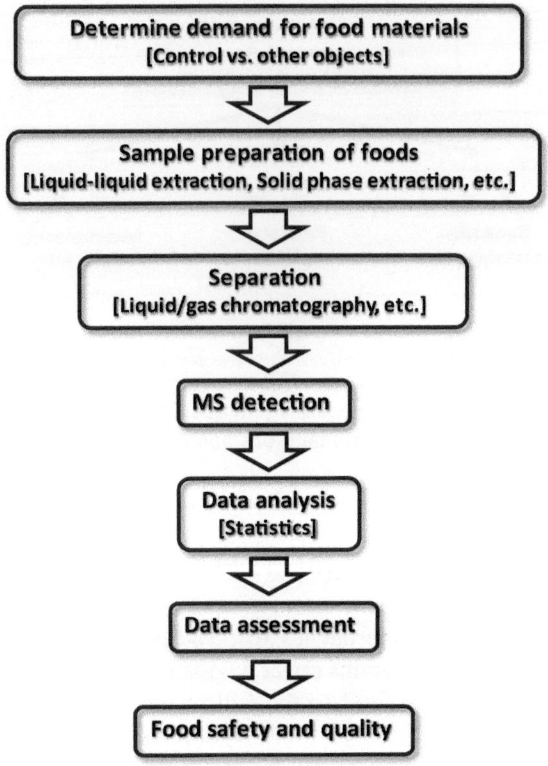

FIGURE 2 General flowchart for the foodomics with mass spectrometry that determine the final goal of food safety and quality.

of foodomics is to achieve valuable goals with the designed flowchart, such as management of the production process, prevention of contamination, and improvement of human nutrition.

Given the structural complexity and concentration diversity of metabolites in any type of food, it is an unviable challenge to measure its whole metabolome without bias. The sample preparation should be generic and should preserve the integrity of the sample metabolome. The use of liquid chromatography (LC), gas chromatography (GC), and capillary electrophoresis (CE) has been combined with an MS detector for the determination of complex food matrices. After acquisition of the MS data, an enormous number of peaks need to be simplified to easy-to-understand results for food safety and quality.

4. FOOD MATRICES

For developing the idea of foodomics, food from a categorical group needs to be selected. For biological studies, the relevant groups to compare are healthy donors and disease patients. On the other hand, the selection of food has a wider range of choices than the metabolomics of biological materials. Thus, foodomics has more to do with the random selection and/or the known destination of food than biological metabolomics. It uses a similar pattern from various foods to evaluate the exhaustive trend of multivariate and comparative statistical analyses. Actually, the transgenic agricultural materials, impact of stress, nutritional status, quality grade, and environmental condition could be simultaneously characterized by MS analyzing hundreds/thousands of metabolites resulting in massive and complex data sets (Table 1) [8].

4.1 Food Metabolites Related to Genes

In the case of genes, recent advances in transgenic agriculture technology have led to concerns about the safety of genetically modified foods for human, animal, and environmental health. The evaluation of the transgenic expression involves the application of integrated genomics, proteomics, transcriptomics, lipidomics, and metabolomics approaches. An analysis of wheat by GC-MS showed that transgenic lines can be distinguished from nontransgenic ones by higher levels of maltose and/or sucrose, and that differences in free amino acids between two types were also apparent [9]. Moreover, genetically modified maize samples and their corresponding nontransgenic parental line grown under identical conditions were analyzed using CE-TOF-MS to identify and quantify

TABLE 1 Background Food Materials for Foodomics and Areas of Application

Background Materials for Foodomics	Anticipated Results	Important Discussions
Gene (cultivar improvement)	Transgenic effects	Environmental and genetic associated perturbations
Impact stress	Response from stress	The time-dependent responses
Nutritional status	Relationship between nutrients and metabolism	Variation of metabolism-induced profiles in health
Quality grade	Taste and/or aroma based on trained specialist's test	The preservation of germplasm-bearing ability and traditional farming/processing
Environmental condition	Evaluation origin or contamination	Rapid detection of bacterial, fermented, metal and unexpected contamination for food safety

the main metabolites in the transgenic organisms [10]. To assess the potential regulatory use of profiling techniques in agricultural biotechnology, metabolomic studies with microarrays and LC-MS were developed for the comparison and quantification of a wide range of metabolites in several tissues from maize varieties [11,12]. These results support and extend previous insights into both environmental and genetic associated perturbations to the metabolome that are not associated with transgenic modification and/or amelioration. All this work shows that high-throughput LC, CE, and GC-MS metabolomics are useful tools for the characterization of transgenic crops. However, researchers will have to take into consideration the impact on the detection and quantitation of a wide range of metabolites on the preliminary design as well as the validation and interpretation of results.

4.2 Food Metabolites Related to the Impact of Stress

In the case of the impact of stress, the reconstruction of metabolic correlation networks has been attempted to evaluate the effectiveness of data mining techniques that apply to comprehensive data sets. A metabolomic study based on LC-MS was carried out for Rambo and Raf tomato cultivars treated with pesticide during a 21-day period [13]. Proteomic and metabolomic studies were used to examine whether the mitochondrial function is altered in soybeans by flooding stress for 2 days [14]. Moreover, the metabolic changes of cultivated and wild soybean samples under salt stress were profiled by GC-MS and LC-MS. Wild soybeans contained higher amounts of disaccharides, sugar alcohols, and acetylated amino acids than cultivated soybeans [15]. In this context, it is increasingly important to understand the time-dependent responses of crops to environmental stress. To support molecular breeding activities, the economic, technical, and statistical feasibility must be assessed using MS methods to evaluate the physiological state under different stress conditions based on time-dependent responses.

4.3 Food Metabolites Related to Nutritional Status

In the case of nutritional status, there has been increased interest in nutritional research using metabolomics and because of the intimate relationship between nutrients and metabolism, there exists a great potential for the use of metabolomics within nutritional research. The metabolite profiling or fingerprinting, which allows the simultaneous monitoring of multiple and dynamic components of biological fluids, may provide metabolic signals indicative of nutritional food intake. For human materials, metabolomics has been successfully applied in pharmacology, toxicology, and medical screening, but nutritional metabolomics is still initiated by minor experimental design. A small number of biomarker for specific food and nutrients has been successfully identified but few targeted and nontargeted methods have been developed. The biomarkers that can achieve a more global characterization of dietary intake have been tested yet. A few reports have described minimal changes of the metabolomics profile in urine and blood samples after cocoa powder consumption and different dietary interventions [16–18]. Basic protocols and/or experimental designs are currently underway in many laboratories to demonstrate the hypothesis that nontargeted and/or exhaustive metabolic fingerprinting using MS can be applied to human samples (blood and urine) to identify dietary lifestyle concerns. These challenges would include the necessity to clearly understand the causes of variation in the metabolic profiling regarding the effects of intestinal microorganisms, aging metabolism, health status, and homeostasis, ultimately resulting in a nutritional status.

4.4 Food Metabolites Related to Quality Grade

In the case of quality grade, the nontargeted and/or exhaustive metabolic fingerprinting using MS was applied to determine the relationship between the sample components and their quality. For example, the evaluation of green tea has traditionally been assessed by highly trained specialists who test it based on the leaves aroma and taste of the brew. Nontargeted metabolomics with MS has been applied to explore the correlation between the quality of green tea and its metabolic fingerprinting regarding the chemical constituents [19,20]. Interestingly, metabolomics using LC-MS and GC-MS reveals chemical differences in the shade grown green tea (tencha) that make it high umami and less astringent [21]. Moreover, a recent metabolomics approach indicated those metabolites having a substantial impact on the quality brands of green tea [22]. Other important application of food metabolomics is for the evaluation of quality through the aromatic profiling of various fruits, such as grape, apple, strawberry, avocado, and melon [23–28]. The metabolic fingerprinting regarding the chemical constituents can reveal specific trends by the principal component analysis (PCA), whereas a correlation analysis demonstrated that specific metabolites correlate directly with the quality traits such as antioxidant activity, total phenols, and total anthocyanin, which are important parameters in the preservation of traditional farming and processing.

4.5 Food Metabolites Related to Environmental Conditions

For the environmental condition, various metabolomics approaches using MS has been applied to investigate the effect of the environment on the variation of many components in agricultural commodities. Discrimination of the origin for traditional herbs, wine, fermented products, and functional foods is important to accurately understand their therapeutic effects, and to appropriately utilize their qualities because different environmental backgrounds can induce diverse metabolic changes from specific plants. Recently, the origin of a plant was differentiated using MS. There are many interesting questions about winemaking that are starting to be answered by its exhaustive metabolic fingerprinting. Recent studies monitored these metabolic chemical fingerprintings and correlate then to the wine sensor property "body," or viscous mouth feeling [29–34]. The impact of the addition of glutathione-enriched inactive dry yeast preparations on the stability of some typical wines stored in an accelerated oxidative background was evaluated using exhaustive CE-MS metabolite fingerprinting [35]. On the other hand, nontargeted metabolomics is a useful approach for the simultaneous analysis of many compounds in herbal products [36]. The geographical origins of *Schisandra chinensis* fruits from Korea and China could be differentiated using metabolomics with GC-MS [37]. The metabolomics applying MS have demonstrated to be a simple and easy to differentiate food sample relating to the environmental conditions. Recently, a generic method to screen for new or unexpected contaminants at ppm levels in orange juice has been developed using LC-MS [38]. For the risk assessment of infant formula, unexpected contamination and degradation were also evaluated by LC-MS [39]. Future studies can probably demonstrated that MS metabolomic approach is useful for the rapid and screening detection of bacteria, biomarkers of fermentation, and several contaminations in various matrices [40–42].

5. SAMPLE PREPARATIONS

Due to the complexity of the food matrices, the first step in most traditional analytical methods used for the determination of compounds in food is the extraction and/or cleanup of the targeted components from the matrix. The foodomics approach is similar to these analytical methods for the targeted components in the food matrices. The sample preparation for pesticides, veterinary medicine, additives, and other compounds has been satisfactorily expanding to develop effective procedures for different foods. For example, several extraction approaches, such as solid-phase extraction (SPE), liquid–liquid extraction (LLE), QuEChERS (acronym of quick, easy, cheap, effective, rugged, and safe), pressurized liquid extraction (PLE), and matrix solid-phase dispersion (MSPD), could be applied [43–47]. For the targeted approach, a variety of sample treatments can be used. However, nontargeted experiment represents a different challenge because a nonselective extraction is required in order to maximize the number of metabolites determined. However, also in nontargeted analysis some selection based on the polar or nonpolar characteristics is applied to choose the extraction methods. The most useful methods for the extraction/cleanup of nontargeted compounds from the unwanted matrix are shown in Figure 3.

5.1 Liquid–Liquid Extraction, Solid-Phase Extraction, and Dispersive Solid-Phase Extraction

LLE is one of the most traditional techniques for polar or moderately polar analytes in food [48]. For instance, in a nontargeted metabolomics

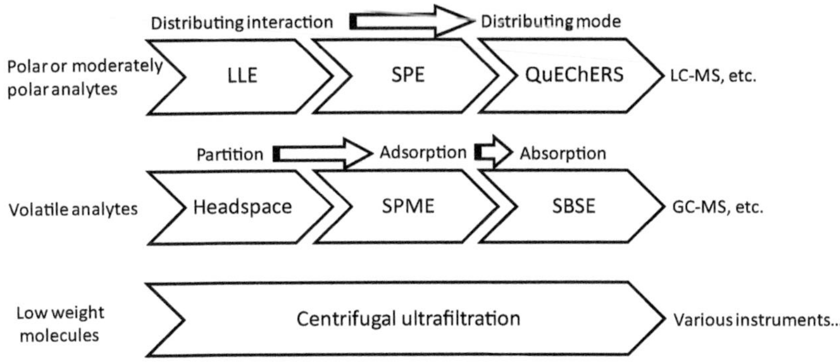

FIGURE 3 Ideal sample preparations for foodomics.

experiment, optimization of solvents, and extraction method is often required to meet the goal of covering as broad a scope of metabolites as possible. Recently, liquid-phase microextraction has increasingly applied for the extraction of both inorganic and organic analytes from different food matrices [49].

SPE appears to be a more adequate method for obtaining accurate and reproducible results from food analysis [50]. Commonsensical C_{18} SPE columns, based on the reversed phase mode that involved a polar or moderately polar sample matrix and a nonpolar stationary phase offer a wide extraction range from hydrophilic to lipophilic compounds. Recently, various phase modes have been obtained. The QuEChERS method was proposed to facilitate the rapid screening of a large number of food and agricultural samples for multipesticide residues [51,52]. The QuEChERS procedure is based on an initial single-phase extraction in a tube for the preparation of solid food samples with acetonitrile. A liquid–liquid partition is carried out by "salting out" adding sodium chlorine and magnesium sulfate. After centrifugation, the acetonitrile layer containing analytes is collected. Thus, an extensive range of QuEChERS are also available in a wide selection of polymer and silica sorbents for the applicable methods of the nontargeted and/or exhaustive analytes based on foodomics.

5.2 SPME and Similar Methods

For GC-MS analysis, current developments in analytical sampling/extracting volatile analytes seem to favor partition and absorption rather than the adsorption concept for efficient and high-recovery results. The traditional headspace technique is most suited to the analysis of the very light volatiles in food samples that can be efficiently partitioned into the headspace gas volume from the liquid or solid matrix sample [53]. Higher boiling volatiles and semivolatiles are not detectable using this technique due to their low partition in the gas headspace. The solid-phase microextraction (SPME) is a relatively renewed sample extraction technique that brings some capabilities to the chromatographic concept for volatile analytes in food matrices [44]. Essentially, SPME has two discrete steps, solute adsorption from the sample matrix into a thick, relative to conventional capillary GC columns, layer of silicone and/or adsorptive material, and transfer of the analytes into a chromatographic instrument. SPME has a significant potential to dramatically reduce solvent consumption and to increase the repeatability and convenience specifications. The SPME coating is selected to have as high distribution constants as possible for the analytes of interest. In fact, four SPME coatings are commercially available; i.e., poly(dimethylsiloxane) (PDMS), poly(dimethylsiloxane/divinylbenzene) (PDMS/DVB), polyacrylate (PA), and Carbowax-templated resin (CW-TPR). The organic PDMS polymers can be easily attached to a sampling device, such as silica fiber, conventional magnetic stirring rod, and injecting syringe, by SPME coupled with GC-MS. One type of SPME is stir bar sorptive extraction (SBSE) using a conventional magnetic stirring rod with PDMS, namely the Twister® system [54]. The main advantage of this process is that it is solvent-free, and is therefore suitable for the detection of low-molecular-weight molecules, generally eluted in the solvent peak [55]. It is possible to change the sorption equilibrium by modulating the pH, temperature, or sodium chloride concentration. Thus, SBSE requires careful optimization and consistent operating conditions for the nontargeted and/or exhaustive volatile low-molecular-weight molecules in the food matrices [56]. Any poorly characterized sampling technique for foodomics has no valid use in analytical laboratories, and the burden of developing these SPME methods included SBSE are no greater than for developing a method for any of the other techniques. SPME has a significant place in multivolatile low-molecular-weight molecules arrays of sample preparation techniques for foodomics with GC-MS.

5.3 Centrifugal Ultrafiltration

Based on the concept from size exclusion, classic centrifugal ultrafiltration is a useful, fast, and easy procedure for the exhaustive sample preparation of various analytes from food materials. Ultrafiltration is a simple process in which a sample solution is filtered through a special filter which only allows passage of molecules of specific molecular weight. Many companies offer a choice of a number of different devices covering sample volumes from 100 μL up to 100 mL with a molecular weight cutoff (<generality less 3000 Da) for separating high-molecular species such as protein, filament, deposition, solid matter, etc., in food materials. Actually, we can prepare the available sample solutions from various matrices such as chicken, fish tissue, and milk based on totally unrelated polar, ionic, and specific species [57]. Ultrafiltration enables applicable coverage of polar metabolites and is recommended as the useful preparation for foodomics with MS. However, ultrafiltration suffers from a significant loss of hydrophobic species that are irreversibly adsorbed on the surface of the filter materials. To improve the metabolic coverage of ultrafiltration, a protein

removal method can also have significant benefits in terms of ensuring the better stability and recovery of exhaustive metabolites from biological samples [58]. Although the overriding requirement is sample preparation in foodomics, it is important to emphasize that the nontargeted MS approach is performed based on an enormous number of ion signals and highly reproducible preparations. The choice of the sample preparation method is very important for the foodomics platform because it affects both the observed food components and indirect markers of the meaningful statistics in the interpretation. An ideal sample preparation method for foodomics with MS should be as nonselective as possible to ensure an adequate and comfortable recovery and cover all components, and be simple, useful, and fast to prevent any nontargeted analyte loss and/or degradation during the reproducible high throughput. Despite its importance, sample preparation is an often downplayed aspect of foodomics. The focus of future studies would be to fix the role, design protocols, and establish trends of sample preparation specifically within the context of global foodomics. Recently, the optimal methods for the sample preparation of a large variety of metabolites in human biological samples using metabolomic approaches are starting to be discussed and investigated [59–65]. Thus, the sample preparation for food materials will be able to establish institutionalized protocols based on these previous reports for biological metabolomics.

6. SEPARATION

For the common analytical technique, separation is needed to provide a sample solution for the MS detection. The most commercial analytical techniques are based on the coupling of the separation concept with chromatography or electrophoresis for the sensitive and selective determination of the nontargeted and/or exhaustive analytes in food materials. Historically, GC was coupled with MS in the 1960s and this proved to be a robust and long-lasting commercial technique for monitoring low-molecular-weight molecules in various samples [66]. Due to advanced technologies in MS, recent years have seen the development of the coupling of LC and/or CE with MS which provides many possible analytical strategies for metabolomics [67,68]. LC-MS is the core technology for the analysis of peptides, proteins, and metabolites in biological samples and it certainly will remain so in the foreseeable future for foodomics [67]. On the other hand, the use of CE, as an alternative separation technique to LC with interfacing to an MS detector, has been characteristically increased in recent years [68]. CE has emerged as a highly efficient and versatile technique, given that it can be used in a number of separation modes such as capillary zone electrophoresis (CZE), capillary gel electrophoresis (CGE), capillary isoelectric focusing (CIEF), capillary electrokinetic chromatography (CEC), micellar electrokinetic chromatography, (MEKC) and isoelectric focusing (IEF) [68]. During the development of separation techniques, the previous papers comparing the advantages and disadvantages of GC-MS, LC-MS, and CE-MS have been extensively discussed for efficient and useful metabolomics. Today, based on the increasing and significant advantage of the MS ability, the exponential development of the last generation of MS coupled to highly efficient and high-performance separation systems has provided better resolution, sensitivity, and reproducibility in a relatively short time. Due to MS's high sensitivity, high mass resolution, and accuracy, it has a directly monitored mass ions with an extremely high scan rate unavailable with any other separation techniques and is ideally suited to studies in metabolomics. On the other hand, the foodomics platform requires to combine several separation techniques prior MS. For this reason, the optimal situation for data analysis is that the majority of detectable signals have independent results with a different eluting time (retention time, migration time, or retention index) to better describe the molecule. As a result multiple, often complementary, analytical technologies are commonly used in metabolomics.

6.1 GC Techniques

In the case of GC separation, GC-MS is the most widely used analytical technique for the metabolomics analysis of volatile organic compounds in foods. Since the targets are volatile organic compounds, the modification of a functional group of a molecule by derivatization is not needed. Thus, nontargeted analytes are readily identified. Databases are presented in simple-to-use tables, summarizing various MS spectral patterns for identified compounds [69]. A recent paper described a method, iMatch2, for compound identification using retention indices in the National Institute of Standard and Technology (NIST11) mass spectra library [70]. It is hard to find studies that use derivatization of the nontargeted analytes in foodomics, however, an easy way to focus on the volatile organic compounds such as aromas in food. For example, the two-dimensional GC-MS method was used to investigate the hop-derived aroma characteristics in beer for the quality of beer [71]. For these separations of analytes, capillary GC columns were conveniently selected regarding the ionic liquid stationary phase, fast or normal chromatography, two-dimensional chromatography, and chiral columns for the food analysis. To maximize the performance for foodomics, an optimized chromatographic separation begins with the selection of the proper column for the nontargeted compounds in food materials.

6.2 LC Techniques

Typically, the most useful and versatile LC-MS method has been applied for the metabolomics approach with various column modes such as reversed phase, hydrophilic interaction, monolithic, and mix modes. A further development is the employ of ultrahigh pressure liquid chromatography (UHPLC) to maintain the chromatographic resolution and peak capacity with shorter elution times than in high-performance LC (HPLC). UHPLC plays an important function in the separation and consequent identification of many analytes in complex food materials, and uses columns packed with sub-2-μm particles with a higher backpressure requirement (>400 bar). The significant development of UHPLC technology has involved a wide variety of stationary phases packed columns with sub-2-μm particles and instruments with maximum pressures ranging between 600 and 1200 bar. Thus, it is realistically possible to speed up the separation by UHPLC compared to the HPLC system. For example, the separation of 12 compounds on a gradient mode by HPLC and UHPLC columns are shown in Figure 4(A) and (B) [72,73]. The analytical time was dramatically reduced

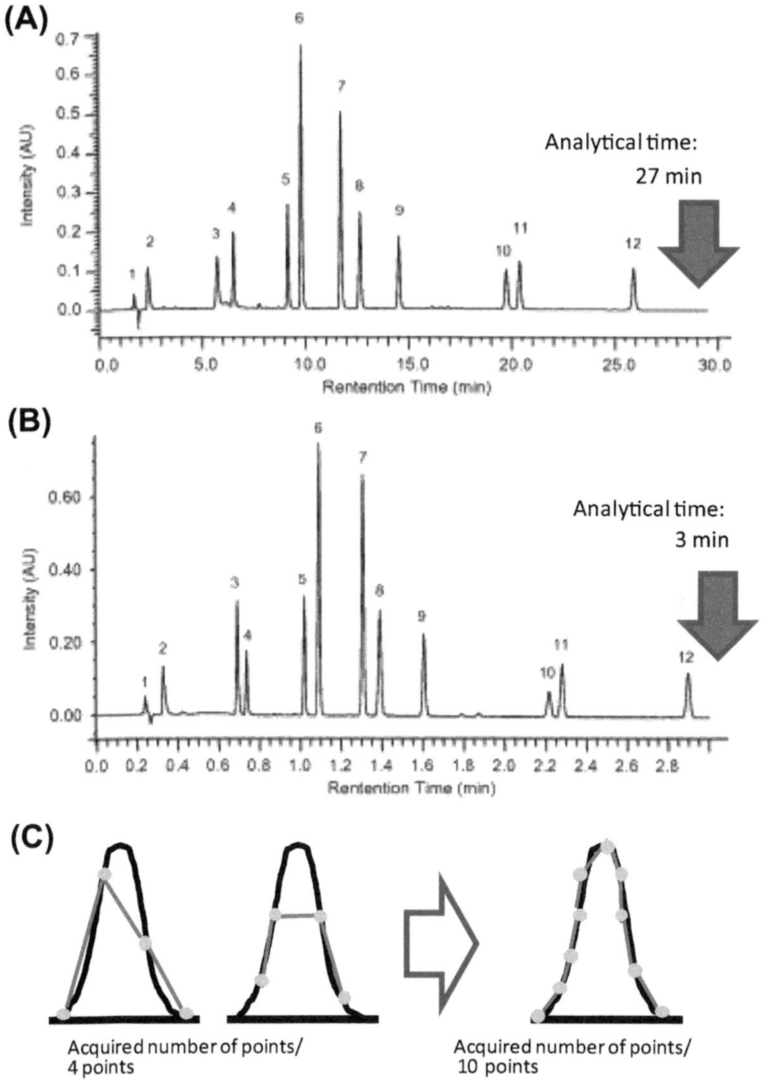

FIGURE 4 UHPLC separation of analytes for MS detection. Separation of a pharmaceutical formulation containing 12 compounds in gradient mode with HPLC (A) and UHPLC (B) systems [60]. MS duty cycle and acquisition rate for MS detection (C).

from HPLC (27 min) by UHPLC (3 min) without any loss in peak capacity and retention of the elution profile in selectivity [73]. However, when a high-speed analysis is conducted using the UHPLC system, the nontargeted peaks entering the conventional MS detector cannot be handled based on the duty cycles, and the acquisition rates reduced the number of points across each peak (Figure 4(C)). The MS instruments expend time to change analysis modes such as positive-/negative-ion switching, monitoring multiple transitions, and scan range of the *m/z* values, which reduces the time available for the real work of acquiring data. On the other hand, recent MS technology using various types of instruments has greatly improved the efficiency, yielding up to a 100-fold gain in sensitivity and full advantage of the higher scan speeds than the normal types. Thus, UHPLC combined with an MS detector can be used for the determination of complex food materials in a short time. Moreover, the Guillarme's review based on these high-resolution LC methods provided useful information for foodomics [74]. Other interesting approach using the LC that could be of aspect in foodomics is the ability to separate chiral analytes. Over the last few years, the reports for chiral metabolomics indicate a growing interest. For instance, a targeted lipidomics approach reported that chiral eicosanoid lipids could be analyzed by LC-MS using a chiral chromatographic column [75]. Moreover, derivatization reagents for chiral metabolomics were synthesized and used to separate the targeted chiral compounds in biological samples [76,77] (Figure 5(A) and (B)). The chiral derivatization for the determination of pesticides in food samples could be achieved using a conventional reversed phase chromatographic column [78] (Figure 5(C)). Thus, future studies would involve chiral foodomics.

6.3 CE Techniques

CE-MS is an ideal analytical technique for almost all *omics* approaches such as proteomics, peptidomics, and metabolomics mainly due to the particular characteristics of this separation. In the case of CE separation, the CE combined with MS for foodomics have already been reviewed and discussed in previous reports [79,80]. According to them, CE-MS provides impressive possibilities as an analytical platform for the foodomics approaches at different levels [79]. Moreover, Ramautar et al. reviewed CE-MS development and applications introduced for the biomedical, clinical, microbial, plant, environmental, and food metabolomics in the period from 2010 to 2012 [81]. Although other advanced separation techniques, such as LC and GC, are also well-established for a metabolomics study, the key advantages of the CE-MS method are the ability to separate a large number of metabolites that have highly polar and ionic behaviors compared to GC-MS and LC-MS. In future studies, it is expected that various solutions, mainly related to the design of new capillary coatings and interfaces combined with cutting-edge methodological

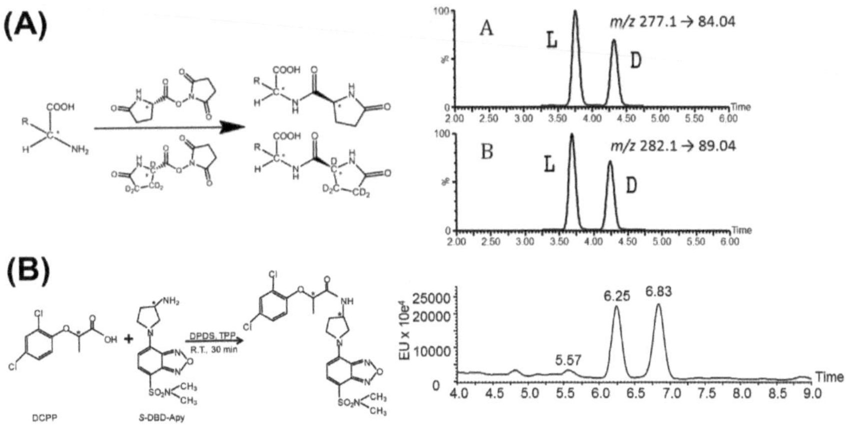

FIGURE 5 **Chiral foodomics approaches using derivatization with LC-MS system.** Novel chiral derivatization was developed for amino acids (A) [63]. Separation of chiral-derivatized analytes in green tea sample was proposed by UHPLC system (B) [65]. Please see these references for detail information of analytical conditions.

advances, will help to overcome these important limitations for nontargeted and/or exhaustive highly polar and ionic compounds in food materials [79]. Thus, we can expect more from the CE-MS ability of specific analytes in food materials that have trouble being monitored in the LC-MS and GC-MS foodomics. For example, CE-MS approaches have been used to detect amino acids, carbohydrates, DNA, vitamins, small organic substances, inorganic ions, and chiral compounds [80].

6.4 Others Separating Techniques

Future trends in foodomics will deal with technological obstacles for the identification of unknown compounds, statistical significant issues, and difference of peak responses in the MS chromatograms. Commonly, the identification of unknown compounds is based on the accurate MS spectra that provide the empirical formula, which is identified by database searches on the Internet. However, the limitations of the MS database for foodomics need consideration, and the implications of such limitations with respect to predicting the identification of significant markers should be explicated. Based on the expertise from the editors of the *Journal of Agricultural and Food Chemistry*, they stated that if marker compounds are pinpointed from such "nontargeted" comparisons, a library search must follow to establish whether a compound is already known, and if the structure is not yet described, the study must then be focused on the identification and quantification of the marker compounds according to standards set forth in previous perspectives [82]. These previous perspectives showed that the standards of the identification require that the determination of the molecular formula of a new compound by high-resolution MS, by combustion analysis, and/or by nuclear magnetic resonance (MNR) [83]. In addition, the latter is preferable as it provides a confirmation of purity. This approach is double work, and should be modified into a more useful and efficient foodomics protocol than the common metabolomics techniques. Thus, the novel separation techniques with capacity of purification will be beneficial in foodomics. These separation techniques are supercritical fluid chromatography (SFC) and high-speed countercurrent chromatography (HSCCC). The SFC and HSCCC are internationally known as purification techniques, and recently, it is possible to connect them to an MS detector for monitoring of analytes in various materials [84,85]. For metabolomics in biological and food materials, these techniques are a very easy and economical way to obtain the large-scale separation and purification of unknown analytes. The HSCCC-MS system was shown in Figure 6.

Also, coupling microfluidic chips (Chip) to MS can greatly expand the potential of metabolomics because it provides faster analysis time, enhanced sensitivity, and throughput. Multiple functions can also be integrated onto one chip, simplifying the operating procedures, and enabling a high-throughput and automated sample preparation for MS profiling. There are few applications of Chip-MS in metabolomic fingerprinting. Given the advantages of the Chip-MS interfaces, such as a high sensitivity and throughput, more applications of Chip-MS in this field can be expected in the near future [86].

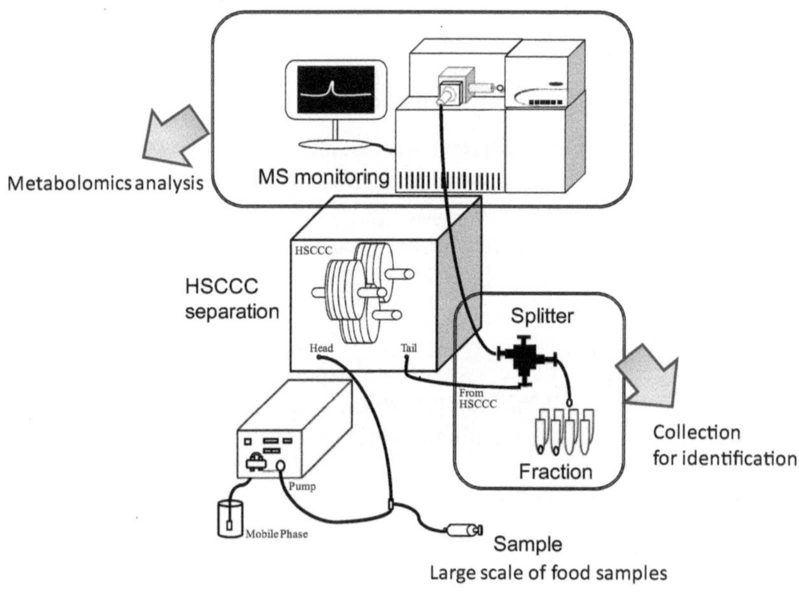

FIGURE 6 HSCCC-MS system for the application of foodomics with large-scale metabolomics analysis and collection for identification [72].

7. MS DETECTION

The goal of foodomics is the exhaustive profiling of all components in food materials. However, because foods vary in configuration, behavior, molecular weight, polarity, and solubility, a method capable of simultaneously detecting all components is realistically unavailable. It is necessary to narrow the number of targeted components and acquire data using analytical instruments. On the other hand, in the metabolomics, the development, characteristics, and application of various analytical instruments that are commonly used worldwide have been discussed [87–90]. Especially, the recent growth of metabolomics has significantly depended on the development of the MS approach, and put special emphasis on this technology today. For the major metabolomics approaches, the higher priority is the global quantitative assessment of metabolites using MS instruments that has played a pivotal role in various fields of science in the postgenomic era. Metabolomics with MS are not only the end product of the gene expression, but also forms part of the regulatory system in an integrated manner. Thus, metabolomics is often considered a powerful tool to provide a specific and comprehensive snapshot of the physiology of a biological phenomenon. The power of metabolomics generates value on the acquisition of analytical data in which metabolites in a biological sample are quantified, and the extraction of the most meaningful elements of the data by using various tools. In this sense, we introduce the latest development of MS techniques for the metabolomics study of food materials, i.e., advanced foodomics with MS.

7.1 MALDI-MS Techniques

Matrix-assisted laser desorption/ionization (MALDI) is an ionization technique in an MS instrument using a specific amount of matrix reagents that is uniformly dispersed with the sample. The targeted sample is ionized by nitrogen laser pulse irradiation of the surface of the mixture with matrix reagents and analytes. TOF is the analyzer most commonly associated with MALDI. In addition, other MALDI-MS systems, such as Fourier transform ion cyclotron resonance (FT-ICR), orbitrap, or the quadrupole time-of-flight (QqTOF), have recently been used to enhance the selectivity for exhaustive profiling. Moreover, a recent advanced method using MALDI-TOF-MS is imaging MS of variety of biomolecules from small metabolites to protein in various samples. The technique can be applied to use biological tissue and cell levels, and provide information regarding the spatial distribution of nontarget molecules. Recent reviews have presented a brief summary of the MALDI-imaging MS technology and its use for the analysis of plant materials [91,92]. Actually, a foodomics approach has been developed by nontargeted MALDI-imaging MS profiling of chemical components in apple samples [93]. However, the MALDI-imaging MS profiling of food has been little applied within foodomics.

7.2 Direct MS Techniques

During the introduction of direct infusion MS (DI-MS), the sample solution is directly introduced into the electrospray ionization (ESI) interface without chromatographic separation using a syringe pump or nanospray chip. The DI-MS technique is a high-throughput method due to the shorter time required for the analysis of one sample in comparison to chromatographic methods. On the other hand, because the intense matrix effect highly observed, quantitative performance is worse than that of LC-MS. Although a stable isotope standard of the whole metabolomic was used to overcome the matrix effect, this method is too complicate and little feasible [94]. Moreover, the molecular selectivity of DI-MS is inferior to LC-MS due to a lack of retention time information for the analytes. To address this limitation, the tandem MS and accurate mass techniques have been used for the DI-MS metabolomics by using FT-ICR and orbitrap systems [95]. Another DI-MS method uses flow injection analysis and an automated sampler without a column for the metabolomics study. The flow injection analysis coupled to MS has been applied in several foodomics approaches [96–98]. However, this method provides better results combined with a short chromatographic column to expand the coverage of nontargeted analytes and reduce the matrix effects. At present, because there is no universal DI-MS instrument capable of measuring various types of components in food, the most appropriate technique for foodomics is selected giving consideration to various factors (resolution, sensitivity, matrix effect, throughput, and cost) based on a shorter analytical time of one run.

7.3 MS Combined with Separation Techniques

Previous foodomics studies have shown that the inclusion of a separation technique is a better approach for the exhaustive profiling of various components in food than DI-MS. The advantages of the chromatography coupled to MS are peak capacity, repeatability of retention time, and readily available MS libraries for identification without using standard compounds. Especially, the chemical identification by GC-MS is relatively easy compared to other analytical platforms because the mass spectrum (with fragment ions) of a specific analyte can be consistently obtained, and can be easily identified using the NIST mass spectral library on electron ionization (EI). On the other hand, LC-MS can analyze a wide range of analytes, from high to low molecular weights and from hydrophilic to hydrophobic character. These versatile abilities can be exploited by selecting the appropriate columns and mobile phases. For LC-MS, atmospheric pressure ESI, a frequent ionization method, can be used for foodomics. However, the ESI technique lacks a quantitative capability due to the matrix effects. For handing this phenomenon, analytical standards labeled with stable isotopes (^{13}C, ^{2}H, and/or ^{15}N compounds) are spiked into the sample. Actually, a nontargeted screening strategy for the detection of novel conjugates of the mycotoxin deoxynivalenol in wheat using stable isotopic labeling and LC-MS was reported [99]. Thus, the future foodomics with LC-MS probably will imply the use of labeled standards.

In most cases, for MS-based foodomics, a minimum of two independent variables are indispensible to identify a molecule, for example, its accurate mass (*m/z*) with response and different eluting times (retention time, migration time, or retention index). As high-resolution MS instruments, such as TOF, FT-ICR, and orbitrap may correctly assign a putative molecular formula. However, at present, it is still impossible to detect all components of food using any of these excellent instruments. Thus, the existing techniques, which aims to find significant changes and validate the data obtained from the food samples, are applied to crucial processes in the foodomics research.

8. DATA ANALYSIS

In the data analysis of MS signals regarding the peak response, the *m/z* values and eluting time, the first step is the extraction of significant signals from the raw data by deconvolution, which utilizes all detected peaks picking an algorithm (Figure 7). The task of an MS peak picking algorithm is the transformation of a profiling spectrum into a list of unknown peaks. For most deconvolution techniques, the profiling spectra from the raw data are merely obtained by the existing digitalization of dependent signals regarding the peak resolution, shape, and isotopic patterns of the overlapping peaks. The resolution of overlapping isotopic peaks is important to evaluate overlapping analytes in a matrix and to identify the distance from the isotopic peaks of the same molecular ion or

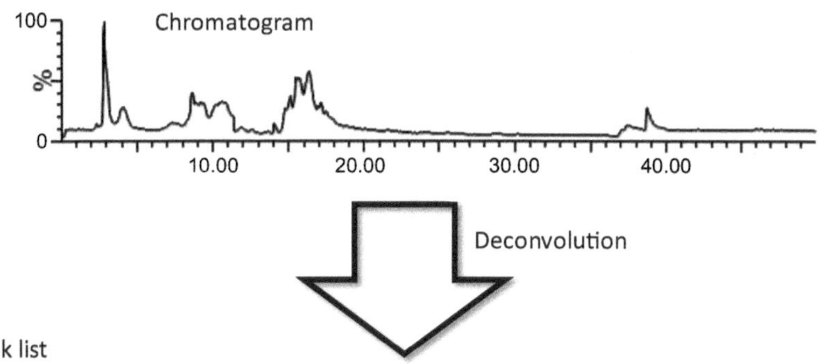

Peak list

Primary ID	Data files	Class	249.206_2.16	266.232_2.17	336.31_2.17	816.638_2.18	496.421_2.18	758.598_2.19	656.525_2.2	665.502_2.2
RT			2.16	2.17	2.17	2.18	2.18	2.19	2.2	2.2
Mass			249.206	266.232	336.31	816.638	496.421	758.598	656.525	665.502
N_1110_1	20140217_	P	0.169865	0.234315	0	0.011269	0	0	0	0.021686
N_1110_2	20140217_	P	0.227174	0.326087	0	0	0.063887	0	0.166607	0
N_1110_3	20140217_	P	0	0.628532	0	0.144382	0	0.592398	0.563284	0
A_1127_1	20140217_	N	0.243024	0.317183	0	0	0.178975	0	0.119119	0
A_1127_2	20140217_	N	0	0.183611	0	0.038046	0	0	0	0
A_1127_3	20140217_	N	0.14512	0.187229	0	0	0	0	0	0

FIGURE 7 Data analysis for foodomics. In the data analysis from MS signals regarding the peak response, the *m/z* values and eluting time, the first step is the extraction of significant signals from the raw chromatogram data by deconvolution.

other ions from differing peaks. Although the peak deconvolution is rather limited, these techniques are still of significant interest especially due to their much better sensitivity, selectivity, and comprehensive fragmentation capabilities. As a matter of fact, this deconvolution has absolute discretion to make a decision of experimental outcome in metabolomics with MS. Thus, a semiautomated strategy using a hierarchical multivariate curve resolution approach was examined for the deconvolution process to improve and automate the data processing from GC-MS metabolomics [100]. Moreover, a recent study showed that an automated data analysis pipeline (ADAP) has been developed for the peak detection, deconvolution, peak alignment, and library search [101]. However, more efficient and reliable deconvolution algorithms are really needed for the digitalization of peak picking, identification of MS spectrum, and quantification of pinpoint markers due to the limitations of existing deconvolution algorithms. Today, newly developed algorithms need to be implemented into user-friendly and high-throughput software tools. These tools should be equipped with visualization capabilities that will allow metabolomics researchers to visually examine the intermediate and final results for verifying the correctness of significant metabolites detected in the data analysis and data interpretation stages [102]. Development of these algorithms and software tools for accurate deconvolution would greatly benefit future foodomics research.

Since the nontargeted MS signals involved an extremely large amount of data, MS users need to focus on the interesting and significant components from the systematization by a deconvolution technique. In this case, multivariable analysis is a commonsense way to extract the visual markers from the MS data. In most cases, a principal component analysis (PCA) is a statistical aggregation algorithm from multivariate distribution and the most frequently used approach in metabolomics studies. PCA is performed to reduce the multidimensional data that can be plotted in a two- or three-dimensional Cartesian coordinate with the axes, named the principal components representing the greatest variations from the MS signals (Figure 8(A)). The popularity of this technique stems from the

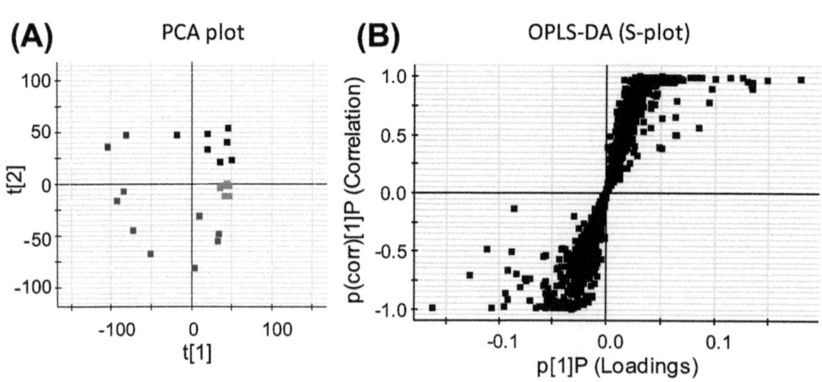

FIGURE 8 Typical PCA and OPLS-DA plots. *This detail information was shown in Ref. [39].*

multivariable data for the easy graphical interpretation based on the class types and trends. The contribution coefficient of individual variables to the distribution of each sample can be identified by the corresponding loading plots which plot the contribution of each variable versus the selected principal components [103]. PCA is followed by a discriminant analysis, such as a partial least squares discriminant analysis (PLS-DA). PLS-DA attempts to maximize the covariance between the independent and dependent variables to discriminate against analytes in the samples, and this method is frequently used for finding a way to identify purposeful markers between sample classes (Figure 8(B)). Moreover, the lack of trend information when determining the principal components in a PCA plot can lead to the discrimination of samples based on nonrelated factors from the PCA. A specialized form of PLS-DA is orthogonal projections to latent structures (OPLS), in which any noncorrelated systematic variance is removed from the model [104]. The interpretative criterion from the exhaustive profiling in food materials by PLS-DA or OPLS-DA is a major property of foodomics research. If foodomics researchers are able to obtain reliable and reproducible results for the targeted approach, they should validate the robustness of the predicted consequence. The predicted consequence must be a statistically significant difference and have beneficial and interesting information for food materials. Cross-validation is frequently used for internal validation of the predicted consequence. After executing a cross-validation, the root mean square error of evaluation (RMSEE) is evaluated for robustness. Moreover, the root mean square error of prediction (RMSEP) can be used as an external validation of the reliability. Foodomics researchers should evaluate the RMSEE and RMSEP values in accordance with the required accuracy and precision. Additionally, in the discriminant analysis, the false positive and false negative rates are routinely evaluated for indicating the degree of precision. A receiver operating characteristic (ROC) curve has been frequently used for this evaluation of the sensitivity and specificity based on the false positive/negative rate. In the ROC curve, the area under curve (AUC) can be calculated for the robustness statistical analysis. Actually, the ROC curve is a fundamental tool for diagnostic test evaluation of medical markers. Using the ROC curve, the true positive rate (sensitivity) is plotted as a function of the false positive rate (=100—specificity, %) for different cutoff points of a targeted marker or parameter. Each point on the ROC curve represents a sensitivity/specificity pair corresponding to a particular decision threshold. In addition, the AUC is a measure of how well a parameter can distinguish between two diagnostic groups of disorder and control. When we evaluate the results of a particular test from two food materials, one food materials is included with a contributory factor, while the other food material has no contributory factor, a perfect separation between the two groups can rarely be observed in the targeted foodomics. Indeed, the distribution of the test results overlaps, as shown in Figure 9. For every possible cutoff point or criterion value, the two groups classified as true positive or negative markers are discriminated for false positive and negative rates. Thus, the final determining factor in the

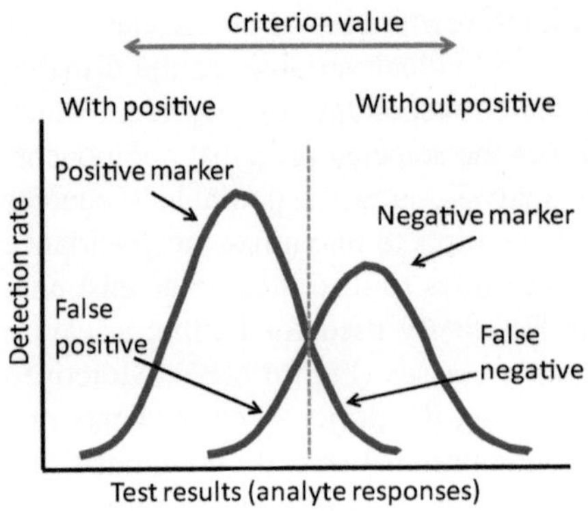

FIGURE 9 Distribution of the test results to overlap between the two groups classified as true positive or negative markers for false positive and negative rates.

foodomics would be decided by using the AUC rate. For current foodomics, a computer is used for the commonsense way to extract the MS data and to process this data set using the multivariate statistical technique. The development of software for the MS data analysis is a very important mission in foodomics.

9. DATA ASSESSMENT

After the process of extracting useful and interesting MS markers from the raw data, unknown compounds in specific food materials need to be identified for the common realization of the causality behavior characterization in food and nutrition sciences. Based on the identification of markers in foodomics, renewed and advanced food science will lead to the development of the assessment, guidelines, follow-up study, toxicity assay, and functional ability of foods. In fact, the identification method has been the key process for metabolomics of human samples. Currently, the free databases are assessed on the Internet based on accurate MS spectra, isotopic patterns, and fragment ions. In the case of human metabolomics MS data, the fundamental access is "The Human Metabolome Database (HMDB)" that is a freely available electronic database containing detailed information about metabolites found in the human body [105]. Since the first HMDB released in 2007, the HMDB has been used to facilitate research for nearly 1000 published metabolomics studies in clinical, biochemical, and biological chemistry. A recently upgraded and enhanced version of the 2012 (version 3.0), is the result of the inclusion both detected and expected metabolites. Moreover, the applied databases for foodomics are the "MassBank," "Golm Metabolome Database," "Metlin," "Fiehn GC-MS Database," and "mzCloud." On the other hand, there are a few specific foodomics databases (such as "FooDB," "DrugBank," and "Data Resources of Plant Metabolomics"). However, it is very difficult to summarize a broad range of food in response to the researcher's varied requests for exploring the application of foodomics. It should be limited to the existing application of the user's database. For example, The "Phenol-Explorer 3.0" is the only available database for the polyphenols content in foods and the in vivo metabolism and pharmacokinetics [106]. An ideal database for the foodomics approach would be related to purpose, compound, group-ability, and food material and specialized for different needs. Given the future situation of specialized databases, the foodomics approach will be modified to fit the useful and hospitable databases, and developed for an easy and clear output for the identification of markers in food materials according to the database.

Finally, foodomics for physiological response monitoring is needed. In the case of human metabolomics, the "Kyoto Encyclopedia of Genes and Genomes (KEGG) from Japan" is a database resource for understanding the high-level functions and utilities of biological systems, such as the cell, the organism, and the ecosystem, from molecular-level information, especially large-scale molecular data sets generated by genome sequencing and other high-throughput experimental technologies [107]. Today, the KEGG database has been improved by international researchers. Actually, the researchers from The Netherlands provide an improved description of the tricarboxylic acid cycle via the community-created database [108]. As other metabolic pathway tools for human metabolomics, the "MetaCyc," "HumanCyc," "BioCyc," and "Reactome" are available databases on the Internet. These key words regarding the database can be investigated by a "Google" search for metabolomics. On the other hand, there are few specific databases for foodomics to use and investigate the dynamic analysis, interactive process, risk and benefit assessment, nutritional intervention, microbiological performance, agricultural productivity and comprehensive promotion, quality-control measure, and human health and well-being. Recently, the review regarding the flatfish aquaculture summarizes the use of comprehensive functional genomics, proteomics, and metabolomics analyses aimed at better identifying the critical genes and molecules that control the traits of commercial interest in aquaculture production, such as growth rates, reproduction, larval development, and disease resistance [109]. This overview described the future generation sequencing platforms, which have drastically transformed the way to address genomic physiological response monitoring for flatfish genomics research. In addition, the determination of the system-wide biochemical effects of diets on an individual's metabolism in nutrition research was reviewed for the interactions in the complex mosaic of both genomic and metagenomic networks [110]. The development and application of advanced database and physiological response monitoring methodologies has contributed to the creation of a better foodomics approach. This is a new approach to replace the old concepts in food and nutrition sciences. In this field, future researchers working in food chemistry, analytical chemistry, nutrition science, biochemistry, microbiology, molecular biology, food technology, information science technology, and renewed generationology can finally work together to reach the main objective of foodomics.

10. ADVANCED FUTURE FOODOMICS APPROACH WITH MS

Based on the foodomics concept, we are facing a wide variety of refined areas which would analyze nontargeted low-molecular-weight molecules in biological and/or food samples with MS. The main focuses of the search related to food and metabolomics are compositional objectives. The perspectives of this approach are the "human responsive by food," "microbial aspect by food and other stress," "quality assessment of food," "safety assessment of food," "investigation of functional food," and other efforts. In this final section, the recent typical and interesting perspectives for advanced future MS foodomics are introduced (Figure 10). These perspectives will be used as references for starting the new research based on foodomics with MS.

10.1 Advanced Foodomics Approach for Human Response by Foods

In the case of human response to foods, changes in metabolic profiling after the intake of different foods can provide an insight into their relativity between the human metabolism and dietary pattern. For example, the profiling of the urinary metabolomics of subjects was reported for evaluating coffee, wine, or cocoa powder consumption [16,111,112]. Moreover, the metabolic profiling for human dietary exposure was evaluated based on an MS analysis of biological samples for the common recognition of nutrition science [18,113]. These procedures would have minimal interferences with the normal daily activities of the subjects. Indeed, by substitution of one component of the standard country intake (i.e., cornflakes, bread, pasta, rice) with different foods deemed to be of high public health significance, we have evidence that metabolic fingerprinting can be efficiently associated with specific dietary components and may be used

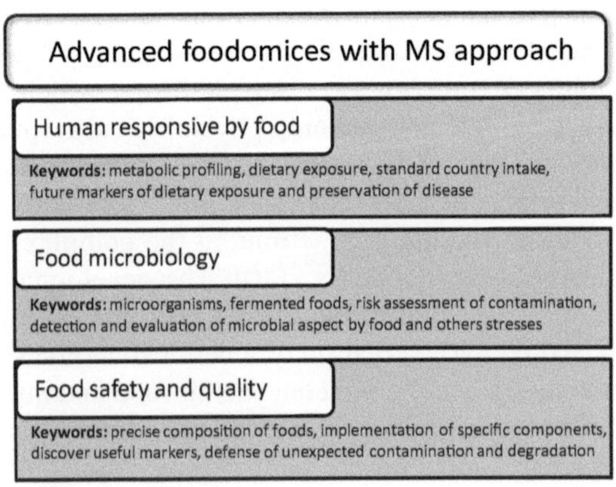

FIGURE 10 Main focuses of foodomics with MS for the following perspective in the food and nutrition sciences.

as targets for the future markers of dietary exposure and preservation of disease. If this metabolic profiling based on human intake is structured and accrued in the future, our dietary environment from various history periods can be digitally preserved by foodomics technology. The accurate and universal digital data based on foodomics technology will show that food style with long histories can be maintained to develop and give birth to interesting ideas based on the earthborn lifestyles of the time.

10.2 Advanced Foodomics Approach for Food Microbiology

The microbiology of foods is the study of the microorganisms used to create food products, control food processes, and/or bring about the negative effect/contamination of foods. Today, these microorganisms are an important essential for the production of foods such as cheese, yogurt, wine, beer, sake, and other fermented foods. On the other hand, fatal contaminants of pathogenic bacteria, viruses, and derived toxins are a serious problem of food safety. Until now, microbiological tests for the detection of these microorganisms have been used for food quality assessment. A few years ago, the polymerase chain reaction (PCR) was developed as a quick and useful method to generate a number of copies of a DNA fragment at a specific band, and used to detect different kinds of viruses or bacteria based on their specific DNA patterns. In today's and future food science, the detection and evaluation of microbial aspect by food and others stresses would be very important to everyone in the world. Using the foodomics approach, it is worth noting that the nontargeted MS analysis of microorganisms or fermented foodstuff is obviously applied to evaluate the advantage of the microbiology for foods. Previously, the metabolic profiling was discussed for the identification and classification of yeasts and filamentous fungi using integration of the MS methodology [114]. Moreover, LC-MS metabolomics was applied to evaluate a Japanese fermented food, namely Miso, of different stages of ripeness by a statistical PCA plot for the classified groups based on the stage of its ripeness [115]. The metabolite changes observed in fast-fermented bean paste may enhance the functional properties of unfermented soybeans, and these results suggest that the metabolomics approaches can provide a more comprehensive understanding of the metabolism involved in the production of fermented foods [116].

10.3 Advanced Foodomics Approach for Food Safety and Quality

In the case of food safety and quality, metabolic fingerprinting with MS techniques can provide valuable and surprising information on the precise composition of foods that can be directly correlated to its quality. Nowadays, the trend in analytical chemistry is toward the creation of multiple methods with extended coverage, enabling the determination of many different classes of compounds in a single analysis in which virtually all classes of different compounds are included in a single run. Actually, comprehensive GC-MS profiling of the volatile components and can, therefore, be easily extended to aroma compounds in various fruits and can therefore be widely used for quality studies in the field of fruit aroma [27]. In addition, the LC-MS approach was used to profile the flavonoid and anthocyanin in grapes samples [117], and the MS metabolic profiling determined the changes during the germination of rice [118]. The most likely use of foodomics is the development of metabolic profiling through the implementation of specific components for the food quality database or as a marker. Interestingly, the Metabolome Tomato Database (MoTo DB) is an open-access metabolome database for tomato fruit [119]. Recent work has reported that orange samples from the Valencia Spanish region and foreign samples (Argentina, South Africa, and Brazil) can be distinguished using citrusin D as biomarker [120].

Other foodomics-based option is the application of stable isotope ratio MS using the $^{13}C/^{12}C$ that can be correlated to different food origins, or even, to the adulterations of products. Besides, the combination of this technique with GC allows the isotopic analysis of each separated compound. This technique was used to develop a novel foodomics approach enabling the simultaneous identification of milk samples either processed with different heating regimens or from different geographical origins [121]. In the investigation and evaluation of food quality using the foodomics approach, we will be able to create a significant area for the renewed perspective of food process, information, and analysis. It is possible to apply this ideal foodomics for novel food quality assessment.

In the first section dealing with the foodomics approach, it is a high-priority problem to preserve many unexpected correspondences in food contamination. Today, we are feeling that a broad vision means not only an application of nutrition and/or functional profiling, but also the mission of resolving any unexpected contamination and degradation of foods. Fortunately, a few foodomics approaches were used to detect any unexpected contamination in food [38,122]. While these present nontargeted foodomics are the limits of every compound, the aim of this method is not to replace each existing targeted screening, but rather complement them to find unexpected contaminants such as very highly concentrated levels, and possible drugs, pesticides and common toxins-laced foods. At the foundation of regular food analyses, a targeted approach is generally used for the purpose of determining ppb (parts per billion) or ppt (parts per trillion) levels of the suspected compounds based on today's food policies aimed to guarantee minimal risks for any toxic effect after lifelong exposure based on food safety concepts. However, it is a very high risk in our lifestyle that highly concentrated levels of food contamination which has contributed to people's suffering on death or severe injury have occurred. If we use a targeted approach for the detection of food contamination, this approach would be based on long periods of harsh measures for the food materials and/or products. On the other hand, the foodomics approach based on MS involves a statistical evaluation of the raw data to indicate single compounds that specifically increase or decrease in foods regarding unexpected contaminants and degradation in any situation. This ideal foodomics approach suggests that nontargeted food materials would be widely used for a routine assay for its broader scope of analytes provided with the possibility to pick up unexpected contaminants during production. To achieve this, foodomics will be a more useful procedure such as the separation, MS detection, statistics, and database and entire protocol. Thus, we cannot wait to develop and apply a future foodomics approach for food safety and quality using the advanced MS technique.

LIST OF ABBREVIATIONS

ADAP	Automated Data Aanalysis Pipeline
APCI	Atmospheric pressure chemical ionization
APPI	Atmospheric pressure photo ionization
AUC	Area under curve
CE	Capillary electrophoresis
CEC	Capillary electrokinetic chromatography
CGE	Capillary gel electrophoresis
Chip	Microfluidic chips
CIEF	Capillary isoelectric focusing
CW-TPR	Carbowax-templated resin
CZE	Capillary zone electrophoresis
DI	Direct infusion
EI	Electron ionization
ESI	Electrospray ionization
FT-ICR	Fourier transform ion cyclotron resonance
GC	Gas chromatography
HMDB	Human Metabolome Database
HPLC	High-performance liquid chromatography
IEF	Isoelectric focusing
KEGG	Kyoto Encyclopedia of Genes and Genomes
LC	Liquid chromatography
LLE	Liquid–liquid extraction
MALDI	Matrix-assisted laser desorption/ionization
MEKC	Micellar electrokinetic chromatography
MoTo DB	Metabolome Tomato Database
MS	Mass spectrometry
MSPD	Matrix solid-phase dispersion
NMR	Nuclear magnetic resonance
OPLS	Orthogonal projections to latent structures
PA	Polyacrylate
PCA	Principal component analysis
PCR	Polymerase chain reaction
PDMS	Poly(dimethylsiloxane)
PDMS/DVB	Poly(dimethylsiloxane/divinylbenzene)
PLE	Pressurized liquid extraction
PLS-DA	Partial least squares discriminant analysis

QuEChERS	Acronym of quick, easy, cheap, effective, rugged, and safe
RMSEE	Root mean square error of evaluation
RMSEP	Root mean square error of prediction
ROC	Receiver operating characteristic
SBSE	Stir bar sorptive extraction
dSPE	Dispersive solid- phase extraction
SPE	Solid-phase extraction
SPME	Solid-phase microextraction
TOF	Time-of-flight
UHPLC	Ultra-high pressure liquid chromatography

REFERENCES

[1] H. Xin, R. Stone, Science 322 (2008) 1310–1311.
[2] Y.C. Tyan, M.H. Yang, S.B. Jong, C.K. Wang, J. Shiea, Anal. Bioanal. Chem. 395 (2009) 729–735.
[3] G. Abernethy, K. Higgs, J. Chromatogr. A 1288 (2013) 10–20.
[4] K. Inoue, T. Sakamoto, J.Z. Min, K. Todoroki, T. Toyo'oka, Food Chem. 156 (2014) 390–393.
[5] A. Cifuentes, J. Chromatogr. A 1216 (2009) 7109.
[6] M. Herrero, C. Simó, V. García-Cañas, E. Ibáñez, A. Cifuentes, Mass Spectr. Rev. 31 (2012) 49–69.
[7] V. García-Cañas, C. Simó, M. Herrero, E. Ibáñez, A. Cifuentes, Anal. Chem. 84 (2012) 10150–10159.
[8] R.A. Dixon, D.R. Gang, A.J. Charlton, O. Fiehn, H.A. Kuiper, T.L. Reynolds, R.S. Tjeerdema, E.H. Jeffery, J.B. German, W.P. Ridley, J.N. Seiber, J. Agric. Food Chem. 54 (2006) 8984–8994.
[9] J.M. Baker, N.D. Hawkins, J.L. Ward, A. Lovegrove, J.A. Napier, P.R. Shewry, M.H. Beale, Plant Biotechnol. J. 4 (2006) 381–392.
[10] T. Levandi, C. Leon, M. Kaljurand, V. Garcia-Cañas, A. Cifuentes, Anal. Chem. 80 (2008) 6329–6335.
[11] W. Zeng, J. Hazebroek, M. Beatty, K. Hayes, C. Ponte, C. Maxwell, C.X. Zhong, J. Agric. Food Chem. 62 (2014) 2997–3009.
[12] H. Baniasadi, C. Vlahakis, J. Hazebroek, C. Zhong, V. Asiago, J. Agric. Food Chem. 62 (2014) 1412–1422.
[13] G.G. Siano, I.S. Pérez, M.D. García, M.M. Galera, H.C. Goicoechea, Talanta 85 (2011) 264–275.
[14] S. Komatsu, A. Yamamoto, T. Nakamura, M.Z. Nouri, Y. Nanjo, K. Nishizawa, K. Furukawa, J. Proteome Res. 10 (2011) 3993–4004.
[15] Y. Lu, H. Lam, E. Pi, Q. Zhan, S. Tsai, C. Wang, Y. Kwan, S. Ngai, J. Agric. Food Chem. 61 (2013) 8711–8721.
[16] R. Llorach-Asunción, O. Jauregui, M. Urpi-Sarda, C. Andres-Lacueva, J. Pharm. Biomed. Anal 51 (2010) 373–381.
[17] C. Legido-Quigley, C. Stella, F. Perez-Jimenez, J. Lopez-Miranda, J. Ordovas, J. Powell, F. van-der-Ouderaa, L. Ware, J.C. Lindon, J.K. Nicholson, E. Holmes, Biomed. Chromatogr. 24 (2010) 737–743.
[18] Ú. Catalán, M.Á. Rodríguez, M.R. Ras, A. Maciá, R. Mallol, M. Vinaixa, S. Fernández-Castillejo, R.M. Valls, A. Pedret, J.L. Griffin, R. Salek, X. Correig, M.J. Motilva, R. Solà, Mol. Biosyst. 9 (2013) 1411–1422.

[19] W. Pongsuwan, E. Fukusaki, T. Bamba, T. Yonetani, T. Yamahara, A. Kobayashi, J. Agric. Food Chem. 55 (2007) 231–236.

[20] W. Pongsuwan, T. Bamba, K. Harada, T. Yonetani, A. Kobayashi, E. Fukusaki, J. Agric. Food Chem. 56 (2008) 10705–10708.

[21] K.M. Ku, J.N. Choi, J. Kim, J.K. Kim, L.G. Yoo, S.J. Lee, Y.S. Hong, C.H. Lee, J. Agric. Food Chem. 58 (2010) 418–426.

[22] T. Ogura, T. Bamba, E. Fukusaki, J. Chromatogr. A 1301 (2013) 73–79.

[23] B. Biais, J.W. Allwood, C. Deborde, Y. Xu, M. Maucourt, B. Beauvoit, W.B. Dunn, D. Jacob, R. Goodacre, D. Rolin, A. Moing, Anal. Chem. 81 (2009) 2884–2894.

[24] J. Zhang, X. Wang, O. Yu, J. Tang, X. Gu, X. Wan, C. Fang, J. Exp. Bot. 62 (2011) 1103–1118.

[25] D. Cuthbertson, P.K. Andrews, J.P. Reganold, N.M. Davies, B.M. Lange, J. Agric. Food Chem. 60 (2012) 8552–8560.

[26] E. Hurtado-Fernández, T. Pacchiarotta, E. Longueira-Suárez, O.A. Mayboroda, A. Fernández-Gutiérrez, A. Carrasco-Pancorbo, J. Chromatogr. A 1313 (2013) 228–244.

[27] U. Vrhovsek, C. Lotti, D. Masuero, S. Carlin, G. Weingart, F. Mattivi, J. Chromatogr. B 966 (2014) 132–139.

[28] J.W. Allwood, W. Cheung, Y. Xu, R. Mumm, R.C. De Vos, C. Deborde, B. Biais, M. Maucourt, Y. Berger, A.A. Schaffer, D. Rolin, A. Moing, R.D. Hall, R. Goodacre, Phytochemistry 99 (2014) 61–72.

[29] K. Skogerson, R. Runnebaum, G. Wohlgemuth, J. de Ropp, H. Heymann, O. Fiehn, J. Agric. Food Chem. 57 (2009) 6899–6907.

[30] J.E. Lee, G.S. Hwang, C.H. Lee, Y.S. Hong, J. Agric. Food Chem. 57 (2009) 10772–10783.

[31] U. Vrhovsek, D. Masuero, M. Gasperotti, P. Franceschi, L. Caputi, R. Viola, F. Mattivi, J. Agric. Food Chem. 60 (2012) 8831–8840.

[32] P. Arapitsas, M. Scholz, U. Vrhovsek, S. Di Blasi, A. Biondi Bartolini, D. Masuero, D. Perenzoni, A. Rigo, F. Mattivi, PLoS One 7 (2012) e37783.

[33] K. Inoue, M. Nishimura, H. Tsutsui, J.Z. Min, K. Todoroki, J.M. Kauffmann, T. Toyo'oka, J. Agric. Food Chem. 61 (2013) 1228–1234.

[34] L.M. Schmidtke, J.W. Blackman, A.C. Clark, P. Grant-Preece, J. Agric. Food Chem. 61 (2013) 11957–11967.

[35] J.J. Rodríguez-Bencomo, I. Andújar-Ortiz, M.V. Moreno-Arribas, C. Simó, J. González, A. Chana, J. Dávalos, M.Á. Pozo-Bayón, J. Agric. Food Chem. 62 (2014) 1373–1383.

[36] M. Commisso, P. Strazzer, K. Toffali, M. Stocchero, F. Guzzo, Comput. Struct. Biotechnol. J. 4 (2013) e201301007.

[37] D.K. Lee, M.H. Yoon, Y.P. Kang, J. Yu, J.H. Park, J. Lee, S.W. Kwon, Food Chem. 141 (2013) 3931–3937.

[38] E. Tengstrand, J. Rosén, K.E. Hellenäs, K.M. Aberg, Anal. Bioanal. Chem. 405 (2013) 1237–1243.

[39] K. Inoue, C. Tanada, T. Sakamoto, H. Tsutsui, T. Akiba, J.Z. Min, K. Todoroki, Y. Yamano, T. Toyo'oka, *Food Chem.* in press. http://dx.doi.org/10.1016/j.foodchem. 2015.02.117.

[40] J.M. Cevallos-Cevallos, M.D. Danyluk, J.I. Reyes-De-Corcuera, J. Food Sci. 76 (2011) M238–M246.

[41] H.J. Namgung, H.J. Park, I.H. Cho, H.K. Choi, D.Y. Kwon, S.M. Shim, Y.S. Kim, J. Sci. Food Agric. 90 (2010) 1926–1935.

[42] Y.K. Kwon, Y.S. Jung, J.C. Park, J. Seo, M.S. Choi, G.S. Hwang, Mar. Pollut. Bull. 64 (2012) 1874–1879.

[43] E. Beceiro-González, M.J. González-Castro, S. Muniategui-Lorenzo, P. López-Mahía, D. Prada-Rodríguez, J. AOAC Int. 95 (2012) 1291–1310.

[44] H.H. Jeleń, M. Majcher, M. Dziadas, Anal. Chim. Acta 738 (2012) 13–26.
[45] V.C. Fernandes, V.F. Domingues, N. Mateus, C. Delerue-Matos, J. Chromatogr. Sci. 49 (2011) 715–730.
[46] S.J. Lehotay, Methods Mol. Biol. 747 (2011) 65–691.
[47] D.A. Lambropoulou, T.A. Albanis, Anal. Bioanal. Chem. 389 (2007) 1663–1683.
[48] C.I. Silvestre, J.L. Santos, J.L. Lima, E.A. Zagatto, Anal. Chim. Acta 652 (2009) 54–65.
[49] M. Asensio-Ramos, L.M. Ravelo-Pérez, M.Á. González-Curbelo, J. Hernández-Borges, J. Chromatogr. A 1218 (2011) 7415–7437.
[50] Y. Picó, M. Fernández, M.J. Ruiz, G. Font, J. Biochem. Biophys. Methods 70 (2007) 117–131.
[51] S.J. Lehotay, Methods Mol. Biol. 747 (2011) 65–91.
[52] B.D. Eitzer, W. Hammack, M. Filigenzi, J. Agric. Food Chem. 62 (2014) 80–87.
[53] C. Bicchi, C. Cordero, E. Liberto, B. Sgorbini, P. Rubiolo, J. Chromatogr. A 1184 (2008) 220–233.
[54] F.M. Lancas, M.E. Queiroz, P. Grossi, I.R. Olivares, J. Sep. Sci. 32 (2009) 813–824.
[55] A. Prieto, O. Basauri, R. Rodil, A. Usobiaga, L.A. Fernández, N. Etxebarria, O. Zuloaga, J. Chromatogr. A 1217 (2010) 2642–2666.
[56] A.A. Argyri, A. Mallouchos, E.Z. Panagou, G.J. Nychas, Int. J. Food Microbiol. 193 (2015) 51–58.
[57] K. Inoue, S. Nitta, T. Hino, H. Oka, J. Chromatogr. B 877 (2009) 461–464.
[58] S. Tiziani, A.H. Emwas, A. Lodi, C. Ludwig, C.M. Bunce, M.R. Viant, U.L. Günther, Anal. Biochem. 377 (2008) 16–23.
[59] Y. Yang, C. Cruickshank, M. Armstrong, S. Mahaffey, R. Reisdorph, N. Reisdorph, J. Chromatogr. A 1300 (2013) 217–226.
[60] D. Vuckovic, J. Pawliszyn, Anal. Chem. 83 (2011) 1944–1954.
[61] A. Rai, S. Umashankar, S. Swarup, Methods Mol. Biol. 1069 (2013) 279–312.
[62] K. Skov, N. Hadrup, J. Smedsgaard, H. Frandsen, J. Chromatogr. B 978–979 (2015) 83–88.
[63] X. Li, D. Long, J. Ji, W. Yang, Z. Zeng, S. Guo, Z. Ji, G. Qi, S. Chen, J. Microbiol. Methods 94 (2013) 61–67.
[64] L.C. Marney, S.C. Kolwicz Jr., R. Tian, R.E. Synovec, Talanta 108 (2013) 123–130.
[65] Z. León, J.C. García-Cañaveras, M.T. Donato, A. Lahoz, Electrophoresis 34 (2013) 2762–2775.
[66] B.A. Goldberger, E.J. Cone, J. Chromatogr. A 674 (1994) 73–86.
[67] S. Becker, L. Kortz, C. Helmschrodt, J. Thiery, U. Ceglarek, J. Chromatogr. B 883–884 (2012) 68–75.
[68] K. Klepárník, Electrophoresis 36 (2015) 159–178.
[69] R. Wehrens, G. Weingart, F. Mattivi, J. Chromatogr. B 966 (2014) 109–116.
[70] I. Koo, X. Shi, S. Kim, X. Zhang, J. Chromatogr. A 1337 (2014) 202–210.
[71] T. Inui, F. Tsuchiya, M. Ishimaru, K. Oka, H. Komura, J. Agric. Food Chem. 61 (2013) 4758–4764.
[72] D. Guillarme, D.T. Nguyen, S. Rudaz, J.L. Veuthey, Eur. J. Pharm. Biopharm. 66 (2007) 475–482.
[73] D. Guillarme, D.T. Nguyen, S. Rudaz, J.L. Veuthey, Eur. J. Pharm. Biopharm. 68 (2008) 430–440.
[74] D. Guillarme, J. Ruta, S. Rudaz, J.L. Veuthey, Anal. Bioanal. Chem. 397 (2010) 1069–1082.
[75] S.H. Lee, I.A. Blair, BMB Rep. 42 (2009) 401–410.
[76] T. Mochizuki, K. Todoroki, K. Inoue, J.Z. Min, T. Toyo'oka, Anal. Chim. Acta 811 (2014) 51–59.

[77] R. Nagao, H. Tsutsui, T. Mochizuki, T. Takayama, T. Kuwabara, J.Z. Min, K. Inoue, K. Todoroki, T. Toyo'oka, J. Chromatogr. A 1296 (2013) 111–118.
[78] K. Inoue, N. Prayoonhan, H. Tsutsui, T. Sakamoto, M. Nishimura, T. Toyo'oka, J. Sep. Sci. 36 (2013) 1356–1361.
[79] C. Ibáñez, C. Simó, V. García-Cañas, A. Cifuentes, M. Castro-Puyana, Anal. Chim. Acta 802 (2013) 1–13.
[80] V. García-Cañas, C. Simó, M. Castro-Puyana, A. Cifuentes, Electrophoresis 35 (2014) 147–169.
[81] R. Ramautar, G.W. Somsen, G.J. de Jong, Electrophoresis 34 (2013) 86–98.
[82] J.N. Seiber, R.J. Molyneux, P. Schieberle, J. Agric. Food Chem. 62 (2014) 22–23.
[83] R.J. Molyneux, P. Schieberle, J. Agric. Food Chem. 55 (2007) 4625–4629.
[84] F. Li, Y. Hsieh, J. Sep. Sci. 31 (2008) 1231–1237.
[85] K. Inoue, Y. Hattori, T. Hino, H. Oka, J. Pharm. Biomed. Anal 51 (2010) 1154–1160.
[86] X. Feng, B.F. Liu, J. Li, X. Liu, Mass Spectrom. Rev. in press. http://dx.doi.org/10.1002/mas.21417.
[87] P.D. Whitfield, A.J. German, P.J. Noble, Br. J. Nutr. 92 (2004) 549–555.
[88] F. van der Kooy, F. Maltese, Y.H. Choi, H.K. Kim, R. Verpoorte, Planta Med. 75 (2009) 763–775.
[89] H.J. Issaq, Q.N. Van, T.J. Waybright, G.M. Muschik, T.D. Veenstra, J. Sep. Sci. 32 (2009) 2183–2199.
[90] A. Zhang, H. Sun, P. Wang, Y. Han, X. Wang, Analyst 137 (2012) 293–300.
[91] S. Kaspar, M. Peukert, A. Svatos, A. Matros, H.P. Mock, Proteomics 11 (2011) 1840–1850.
[92] Y.J. Lee, D.C. Perdian, Z. Song, E.S. Yeung, B.J. Nikolau, Plant J. 70 (2012) 81–95.
[93] P. Franceschi, Y. Dong, K. Strupat, U. Vrhovsek, F. Mattivi, J. Exp. Bot. 63 (2012) 1123–1133.
[94] P. Giavalisco, J. Hummel, J. Lisec, A.C. Inostroza, G. Catchpole, L. Willmitzer, Anal. Chem. 80 (2008) 9417–9425.
[95] R.J. Weber, A.D. Southam, U. Sommer, M.R. Viant, Anal. Chem. 83 (2011) 3737–3743.
[96] S.C. Nanita, J.J. Stry, A.M. Pentz, J.P. McClory, J.H. May, J. Agric. Food Chem. 59 (2011) 7557–7568.
[97] S.C. Nanita, N.L. Padivitage, Anal. Chim. Acta 768 (2013) 1–11.
[98] B. Gao, Y. Lu, Y. Sheng, P. Chen, L.L. Yu, J. Agric. Food Chem. 61 (2013) 2957–2963.
[99] B. Kluger, C. Bueschl, M. Lemmens, F. Berthiller, G. Häubl, G. Jaunecker, G. Adam, R. Krska, R. Schuhmacher, Anal. Bioanal. Chem. 405 (2013) 5031–5036.
[100] P. Jonsson, A.I. Johansson, J. Gullberg, J. Trygg, J. A, B. Grung, S. Marklund, M. Sjöström, H. Antti, T. Moritz, Anal. Chem. 77 (2005) 5635–5642.
[101] W. Jiang, Y. Qiu, Y. Ni, M. Su, W. Jia, X. Du, J. Proteome Res. 9 (2010) 5974–5981.
[102] X. Du, S.H. Zeisel, Comput. Struct. Biotechnol. J. 4 (2013) e201301013.
[103] G. Schlotterbeck, A. Ross, F. Dieterle, H. Senn, Pharmacogenomics 7 (2006) 1055–1075.
[104] J. Trygg, S. Wold, J. Chemom. 16 (2002) 119–128.
[105] D.S. Wishart, T. Jewison, A.C. Guo, M. Wilson, C. Knox, Y. Liu, Y. Djoumbou, R. Mandal, F. Aziat, E. Dong, S. Bouatra, I. Sinelnikov, D. Arndt, J. Xia, P. Liu, F. Yallou, T. Bjorndahl, R. Perez-Pineiro, R. Eisner, F. Allen, V. Neveu, R. Greiner, A. Scalbert, Nucleic Acids Res. 41 (2013) D801–D807.
[106] J.A. Rothwell, J. Perez-Jimenez, V. Neveu, A. Medina-Remón, N. M'hiri, P. García-Lobato, C. Manach, C. Knox, R. Eisner, D.S. Wishart, A. Scalbert, Database 2013 (2013) bat070.

[107] H. Ogata, S. Goto, K. Sato, W. Fujibuchi, H. Bono, M. Kanehisa, Nucleic Acids Res. 27 (1999) 29–34.

[108] M.D. Stobbe, S.M. Houten, A.H. van Kampen, R.J. Wanders, P.D. Moerland, FASEB J. 26 (2012) 3625–3636.

[109] J. Cerdà, M. Manchado, Genes Nutr. 8 (2013) 5–17.

[110] I. Montoliu, U. Genick, M. Ledda, S. Collino, F.P. Martin, J. le Coutre, S. Rezzi, Genes Nutr. 8 (2013) 19–27.

[111] J.A. Rothwell, Y. Fillâtre, J.F. Martin, B. Lyan, E. Pujos-Guillot, L. Fezeu, S. Hercberg, B. Comte, P. Galan, M. Touvier, C. Manach, PLoS One 9 (2014) e93474.

[112] J. Regueiro, A. Vallverdú-Queralt, J. Simal-Gándara, R. Estruch, R. Lamuela-Raventós, J. Agric. Food Chem. 61 (2013) 6763–6768.

[113] G. Favé, M. Beckmann, A.J. Lloyd, S. Zhou, G. Harold, W. Lin, K. Tailliart, L. Xie, J. Draper, J.C. Mathers, Metabolomics 7 (2011) 469–484.

[114] J. Smedsgaard, J. Nielsen, J. Exp. Bot. 56 (2005) 273–286.

[115] H. Yoshida, J. Yamazaki, S. Ozawa, T. Mizukoshi, H. Miyano, J. Agric. Food Chem. 57 (2009) 1119–1126.

[116] J. Kim, J.N. Choi, K.M. John, M. Kusano, A. Oikawa, K. Saito, C.H. Lee, J. Agric. Food Chem. 60 (2012) 9746–9753.

[117] F. Mattivi, R. Guzzon, U. Vrhovsek, M. Stefanini, R. Velasco, J. Agric. Food Chem. 54 (2006) 7692–7702.

[118] X.L. Shu, T. Frank, Q.Y. Shu, K.H. Engel, J. Agric. Food Chem. 56 (2008) 11612–11620.

[119] A.K. Grennan, Plant Physiol. 151 (2009) 1701–1702.

[120] R. Díaz, O.J. Pozo, J.V. Sancho, F. Hernández, Food Chem. 157 (2014) 84–93.

[121] M. Scampicchio, T. Mimmo, C. Capici, C. Huck, N. Innocente, S. Drusch, S. Cesco, J. Agric. Food Chem. 60 (2012) 11268–11273.

[122] B. Gilbert-López, J.F. García-Reyes, C. Meyer, A. Michels, J. Franzke, A. Molina-Díaz, H. Hayen, Analyst 137 (2012) 5403–5410.

Index

Note: Page numbers followed by "f" and "t" indicate figures and tables, respectively.

A

AA. *See* Amino acid
AAS. *See* Atomic Absorption Spectrometry
AB. *See* Arsenobetaine
Abiotic stress, 327–329. *See also* Biotic stress
  tolerance improvement in crops, 347f
ABM. *See* "Aceto balsamico di Modena"
Absorption, distribution, metabolism, and excretion (ADME), 345–346
Abundance sensitivity, 147
AC. *See* Alternating current; Arsenocholine
Acacia honey, 458
Accelerated solvent extraction (ASE), 412
Accurate mass tag (AMT), 320
Acetaia, 450
"Aceto balsamico di Modena" (ABM), 218, 450
ACP. *See* Acyl carrier protein
Active food packaging, 584, 586
Acyl carrier protein (ACP), 401
ADAP. *See* Automated data analysis pipeline
Adequate intake (AI), 138–139
ADME. *See* Absorption, distribution, metabolism, and excretion
Adulteration
  buffalo mozzarella, 41–42
  food, 13–14, 214–216, 444
  of milk, 338
Aflatoxin (AF), 302, 520
Aflatoxin B1 (AFB1), 302
Aflatoxin B2 (AFB2), 302
Ag-HPLC. *See* Silver ion high-performance liquid chromatography
AgNP. *See* Silver nanoparticle
AI. *See* Adequate intake
All-ion fragmentation (AIF), 109–111
Allergenic proteins, 367f
α-lactalbumin, 369
α-tocopherol, 407f
Alternating current (AC), 147–148
Ambient ionization MS techniques
  application, 259
    authentication and quality control, 264–266
    DESI, 261–262
    EESI-QTOF mass spectra of strawberries, 266f
    in food analysis, 259, 259t–260t, 261
    imaging analysis of imprinted potato sprout, 262f
    maturity, 266–267
    novel ambient air techniques, 264
    pesticides in fruits and vegetables, 262–264
    PGDI, 267
    rapid screening of food frauds, 267
  principles and classes, 248–249
    laser ablation and ionization, 257–258
    solid–liquid extraction ambient ionization techniques, 249–253
    thermal desorption and chemical ionization, 254–257
Ambient ionization techniques, 517. *See also* Solid–liquid extraction ambient ionization techniques
Ambient ionization–accurate mass spectrometry (AMI–AMS), 264
Ambient pressure laser desorption (APLD), 298
AMDIS. *See* Automated mass spectral deconvolution and identification system
AMI–AMS. *See* Ambient ionization–accurate mass spectrometry
Amino acid (AA), 290, 291t, 313
(2S,3R,4E)-2-amino-4-octadecen-1,3-diol. *See* Sphingosine

Amnesic shellfish poisoning (ASP), 520–521
AMT. *See* Accurate mass tag
Animal farming, 330–331
Animal products, 457–458, 473–474, 485, 497
ANN. *See* Artificial neural network
Anodic stripping voltammetry (ASV), 476
Antigen-presenting cell (APC), 361–362
APC. *See* Antigen-presenting cell
APCI. *See* Atmospheric pressure chemical ionization
APGDI. *See* Atmospheric pressure glow discharge ionization
API. *See* Atmospheric pressure ionization
APLD. *See* Ambient pressure laser desorption
APPI. *See* Atmospheric pressure photoionization
Aptamers, 301
APTDI. *See* Atmospheric pressure thermal desorption/ionization
Aquaculture, 331–332
Aquanova®, 590–591
Aquatic invertebrates, 338–341
AR. *See* Average requirements
Arabidopsis Information Resource (TAIR), 326
Area under curve (AUC), 672–674
Argentation chromatography, 423
Arsenic species, 201
Arsenobetaine (AB), 201
Arsenocholine (AC), 201
Artificial coloring reagents, 293–294
Artificial neural network (ANN), 645–646
ASAP. *See* Atmospheric pressure solids analysis probe
ASE. *See* Accelerated solvent extraction
AsFFFF. *See* Asymmetrical flow field-flow fractionation
ASP. *See* Amnesic shellfish poisoning
Aspartame, 292–293
ASV. *See* Anodic stripping voltammetry
Asymmetrical flow field-flow fractionation (AsFFFF), 22, 26f
Atmospheric pressure chemical ionization (APCI), 39, 94, 413–414, 416, 489, 517, 538
  ambient ionization MS systems, 248–249
  for HPLC/MS determination of lipids, 426–427
Atmospheric pressure glow discharge ionization (APGDI), 249, 256
Atmospheric pressure ionization (API), 5, 246–247, 489, 516–517
Atmospheric pressure photoionization (APPI), 254, 413–414, 489, 545
Atmospheric pressure solids analysis probe (ASAP), 249, 256, 517, 565
Atmospheric pressure thermal desorption/ionization (APTDI), 249, 257
Atomic Absorption Spectrometry (AAS), 172
AUC. *See* Area under curve
Authentication, 445, 446f
  food, 205
  ICP-MS, 206–207
  trace element availability to plants, 206
Authenticity, 22, 41–42, 205–206
Automated data analysis pipeline (ADAP), 671–672
Automated mass spectral deconvolution and identification system (AMDIS), 48–55
Average requirements (AR), 138–139
Azaspiracid shellfish poisoning (AZP), 520–521

B

Bacterial identification, 624
Bacterial source tracking method (BST method), 632
BAMS. *See* Bioaerosol mass spectrometry
Basic local alignment search tool (BLASTing), 372–374
BBM. *See* Brush border membrane
Beef cattle meat, 333–334
Bellows system, 179–180
Benzene, toluene, ethylbenzene, and xylene (BTEX), 48–55
β-lactoglobulin, 369
Beverages, 449–450
Bioactive compound protection, 592
Bioactive peptides (BP), 341
  classical, bioinformatics, and integrated approaches, 343f
  effects, 341–342
  encryption in protein amino acid sequence, 342–344
  food-derived, 341
  sources, 345–346
  in vitro and in vivo BP production, 344–345
Bioaerosol mass spectrometry (BAMS), 640–641
Bioavailability improvement, 591–592
Biodegradable packaging system, 588
Biogenic amines, 286–288

Bioinformatics
software, 343
tools, 318, 321–322
Biological approaches, 246
Biomarker-selective algorithm, 645–646
BIOPEP, 343
Bioral™, 590–591
Biotic stress, 329–330. *See also* Abiotic stress
Biotoxins, 520
N,O-Bis(trimethylsilyl)trifluoroacetamide (BSTFA), 528
BLASTing. *See* Basic local alignment search tool
Bottom-up proteomics, 313, 314f, 316
analytical steps and techniques, 316f
in-gel bottom-up proteomics, 316–318
shotgun proteomics, 318
mass spectrometry, 320–321
peptide separation, 319–320
proteome informatics tools, 321–325
quantitative proteomics, 321–325
sample preparation, 318–319
"Bottom-up" manufacturing, 583
Bovine serum albumin (BSA), 367–369
Bovine spongiform encephalopathy (BSE), 453
BP. *See* Bioactive peptides
Brush border membrane (BBM), 361–362
BSA. *See* Bovine serum albumin
BSE. *See* Bovine spongiform encephalopathy
BSIA. *See* Bulk Stable Isotope Analysis
BST method. *See* Bacterial source tracking method
BSTFA. *See* N,O-Bis(trimethylsilyl) trifluoroacetamide
BTEX. *See* Benzene, toluene, ethylbenzene, and xylene
Bulk Stable Isotope Analysis (BSIA), 180. *See also* Compound Specific Isotope Analysis (CSIA)

C

^{13}C amino acids fingerprints, 219–220
C_{18} phase, 424–425
CA. *See* Cluster analysis
Caco-2 cells, 369
CAD. *See* Collision-assisted dissociation
Caffeine, 290–292
Calcium, 138
CAM. *See* Crassulacean acid metabolism
Campylobacter species, 635–636
Canola Active oil, 590–591
Canonical discriminant analysis (CDA), 468–469
Capillary electrochromatography (CEC), 420–421, 663–664
Capillary electrophoresis (CE), 77, 164, 246–247, 489, 563–565, 656
coupling with ICP-MS, 170
in lipidomics, 397
Capillary electrophoresis inductively coupled plasma mass spectrometry (CE-ICP-MS), 170–171
Capillary electrophoresis–mass spectrometry (CE-MS), 4, 61, 642, 666–667
CE-MS/MS EIE, 62f
food safety and quality, 61, 63t–64t
Capillary gel electrophoresis (CGE), 663–664
Capillary isoelectric focusing (CIEF), 663–664
Capillary zone electrophoresis (CZE), 171, 420–421, 663–664
Capillary-liquid chromatography (CLC), 563–565
Carbon dioxide (CO_2), 411
Carbon number (CN), 423
Carbowax-templated resin (CW-TPR), 661–662
Cardiolipin. *See* Diphosphatidylglycerol
Caseins, 337
Caviar, 485
CCC. *See* Counter-current chromatography
CD. *See* Celiac disease; Corona discharge
CDA. *See* Canonical discriminant analysis
CE. *See* Capillary electrophoresis
CE-ICP-MS. *See* Capillary electrophoresis inductively coupled plasma mass spectrometry
CE-MS. *See* Capillary electrophoresis–mass spectrometry
CEC. *See* Capillary electrochromatography
Celiac disease (CD), 363
binding of gliadin fragments to HLA-DQ2/DQ8 molecules, 364–365
genetic factors, 363
gliadin peptides, 365
Gluten proteins, 364
wheat proteins distribution, 364f
Centrifugal ultrafiltration, 662–663
Cephalin. *See* Phosphatidylethanolamine (PE)
Ceramides, 405, 406f
Cereals products, 208, 458–459, 474, 486, 497

Cesium (Cs), 206
CF-EDESI. *See* Continuous flow extractive desorption electrospray ionization
CF-IRMS. *See* Continuous Flow Isotope Ratio Mass Spectrometry
CFP. *See* Ciguatera fish poisoning
cfu. *See* colony-forming unit
CGE. *See* Capillary gel electrophoresis
Changeover valve, 179–180
Charge-residue model (CRM), 414–415
Chemical element, 174
Chemical ionization (CI), 413–414, 417–418, 538
Chemical markers
  foodomics, 488f
  isotope ratios as, 461–475
  molecular ions as, 486–498
  trace elements as, 476–486
Chemometrics, 65
  in food forensics, 448–449
Chernobyl, 481–482
Chinese liquors, 267
Cholesterol, 407f
Chromatographic fingerprinting, 60
Chromium, 202
CI. *See* Chemical ionization
CID. *See* Collision-induced dissociation
CIEF. *See* Capillary isoelectric focusing
Ciguatera fish poisoning (CFP), 520–521
CII. *See* Compressed ion injection
Cis-9-octadecenoic acid. *See* Linoleic acid
CJD. *See* Creutzfeldt–Jakob disease
Classical dry ashing procedures, 21–22
Classification trees (CTs), 208
CLC. *See* Capillary-liquid chromatography
Cluster analysis (CA), 209, 448–449
CN. *See* Carbon number
Coagulase-negative *Staphylococci* (CoNS), 632–633
Cobalamin, 136
Cobalt, 136
Codex Alimentarius Commission (CODEX), 135
Coenzyme Q10 (CoQ10), 294f
Coffee, 209, 283, 473, 484
Cold Vapor-Atomic Absorption Spectrometry (CV-AAS), 172
Collision-assisted dissociation (CAD), 290
Collision-induced dissociation (CID), 92, 372–374, 425–426, 526, 558–559
Collision/reaction cell technology, 148
  DRC, 148–149, 149f
  KED, 149
  linear RF-driven multipole ion guide, 148
colony-forming unit (cfu), 633
Compound lipoids, 397
Compound-specific isotope analysis (CSIA), 57, 180, 462. *See also* Bulk Stable Isotope Analysis (BSIA)
  GC-IRMS, 181
  LC-IRMS, 181–182
Comprehensive GC (GC × GC), 543–545
Comprehensive LC (LC × LC), 561–563
  two-dimensional separation of HBCD, 564f
Compressed ion injection (CII), 92
CoNS. *See* Coagulase-negative *Staphylococci*
Continuous flow extractive desorption electrospray ionization (CF-EDESI), 418
Continuous Flow Isotope Ratio Mass Spectrometry (CF-IRMS), 179–180
  BSIA, 180
  CSIA, 181
    GC-IRMS, 181
    LC-IRMS, 181–182
Copper, 137
CoQ10. *See* Coenzyme Q10
Corona discharge (CD), 279
Counter-current chromatography (CCC), 523–524
Crassulacean acid metabolism (CAM), 177, 462–464
Creutzfeldt–Jakob disease (CJD), 453
CRM. *See* Charge-residue model
Crop yield improvement, 310–311
CSIA. *See* Compound-specific isotope analysis
CTs. *See* Classification trees
CV-AAS. *See* Cold Vapor-Atomic Absorption Spectrometry
CW-TPR. *See* Carbowax-templated resin
Cyclo-1,3,5-trimethylene trinitramine (RDX)3, 264
Cyclo-1,3,5,7-tetramethylene tetranitrate (HMX), 264
CZE. *See* Capillary zone electrophoresis

D

D-erythro-sphingosine. *See* Sphingosine
DA. *See* Discriminant analysis
Dairy products, 208, 337–338
DAPCI. *See* Desorption atmospheric pressure chemical ionization
DAPPI. *See* Desorption atmospheric pressure photoionization
DART. *See* Direct analysis in real time

Data analysis, 671–674
for foodomics, 671f
methods, 645–646
Data dependent acquisition (DDA), 9–11, 105–108, 321
Data-independent acquisition (DIA), 108–109
AIF, 109–111
fragmentation of flavonoid mono-O-glycosides in MSE, 110f
MS^E approach, 109
Data matrix, 12
DBDI. *See* Dielectric barrier discharge ionization
DBN. *See* Number of double bonds
DBs. *See* Double bonds
DC. *See* Direct current
DCBI. *See* Desorption corona beam ionization
DCHP. *See* Dicyclohexyl phthalate
DDA. *See* Data dependent acquisition
Defatted dry mass (DDM), 469–470
δ-notation, 174–175
Deoxynivalenol (DON), 520
Derivatization, 278, 286
Derived lipoids, 397
DESI. *See* Desorption electrospray ionization
Desorption atmospheric pressure chemical ionization (DAPCI), 12–13, 249, 256
Desorption atmospheric pressure photoionization (DAPPI), 12–13, 249, 257
Desorption corona beam ionization (DCBI), 249, 256
Desorption electrospray ionization (DESI), 12–13, 248, 250, 279, 418, 489, 565–567, 623, 637
Desorption ionization by charge exchange (DICE), 249, 253
Desorption ionization on porous silicon (DIOS), 19
Detection limits (DLs), 139–140
Determination of origin, 216–218
DFA. *See* Discriminant function analysis
DHA. *See* Docosahexaenoic acid
DIA. *See* Data-independent acquisition
DI-IRMS. *See* Dual-Inlet Isotope Ratio Mass Spectrometry
DI-MS. *See* Direct infusion mass spectrometry
1,2-diacyl-*sn*-glycerol-3-phosphate. *See* Phosphatidic acid
Diarrheic shellfish poisoning (DSP), 520–521
DICE. *See* Desorption ionization by charge exchange
Dicyclohexyl phthalate (DCHP), 296–297
Dielectric barrier discharge ionization (DBDI), 249, 256
Differential in-gel electrophoresis (DIGE), 317
Differential mobility spectrometry (DMS), 276–277, 277f
Diffusive fractionation, 176
DIGE. *See* Differential in-gel electrophoresis
"Dilute-and-shoot" extraction, 559
Dimethyl arsonous acid (DMAA-As(III)), 201
Dimethyl diselenide (DMDSe), 202
Dimethyl phthalate (DMP), 297
Dimethyl selenide (DMSe), 202
Dimethylarsinic acid (DMAA), 201
DIOS. *See* Desorption ionization on porous silicon
Dip Pen Nanolithography, 587
Diphosphatidylglycerol, 402, 403f
Direct analysis in real time (DART), 12–13, 248, 254, 279, 489, 517, 637
Direct current (DC), 147–148
Direct infusion mass spectrometry (DI-MS), 670
Direct solid sampling analysis, 161
food commodities, 162
laser, 161–162
Direct swabbing and analysis, 277–278
Direct thermal desorption (DTD), 282–283
Directspray ESI source, 296, 296f
Discrete dynode detector, 154
Discriminant analysis (DA), 208, 448–449
Discriminant function analysis (DFA), 497
Dispersive SPE (dSPE), 521–522
Divinylbenzene (DVB), 661–662
DLs. *See* Detection limits
DLS. *See* Dynamic light scattering
DMAA. *See* Dimethylarsinic acid
DMAA-As(III). *See* Dimethyl arsonous acid
DMDSe. *See* Dimethyl diselenide
DMP. *See* Dimethyl phthalate
DMS. *See* Differential mobility spectrometry
DMSe. *See* Dimethyl selenide
Docosahexaenoic acid (DHA), 284–285
DON. *See* Deoxynivalenol
Double bonds (DBs), 399
Double-sector instruments, 82–83
DRC. *See* Dynamic reaction cell
Drift tube ion mobility spectrometry (DT-IMS), 276–277, 277f
Drought, 328
DRV. *See* European Dietary Reference Values
DRVs. *See* US Dietary reference values
Dry-cured ham, 334–335
DSP. *See* Diarrheic shellfish poisoning

dSPE. *See* Dispersive SPE
DT-IMS. *See* Drift tube ion mobility spectrometry
DTD. *See* Direct thermal desorption
Dual-Inlet Isotope Ratio Mass Spectrometry (DI-IRMS), 179–180
DVB. *See* Divinylbenzene
Dynamic light scattering (DLS), 596–601
Dynamic reaction cell (DRC), 21–22, 148, 149f, 157

E

e-nose. *See* Electronic nose
E. coli. *See Escherichia coli*
EA-IRMS. *See* Elemental analyzer coupled with isotope ratio mass spectrometer
EADESI. *See* Electrode-assisted desorption electrospray ionization
Easy ambient sonic spray ionization (EASI), 249–253
EC. *See* European Commission
ECN. *See* Equivalent carbon number
ED. *See* Erectile dysfunction
Edible coatings, 588
EDX. *See* Energy dispersive X-ray analysis
EFSA. *See* European Food Safety Authority
EGCG. *See* Epigallocatechin gallate
Eggs, 337–338
EI. *See* Electron ionization
EIA. *See* Enzyme immunoassay
EIC. *See* Extracted ion chromatogram
Eicosapentaenoic acid (EPA), 284–285
ELDI. *See* Electrospray laser desorption ionization
Electrode-assisted desorption electrospray ionization (EADESI), 249–253
Electron impact ionization. *See* Electron ionization (EI)
Electron ionization (EI), 5, 48, 413–414, 417–418, 538, 670–671
Electron microscopy (EM), 596–598
Electron transfer dissociation (ETD), 97–100
Electronic nose (*e-nose*), 283
  devices, 283
  system, 490, 493
Electrospray ionization (ESI), 5, 168–169, 248–250, 278, 413–415, 487–496, 618, 623
Electrospray laser desorption ionization (ELDI), 112–113, 249, 257–258
Electrostatic analyzer (ESA), 150
ElectroThermal Atomic Absorption Spectrometry (ETAAS), 172
Electrothermal vaporization (ETV), 161
Elemental analyzer coupled with isotope ratio mass spectrometer (EA-IRMS), 180
Elemental mass spectrometry. *See also* Isotopic mass spectrometry
  ICP-MS, 139–172
Elemental speciation, 200
ELISA. *See* Enzyme-linked immunosorbent assay
EM. *See* Electron microscopy
Emerging contaminants, 515–516
  acrylamides, 521
  analytical methods using for separation and detection, 516f
  DESI, 565–567
  extraction techniques, 521–524
  food analysis, 563–565
  GC-MS, 528–545
  IMS, 517, 568
  LC-MS, 545–563
  LESA automation steps, 567f
  mass spectrometry analyzers, 524–528
  UHPLC, 517
Encapsulation, 592
Energy dispersive X-ray analysis (EDX), 596–598
Engineered Nanomaterials (ENM), 582–583, 595. *See also* Nanomaterials
Enhanced product ion (EPI), 525–526
ENM. *See* Engineered Nanomaterials
Environmental condition, 659–660
Enzyme immunoassay (EIA), 620
Enzyme-linked immunosorbent assay (ELISA), 372, 620
EPA. *See* Eicosapentaenoic acid
EPI. *See* Enhanced product ion
Epigallocatechin gallate (EGCG), 294
Epitope mapping, 369
Equilibrium fractionation, 176
Equivalent carbon number (ECN), 423
Erectile dysfunction (ED), 295–296
ESA. *See* Electrostatic analyzer
Escherichia coli (E. coli), 618
ESI. *See* Electrospray ionization
EST. *See* Expressed sequence tag
ETAAS. *See* ElectroThermal Atomic Absorption Spectrometry
ETD. *See* Electron transfer dissociation
ETV. *See* Electrothermal vaporization
EU. *See* European Union

European Commission (EC), 135
European Dietary Reference Values (DRV), 138–139
European Food Safety Authority (EFSA), 138–139, 594–596
European Union (EU), 297, 332, 360, 515–516
EVOO. *See* Extra virgin olive oil
Expressed sequence tag (EST), 47, 326
Extra virgin olive oil (EVOO), 289
Extracted ion chromatogram (EIC), 322–325, 524–525
Extraction techniques, 521
  ion suppression/enhancement, 523
  ion-pair extraction, 522–523
  miniaturized extraction techniques, 524
  MIPs, 523
  QuEChERS, 521–522

F

^{4}F. *See* Asymmetrical flow field-flow fractionation (AsFFFF)
FA. *See* Fatty acids; Food allergy
FA methyl esters (FAME), 431
FA-APGDI. *See* Flowing afterglow/atmospheric pressure glow discharge ionization
FAB. *See* Fast atom bombardment
Fabrication of nanomaterials, 583
FAIMS. *See* Field asymmetric ion mobility spectrometry
FAME. *See* FA methyl esters
FAO. *See* Food and Agriculture Organization
Faraday cups (FC), 183
Farm animals, 335–336
  milk, 337–338
Fast atom bombardment (FAB), 5, 413–414, 418
Fatty acids (FA), 399, 400t
  ACP, 401
  conjugated dienoic, 401
  essential, 399–401
  monoenoic, 399
  octadecaenoic acid, 399
FBs. *See* Fumonisins
FC. *See* Faraday cups
FCIA. *See* Flow cytometric immunoassay
FDA. *See* Food and Drug Administration
Fermented drinks, 449–450, 465–468, 479–480, 490–493
FFA. *See* Free fatty acids
FFF. *See* Field-flow fractionation
FFF-ICP-MS. *See* Field-flow fractionation inductively coupled plasma mass spectrometry
FI-ESI-MS. *See* Flow-injection electrospray ionization mass spectrometry
Field asymmetric ion mobility spectrometry (FAIMS), 276–277, 281–282
Field-flow fractionation (FFF), 164, 169, 600
Field-flow fractionation inductively coupled plasma mass spectrometry (FFF-ICP-MS), 169–170, 599–600
  AF4-ICP-MS, 602
  asymmetric flow FFF, 600–601, 601f
  ICP-MS, 601–602
Fingerprint, 47–48
Fingerprint approach, 624
  MALDI-TOF-MS, 626
Fish, 331–332, 338–341, 345, 454–455, 470–471, 482, 495–496
Fish products, 209
Flavonoids, 290
Flavor-related compounds, 288
  analytical properties, 289
  IMS, 289
  organoleptic deviations in wines, 288
  2,4,6-TCA, 288–289
Flow cytometric immunoassay (FCIA), 565
Flow-injection electrospray ionization mass spectrometry (FI-ESI-MS), 637–638
Flowing afterglow/atmospheric pressure glow discharge ionization (FA-APGDI), 256
Fluoroquinolone antibiotic enrofloxacin, 264
FMASE. *See* Focused microwave-assisted solvent extraction
Focused microwave-assisted solvent extraction (FMASE), 412
Food
  additives, 292–293
  analysis, 563–565
  authentication, 194, 205–210, 446f
  authenticity, 205–206, 311–312, 312f
  contact materials, 584
    coatings for kitchen utensils and processing equipment, 588
    nanofiltration membranes, 589
  contaminant, 297
  edibility, 311–312
  frauds, 442, 444
  hypersensitivities, 361f
  ingredients, 209
  microbiology, 677

Food (*Continued*)
nanoparticles in, 211–212, 595–596
EFSA, 596
EM, 596–598
FFF-ICP-MS, 600–602
nanoparticle detection and characterization, 599–600
NIST, 596
sp-ICP-MS, 602–605
nanotechnology, 581–582
packaging system, 584
active packaging, 586
applications of nanomaterials, 584
barrier properties, 584–585
biodegradable packaging system, 588
edible coatings, 588
heat resistance, 586
intelligent/smart packaging, 587
mechanical strength, 586
tracking, tracing, and brand protection, 587
proteins triggering allergic reactions, 362–363
quality, 132, 311–312, 654, 677–679
safety, 3, 132, 311–312, 312f, 516, 654, 677–679
sensitivity, 360–361
structuring, 592–593
traceability, 311–312
Food adulterant detection
dietary supplements and other food, 294–295
PDE-5 inhibitors, 295–296
phthalate compounds, 297
phthalate ester, 296–297
portable IMS, 295–296
weight loss supplements, 295
Food allergy (FA), 359
advances in MS methods
intact food allergens detection, 375
peptides tracing detection, 376–386
CD, 363
binding of gliadin fragments to HLA-DQ2/DQ8 molecules, 364–365
genetic factors, 363
gliadin peptides, 365
Gluten proteins, 364
wheat proteins distribution, 364f
FA-related diseases, 360
food proteins triggering allergic reactions, 362–363
IgE *vs*. non-IgE-mediated FA, 360–362
mass spectrometry role applied to
allergen characterization, 367–370
in celiac disease, 370–372
features, 372–374
proteomic-based approaches, 368f, 373f
prevalence and Legislative Frame Worldwide, 365
allergenic proteins, 367f
EU, 366
International Codex Alimentarius Commission, 365–366
regulatory framework, 366–367
Food and Agriculture Organization (FAO), 310–311, 362
Food and Drug Administration (FDA), 135, 454
Food flavor compounds, 282–283
coffee, 283
"electronic nose" devices, 283
ion mobility spectra for fresh and aged coffee, 284f
ion mobility-based electronic nose, 283
omega-3 acids, 284–285
portable UV-IMS instrument, 283–284
Food forensics, 442
chemometrics, 448–449
crooked traders, 443
food authentication, 445, 446f
food production chains, 447f
food traceability, 446f
MS analysis
analytical methods in, 460
chemical markers in food classification, 461f
isotope ratios as chemical markers, 461–475
molecular ions as chemical markers, 486–498
trace elements as chemical markers, 476–486
principal foods subjected to fraud, 449
animal products, 457–458
beverages, 449–450
fermented drinks, 449–450
fish, 454–455
fruits, 455–457
meats, 453–454
milk and dairy products, 451–452
organic food, 459–460
products from cereals, 458–459
vegetable oils, 452
vegetables, 455–457
wine, 449–450

types of frauds, 444
verifying foodstuff, 445
Food matrices, 656–657
food metabolites related to
environmental condition, 659–660
genes, 657–658
nutritional status, 658–659
quality grade, 659
stress impact, 658
Food pathogens, 618
foodborne pathogens, 618–620
identification by MS, 624
BAMS, 640–641
CE-MS, 642
ESI-MS, 636–638
GC-MS, 641
LC-MS, 642–644
MALDI-TOF-MS, 624–636
MS data analysis methods, 645–646
PCR-ESI-MS, 638–640
pyrolysis–GC-MS, 641
SELDI-MS, 644–645
microorganism diagnosis, 620
MS in microbial analysis, 620–624
Food peptidomics and digestomics, 341
bioactive peptides, 341
classical, bioinformatics, and integrated approaches, 343f
effects, 341–342
encryption in protein amino acid sequence, 342–344
food-derived, 341
sources, 345–346
in vitro and in vivo BP production, 344–345
Food safety and quality
IMS applications in, 114t
Orbitrap analyzer applications in, 101t–104t
TOF analyzer applications in, 85t–92t
Food safety, quality, and authentication, 332
food processing and quality control, 332–333
farm animal milk, dairy products, eggs, 337–338
fish and aquatic invertebrates, 338–341
meat, 333–337
GMOs, 332
Food security, 310, 312f, 326. *See also* Proteomics
livestock, 330–331
aquaculture, 331–332
factors related to animal agricultural sustainability, 331f
plant proteomics, 326–327
stress effects on gene expression, 327
abiotic stress, 327–329
biotic stress, 329–330
postharvest proteomics, 330
Foodborne
microbial detection
HS MCC-IMS, 286, 287f
with ion mobility spectrometry, 285
PCA, 285
pathogens
pathogenic bacteria, 619–620
spoilage bacteria, 618–619
Foodomics, 247, 311, 488f, 654–655
advanced approach with ms, 676
for food microbiology, 677
for food safety and quality, 677–679
for human response by foods, 676–677
data
analysis, 671–674, 671f
assessment, 674–675
food materials, 657t
food matrices, 656–660
with MS, 655–656, 655f–656f
MS detection, 669–671
sample preparations, 660–663
separation, 663–668
Fourier transform infrared (FTIR), 431
Fourier transform ion cyclotron resonance mass spectrometry (FT-ICR-MS), 95–97, 247–248, 425–426, 492–493, 527–528, 669
operating principles, 96f
Frauds in meat, 336–337
Free biogenic amines, 286
Free fatty acids (FFA), 40, 264–266
Free induction decay, 95–96
Fruits, 455–457, 471–473, 482–484, 496–497
juices, 208
FT-ICR MS. *See* Fourier transform ion cyclotron resonance mass spectrometry
FT-Infrared spectrophotometry (FT-IR), 460
FTIR. *See* Fourier transform infrared
Full width at half maximum (FWHM), 146–147
Fumonisins (FBs), 520
Fungi, 329
Fuzzy rule-building expert system (FuRES), 286
FWHM. *See* Full width at half maximum

G

Galician mussel (*Mytilus galloprovincialis*), 482
GALT. *See* Gut associated lymphoid tissue
Gas chromatography (GC), 22, 77, 164, 429–430, 516–517, 622, 656, 664
  derivatization, 278
  FA separation, 420–421
  FAMEs separation, 431
  free FAs, 431
  GC-FTIR, 432
  HTGC, 432
  lipid samples, 430–431
  multidimensional GC system, 432
  plasma FAMEs GC × GC-FID analysis, 433f
Gas chromatography combustion isotope ratio mass spectrometry (GCC-IRMS), 57
Gas chromatography-differential ion mobility spectrometry (GC-DMS), 288
Gas chromatography-mass spectrometry (GC-MS), 48, 56f, 528, 618, 641
  band broadening, 538
  comprehensive GC, 543–545
  emerging contaminants and analytical techniques, 529t–537t
  food quality and safety, 48, 49t–54t
  GC-MS/MS, 539
    ^{3}DIT application, 539
    GC-QqQ-MS/MS, 540
    QqTOF, 540–542
    SRM chromatograms, 541f
  MS detectors, 538–539
  pyrolysis–GC-MS, 641
  target and nontarget metabolomics, 57–58
    GC-QqQ/MS in SRM, 58
    novel approach, 58
  target compounds analysis, 48
    conventional applications, 48–55
    CSIA, 57
    ICP-MS, 57
    innovation strategies, 55
    speed of analysis, 55
    two-dimensional gas chromatography, 57
  TOF-MS types, 539
  volatile compounds profiling, 58–59
    GC×GC-qMS plot of Italian hazelnuts, 60f
    kiwifruit processing, 59–60
    quality control of cocoa beans, 59
    SBSE, 60–61
    volatile fractions of roasted hazelnuts, 60
Gas chromatography-tandem mass spectrometry (GC-MS/MS), 528
Gas chromatography–high-resolution mass spectrometry (GC–HRMS), 9
Gas chromatography–quadrupole mass spectrometry (GC–qMS), 9
Gastrointestinal tract (GIT), 313, 590
GC. *See* Gas chromatography
GC × GC. *See* Comprehensive GC
GC × GC-MS. *See* Two-dimensional gas chromatography-mass spectrometry
GC-DMS. *See* Gas chromatography-differential ion mobility spectrometry
GC-MS. *See* Gas chromatography-mass spectrometry
GC-MS/MS. *See* Gas chromatography-tandem mass spectrometry
GCB. *See* Graphitized carbon black
GCC-IRMS. *See* Gas chromatography combustion isotope ratio mass spectrometry
GC–HRMS. *See* Gas chromatography–high-resolution mass spectrometry
GC–qMS. *See* Gas chromatography–quadrupole mass spectrometry
GE. *See* Gel electrophoresis
GE-ICP-MS. *See* Gel electrophoresis inductively coupled plasma mass spectrometry
Gel electrophoresis (GE), 170
Gel electrophoresis inductively coupled plasma mass spectrometry (GE-ICP-MS), 170–171
Gel-filtration chromatography, 425
Gel-permeation chromatography (GPC), 425, 522–524
Gene expression, stress effects on, 327
  abiotic stress, 327–329
  biotic stress, 329–330
  postharvest proteomics, 330
General Food Law, 366–367
Genetically modified organisms (GMO), 310–311, 332, 498
Genomics, 309–310
Geographical origin, 205–206
Geometry-independent desorption electrospray ionization (GI-DESI), 249–253
Geosmin, 288, 288f
GF-AAS. *See* Graphite furnace atomic absorption spectrometry
GI-DESI. *See* Geometry-independent desorption electrospray ionization
GIT. *See* Gastrointestinal tract
GL. *See* Glycerolipids
Gliadins, 364

Glucosylceramides, 406
Glutathione-S-transferase (GST), 327–328
Gluten proteins, 364
Gluten-free diet, 371
Glutenins (GS), 364
Glyceroglycolipids, 402
Glycerolipids (GL), 401–402
Glycerophospholipids (GP), 402
Glycosphingolipids, 405
GMO. *See* Genetically modified organisms
GP. *See* Glycerophospholipids
GPC. *See* Gel-permeation chromatography
Graphite furnace atomic absorption spectrometry (GF-AAS), 21, 172, 476
Graphitized carbon black (GCB), 523, 543
Gray mold (*Botrytis cinerea*), 286
Green packaging, 584
GS. *See* Glutenins
GST. *See* Glutathione-S-transferase
Gut associated lymphoid tissue (GALT), 361–362

H

H/D exchange. *See* Hydrogen–deuterium exchange
Halogenated naturally produced compound (HNP), 543
HAPGDI. *See* Helium atmospheric pressure glow discharge ionization
HBCD. *See* Hexabromocyclododecane
HCD. *See* Higher energy collision dissociation
Head space isolation with solid phase microextraction (HS-SPME), 58–60
Head space stir-bar sorptive extraction (HS-SBSE), 58–59
Headspace (HS), 246
Headspace mass spectrometer (HS-MS), 490
Helium atmospheric pressure glow discharge ionization (HAPGDI), 249, 256
Herbicides, 300–301
Hexabromocyclododecane (HBCD), 520, 563
HF. *See* Hollow-fiber
HF-LPME. *See* Hollow fiber liquid phase microextraction
HFCS. *See* High-fructose corn syrup
HG-AAS. *See* Hydride Generation-Atomic Absorption Spectrometry
Hidden allergens, 366–367
High molecular weight glutenin subunits (HMW-GS), 364
High resolution electrospray ionization mass spectrometry (HR-ESI-MS), 638–640
High-resolution mass spectrometry (HRMS), 9–11, 78, 378–386, 524, 555
High temperature gas chromatography (HTGC), 432
High-fructose corn syrup (HFCS), 474
High-performance (HP), 422
  HP TLC, 422
High-performance liquid chromatography (HPLC), 372, 420–422, 664–666
  Ag-HPLC, 423–424
  NARP-HPLC, 424–425
  NP-HPLC, 422–423
High-resolution LC (HRLC), 558–559
High-resolution time-of-flight mass spectrometry (HR-TOF-MS), 539
High-speed countercurrent chromatography (HSCCC), 667–668
Higher energy collision dissociation (HCD), 7–8, 41, 97–100, 109–111
HILIC. *See* Hydrophilic interaction liquid chromatography
HLB. *See* Hydrophilic–lipophilic balance
HMDB. *See* Human Metabolome Database
HMO. *See* Human milk oligosaccharides
HMW-GS. *See* High molecular weight glutenin subunits
HMX. *See* Cyclo-1,3,5,7-tetramethylene tetranitrate
HNP. *See* Halogenated naturally produced compound
Hollow fiber liquid phase microextraction (HF-LPME), 300
Hollow-fiber (HF), 524
Homemade-inlet systems, 178–179
Honeys, 209, 458, 474
Horsegate, 442
HPLC. *See* High-performance liquid chromatography
HR-ESI-MS. *See* High resolution electrospray ionization mass spectrometry
HR-TOF-MS; High-resolution time-of-flight mass spectrometry
HRLC. *See* High-resolution LC
HRMS. *See* High-resolution mass spectrometry
HS. *See* Headspace
HS-MS. *See* Headspace mass spectrometer
HS-SBSE. *See* Head space stir-bar sorptive extraction
HS-SPME. *See* Head space isolation with solid phase microextraction

HSCCC. *See* High-speed countercurrent chromatography
HTGC. *See* High temperature gas chromatography
Human digestion, 344
Human health, 515–516
Human Metabolome Database (HMDB), 674–675
Human milk oligosaccharides (HMO), 220
Human response by foods, 676–677
Hydride Generation-Atomic Absorption Spectrometry (HG-AAS), 172
Hydrogen–deuterium exchange (H/D exchange), 11
Hydrophilic interaction liquid chromatography (HILIC), 26–39, 246–247, 425, 545
Hydrophilic–lipophilic balance (HLB), 523
Hydroxyceramides, 406
Hyphenated system, 489
Hyphenated techniques, 164
  CE-ICP-MS and GE-ICP-MS, 170–171
  conventional liquid phase separations, 164–165
  FFF-ICP-MS, 169–170
  GC-ICP-MS, 167–168, 168f
  ICP-MS, 165
  LC–ICP–MS, 165–166, 166f
    feasibility of coupling, 166
    low tolerance, 166–167
    robustness of ICP-MS instruments, 167
    semiquantitative results, 167
  multi-MS, 168–169
  separation techniques, 247

I

IAC. *See* Immunoaffinity columns
IC. *See* Ion chromatography
IC-ICP-MS. *See* Ion chromatography inductively coupled plasma mass spectrometry
IC-MS. *See* Intact cell mass spectrometry
ICAT. *See* Isotopic coding with affinity tag
ICG. *See* Immunochromatography
ICP-AES. *See* Inductively coupled plasma atomic emission spectroscopy
ICP-MS. *See* Inductively coupled plasma mass spectrometry
ICP-SF-MS. *See* Sector field inductively coupled plasma mass spectrometry (SF-ICP-MS)
ID. *See* Isotope dilution
IDA. *See* Information-dependent acquisition
IDM. *See* Ion desorption model
IEF. *See* Isoelectric focusing
IgE-mediated food allergies, 361–362
IM. *See* Ion mobility
Imaging mass spectrometry (IMS), 12–13, 19–21, 78, 112, 113f, 419, 517, 568. *See also* Mass spectrometry (MS)
  in food safety and quality, 114t
  ionization methods, 112–113
  in medical and biomedical field, 113
IMMS. *See* Ion mobility spectrometry and mass spectrometry
Immunoaffinity columns (IAC), 523
Immunochromatography (ICG), 620
IMS. *See* Imaging mass spectrometry; Ion mobility mass spectrometry; Ion mobility spectrometry
In-gel bottom-up proteomics, 316. *See also* Shotgun proteomics
  bioinformatics tools, 318
  PMF, 318
  problem with IEF, 316–317
  sample preparation, 317–318
  2D-PAGE, 316–317
INAO. *See* Institut National des Appellations d'Origine
Inductively coupled plasma atomic emission spectroscopy (ICP-AES), 21, 172, 476
Inductively coupled plasma mass spectrometry (ICP-MS), 21, 57, 133, 139–140, 466
  authentication, 206–207
    authenticity, 205–206
    European Union legislation, 206
    food types, 207–210
    trace element availability to plants, 206
  components, 140f
  conclusion, strengths and limitations, 171–172
  detection limits of contemporary, 143f
  direct solid sampling analysis, 161
    food commodities, 162
    laser, 161–162
  features and operation principles, 140
  food laboratory analysis, 21
  for food safety and quality, 23t–25t
  historical overview, 139–140
  hyphenated techniques, 164–171
  interferences, 154–155, 155t
    ICP-MS comparison with spectroscopic techniques, 173t
    nonspectroscopic interferences, 157–159
    spectroscopic interferences, 155–157

ion detection systems, 154
ion formation, 141–142
capacitive coupling, 144
plasma torch, 142
positively charged ions, 142–144
ion optics, 144–145
ion source, 141, 142f
limitations, 21–22
mass analyzer, 145–146
mass resolution and abundance sensitivity, 146, 146f
FWHM, 146–147
IT mass analyzer, 153–154, 153f
quadrupole, 147–149
SF, 149–151
TOF analyzer, 151–152
migration studies, 210–211
dietary intake studies, 212–213
metal elements, 211
nanoparticle, 211–212
MW-assisted digestion, 161
other techniques *vs*, 172
percentage distribution, 194f
remarkable types of applications, 22
sample introduction, 140–141
sample preparation, 159
liquid samples analysis, 159–160
nitric acid, 160
solid foods, 159–160
wet digestion methods, 160
separation techniques, 22
SP-ICP-MS, 22
time evolution, 195f
toxic and nutritive elements, 195
article deals, 196–197
dietary exposure estimates, 196
intake, 195
metallomics, 204–205
mineral content and levels, 198
rapid fusion method, 198–199
speciation, 199–204
trace elements, 197
trends, 194
Inductively coupled plasma optical emission spectrometry (ICP-OES). *See* Inductively Coupled Plasma Atomic Emission Spectroscopy (ICP-AES)
Information-dependent acquisition (IDA), 7–11, 105–108, 557–558
Infrared laser ablation metastable-induced chemical ionization (IR-LAMICI), 249
Infrared laser-assisted desorption electrospray ionization (IR-LADESI), 249, 258
Innovation strategies, 55
Insect infestation, 330
Institut National des Appellations d'Origine (INAO), 205–206
Intact cell mass spectrometry (IC-MS), 624–625
Intact food allergens detection, 375
Integrated bioinformatics process, 343–344
Intelligent food packaging, 584, 587
Inter alia, 313
Interferences, 154–155, 155t
ICP-MS comparison with spectroscopic techniques, 173t
nonspectroscopic interferences, 157–159
spectroscopic interferences, 155–157
Internal standardization, 158
Internal standards (ISs), 325
International Codex Alimentarius Commission, 365–366
International Union of Biochemistry and Molecular Biology (IUPAC-IUBMB), 398
International Union of Immunological Societies (IUS), 362
International Union of Pure and Applied Chemistry (IUPAC), 79
Iodine, 137
Ion
detection systems, 154
detector, 624
formation, 141–142
capacitive coupling, 144
plasma torch, 142
positively charged ions, 142–144
lenses, 145
optics, 144–145
source, 141, 142f
suppression/enhancement, 523
Ion chromatography (IC), 164, 302
Ion chromatography inductively coupled plasma mass spectrometry (IC-ICP-MS), 166
Ion desorption model (IDM), 414–415
Ion mobility (IM), 94
Ion mobility mass spectrometry (IMMS), 12–13, 280
Ion mobility spectrometry (IMS), 276, 517
components, 281t
CoQ10 sample, 294f
food adulterant detection
dietary supplements and other food with, 294–295
ESI, 296
PDE-5 inhibitors, 295–296

Ion mobility spectrometry (IMS) *(Continued)*
phthalate compounds, 297
phthalate ester, 296–297
portable IMS, 296
portable ion mobility spectrometers, 295
weight loss supplements, 295
food composition determination and food additives
amino acids, 290, 291t
artificial coloring and sweetening reagents, 293–294
caffeine, 290–292
flavonoids, 290
food additives, 292–293
mobility and mass spectra, 293f
nutritional supplements, 294
protonated aglycone isomers, quercetin and hesperetin separation, 292f
theophylline, 290–292
food contaminant analysis and residue, 297
microbial toxins analysis in food, 302–303
nitrite and nitrate analysis in food, 302
OTA, 302–303
pesticides detection, 297–301
veterinary drugs analysis, 301
food flavors analysis and composition
biogenic amines, 286–288
flavor-related compounds, 288–289
food flavor compounds, 282–285
foodborne microbial detection, 285–286
hyphenation, 280–282
mobility coefficient, 276
performance, 279
resolving power measurement in, 280f
sample introduction and ionization method, 277
CD sources, 279
DART ionization source, 279
derivatization, 278
ionization sources, 279
less-volatile samples, 277–278
^{63}Ni source, 279
Ion trap (IT), 55, 518, 525–526, 538–539
mass analyzer, 153–154, 153f
Ion trap time-of-flight hybrid mass spectrometer (qIT-TOF-MS), 645–646
Ion trap-time-of-flight (IT-ToF), 426
Ionization modes, 413–414
APCI, 416
APPI, 416
CI, 417–418
DESI, 418
EI, 417–418
ESI, 414–415
FAB, 418
MALDI, 416–417
Ionization potential (IP), 167
Ionization source, 622–623
IP. *See* Ionization potential
IPA. *See* Isopropanol
IR. *See* Isotope ration
IR-LADESI. *See* Infrared laser-assisted desorption electrospray ionization
IR-LAMICI. *See* Infrared laser ablation metastable-induced chemical ionization
IRMS. *See* Isotope ratio mass spectrometry
Iron, 136
Isobaric interferences, 184
Isobaric tags for relative and absolute quantification (iTRAQ), 322, 376
Isocratic elution, 423
Isoelectric focusing (IEF), 316, 663–664
Isopropanol (IPA), 423
Isotope dilution (ID), 164, 191
advantage, 192
critical stages, 192
procedure, 191–192
sources of error, 192
Isotope ratio mass spectrometry (IRMS), 133, 178, 184, 461–462. *See also* Inductively coupled plasma mass spectrometry (ICP-MS); Liquid chromatography-mass spectrometry (LC-MS); Multicollector inductively coupled plasma mass spectrometry (MC-ICP-MS); Thermal ionization mass spectrometry (TIMS)
authentication
adulteration, 214–216
determination of origin, 216–218
nutrient intake and bioavailability, 218–220
causes of stable isotopes variations, 176
electron impact ion production, 182
inlet system, 178
CF-IRMS, 180–182
DI-IRMS, 179–180
homemade-inlet systems, 178–179
instrumentation, 178
internationally recognized standards, 176t
isobaric interferences, 184
isotopes, 174
isotopic variations in

nature, 176–177
plants and animals, 177–178
percentage distribution, 213f
separation and ions detection
magnetic sectors, 182–183
multiple Faraday cup detectors, 183
stable isotope abundances expression and δ-notation, 174–175, 175t
time evolution, 214f
trends, 213
Isotope ration (IR), 22
Isotope ratios as chemical markers, 460–462
animal products, 473–474
beverages, 465–468
CSIA, 462
fermented drinks, 465–468
fish, 470–471
in food forensics, 464t
fractionation mechanism, 464–465
fruits and vegetables, 471–473
IRMS, 461–462
meats, 469–470
milk and dairy products, 468
organic food, 474–475
products from cereals, 474
vegetable oils, 468–469
wine, 465–468
Isotopes, 174
Isotopic coding with affinity tag (ICAT), 322, 376
Isotopic mass spectrometry, 174. *See also* Elemental mass spectrometry
IRMS, 174–184
MC-ICP-MS, 187–192
TIMS, 184–187
ISs. *See* Internal standards
IT. *See* Ion trap
IT-ToF. *See* Ion trap-time-of-flight
iTRAQ. *See* Isobaric tags for relative and absolute quantification
IUPAC. *See* International Union of Pure and Applied Chemistry
IUPAC-IUBMB. *See* International Union of Biochemistry and Molecular Biology
IUS. *See* International Union of Immunological Societies

K

K-nearest neighbors (KNN), 207–208
KED. *See* Kinetic energy discrimination
Kinetic energy discrimination (KED), 149
Kinetic fractionation, 176
Kitchen utensils, coatings for, 588
KNN. *See* K-nearest neighbors

L

LA. *See* Laser ablation
LA-FAPA. *See* Laser ablation flowing atmospheric pressure afterglow
Lactoglobulins, 375
LADESI. *See* Laser-assisted desorption electrospray ionization
LAESI. *See* Laser ablation electrospray ionization
LASER. *See* Light Amplification by Stimulated Emission of Radiation
Laser ablation (LA), 161
ELDI, 257–258
ESI, 257
LAESI, 258
MALDESI, 258
Laser ablation electrospray ionization (LAESI), 249, 258
Laser ablation flowing atmospheric pressure afterglow (LA-FAPA), 112–113
Laser desorption atmospheric pressure chemical ionization (LD-APCI), 249, 258
Laser desorption/ionization (LDI), 112–113
Laser desorption spray postionization (LDSPI), 249, 258
Laser diode thermal desorption (LDTD), 249, 258
Laser electrospray mass spectrometry (LEMS), 249, 258
Laser Spray Ionization (LSI), 249, 257
Laser-assisted desorption electrospray ionization (LADESI), 249
LC. *See* Liquid chromatography
LC-HRMS. *See* Liquid chromatography-High Resolution Mass Spectrometry
LC-MS. *See* Liquid chromatography-mass spectrometry
LC–ICP–MS. *See* Liquid chromatography coupled to inductively coupled plasma mass spectrometry
LD-APCI. *See* Laser desorption atmospheric pressure chemical ionization
LDA. *See* Linear discriminant analysis
LDI. *See* Laser desorption/ionization
LDSPI. *See* Laser desorption spray postionization
LDTD. *See* Laser diode thermal desorption
Lecithin. *See* Phosphatidylcholine (PC)

LEMS. *See* Laser electrospray mass spectrometry
LESA. *See* Liquid extraction surface analysis
Leucine, 220
Ligands, 200
Light Amplification by Stimulated Emission of Radiation (LASER), 161–162
Limit of detection (LOD), 39, 375, 527
Limit of quantification (LOQ), 39
Linear discriminant analysis (LDA), 208–209, 471
Linear ion trap (LIT), 264, 372–374, 525–526
Linear ion trap (LTQ), 5–7, 9
Linear quadrupole Orbitrap (LTQ–Orbitrap), 524
Linoleic acid, 400f
LIPID MAPS Structure Database (LMSD), 398
Lipidomics, 396, 414f. *See also* Lipids
  chromatographic separations, 420–421
    GC, 429–432
    HP TLC, 422
    HPLC, 422–425
    MDLC, 428–429
    TLC, 421
    two-dimensional TLC, 422
  mass spectrometry in, 413
    advances in, 419–420
    DESI, 420
    IMS, 419
    ionization modes, 413–419
    N_2 UV laser beams, 420
    SIMS, 420
Lipids, 397
  analysis, 396–397
  extraction methods, 408
    ASE, 412
    MAE, 412
    method of Bligh and Dyer, 410–411
    method of Folch, Lees, and Stanley, 410
    sample preparation, 408–409
    sampling, 408–409
    SFE, 411–412
    Soxhlet extraction, 410
    SPE, 412–413
  fatty acids and derivatives, 398
  LIPID MAPS Lipid Classification System, 398
  structures, functions, and occurrence, 398–399
    FA, 399–401
    GL, 401–402
    GP, 402–404
    PK, 407
    PR, 406–407
    SL, 407, 408f
    SP, 404–406
    ST, 406
    triacyl-*sn*-glycerol, 402f
  system-based lipid approach, 398
Lipoids. *See* Lipids
Liquid chromatography (LC), 22, 77, 93, 164, 517, 622
  applications in food analysis, 246–247
  coupling MS, 516–517
  techniques, 664–666
  use with MS detector, 656
Liquid chromatography coupled to inductively coupled plasma mass spectrometry (LC–ICP–MS), 26–39, 165–166, 166f
  feasibility of coupling, 166
  low tolerance, 166–167
  robustness of ICP-MS instruments, 167
  semiquantitative results, 167
Liquid chromatography-High Resolution Mass Spectrometry (LC-HRMS), 558
  emerging contaminants, 561
  HRLC, 558–559
  PRO detection, 560f
  UHPLC applications, 559
  UHPLC peak, 562f
Liquid chromatography-mass spectrometry (LC-MS), 4, 26, 545, 599, 618, 642–643. *See also* Gas chromatography-mass spectrometry (GC-MS)
  combination of API-MS/MS, 545
  comprehensive LC, 561–563
  emerging contaminants, analytical techniques, 546t–554t
  ESI, 39
  for food safety and quality, 27t–38t
  GC × GC-TOF-MS chromatogram, 544f
  LC-HRMS, 558–561
  LC-MS/MS, 555
    chromatograms of standard solution, 556f
    emerging contaminants determination, 557
    IDA, 557–558
    QqQ and QqLIT, 555
    on QqQ instruments in SRM, 555–557
  matrix effect, 39
  nontarget analysis, 44
    fingerprint, 47–48
    foodomics platform, 46f
    metabolites identification, 45
    metabolomics, 44–45

phytochemical compounds, 45
traditional shotgun proteomics, 47
volatile and aminothiols in wine, 46–47
SNP analyses, 644
target analysis, 40
agnuside identification, 43f
biomarkers, 42–44
for food control and analysis, 40
HRMS, 41
MRM method, 41–42
multitargeted screening, 40–41
"omic" sciences, 41
PCA, 42
UHPLC, 545–555
Liquid extraction surface analysis (LESA), 249, 253, 567
Liquid-phase microextraction techniques (LPME), 524
Liquid–liquid extraction (LLE), 522, 660–661
LIT. *See* Linear ion trap
LIT-FTICR instrument, 315
Livestock, 330–331
aquaculture, 331–332
factors related to animal agricultural sustainability, 331f
LLE. *See* Liquid–liquid extraction
LMSD. *See* LIPID MAPS Structure Database
LMW-GS. *See* Low molecular weight glutenin subunits
LOD. *See* Limit of detection
LOQ. *See* Limit of quantification
Low molecular weight glutenin subunits (LMW-GS), 364
Low-abundance proteins, 319
Low-pressure gas chromatography (LP-GC), 538
Low-temperature plasma ionization (LTP), 517
Lowest threshold intake (LTI), 138–139, 196
LP-GC. *See* Low-pressure gas chromatography
LPE. *See* Lysophosphatidylethanolamine
LPME. *See* Liquid-phase microextraction techniques
LSI. *See* Laser Spray Ionization
LTI. *See* Lowest threshold intake
LTP. *See* Low-temperature plasma ionization
LTQ. *See* Linear ion trap; Quadruple linear ion trap
LTQ–Orbitrap. *See* Linear quadrupole Orbitrap
Lysophosphatidylcholine (Lys-PC), 402
Lysophosphatidylethanolamine (LPE), 264–266
Lysophosphatidylserine (Lys-PS), 402

M

MA(L)LS. *See* Multiangle (laser) light scattering
MAE. *See* Microwave-assisted extraction
MAG. *See* Monoacylglycerols
Magnesium, 137–138
Magnetic sectors, 82, 182–183
Mahalanobis distance (MD), 208
MALDESI. *See* Matrix-assisted laser desorption electrospray ionization
MALDI. *See* Matrix-assisted laser desorption/ionization
MALDI IMS. *See* Matrix-assisted laser desorption/ionization imaging mass spectrometry
MALDI-TOF. *See* Matrix-assisted laser ionization time-of-flight
MALDI-TOF-MS. *See* Matrix-assisted laser desorption ionization time-of-flight mass spectrometry
MAMP. *See* Microbe-associated molecular pattern
Manganese, 136
Mascot search engine, 321–322
Masking of off-flavors, 592
Mass analyzer, 145–146, 623
Mass bias, 187
Mass spectrometry (MS), 5, 9, 77, 132–133, 280–282, 313, 320–321, 397, 460, 516–517, 618, 654. *See also* Food forensics; Food pathogens; Imaging mass spectrometry (IMS)
analyzers, 524
features, 524
food safety, 528
high-speed GC-TOF-MS, 526–527
IT and LIT, 525–526
orbitrap mass spectrometer, 527–528
QqQ, 524–525
QqTOF, 527
TOF, 526
working mode and, 525f
applications, 133
current instrumentation and operation, 5
bioinformatics tools, 12
data matrix, 12
EI, 5
food safety and quality analysis platforms, 9
full-scan MS spectra, 11
future trends, 12–13
ionization sources, 6f
MS-based techniques, 9

Mass spectrometry (MS) *(Continued)*
nontarget analysis, 9–11
"omic", 11–12
orbitrap, 5–7, 7f
TOF-MS, 8–9
trapped-ion mass analyzers, 7–8
detection, 669
direct MS techniques, 670
MALDI-MS techniques, 669
elemental and isotope content, 133
foodomics, 655–656, 655f–656f
instruments, 78
in lipidomics, 413–420
in microbial analysis, 620
ion detector, 624
ionization source, 622–623
mass analyzer, 623
obtaining and processing information, 100–104
data analysis methodology, 111f
DDA, 105–108
DIA, 108–111
emerging data analysis/HSM
IDA, 105–108
MS system configuration, 104–105
MS/MS spectrum matches, 107f
preprocessing and postprocessing steps, 111–112
UHPLC-LTQ-orbitrap MS/dd-MS^2 chromatogram, 108f
resolution and accuracy, 78–79
FT-ICR, 95–97
HRMS, 79, 80t–81t
magnetic sector instruments, 82–84
mass resolving power calculation, 82f
Orbitrap, 97–100
TOF analyzers in food safety and quality, 85t–92t
TOF instruments, 84–95, 84f
Mass/charge ratio (m/z ratio), 524, 623–624
Mastitis, 338
Matrix effect, 39, 523
Matrix solid-phase dispersion (MSPD), 524, 660
Matrix-assisted laser desorption electrospray ionization (MALDESI), 249, 258
Matrix-assisted laser desorption ionization time-of-flight mass spectrometry (MALDI-TOF-MS), 4, 13–14, 489, 618, 624
bacteria identification, 646–648
capacity, 14
data acquisition with, 626
disadvantage, 626–627
DNA determination, 14
fingerprinting, 626
for food safety and quality, 15t–18t
in foodborne pathogens, 629–630
BST method, 632
with Internet-based proteome database, 630
L. monocytogenes primary detection, 633
microorganisms identification, 630–631
MR-CNS, 632
proteomic approach, 633–635
SCMV, 636
Web-application SPECLUST, 635–636
imaging capability, 19–21
MALDI Biotyper 3. 0, 627–628
MALDI-TOFSNP mass spectrum, 19f
microorganisms detection and characterization, 625
optical images of eggplant, 20f
peak-matching techniques, 628–629
phyloproteomics, 628–629
SELDI-TOF-MS, 19
Matrix-assisted laser desorption/ionization (MALDI), 5, 112–113, 258, 279, 315, 320, 372, 413–414, 416–417, 426–427, 622, 669
Matrix-assisted laser desorption/ionization imaging mass spectrometry (MALDI IMS), 568
Matrix-assisted laser ionization time-of-flight (MALDI-TOF), 369
Matrix-induced interferences, 158
Maturity, 266–267
MC-ICP-MS. *See* Multicollector inductively coupled plasma mass spectrometry
MCC. *See* Multicapillary column
MCC-IMS. *See* Multicapillary column-ion mobility spectrometry
MDLC. *See* Multidimensional liquid chromatography
MDMS-SL. *See* Multidimensional mass spectrometry-based shotgun lipidomics
3-MCPD. *See* 3-monochloropropane-1,2-diol
ME. *See* Microextraction techniques
Meat(s), 208–209, 333, 453–454, 469–470, 481–482, 494–495
beef cattle, 333–334
farm animals, 335–336
processed meat and fraud, 336–337
swine, 334–335
Mechanically separated meat (MSM), 454

MEKC. *See* Micellar electrokinetic chromatography
MeO-PBDE. *See* Methoxylated-polybrominated diphenyl ether
Metabolic fingerprinting, 487–489
Metabolic profiling, 487–489
Metabolites identification, 45
Metabolome Tomato Database (MoTo DB), 677–678
Metabolomics, 11–12, 44–45, 487–489
Metallomics, 204–205
Methicillin-resistant coagulase-negative *Staphylococci* (MR-CNS), 632
Methicillin-resistant S. *aureus* (MRSA), 632
Methoxylated-polybrominated diphenyl ether (MeO-PBDE), 543
Methylmercury, 201
MFGM. *See* Milk fat globule membrane
MHC-1. *See* Mixed halogenated compound
Micellar electrokinetic chromatography (MEKC), 420–421, 663–664
Microbe-associated molecular pattern (MAMP), 648
Microbial analysis
  MS in, 620–624
  toxin analysis, 302–303
Microbiota, 313
Microextraction techniques (ME), 246
Microorganism diagnosis, 620
Microwave-assisted digestion (MW-assisted digestion), 159–161
Microwave-assisted extraction (MAE), 246, 412, 522
Milk and dairy products, 451–452, 468, 480, 493–494
Milk fat globule membrane (MFGM), 337
MIP. *See* Molecularly imprinted polymer
MIP-SPE. *See* Molecularly imprinted polymer-solid phase extraction
Mixed halogenated compound (MHC-1), 543
MLP. *See* Multilayer Perceptron
MMAA. *See* Monomethyl arsonic acid
Mobility coefficient, 276
Modern nutritional research, 312–313
Molecular analysis strategies, 486–489
Molecular ions as chemical markers
  animal products, 497
  beverages, 490–493
  fermented drinks, 490–493
  fish, 495–496
  fruits and vegetables, 496–497
  hyphenated system, 489
  meats, 494–495
  milk and dairy products, 493–494
  molecular analysis strategies, 486–489
  organic food, 497–498
  products from cereals, 497
  stand-alone system, 489–490
  vegetable oils, 494
  wine, 490–493
Molecular profiling, 490
Molecularly imprinted polymer (MIP), 246, 290–292, 523
Molecularly imprinted polymer-solid phase extraction (MIP-SPE), 300
Molybdenum, 136
Monoacylglycerols (MAG), 40
3-monochloropropane-1, 2-diol (3-MCPD), 540
Monomethyl arsonic acid (MMAA), 201
Monomethyl arsonous acid (MMAA-As(III)), 201
Monounsaturated FAs (MUFA), 399
MoTo DB. *See* Metabolome Tomato Database
Mozzarella cheese, 468
MR-CNS. *See* Methicillin-resistant coagulase-negative *Staphylococci*
MRSA. *See* Methicillin-resistant S. *aureus*
MS. *See* Mass spectrometry
MS/MS. *See* Tandem mass spectrometry
MS^E acquisition mode, 109, 425–426
MSM. *See* Mechanically separated meat
MSPD. *See* Matrix solid-phase dispersion
MudPIT. *See* Multidimensional protein identification technology
MUFA. *See* Monounsaturated FAs
Multi-MS, 168–169
Multi-target analysis, 376–378
Multiangle (laser) light scattering (MA(L)LS), 600–601
Multicapillary column (MCC), 289
Multicapillary column-ion mobility spectrometry (MCC-IMS), 286
Multicollector inductively coupled plasma mass spectrometry (MC-ICP-MS), 133, 187, 188f. *See also* Inductively coupled plasma mass spectrometry (ICP-MS); Isotope ratio mass spectrometry (IRMS); Thermal ionization mass spectrometry (TIMS)
  fundamentals, 190
  ID analysis, 191
    advantage, 192
    critical stages, 192
    procedure, 191–192
    sources of error, 192

Multicollector inductively coupled plasma mass spectrometry (MC-ICP-MS) *(Continued)*
instrument, 187–188
Ar ICP-source, 188
magnetic sector mass analyzers, 189
multiple Faraday cups, 189
optimization, 189–190
quadrupole ICP-MS, 188
single-collector instruments, 189
limitations, 191
strengths, 190–191
Multidimensional liquid chromatography (MDLC), 428–429
Multidimensional mass spectrometry-based shotgun lipidomics (MDMS-SL), 419
Multidimensional protein identification technology (MudPIT), 319–320
Multilayer Perceptron (MLP), 208–210
Multiple Faraday cup detectors, 183
Multitargeted screening, 40–41
Multivariate analysis (MVA), 645–646
MVA. *See* Multivariate analysis
MW. *See* Molecular weight
MW-assisted digestion. *See* Microwave-assisted digestion
Mycotoxins, 520, 566

N

Naïve Bayes (NB), 209–210
NALDI™. *See* Nano-assisted laser desorption ionization
Nano-assisted laser desorption ionization (NALDI™), 19
nano-DESI. *See* Nanospray desorption electrospray ionization
Nanoencapsulation, 590
nanoESI-MS/MS, 320
Nanofiltration
membranes, 589
technology, 589
Nano–initiator mass spectrometry (NIMS), 19
Nanomaterials, 580–581
derived from conventional food processing, 582–583
fabrication of, 583
food contact materials, 584–589
in food sector, 581
legal obligation, 595
nanoparticles in food, 595–605
nanosensors, 593
nanosized delivery system, 590–592
nanosized food ingredients, 589–590
parameters and analysis methods for characterization of, 597t–598t
safety aspects in food, 593–594
EFSA, 594–595
risk assessment, 593–594
structuring of foods, 592–593
Nanoparticles, 211–212, 595–596
EFSA, 596
EM, 596–598
FFF-ICP-MS, 600–602
nanoparticle detection and characterization, 599–600
NIST, 596
sp-ICP-MS, 602–605
Nanoscience, 580
Nanosized delivery system, 590
bioactive compound protection, 592
bioavailability improvement, 591–592
masking of off-flavors, 592
nanoencapsulation, 590
Nanosized food ingredients, 589–590
Nanospray desorption electrospray ionization (nano-DESI), 249, 253
Nanotechnology, 580, 606. *See also* Food nanotechnology; Nanomaterials
NARP-HPLC. *See* Nonaqueous reversed-phase high-performance liquid chromatography
National Center for Biotechnology Information (NCBI), 326
National Institute of Standards and Technology (NIST), 596, 664
NB. *See* Naïve Bayes
NBFR. *See* Novel brominated flame retardants
NCBI. *See* National Center for Biotechnology Information
NCI. *See* Negative chemical ionization
Negative chemical ionization (NCI), 48
Neurologic shellfish poisoning (NSP), 520–521
^{63}Ni source, 279
Nickel, 136
NIMS. *See* Nano-initiator mass spectrometry
NIST. *See* National Institute of Standards and Technology
Nitrate analysis, 302
Nitrite analysis, 302
NMR spectroscopy. *See* Nuclear magnetic resonance spectroscopy
Nonaqueous reversed-phase high-performance liquid chromatography (NARP-HPLC), 424–425

nonspecific lipid transfer proteins (nsLTPs), 363
Nonspectroscopic interferences, 157–158. *See also* Spectroscopic interferences
internal standardization, 158
magnitude of signal suppression, 158–159
matrix-induced interferences, 158
physical interferences, 158
Nonsteroidal antiinflammatory drugs (NSAIDs), 563–565
Nontarget analysis, 9–11, 44. *See also* Target analysis
fingerprint, 47–48
foodomics platform, 46f
metabolites identification, 45
metabolomics, 44–45
phytochemical compounds, 45
traditional shotgun proteomics, 47
volatile and aminothiols in wine, 46–47
Normal-phase high-performance liquid chromatography (NP-HPLC), 422–423
NovaSOL®, 590–591
Novel ambient air techniques, 264
Novel brominated flame retardants (NBFR), 543
NP-HPLC. *See* Normal-phase high-performance liquid chromatography
NSAIDs. *See* Nonsteroidal antiinflammatory drugs
nsLTPs. *See* nonspecific lipid transfer proteins
NSP. *See* Neurologic shellfish poisoning
Nuclear magnetic resonance spectroscopy (NMR spectroscopy), 45, 419, 460, 638–640
Number of double bonds (DBN), 423
NutraLease™, 590–591
Nutriproteomics, 312–313
Nutritional status, 658–659
Nutritional supplements, 294, 295t
Nutritive elements. *See also* Toxic elements
biological function, deficiency problems and bioavailability, 135
calcium, 138
cobalt, 136
copper, 137
inorganic nutrients, 138
iodine, 137
iron, 136
magnesium, 137–138
manganese, 136
molybdenum, 136
selenium, 137
Worldwide situation, guidelines, 138
nutrient values, 139
risk of chronic diseases, 138–139
WHO/FAO and USDA, 139

O

oa-ToF. *See* Orthogonal accelerated time of flight
OB. *See* Organobromine
Ochratoxin (OT), 520
Ochratoxin A (OTA), 302–303
Oddo–Harkins rule, 478
OECD. *See* Organisation for Economic Co-operation and Development
OIV. *See* Organisation International de la Vigne et du Vin
Olive oil (OO), 289, 289f, 445, 468–469
Omega-3 acids, 284–285
"Omic" sciences, 11–12, 41
science evolution, 310
"Omic" technologies, 78, 654
OO. *See* Olive oil
OPA. *See* Orthogonal projection analysis
Open access PeptideRanker server, 343–344
OPLS. *See* Orthogonal projections to latent structures
Optional quality terms (OQT), 206
Orbitrap, 97, 247–248, 321
analyzers in food safety and quality, 101t–104t
high resolution mass analyzers, 100
mass analyzers, 5–7, 7f, 247–248
mass spectrometer, 527–528
minicake black sample, 99f
MS incorporating, 98f
Q-Orbitrap, 100
Organic foods, 209–210, 459–460, 474–475, 486, 497–498
Organisation for Economic Co-operation and Development (OECD), 332, 580
Organisation International de la Vigne et du Vin (OIV), 462
Organobromine (OB), 543
Organoleptic deviations, 288
Orthogonal accelerated time of flight (oa-ToF), 426
Orthogonal projection analysis (OPA), 208
Orthogonal projections to latent structures (OPLS), 672–674
OT. *See* Ochratoxin
OTA. *See* Ochratoxin A
Ozone electrospray (OzESI), 415

P

PA. *See* Polyacrylate
PADI. *See* Plasma-assisted desorption ionization
PAH. *See* Polycyclic aromatic hydrocarbon
Paralytic shellfish poisoning (PSP), 520–521
Parathyroid hormone (PTH), 137–138
Partial least square (PLS), 283
Partial least squares-discriminant analysis (PLS-DA), 44–45, 57–58, 208, 482, 672–674
Particle beam (PB), 5
Partition number (PN), 423
Pathogenic bacteria, 619–620
Pattern recognition methods, 448–449
Paul ion trap, 320
PB. *See* Particle beam
PBB. *See* Polybrominated biphenyl
PBDE. *See* Polybrominated diphenyl ether
PBFR. *See* Polybrominated flame retardant
PBHD. *See* Polybrominated hexahydroxanthene derivative
PC. *See* Phosphatidylcholine
PCA. *See* Principal component analysis
PCB. *See* Polychlorinated biphenyl
PCDD. *See* Polychlorinated dibenzo-p-dioxin
PCDF. *See* Polychlorinated dibenzofuran
PCI. *See* Positive chemical ionization
PCR. *See* Polymerase chain reaction
PCR-electrospray ionization mass spectrometry (PCR-ESI-MS), 638–640
PCR-ESI-MS. *See* PCR-electrospray ionization mass spectrometry
PCR/ESI-MS. *See* Polymerase chain reaction coupled to mass spectrometry using electrospray ionization
PDE-5 inhibitors. *See* Type-5 phosphodiesterase inhibitors
PDMS. *See* Poly(dimethylsiloxane)
PDO. *See* Protected denominations of origin; Protected designation of origin
PE. *See* Phosphatidylethanolamine
Peak capacity, 279–280
Peak tails, 147
Peptide fragmentation fingerprint (PFF), 321–322
Peptide mass fingerprinting (PMF), 318
Peptide separation, 319–320
Peptide detection, 376
- criteria for proper protein and peptide identification, 376t
- experimental workflow and double MS acquisition scheme, 379f
- HRMS, 378–386
- LC-MS/MS quantitative methods, 380t–385t
- workflow underpinning bottom-up strategy, 376–378

Peptidomics, 487–489, 654–655
Per mil values, 175
Perfluoroalkyl substance (PFAS), 516, 520
Perfluorooctane sulfonate (PFOS), 557
Perfluorooctanoic acid (PFOA), 557
Persistent organic pollutants (POPs), 83–84
Personalized nutrition, 312–313
Pervaporation (PV), 59–60
PESI. *See* Prove electrospray ionization
Pesticides, 519
- detection, 297–298, 299f
  - direct analysis of complex samples, 300
  - herbicides, 300–301
  - HF-LPME, 300
  - HP-IMS, 298
- intoxications, 298

PFAS. *See* Perfluoroalkyl substance
PFF. *See* Peptide fragmentation fingerprint
PFOA. *See* Perfluorooctanoic acid
PFOS. *See* Perfluorooctane sulfonate
PGI. *See* Protected geographical indication
Phosphatidic acid, 402, 403f
Phosphatidylcholine (PC), 402–403, 403f
Phosphatidylethanolamine (PE), 402–403, 404f
Phosphatidylinositol (PI), 402–403, 404f
Phosphatidylserine (PS), 402–403, 404f
Phosphocreatine, 333–334
Phospholipids. *See* Glycerophospholipids (GP)
Phthalate ester, 296–297
Phycotoxins, 520–521
Phyloproteomics, 628–629
PI. *See* Phosphatidylinositol; Polydispersity index
PIF. *See* Powdered infant formula
PK. *See* Polyketides
Plant proteomics, 326–327. *See also* Postharvest proteomics; Proteomics
Plant–pathogen interactions, 329
Plasma-assisted desorption ionization (PADI), 249, 256
PLE. *See* Pressurized liquid extraction
PLS. *See* Partial least square; Projection to latent structure
PLS-DA. *See* Partial least squares-discriminant analysis
Plutonium isotopes, 198–199
PMF. *See* Peptide mass fingerprinting
PN. *See* Partition number

PNN. *See* Probabilistic neural network
Poly(dimethylsiloxane) (PDMS), 661–662
Polyacrylate (PA), 661–662
Polyatomic ion interferences, 156
Polybrominated biphenyl (PBB), 520, 543
Polybrominated diphenyl ether (PBDE), 516, 520, 543
Polybrominated flame retardant (PBFR), 516, 520
Polybrominated hexahydroxanthene derivative (PBHD), 543
Polychlorinated biphenyl (PCB), 48–55, 83–84
Polychlorinated dibenzo-p-dioxin (PCDD), 83–84
Polychlorinated dibenzofuran (PCDF), 83–84
Polycyclic aromatic hydrocarbon (PAH), 48–55, 516, 520
Polydispersity index (PI), 598–599
Polyketides (PK), 407, 408f
Polymerase chain reaction (PCR), 677
Polymerase chain reaction coupled to mass spectrometry using electrospray ionization (PCR/ESI-MS), 618
Polymer–clay nanocomposites, 586
Polystyrene–divinylbenzene (PSDVB), 523
Polyunsaturated fatty acid (PUFA), 284–285, 399–401
Pomace olive oil (POO), 289
POPs. *See* Persistent organic pollutants
Population reference intakes (PRI), 138–139
Portable ion mobility spectrometers, 295
Portable UV-IMS instrument, 283–284
Positive chemical ionization (PCI), 48
Postharvest proteomics, 330
Postmortem proteolytic tenderization, 340–341
Posttranslational modification (PTM), 309–310, 319, 644
Powdered infant formula (PIF), 635–636
PR. *See* Prenol lipids
Prenol lipids (PR), 406–407
Pressurized hot water extraction, 522
Pressurized liquid extraction (PLE), 246, 412, 522, 660
PRI. *See* Population reference intakes
Primary geographic indicators, 476
Primary/secondary amine silica gel (PSA), 543
Principal component analysis (PCA), 42, 109, 207, 209, 266–267, 285, 335, 448–449, 628–629, 659, 672–674
PRO. *See* Propamocarb
Probabilistic neural network (PNN), 209, 482
Probe electrospray ionization (PESI), 112–113
Processed meat, 336–337
Programmed temperature-vaporizing (PTV), 538
Projection to latent structure (PLS), 57–58
Propamocarb (PRO), 560f
Protected denominations of origin (PDO), 206
Protected designation of origin (PDO), 468
Protected geographical indication (PGI), 206, 483–484
Proteins, 313
  digestomics, 313
  profiling, 493–494
Proteome, 312–313
  informatics tools, 321–325
Proteomics, 309–310, 312f, 313, 487–489, 648, 654–655. *See also* Food security; Plant proteomics
  bottom-up proteomics, 316
    analytical steps and techniques, 316f
    in-gel bottom-up proteomics, 316–318
    shotgun proteomics, 318–325
  top-down proteomics, 315
Proton transfer reaction ionization quadrupole mass analyzers (PTR-Quad-MS), 94–95
Proton transfer reaction time of flight mass spectrometry (PTR-ToF-MS), 94–95, 489, 495
Prove electrospray ionization (PESI), 267
Provisional tolerable weekly intake (PTWI), 195–196
PS. *See* Phosphatidylserine
PSA. *See* Primary/secondary amine silica gel
PSDVB. *See* Polystyrene–divinylbenzene
PSP. *See* Paralytic shellfish poisoning
PTH. *See* Parathyroid hormone
PTM. *See* Posttranslational modification
PTR-Quad-MS. *See* Proton transfer reaction ionization quadrupole mass analyzers
PTR-ToF-MS. *See* Proton transfer reaction time of flight mass spectrometry
PTV. *See* Programmed temperature-vaporizing
PTWI. *See* Provisional tolerable weekly intake
PUFA. *See* Polyunsaturated fatty acid
PV. *See* Pervaporation
Pyrolysis, 641
  pyrolysis–GC-MS, 641

Q

Q-Orbitrap. *See* quadrupole with Orbitrap
QC. *See* Quality control
QDA. *See* Quantitative descriptive analysis
QIT. *See* Quadrupole ion traps

qIT-TOF-MS. *See* Ion trap time-of-flight hybrid mass spectrometer
qPCR. *See* Quantitative polymerase chain reaction
QqLIT. *See* Quadrupole linear ion trap
QqQ. *See* Triple quadrupole
QqTOF. *See* Quadrupole time of flight
QTRAP. *See* Quadrupole linear ion trap (QqLIT)
Quadruple linear ion trap (LTQ), 524
Quadrupole ion traps (QIT), 524
Quadrupole linear ion trap (QqLIT), 247–248, 518, 623
Quadrupole time of flight (QqTOF), 5–9, 247–248, 315, 320, 372, 426, 518, 524, 527, 669
Quadrupole with Orbitrap (Q-Orbitrap), 100, 524
Quadrupoles, 5–7, 147–148
  collision/reaction cell technology, 148
    DRC, 148–149, 149f
    KED, 149
    linear RF-driven multipole ion guide, 148
  ICP-MS, 147
  low-resolution instruments, 147
  scan rates, 147–148
Quality control (QC), 648
Quality grade, 659
Quality of food, 515–516
Quantitative descriptive analysis (QDA), 57–58
Quantitative polymerase chain reaction (qPCR), 643
Quantitative proteomics, 321–325
Quantitative relative proteomics, 325
Quick, Easy, Cheap, Effective, Rugged, Safe (QuEChERS), 521–522, 660–661
  QuEChERS-like extraction method, 555–557

R

Rabbit (*Oryctolagus cuniculus*), 335–336
Radial Basis Function (RBF), 208
Radio frequency (RF), 97, 141–142
  RF field, 7–8
Radio frequency identification display (RFID), 587
Radioactive isotopes, 174
RAM. *See* Restricted access materials
Rapid fusion method, 198–199
Rare earth elements (REE), 206, 479
RBF. *See* Radial Basis Function
RDA. *See* Recommended Dietary Allowances
RDX. *See* Cyclo-1,3,5-trimethylene trinitramine
Reactive desorption electrospray ionization (Reactive DESI), 249–253
Reactive oxygen species (ROS), 327
Receiver operating characteristic (ROC), 672–674
Recommended Dietary Allowances (RDA), 138–139
REE. *See* Rare earth elements
Reference intake (RI), 138–139
Reference material (RM), 174–175
Refractive index (RI), 600–601
Relative quantitative proteomics, 327
Relative standard deviation (RSD), 302–303
Repetitive sequence-based polymerase chain reaction (Rep-PCR), 632
Resolving power, 279–280
Restricted access materials (RAM), 246
Retention time locking (RTL), 48–55
Reverse Nier–Johnson spectrometer, 150, 150f
Reversed phase (RP), 319–320, 545
Reversed-phase chromatography (RPC), 166
Reversed-phase ion-pairing chromatography (RPIPC), 166
Reversed-phase liquid chromatography (RP-HPLC), 424
Reversed-phase Liquid chromatography (RP-LC), 26–39
Reversed-phase octadecyl silica (RP-C18), 523
RF. *See* Radio frequency
RFID. *See* Radio frequency identification display
RI. *See* Reference intake; Refractive index
RM. *See* Reference material
ROC. *See* Receiver operating characteristic
Root mean square error of evaluation (RMSEE), 672–674
Root mean square error of prediction (RMSEP), 672–674
ROS. *See* Reactive oxygen species
RP. *See* Reversed phase
RP-C18. *See* Reversed-phase octadecyl silica
RP-HPLC. *See* Reversed-phase liquid chromatography
RP-LC. *See* Reversed-phase Liquid chromatography
RPC. *See* Reversed-phase chromatography
RPIPC. *See* Reversed-phase ion-pairing chromatography
RSD. *See* Relative standard deviation

RTL. *See* Retention time locking
Rubidium (Rb), 206

S

S/B ratios. *See* Signal-to-background ratios
Saccharolipids (SL), 407, 408f
Salinity, 328
Salmon, 470–471
Sample preparation, 408–409, 660
 centrifugal ultrafiltration, 662–663
 dispersive SPE, 660–661
 for foodomics, 661f
 LLE, 660–661
 in shotgun proteomics, 318–319
 SPE, 660–661
 SPME, 661–662
Sampling, 408–409
SAWN. *See* Surface acoustic wave nebulization
SAX. *See* Strong anion exchange
SBS. *See* Styrene butadiene co-styrene
SBSE. *See* Stir bar sorptive extraction
SCMV. *See* Sugarcane mosaic virus
SCX. *See* Strong cation exchange
SDAPCI. *See* Surface desorption atmospheric pressure chemical ionization
SdFFF. *See* Sedimentation field-flow fractionation
SDME. *See* Single-drop microextraction
SDS-PAGE. *See* Sodium dodecyl sulfate-PAGE
SE. *See* Solvent extraction
SEC. *See* Size exclusion chromatography
SEC-ICP-MS. *See* Size exclusion chromatography coupled to ICP-MS
Secondary electrospray ionization (SESI-MS), 489
Secondary geographic indicators, 476
Secondary ion mass spectrometry (SIMS), 112–113, 418
Sector field (SF), 149–150
 double-focusing magnetic-sector instrument, 150–151
 high-throughput trace analysis, 151
 reverse Nier–Johnson spectrometer, 150, 150f
Sector field inductively coupled plasma mass spectrometry (SF-ICP-MS), 211–212, 605
SeCys. *See* Selenocystine
Sedimentation field-flow fractionation (SdFFF), 600
SeEt. *See* Selenoethionine
SELDI. *See* Surface enhanced laser desorption ionization
SELDI-MS. *See* Surface enhanced laser desorption ionization-mass spectrometry
SELDI-TOF-MS. *See* Surface-enhanced laser desorption/ionization time-of-flight mass spectrometry
Selected ion monitoring (SIM), 48–55, 83–84, 376–378, 528
Selected reaction monitoring (SRM), 9, 325, 376–378, 524–525, 555, 641
Selective proteotypicpeptide analysis (SPA), 642–643
Selenium, 137, 202–203
Selenocystine (SeCys), 203–204
Selenoethionine (SeEt), 203–204
Selenomethionine (SeMet), 137, 202–204
SeMet. *See* Selenomethionine
Senescence, 330
Separation, 663–664. *See also* Sample preparation
 CE techniques, 666–667
 coupling microfluidic chips, 668
 GC techniques, 664
 HSCCC-MS system, 668f
 LC techniques, 664–666
 MS combination, 670–671
 MS spectra, 667–668
 UHPLC separation of analytes, 665f
Sequenom® MassARRAY® MALDI-TOF-MS, 14
SEQUEST search engine, 321–322
SESI-MS. *See* Secondary electrospray ionization
Setoff, 264
SF. *See* Sector field
SF-ICP-MS. *See* Sector field inductively coupled plasma mass spectrometry
SFC. *See* Supercritical fluid chromatography
SFE. *See* Supercritical fluid extraction
Shotgun lipidomics, 413, 415
Shotgun proteomics, 318. *See also* In-gel bottom-up proteomics
 information flow during data processing, 325f
 mass spectrometry, 320–321
 MS-based quantitative proteomics, 324f
 peptide identification, 323f
 peptide separation, 319–320
 proteome informatics tools, 321–325
 quantitative proteomics, 321–325
 sample preparation, 318–319

"Shotgun" approach, 376
SIA. *See* Stable isotopes analysis
Sibutramine, 295
SID. *See* Stable isotope dilution
Signal-to-background ratios (S/B ratios), 281–282
SILAC. *See* Stable isotope labeling with amino acids in cell culture
Silica plate imprinting (SPI), 113
Silicon, 137
  dioxide, 589–590
Silver ion high-performance liquid chromatography (Ag-HPLC), 423–424
Silver nanoparticle (AgNP), 210–212
Silver nanoparticles, 586
SIM. *See* Selected ion monitoring
SIMCA. *See* Soft independent modeling of class analogy
Simple isoprenoids, 406
Simple lipoids, 397
SIMS. *See* Secondary ion mass spectrometry
Single nucleotide polymorphism (SNP), 14, 497
Single particle Inductively Coupled Plasma Mass Spectrometry (SP-ICP-MS), 22
Single-drop microextraction (SDME), 288–289
Single-nucleotide polymorphism analyses (SNP analyses), 644
Single-particle ICP-MS (sp-ICP-MS), 602–603
  AF4-ICP-MS *vs*., 606t
  dwell time, 603
  hydrodynamic chromatography with MALDI-TOF, 605
  measurement of ionic and nanoparticulate silver mixture, 603f
  multielemental analysis, 605
  time-resolved data acquisition, 604f
  validation, 605
Size exclusion chromatography (SEC), 203–204, 320
Size exclusion chromatography coupled to ICP-MS (SEC-ICP-MS), 204–205
SL. *See* Saccharolipids
SLDA. *See* Stepwise linear discriminant analysis
Slim Shake chocolate, 590–591
Smart food packaging. *See* Intelligent food packaging
SNP. *See* Single nucleotide polymorphism
SNP analyses. *See* Single-nucleotide polymorphism analyses
Sodium dodecyl sulfate-PAGE (SDS-PAGE), 316
Soft independent modeling of class analogy (SIMCA), 209, 448–449
Solid phase microextraction (SPME), 277–278, 490–491, 522
Solid-phase extraction (SPE), 246, 277–278, 301, 409, 412–413, 521, 523, 660–661
Solid-phase microextraction (SPME), 524, 661–662
Solid–liquid extraction ambient ionization techniques, 249–250, 252f
  DART, 252f
  DESI, 250, 252f
  DICE, 253
  reactive DESI, 250–253
Solvent extraction (SE), 521
Sonic spray ionization (SSI), 250–253
Sophistication, 444
Soxhlet extraction, 246, 410
SP. *See* Sphingolipids
SP-ICP-MS. *See* Single particle Inductively Coupled Plasma Mass Spectrometry
sp-ICP-MS. *See* Single-particle ICP-MS
SPA. *See* Selective proteotypicpeptide analysis
SPE. *See* Solid-phase extraction
Speciation, 199, 204
  arsenic, 201
  chromium, 202
  elemental, 200
  forms, 201
  hyphenated ICP-MS, 199
  ICP-MS in, 201
  organic forms, 201–202
  SEC, 203–204
  selenium, 202–203
  SPME-enantioselective gas chromatography, 204
  toxic elements, 200
  volatile species, 202
Spectroscopic interferences, 155–156. *See also* Nonspectroscopic interferences
  adjacent lower abundant peak, 156
  molecular or polyatomic ion interferences, 156–157
  multielement analysis of food, 157
  polyisotopic elements, 157
Sphing-4-enine. *See* Sphingosine
Sphingolipids (SP), 404–405
  ceramides, 405, 406f
  glycosphingolipids, 405
  sphingoid bases, 405

Sphingomyelins, 405
Sphingosine, 405
SPI. *See* Silica plate imprinting
Spices, 471–472
SPME. *See* Solid phase microextraction
Spoilage bacteria, 618–619
SPR. *See* Surface-plasmon resonance
$^{87}Sr/^{86}Sr$ ratio, 177
SRM. *See* Selected reaction monitoring
SSI. *See* Sonic spray ionization
ST. *See* Sterol lipids
Stable isotope dilution (SID), 376–378
Stable isotope labeling with amino acids in cell culture (SILAC), 376
Stable isotopes, 174
  abundance expression, 174–175, 175t
Stable isotopes analysis (SIA), 461–462
Stand-alone system, 489–490
Staple crops, 330
Stepwise linear discriminant analysis (SLDA), 496
Steroids, 406
Sterol lipids (ST), 406
Stir bar sorptive extraction (SBSE), 60–61, 524
Strong anion exchange (SAX), 320
Strong cation exchange (SCX), 320
Strontium (Sr), 206–207
Styrene butadiene co-styrene (SBS), 59–60
Substantial equivalence, 332
Sugarcane mosaic virus (SCMV), 636
Supercritical fluid chromatography (SFC), 397, 420–421, 667–668
Supercritical fluid extraction (SFE), 246, 411–412
Support vector machine (SVM), 209–210
Support vector machine discriminant analysis (SVM-DA), 207
Supramolecular solvents (SUPRAS), 524
SUPRAS. *See* Supramolecular solvents
Surface acoustic wave nebulization (SAWN), 418–419
Surface desorption atmospheric pressure chemical ionization (SDAPCI), 112–113
Surface enhanced laser desorption ionization (SELDI), 623
Surface enhanced laser desorption ionization-mass spectrometry (SELDI-MS), 644–6453
Surface-enhanced laser desorption/ionization time-of-flight mass spectrometry (SELDI-TOF-MS), 19, 93
Surface-plasmon resonance (SPR), 264
SVM. *See* Support vector machine
SVM-DA. *See* Support vector machine discriminant analysis
Sweetening reagents, 293–294
Swine meat, 334–335
Szegedi Füszerpaprika, 471–472

T

T-2 toxin (T-2), 520
T-cell epitopes, 361–362
TAG. *See* Triacylglycerols
TAIR. *See* Arabidopsis Information Resource
TALs. *See* Tetraalkyl leads
Tandem "in-time", 524
Tandem mass spectrometry (MS/MS), 425–426, 524, 636–637
Target analysis, 40, 487–489. *See also* Nontarget analysis
  agnuside identification, 43f
  biomarkers, 42–44
  for food control and analysis, 40
  HRMS, 41
  MRM method, 41–42
  multitargeted screening, 40–41
  "omic" sciences, 41
  PCA, 42
TBBPA. *See* Tetrabromobisphenol A
2,4,6-TCA. *See* 2,4,6-Trichloroanisole
TCA. *See* Trichloroacetic acid
Tea, 473, 484
  samples, 209
TEM. *See* Transmission electron microscopy
Tetraalkyl leads (TALs), 201–202
Tetrabromobisphenol A (TBBPA), 520
Tetracyclin, 301
Tetramethylammonium hydroxide (TMAH), 204
TFC. *See* Turbulent flow chromatography
TG2. *See* Transglutaminase
TGL. *See* "Tonda Gentile delle Langhe"
Theophylline, 290–292
Thermal desorption and chemical ionization, 254
  APCI and ionization techniques, 256
  APTDI, 257
  ASAP, 256
  DAPCI, 256
  DAPPI, 257
  DART, 254
  DESI, 254–256
  matrix-matched standards and ID, 255f

Thermal ionization mass spectrometry (TIMS), 133, 184. *See also* Inductively coupled plasma mass spectrometry (ICP-MS); Isotope ratio mass spectrometry (IRMS); Multicollector inductively coupled plasma mass spectrometry (MC-ICP-MS)
  applications, 220
  comparison with MC-ICP-MS, 186–187
  instrumentation, 184
    lighter isotopes, 186
    magnetic sector mass spectrometer, 184–185, 185f
    mass-resolved beams, 185
  limitations, 186
  strengths, 186
Thermal processes, 332–333
Thin-layer chromatography (TLC), 397, 417, 420–421
33-meris, 371
Thousand flower honey, 458
Three-dimensional ion traps (^{3}DIT), 5–7
TIC. *See* Total ion chromatogram
Time-of-flight (TOF), 5–7, 55, 78–79, 108–109, 518, 526, 623
  analyzers, 84f, 151–152, 425–426
    advantages, 152
    in food safety and quality, 85t–92t
    ICP-TOF-MS systems, 152, 152f
  within food analysis, 95
  GC-TOF-MS, 94
  ICP-TOF mass spectrometers, 95
  instruments, 84
  LC-TOF-HRMS method, 93–94
  MALDI-TOF, 93
  MS detectors, 538–539
  TOF-based instruments, 145–146
  TOF-TOF-MS, 92
Time-of-flight mass spectrometry (TOF-MS), 8–9, 262–264
Time-of-flight secondary ion mass spectrometry (TOF-SIMS), 92–93
TIMS. *See* Thermal ionization mass spectrometry
Titanium dioxide (TiO_2), 212, 589–590
TLC. *See* Thin-layer chromatography
TM-DESI. *See* Transmission mode desorption electrospray ionization
TMA. *See* Trimethylamine
TMAH. *See* Tetramethylammonium hydroxide
TOF. *See* Time-of-flight
TOF-MS. *See* Time-of-flight mass spectrometry
TOF-SIMS. *See* Time-of-flight secondary ion mass spectrometry
"Tonda Gentile delle Langhe" (TGL), 208
Top-down proteomics, 313, 314f
"Top-down" approach, 375
  lipidomic approach, 397
"Top-down" manufacturing, 583
Total ion chromatogram (TIC), 42–44
Toxic elements. *See also* Nutritive elements
  contamination, 133–134
  health problems, 134
  legislation on, 135
  toxicity between chemical forms, 134
Trace elements as chemical markers, 476
  animal products, 485
  beverages, 479–480
  with elemental techniques, 476
  elemental techniques analytical features, 477t–478t
  fermented drinks, 479–480
  fish, 482
  fruits and vegetables, 482–484
  meats, 481–482
  milk and dairy products, 480
  organic food, 486
  products from cereals, 486
  traceability of foodstuffs, 478
  vegetable oils, 480–481
  wine, 479–480
Traceability, 445, 446f
Traditional shotgun proteomics, 47
Traditional specialities guaranteed (TSG), 206
Transgenic foods, 498
Transglutaminase (TG2), 364–365
Transmission electron microscopy (TEM), 602
Transmission mode desorption electrospray ionization (TM-DESI), 249–253
Trapped-ion mass analyzers, 7–8
Triacyl-*sn*-glycerol, 402f
Triacylglycerol. *See* Triacyl-*sn*-glycerol
Triacylglycerols (TAG), 264–266, 398
Trichloroacetic acid (TCA), 319
2,4,6-Trichloroanisole (2,4,6-TCA), 288–289
Triglycerides, 401
Trimethylamine (TMA), 286
Triple quadrupole (QqQ), 55, 247–248, 325, 518, 524–525, 538–539
Triple quadrupole mass spectrometer, 5–7
"Trojan Horse Effect", 593–594

Truffle, 455–456
TSG. *See* Traditional specialities guaranteed
Turbulent flow chromatography (TFC), 557
Two-dimensional gas chromatography (GCxGC), 641
Two-dimensional gas chromatography-mass spectrometry (GC × GC-MS), 528
Two-dimensional gel electrophoresis (2-DE), 493–494
2D polyacrylamide gel electrophoresis (2D-PAGE), 316–317, 327
Two-dimensional thin-layer chromatography (2D TLC), 422
Two-sector instruments. *See* Double-sector instruments
"Type I" hypersensitivity reactions. *See* IgE-mediated food allergies
Type-5 phosphodiesterase inhibitors (PDE-5 inhibitors), 295–296

U

UAE. *See* Ultrasonic-assisted extraction
UHPLC. *See* Ultra-high performance liquid chromatography
UL. *See* Upper intake levels
Ultra-high performance liquid chromatography (UHPLC), 26–39, 77–78, 517, 545–555, 664–666
Ultra-performance liquid chromatography (UPLC), 246–247
Ultrafiltration, 662–663
Ultrasonic extraction, 161
Ultrasonic-assisted extraction (UAE), 246
Ultrasound-assisted emulsification microextraction (USAEME), 524
Uniform Resource Locators (URLs), 586
UniProt. *See* Universal Protein Resource
United States Department of Agriculture (USDA), 139
Universal Protein Resource (UniProt), 326
UPLC. *See* Ultra-performance liquid chromatography
Upper intake levels (UL), 139
Urea, 316–317
URLs. *See* Uniform Resource Locators
US Dietary reference values (DRVs), 138–139
USAEME. *See* Ultrasound-assisted emulsification microextraction
USDA. *See* United States Department of Agriculture

V

Vegetable oils, 208, 445, 452, 468–469, 480–481, 494
Vegetables, 207–208, 455–457, 471–473, 482–484, 496–497
Venturi-Easy ambient sonic spray ionization (V-EASI), 253
Veterinary drugs, 516–517, 519
  analysis, 301
Vienna Pee Dee Belemnite (VPDB), 174–175
Viruses, 329
Vitamin B_{12}. *See* Cobalamin
Vitamin E, 407f
Volatile organic compound (VOC), 94–95, 288

W

Web-application, 628–629
Western blotting (WB), 367–369
Wet digestion
  methods, 160
  procedures, 21–22
Wheat proteins distribution, 364f
Whey milk proteins, 337
Whisky, 467
Wines, 207, 449–450, 465–468, 479–480, 490–493
World Health Organization (WHO), 3, 138

X

XIC. *See* Extracted ion chromatogram (EIC)

Y

Yerba mate (*Ilex paraguariensis*), 207

Z

Zearalenone (ZEN), 520
Zinc, 136

FSC
www.fsc.org
MIX
Papier | Fördert
gute Waldnutzung
FSC® C083411

Kolibri 360 GmbH
Ferdinand-Jühlke-Straße 7
99095 Erfurt, Deutschland
produktsicherheit@kolibri360.de